Mathematical Methods and Models in Composites

Second Edition

Computational and Experimental Methods in Structures

ISSN: 2044-9283

Series Editor: Ferri M. H. Aliabadi *(Imperial College London, UK)*

This series will include books on state-of-the-art developments in computational and experimental methods in structures, and as such it will comprise several volumes covering the latest developments. Each volume will consist of single-authored work or several chapters written by the leading researchers in the field. The aim will be to provide the fundamental concepts of experimental and computational methods as well as their relevance to real world problems.

The scope of the series covers the entire spectrum of structures in engineering. As such it will cover both classical topics in mechanics, as well as emerging scientific and engineering disciplines, such as: smart structures, nanoscience and nanotechnology; NEMS and MEMS; micro- and nano-device modelling; functional and smart material systems.

Published:

Vol. 13 *Mathematical Methods and Models in Composites (Second Edition)*
edited by V. Mantič (University of Seville, Spain)

Vol. 12 *Wear in Advanced Engineering Applications and Materials*
edited by Luis Rodríguez-Tembleque (Universidad de Sevilla, Spain), Jesús Vázquez (Universidad de Sevilla, Spain) and M. H. Ferri Aliabadi (Imperial College London, UK)

Vol. 11 *Advances in Computational Coupling and Contact Mechanics*
edited by Luis Rodríguez-Tembleque (Universidad de Sevilla, Spain) and M. H. Ferri Aliabadi (Imperial College London, UK)

Vol. 10 *Vibration-Based Techniques for Damage Detection and Localization in Engineering Structures*
edited by Ali S. Nobari (Imperial College London, UK & Amirkabir University of Technology, Iran) and M. H. Ferri Aliabadi (Imperial College London, UK)

More information on this series can also be found at http://www.worldscientific.com/series/cems

(Continued at end of book)

Computational and Experimental Methods in Structures – Vol. 13

Mathematical Methods and Models in Composites

Second Edition

Editor

Vladislav Mantič

University of Seville, Spain

World Scientific

NEW JERSEY • LONDON • SINGAPORE • BEIJING • SHANGHAI • HONG KONG • TAIPEI • CHENNAI • TOKYO

Published by

World Scientific Publishing Europe Ltd.

57 Shelton Street, Covent Garden, London WC2H 9HE

Head office: 5 Toh Tuck Link, Singapore 596224

USA office: 27 Warren Street, Suite 401-402, Hackensack, NJ 07601

Library of Congress Cataloging-in-Publication Data
Names: Mantič, Vladislav, editor.
Title: Mathematical methods and models in composites / editor,
Vladislav Mantič, University of Seville, Spain.
Description: Second edition. | New Jersey : World Scientific, [2023] |
Series: Computational and experimental methods in structures, 2044-9283 ; vol. 13 |
Includes bibliographical references and index.
Identifiers: LCCN 2022010505 | ISBN 9781800611870 (hardcover) |
ISBN 9781800611887 (ebook for institutions) | ISBN 9781800611894 (ebook for individuals)
Subjects: LCSH: Composite materials--Mathematical models. | Computersimulation.
Classification: LCC TA418.9.C6 M324 2023 | DDC 620.1/18--dc23/eng/20220307
LC record available at https://lccn.loc.gov/2022010505

British Library Cataloguing-in-Publication Data
A catalogue record for this book is available from the British Library.

For any available supplementary material, please visit
https://www.worldscientific.com/worldscibooks/10.1142/Q0350#t=suppl

Desk Editors: Soundararajan Raghuraman/Shi Ying Koe

Typeset by Stallion Press
Email: enquiries@stallionpress.com

This book is dedicated to Professor Federico París
on the occasion of his 70th birthday

https://doi.org/10.1142/9781800611887_fmatter

Preface

The range of engineering applications of composite materials and structures is growing significantly, leading to an increasing demand for their modeling. The purpose of this book is to present a variety of mathematical methods and models that can be useful in the design, certification, analysis, and characterization of composite materials, the simulation of composite materials manufacturing, and in the prediction of damage and failure of composite materials in service. The chapters written by well-known scholars alongside promising young researchers — engineers, mathematicians, and physicists — present up-to-date reviews of such mathematical methods and models and their applications to the solutions of challenging engineering problems in composites.

Since 2014, when the very successful first edition of this book appeared, this topic has evolved rapidly, so the second extended edition containing 13 chapters is an opportunity to address several relevant issues not covered by the first edition. The chapters of the first edition included in the second edition have been revised and updated, and some of them have also been substantially extended (Chapters 1, 4, 8, 10–13). Moreover, the second edition includes three new chapters, Chapters 5, 7, and 9. Thus, this second edition includes a broad spectrum of mathematical methods and models currently applied to composite materials and structures.

The chapters are organized and grouped by topic. Chapters 1 and 2 present and analyze homogenization methods for composite materials and structures. Several equivalent single-layer and layer-wise models for multilayered plates and shells are reviewed and applied to a variety of problems in Chapters 3–6. Instabilities in nonlinearly elastic multilayers and fiber-reinforced solids are analyzed in Chapters 6 and 7. The Stroh formalism is applied to the analysis of wave-propagation in anisotropic homogenized composites in Chapter 8. Advanced mathematical models and efficient numerical simulations of composite processes, reviewed in Chapter 9, are based on the Proper Generalized Decomposition (PGD), a powerful model reduction technique. The aim of different mathematical models and computational tools presented in Chapters 10–13 is to predict damage and fracture initiation and propagation in composites at micro-, meso-, and macroscales. Comprehensive explanations and references make all chapters self-contained.

Chapter 1 by A.L. Kalamkarov reviews the multiscale asymptotic homogenization method and some of its applications to composite materials and three-dimensional thin-walled composite reinforced structures, including smart composite materials, plates, and shells. Analytical formulae for the effective elastic properties of an equivalent homogeneous material are derived through the analytical solution of the corresponding unit-cell problems. Examples of the application of the method to rib- and wafer-reinforced shells, orthotropic grid-reinforced composite plates and shells, sandwich composite shells with cellular cores, and carbon nanotubes- are presented.

Chapter 2 by M. Ostoja-Starzewski and S.I. Ranganathan considers the scale-dependent properties of random microstructures employing the framework of stochastic (micro)mechanics consistent with the mathematical statement of the Hill–Mandel condition. The scaling from a statistical toward a representative volume element is analyzed and the scale-dependent bounds and scaling laws in planar conductivity, linear and nonlinear (thermo)elasticity, plasticity, and Darcy permeability are studied and illustrated by examples.

Chapter 3 by C. Hwu presents the Stroh-like formalism for thin anisotropic plates with stretching-bending coupling. This formalism is a powerful complex-variable method for the analysis of general symmetric and unsymmetric composite laminates. Extensions of this formalism to hygrothermal and electro-elastic problems in composite laminates are also introduced. Analytical solutions of several problems for laminates with holes and cracks are shown.

Chapter 4 by E. Carrera, M. Cinefra, and M. Petrolo first provides a comprehensive review of available classical, refined, zig-zag, and layer-wise models for multilayered plates and shells, paying particular attention to their historical origins and evolution and to the key relations between them. The best theory diagram is introduced as a tool to evaluate the accuracy of structural models against reference solutions. Then, a refined finite element formulation for multilayered shell structures based on the first author's unified formulation is developed and numerically tested on classical discriminating problems showing good convergence and robustness. Finally, the best theory diagrams for the plate and shell models considered provide recommendations for a suitable development of refined structural theories.

Chapter 5 by J. Reinoso, M. Paggi, and A. Blázquez reviews various nonlinear continuum-based shell parametrizations with the corresponding extension for laminated composite shells, and a nonlinear cohesive interface element including nonlinear geometric and material effects. A variational formulation and the corresponding finite element discretizations are presented. These computational strategies are then used for the nonlinear analysis of different composite structures: a postbuckling analysis of a composite stiffened panel, and a wrinkling–delamination analysis of a layer–substrate composite system predicting a postbuckling deformation pattern and delamination. Comparisons with previous semi-analytical investigations and experimental data demonstrate the reliability of the proposed methodology for the analysis of complex thin-walled composite structures involving instabilities.

Chapter 6 by D. Bigoni, M. Gei, and S. Roccabianca applies an incremental bifurcation theory of prestressed elastic solids to analyze instabilities that often lead to delamination in multilayers, for different deformation paths including finite tension/compression and finite bending. The interlaminar contact is described by introducing linear imperfect interfaces. It is shown that several instabilities such as Euler buckling, necking, surface instability, and several wavelike modes may occur in a multilayer.

Chapter 7 by J. Merodio and R.W. Ogden analyzes several instability mechanisms in fiber-reinforced materials, fiber kinking and fiber splitting in compression, and fiber debonding and matrix failure under tension. The nonlinearly elastic composite material is modeled by a constitutive law consisting of an isotropic contribution by a matrix material, and a transversely isotropic contribution associated with the fiber direction. The studied loss of ellipticity of the governing equations of equilibrium is associated with surfaces of discontinuity of the deformation gradient tensor (or of its gradient) in the material at different orientations relative to the fiber direction. It is shown how the loss of ellipticity can be related to different failure mechanisms in fiber-reinforced materials.

Chapter 8 by K. Tanuma and C.-S. Man introduces the Stroh formalism for anisotropic elastodynamics and shows some of its applications. A perturbation formula for the phase velocities and polarization ratio of Rayleigh waves that propagate along the free surface of a prestressed orthotropic or transversely isotropic half-space is derived. The possibility of determination of the prestress in homogenized orthotropic or transversely isotropic composites by boundary measurement of angular dependence of phase velocities or polarization ratio of Rayleigh waves is examined. The properties of the polarization ratio make it advantageous for use in the non-destructive evaluation of the prestress.

Chapter 9 by E. Abisset-Chavanne, A. Barasinski, and F. Chinesta, after a review of the latest developments in reinforced polymer flow, focuses on the efficient modeling and simulation of composite manufacturing processes of laminates, requiring very fine meshes in the thickness direction because of the complexity of the involved physics. Special attention is paid to the appropriate representation of the physics at the interfaces where thermal resistance, molecular diffusion, and squeeze flow during consolidation of composite laminates occur. Often such complex models, where hypotheses for reducing the dimensionality of the model from 3D to 2D is not possible, should be solved many times because of the history-dependent thermomechanical behavior. To circumvent the curse of dimensionality in solving such complex problems in the fabrication of composite structures, the necessary application of a model reduction technique, the proper generalized decomposition, is discussed.

Chapter 10 by C.G. Dávila, C.A. Rose, E.V. Iarve, and F.A. Leone re-examines the most common computational modeling strategies for the prediction of localized and distributed damage in laminated composites with an emphasis on the scale of the damage idealization and size effects. The capabilities of cohesive crack models to represent crack initiation and propagation and the intrinsic limitations of continuum damage models for modeling laminated composites are analyzed. An x-FEM technique modeling damage propagation by inserting cohesive cracks in arbitrary directions is introduced and its capabilities for avoiding some of the limitations of continuum damage models are shown.

Chapter 11 by T. Roubíček, M. Kružík, J. Zeman, C.G. Panagiotopoulos, R. Vodička, and V. Mantič describes a general and rigorous mathematical framework based on various concepts of solutions, such as the energetic or stress-driven type of solutions, for quasistatic and rate-independent processes of delamination and debonding in composites, possibly combined with interface plasticity or friction. Nevertheless, these approaches also allow consideration of other inelastic or rate-dependent processes in the bulk, such as plasticity, viscosity, inertia, and thermal processes. The process evolution is assumed to be governed by the minimum-energy and minimum dissipation-potential principles written in terms of a time-dependent stored energy functional and a dissipation potential. This ultimately leads to the solution of a time-incremental variational problem. The methodology developed is applied to several macroscopic delamination/debonding and microscopic fiber-matrix debonding problems.

Chapter 12 by D. Leguillon and E. Martin presents the coupled criterion in the framework of finite fracture mechanics, which allows to predict the nucleation of a crack of finite length at stress concentrations or stress singularities. This criterion uses the conditions for energy and tensile stress and involves both the fracture toughness and tensile strength of the material. It is applied to semi-analytical studies of the following damage mechanisms: a transverse crack in a layer approaching the interface with an adjacent layer, where this crack can stop, deflect originating a delamination crack, or cross the interface; the mechanism of crack kinking out of an interface by considering the T-stress; and the initiation of damage at the corner of an interface.

Chapter 13 by V. Mantič, A. Barroso, and F. París first introduces a general semi-analytic matrix formalism for the evaluation of singular stresses in anisotropic elastic multi-material corners, covering also the case of sliding frictional contact surfaces. A least-squares fitting technique for extracting generalized stress intensity factors in such corners from FEM and BEM results is then presented and used for the singularity analysis of a crack terminating at a ply interface in a laminate, and a bimaterial corner in a double-lap joint. A criterion for the failure initiation at a multi-material corner tip is proposed, and a new experimental procedure is introduced using a modified Brazilian disc specimen for the determination of the corresponding failure envelope for the bimaterial corner in a double-lap joint, and tested. Finally, examples of application of the developed procedure to the elimination of stress singularities in some bimaterial joints are presented.

I am greatly indebted to all the contributors to this book for their efforts in producing chapters of high quality and originality, and for their patience in waiting for the book to be published. I sincerely apologize to the contributors for the long delay in the publication of the book, especially on my part. I thank all the reviewers who recommended numerous improvements to the original manuscripts.

I also thank Professor Tomáš Roubíček (Charles University in Prague) for his practical advice on the book layout, and for his constant motivation to push this second edition forward.

This book is dedicated to Professor Federico París (University of Seville) with whom I have worked for the last 30 years. I am very grateful to him not only for his professional advice and guidance and for his important contribution to my professional career, but also for his personal support to me and my family, especially in the difficult times when we needed it most.

I would like to express my special gratitude to Prof. Ferri Aliabadi (Imperial College London) and the editorial team at World Scientific for inviting me to edit this second extended edition of the book, and for their very professional advice and support throughout the preparation of the book.

Vladislav Mantič
Seville
December, 2022

https://doi.org/10.1142/9781800611887_fmatter

About the Editor

Vladislav Mantič is Professor of Continuum Mechanics and Structural Analysis at the University of Seville, where he has been working since 1994. Previously he worked at the Technical University of Košice and at the Institute of Materials and Machine Mechanics of the Slovak Academy of Sciences in Košice. He graduated in Mathematical Engineering from the Faculty of Nuclear Sciences and Physical Engineering of the Czech Technical University in Prague with the Rector's Award in 1984 and received his doctorate from the Institute of Computer Science of the Czech Academy of Sciences in Prague in 1993. For six years he served as the director of the Doctoral Program in Mechanical Engineering and Industrial Management at the University of Seville.

Currently he coordinates the European research project NewFrac focused on the Phase Field models of fracture and the Coupled Criterion of Finite Fracture Mechanics. He is working in the singularity analysis of corners and cracks with spring boundary or interface conditions, in the prediction of crack onset and propagation in composite materials, and on the development of the Coupled Criterion, especially in its computational implementation. Some of his achievements in the field of boundary integral equations and fracture mechanics are mentioned here. In 1993 he deduced a general closed-form expression for the coefficient-tensor of the free term in the 3D Somigliana identity for linear elastic isotropic solids. In 2004 he derived a relationship between the fracture-mode-mixity angles based on the Stress Intensity Factors and the Energy Release Rates in interface cracks between dissimilar linear elastic isotropic adherents. He introduced two alternative formulations of the Coupled Criterion, the first based on the minimization of the envelope of the stress criterion and energy criterion functions in 2009, and the second based on the minimization of the total energy functional subjected to a stress condition in a load stepping scheme in 2014. He also proposed a general and semi-analytic matrix formalism for singularity analysis of anisotropic linear elastic multi-material corners considering many types of boundary and interface conditions and covering also frictional sliding contact in 2014.

https://doi.org/10.1142/9781800611887_fmatter

Contents

https://doi.org/10.1142/9781800611887_0001

Chapter 1

Micromechanical Modeling of Advanced Composites and Smart Composite Structures Using the Asymptotic Homogenization Method

Alexander L. Kalamkarov
Department of Mechanical Engineering, Dalhousie University, Halifax, Nova Scotia, Canada
alex.kalamkarov@dal.ca

Abstract

The basics of the multi-scale asymptotic homogenization method and its applications to the analysis of the advanced composite materials, thin-walled composite structures and smart composite materials and structures are presented. Asymptotic homogenization is a powerful mathematically rigorous technique for analyzing composite materials and structures. The proof of the possibility of homogenizing a composite material of a regular structure, i.e., of examining an equivalent homogeneous material instead of the original inhomogeneous composite material, is one of the principal results of this theory. Method of asymptotic homogenization has also indicated a procedure of transition from the original problem for the inhomogeneous composite medium to a problem for a homogeneous medium. The effective properties of this equivalent homogeneous medium are determined through the solution of the unit-cell problems formulated within a single unit cell of the material and derived in the process of asymptotic homogenization. The asymptotic homogenization technique is applied to the analysis of three-dimensional (3D) composite materials, thin-walled composite reinforced structures, and smart composite materials and structures. The analytical solution of the corresponding unit cell problems is obtained and the explicit analytical formulae for the effective elastic properties of 3D grid-reinforced composites of various structures are derived. Asymptotic homogenization of 3D thin-walled composite reinforced structures is presented, and the general homogenization composite shell model is introduced. It is applied to the analysis of practically important composite reinforced shells and plates, including rib- and wafer-reinforced shells, orthotropic grid-reinforced composite shells and plates, and sandwich composite shells with cellular cores of different geometrical configuration. In particular, one of considered examples represents micromechanical modeling of the carbon nanotubes. The analytical expressions for the effective stiffness moduli of these composite reinforced shells and plates are presented. For many problems presented in this chapter, the asymptotic homogenization is much more effective and mathematically rigorous method than any other analytical or numerical approach.

Contents

1.1. Introduction

The rapidly increasing popularity of composite materials and structures in recent years has been seen through their incorporation in the wide-ranging engineering applications. In particular, the advanced composites are used to reinforce and monitor components in the civil and structural engineering (see, e.g., [1, 2]), in the aerospace, automotive and marine engineering components of all sizes, medical prosthetic devices, sports and recreational goods and others. Success in practical application of composites largely depends on a possibility to predict their mechanical properties and behavior through the development of the appropriate mechanical models. The micromechanical modeling of composite structures however, can be rather complicated as a result of the distribution and orientation of the multiple inclusions and reinforcements within the matrix, and their mechanical interactions on a local (micro-)level. Therefore, it is important to establish such micromechanical models that are neither too complicated to be developed and applied, nor too simple to reflect the real mechanical properties and behavior of the composite materials and structures.

At present, asymptotic techniques are applied in many cases in micromechanics of composites. Various asymptotic approaches to the analysis of composite materials have apparently reached their conclusion within the framework of the mathematical theory of asymptotic homogenization. Indeed, the proof of the possibility of homogenizing a composite material of a regular structure, i.e., of examining an equivalent homogeneous solid instead of the original inhomogeneous composite solid, is one of the principal results of this theory. Theory of homogenization has also indicated a method of transition from the original problem (which contains in its formulation a small parameter related to the small dimensions of the constituents of the composite) to a problem for a homogeneous solid. The effective properties of this equivalent homogeneous material are determined through the solution of so-called local problems formulated on the unit cell of the composite material. These solutions also enable calculation of local stresses and strains in the composite material.

In the present chapter, we will review the basics of the asymptotic homogenization in Section 1.2; consider the formulation of the unit-cell problems in Section 1.3. In Section 1.4, the asymptotic homogenization technique is applied to the analysis of the three-dimensional (3D) grid-reinforced composites with generally orthotropic reinforcement materials. The analytical solution of the corresponding unit-cell problems is obtained and the explicit analytical formulas for the effective elastic properties of 3D grid-reinforced composites of various structures are derived. The general homogenization composite shell model is presented in Section 1.5. It is applied to the analysis of thin-walled composite structures, including rib- and wafer-reinforced shells, and sandwich composite shells with honeycomb fillers. The analytical expressions for the effective stiffness moduli of these composite reinforced shells are obtained. Section 1.6 is devoted to the application of the general homogenization composite shell model to the analysis of generally orthotropic grid-reinforced composite shells. The analytical solutions of the pertinent unit-cell problems are obtained, and used in Section 1.7 to derive formulas for the effective stiffness moduli of practically important types of grid-reinforced composite shells with orthotropic reinforcements. In particular, one of examples considered in Section 1.7 represents analytical modeling of the mechanical behavior of the carbon nanotubes. Section 1.8 presents application of the general homogenization composite shell model to the analysis of the sandwich composite shells with cellular cores of a different geometrical configuration. Finally, the introduction into the smart materials is presented in Section 1.9. The asymptotic homogenization is applied in this section to the smart composite materials and smart composite reinforced shells and plates. The homogenization composite shell model presented in Section 1.5 is generalized in Section 1.9 to the case of a smart composite materials.

The present chapter is largely based on research results obtained by the author and his graduate students.

1.2. Asymptotic Homogenization Method

For the past 25 years, asymptotic homogenization methods have proven to be powerful techniques for the study of heterogeneous media. Some of these classical tools today include multiscale expansions [3–8], G- and Γ-convergence [9, 10] and energy methods [11, 12].

An approach based on Fourier analysis has been proposed in [13, 14]. This method works in the following way. First, the original operator is transformed into an equivalent operator in the Fourier space. Standard Fourier series are used to expand the coefficients of the operator and a Fourier transform is used to decompose the integrals. Next, the Fourier transforms of the integrals are expanded using a suitable two-scale expansion and the homogenized problem is finally derived by merely neglecting high-order terms in the above expansions when moving to the limit as the period tends to zero.

The method of orientational averaging was proposed in [15]. It is based on the following assumptions: a characteristic volume (repeated throughout the bulk of the composite) is isolated from the composite medium. The properties of the composite as a whole are assumed to be the same as those of this characteristic volume. In the case of ideally straight fibers, the set of fibers is represented in the form of an array of unidirectional reinforced cylinders. The papers on homogenization using wavelet approximations [16] and non-smooth transformations [17] should also be mentioned.

In this section, we describe a variant of the asymptotic homogenization approach that will be used later. For simplicity, we will start with a two-dimensional (2D) heat conduction problem. However, these results will remain valid for other kinds of transport coefficients such as electrical conductivity, diffusion, magnetic permeability, etc. Due to the well-known

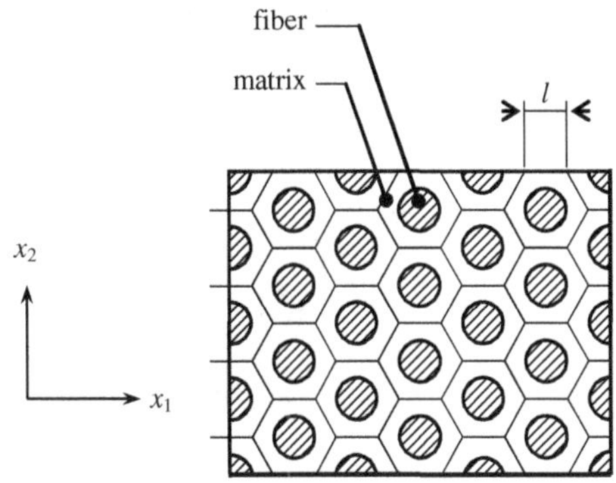

Fig. 1.1. Composite material with hexagonal array of cylindrical fibers.

longitudinal shear–transverse conduction analogy, see [18], the elastic antiplane shear deformation can also be evaluated in a similar mathematical way. This will be followed by a summary of asymptotic homogenization applied to an elasticity problem for a 3D composite solid. An analogous asymptotic homogenization technique has been developed for a number of more complicated nonlinear models, see [3, 5, 11].

Let us consider a transverse transport process through a periodic composite structure, when the fibers are arranged in a periodic square lattice, as shown in Fig. 1.1.

The characteristic size l of the inhomogeneities is assumed to be much smaller than the global size L of the whole structure: $l \ll L$. Assuming perfect bonding conditions on the interface $\partial\Omega$ between the constituents, the governing boundary-value problem can be written as follows:

$$k^a\left(\frac{\partial^2 u^a}{\partial x_1^2}+\frac{\partial^2 u^a}{\partial x_2^2}\right)=-f^a \quad \text{in } \Omega_a,\ \ u^m=u^f,$$

$$k^m\frac{\partial u^m}{\partial \mathrm{n}}=k^f\frac{\partial u^f}{\partial \mathrm{n}} \quad \text{on } \partial\Omega. \tag{1.1}$$

Here and in the following, variables indexed by m correspond to the matrix, and those indexed by f correspond to the fibers, index a takes both of these references: $a = m$ or $a = f$. Generally, the boundary-value problem (1.1) has a number of different physical interpretations, but here it is discussed with reference to heat conduction. Then, in the above expressions, k^a are the heat conductivities of the constituents, u^a is a temperature distribution, f^a is a density of heat sources and $\partial/\partial\mathrm{n}$ is a derivative in the normal direction to the interface $\partial\Omega$. Let us now consider the governing boundary-value problem (1.1) using the asymptotic homogenization method [3–8]. We will introduce a dimensionless small parameter $\varepsilon = l/L$, $\varepsilon \ll 1$, characterizing the rate of heterogeneity of the composite structure.

In order to separate the micro- and macroscale components of the solution, we introduce the so-called slow ($\mathbf{x}$) and fast ($\mathbf{y}$) coordinates

$$y_1 = \frac{x_1}{\varepsilon}, \quad y_2 = \frac{x_2}{\varepsilon} \tag{1.2}$$

and we express the temperature field in the form of an asymptotic expansion:

$$u^a = u_0(\mathbf{x}) + \varepsilon u_1^a(\mathbf{x}, \mathbf{y}) + \varepsilon^2 u_2^a(\mathbf{x}, \mathbf{y}) + \cdots, \tag{1.3}$$

where $\mathbf{x} = x_1\mathbf{e}_1 + x_2\mathbf{e}_2$, $\mathbf{y} = y_1\mathbf{e}_1 + y_2\mathbf{e}_2$, and $\mathbf{e}_1$ and $\mathbf{e}_2$ are the Cartesian unit vectors. The first term $u_0(\mathbf{x})$ of expansion (1.3) represents the homogeneous part of the solution; it changes slowly within the whole domain of the material and does not depend on the fast coordinates. All the further terms $u_i^a(\mathbf{x}, \mathbf{y})$, $i = 1, 2, 3, \ldots$, describe the local variation of the temperature field on the scale of the heterogeneities. In the perfectly regular case, the periodicity of the medium induces the same periodicity for $u_i^a(\mathbf{x}, \mathbf{y})$ with respect to the fast variables:

$$u_k^a(\mathbf{x}, \mathbf{y}) = u_k^a(\mathbf{x}, \mathbf{y} + \mathbf{L}_p), \tag{1.4}$$

where $\mathbf{L}_p = \varepsilon^{-1}\mathbf{l}_p$, $\mathbf{l}_p = p_1\mathbf{l}_1 + p_2\mathbf{l}_2$, $p_s = 0, \pm1, \pm2, \ldots$, and $\mathbf{l}_1$ and $\mathbf{l}_2$ are the fundamental translation vectors of the square lattice.

The spatial derivatives are defined as follows:

$$\frac{\partial}{\partial x_1} \to \frac{\partial}{\partial x_1} + \varepsilon^{-1}\frac{\partial}{\partial y_1}, \quad \frac{\partial}{\partial x_2} \to \frac{\partial}{\partial x_2} + \varepsilon^{-1}\frac{\partial}{\partial y_2}. \tag{1.5}$$

Substituting expressions (1.2), (1.3) and (1.5) into the governing boundary-value problem (1.1) and splitting it with respect to equal powers of ε one comes to a recurrent sequence of problems:

$$\begin{gathered} \frac{\partial^2 u_1^a}{\partial y_1^2} + \frac{\partial^2 u_1^a}{\partial y_2^2} = 0 \quad \text{in } \Omega, \quad [u_1^m = u_1^f]|_{\partial\Omega}, \\ \left[k^m \frac{\partial u_1^m}{\partial \mathrm{m}} - k^f \frac{\partial u_1^f}{\partial \mathrm{m}} = (k^f - k^m)\frac{\partial u_0}{\partial \mathrm{n}} \right]\Bigg|_{\partial\Omega}; \end{gathered} \tag{1.6}$$

$$k^a \left(\frac{\partial^2 u_0}{\partial x_1^2} + \frac{\partial^2 u_0}{\partial x_2^2} + 2\frac{\partial^2 u_1^a}{\partial x_1 \partial y_1} + 2\frac{\partial^2 u_1^a}{\partial x_2 \partial y_2} + \frac{\partial^2 u_2^a}{\partial y_1^2} + \frac{\partial^2 u_2^a}{\partial y_2^2} \right)$$

$$= -f^a \quad \text{in } \Omega, \tag{1.7}$$

$$[u_2^m = u_2^f]|_{\partial\Omega}, \quad \left[k^m \frac{\partial u_2^m}{\partial \mathrm{m}} - k^f \frac{\partial u_2^f}{\partial \mathrm{m}} = k^f \frac{\partial u_1^f}{\partial \mathrm{n}} - k^m \frac{\partial u_1^m}{\partial \mathrm{n}} \right] \Bigg|_{\partial\Omega};$$

and so on.

Here $\partial/\partial\mathrm{m}$ is a derivative in the normal direction to the interface $\partial\Omega$ in the fast coordinates y_1, y_2.

The boundary-value problem (1.6) allows evaluation of the higher-order component $u_i^a(\mathbf{x}, \mathbf{y})$ of the temperature field; owing to the periodicity condition (1.4) it can be considered within only one periodically repeated unit cell. It follows from the boundary-value problem (1.6) that the variables $\mathbf{x}$ and $\mathbf{y}$ can be separated in $u_1(\mathbf{x}, \mathbf{y})$ by assuming

$$u_1(\mathbf{x}, \mathbf{y}) = \frac{\partial u_0(\mathbf{x})}{\partial x_1} U_1(\mathbf{y}) + \frac{\partial u_0(\mathbf{x})}{\partial x_2} U_2(\mathbf{y}), \tag{1.8}$$

where $U_1(\mathbf{y})$ and $U_2(\mathbf{y})$ are local functions for which problem (1.6) yields the following unit-cell problems:

$$\begin{gathered}
\frac{\partial^2 U_1(\mathbf{y})}{\partial y_1^2} + \frac{\partial^2 U_1(\mathbf{y})}{\partial y_2^2} = 0, \quad \frac{\partial^2 U_2(\mathbf{y})}{\partial y_1^2} + \frac{\partial^2 U_2(\mathbf{y})}{\partial y_2^2} = 0 \quad \text{in } \Omega, \\
U_1^m(\mathbf{y}) = U_1^f(\mathbf{y}), \quad U_2^m(\mathbf{y}) = U_2^f(\mathbf{y}) \quad \text{on } \partial\Omega, \\
k^m \frac{\partial U_1^m(\mathbf{y})}{\partial \mathrm{m}} - k^f \frac{\partial U_1^f(\mathbf{y})}{\partial \mathrm{m}} = (k^f - k^m) n_1, \\
k^m \frac{\partial U_2^m(\mathbf{y})}{\partial \mathrm{m}} - k^f \frac{\partial U_2^f(\mathbf{y})}{\partial \mathrm{m}} = (k^f - k^m) n_2 \quad \text{on } \partial\Omega.
\end{gathered} \tag{1.9}$$

The effective heat conductivity can be determined from the boundary-value problem (1.7). The following homogenization operator over the unit-cell area Ω_0 will be applied to Eq. (1.7):

$$\left[\iint_{\Omega_0^m} (\cdots) dy_1 dy_2 + \iint_{\Omega_0^{in}} (\cdots) dy_1 dy_2 \right] L^{-2},$$

where Ω_0^m and Ω_0^{in} denote unit-cell areas occupied by the matrix and inclusion, respectively.

Terms containing u_2^a will be eliminated by means of Green's theorem and taking into account the boundary conditions (1.7) and the periodicity condition (1.7), which yields:

$$
\begin{aligned}
&[(1-c)k^m + ck^f]\left(\frac{\partial^2 u_0}{\partial x_1^2} + \frac{\partial^2 u_0}{\partial x_2^2}\right) \\
&\quad + \frac{k^m}{L^2}\iint_{\Omega_0^m}\left(\frac{\partial^2 u_1^m}{\partial x_1 \partial y_1} + \frac{\partial^2 u_1^m}{\partial x_2 \partial y_2}\right) dy_1 dy_2 \\
&\quad + \frac{k^f}{L^2}\iint_{\Omega_0^{in}}\left(\frac{\partial^2 u_1^f}{\partial x_1 \partial y_1} + \frac{\partial^2 u_1^f}{\partial x_2 \partial y_2}\right) dy_1 dy_2 \\
&= -[(1-c)f^m + cf^f], \qquad (1.10)
\end{aligned}
$$

where c is the fiber volume fraction.

The homogenized heat conduction equation can be obtained by substituting expression (1.8) for $u_1(\mathbf{x}, \mathbf{y})$ into Eq. (1.10), which yields

$$
\langle k_{ij} \rangle \frac{\partial u_0^2(\mathbf{x})}{\partial x_i \partial x_j} = -\langle f \rangle, \qquad (1.11)
$$

$$
\begin{aligned}
\langle k_{ij} \rangle &= [(1-c)k^m + ck^f]\delta_{ij} + \frac{k^m}{L^2}\iint_{\Omega_0^m} \frac{\partial U_j^m}{\partial y_i} dy_1 dy_2 \\
&\quad + \frac{k^f}{L^2}\iint_{\Omega_0^{in}} \frac{\partial U_j^f}{\partial y_i} dy_1 dy_2, \qquad (1.12)
\end{aligned}
$$

where $\langle f \rangle = (1-c)f^m + cf^f$ is the effective density of heat sources; δ_{ij} is Kronecker's delta; indexes $i, j, l = 1, 2$; and the summation over the repeated indexes is implied.

Note that in general the homogenized material will be anisotropic, and $\langle k_{ij} \rangle$ in Eq. (1.11) is a tensor of effective coefficients of heat conductivity. Tensor $\langle k_{ij} \rangle$ is defined by the expression (1.12), and it can be readily calculated as soon as the unit-cell problems (1.9) are solved and the local functions $U_1(\mathbf{y})$ and $U_2(\mathbf{y})$ are found. The unit-cell problems (1.9) can be solved analytically or numerically. The approximate methods of their analytical solution will be presented below in a number of practically important cases.

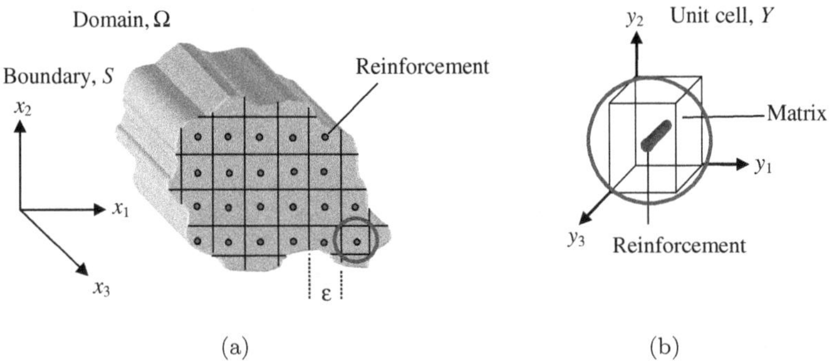

Fig. 1.2. (a) Three-dimensional periodic composite structure, (b) unit cell, Y.

Let us now consider the asymptotic homogenization of an elasticity problem for a 3D periodic composite material occupying region Ω with a boundary S; see Fig. 1.2.

We assume that the region Ω is made up by the periodic repetition of the unit cell Y in the form of a parallelepiped with dimensions εY_i, $i = 1, 2, 3$. The elastic deformation of this composite solid is described by the following boundary-value problem:

$$\frac{\partial \sigma_{ij}^{\varepsilon}}{\partial x_j} = f_i \quad \text{in } \Omega, \quad u_i^{\varepsilon}(\mathbf{x}) = 0 \quad \text{on } S, \tag{1.13}$$

$$\sigma_{ij}^{\varepsilon} = C_{ijkl} e_{kl}^{\varepsilon}, \quad e_{ij}^{\varepsilon} = \frac{1}{2}\left(\frac{\partial u_i^{\varepsilon}}{\partial x_j} + \frac{\partial u_j^{\varepsilon}}{\partial x_i}\right) \quad \text{in } \Omega, \tag{1.14}$$

where C_{ijkl}, is a tensor of elastic coefficients. The coefficients C_{ijkl} are assumed to be periodic functions with a unit cell Y. Here and in the following all Latin indexes assume values $1, 2, 3$, and repeated indexes are summed.

The introduction of the fast variables $y_i = \frac{x_i}{\varepsilon}$, $i = 1, 2, 3$, similar to Eq. (1.2), into Eqs. (1.13) and (1.14) and the rule of differentiation (1.5) leads to the following boundary-value problem:

$$\frac{\partial \sigma_{ij}^{\varepsilon}}{\partial x_j} + \frac{1}{\varepsilon}\frac{\partial \sigma_{ij}^{\varepsilon}}{\partial y_j} = f_i \quad \text{in } \Omega, \quad u_i^{\varepsilon}(\mathbf{x}, \mathbf{y}) = 0 \quad \text{on } S, \tag{1.15}$$

$$\sigma_{ij}^{\varepsilon}(\mathbf{x}, \mathbf{y}) = C_{ijkl}(\mathbf{y}) \frac{\partial u_k^{\varepsilon}}{\partial x_l}(\mathbf{x}, \mathbf{y}) \quad \text{in } \Omega. \tag{1.16}$$

The next step is to expand the displacements and as a result the stresses into the asymptotic expansions in powers of the small parameter ε, as for expansion (1.3):

$$u_i^{\varepsilon}(\mathbf{x},\mathbf{y}) = u_i^{(0)}(\mathbf{x},\mathbf{y}) + \varepsilon u_i^{(1)}(\mathbf{x},\mathbf{y}) + \varepsilon^2 u_i^{(2)}(\mathbf{x},\mathbf{y}) + \cdots\,, \tag{1.17}$$

$$\sigma_{ij}^{\varepsilon}(\mathbf{x},\mathbf{y}) = \sigma_{ij}^{(0)}(\mathbf{x},\mathbf{y}) + \varepsilon \sigma_{ij}^{(1)}(\mathbf{x},\mathbf{y}) + \varepsilon^2 \sigma_{ij}^{(2)}(\mathbf{x},\mathbf{y}) + \cdots\,, \tag{1.18}$$

where all the above functions are periodic in $\mathbf{y}$ with the unit cell Y. Substituting Eqs. (1.17) and (1.18) into Eqs. (1.15) and (1.16), while considering at the same time the periodicity of $\mathbf{u}^{(i)}$ in $\mathbf{y}$, reveals that $\mathbf{u}^{(0)}$ is independent of the fast variable $\mathbf{y}$; see [5] for details. Subsequently, equating terms with similar powers of ε results in the following set of equations:

$$\frac{\partial \sigma_{ij}^{(0)}(\mathbf{x},\mathbf{y})}{\partial y_j} = 0, \tag{1.19}$$

$$\frac{\partial \sigma_{ij}^{(1)}(\mathbf{x},\mathbf{y})}{\partial y_j} + \frac{\partial \sigma_{ij}^{(0)}(\mathbf{x},\mathbf{y})}{\partial x_j} = f_{\mathrm{i}}, \tag{1.20}$$

where

$$\sigma_{ij}^{(0)} = C_{ijkl}\left(\frac{\partial u_k^{(0)}}{\partial x_l} + \frac{\partial u_k^{(1)}}{\partial y_l}\right), \tag{1.21}$$

$$\sigma_{ij}^{(1)} = C_{ijkl}\left(\frac{\partial u_k^{(1)}}{\partial x_l} + \frac{\partial u_k^{(2)}}{\partial y_l}\right). \tag{1.22}$$

Substitution of Eq. (1.21) into Eq. (1.19) yields:

$$\frac{\partial}{\partial y_j}\left(C_{ijkl}\frac{\partial u_k^{(1)}(\mathbf{x},\mathbf{y})}{\partial y_l}\right) = -\frac{\partial C_{ijkl}(\mathbf{y})}{\partial y_j}\frac{\partial u_k^{(0)}(\mathbf{x})}{\partial x_l}. \tag{1.23}$$

Due to the separation of variables in the right-hand side of Eq. (1.23) the solution of Eq. (1.23) can be written as follows, as with Eq. (1.8):

$$u_n^{(1)}(\mathbf{x},\mathbf{y}) = \frac{\partial u_k^{(0)}(\mathbf{x})}{\partial x_l} N_n^{kl}(\mathbf{y}), \tag{1.24}$$

where $N_n^{kl}(\mathbf{y})$ $(n,k,l = 1,2,3)$ are periodic functions with a unit cell Y satisfying the following equation:

$$\frac{\partial}{\partial y_j}\left(C_{ijmn}(\mathbf{y})\frac{\partial N_m^{kl}(\mathbf{y})}{\partial y_n}\right) = -\frac{\partial C_{ijkl}}{\partial y_j}. \tag{1.25}$$

It is observed that Eq. (1.25) depends only on the fast variable $\mathbf{y}$ and it is entirely formulated within the unit cell Y. Thus, the problem (1.25) is appropriately called an elastic unit-cell problem. Note that instead of boundary conditions, this problem has a condition of a periodic continuation of functions $N_m^{kl}(\mathbf{y})$.

If inclusions are perfectly bonded to the matrix on the interfaces of the composite material, then the functions $N_m^{kl}(\mathbf{y})$ together with the expressions $[(C_{ijkl} + C_{ijmn}(\mathbf{y})\frac{\partial N_m^{kl}(\mathbf{y})}{\partial y_n})n_j^{(c)}]$, $i=1,2,3$, must be continuous on the interfaces. Here, $n_j^{(c)}$ are the components of the unit normal to the interface.

The next important step in the homogenization process is achieved by substituting Eq. (1.24) into Eqs. (1.21) and (1.22), and the resulting expression into Eq. (1.20). The result is then integrated over the domain Y of the unit cell (with volume $|Y|$), remembering to treat $\mathbf{x}$ as a parameter as far as integration with respect to $\mathbf{y}$ is concerned. After canceling out terms that vanish due to the periodicity, we obtain the homogenized global problem

$$\tilde{C}_{ijkl}\frac{\partial^2 u_k^{(0)}(\mathbf{x})}{\partial x_j \partial x_l} = f_i \quad \text{in } \Omega, \quad u_i^{(0)}(\mathbf{x}) = 0 \quad \text{on } S, \tag{1.26}$$

where the following notation is introduced:

$$\tilde{C}_{ijkl} = \frac{1}{|Y|}\int_Y \left(C_{ijkl}(\mathbf{y}) + C_{ijmn}(\mathbf{y})\frac{\partial N_m^{kl}}{\partial y_n}\right)dv. \tag{1.27}$$

Similarly, substitution of Eq. (1.24) into Eq. (1.21) and then integrating the resulting expression over the domain of the unit cell Y yields:

$$\langle \sigma_{ij}^{(0)} \rangle = \frac{1}{|Y|}\int_Y \sigma_{ij}^{(0)}(\mathbf{y})dv = \tilde{C}_{ijkl}\frac{\partial u_k^{(0)}}{\partial x_l}. \tag{1.28}$$

Equations (1.26) and (1.28) represent the homogenized elasticity boundary-value problem. The coefficients $\tilde{C}_{ijkl}$ given by Eq. (1.27) are the effective elastic coefficients of the homogenized material. They are readily determined as soon as the unit-cell problem (1.25) is solved and the functions $N_m^{kl}(\mathbf{y})$ are found. It is observed that these effective coefficients are free from the complications that characterize the original rapidly varying elastic coefficients $C_{ijkl}(\mathbf{y})$. They are universal for a composite material under study, and can be used to solve a wide variety of boundary-value problems associated with the given composite material.

It should be noted that while asymptotic homogenization leads to a much simpler problem for an equivalent homogeneous material with certain effective properties, the construction of a solution in the vicinity of the boundary S of the composite solid, i.e., at the distances of the order of ε, remains beyond the capabilities of classical homogenization. In order to determine the stresses and strains near the boundary, a boundary-layer problem should be considered as an extension to the asymptotic homogenization. A boundary-layer method in asymptotic homogenization was developed by Kalamkarov [5]. This approach was further developed by Kalamkarov and Georgiades [19] in the asymptotic homogenization of smart periodic composites. The exponential decaying of boundary layers was proved in [10] for problems with a simple geometry.

New generalized integral transforms for the analytical solution of the boundary-value problems for composite materials have been developed by Kalamkarov [5, 20], and Kalamkarov *et al.* [21].

The properties of boundary layers in periodic homogenization in rectangular domains that are either fixed or have an oscillating boundary are investigated in [22]. Such boundary layers are highly oscillating near the boundary and decay exponentially fast in the interior to a non-zero limit, which the authors called a boundary-layer tail. It is shown that these boundary-layer tails can be incorporated into the homogenized equation by adding dispersive terms and a Fourier boundary condition. Although finding the explicit analytical solutions of boundary-layer problems in the theory of homogenization still remains an open problem, the effective numerical procedures have been proposed in [23, 24].

1.3. Unit-Cell Problems

As we have seen in Section 1.2, the derivation of the homogenized equations for the periodic composites includes solution of the unit-cell problems (1.9) or (1.25). In some particular cases, these problems can be solved analytically producing exact solutions, for example for laminated composites and grid-reinforced structures; see [5, 25–27]. The explicit formulas for the effective moduli are very useful, especially for the design and optimization of composite materials and structures [27, 28]. But in the general case, the unit-cell problems cannot be solved analytically and therefore numerical methods should be used. In some cases, approximate analytical solutions of the unit-cell problems can be found, and explicit formulas for the effective coefficients can be obtained due to the presence of additional small parameters within the unit cell, not to be confused with the small parameter of inhomogeneity.

For a small volume fraction of inclusions, $c \ll c_{\max}$, one can use the three-phase model [29–31]. It is based on the following assumption: the periodically heterogeneous composite structure is approximately replaced by a three-phase medium consisting of a single inclusion, a matrix layer and an infinite effective medium with homogenized mechanical properties. An asymptotic justification of the three-phase composite model is given in [29].

For laminated composite materials, the unit-cell problems (1.9) and (1.25) are one-dimensional and they can be solved analytically. Using this analytical solution, the effective properties of laminated composites can be obtained in an explicit analytical form from Eqs. (1.12) and (1.27); see [5, 27]. In the more complicated case of generally anisotropic constituent materials, explicit formulas for the effective elastic, actuation, thermal conductivity and hygroscopic absorption properties of laminated smart composites have been derived by Kalamkarov and Georgiades [32]. In particular, the following explicit formula for the effective elastic coefficients of a laminated composite in the case of generally anisotropic constituent materials is derived in [32]:

$$\tilde{C}_{ijkl} = \langle C_{ijkl}\rangle - \langle C_{ijm3}C^{-1}_{m3q3}C_{q3kl}\rangle + \langle C_{ijm3}C^{-1}_{m3q3}\rangle\langle C^{-1}_{q3p3}\rangle^{-1}\langle C^{-1}_{p3n3}C_{n3kl}\rangle, \tag{1.29}$$

where the angle brackets denote a rule of mixture, and as earlier indicated all Latin indexes assume values $1, 2, 3$, and repeated indexes are summed.

For fiber-reinforced periodic composites the unit-cell problem (1.25) becomes 2D, and it can be solved analytically for some simple geometries, or numerically; see [5, 33, 34].

1.4. Three-Dimensional Grid-Reinforced Composites

In this section, we will apply the above described asymptotic homogenization technique to the analysis of a 3D composite structure reinforced with N families of reinforcements. An example of such a structure with three families of mutually perpendicular reinforcements is shown in Fig. 1.3 [35]. We assume that the members of each family are made of individual, generally orthotropic materials and have relative orientation angles $\theta_1^n, \theta_2^n, \theta_3^n$ (where $n = 1, 2, \ldots, N$) with the y_1, y_2, y_3 axes respectively.

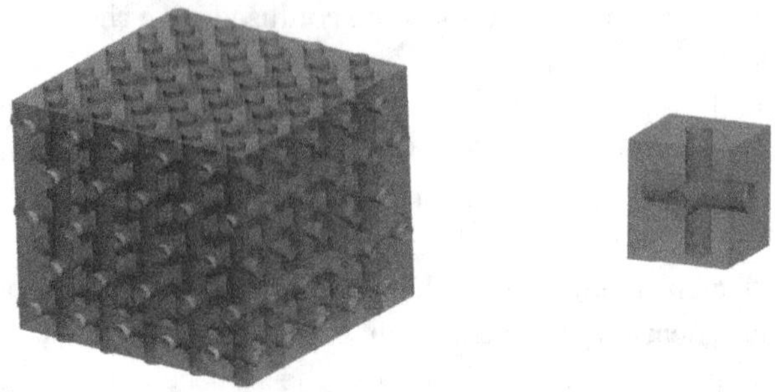

(a) (b)

Fig. 1.3. (a) Cubic grid-reinforced structure and (b) its unit cell.

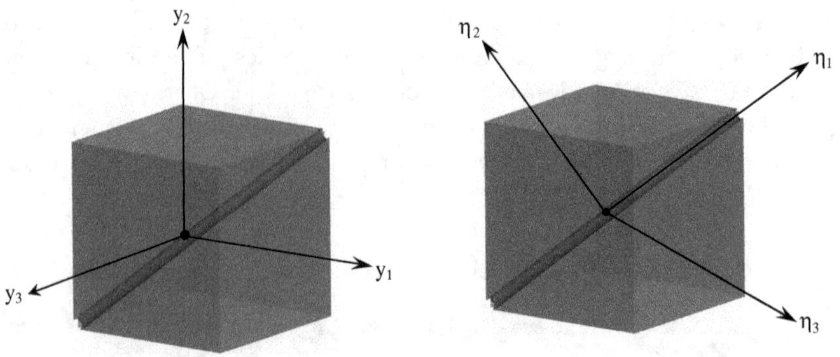

(a) (b)

Fig. 1.4. Unit cell for a single reinforcement family in original (a) and rotated (b) microscopic coordinates.

It is further assumed that the orthotropic reinforcements have significantly higher elasticity moduli than the matrix material, so we are justified in neglecting the contribution of the matrix phase in the analytical treatment. Clearly, for the particular case of framework or lattice network structures the surrounding matrix is absent and this is modeled by assuming zero matrix rigidity.

The nature of the network structure of Fig. 1.3 is such that it would be more efficient if we first considered a simpler type of unit cell made of only a single reinforcement as shown in Fig. 1.4. Having solved this, the effective elastic coefficients of more general structures with several families of reinforcements can be determined by the superposition of the found solution for each of them separately. In following this procedure, one must naturally accept the error incurred at the regions of intersection between the reinforcements. However, our approximation will be quite accurate because these regions of intersection are highly localized and do not contribute significantly to the integral over the entire volume of the unit cell. A mathematical justification for this argument in the form of the so-called principle of the split homogenized operator can be found in [4].

In order to calculate the effective coefficients of the simpler structure of Fig. 1.4, the unit-cell problem given by Eq. (1.25) must be solved and subsequently Eq. (1.27) must be applied. The problem formulation for the structure shown in Fig. 1.4 begins with the introduction of the following notation [5, 35]:

$$b_{ij}^{kl} = C_{ijmn}(\mathbf{y})\frac{\partial N_m^{kl}(y)}{\partial y_n} + C_{ijkl}. \tag{1.30}$$

With this definition in mind, the unit cell of the problem given by Eq. (1.25) can be solved as:

$$b_{mm}^{mm} = C_{mmmm} + \begin{bmatrix} \lambda_1^{mm}\{C_{m1}q_{21} + C_{m6}q_{22} + C_{m5}q_{23}\} \\ +\lambda_2^{mm}\{C_{m1}q_{31} + C_{m6}q_{32} + C_{m5}q_{33}\} \\ +\lambda_3^{mm}\{C_{m6}q_{21} + C_{m2}q_{22} + C_{m4}q_{23}\} \\ +\lambda_4^{mm}\{C_{m6}q_{31} + C_{m2}q_{32} + C_{m4}q_{33}\} \\ +\lambda_5^{mm}\{C_{m5}q_{21} + C_{m4}q_{22} + C_{m3}q_{23}\} \\ +\lambda_6^{mm}\{C_{m5}q_{31} + C_{m4}q_{32} + C_{m3}q_{33}\} \end{bmatrix}, \tag{1.31}$$

$$b_{mn}^{mn} = C_{mnmn} + \begin{bmatrix} \lambda_1^{mm}\{C_{mn11}q_{21} + C_{mn12}q_{22} + C_{mn13}q_{23}\} \\ +\lambda_2^{mn}\{C_{mn11}q_{31} + C_{mn12}q_{32} + C_{mn13}q_{33}\} \\ +\lambda_3^{mm}\{C_{mn12}q_{21} + C_{mn22}q_{22} + C_{mn23}q_{23}\} \\ +\lambda_4^{mn}\{C_{mn12}q_{31} + C_{mn22}q_{32} + C_{mn23}q_{33}\} \\ +\lambda_5^{mm}\{C_{mn13}q_{21} + C_{mn23}q_{22} + C_{mn33}q_{23}\} \\ +\lambda_6^{mm}\{C_{mn13}q_{31} + C_{mn23}q_{32} + C_{mn33}q_{33}\} \end{bmatrix}, \tag{1.32}$$

where there is no summation on either index m or n. The C_{IJ} $(I, J = 1, 2, 3, \ldots, 6)$ in Eq. (1.31) are the elastic coefficients of the orthotropic reinforcements in the contracted notation; see, e.g., [5, 36]. These components are obtained from C_{ijkl} by the following replacement of subscripts: $11 \rightarrow 1$, $22 \rightarrow 2$, $33 \rightarrow 3$, $23 \rightarrow 4$, $13 \rightarrow 5$, $12 \rightarrow 6$. The resulting C_{IJ}

are symmetric, i.e., $C_{IJ} = C_{JI}$. The coefficients q_{ij} in Eqs. (1.31) and (1.32) represent the components of the matrix of direction cosines characterizing the axes of rotation in Fig. 1.4. The constants λ_i^{kl}in Eqs. (1.31) and (1.32) satisfy the following linear algebraic equations:

$$\begin{aligned}
A_1\lambda_1^{kl} + A_2\lambda_2^{kl} + A_3\lambda_3^{kl} + A_4\lambda_4^{kl} + A_5\lambda_5^{kl} + A_6\lambda_6^{kl} + A_7^{kl} &= 0,\\
A_8\lambda_1^{kl} + A_9\lambda_2^{kl} + A_{10}\lambda_3^{kl} + A_{11}\lambda_4^{kl} + A_{12}\lambda_5^{kl} + A_{13}\lambda_6^{kl} + A_{14}^{kl} &= 0,\\
A_{15}\lambda_1^{kl} + A_{16}\lambda_2^{kl} + A_{17}\lambda_3^{kl} + A_{18}\lambda_4^{kl} + A_{19}\lambda_5^{kl} + A_{20}\lambda_6^{kl} + A_{21}^{kl} &= 0,\\
A_{22}\lambda_1^{kl} + A_{23}\lambda_2^{kl} + A_{24}\lambda_3^{kl} + A_{25}\lambda_4^{kl} + A_{26}\lambda_5^{kl} + A_{27}\lambda_6^{kl} + A_{28}^{kl} &= 0,\\
A_{29}\lambda_1^{kl} + A_{30}\lambda_2^{kl} + A_{31}\lambda_3^{kl} + A_{32}\lambda_4^{kl} + A_{33}\lambda_5^{kl} + A_{34}\lambda_6^{kl} + A_{35}^{kl} &= 0,\\
A_{36}\lambda_1^{kl} + A_{37}\lambda_2^{kl} + A_{38}\lambda_3^{kl} + A_{39}\lambda_4^{kl} + A_{40}\lambda_5^{kl} + A_{41}\lambda_6^{kl} + A_{42}^{kl} &= 0,
\end{aligned} \tag{1.33}$$

where A_i^{kl} are constants that depend on the geometric parameters of the unit cell and the material properties of the reinforcement. The explicit expressions for these constants can be found in [35]. Once the system in Eq. (1.33) is solved, the obtained coefficients λ_i^{kl} are substituted back into Eqs. (1.31) and (1.32) to determine the b_{ij}^{kl} coefficients. In turn these are used to calculate the effective elastic coefficients of the 3D grid-reinforced composite structures by integrating over the volume of the unit cell.

The effective elastic moduli of the 3D grid-reinforced composite with generally orthotropic reinforcements with a unit cell shown in Fig. 1.4 are obtained on the basis of Eq. (1.27), which, on account of notation (1.30), becomes:

$$\tilde{C}_{ijkl} = \frac{1}{|Y|}\int_Y b_{ij}^{kl} dv. \tag{1.34}$$

Noting that the b_{ij}^{kl} are constants in the considered case, and denoting the length and cross-sectional area of the reinforcement (in coordinates y_1, y_2, y_3) by L and A respectively, and the volume of the unit cell by V, the effective elastic coefficients become

$$\tilde{C}_{ijkl} = \frac{AL}{V} b_{ij}^{kl} = V_f b_{ij}^{kl}, \tag{1.35}$$

where V_f is the volume fraction of the reinforcement within the unit cell. For structures with more than one family of reinforcements (a particular case of which is shown in Fig. 1.3) the effective moduli can be obtained by superposition. The influence of a fiber coating on the mechanical properties of fiber-reinforced composites was analyzed in [37].

1.4.1. Examples of 3D grid-reinforced composite structures

Let us now apply the above asymptotic homogenization model to calculate the effective elastic coefficients for three different examples of 3D grid-reinforced composite structures. We will also compare the analytical (asymptotic homogenization) results with numerical (finite element) calculations. We will consider consequently a grid-reinforced structure with two families of mutually perpendicular reinforcements (structure A_1 shown in Fig. 1.5); a 3D model with three mutually perpendicular reinforcements oriented along the three coordinate axes (structure A_2 shown in Fig. 1.3); and, finally, a 3D structure with a rhombic arrangement of orthotropic reinforcements: two reinforcements are oriented in the y_1–y_2 plane at 45° to one another with a third reinforcement oriented along the y_3 axis (structure A_3 shown in Fig. 1.6).

The properties of the orthotropic reinforcement and isotropic matrix materials are listed in Table 1.1.

The results are shown below in Figs. 1.7–1.9. Figure 1.7 shows the variation of the effective elastic coefficient $\tilde{E}_1 = \tilde{E}_3$ for the structure A_1 vs. the total reinforcement volume fraction. Three different lines are shown in this figure. The first line represents the asymptotic homogenization (AHM) results, the second line represents the finite element (FEM) results considering the reinforcement contribution only (i.e., neglecting the matrix), and the third line represents the FEM results with both the reinforcement and matrix contributions; see [38] for details. Certain interesting observations are apparent from Fig. 1.7. First of all the high

(a) (b)

Fig. 1.5. (a) Grid-reinforced composite structure A_1, with reinforcements oriented along the y_1 and y_2 directions. (b) Unit cell of structure A_1.

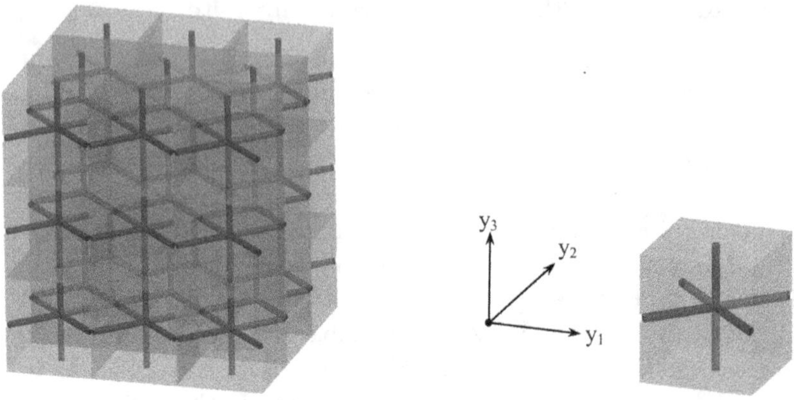

(a) (b)

Fig. 1.6. (a) 3D grid-reinforced composite structure A_3, with reinforcements arranged in a rhombic fashion. (b) Unit cell of structure A_3.

Table 1.1. Material properties.

(a) Material properties of carbon reinforcement

E_1	E_2	E_3	G_{12}	G_{13}	G_{23}	ν_{12}	ν_{13}	ν_{23}
173.0 GPa	33.1 GPa	5.2 GPa	9.4 GPa	8.3 GPa	3.2 GPa	0.036	0.25	0.171

(b) Material properties of epoxy matrix

E	ν
3.6 GPa	0.35

degree of conformity between the first and second lines validates the accuracy of the asymptotic homogenization model in the case when the matrix contribution is neglected. We recall that we have assumed that the reinforcements were much stiffer than the matrix and we consequently neglected the contribution of the latter. The discrepancy between the first and third lines is due to the contribution of the matrix. Figure 1.7 also validates another assumption of the asymptotic homogenization model. In using superposition to determine the effective properties of structures with two or more families of reinforcements, an error will be incurred at the region of overlap between the reinforcements. However, we assume that for the practical purposes this error will not contribute significantly to the integral in Eq. (1.34) and thus will not appreciably affect the effective coefficients. This assumption is confirmed by the excellent agreement between the first and second lines in Fig. 1.7. Of course we expect that in

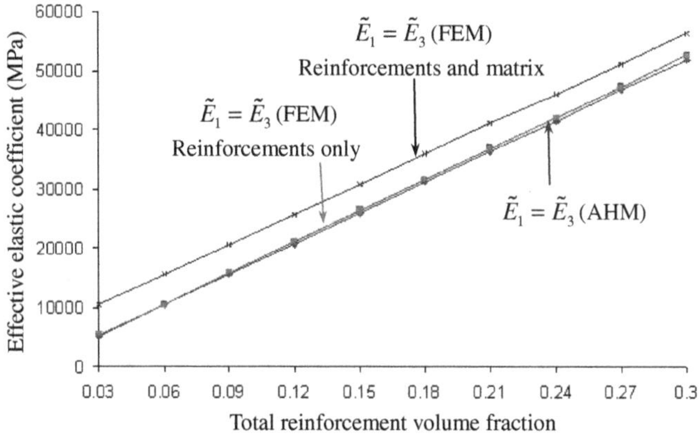

Fig. 1.7. Variation of the effective stiffness moduli $\tilde{E}_1$ (or $\tilde{E}_3$) for structure A_1 (shown in Fig. 1.5).

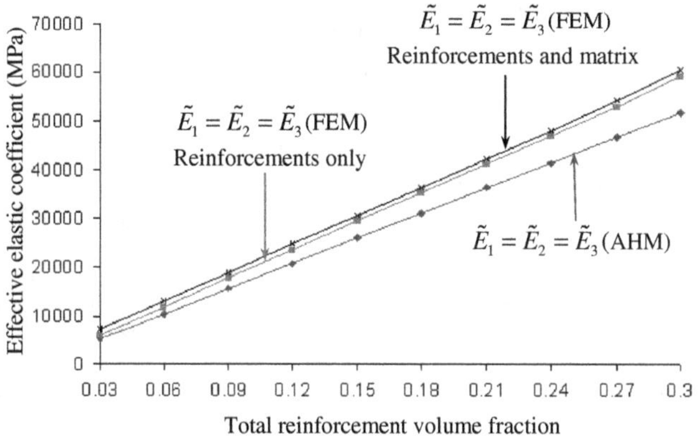

Fig. 1.8. Variation of the effective stiffness moduli $\tilde{E}_1 = \tilde{E}_2 = \tilde{E}_3$ for structure A_2 (shown in Fig. 1.3).

more complex unit-cell structures with a larger extent of overlap between reinforcements this error could be more pronounced. This will be illustrated in subsequent examples.

We now turn our attention to structure A_2 shown in Fig. 1.3 for which the variation of the effective elastic moduli vs. the reinforcement volume fraction is shown in Fig. 1.8. Again, three lines are plotted corresponding to the AHM results, the FEM (contribution of reinforcements only) and FEM (contribution of reinforcements and matrix) results. It should be expected that the discrepancy between the AHM results and the FEM results (considering only the reinforcements) is higher for structure A_2 than

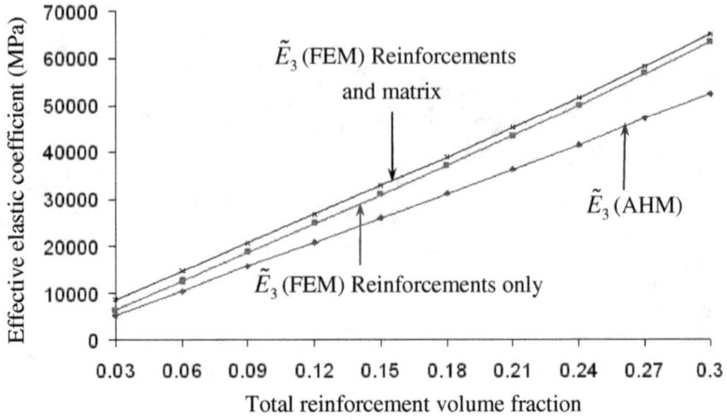

Fig. 1.9. Variation of the effective stiffness modulus $\tilde{E}_3$ for structure A_3 (shown in Fig. 1.6).

for structure A_1. This is attributed to the larger volume of overlap between the various reinforcements in the unit cell of structure A_2.

The final structure to be considered is structure A_3 shown in Fig. 1.6, for which the variation of $\tilde{E}_3$ vs. the reinforcement volume fraction is shown in Fig. 1.9. As with the previous example, the discrepancy between the lower two lines is attributed to the regions of overlap between the different reinforcement families. The difference between the upper two lines is the contribution of the matrix on the effective elastic coefficients.

1.5. Asymptotic Homogenization of Thin-Walled Composite Reinforced Structures

In numerous engineering applications, composite materials are used in the form of thin-walled structural members such as shells and plates. Their stiffness and strength combined with the reduced weight and associated material savings offer very impressive possibilities. It is very common that the reinforcing elements such as embedded fibers or surface ribs form a regular array with a period much smaller than the characteristic dimensions of the whole composite structure. Consequently, the asymptotic homogenization analysis becomes applicable.

An asymptotic homogenized model for plates with periodic inhomogeneities in tangential directions was developed for the first time by Duvaut [39, 40]. In these works the asymptotic homogenization procedure was applied directly to a 2D plate problem. Evidently, the asymptotic homogenization method cannot be applied directly to 3D thin composite layers if their small thickness (in the direction in which there is no periodicity) is comparable with the small dimensions of the periodicity cell (in the two tangential directions). To deal with the 3D problem for a thin composite layer, a modified asymptotic homogenization approach was proposed by Caillerie [41, 42] in heat conduction studies. It consists of applying a two-scale asymptotic formalism directly to the 3D problem for a thin inhomogeneous layer with the following modification. Two sets of 'rapid' coordinates are introduced. Two tangential coordinates are associated with the rapid periodic variation in the composite properties. The third one is in the transverse direction and it is associated with the small thickness of the layer. It takes into account that there is no periodicity in this transverse direction. There are two small parameters: one is a measure of the periodic variation in the two tangential directions and the other is a measure of the small thickness. Generally, these two parameters may or may not be of the same order of magnitude, although in practical applications, they are commonly small values of the same order. Kohn and Vogelius [43–45] adopted this approach in their study of the pure bending of a thin, linearly elastic homogeneous plate with wavy surfaces.

A generalization of this approach to the most comprehensive case of a thin 3D composite layer with wavy surfaces (which model the surface reinforcements) was offered by Kalamkarov [5, 25, 26]; see also [27]. In these works the general asymptotic homogenization model for a composite shell was developed by applying the modified two-scale asymptotic technique directly to 3D elastic and thermoelastic problems for a thin curvilinear composite layer with wavy surfaces; see Fig. 1.10. Homogenization models were also developed for nonlinear problems for composite shells; see [46, 47].

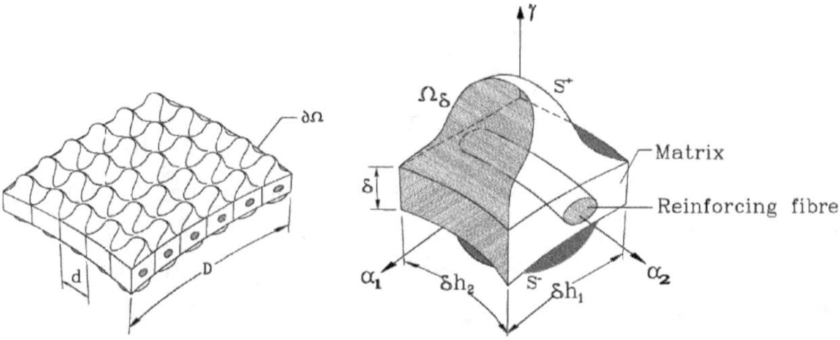

(a) (b)

Fig. 1.10. Thin 3D curvilinear composite layer (a) with a periodicity cell Ω_δ (b).

The homogenization models developed for a composite shell were applied in the design and optimization of composite and reinforced shells [27, 28]. Most recently, this technique was adopted in modeling smart composite shells and plates [19, 48–57]. The general homogenization model for a composite shell has found numerous applications in the analysis of various practically important composite structures. Grid-reinforced and network-thin generally orthotropic composite shells, as well as 3D network-reinforced composite structures are studied in [58–61]. Sandwich composite shells and in particular honeycomb sandwich composite shells made of generally orthotropic materials are analysed in [62–64]. Asymptotic homogenization was also applied by Kalamkarov *et al.* [65–67] to calculate the effective properties of carbon nanotubes and carbon nanotube-reinforced structures.

Let us now summarize the above general homogenization model for a composite shell; see [5, 27] for details. Consider a general thin 3D composite layer of a periodic structure with the unit cell Ω_δ shown in Fig. 1.10. In this figure, α_1, α_2 and γ are the orthogonal curvilinear coordinates, such that the coordinate lines α_1 and α_2 coincide with the main curvature lines of the mid-surface of the carrier layer, and coordinate line γ is normal to its mid-surface (at which $\gamma = 0$).

The thickness of the layer and the dimensions of the unit cell of the composite material (which define the scale of the composite material's inhomogeneity) are assumed to be small compared with the dimensions of the whole structure. These small dimensions of the periodicity cell are characterized by a small parameter δ.

The unit cell Ω_δ shown in Fig. 1.10(b) is defined by the following relations:

$$-\frac{\delta h_1}{2} < \alpha_1 < \frac{\delta h_1}{2}, \quad -\frac{\delta h_2}{2} < \alpha_2 < \frac{\delta h_2}{2}, \quad \gamma^- < \gamma < \gamma^+,$$
$$\gamma^\pm = \pm\frac{\delta}{2} \pm \delta F^\pm\left(\frac{\alpha_1}{\delta h_1}, \frac{\alpha_2}{\delta h_2}\right). \tag{1.36}$$

Here, δ is the thickness of the layer, δh_1 and δh_2 are the tangential dimensions of the periodicity cell Ω_δ. The functions $F^\pm$ in Eq. (1.36) define the geometry of the upper (S^+) and lower (S^-) reinforcing elements, for example, the ribs or stiffeners; see Figs. 1.10 and 1.11. If there are no reinforcing elements then $F^+ = F^- = 0$ and the composite layer has a uniform thickness of the order of δ as for example in the case shown in Fig. 1.12.

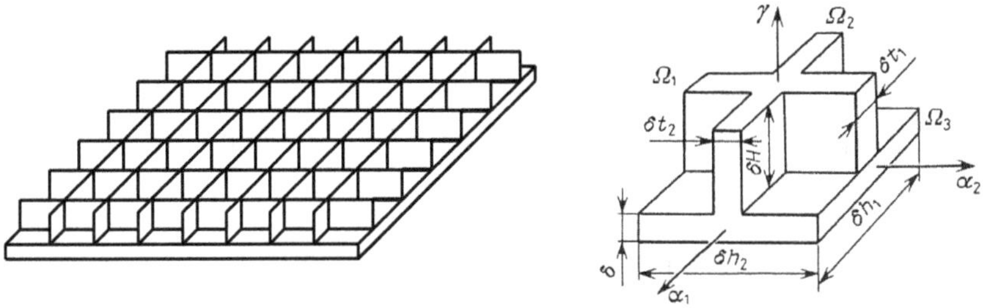

(a) (b)

Fig. 1.11. (a) Wafer-reinforced shell and (b) its unit cell.

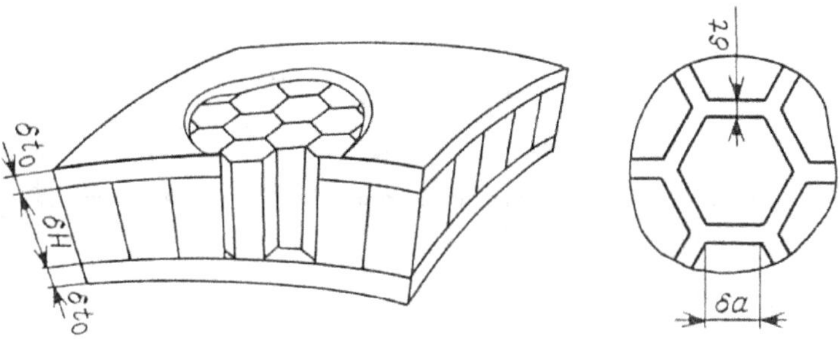

Fig. 1.12. Sandwich composite shell with honeycomb filler.

The periodic inhomogeneity of the composite material is modeled by the assumption that the elastic coefficients $C_{ijkl}(\alpha_1, \alpha_2, \gamma)$ are periodic functions of the variables α_1 and α_2 with a unit cell Ω_δ.

The elasticity problem for the above 3D thin composite layer is formulated as follows:

$$\frac{\partial \sigma_{ij}}{\partial \alpha_j} = f_i,$$

$$\sigma_{ij} = C_{ijkl}(\alpha_1, \alpha_2, \gamma) e_{kl}, \quad e_{kl} = \frac{1}{2}\left(\frac{\partial u_k}{\partial \alpha_l} + \frac{\partial u_l}{\partial \alpha_k}\right), \tag{1.37}$$

$$\sigma_{ij} n_j^{\pm} = p_i^{\pm}.$$

Here f_i, $p_i^{\pm}$ and u_k represent body forces, surface tractions and the displacement field, respectively; $n_j^{\pm}$ is the unit normal to the upper and lower wavy surfaces $\gamma^{\pm}(\alpha_1, \alpha_2)$ and is given by

$$\mathbf{n}^{\pm} = \left\{-\frac{\partial \gamma^{\pm}}{\partial \alpha_1}, -\frac{\partial \gamma^{\pm}}{\partial \alpha_2}, 1\right\}\left(\frac{1}{H_1^2}\left(\frac{\partial \gamma^{\pm}}{\partial \alpha_1}\right)^2 + \frac{1}{H_2^2}\left(\frac{\partial \gamma^{\pm}}{\partial \alpha_2}\right)^2 + 1\right)^{-1/2}, \tag{1.38}$$

where H_1 and H_2 are the Lamé coefficients defined by

$$H_1 = A_1(1 + \kappa_1 \gamma), \quad H_2 = A_2(1 + \kappa_2 \gamma), \tag{1.39}$$

where $A_1(\alpha_1, \alpha_2)$ and $A_2(\alpha_1, \alpha_2)$ are the coefficients of the first quadratic form and κ_1 and κ_2 are the main curvatures of the mid-surface of the carrier layer ($\gamma = 0$).

We introduce the following fast variables, $\xi = (\xi_1, \xi_2)$, and z:

$$\xi_1 = \frac{\alpha_1 A_1}{\delta h_1}, \quad \xi_2 = \frac{\alpha_2 A_2}{\delta h_2}, \quad z = \frac{\gamma}{\delta}. \tag{1.40}$$

The displacements and stresses are expressed in the form of the following two-scale asymptotic expansions:

$$\begin{aligned} u_i(\alpha, \xi, z) &= u_i^{(0)}(\alpha) + \delta u_i^{(1)}(\alpha, \xi, z) + \delta^2 u_i^{(2)}(\alpha, \xi, z) + \cdots, \\ \sigma_{ij}(\alpha, \xi, z) &= \sigma_{ij}^{(0)}(\alpha, \xi, z) + \delta \sigma_{ij}^{(1)}(\alpha, \xi, z) + \delta^2 \sigma_{ij}^{(2)}(\alpha, \xi, z) + \cdots. \end{aligned} \tag{1.41}$$

As a result of the asymptotic homogenization procedure, see [5, 27] for details, the following relations for the displacements and stresses are derived:

$$\begin{cases} u_1 = v_1(\alpha) - \delta \dfrac{z}{A_1} \dfrac{\partial w(\alpha)}{\partial \alpha_1} + \delta U_1^{\mu\nu} e_{\mu\nu} + \delta^2 V_1^{\mu\nu} \tau_{\mu\nu} + \mathrm{O}(\delta^3), \\[2ex] u_2 = v_2(\alpha) - \delta \dfrac{z}{A_2} \dfrac{\partial w(\alpha)}{\partial \alpha_2} + \delta U_2^{\mu\nu} e_{\mu\nu} + \delta^2 V_2^{\mu\nu} \tau_{\mu\nu} + \mathrm{O}(\delta^3), \\[2ex] u_3 = w(\alpha) + \delta U_3^{\mu\nu} e_{\mu\nu} + \delta^2 V_3^{\mu\nu} \tau_{\mu\nu} + \mathrm{O}(\delta^3), \end{cases} \tag{1.42}$$

$$\sigma_{ij} = b_{ij}^{\mu\nu} e_{\mu\nu} + \delta b_{ij}^{*\mu\nu} \tau_{\mu\nu}. \tag{1.43}$$

Here and in the following Latin indexes assume values 1, 2, 3; Greek indexes 1, 2; and repeated indexes are summed; the mid-surface strains are denoted as follows: $e_{11} = e_1, e_{22} = e_2$ (elongations), $e_{12} = e_{21} = \omega/2$ (shear), $\tau_{11} = k_1$, $\tau_{22} = k_2$ (bending) and $\tau_{12} = \tau_{21} = \tau$ (twisting).

The following notation is used in Eq. (1.43):

$$b_{ij}^{lm} = \frac{1}{h_\beta} C_{ijn\beta} \frac{\partial U_n^{lm}}{\partial \xi_\beta} + C_{ijn3} \frac{\partial U_n^{lm}}{\partial z} + C_{ijlm}, \tag{1.44}$$

$$b_{ij}^{*lm} = \frac{1}{h_\beta} C_{ijn\beta} \frac{\partial V_n^{lm}}{\partial \xi_\beta} + C_{ijn3} \frac{\partial V_n^{lm}}{\partial z} + z C_{ijlm}. \tag{1.45}$$

The functions $U_n^{lm}(\xi_1, \xi_2, z)$ and $V_n^{lm}(\xi_1, \xi_2, z)$ in Eqs. (1.42), (1.44) and (1.45) are solutions of the unit-cell problems. Note that all the above functions are periodic in the variables ξ_1 and ξ_2 with periods A_1 and A_2, respectively. The above unit-cell problems are formulated as follows:

$$\begin{cases} \dfrac{1}{h_\beta}\dfrac{\partial b_{i\beta}^{\lambda\mu}}{\partial \xi_\beta} + \dfrac{\partial b_{i3}^{\lambda\mu}}{\partial z} = 0, \\ \dfrac{1}{h_\beta} n_\beta^{\pm} b_{i\beta}^{\lambda\mu} + n_3^{\pm} b_{i3}^{\lambda\mu} = 0 \quad \text{at } z = z^{\pm}, \end{cases} \tag{1.46}$$

$$\begin{cases} \dfrac{1}{h_\beta}\dfrac{\partial b_{i\beta}^{*\lambda\mu}}{\partial \xi_\beta} + \dfrac{\partial b_{i3}^{*\lambda\mu}}{\partial z} = 0, \\ \dfrac{1}{h_\beta} n_\beta^{\pm} b_{i\beta}^{*\lambda\mu} + n_3^{\pm} b_{i3}^{*\lambda\mu} = 0 \quad \text{at } z = z^{\pm}, \end{cases} \tag{1.47}$$

where n_i^+ and n_i^- are components of the normal unit vector to the upper $(z = z^+)$ and lower $(z = z^-)$ surfaces of the unit cell, respectively, defined in the coordinate system ξ_1, ξ_2, z.

If inclusions are perfectly bonded to the matrix on the interfaces of the composite material then the functions U_n^{lm} and V_n^{lm} together with the expressions $[\frac{1}{h_\beta} n_\beta^{(c)} b_{i\beta}^{\lambda\mu} + n_3^{(c)} b_{i3}^{\lambda\mu}]$ and $[\frac{1}{h_\beta} n_\beta^{(c)} b_{i\beta}^{*\lambda\mu} + n_3^{(c)} b_{i3}^{*\lambda\mu}]$ must be continuous on the interfaces. Here $n_i^{(c)}$ are the components of the unit normal to the interface.

It should be noted that unlike the unit-cell problems of 'classical' homogenization models, e.g., Eqs. (1.9) and (1.25), those set by Eqs. (1.46) and (1.47) depend on the boundary conditions at $z = z^{\pm}$ in the z direction rather than on the periodicity.

After the local functions $U_n^{lm}(\xi_1, \xi_2, z)$ and $V_n^{lm}(\xi_1, \xi_2, z)$ are found from the unit-cell problems given by Eqs. (1.44)–(1.47), the functions $b_{ij}^{lm}(\xi_1, \xi_2, z)$ and $b_{ij}^{*lm}(\xi_1, \xi_2, z)$ given by Eqs. (1.44) and (1.45) can be calculated. These local functions define the stress σ_{ij} as in from Eq. (1.43). They also define the effective stiffness moduli of the homogenized shell. Indeed the constitutive relations of the equivalent anisotropic homogeneous shell, that is between the stress resultants N_{11}, N_{22} (normal), N_{12} (shear) and moment resultants M_{11}, M_{22} (bending), M_{12} (twisting) on the one hand, and the mid-surface strains $e_{11} = e_1, e_{22} = e_2$ (elongations), $e_{12} = e_{21} = \omega/2$ (shear), $\tau_{11} = k_1$, $\tau_{22} = k_2$ (bending), $\tau_{12} = \tau_{21} = \tau$ (twisting) on the other, can be represented as follows (see [5, 27] for details):

$$\begin{cases} N_{\alpha\beta} = \delta\langle b_{\alpha\beta}^{\lambda\mu}\rangle e_{\lambda\mu} + \delta^2\langle b_{\alpha\beta}^{*\lambda\mu}\rangle \tau_{\lambda\mu}, \\ M_{\alpha\beta} = \delta^2\langle z b_{\alpha\beta}^{\lambda\mu}\rangle e_{\lambda\mu} + \delta^3\langle z b_{\alpha\beta}^{*\lambda\mu}\rangle \tau_{\lambda\mu}. \end{cases} \tag{1.48}$$

The angle brackets in Eq. (1.48) denote averaging by the integration over the volume of the 3D unit cell:

$$\langle f(\xi_1, \xi_2, z)\rangle = \frac{1}{|\Omega|}\int_\Omega f(\xi_1, \xi_2, z)\, d\xi_1 d\xi_2 dz.$$

The coefficients in the constitutive relations Eq. (1.48) $\langle b_{\alpha\beta}^{\lambda\mu}\rangle$, $\langle b_{\alpha\beta}^{*\lambda\mu}\rangle$, $\langle zb_{\alpha\beta}^{\lambda\mu}\rangle$ and $\langle zb_{\alpha\beta}^{*\lambda\mu}\rangle$ are the effective stiffness moduli of the homogenized shell.

The following symmetry relationships for the effective stiffness moduli have been proved in [5]:

$$\langle b_{ij}^{mn}\rangle = \langle b_{mn}^{ij}\rangle, \quad \langle zb_{ij}^{mn}\rangle = \langle b_{mn}^{*ij}\rangle, \quad \langle zb_{ij}^{*mn}\rangle = \langle zb_{mn}^{*ij}\rangle. \tag{1.49}$$

The mid-surface strains $e_{\lambda\mu}(\alpha_1, \alpha_2)$ and $\tau_{\lambda\mu}(\alpha_1, \alpha_2)$ can be determined by solving a global boundary-value problem for the homogenized anisotropic shell with the constitutive relations (1.48); see [5, 27] for details. It should be noted that, as can be observed from Eq. (1.48), there is a following one-to-one correspondence between the effective stiffness moduli and the extensional [A] coupling [B] and bending [D] stiffnesses familiar from classical composite laminate theory (see, e.g., [36]):

$$\begin{bmatrix} A & B \\ B & D \end{bmatrix} = \begin{bmatrix} \delta\langle b_{11}^{11}\rangle & \delta\langle b_{11}^{22}\rangle & \delta\langle b_{11}^{12}\rangle & \delta^2\langle zb_{11}^{11}\rangle & \delta^2\langle zb_{11}^{22}\rangle & \delta^2\langle zb_{11}^{12}\rangle \\ \delta\langle b_{11}^{22}\rangle & \delta\langle b_{22}^{22}\rangle & \delta\langle b_{22}^{12}\rangle & \delta^2\langle zb_{11}^{22}\rangle & \delta^2\langle zb_{22}^{22}\rangle & \delta^2\langle zb_{22}^{12}\rangle \\ \delta\langle b_{11}^{12}\rangle & \delta\langle b_{22}^{12}\rangle & \delta\langle b_{12}^{12}\rangle & \delta^2\langle zb_{11}^{12}\rangle & \delta^2\langle zb_{22}^{12}\rangle & \delta^2\langle zb_{12}^{12}\rangle \\ \delta^2\langle b_{11}^{*11}\rangle & \delta^2\langle b_{11}^{*22}\rangle & \delta^2\langle b_{11}^{*12}\rangle & \delta^3\langle zb_{11}^{*11}\rangle & \delta^3\langle zb_{11}^{*22}\rangle & \delta^3\langle zb_{11}^{*12}\rangle \\ \delta^2\langle b_{11}^{*22}\rangle & \delta^2\langle b_{22}^{*22}\rangle & \delta^2\langle b_{22}^{*12}\rangle & \delta^3\langle zb_{11}^{*22}\rangle & \delta^3\langle zb_{22}^{*22}\rangle & \delta^3\langle zb_{22}^{*12}\rangle \\ \delta^2\langle b_{11}^{*12}\rangle & \delta^2\langle b_{22}^{*12}\rangle & \delta^2\langle b_{12}^{*12}\rangle & \delta^3\langle zb_{11}^{*12}\rangle & \delta^3\langle zb_{22}^{*12}\rangle & \delta^3\langle zb_{12}^{*12}\rangle \end{bmatrix}. \tag{1.50}$$

It is worth mentioning at this point that the coordinates ξ_1 and ξ_2, defined by Eq. (1.40) in terms of the functions $A_1(\alpha_1, \alpha_2)$ and $A_2(\alpha_1, \alpha_2)$, are involved in the local problems. Functions $A_1(\alpha_1, \alpha_2)$ and $A_2(\alpha_1, \alpha_2)$ are the coefficients of the first quadratic form of the mid-surface of the carrier layer. This means that if the mid-surface is not a developing surface, so that these functions are not constant, the effective stiffness coefficients will also depend on the macroscopic coordinates α_1 and α_2 through these functions. Therefore, even in the case of the originally homogeneous material, we may find structural inhomogeneity after the homogenization process.

The unit-cell problems given by Eqs. (1.44), (1.46) and (1.45), (1.47) have been solved analytically for a number of structures of practical interest, and the explicit analytical formulas for the effective stiffness moduli have been obtained for the following types of composite and reinforced shells and plates: angle-ply fiber-reinforced shells, grid-reinforced and network shells [5, 25, 27, 55, 56, 58, 59]; rib- and wafer-reinforced shells [5, 26, 27, 49, 50, 54]; sandwich composite shells, in particular, honeycomb sandwich composite shells made of generally orthotropic materials [5, 27, 62–64] and carbon nanotubes [65–67].

As examples of these results, we will present here the analytical formulae for the effective stiffness moduli of a wafer-reinforced shell (Fig. 1.11) and a sandwich composite shell with a honeycomb filler (Fig. 1.12).

All the non-zero effective stiffness moduli of the wafer-reinforced shell shown in Fig. 1.11 are obtained as follows, (see [5, 26, 27, 50] for details):

$$
\begin{aligned}
&\langle b_{11}^{11}\rangle = \frac{E_1^{(3)}}{1-\nu_{12}^{(3)}\nu_{21}^{(3)}} + E_1^{(2)}F_2^{(w)}, \quad \langle b_{22}^{22}\rangle = \frac{E_2^{(3)}}{1-\nu_{12}^{(3)}\nu_{21}^{(3)}} + E_2^{(1)}F_1^{(w)},\\
&\langle b_{22}^{11}\rangle = \langle b_{11}^{22}\rangle = \frac{\nu_{12}^{(3)}E_1^{(3)}}{1-\nu_{12}^{(3)}\nu_{21}^{(3)}}, \quad \langle b_{12}^{12}\rangle = G_{12}^{(3)},\\
&\langle zb_{11}^{11}\rangle = \langle b_{11}^{*11}\rangle = E_1^{(2)}S_2^{(w)}, \quad \langle zb_{22}^{22}\rangle = \langle b_{22}^{*22}\rangle = E_2^{(1)}S_1^{(w)},\\
&\langle zb_{11}^{*11}\rangle = \frac{E_1^{(3)}}{12\big(1-\nu_{12}^{(3)}\nu_{21}^{(3)}\big)} + E_1^{(2)}J_2^{(w)}, \qquad\qquad (1.51)\\
&\langle zb_{22}^{*22}\rangle = \frac{E_2^{(3)}}{12\big(1-\nu_{12}^{(3)}\nu_{21}^{(3)}\big)} + E_2^{(1)}J_1^{(w)},\\
&\langle zb_{22}^{*11}\rangle = \langle zb_{11}^{*22}\rangle = \frac{\nu_{12}^{(3)}E_1^{(3)}}{12(1-\nu_{12}^{(3)}\nu_{21}^{(3)})},\\
&\langle zb_{12}^{*12}\rangle = \frac{G_{12}^{(3)}}{12} + \frac{G_{12}^{(1)}}{12}\left(\frac{H^3t_1}{h_1} - K_1\right) + \frac{G_{12}^{(2)}}{12}\left(\frac{H^3t_2}{h_2} - K_2\right),
\end{aligned}
$$

where

$$K_1 = \frac{96H^4}{\pi^5 A_1 h_1}\sqrt{\frac{G_{12}^{(1)}}{G_{23}^{(1)}}}\sum_{n=1}^{\infty}\frac{[1-(-1)^n]}{n^5}\tanh\left(\sqrt{\frac{G_{23}^{(1)}}{G_{12}^{(1)}}}\frac{n\pi A_1 t_1}{2H}\right),$$
$$K_2 = \frac{96H^4}{\pi^5 A_2 h_2}\sqrt{\frac{G_{12}^{(2)}}{G_{13}^{(2)}}}\sum_{n=1}^{\infty}\frac{[1-(-1)^n]}{n^5}\tanh\left(\sqrt{\frac{G_{13}^{(2)}}{G_{12}^{(2)}}}\frac{n\pi A_2 t_2}{2H}\right). \qquad (1.52)$$

Here the superscripts indicate the elements of the unit cell Ω_1, Ω_2 and Ω_3; see Fig. 1.11(b); A_1 and A_2 are the coefficients of the first quadratic form of the mid-surface of a carrier layer; $F_1^{(w)}, F_2^{(w)}$, $S_1^{(w)}, S_2^{(w)}$ and $J_1^{(w)}, J_2^{(w)}$ are defined as follows:

$$F_1^{(w)} = \frac{Ht_1}{h_1}, \quad F_2^{(w)} = \frac{Ht_2}{h_2}, \quad S_1^{(w)} = \frac{(H^2+H)t_1}{2h_1},$$
$$S_2^{(w)} = \frac{(H^2+H)t_2}{2h_2}, \quad J_1^{(w)} = \frac{(4H^3+6H^2+3H)t_1}{12h_1}, \qquad (1.53)$$
$$J_2^{(w)} = \frac{(4H^3+6H^2+3H)t_2}{12h_2}.$$

All the non-zero effective stiffness moduli of the sandwich composite shell with a honeycomb filler shown in Fig. 1.12 are obtained as follows (see [5, 27] for details):

$$\langle b_{11}^{11}\rangle = \langle b_{22}^{22}\rangle = \frac{2E_0t_0}{1-\nu_0^2} + \frac{\sqrt{3}}{4}\frac{EHt}{a},$$
$$\langle b_{12}^{12}\rangle = \frac{E_0t_0}{(1+\nu_0)} + \frac{\sqrt{3}}{12}\frac{EHt}{a},$$
$$\langle b_{22}^{11}\rangle = \langle b_{11}^{22}\rangle = \frac{2\nu_0E_0t_0}{1-\nu_0^2} + \frac{\sqrt{3}}{12}\frac{EHt}{a},$$
$$\langle zb_{11}^{*11}\rangle = \langle zb_{22}^{*22}\rangle = \frac{E_0}{1-\nu_0^2}\left(\frac{H^2t_0}{2} + Ht_0^2 + \frac{2t_0^3}{3}\right) + \frac{\sqrt{3}}{48}\frac{EH^3t}{a}, \qquad (1.54)$$
$$\langle zb_{22}^{*11}\rangle = \langle zb_{11}^{*22}\rangle = \frac{\nu_0E_0}{1-\nu_0^2}\left(\frac{H^2t_0}{2} + Ht_0^2 + \frac{2t_0^3}{3}\right) + \frac{\sqrt{3}}{144}\frac{EH^3t}{a},$$
$$\langle zb_{12}^{*12}\rangle = \frac{E_0}{2(1+\nu_0)}\left(\frac{H^2t_0}{2} + Ht_0^2 + \frac{2t_0^3}{3}\right) + \frac{EH^3t}{12(1+\nu)a}$$
$$\times\left[\frac{3+\nu}{4\sqrt{3}} - \frac{128H}{(\sqrt{3}\pi^5At)}\sum_{n=1}^{\infty}\frac{\tanh(\pi(2n-1)At/(2H))}{(2n-1)^5}\right].$$

The first terms in Eq. (1.54) define the contribution from the top and bottom carrier layers of the sandwich shell while the latter terms represent the contribution from the honeycomb filler. E_0 and ν_0 are the properties of the material of the carrier layers, and E and ν of the honeycomb foil material. We have confined our attention here to the case of equal coefficients of the first quadratic form of the mid-surface of the shell, i.e., $A_1 = A_2 = A$. The details of the derivation of Eq. (1.54) and more complicated cases of composite sandwich shells with generally orthotropic constituent materials will be presented below in Section 1.8.

1.6. Generally Orthotropic Grid-Reinforced Composite Shell

In this section, we will apply the asymptotic homogenization model to a composite shell reinforced with a number of families of parallel reinforcing elements. An example of a shell with two mutually perpendicular reinforcement families is shown in Fig. 1.13. We assume that the reinforcements are made of generally orthotropic materials and that they are much stiffer than the surrounding matrix material. As such, we may neglect the contribution of the matrix in the ensuing analysis.

As in the approach we used above in Section 1.4, we will first consider a simpler type of shell with only one family of reinforcements. The effective elastic coefficients of more general structures with several reinforcement families will be determined by superposition.

Consider the unit cell of Fig. 1.14 shown both before and after the introduction of the microscopic variables ξ_1, ξ_2 and z, defined in Eq. (1.40). Note that the matrix $[C]$ of elastic coefficients of an orthotropic material referenced to a coordinate system that has been rotated by the angle φ of the reinforcing grid orientation (in the ξ_1–ξ_2 plane) with respect to the

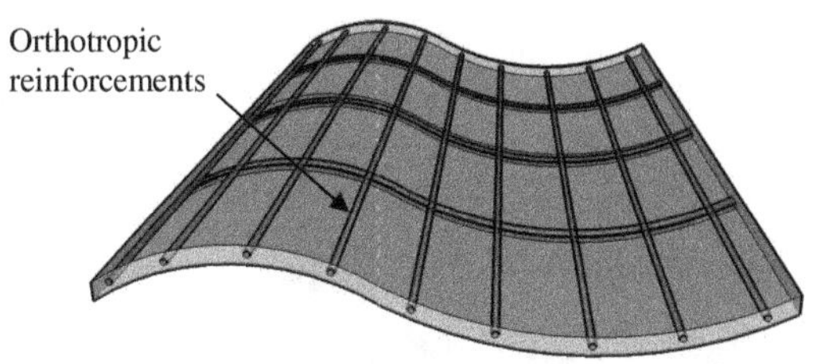

Fig. 1.13. Composite shell with two families of orthotropic reinforcements.

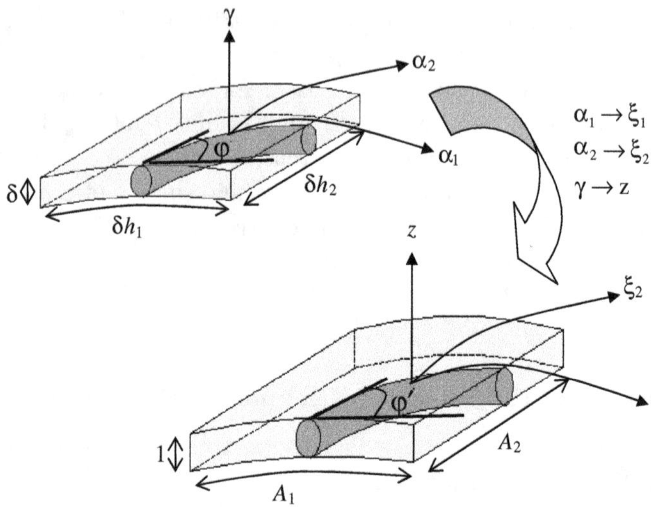

Fig. 1.14. Unit cell in microscopic (ξ_1, ξ_2, z) and macroscopic $(\alpha_1, \alpha_2, \gamma)$ variables.

principal material coordinate system coincides with that of a monoclinic material and has the following form:

$$[C] = \begin{bmatrix} C_{11} & C_{12} & C_{13} & 0 & 0 & C_{16} \\ C_{12} & C_{22} & C_{23} & 0 & 0 & C_{26} \\ C_{13} & C_{23} & C_{33} & 0 & 0 & C_{36} \\ 0 & 0 & 0 & C_{44} & C_{45} & 0 \\ 0 & 0 & 0 & C_{45} & C_{55} & 0 \\ C_{16} & C_{26} & C_{36} & 0 & 0 & C_{66} \end{bmatrix}. \tag{1.55}$$

After this coordinate transformation, the shape of the unit cell changes and the angle between the reinforcement and the ξ_1-axis changes to angle φ' (see Fig. 1.14) through the following relationship $\varphi' = \arctan(\frac{A_2 h_1}{A_1 h_2} \tan\varphi)$.

We begin by first solving for $b_{ij}^{\lambda\mu}$ from Eq. (1.44). These local functions are given as follows for an orthotropic material of reinforcement:

$$\begin{aligned} b_{11}^{\lambda\mu} &= \frac{1}{h_1} C_{11} \frac{\partial U_1^{\lambda\mu}}{\partial \xi_1} + \frac{1}{h_2} C_{12} \frac{\partial U_2^{\lambda\mu}}{\partial \xi_2} + C_{13} \frac{\partial U_3^{\lambda\mu}}{\partial z} \\ &\quad + C_{16} \left[\frac{1}{h_1} \frac{\partial U_2^{\lambda\mu}}{\partial \xi_1} + \frac{1}{h_2} \frac{\partial U_1^{\lambda\mu}}{\partial \xi_2} \right] + C_{11\lambda\mu}, \end{aligned} \tag{1.56}$$

$$b_{22}^{\lambda\mu} = \frac{1}{h_1}C_{12}\frac{\partial U_1^{\lambda\mu}}{\partial \xi_1} + \frac{1}{h_2}C_{22}\frac{\partial U_2^{\lambda\mu}}{\partial \xi_2} + C_{23}\frac{\partial U_3^{\lambda\mu}}{\partial z}$$

$$+ C_{26}\left[\frac{1}{h_1}\frac{\partial U_2^{\lambda\mu}}{\partial \xi_1} + \frac{1}{h_2}\frac{\partial U_1^{\lambda\mu}}{\partial \xi_2}\right] + C_{22\lambda\mu}, \tag{1.57}$$

$$b_{33}^{\lambda\mu} = \frac{1}{h_1}C_{13}\frac{\partial U_1^{\lambda\mu}}{\partial \xi_1} + \frac{1}{h_2}C_{23}\frac{\partial U_2^{\lambda\mu}}{\partial \xi_2} + C_{33}\frac{\partial U_3^{\lambda\mu}}{\partial z}$$

$$+ C_{36}\left[\frac{1}{h_1}\frac{\partial U_2^{\lambda\mu}}{\partial \xi_1} + \frac{1}{h_2}\frac{\partial U_1^{\lambda\mu}}{\partial \xi_2}\right] + C_{33\lambda\mu}, \tag{1.58}$$

$$b_{12}^{\lambda\mu} = \frac{1}{h_1}C_{16}\frac{\partial U_1^{\lambda\mu}}{\partial \xi_1} + \frac{1}{h_2}C_{26}\frac{\partial U_2^{\lambda\mu}}{\partial \xi_2} + C_{36}\frac{\partial U_3^{\lambda\mu}}{\partial z}$$

$$+ C_{66}\left[\frac{1}{h_1}\frac{\partial U_2^{\lambda\mu}}{\partial \xi_1} + \frac{1}{h_2}\frac{\partial U_1^{\lambda\mu}}{\partial \xi_2}\right] + C_{12\lambda\mu}, \tag{1.59}$$

$$b_{13}^{\lambda\mu} = C_{55}\left[\frac{1}{h_1}\frac{\partial U_3^{\lambda\mu}}{\partial \xi_1} + \frac{\partial U_1^{\lambda\mu}}{\partial z}\right] + C_{45}\left[\frac{1}{h_2}\frac{\partial U_3^{\lambda\mu}}{\partial \xi_2} + \frac{\partial U_2^{\lambda\mu}}{\partial z}\right] + C_{13\lambda\mu}, \tag{1.60}$$

$$b_{23}^{\lambda\mu} = C_{45}\left[\frac{1}{h_1}\frac{\partial U_3^{\lambda\mu}}{\partial \xi_1} + \frac{\partial U_1^{\lambda\mu}}{\partial z}\right] + C_{44}\left[\frac{1}{h_2}\frac{\partial U_3^{\lambda\mu}}{\partial \xi_2} + \frac{\partial U_2^{\lambda\mu}}{\partial z}\right] + C_{23\lambda\mu}. \tag{1.61}$$

In order to reduce the complexity of the associated problems, we introduce a new coordinate system $\{\eta_1, \eta_2, z\}$ obtained via rotation through an angle φ' around the z-axis, such that the η_1-coordinate axis coincides with the longitudinal axis of the reinforcing element and the η_2-coordinate axis is perpendicular to it; see Fig. 1.15. With this transformation the problem at hand is now independent of the η_1 coordinate and will only

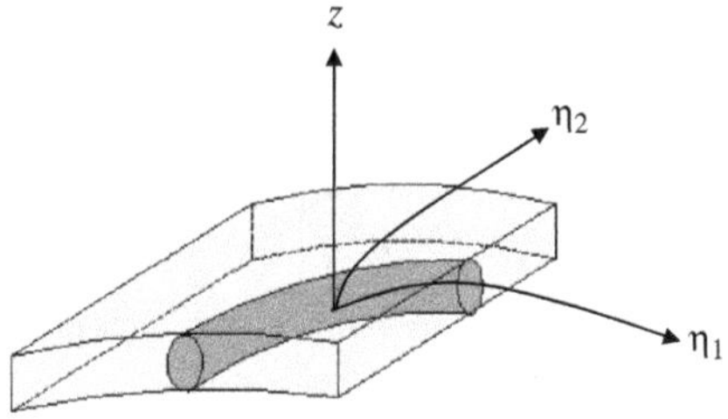

Fig. 1.15. Coordinate transformation to the microscopic coordinates (η_1, η_2, z).

depend on η_2 and z. Consequently, the order of the differential equations is reduced by one and the analysis of the problem is simplified. Thus the $b_{ij}^{\lambda\mu}$ functions from Eqs. (1.56)–(1.61) can be written as follows:

$$b_{11}^{\lambda\mu} = -\frac{1}{h_1}C_{11}\sin\varphi'\frac{\partial U_1^{\lambda\mu}}{\partial\eta_2} + \frac{1}{h_2}C_{12}\cos\varphi'\frac{\partial U_2^{\lambda\mu}}{\partial\eta_2} + C_{13}\frac{\partial U_3^{\lambda\mu}}{\partial z}$$

$$+ C_{16}\left[-\frac{1}{h_1}\sin\varphi'\frac{\partial U_2^{\lambda\mu}}{\partial\eta_2} + \frac{1}{h_2}\cos\varphi'\frac{\partial U_1^{\lambda\mu}}{\partial\eta_2}\right] + C_{11\lambda\mu}, \tag{1.62}$$

$$b_{22}^{\lambda\mu} = -\frac{1}{h_1}C_{12}\sin\varphi'\frac{\partial U_1^{\lambda\mu}}{\partial\eta_2} + \frac{1}{h_2}C_{22}\cos\varphi'\frac{\partial U_2^{\lambda\mu}}{\partial\eta_2} + C_{23}\frac{\partial U_3^{\lambda\mu}}{\partial z}$$

$$+ C_{26}\left[-\frac{1}{h_1}\sin\varphi'\frac{\partial U_2^{\lambda\mu}}{\partial\eta_2} + \frac{1}{h_2}\cos\varphi'\frac{\partial U_1^{\lambda\mu}}{\partial\eta_2}\right] + C_{22\lambda\mu}, \tag{1.63}$$

$$b_{33}^{\lambda\mu} = -\frac{1}{h_1}C_{13}\sin\varphi'\frac{\partial U_1^{\lambda\mu}}{\partial\eta_2} + \frac{1}{h_2}C_{23}\cos\varphi'\frac{\partial U_2^{\lambda\mu}}{\partial\eta_2} + C_{33}\frac{\partial U_3^{\lambda\mu}}{\partial z}$$

$$+ C_{36}\left[-\frac{1}{h_1}\sin\varphi'\frac{\partial U_2^{\lambda\mu}}{\partial\eta_2} + \frac{1}{h_2}\cos\varphi'\frac{\partial U_1^{\lambda\mu}}{\partial\eta_2}\right] + C_{33\lambda\mu}, \tag{1.64}$$

$$b_{12}^{\lambda\mu} = -\frac{1}{h_1}C_{16}\sin\varphi'\frac{\partial U_1^{\lambda\mu}}{\partial\eta_2} + \frac{1}{h_2}C_{26}\cos\varphi'\frac{\partial U_2^{\lambda\mu}}{\partial\eta_2} + C_{36}\frac{\partial U_3^{\lambda\mu}}{\partial z}$$

$$+ C_{66}\left[-\frac{1}{h_1}\sin\varphi'\frac{\partial U_2^{\lambda\mu}}{\partial\eta_2} + \frac{1}{h_2}\cos\varphi'\frac{\partial U_1^{\lambda\mu}}{\partial\eta_2}\right] + C_{12\lambda\mu}, \tag{1.65}$$

$$b_{13}^{\lambda\mu} = C_{55}\left[-\frac{1}{h_1}\sin\varphi'\frac{\partial U_3^{\lambda\mu}}{\partial\eta_2} + \frac{\partial U_1^{\lambda\mu}}{\partial z}\right]$$

$$+ C_{45}\left[\frac{1}{h_2}\cos\varphi'\frac{\partial U_3^{\lambda\mu}}{\partial\eta_2} + \frac{\partial U_2^{\lambda\mu}}{\partial z}\right] + C_{13\lambda\mu}, \tag{1.66}$$

$$b_{23}^{\lambda\mu} = C_{45}\left[-\frac{1}{h_1}\sin\varphi'\frac{\partial U_3^{\lambda\mu}}{\partial\eta_2} + \frac{\partial U_1^{\lambda\mu}}{\partial z}\right]$$

$$+ C_{44}\left[\frac{1}{h_2}\cos\varphi'\frac{\partial U_3^{\lambda\mu}}{\partial\eta_2} + \frac{\partial U_2^{\lambda\mu}}{\partial z}\right] + C_{23\lambda\mu}, \tag{1.67}$$

and the unit-cell problem and associated boundary condition (1.46) can be rewritten in terms of the coordinates η_2 and z as follows:

$$-\frac{\sin\varphi'}{h_1}\frac{\partial}{\partial\eta_2}b_{i1}^{\lambda\mu}+\frac{\cos\varphi'}{h_2}\frac{\partial}{\partial\eta_2}b_{i2}^{\lambda\mu}+\frac{\partial}{\partial z}b_{i3}^{\lambda\mu}=0, \tag{1.68}$$

$$\left[n_2'\left(-\frac{\sin\varphi'}{h_1}b_{i1}^{\lambda\mu}+\frac{\cos\varphi'}{h_2}b_{i2}^{\lambda\mu}\right)+n_3'b_{i3}^{\lambda\mu}\right]\bigg|_{\Im}=0, \tag{1.69}$$

where n_2' and n_3' are the components of the unit vector normal to the lateral surface of the reinforcement with respect to the $\{\eta_1,\eta_2,z\}$ coordinate system, and the suffix $\Im$ stands for the matrix/reinforcement interface. We will now solve the system defined by Eqs. (1.62)–(1.68) and associated boundary condition (1.69) by assuming that the local functions $U_1^{\lambda\mu}$ and $U_2^{\lambda\mu}$ are linear in η_2 and are independent of z, whereas $U_3^{\lambda\mu}$ is linear in z and independent of η_2. That is, the solution can be found as follows:

$$U_1^{\lambda\mu}=A^{\lambda\mu}\eta_2,\quad U_2^{\lambda\mu}=B^{\lambda\mu}\eta_2,\quad U_3^{\lambda\mu}=C^{\lambda\mu}z, \tag{1.70}$$

where $A^{\lambda\mu}$, $B^{\lambda\mu}$ and $C^{\lambda\mu}$ are constants to be determined. Equation (1.70) is substituted into the expressions (1.62)–(1.67), which allows the calculation of the aforementioned constants in conjunction with Eq. (1.69). After solving the pertinent system of algebraic equations the results are then back-substituted into Eqs. (1.62)–(1.67) to yield the following formulas for all the non-zero local functions b_{ij}^{kl}:

$$b_{11}^{\lambda\mu}=\frac{C_{12\lambda\mu}[\Lambda_4\Lambda_7+\Lambda_8\Lambda_3]+C_{11\lambda\mu}[\Lambda_5\Lambda_7-\Lambda_9\Lambda_3]+C_{22\lambda\mu}[\Lambda_6\Lambda_7]+C_{33\lambda\mu}\Lambda_3}{\Lambda_7\left[\frac{A_2\tan\phi}{A_1}\Lambda_4+\Lambda_5+\frac{A_2^2\tan^2\phi}{A_1^2}\Lambda_6\right]+\frac{A_2\tan\phi}{A_1}[\Lambda_8\Lambda_3]-\Lambda_9\Lambda_3}, \tag{1.71}$$

$$b_{22}^{\lambda\mu}=\frac{C_{12\lambda\mu}[\Lambda_4\Lambda_7+\Lambda_8\Lambda_3]+C_{11\lambda\mu}[\Lambda_5\Lambda_7-\Lambda_9\Lambda_3]+C_{22\lambda\mu}[\Lambda_6\Lambda_7]+C_{33\lambda\mu}\Lambda_3}{\Lambda_7\left[\frac{A_1}{A_2\tan\phi}\Lambda_4+\frac{A_1^2}{A_2^2\tan^2\phi}\Lambda_5+\Lambda_6\right]+\frac{A_1}{A_2\tan\phi}[\Lambda_8\Lambda_3]-\frac{A_1^2}{A_2^2\tan^2\phi}\Lambda_9\Lambda_3}, \tag{1.72}$$

$$b_{12}^{\lambda\mu}=\frac{C_{12\lambda\mu}[\Lambda_4\Lambda_7+\Lambda_8\Lambda_3]+C_{11\lambda\mu}[\Lambda_5\Lambda_7-\Lambda_9\Lambda_3]+C_{22\lambda\mu}[\Lambda_6\Lambda_7]+C_{33\lambda\mu}\Lambda_3}{\Lambda_7\left[\Lambda_4+\frac{A_1}{A_2\tan\phi}\Lambda_5+\frac{A_2\tan\phi}{A_1}\Lambda_6\right]+[\Lambda_8\Lambda_3]-\frac{A_1}{A_2\tan\phi}\Lambda_9\Lambda_3}, \tag{1.73}$$

where the quantities $\Lambda_1,\Lambda_2,\ldots,\Lambda_9$ can be found in [60].

We now turn our attention to the local functions b_{ij}^{*kl}. We begin by expanding Eq. (1.45) keeping Eq. (1.55) in mind as well as the coordinate transformation defined by Figs. 1.14 and 1.15. The resulting expressions are as follows:

$$\begin{aligned} b_{11}^{*\lambda\mu} &= -\frac{1}{h_1} C_{11} \sin\varphi' \frac{\partial V_1^{\lambda\mu}}{\partial \eta_2} + \frac{1}{h_2} C_{12} \cos\varphi' \frac{\partial V_2^{\lambda\mu}}{\partial \eta_2} + C_{13} \frac{\partial V_3^{\lambda\mu}}{\partial z} \\ &\quad + C_{16} \left[-\frac{1}{h_1} \sin\varphi' \frac{\partial V_2^{\lambda\mu}}{\partial \eta_2} + \frac{1}{h_2} \cos\varphi' \frac{\partial V_1^{\lambda\mu}}{\partial \eta_2} \right] + z C_{11\lambda\mu}, \end{aligned} \tag{1.74}$$

$$\begin{aligned} b_{22}^{*\lambda\mu} &= -\frac{1}{h_1} C_{12} \sin\varphi' \frac{\partial V_1^{\lambda\mu}}{\partial \eta_2} + \frac{1}{h_2} C_{22} \cos\varphi' \frac{\partial V_2^{\lambda\mu}}{\partial \eta_2} + C_{23} \frac{\partial V_3^{\lambda\mu}}{\partial z} \\ &\quad + C_{26} \left[-\frac{1}{h_1} \sin\varphi' \frac{\partial V_2^{\lambda\mu}}{\partial \eta_2} + \frac{1}{h_2} \cos\varphi' \frac{\partial V_1^{\lambda\mu}}{\partial \eta_2} \right] + z C_{22\lambda\mu}, \end{aligned} \tag{1.75}$$

$$\begin{aligned} b_{33}^{*\lambda\mu} &= -\frac{1}{h_1} C_{13} \sin\varphi' \frac{\partial V_1^{\lambda\mu}}{\partial \eta_2} + \frac{1}{h_2} C_{23} \cos\varphi' \frac{\partial V_2^{\lambda\mu}}{\partial \eta_2} + C_{33} \frac{\partial V_3^{\lambda\mu}}{\partial z} \\ &\quad + C_{36} \left[-\frac{1}{h_1} \sin\varphi' \frac{\partial V_2^{\lambda\mu}}{\partial \eta_2} + \frac{1}{h_2} \cos\varphi' \frac{\partial V_1^{\lambda\mu}}{\partial \eta_2} \right] + z C_{33\lambda\mu}, \end{aligned} \tag{1.76}$$

$$\begin{aligned} b_{12}^{*\lambda\mu} &= -\frac{1}{h_1} C_{16} \sin\varphi' \frac{\partial V_1^{\lambda\mu}}{\partial \eta_2} + \frac{1}{h_2} C_{26} \cos\varphi' \frac{\partial V_2^{\lambda\mu}}{\partial \eta_2} + C_{36} \frac{\partial V_3^{\lambda\mu}}{\partial z} \\ &\quad + C_{66} \left[-\frac{1}{h_1} \sin\varphi' \frac{\partial V_2^{\lambda\mu}}{\partial \eta_2} + \frac{1}{h_2} \cos\varphi' \frac{\partial V_1^{\lambda\mu}}{\partial \eta_2} \right] + z C_{12\lambda\mu}, \end{aligned} \tag{1.77}$$

$$\begin{aligned} b_{13}^{*\lambda\mu} &= C_{55} \left[-\frac{1}{h_1} \sin\varphi' \frac{\partial V_3^{\lambda\mu}}{\partial \eta_2} + \frac{\partial V_1^{\lambda\mu}}{\partial z} \right] \\ &\quad + C_{45} \left[\frac{1}{h_2} \cos\varphi' \frac{\partial V_3^{\lambda\mu}}{\partial \eta_2} + \frac{\partial V_2^{\lambda\mu}}{\partial z} \right] + z C_{13\lambda\mu}, \end{aligned} \tag{1.78}$$

$$\begin{aligned} b_{23}^{*\lambda\mu} &= C_{45} \left[-\frac{1}{h_1} \sin\varphi' \frac{\partial V_3^{\lambda\mu}}{\partial \eta_2} + \frac{\partial V_1^{\lambda\mu}}{\partial z} \right] \\ &\quad + C_{44} \left[\frac{1}{h_2} \cos\varphi' \frac{\partial V_3^{\lambda\mu}}{\partial \eta_2} + \frac{\partial V_2^{\lambda\mu}}{\partial z} \right] + z C_{23\lambda\mu}. \end{aligned} \tag{1.79}$$

Similarly, the unit-cell problem (1.47) becomes:

$$-\frac{\sin\varphi'}{h_1}\frac{\partial}{\partial\eta_2}b_{i1}^{*\lambda\mu}+\frac{\cos\varphi'}{h_2}\frac{\partial}{\partial\eta_2}b_{i2}^{*\lambda\mu}+\frac{\partial}{\partial z}b_{i3}^{*\lambda\mu}=0, \qquad (1.80)$$

$$\left[n_2'\left(-\frac{\sin\varphi'}{h_1}b_{i1}^{*\lambda\mu}+\frac{\cos\varphi'}{h_2}b_{i2}^{*\lambda\mu}\right)+n_3'b_{i3}^{*\lambda\mu}\right]\Big|_{\Im}=0. \qquad (1.81)$$

Since the local functions b_{ij}^{*kl} are related to the bending deformations it is expected that the pertinent solution will depend on the shape of the reinforcing elements (unlike the corresponding b_{ij}^{kl} coefficients). Indeed the presence of the z coordinates in Eqs. (1.74)–(1.79) implies exactly that. From the practical viewpoint, let us assume that the reinforcing elements have a circular cross-section. From the coordinate transformation from α_1, α_2 to ξ_1, ξ_2, defined by Eq. (1.40), we note that the cross-section will change from circular to elliptical with the eccentricity e' given by the following formula:

$$e'=\left(1-\frac{A_1^2A_2^2}{A_2^2h_1^2\sin^2\varphi+A_1^2h_2^2\cos^2\varphi}\right)^{1/2}. \qquad (1.82)$$

Additionally, the components n_2' and n_3' (clearly $n_1'=0$) of the unit vector normal to the surface of the reinforcing element are as follows:

$$n_2'=\eta_2[1-(e')^2]^{-1} \quad \text{and} \quad n_3'=z. \qquad (1.83)$$

It is possible to satisfy the differential equation (1.81) and boundary conditions (1.81) by assuming that the functions $V_i^{\lambda\mu}$ have the following general functional form:

$$V_i^{\lambda\mu}=W_{i1}^{\lambda\mu}\eta_2z+W_{i2}^{\lambda\mu}\frac{\eta_2^2}{2}+W_{i3}^{\lambda\mu}\frac{z^2}{2}, \qquad (1.84)$$

where $W_{ij}^{\lambda\mu}$ are constants to be determined. The determination of the local functions $b_{ij}^{*\lambda\mu}$ follows in a straightforward, albeit algebraically tedious manner. Keeping Eqs. (1.82) and (1.83) in mind, we first substitute Eq. (1.84) into Eqs. (1.80) and (1.81) and calculate the constants $W_{ij}^{\lambda\mu}$ by comparing terms with like powers of η_2 and z. Once the $W_{ij}^{\lambda\mu}$ functions are determined they are substituted into Eq. (1.84) and the resulting expressions are back-substituted into Eqs. (1.74)–(1.79) to obtain the desired local functions $b_{ij}^{*\lambda\mu}$. As a result, the following expressions for the functions $b_{11}^{*\lambda\mu}$, $b_{22}^{*\lambda\mu}$ and $b_{12}^{*\lambda\mu}$ are found:

$$b_{11}^{*\lambda\mu} = zB_{11}^{\lambda\mu}; \quad b_{22}^{*\lambda\mu} = zB_{22}^{\lambda\mu}; \quad b_{12}^{*\lambda\mu} = zB_{12}^{\lambda\mu}, \tag{1.85}$$

where

$$\begin{gathered} B_{11}^{\lambda\mu} = \frac{\sum_5 \sum_6 - \sum_2 \sum_3}{\sum_1 \sum_2 - \sum_4 \sum_5}, \quad B_{22}^{\lambda\mu} = \frac{\sum_3 \sum_4 - \sum_1 \sum_6}{\sum_1 \sum_2 - \sum_4 \sum_5}, \\ B_{12}^{\lambda\mu} = \frac{A_2 \tan\varphi}{2A_1} B_{11}^{\lambda\mu} + \frac{A_1}{2A_2 \tan\varphi} B_{22}^{\lambda\mu}. \end{gathered} \tag{1.86}$$

Explicit expressions for $\sum_1, \sum_2, \ldots, \sum_6$, which depend on the geometric parameters of the unit cell and the material properties of the reinforcements, can be found in [60].

1.6.1. Calculation of the effective elastic coefficients

The effective elastic coefficients for the reinforced generally orthotropic composite shell of Fig. 1.14 can be calculated by means of expressions (1.71)–(1.73), (1.85) and (1.86). Let us denote the volume of one reinforcing element within the unit cell of Fig. 1.14 by $\delta^3 V$. Then, the effective elastic coefficients are given by

$$\begin{gathered} \langle b_{ij}^{\lambda\mu} \rangle = \frac{1}{|\Omega|} \int_\Omega b_{ij}^{\lambda\mu} dv = \frac{V}{h_1 h_2} b_{ij}^{\lambda\mu}, \quad \langle z b_{ij}^{\lambda\mu} \rangle = \frac{1}{|\Omega|} \int_\Omega z b_{ij}^{\lambda\mu} dv = 0, \\ \langle b_{ij}^{*\lambda\mu} \rangle = 0, \quad \langle z b_{ij}^{*\lambda\mu} \rangle = \frac{V}{16 h_1 h_2} B_{ij}^{\lambda\mu}. \end{gathered} \tag{1.87}$$

The corresponding results for composite shells reinforced by more than one family of orthotropic reinforcements can be obtained from Eq. (1.87) by superposition. In doing so, we accept an error incurred due to stress variations at the regions of overlap of the reinforcements. However, this error is small and will not contribute significantly to the integral over the volume of the unit cell as discussed earlier in Section 1.4.

1.7. Examples of Grid-Reinforced Composite Shells with Orthotropic Reinforcements

The mathematical model and methodology presented in Section 1.6 can be used in analysis and design to tailor the effective elastic properties of the above reinforced composite shells to meet the criteria of a particular application, by selecting the appropriate shape of the shells as well as the type, number, orientation and geometric characteristics of the reinforcements. In this section, we will apply our general solution to different composite shells and plates. In the first example, we will consider a general composite shell reinforced with isotropic reinforcements. In the second example, the special case of a cylindrical shell will be considered. In the third example, we will obtain the closed-form expressions for the effective elastic properties of single-walled carbon nanotubes. In the fourth example, we will consider general multilayered composite shells and illustrate our results with a typical three-layer cylindrical shell. Finally, we will observe how our model can be used to derive the effective elastic coefficients of grid-reinforced composite plates. Without loss of generality, we will assume that in the considered grid-reinforced structures all reinforcements have similar cross-sectional areas and are made of the same material. If desired, however, the model allows for each family of reinforcements to have specific geometric and material properties.

Example 1.1: Composite shell reinforced with isotropic reinforcements

In the case of isotropic reinforcements, all the non-zero effective stiffness moduli are as follows [5, 25, 27]:

$$
\begin{aligned}
\langle b_{11}^{11}\rangle &= \frac{VA_1^4}{h_1h_2\Theta^4}E\cos^4\phi, \quad \langle b_{22}^{22}\rangle = \frac{VA_2^4}{h_1h_2\Theta^4}E\sin^4\phi,\\
\langle b_{11}^{12}\rangle &= \frac{VA_1^3A_2}{h_1h_2\Theta^4}E\cos^3\phi\sin\phi,\\
\langle b_{22}^{12}\rangle &= \frac{VA_1A_2^3}{h_1h_2\Theta^4}E\cos\phi\sin^3\phi,\\
\langle b_{11}^{22}\rangle &= \langle b_{12}^{12}\rangle = \frac{VA_1^2A_2^2}{h_1h_2\Theta^4}E\cos^2\phi\sin^2\phi;
\end{aligned}
\tag{1.88}
$$

$$
\begin{aligned}
\langle zb_{11}^{*11}\rangle &= \frac{V}{16h_1h_2}\frac{EA_1^4}{(1+\nu)\Theta^4}\cos^2\phi[2A_2^4\Psi\sin^2\phi+\cos^2\phi(1+\nu)],\\
\langle zb_{11}^{*22}\rangle &= \frac{V}{16h_1h_2}\frac{EA_1^2A_2^2}{(1+\nu)\Theta^4}\cos^2\phi\sin^2\phi[-2A_1^2A_2^2\Psi+1+\nu],\\
\langle zb_{11}^{*12}\rangle &= \frac{V}{16h_1h_2}\frac{EA_1^3A_2}{(1+\nu)\Theta^4}\cos\phi\sin\phi\\
&\quad\times[A_2^2\Psi(A_2^2\sin^2\phi-A_1^2\cos^2\phi)+\cos^2\phi(1+\nu)],
\end{aligned}
$$

$$\langle zb_{12}^{*12}\rangle = \frac{V}{16h_1h_2}\frac{EA_1^2A_2^2}{2(1+\nu)\Theta^4} \times [(A_1^2\cos^2\phi - A_2^2\sin^2\phi)\Psi + 2\cos^2\phi\sin^2\phi(1+\nu)], \tag{1.89}$$

$$\langle zb_{22}^{*12}\rangle = \frac{V}{16h_1h_2}\frac{EA_1A_2^3}{(1+\nu)\Theta^4}\cos\phi\sin\phi \times [A_1^2\Psi(A_1^2\cos^2\phi - A_2^2\sin^2\phi) + \sin^2\phi(1+\nu)],$$

$$\langle zb_{22}^{*22}\rangle = \frac{V}{16h_1h_2}\frac{EA_2^4}{(1+\nu)\Theta^4}\sin^2\phi[2A_1^4\Psi\cos^2\phi + \sin^2\phi(1+\nu)].$$

In Eqs. (1.72) and (1.73), E and ν are Young's modulus and Poisson's ratio of the reinforcement and

$$\Theta = A_1^2\cos^2\varphi + A_2^2\sin^2\varphi, \quad \Psi = \sqrt{\Theta^2 + A_1^2A_2^2}. \tag{1.90}$$

Example 1.2: Thin cylindrical shell

The second example represents a cylindrical composite shell (i.e., we can assume that $A_1 = A_2 = 1$) reinforced with a single family of reinforcements parallel to the longitudinal axis of the shell ($\varphi = 0^0$) as shown in Fig. 1.16. The effective elastic coefficients of this structure can readily be determined from Eq. (1.87) with the use of the solutions (1.71)–(1.73) and (1.86). Although the resulting expressions are too lengthy to be reproduced here, typical coefficients will be presented graphically for the reinforcement material properties given in Table 1.2.

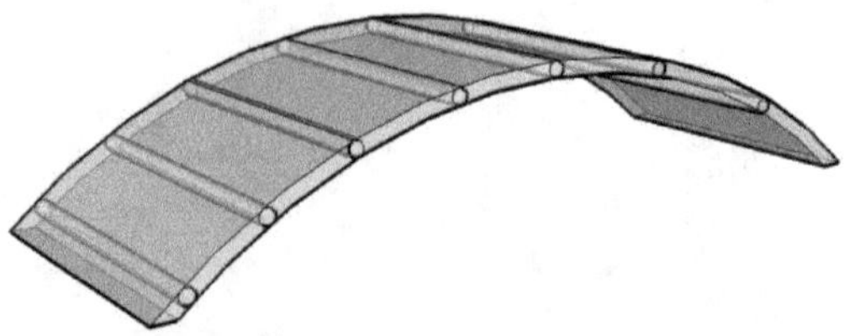

Fig. 1.16. Cylindrical composite shell with a single family of orthotropic reinforcements.

Table 1.2. Reinforcement material properties.

Property	E_1	$E_2 = E_3$	$G_{12} = G_{13}$	G_{23}	$\nu_{12} = \nu_{13} = \nu_{23}$
Value	152.0 GPa	4.1 GPa	2.9 GPa	1.5 GPa	0.35

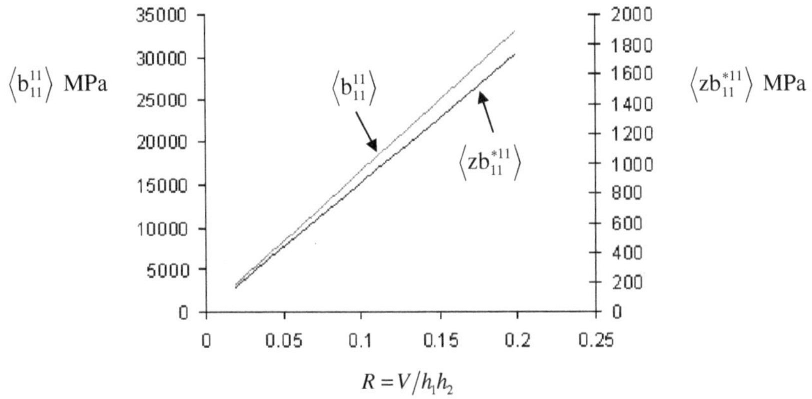

Fig. 1.17. Plot of $\langle b_{11}^{11} \rangle$ and $\langle zb_{11}^{*11} \rangle$ vs. R, volume fraction of reinforcement for a composite shell reinforced with a single family of orthotropic reinforcements (shown in Fig. 1.16).

Figure 1.17 shows a typical plot for the variation of $\langle b_{11}^{11} \rangle$ and $\langle zb_{12}^{*12} \rangle$ vs. R for the reinforced shell of Fig. 1.16, where R is the ratio of the volume of one reinforcing element within the unit cell to the volume of the entire unit cell. In other words, R is the volume fraction of the reinforcements and can be expressed as:

$$R = V/(h_1 h_2). \tag{1.91}$$

As expected, both the bending and extensional stiffnesses in the direction of the reinforcements increase with an increase in the volume fraction. Clearly, all the effective coefficients can be modified to fit different requirements by changing either the geometrical characteristics of the shell and reinforcements or by changing the type and number of reinforcement families.

Example 1.3: Single-walled carbon nanotube

Of particular interest in the context of cylindrical grid-reinforced shells is the case of a single-walled carbon nanotube (SWCNT). Carbon nanotubes are a recently discovered allotrope of carbon comprising long-chained molecules of carbon with carbon atoms arranged in a hexagonal network to form a tubular structure. They are classified as single- or multi-walled depending on the number of walls. Typically, the nanotubes are about 20 to 150 Å in diameter and about 1000 to 2000 Å in length. And they demonstrate remarkable strength and stiffness properties: Young's modulus

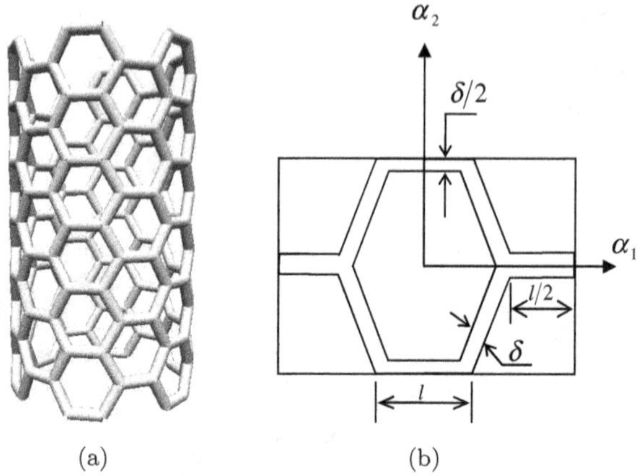

Fig. 1.18. (a) Schematic representation of an SWCNT; (b) unit cell of an SWCNT.

of 1.3 ± 0.5 TPa and tensile strength of 150 GPa for an SWCNT. Carbon nanotubes can be made by rolling up a carbon-sheet in various ways. Figure 1.18(b) shows the unit cell of an SWCNT in the so-called 'armchair' configuration.

As Fig. 1.18 demonstrates, the periodic nature of SWCNTs makes them particularly amenable to study by asymptotic homogenization techniques and the micromechanical model developed in Section 1.6; see Kalamkarov *et al.* [65, 66] for details. In this micromechanical model the C–C bonds are modeled by bars as shown in Fig. 1.18(b). It is assumed that these bars are of circular cross-section with the material properties E and ν. Equations (1.88)–(1.90) can be applied to the geometry of the unit cell of the SWCNT shown in Fig. 1.18(b), and all the effective stiffnesses moduli entering the constitutive relations (1.48) as coefficients can be calculated. As a result, the following constitutive relations of the homogenized SWCNT were obtained [65, 66]:

$$
\begin{aligned}
N_{11} &= \delta^2 \frac{E}{l} \frac{\pi}{16\sqrt{3}} (3\varepsilon_{11} + \varepsilon_{22}), \\
N_{22} &= \delta^2 \frac{E}{l} \frac{\pi}{16\sqrt{3}} (\varepsilon_{11} + 3\varepsilon_{22}), \\
N_{12} &= \delta^2 \frac{E}{l} \frac{\pi}{16\sqrt{3}} \varepsilon_{12},
\end{aligned}
\tag{1.92}
$$

$$
\begin{aligned}
M_{11} &= \delta^3 \frac{E}{(1+\nu)l} \frac{\pi\sqrt{3}}{768}[(4+3\nu)k_{11} + \nu k_{22}], \\
M_{22} &= \delta^3 \frac{E}{(1+\nu)l} \frac{\pi\sqrt{3}}{768}[\nu k_{11} + (4+3\nu)k_{22}], \qquad (1.93) \\
M_{12} &= \delta^3 \frac{E}{l} \frac{\pi\sqrt{3}}{768} k_{12}.
\end{aligned}
$$

The constitutive relations given by Eqs. (1.92) and (1.93) can be further applied to derive the analytical formulas for the engineering constants of SWCNTs. In particular, Eq. (1.76) yields the following formula for the effective Young's moduli E_{11} and E_{22} of SWCNTs:

$$
E_{11} = E_{22} = E_{\mathrm{SWCNT}} = \frac{\pi}{6\sqrt{3}} \frac{\delta E}{l}. \qquad (1.94)
$$

Using Eq. (1.76) we also obtain $\nu_{12} = 0.33$ and the following formula for the effective shear modulus G_{12} of an SWCNT:

$$
G_{12} = \frac{\pi}{32\sqrt{3}} \frac{\delta E}{l}. \qquad (1.95)
$$

It can be observed from Eq. (1.78) that the Young's modulus of a carbon nanotube increases with increasing bar diameter δ and with decreasing C–C link length l; see Fig. 1.18(b). In other words, this can be interpreted by saying that the Young's modulus of an SWCNT increases with decreasing tube diameter d. This dependency of Young's modulus on the tube diameter of an SWCNT is consistent with experimental observations. Using typical values of $E = 5.488 \times 10^{-6}\,\mathrm{N/nm}^2$, $\delta = 0.147\,\mathrm{nm}$ and $l = 0.142\,\mathrm{nm}$, the effective Young's and shear moduli of SWCNTs were determined from Eqs. (1.94) and (1.95) to be 1.71 TPa and 0.32 TPa, respectively [66]. These results compare favorably with the results of other researchers who used experimental or numerical techniques in their analyses; see, e.g., [68, 69].

Example 1.4: Laminated grid-reinforced composite shell with generally orthotropic reinforcements

In this example, we will analyze a laminated composite shell formed by N layers, each layer reinforced with a single family of orthotropic reinforcements; see Fig. 1.19. We assume that the family of reinforcements in the jth layer of the shell makes an angle φ_j with the coordinate line α_1.

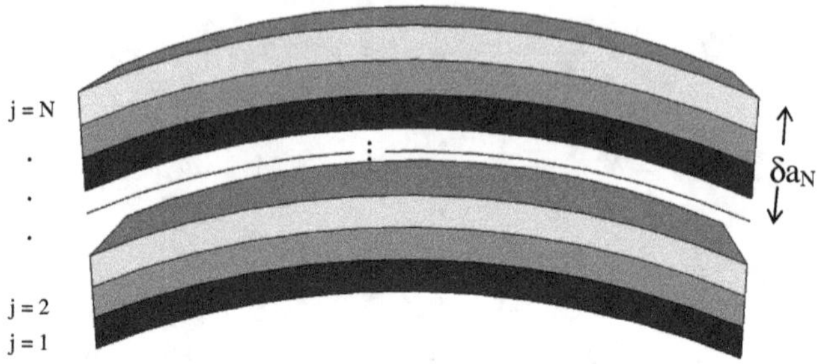

Fig. 1.19. Composite N-layered reinforced shell with each layer reinforced with a family of orthotropic reinforcements.

The distance between the axis of the jth reinforcement from the shell's mid-surface is denoted by δa_j (in the $\{\alpha_1, \alpha_2, \gamma\}$ coordinate system) as shown in Fig. 1.19.

One may derive expressions for the effective properties in a similar way as demonstrated in Section 1.6 after modifying the unit-cell problems in Eqs. (1.44) and (1.45) by replacing z with $(z' + a_j)$. The procedure, though algebraically tedious, is straightforward. The final results show that the local functions b_{ij}^{kl} remain as in Eqs. (1.71)–(1.73) while the local functions b_{ij}^{*kl} become

$$b_{ij}^{*\lambda\mu} = z' B_{ij}^{\lambda\mu} + a_j b_{ij}^{\lambda\mu}, \tag{1.96}$$

where $b_{ij}^{\lambda\mu}$ and $B_{ij}^{\lambda\mu}$ are given by Eqs. (1.71)–(1.73) and (1.86) after replacing $\tan\varphi$ with $\tan\varphi_j$ and $C_{mn}^{\lambda\mu}$ with $C_{mn(j)}^{\lambda\mu}$. Finally, the effective properties as calculated by summation over all N layers are as follows:

$$\begin{gathered} \langle b_{ij}^{\lambda\mu} \rangle = \sum_{j=1}^{N} b_{ij}^{\lambda\mu} \gamma_j, \quad \langle z b_{ij}^{\lambda\mu} \rangle = \langle b_{ij}^{*\lambda\mu} \rangle = \sum_{j=1}^{N} b_{ij}^{\lambda\mu} \gamma_j a_j, \\ \langle z b_{ij}^{*\lambda\mu} \rangle = \textstyle\sum_{j=1}^{N} \left(\frac{B_{ij}^{\lambda\mu} \gamma_j}{16} + a_j^2 b_{ij}^{\lambda\mu} \gamma_j \right), \end{gathered} \tag{1.97}$$

where γ_j is the volume fraction of reinforcements in the jth layer and is given by

$$\gamma_j = \frac{V_j}{h_1 h_2}. \tag{1.98}$$

We will now illustrate Example 1.4 by considering a three-layer composite shell with orthotropic reinforcements oriented at $\varphi = 60°$, $\varphi = 90°$ and $\varphi = 120°$ as shown in Fig. 1.20. The effective stiffness moduli

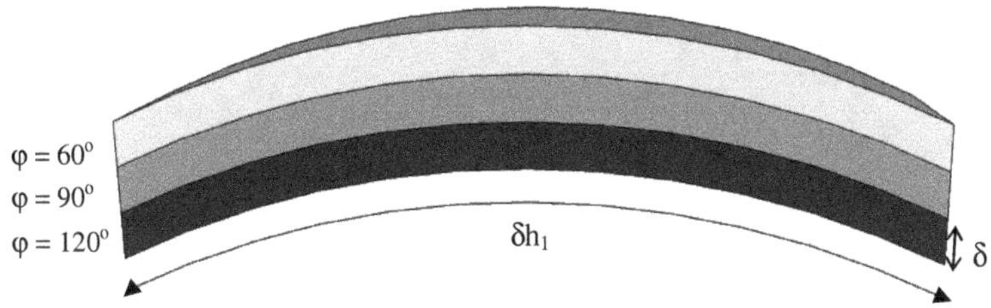

Fig. 1.20. Unit cell for three-layer composite shell.

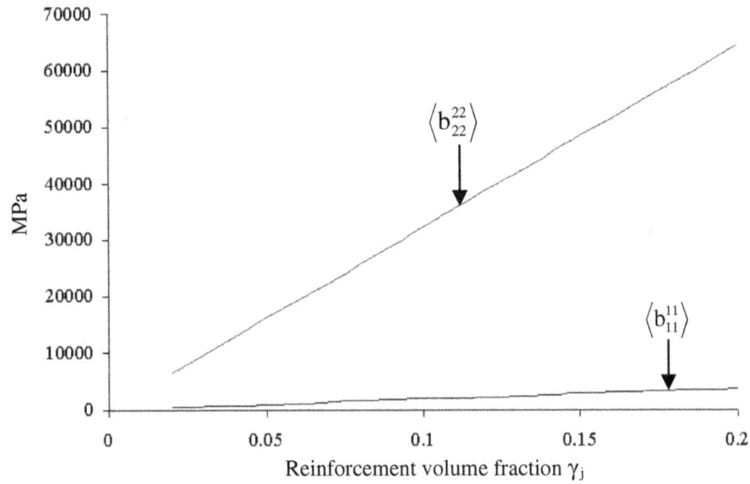

Fig. 1.21. Plot of typical extensional effective stiffness moduli vs. reinforcement volume fraction per layer γ_j.

are readily obtained from Eq. (1.97) and although the resulting expressions are too lengthy to be reproduced here, some of the effective stiffnesses will be presented graphically. We will assume that the shell layers are cylindrical and that the reinforcements have the properties given in Table 1.2.

Figure 1.21 shows the variation of effective stiffnesses $\langle b_{11}^{11} \rangle$ and $\langle b_{22}^{22} \rangle$ vs. γ_j. Without loss of generality, we assume that the reinforcement volume fraction is the same in each of the three layers. As expected, the extensional stiffness for the shell is larger in the α_2 direction than in the α_1 direction because there are more reinforcements either entirely (middle layer) or partially (top and bottom layers) oriented in the α_2 direction. For the same reason, the bending stiffness in the α_2 direction is larger than its counterpart in the α_1 direction as shown in Fig. 1.22.

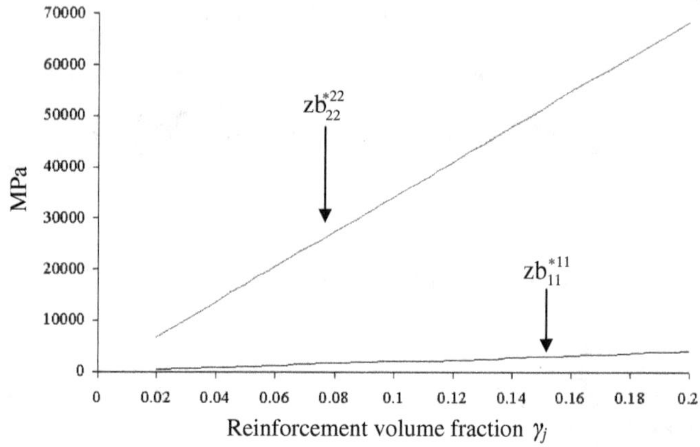

Fig. 1.22. Plot of typical bending effective elastic stiffness vs. reinforcement volume fraction per layer γ_j.

Example 1.5: Composite plates with generally orthotropic reinforcements

As a final example, we will now apply the obtained general results for the case of a thin plate reinforced with a grid of generally orthotropic reinforcements, shown in Fig. 1.23.

The effective stiffness moduli for the grid-reinforced plate can readily be obtained from Eqs. (1.71)–(1.73), (1.85) and (1.86) by letting $A_1 = A_2 = 1$. The results are too lengthy to reproduce here, and for purposes of illustration we will compare graphically some of the effective stiffnesses pertaining to the two structures shown in Figs. 1.24 and 1.25. The structure of Fig. 1.24 consists of two mutually perpendicular families of orthotropic reinforcements ($\varphi = 0°$ and $\varphi = 90°$) forming a rectangular reinforcing grid. This structure will be referred to in the following as S_1.

The structure of Fig. 1.25 has three families of orthotropic reinforcements oriented at $\varphi = 45°$, $\varphi = 90°$ and $\varphi = 135°$ forming a triangular reinforcing grid. This structure will be referred to as S_2. The unit cells of S_1 and S_2 are also shown in Figs. 1.24 and 1.25. In the ensuing plots we will assume that the reinforcements have the elastic properties given in Table 1.2. The different effective stiffnesses are plotted vs. the volume fraction of the reinforcements R defined in Eq. (1.91).

Figure 1.26 shows the variation of $\langle b_{11}^{11} \rangle$ vs. R for the two structures S_1 and S_2. It can be observed that the stiffness in the ξ_1 direction is larger for S_1 than for S_2 because S_1 has more reinforcements oriented in the ξ_1 direction.

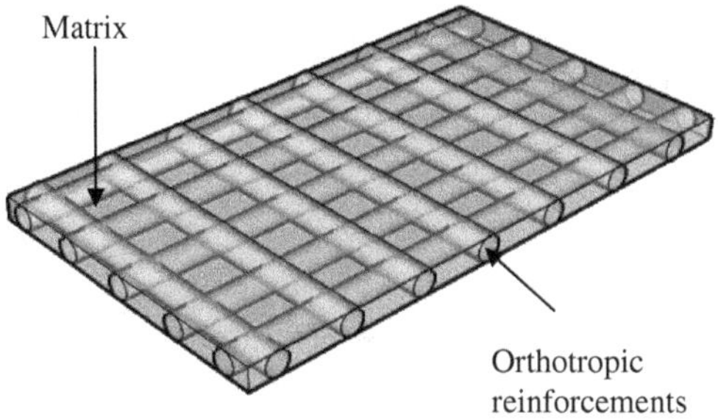

Fig. 1.23. Composite plate reinforced with a network of orthotropic bars.

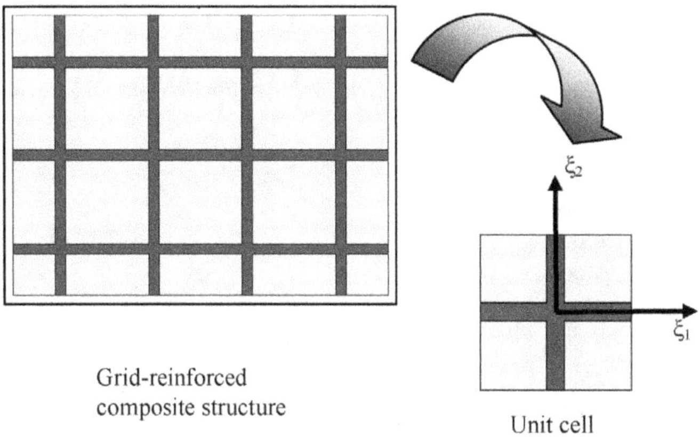

Fig. 1.24. Grid-reinforced plate (structure S_1) with reinforcements arranged at angles $\varphi = 0°$ and $\varphi = 90°$.

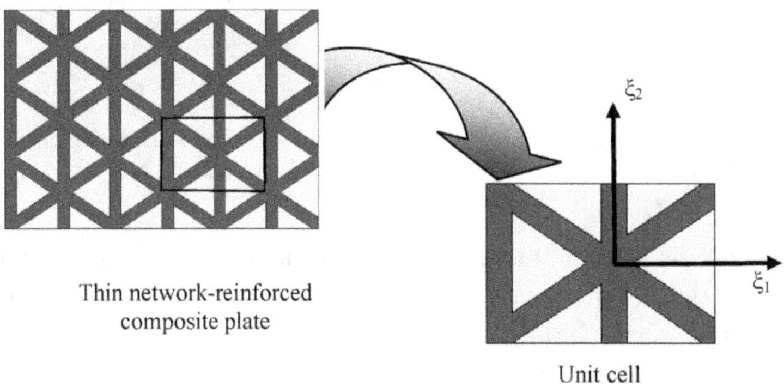

Fig. 1.25. Grid-reinforced plate (structure S_2) with reinforcements arranged at angles $\varphi = 45°$, $\varphi = 90°$ and $\varphi = 135°$.

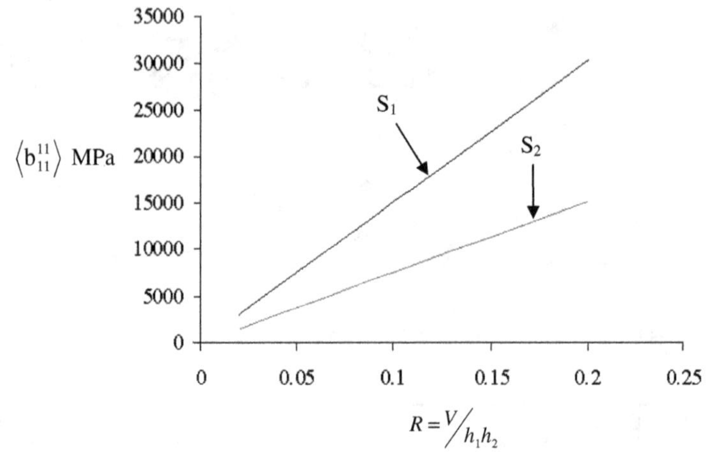

Fig. 1.26. Plot of elastic stiffness $\langle b_{11}^{11} \rangle$ vs. volume fraction of reinforcements R for structures S_1 and S_2.

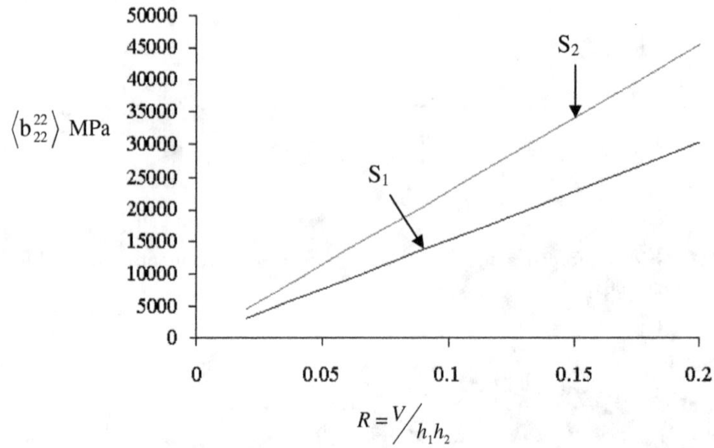

Fig. 1.27. Plot of elastic stiffness $\langle b_{22}^{22} \rangle$ vs. volume fraction of reinforcements R for structures S_1 and S_2.

S_2 has an overall larger number of reinforcements but one of them is oriented entirely in the ξ_2 direction and therefore makes no contribution to the stiffness in the ξ_1 direction while the other two are oriented at an angle to the ξ_1 axis and therefore only partially contribute to the value of $\langle b_{11}^{11} \rangle$. For the same reason, we expect that the trend in the ξ_2 direction will be reversed and that S_2 should be stiffer. Indeed Fig. 1.27 shows precisely that.

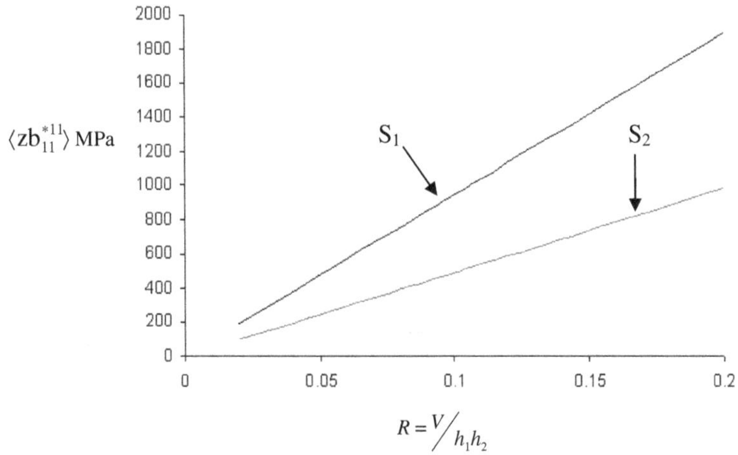

Fig. 1.28. Plot of $\langle zb_{11}^{*11} \rangle$ vs. volume fraction of reinforcements R for structures S_1 and S_2.

Similar considerations hold for the remaining effective stiffness moduli. Figure 1.28 shows the variation of $\langle zb_{11}^{*11} \rangle$ vs. R for S_1 and S_2. We note that this coefficient characterizes the bending stiffness of the composite plate in the ξ_1–z plane. On the basis of the arguments given above the value of the $\langle zb_{11}^{*11} \rangle$ coefficient is higher for S_1 than S_2. It is important to note, however, that all of these trends and characteristics can be easily modified by changing the size, type, angular orientation, etc. so that the desirable elastic coefficients are obtained to conform to a particular application.

1.8. Sandwich Composite Shells with Cellular Cores

Composite sandwich structures with cellular cores have found numerous engineering applications; see, e.g., [70, 71]. In many cases, the methods of synthesis and fabrication of these structures are controlled by the phase separation processes with surface tension being the controlling physical factor; see [72]. Therefore, the hexagonal cell that has the minimal surface area will have a big advantage in relation to the thermoelastic performance of sandwich structures in engineering applications.

We considered earlier in Section 1.5 an example of a three-layered sandwich shell with a hexagonal honeycomb filler; see Fig. 1.12. Now we will present the details of the derivation of the effective stiffness moduli given earlier in Eq. (1.41) in the case of isotropic constituent materials. We will apply the above-introduced asymptotic homogenization technique

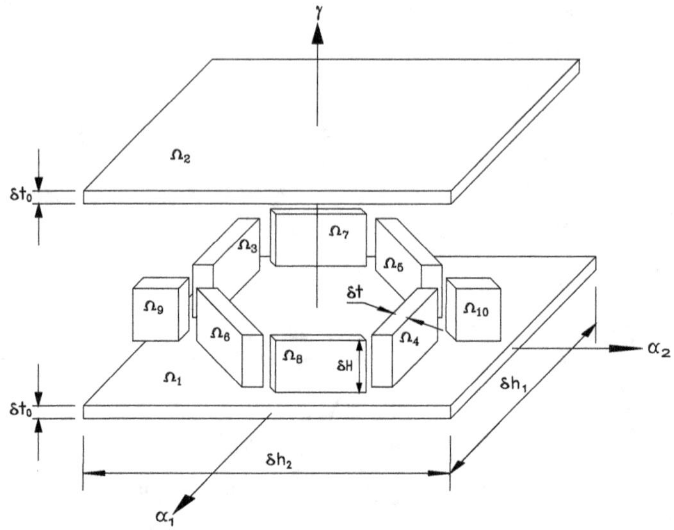

Fig. 1.29. Unit cell of a three-layered composite sandwich with honeycomb filler.

to examine the composite sandwich shell with hexagonal honeycomb filler with a periodicity cell consisting of ten individual elements as shown in Fig. 1.29. We assume that $A_1 = A_2 = 1$ and that all elements of the sandwich structure shown in Fig. 1.29 are made of different generally orthotropic materials.

As noted above, the matrix $[C]$ of elastic coefficients of an orthotropic material referenced to a coordinate system that has been rotated by an angle φ with respect to its principal material coordinate system coincides with that of a monoclinic material and has the form given in Eq. (1.55). The analytical solution of local problems (1.44), (1.46), and (1.45), (1.47) in the considered case can be found with the assumption that the thickness of each of the cell elements is small in comparison to the other dimensions, i.e., under the conditions $t \ll h_1, h_2, H$, $t_0 \ll H$ and $H \sim h_1, h_2$. This assumption is very appropriate for sandwich structures with cellular cores used in engineering applications. On the basis of this analytical solution for all non-zero effective stiffness moduli we obtain (see [63, 64] for details):

$$\langle b_{11}^{11} \rangle = \frac{E_2^{(1)} F_1}{1 - \nu_{12}^{(1)} \nu_{21}^{(1)}} + \frac{E_2^{(2)} F_2}{1 - \nu_{12}^{(2)} \nu_{21}^{(2)}}$$

$$+ \sum_{i=3}^{10} E^{(i)} F_i S_{(i)}^4 h_2^{-4} \left(C_{(i)}^2 h_1^{-2} + S_{(i)}^2 h_2^{-2} \right)^{-2},$$

$$\langle b_{22}^{22}\rangle = \frac{E_1^{(1)}F_1}{1-\nu_{12}^{(1)}\nu_{21}^{(1)}} + \frac{E_1^{(2)}F_2}{1-\nu_{12}^{(2)}\nu_{21}^{(2)}} + \sum_{i=3}^{10} E^{(i)}F_i C_{(i)}^4 h_2^{-4}\left(C_{(i)}^2 h_1^{-2} + S_{(i)}^2 h_2^{-2}\right)^{-2},$$

$$\langle b_{22}^{11}\rangle = \langle b_{11}^{22}\rangle = \frac{\nu_{12}^{(1)}E_1^{(1)}F_1}{1-\nu_{12}^{(1)}\nu_{21}^{(1)}} + \frac{\nu_{12}^{(2)}E_1^{(2)}F_2}{1-\nu_{12}^{(2)}\nu_{21}^{(2)}} + \sum_{i=3}^{10} E^{(i)}F_i C_{(i)}^2 S_{(i)}^2 h_1^{-2}h_2^{-2}\left(C_{(i)}^2 h_1^{-2} + S_{(i)}^2 h_2^{-2}\right)^{-2},$$

$$\langle b_{12}^{12}\rangle = G_{12}^{(1)}F_1 + G_{12}^{(2)}F_2 + \sum_{i=3}^{10} E^{(i)}F_i C_{(i)}^2 S_{(i)}^2 h_1^{-2}h_2^{-2}\left(C_{(i)}^2 h_1^{-2} + S_{(i)}^2 h_2^{-2}\right)^{-2},$$

$$\langle zb_{11}^{*11}\rangle = \frac{E_1^{(1)}J_1}{1-\nu_{12}^{(1)}\nu_{21}^{(1)}} + \frac{E_1^{(2)}J_2}{1-\nu_{12}^{(1)}\nu_{21}^{(1)}} + \sum_{i=3}^{10} E^{(i)}J_i S_{(i)}^4 h_2^{-4}\left(C_{(i)}^2 h_1^{-2} + S_{(i)}^2 h_2^{-2}\right)^{-2},$$

$$\langle zb_{22}^{*22}\rangle = \frac{E_2^{(1)}J_1}{1-\nu_{12}^{(1)}\nu_{21}^{(1)}} + \frac{E_2^{(2)}J_2}{1-\nu_{12}^{(1)}\nu_{21}^{(1)}} + \sum_{i=3}^{10} E^{(i)}J_i C_{(i)}^4 h_1^{-4}\left(C_{(i)}^2 h_1^{-2} + S_{(i)}^2 h_2^{-2}\right)^{-2},$$

$$\langle zb_{22}^{*11}\rangle = \langle zb_{11}^{*22}\rangle = \frac{\nu_{12}^{(1)}E_1^{(1)}J_1}{1-\nu_{12}^{(1)}\nu_{21}^{(1)}} + \frac{\nu_{12}^{(2)}E_1^{(2)}J_2}{1-\nu_{12}^{(2)}\nu_{21}^{(2)}} + \sum_{i=3}^{10} E^{(i)}J_i C_{(i)}^2 S_{(i)}^2 h_1^{-2}h_2^{-2}\left(C_{(i)}^2 h_1^{-2} + S_{(i)}^2 h_2^{-2}\right)^{-2},$$

$$\langle zb_{12}^{*12}\rangle = G_{12}^{(1)}J_1 + G_{12}^{(2)}J_2 + \sum_{i=3}^{10} G_{12}^{(i)}J_i\left(\frac{C_{(i)}^2 S_{(i)}^2 h_1^{-2}h_2^{-2}}{\left(C_{(i)}^2 h_1^{-2} + S_{(i)}^2 h_2^{-2}\right)^2} - \frac{8H}{3\pi^5 h_1}\sum_{n=1}^{\infty}\frac{\tanh[\pi(2n-1)t/(2H)]}{(2n-1)^5}\right). \tag{1.99}$$

In the above formulas, the superscripts (1), (2) and $(i), i = 3, 4, \ldots, 10$, refer to the corresponding elements of the unit cell $\Omega_1, \Omega_2, \Omega_3, \ldots, \Omega_{10}$, shown in Fig. 1.29, and indicate the material properties of the corresponding structural element; $C_{(i)}$ and $S_{(i)}$ stand for $\cos\varphi_i$ and $\sin\varphi_i$, respectively, where φ_i is an angle that element Ω_i makes with the α_2-axis, $i = 3, 4, \ldots, 10$; and the quantities $F_1, F_2, \ldots, F_{10}$ and $J_1, J_2, \ldots, J_{10}$ are defined as follows:

$$
\begin{aligned}
&F_i = \langle 1 \rangle_{\Omega_i} = t_0; \quad J_i = \langle z^2 \rangle_{\Omega_i} = \frac{t_0}{3}\left(\frac{3\,H^2}{4} + \frac{3\,H\,t_0}{2} + t_0^2\right), \quad \text{for } i = 1, 2, \\
&F_i = \langle 1 \rangle_{\Omega_i} = \frac{\sqrt{3}\,Ht}{9\ \ a}; \quad J_i = \langle z^2 \rangle_{\Omega_i} = \frac{\sqrt{3}\ H^3\,t}{108\ \ a}, \quad \text{for } i = 3, 4, 5, 6, \qquad (1.100) \\
&F_i = \langle 1 \rangle_{\Omega_i} = \frac{\sqrt{3}\,Ht}{18\ \ a}; \quad J_i = \langle z^2 \rangle_{\Omega_i} = \frac{\sqrt{3}\ H^3\,t}{216\ \ a}, \quad \text{for } i = 7, 8, 9, 10.
\end{aligned}
$$

Here $F_1, F_2, \ldots, F_{10}$ are the cross-sectional areas and $J_1, J_2, \ldots, J_{10}$ are the moments of inertia of the cross-sections of the corresponding elements $\Omega_1, \Omega_2, \ldots, \Omega_{10}$ relative to the middle surface of the shell, and are calculated in the coordinate system ξ_1, ξ_2, z.

It is seen in Eq. (1.99) that the first terms represent the contribution from the top and bottom face carriers and the latter terms describe the contribution of the sandwich core. If we now consider a particular case when the face carriers and honeycomb core are made of similar orthotropic material then Eq. (1.99) reduces to the following formulas [64]:

$$
\begin{aligned}
\langle b_{11}^{11} \rangle &= 2\,E_1 t_0/(1 - \nu_{12}\nu_{21}) + 1.1732\,E_1\nu_{21}Ht/(\nu_{12}a), \\
\langle b_{22}^{22} \rangle &= 2\,E_2 t_0/(1 - \nu_{12}\nu_{21}) + 0.5152\,E_2 Ht/(a), \\
\langle b_{22}^{11} \rangle &= \langle b_{11}^{22} \rangle = 2\nu_{12}E_1 t_0/(1 - \nu_{12}\nu_{21}) + 0.3908\,E_1\nu_{21}Ht/(\nu_{12}a), \\
\langle b_{12}^{12} \rangle &= 2G_{12}t_0 + 0.3908 E_2 H\,t/a, \\
\langle z b_{11}^{*11} \rangle &= E_1 t_0(0.5\,H^2 + Ht_0 + 0.6667 t_0^2)/(1 - \nu_{12}\nu_{21}) \qquad (1.101) \\
&\quad + 0.0976\,E_1\nu_{21}H^3 t/(\nu_{12}a), \\
\langle z b_{22}^{*22} \rangle &= E_2 t_0(0.5H^2 + Ht_0 + 0.6667 t_0^2)/(1 - \nu_{12}\nu_{21}) \\
&\quad + 0.0429336\,E_2 H^3 t/a,
\end{aligned}
$$

$$\left\langle zb_{22}^{*11}\right\rangle = \left\langle zb_{11}^{*22}\right\rangle = \nu_{12}E_1t_0(0.5H^2 + H\,t_0 + 0.6667t_0^2)/(1-\nu_{12}\nu_{21})$$
$$+\ 0.03256\,E_1\nu_{21}H^3t/(\nu_{12}a),$$
$$\left\langle zb_{12}^{*12}\right\rangle = G_{12}t_0(0.5\,H^2 + Ht_0 + 0.6667t_0^2) + 0.03256\,E_2H^3t/a.$$

Note that Eqs. (1.99)–(1.101) generalize the earlier given Eq. (1.54) for much more complicated cases of generally orthotropic constituent materials.

1.8.1. Examples of sandwich shells

The effective properties of sandwich composite structures can be tailored to meet the requirements of a particular application by changing the geometric parameters including the thickness of the face carriers, the width and cross-sectional areas of the core elements or the relative height of the core, the angular orientation of the elements and by changing the materials of the core and face carriers. This is demonstrated below where a comparison of the effective elastic properties of two different cellular cores is made; see Fig. 1.30. The unit cells of the analysed structures are shown in Figs. 1.31 and 1.32. Both structures have hexagonal symmetry beginning with the triangular cell and maintaining the basic 60°, 120° angular architecture.

The analytical formulas for all the effective stiffness moduli of these sandwich structures have been derived in the case when the structural elements are made of generally orthotropic materials. All the non-zero effective stiffness moduli are given in Table 1.3 [64].

Figures 1.33–1.37 show the variation of the effective stiffness moduli of the hexagonal-triangular and star-hexagonal cores vs. core height H. A graphite/epoxy core material is considered with the following properties: $C_{11} = 183.443\,\text{GPa}$, $C_{12} = C_{13} = 4.363\,\text{GPa}$, $C_{22} = C_{33} = 11.662\,\text{GPa}$, $C_{23} = 3.918\,\text{GPa}$, $C_{44} = 2.87\,\text{GPa}$, $C_{55} = C_{66} = 7.17\,\text{GPa}$, $\nu_{12} = \nu_{13} = 0.01593$, $\nu_{21} = \nu_{31} = 0.28003$ and $\nu_{23} = \nu_{32} = 0.33$.

Figure 1.33 shows that the stiffness $\left\langle b_{11}^{11}\right\rangle$ in the ξ_1 direction is significantly larger for the hexagonal-triangular core than its star-hexagonal counterpart because the former has more reinforcements oriented in the ξ_1 direction. Although the latter structure has an overall similar number of reinforcements to those of the hexagonal-triangular core, the horizontal component of the reinforcements of that structure is much shorter in length in the ξ_1 direction and therefore makes a significant difference in the homogenized stiffness in the ξ_1 direction, while making almost no contribution in the elastic properties in the ξ_2 direction. In other words, it

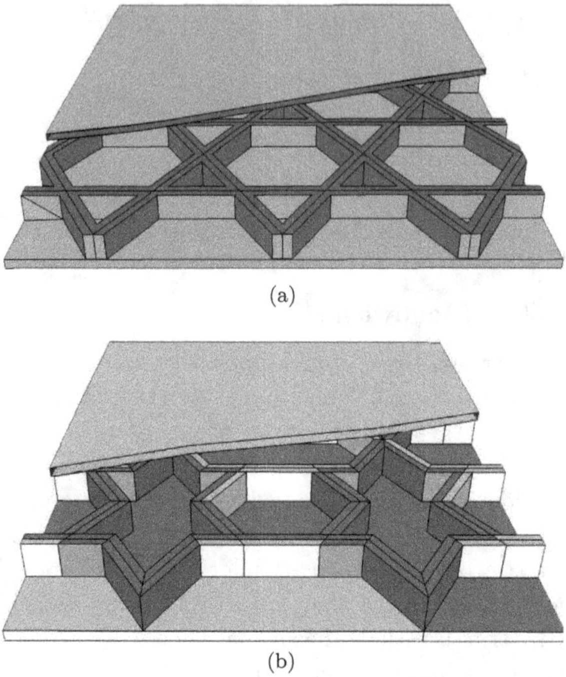

Fig. 1.30. (a) Hexagonal-triangular and (b) star-hexagonal cored composite sandwich structures.

is expected that the stiffness properties in the ξ_2 direction for the two cores remain almost unchanged. Indeed, Fig. 1.34 proves this trend precisely.

Figure 1.35 shows the variation of the effective stiffness $\langle zb_{11}^{*11} \rangle$ vs. core height H. It is obvious that the presence of additional members in the hexagonal-triangular core has contributed to making the structure stiffer than its star-hexagonal counterpart. In Fig. 1.36, the effective stiffnesses of the two cores are very close due to the same reason as explained for Fig. 1.34. Finally, Fig. 1.37 shows the variation of effective stiffnesses $\langle b_{12}^{12} \rangle$ and $\langle zb_{12}^{*12} \rangle$ vs. core height H. As expected both the extensional and torsional stiffnesses in the direction of the reinforcements increase with the height of the core.

1.9. Smart Composite Materials and Structures

The high maintenance cost and limited service life condition often associated with traditional structural materials like concrete and steel can be significantly offset by the application of composites in the areas of civil

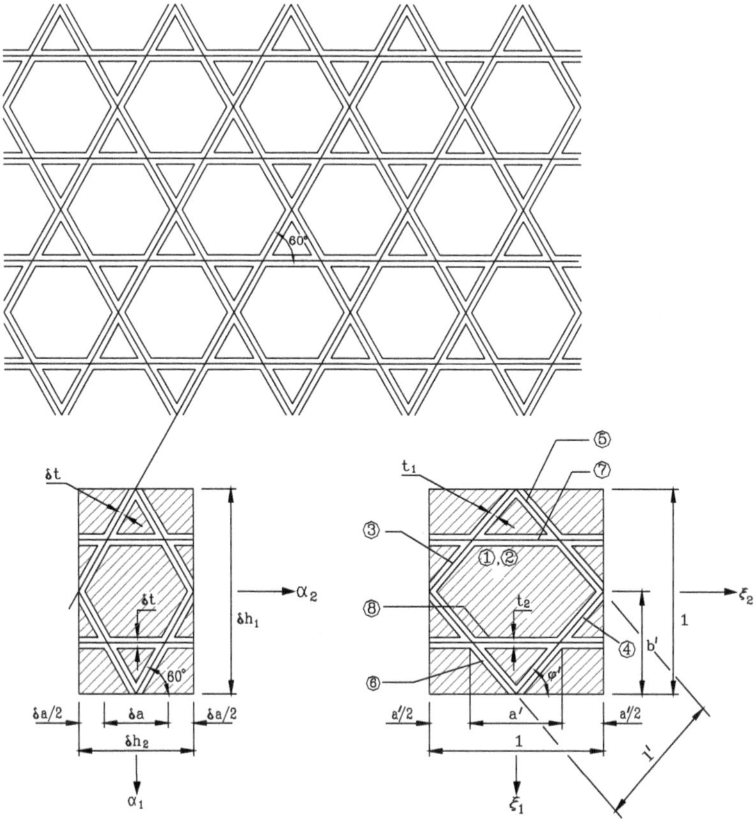

Fig. 1.31. Unit cell of hexagonal-triangular core in global $(\alpha_1, \alpha_2, \gamma)$ and local (ξ_1, ξ_2, z) coordinates.

engineering, aerospace, transportation industry, oil and gas, and marine engineering. At the same time, new technologies permitted the growth of other fields such as new actuator materials, advancements in fiber-optics, MEMS and telecommunications, significantly facilitates the development of new and highly effective sensors and actuators that now became available at reasonable prices. Their merge with the field of composites gave birth to the so-called smart materials and structures. Smart materials have the ability to respond adaptively in a pre-designed useful and efficient manner to changes in environmental conditions, including certain changes in their own state. Smart structures incorporate sensors and actuators made of smart materials and they can perform self-adjustment or self-repair as conditions change,

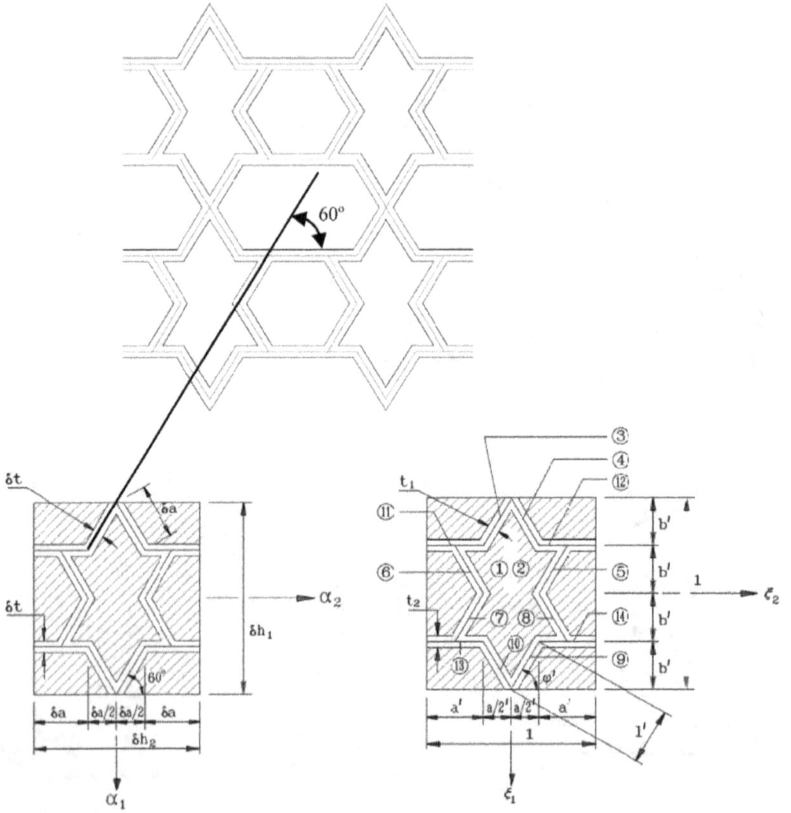

Fig. 1.32. Unit cell of star-hexagonal core in global $(\alpha_1, \alpha_2, \gamma)$ and local (ξ_1, ξ_2, z) coordinates.

as shown in Fig. 1.38. Ideally, they demonstrate optimum performance under a variety of environmental conditions.

In general, depending on their type, smart materials and structures can be classified as passive or actively controlled. Passive smart materials incorporate sensors that provide information on their state and integrity, while actively controlled smart materials incorporate both sensors and actuators. Two basic types of control can be specified: passive control, with the use of auxiliary non-adaptable elements; and active control, with the use of adaptable materials or mechanisms, such as electromechanical, piezoelectric, magnetostrictive, electro- or magnetorheological, or actuators using the shape memory effect, etc. The most popular material systems being used for sensors and actuators are as follows: (1) piezoelectric

Table 1.3. Effective stiffness moduli of hexagonal-triangular and star-hexagonal cores ($A_1 = A_2 = 1$, $a = 2.5$).

Effective stiffness moduli	Hexagonal-triangular core shown in Figs. 1.30(a) and 1.31	Star-hexagonal core shown in Figs. 1.30(b) and 1.32
$\langle b_{11}^{11} \rangle$	$16.6276 E_1 \nu_{21} Ht/(\nu_{12} a)$	$3.3366 E_1 \nu_{21} Ht/(\nu_{12} a)$
$\langle b_{22}^{22} \rangle$	$2.4250\, E_2 Ht/(a)$	$2.3287\, E_2 Ht/(a)$
$\langle b_{22}^{11} \rangle = \langle b_{11}^{22} \rangle$	$5.5424 E_1 \nu_{21} Ht/(\nu_{12} a)$	$1.1122 E_1 \nu_{21} Ht/(\nu_{12} a)$
$\langle b_{12}^{12} \rangle$	$5.5424 E_2 Ht/a$	$1.1122 E_2 Ht/a$
$\langle z b_{11}^{*11} \rangle$	$1.3856\, E_1 \nu_{21} H^3 t/(\nu_{12} a)$	$0.2780\, E_1 \nu_{21} H^3 t/(\nu_{12} a)$
$\langle z b_{22}^{*22} \rangle$	$0.2022 E_2 H^3 t/a$	$0.1941 E_2 H^3 t/a$
$\langle z b_{22}^{*11} \rangle = \langle z b_{11}^{*22} \rangle$	$0.4620\, E_1 \nu_{21} H^3 t/(\nu_{12} a)$	$0.0927\, E_1 \nu_{21} H^3 t/(\nu_{12} a)$
$\langle z b_{12}^{*12} \rangle$	$0.4620\, E_2 H^3 t/a$	$0.0927\, E_2 H^3 t/a$

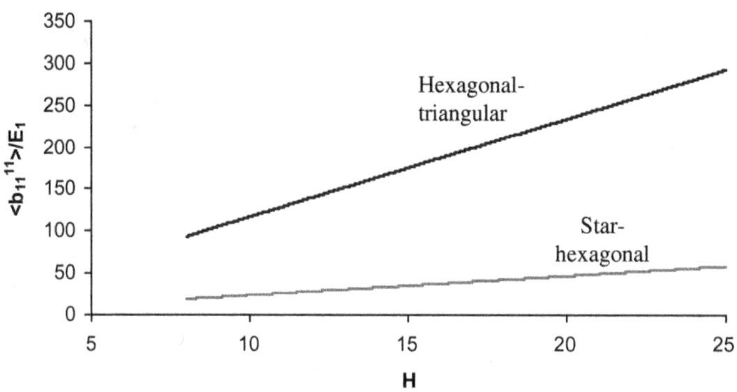

Fig. 1.33. Effective stiffness $\langle b_{11}^{11} \rangle / E_1$ vs. core height H of hexagonal-triangular and star-hexagonal sandwich cores.

materials, (2) magnetostrictive and electrostrictive materials, (3) shape memory alloys, (4) electrorheological and magnetorheological fluids, (5) carbon nanotubes, (6) optical fibers, (7) electrochromic materials, and (8) smart gels.

Piezoelectric materials, e.g., lead zirconate titanate (PZT) and barium titanate, by virtue of their unique electromechanical coupling characteristics, low power requirements, and relatively high generative forces, play a prominent role in the modern electro-ceramic industry. In conjunction with efforts to develop monolithic materials with improved piezoelectric properties, there have been a number of efforts to develop

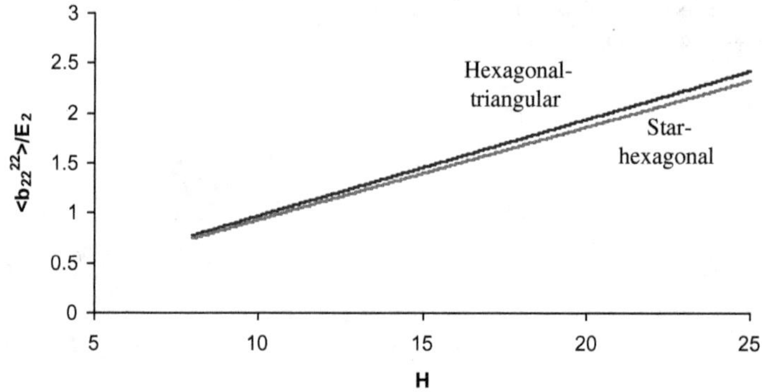

Fig. 1.34. Effective stiffness $\langle b_{22}^{22} \rangle / E_2$ vs. core height H of hexagonal-triangular and star-hexagonal sandwich cores.

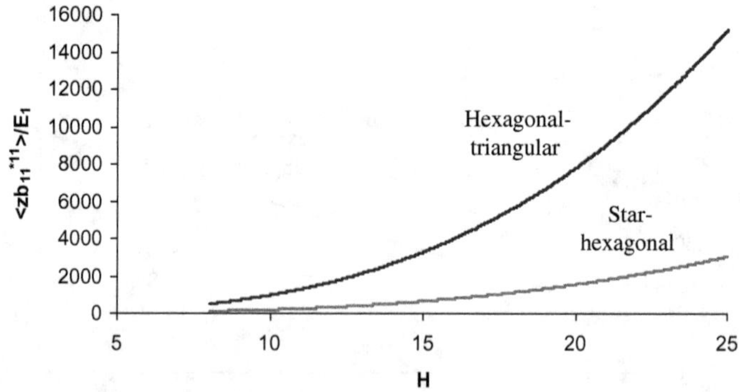

Fig. 1.35. Effective stiffness $\langle zb_{11}^{*11} \rangle / E_1$ vs. core height H of hexagonal-triangular and star-hexagonal sandwich cores.

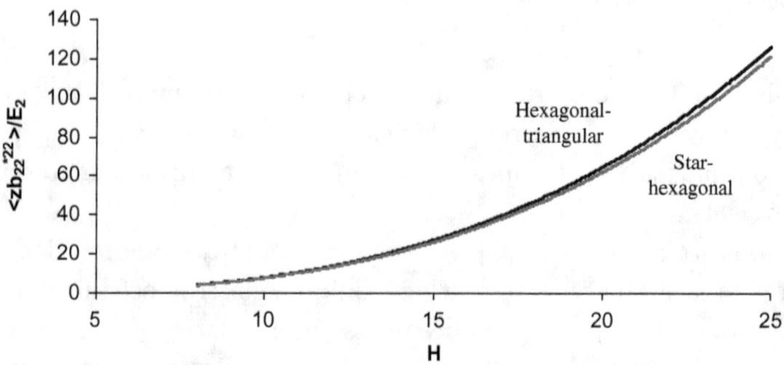

Fig. 1.36. Effective stiffness $\langle zb_{22}^{*22} \rangle / E_2$ vs. core height H of hexagonal-triangular and star-hexagonal sandwich cores.

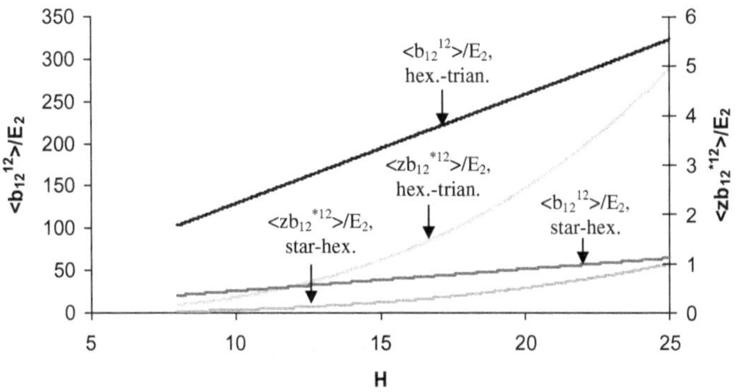

Fig. 1.37. Effective stiffness moduli $\langle b_{12}^{12}\rangle/E_2$ and $\langle zb_{12}^{*12}\rangle/E_2$ vs. core height H of hexagonal-triangular and star-hexagonal sandwich cores.

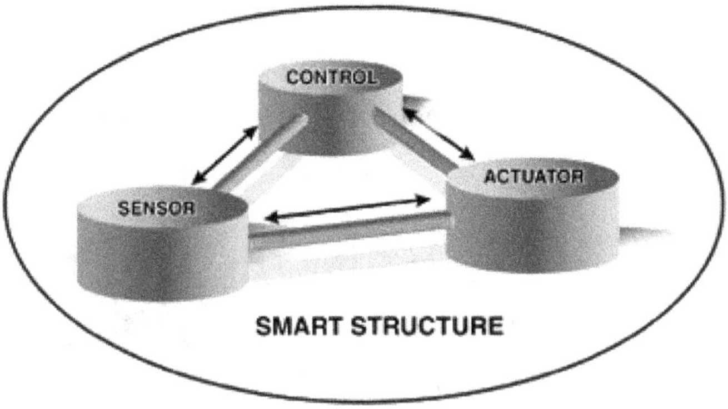

Fig. 1.38. Three basic constituents of smart structure: (1) Sensor — data acquisition. (2) Actuator — action triggered by control. (3) Control — analyzing data and reaching decision.

composite piezoelectric materials/structures with enhanced and unique mechanical and piezoelectric properties as well. Overall, in order to facilitate the continuous integration of piezoelectric smart composite structures in new engineering platforms, the effective homogenized mechanical, thermal and actuation properties of such composites structures need to be characterized accurately. The actuation coefficients characterize the intrinsic transducer nature of active smart composites that can be used to induce strains and stresses in a coordinated fashion.

In the present section, we will develop a general asymptotic homogenization model to determine the effective mechanical, thermal and actuation properties of (i) 3D smart composite materials, and (ii) smart composite shells and plates.

1.9.1. Asymptotic homogenization of 3D smart composite materials

Consider a general smart composite structure representing an inhomogeneous solid occupying domain Ω with boundary S that contains a large number of periodically arranged reinforcements and actuators as shown in Fig. 1.39(a). It can be observed that this periodic structure is obtained by repeating a small unit cell Y in the domain Ω, see Fig. 1.39(b).

The elastic deformation of this structure can be described by means of the following boundary-value problem (cf. Eqs. (1.13)–(1.16)):

$$\frac{\partial \sigma_{ij}\left(\mathbf{x}, \frac{\mathbf{x}}{\varepsilon}\right)}{\partial x_j} = F_i \quad \text{in } \Omega, \tag{1.102}$$

$$u_i\left(\mathbf{x}, \frac{\mathbf{x}}{\varepsilon}\right) = 0 \quad \text{on } S, \tag{1.103}$$

where

$$\begin{aligned} \sigma_{ij}\left(\mathbf{x}, \frac{\mathbf{x}}{\varepsilon}\right) &= C_{ijkl}\left(\frac{\mathbf{x}}{\varepsilon}\right)\left\{e_{kl}\left(\mathbf{x}, \frac{\mathbf{x}}{\varepsilon}\right)\right\} \\ &\quad - P_{ijkl}\left(\frac{\mathbf{x}}{\varepsilon}\right) R_k(\mathbf{x}) - \Theta_{ij}\left(\frac{\mathbf{x}}{\varepsilon}\right) T(\mathbf{x}), \end{aligned} \tag{1.104}$$

$$\varepsilon_{ij}\left(\mathbf{x}, \frac{\mathbf{x}}{\varepsilon}\right) = \frac{1}{2}\left\{\frac{\partial u_i}{\partial x_j}\left(\mathbf{x}, \frac{\mathbf{x}}{\varepsilon}\right) + \frac{\partial u_j}{\partial x_i}\left(\mathbf{x}, \frac{\mathbf{x}}{\varepsilon}\right)\right\}. \tag{1.105}$$

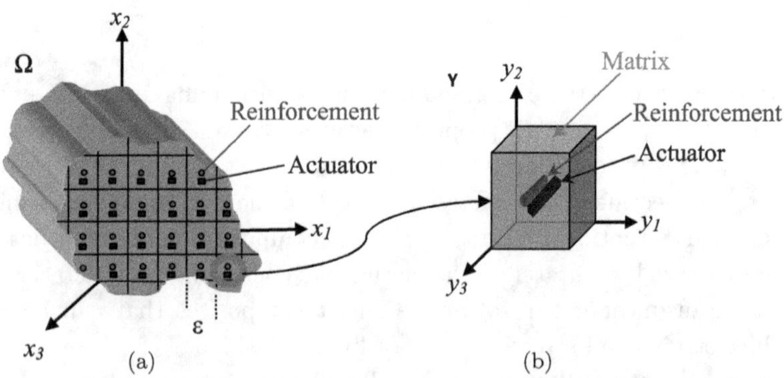

Fig. 1.39. (a) 3D periodic smart composite solid, (b) unit cell Y.

In Eqs. (1.102)–(1.105), in addition to earlier defined notations, P_{ijk} is a tensor of actuation coefficients describing the effect of a control signal $\mathbf{R}$ on the stress field σ_{ij}, Θ_{ij} is the thermal expansion tensor, and T represents change in temperature with respect to a reference state. In case of piezoelectric actuators, $\mathbf{R}$ is the electric field, and P_{ijk} are the piezoelectric coefficients. It is assumed that the elastic, actuation (piezoelectric in case of piezoelectric actuators) and thermal expansion coefficients are all periodic with a unit cell Y of characteristic dimension ε. Consequently, the periodic smart composite structure in Fig. 1.39(a) is seen to be made up of a large number of unit cells periodically arranged within the domain Ω. It is noteworthy to consider at this point that if the boundary conditions in Eq. (1.103) were made non-zero to examine a very general model, then boundary-layer type solutions can be developed to satisfy such inhomogeneous boundary conditions; see [19]. However, the obtained effective coefficients will not be altered in any way. The development of asymptotic homogenization model for the 3D smart composite structures can be found in [19, 32, 73–75]. The application of asymptotic homogenization method for the micromechanical analysis of quantum dot-embedded smart nanocomposite materials can be found in [88].

Similar to the asymptotic homogenization model developed for 3D bulk composite structures, the first step is to define the fast variables according to (cf. Eq. (1.2)):

$$y_i = \frac{x_i}{\varepsilon}, \quad i = 1, 2, 3. \tag{1.106}$$

As a consequence of introducing y, the chain rule of differentiation mandates that the derivatives must be transformed according to (cf. Eq. (1.5))

$$\frac{\partial}{\partial x_i} \rightarrow \frac{\partial}{\partial x_i} + \frac{1}{\varepsilon}\frac{\partial}{\partial y_i}. \tag{1.107}$$

Introduction of $\mathbf{y}$ necessitates the transformation of Eqs. (1.102)–(1.104) into the following expressions:

$$\frac{\partial \sigma_{ij}(\mathbf{x}, \mathbf{y})}{\partial x_j} + \frac{1}{\varepsilon}\frac{\partial \sigma_{ij}(\mathbf{x}, \mathbf{y})}{\partial y_i} = F_i \quad \text{in } \Omega, \tag{1.108}$$

$$u_i(\mathbf{x}, \mathbf{y}) = 0 \quad \text{on } S, \tag{1.109}$$

and

$$\sigma_{ij}(\mathbf{x}, \mathbf{y}) = C_{ijkl}(\mathbf{y})\left\{\frac{\partial u_k}{\partial x_l}(\mathbf{x}, \mathbf{y})\right\} - P_{ijk}(\mathbf{y})R_k(\mathbf{x}) - \Theta_{ij}(\mathbf{y})T(\mathbf{x}). \tag{1.110}$$

The next step in the model development is to asymptotically expand the stress and displacement fields into infinite series in terms of the small parameter ε as shown:

(i) *Asymptotic expansion for the displacement field*:

$$u_i(\mathbf{x},\mathbf{y}) = u_i^{(0)}(\mathbf{x},\mathbf{y}) + \varepsilon u_i^{(1)}(\mathbf{x},\mathbf{y}) + \varepsilon^2 u_i^{(2)}(\mathbf{x},\mathbf{y}) + \cdots . \tag{1.111}$$

(ii) *Asymptotic expansion for the stress field*:

$$\sigma_{ij}(\mathbf{x},\mathbf{y}) = \sigma_{ij}^{(0)}(\mathbf{x},\mathbf{y}) + \varepsilon \sigma_{ij}^{(1)}(\mathbf{x},\mathbf{y}) + \varepsilon^2 \sigma_{ij}^{(2)}(\mathbf{x},\mathbf{y}) + \cdots . \tag{1.112}$$

Substituting Eqs. (1.106), (1.107) and (1.110) into Eq. (1.108) and considering at the same time the periodicity of $u^{(i)}$ in y_j one can readily eliminate the microscopic variable y from the first term $u^{(0)}$ in the asymptotic displacement field expansion thus showing that it depends only on the macroscopic variable x. Subsequently, by substituting Eq. (1.112) into Eq. (1.108) and separating terms with like powers of ε one obtains a series of differential equations the first two of which are as follows:

$$\frac{\partial \sigma_{ij}^{(0)}}{\partial y_j} = 0, \tag{1.113}$$

$$\frac{\partial \sigma_{ij}^{(1)}}{\partial y_j} + \frac{\partial \sigma_{ij}^{(0)}}{\partial x_j} = f_i, \tag{1.114}$$

where

$$\sigma_{ij}^{(0)} = C_{ijkl}\left(\frac{\partial u_k^{(0)}}{\partial x_l} + \frac{\partial u_k^{(1)}}{\partial y_l}\right) - P_{ijk}R_k - \Theta_{ij}T, \tag{1.115}$$

$$\sigma_{ij}^{(1)} = C_{ijkl}\left(\frac{\partial u_k^{(1)}}{\partial x_l} + \frac{\partial u_k^{(2)}}{\partial y_l}\right). \tag{1.116}$$

From Eq. (1.113) and Eq. (1.115) one obtains:

$$\frac{\partial}{\partial y_j}\left(C_{ijkl}\frac{\partial u_k^{(1)}(\mathbf{x},\mathbf{y})}{\partial y_l}\right) = \frac{\partial P_{ijk}(\mathbf{y})}{\partial y_j}R_k(\mathbf{x}) + \frac{\partial \Theta_{ij}(\mathbf{y})}{\partial y_j}T(\mathbf{x}) - \frac{\partial C_{ijkl}(\mathbf{y})}{\partial y_j}\frac{\partial u_k^{(0)}(\mathbf{x})}{\partial x_l}. \tag{1.117}$$

The separation of variables on the right-hand side of Eq. (1.117) prompts us to write down the solution for $u^{(1)}$ as follows:

$$u_n^{(1)}(\mathbf{x},\mathbf{y}) = R_k(\mathbf{x})N_n^k(\mathbf{y}) + T(\mathbf{x})N_n(\mathbf{y}) + \frac{\partial u_k^{(1)}(\mathbf{x})}{\partial x_l}N_n^{kl}(\mathbf{y}), \tag{1.118}$$

where the auxiliary functions N_n^{kl}, N_n^k and N_n are periodic in $\mathbf{y}$ and satisfy (cf. Eq. (1.25)):

$$\frac{\partial}{\partial y_j}\left(C_{ijmn}(\mathbf{y})\frac{\partial N_m^{kl}(\mathbf{y})}{\partial y_n}\right) = -\frac{\partial C_{ijkl}}{\partial y_j}, \tag{1.119}$$

$$\frac{\partial}{\partial y_j}\left(C_{ijmn}(\mathbf{y})\frac{\partial N_m^{k}(\mathbf{y})}{\partial y_n}\right) = \frac{\partial P_{ijk}}{\partial y_j}, \tag{1.120}$$

$$\frac{\partial}{\partial y_j}\left(C_{ijmn}(\mathbf{y})\frac{\partial N_m(\mathbf{y})}{\partial y_n}\right) = \frac{\partial \Theta_{ij}}{\partial y_j}. \tag{1.121}$$

Equations (1.119)–(1.121) depend entirely on the fast variable $\mathbf{y}$ and are thus solved in the domain Y of the unit cell, remembering at the same time that all of C_{ijkl}, P_{ijk}, Θ_{ij} and N_m^{kl}, N_m^k, and N_m are Y-periodic in $\mathbf{y}$. They are appropriately called as unit-cell problems.

The next important step in the model development is the homogenization procedure. This is carried out by first substituting Eq. (1.118) into Eq. (1.115), and combining the result with Eq. (1.114). The resulting expressions are then integrated over the domain Y of the unit cell (with volume $|Y|$) remembering to treat x_i as a parameter as far as integration with respect to y_j is concerned. Canceling out terms that vanish due to periodicity considerations yields

$$\tilde{C}_{ijkl}\frac{\partial^2 u_k^{(0)}(\mathbf{x})}{\partial x_j \partial x_l} - \tilde{P}_{ijk}\frac{\partial R_k(\mathbf{x})}{\partial x_j} - \tilde{\Theta}_{ij}\frac{\partial T(\mathbf{x})}{\partial x_j} = F_i, \tag{1.122}$$

where the following definitions are introduced (cf. Eq. (1.27)):

$$\tilde{C}_{ijkl} = \frac{1}{|Y|}\int_Y \left(C_{ijkl}(\mathbf{y}) + C_{ijmn}(\mathbf{y})\frac{\partial N_m^{kl}(\mathbf{y})}{\partial y_n}\right)dv, \tag{1.123}$$

$$\tilde{P}_{ijk} = \frac{1}{|Y|}\int_Y \left(P_{ijk}(\mathbf{y}) - C_{ijmn}(\mathbf{y})\frac{\partial N_m^{kl}(\mathbf{y})}{\partial y_n}\right)dv, \tag{1.124}$$

$$\tilde{\Theta}_{ij} = \frac{1}{|Y|}\int_Y \left(K_{ij}(\mathbf{y}) - C_{ijmn}(\mathbf{y})\frac{\partial N_m(\mathbf{y})}{\partial y_n}\right)dv. \tag{1.125}$$

The coefficients $\tilde{C}_{ijkl}$, $\tilde{P}_{ijk}$, $\tilde{\Theta}_{ij}$ are called, respectively, the effective elastic, actuation (piezoelectric in case of piezoelectric actuators) and thermal expansion coefficients of the general 3D smart composite structure. As it was mentioned above, all the effective coefficients are free from the inhomogeneity complications that characterize their actual rapidly varying material counterparts C_{ijkl}, P_{ijk}, Θ_{ij}, and as such, are more amenable to analytical and numerical treatment. They are universal in nature and can be used to study a wide variety of boundary value problems associated with a given composite structure. Note that for a general 3D composite material, the elastic unit cell problem is given by Eq. (1.25) and the formulae for the effective elastic coefficient is given by Eq. (1.27).

In summary, Eqs. (1.119)–(1.125) represent the governing equations of the homogenized model of a smart composite structure with periodically arranged reinforcements and actuators. Equations (1.119)–(1.121) represent the unit cell problems, formulae (1.123)–(1.125) define the effective properties, and expression (1.118) provides an asymptotic formula for the local displacement field.

1.9.2. Asymptotic homogenization of smart composite shells and plates

In the present section, a general 3D micromechanical model pertaining to thin composite layers with wavy boundaries is applied to the case of smart shells and plates. Similar to Section 1.5, the asymptotic homogenization method is used, which reduces the original boundary value problem into a set of three decoupled problems, each problem characterized by two differential equations. These three sets of differential equations, referred to as unit-cell problems, deal, independently, with the elastic, piezoelectric, and thermal expansion behavior of the smart composite shells and plates. The solution of the unit-cell problems yields expressions for effective elastic, piezoelectric and thermal expansion coefficients which, as a consequence of their universal nature, can be used to study a wide variety of boundary value problems associated with a smart structure of a given geometry. These formulas for the effective properties can readily be used to tailor their values to meet the requirements of a particular application by changing certain material or geometric parameters such as the size or properties of the reinforcements.

The general 3D micromechanical models pertaining to smart composite layers with wavy boundaries can be found in [48, 49, 51, 52, 76, 77, 82–87].

The general homogenization model for smart composite shell has found numerous applications in the analysis of various practically important composite structures. Georgiades *et al.* [55] and Challagulla *et al.* [53, 60] studied generally orthotropic grid-reinforced and network smart composite plates and shells. Saha *et al.* [62, 63] and Saha and Kalamkarov [64] analyzed the smart sandwich composite shells, and, in particular, the honeycomb sandwich composite shells made of generally orthotropic materials.

Consider a 3D inhomogeneous thin layer with wavy surfaces and with a large number of embedded and periodically arranged reinforcements/actuators; see Fig. 1.10. This solid can be constructed by repeating a certain small unit cell Ω_δ in the α_1–α_2 plane (Fig. 1.10). All three coordinates (α_1, α_2, γ) in Fig. 1.10 are made dimensionless by dividing them with a certain characteristic dimension of the solid, D. Furthermore, δ is a small thickness of the smart shell and δh_1, δh_2 are the tangential dimensions of the unit cell.

Thus, the unit cell Ω_δ of the structure under consideration is characterized by the inequalities (1.36). The elastic deformation of this periodic structure can be represented by means of following expressions (cf. Eq. (1.37)):

$$\frac{\partial \sigma_{ij}}{\partial \alpha_j} = f_i, \tag{1.126}$$

$$\sigma_{ij} = C_{ijkl}\left\{e_{kl} - d^{(\mathrm{r})}_{klm}R_m - \theta^{(\mathrm{t})}_{kl}T\right\} \quad \text{and} \quad e_{ij} = \frac{1}{2}\left(\frac{\partial u_i}{\partial \alpha_j} + \frac{\partial u_j}{\partial \alpha_i}\right). \tag{1.127}$$

Here, in addition to earlier defined notations, $d^{(\mathrm{r})}_{ijk}$ is a tensor of actuation coefficients describing the effect of a control signal $\mathbf{R}$ on the stress field σ_{ij}, and $\theta^{(\mathrm{t})}_{ij}$ is the thermal expansion tensor (relating strain with temperature change). In case of piezoelectric actuators, $\mathbf{R}$ is the electric field, and $d^{(\mathrm{r})}_{ijk}$ are the piezoelectric coefficients. All indices take on values $1, 2, 3$ with $\alpha_3 = \gamma$. The inhomogeneity of composite material is modeled by assuming that the elastic, piezoelectric, and thermal expansion coefficients are the functions in spatial coordinates α_1, α_2, γ, periodic in tangential coordinates α_1 and α_2 with periodicity cell Ω_δ. Assume also that the top and bottom surfaces of the layer, $S^\pm$, are subjected to surface tractions p_i which are related to stresses as $\sigma_{ij}n_j = p_i$, where $\mathbf{n}$ is the unit vector normal to the surfaces $\gamma^\pm$ and is given by Eq. (1.38).

The fast variables, $\{\xi_1, \xi_2, z\}$ are introduced according to Eq. (1.40), and the asymptotic expansions for the displacement and stress fields given by Eq. (1.41) are subsequently assumed.

The solution of the pertinent 3D problem is obtained from Eqs. (1.126), (1.127), (1.40), (1.41) and results in the homogenized smart composite shell model. The constitutive relations of the equivalent (homogenized) smart shell are obtained in terms of the stress resultants ($N_{\alpha\beta}$), moment resultants ($M_{\alpha\beta}$), averaged (homogenized) stresses $\left\langle \sigma_{\alpha\beta}^{(1)} \right\rangle$ and $\left\langle z\sigma_{\alpha\beta}^{(1)} \right\rangle$, and the mid-surface strains ($\varepsilon_{\alpha\beta}$) and curvatures ($k_{\alpha\beta}$) as follows (cf. (1.48)):

$$N_{\alpha\beta} = \delta \left\langle b_{\alpha\beta}^{\lambda\mu} \right\rangle \varepsilon_{\lambda\mu} + \delta^2 \left\langle b_{\alpha\beta}^{*\lambda\mu} \right\rangle \kappa_{\lambda\mu} - \delta^2 \left\langle d_{\alpha\beta}^{k} \right\rangle R_k^{(0)} - \delta^2 \left\langle d_{\alpha\beta}^{*k} \right\rangle R_k^{(1)}$$
$$-\delta^2 \left\langle \theta_{\alpha\beta} \right\rangle T^{(0)} - \delta^2 \left\langle \theta_{\alpha\beta}^{*} \right\rangle T^{(1)}, \tag{1.128}$$

$$M_{\alpha\beta} = \delta \left\langle z b_{\alpha\beta}^{\lambda\mu} \right\rangle \varepsilon_{\lambda\mu} + \delta^2 \left\langle z b_{\alpha\beta}^{*\lambda\mu} \right\rangle \tau_{\lambda\mu} - \delta^2 \left\langle z d_{\alpha\beta}^{k} \right\rangle R_k^{(0)} - \delta^2 \left\langle z d_{\alpha\beta}^{*k} \right\rangle R_k^{(1)}$$
$$-\delta^2 \left\langle z\theta_{\alpha\beta} \right\rangle T^{(0)} - \delta^2 \left\langle z\theta_{\alpha\beta}^{*} \right\rangle T^{(1)}, \tag{1.129}$$

$$\left\langle \sigma_{\alpha\beta}^{(1)} \right\rangle = \left\langle b_{\alpha\beta}^{\lambda\mu} \right\rangle \varepsilon_{\lambda\mu}^{(1)} + \left\langle b_{\alpha\beta}^{*\lambda\mu} \right\rangle \kappa_{\lambda\mu} - \left\langle d_{\alpha\beta}^{k} \right\rangle R_k^{(0)} - \left\langle d_{\alpha\beta}^{*k} \right\rangle R_k^{(1)}$$
$$- \left\langle \theta_{\alpha\beta} \right\rangle T^{(0)} - \left\langle \theta_{\alpha\beta}^{*} \right\rangle T^{(1)}, \tag{1.130}$$

$$\left\langle z\sigma_{\alpha\beta}^{(1)} \right\rangle = \left\langle z b_{\alpha\beta}^{\lambda\mu} \right\rangle \varepsilon_{\lambda\mu}^{(1)} + \left\langle z b_{\alpha\beta}^{*\lambda\mu} \right\rangle \kappa_{\lambda\mu} - \left\langle z d_{\alpha\beta}^{k} \right\rangle R_k^{(0)} - \left\langle z d_{\alpha\beta}^{*k} \right\rangle R_k^{(1)}$$
$$- \left\langle z\theta_{\alpha\beta} \right\rangle T^{(0)} - \left\langle z\theta_{\alpha\beta}^{*} \right\rangle T^{(1)}, \tag{1.131}$$

where the temperature T and control signal $\mathbf{R}$ are assumed to follow linear through-the-thickness relationships. Throughout this work, it is assumed that Greek indices α, β, γ, etc. take values 1 and 2, whereas Latin indices, i, j, k, etc. vary from 1 to 3. In addition to earlier defined effective elastic coefficients $\langle b_{\alpha\beta}^{\lambda\mu} \rangle$, $\langle b_{\alpha\beta}^{*\lambda\mu} \rangle$, $\langle z b_{\alpha\beta}^{\lambda\mu} \rangle$, and $\langle z b_{\alpha\beta}^{*\lambda\mu} \rangle$ the quantities $\langle d_{\alpha\beta}^{k} \rangle$, $\langle d_{\alpha\beta}^{*k} \rangle$ are called the effective actuation coefficients, and $\langle \theta_{\alpha\beta} \rangle$, $\langle \theta_{\alpha\beta}^{*} \rangle$ are the effective thermal expansion coefficients of the homogenized smart shell. The effective coefficients are obtained through integration of local functions $b_{\alpha\beta}^{\lambda\mu}$, $b_{\alpha\beta}^{*\lambda\mu}$, $d_{\alpha\beta}^{k}$, $d_{\alpha\beta}^{*k}$, $\theta_{\alpha\beta}$, and $\theta_{\alpha\beta}^{*}$ over the 3D unit cell. In case of piezoelectric actuators, $\langle d_{\alpha\beta}^{k} \rangle$, $\langle d_{\alpha\beta}^{*k} \rangle$ are the effective piezoelectric coefficients.

The above introduced local functions are periodic in tangential coordinates ξ_1 and ξ_2 with periodicity cell Ω, but, evidently, they are not periodic in the z-direction and as such differ from classical homogenization schemes, see [5]. They are defined in terms of pertinent elastic C_{ijkl}, actuation (in case of piezoelectric actuators, piezoelectric) P_{ijk}, and thermal expansion K_{ij} material coefficients and the yet unknown local functions $U_n^{lm}(\xi_1, \xi_2, z)$, $V_n^{lm}(\xi_1, \xi_2, z)$, $U_n^l(\xi_1, \xi_2, z)$, etc. as follows:

$$\begin{aligned} b_{ij}^{lm} &= \frac{1}{h_\beta} C_{ijn\beta} \frac{\partial U_n^{lm}}{\partial \xi_\beta} + C_{ijn3} \frac{\partial U_n^{lm}}{\partial z} + C_{ijlm}, \\ b_{ij}^{*lm} &= \frac{1}{h_\beta} C_{ijn\beta} \frac{\partial V_n^{lm}}{\partial \xi_\beta} + C_{ijn3} \frac{\partial V_n^{lm}}{\partial z} + zC_{ijlm}, \end{aligned} \tag{1.132}$$

$$\begin{aligned} d_{ij}^{k} &= P_{ijk} - \frac{1}{h_\beta} C_{ijn\beta} \frac{\partial U_n^{k}}{\partial \xi_\beta} - C_{ijm3} \frac{\partial U_n^{k}}{\partial z}, \\ d_{ij}^{*k} &= zP_{ijk} - \frac{1}{h_\beta} C_{ijn\beta} \frac{\partial V_n^{k}}{\partial \xi_\beta} - C_{ijm3} \frac{\partial V_n^{k}}{\partial z}, \end{aligned} \tag{1.133}$$

$$\begin{aligned} \theta_{ij} &= K_{ij} - \frac{1}{h_\beta} C_{ijn\beta} \frac{\partial U_n}{\partial \xi_\beta} - C_{ijm3} \frac{\partial U_n}{\partial z}, \\ \theta_{ij}^{*k} &= zK_{ij} - \frac{1}{h_\beta} C_{ijn\beta} \frac{\partial V_n}{\partial \xi_\beta} - C_{ijm3} \frac{\partial V_n}{\partial z}, \end{aligned} \tag{1.134}$$

where

$$P_{ijm} = C_{ijkl} d_{klm}^{(r)}, \quad K_{ij} = C_{ijkl} \theta_{kl}^{(t)}. \tag{1.135}$$

Definitions given in Eqs. (1.132)–(1.134) are then substituted in the following unit-cell local problems to determine the local functions $U_n^{lm}(\xi_1, \xi_2, z)$, $V_n^{lm}(\xi_1, \xi_2, z)$, $U_n^l(\xi_1, \xi_2, z)$, etc., all of them periodic in ξ_1, ξ_2 with periodicity cell Ω (cf. (1.46) and (1.47)),

$$\frac{1}{h_\beta} \frac{\partial}{\partial \xi_\beta} b_{i\beta}^{\lambda\mu} + \frac{\partial}{\partial z} b_{i3}^{\lambda\mu} = 0, \qquad \frac{1}{h_\beta} n_\beta^{\pm} b_{i\beta}^{\lambda\mu} + n_3^{\pm} b_{i3}^{\lambda\mu} = 0 \text{ at } z = z^{\pm}, \tag{1.136}$$

$$\frac{1}{h_\beta} \frac{\partial}{\partial \xi_\beta} b_{i\beta}^{*\lambda\mu} + \frac{\partial}{\partial z} b_{i3}^{*\lambda\mu} = 0, \qquad \frac{1}{h_\beta} n_\beta^{\pm} b_{i\beta}^{*\lambda\mu} + n_3^{\pm} b_{i3}^{*\lambda\mu} = 0 \text{ at } z = z^{\pm}, \tag{1.137}$$

$$\frac{1}{h_\beta} \frac{\partial}{\partial \xi_\beta} d_{i\beta}^{k} + \frac{\partial}{\partial z} d_{i3}^{k} = 0, \qquad \frac{1}{h_\beta} n_\beta^{\pm} d_{i\beta}^{k} + n_3^{\pm} d_{i3}^{k} = 0 \text{ at } z = z^{\pm}, \tag{1.138}$$

$$\frac{1}{h_\beta} \frac{\partial}{\partial \xi_\beta} d_{i\beta}^{*k} + \frac{\partial}{\partial z} d_{i3}^{*k} = 0, \qquad \frac{1}{h_\beta} n_\beta^{\pm} d_{i\beta}^{*k} + n_3^{\pm} d_{i3}^{*k} = 0 \text{ at } z = z^{\pm}, \tag{1.139}$$

$$\frac{1}{h_\beta} \frac{\partial}{\partial \xi_\beta} \theta_{i\beta} + \frac{\partial}{\partial z} \theta_{i3} = 0, \qquad \frac{1}{h_\beta} n_\beta^{\pm} \theta_{i\beta} + n_3^{\pm} \theta_{i3} = 0 \text{ at } z = z^{\pm}, \tag{1.140}$$

$$\frac{1}{h_\beta} \frac{\partial}{\partial \xi_\beta} \theta_{i\beta}^{*} + \frac{\partial}{\partial z} \theta_{i3}^{*} = 0, \qquad \frac{1}{h_\beta} n_\beta^{\pm} \theta_{i\beta}^{*} + n_\beta^{\pm} \theta_{i3}^{*} = 0 \text{ at } z = z^{\pm}. \tag{1.141}$$

Here, $z^{\pm}$ represent the profiles of the upper and lower surfaces of the composite layer in terms of the macroscopic variables. The differential equations and pertinent boundary conditions in Eqs. (1.138)–(1.141) are formulated and solved entirely on the domain of the unit cell and are independent of the global formulation of the problem. Once functions $U_n^{lm}(\xi_1,\xi_2,z)$, $V_n^{lm}(\xi_1,\xi_2,z)$, $U_n^l(\xi_1,\xi_2,z)$, etc. are determined, they are back-substituted into Eqs. (1.132)–(1.134) to obtain the local functions $b_{\alpha\beta}^{\lambda\mu}$, $b_{\alpha\beta}^{*\lambda\mu}$, $d_{\alpha\beta}^k$, $d_{\alpha\beta}^{*k}$, $\theta_{\alpha\beta}$, and $\theta_{\alpha\beta}^*$ and finally these are used to determine the effective coefficients by averaging over the volume of the 3D unit cell Ω as follows:

$$\langle f(\xi_1,\xi_2,z)\rangle = \frac{1}{|\Omega|}\int_\Omega f(\xi_1,\xi_2,z)d\xi_1 d\xi_2 dz. \tag{1.142}$$

In addition to the above given symmetry properties of the effective coefficients given by the Eq. (1.49), the following new symmetry properties have been proved by Kalamkarov and Georgiades [49], Kalamkarov *et al.* [51, 52]:

$$\begin{aligned} &\delta\left\langle d_{mn}^k\right\rangle = \left\langle d_{ijk}^{(r)} b_{ij}^{mn}\right\rangle, \quad \delta\left\langle z d_{mn}^k\right\rangle = \left\langle d_{ijk}^{(r)} b_{ij}^{*mn}\right\rangle,\\ &\delta\left\langle d_{mn}^{*k}\right\rangle = \left\langle z d_{ijk}^{(r)} b_{ij}^{mn}\right\rangle, \quad \delta\left\langle z d_{mn}^{*k}\right\rangle = \left\langle z d_{ijk}^{(r)} b_{ij}^{*mn}\right\rangle, \end{aligned} \tag{1.143}$$

$$\begin{aligned} &\delta\left\langle \theta_{mn}\right\rangle = \left\langle \alpha_{ij}^{(\theta)} b_{ij}^{mn}\right\rangle, \quad \delta\left\langle z\theta_{mn}\right\rangle = \left\langle \alpha_{ij}^{(\theta)} b_{ij}^{*mn}\right\rangle,\\ &\delta\left\langle \theta_{mn}^*\right\rangle = \left\langle z\alpha_{ij}^{(\theta)} b_{ij}^{mn}\right\rangle, \quad \delta\left\langle z\theta_{mn}^*\right\rangle = \left\langle z\alpha_{ij}^{(\theta)} b_{ij}^{*mn}\right\rangle. \end{aligned} \tag{1.144}$$

The unit-cell problems given by Eqs. (1.132)–(1.134) and (1.136)–(1.141) have been solved analytically for a number of structures of a practical interest, and the explicit analytical formulas for the effective stiffness moduli have been obtained for the following types of smart composite and reinforced shells and plates: angle-ply fiber-reinforced shells, grid-reinforced and network shells (Georgiades *et al.* [54], Challagulla *et al.* [53, 58, 60], Hadjiloizi *et al.* [77]), rib- and wafer-like reinforced shells (Kalamkarov and Georgiades [49], Kalamkarov *et al.* [52], Georgiades and Kalamkarov [50], Hadjiloizi *et al.* [76, 77, 82–86]), sandwich composite shells, in particular, the honeycomb sandwich composite shells made of generally orthotropic materials (Kalamkarov *et al.* [51], Saha *et al.* [62, 63], Saha and Kalamkarov [64]).

Asymptotic analysis of perforated plates, shells and membranes is developed in [78, 79]. The methods of three-phase composite model, boundary shape perturbation technique, and Padé approximants are applied to the solution of the unit-cell problems for composites in [80, 81, 89]. Asymptotic homogenization model is generalized for the geometrically nonlinear elastic composite plates with wavy surfaces by Kalamkarov *et al.* [90]. Comparative analysis of micromechanical models for the elastic composite laminae and the analysis of longitudinal, shear and transversal strength of composite laminae is conduced in [91–94] Theoretical models are compared with the numerous experimental data. It is shown, in particular that the asymptotic homogenization method presents the best predictions among the elasticity-based models.

The general asymptotic homogenization model pertaining to smart magnetoelectric composite shells composed of reinforcements and/or actuators attached to the surface of a shell or sandwiched between two carrier layers is developed in [95]. This paper highlights the strong influence of the curvature of the shell's middle surface on the effective properties. That is caused by entrance of the coefficients $A_1(\alpha_1, \alpha_2), A_2(\alpha_1, \alpha_2)$ of the first quadratic form of mid-surface into the Eq. (1.40), as explained on pages 25 and 26. After homogenization the effective shell can be structurally inhomogeneous even if the material of original shell was homogeneous. The general model developed in [95] is illustrated by several practically important examples which include spherical, paraboloidal, ellipsoidal and cylindrical shells reinforced with ribs, wafers, or triangularly arranged reinforcements. It is also shown in [95] that in case of shell reinforced with piezoelectric and piezomagnetic constituents, the effective elastic and most other coefficients are functions of not only elastic properties but also of the electric and magnetic properties of the constituents.

1.10. Conclusion

Asymptotic homogenization is a mathematically rigorous powerful tool for analyzing composite materials and structures. The proof of the possibility of homogenizing a composite material of a regular structure, i.e., of examining an equivalent homogeneous solid instead of the original inhomogeneous composite solid, is one of the principal results of this theory. Method of asymptotic homogenization has also indicated a procedure of transition from the original problem (which contains in its formulation a small parameter related to the small dimensions of the constituents of the composite) to a problem for a homogeneous solid. The effective properties of this equivalent homogeneous material are determined through the solution of the unit-cell problems. Important advantage of the asymptotic homogenization is that, in addition to the effective properties, it allows to determine with a high accuracy the local stress and strain distributions defined by the microstructure of composite materials.

The present chapter reviews the basics of the asymptotic homogenization of advanced composite materials, thin-walled composite structures and smart composite materials and structures.

The explicit analytical formulae for the effective elastic properties of the generally orthotropic 3D grid-reinforced composites of various structures are derived. The asymptotic homogenization of 3D thin-walled composite reinforced structures is presented, and the general homogenization smart composite shell model is introduced. Micromechanical models are derived and applied to obtain the analytical formulas for the effective stiffnesses of the generally orthotropic composite shells and plates, rib- and wafer-reinforced shells, network and grid-reinforced shells as well as sandwich composite shells with cellular cores of different geometrical configuration. In particular, one of considered examples represents micromechanical modeling of the carbon nanotubes.

For many problems that are presented in the present chapter, any other analytical or numerical approaches are not as effective as the asymptotic homogenization method.

References

[1] Kalamkarov A.L., Fitzgerald S., MacDonald D., Georgiades A., 2000. Mechanical performance of pultruded composites rods with embedded fiber optic sensors, *J. Comp. Sci Tech.*, 60, 1171–1179.

[2] Kalamkarov A.L., Fitzgerald S., MacDonald D., Georgiades A., 1999. On the processing and evaluation of pultruded smart composites, *Comp. Part B: Eng.*, 30(2), 167–175.

[3] Kalamkarov A.L., Andrianov I.V., Danishevs'kyy V.V., 2009. Asymptotic homogenization of composite materials and structures, *Trans. ASME, Appl. Mech. Rev.*, 62(3), 030802-1–030802-20.

[4] Bakhvalov N.S., Panasenko G.P., 1989. *Homogenisation: Averaging Processes in Periodic Media. Mathematical Problems in Mechanics of Composite Materials*, Kluwer, Dordrecht.

[5] Kalamkarov A.L., 1992. *Composite and Reinforced Elements of Construction*, Wiley, Chichester, NY.

[6] Bensoussan A., Lions J.-L., Papanicolaou G., 1978. *Asymptotic Methods in Periodic Structures*, North-Holland, Amsterdam.
[7] Panasenko G.P., 2005. *Multi-Scale Modeling for Structures and Composites*, Springer, Berlin.
[8] Sanchez-Palencia E., 1980. *Non-homogeneous Media and Vibration Theory*, Springer, Berlin, New York.
[9] Dal Maso G., 1993. *An Introduction to Γ-Convergence*, Birkhäuser, Boston.
[10] Oleynik O.A., Shamaev A.S., Yosifian G.A., 1992. *Mathematical Problems in Elasticity and Homogenization*, North-Holland, Amsterdam.
[11] Berdichevsky V., Jikov V., Papanicolaou G., (eds) 1999. *Homogenization*, World Scientific, Singapore.
[12] Tartar L., 1990. H-measure, a new approach for studying homogenization, oscillation and concentration effects in partial differential equations, *Proc. R. Soc. Edinburg A*, 115, 193–230.
[13] Allaire G., Conca C., 1998. Boundary layers in the homogenization of a spectral problem in fluid-solid structures, *SIAM J. Math. Anal.*, 29(2), 343–379.
[14] Conca C., Lund F., 1999. Fourier homogenization method and the propagation of acoustic waves through a periodic vortex array, *SIAM J. Appl. Math.*, 59(5), 1573–1581.
[15] Lagzdins A.Zh., Tamuzh V.P., Teters G.A., Kregers A.F., 1992. *Orientational Averaging in Mechanics of Solids*, Longman Scientific & Technical, Harlow, UK, Wiley, NY.
[16] Dorobantu M., Enquist B., 1998. Wavelet-based numerical homogenization, *SIAM J. Math. Anal.*, 35(2), 540–559.
[17] Pilipchuk V.N., Starushenko G.A., 1997. A version of non-smooth transformations of variables for one-dimensional elastic systems of periodic structures, *J. Appl. Math. Mech.*, 61(2), 265–274.
[18] Milton G.W., 2002. *The Theory of Composites*, Cambridge University Press, Cambridge.
[19] Kalamkarov A.L., Georgiades A., 2002. Micromechanical modeling of smart composite structures, *Smart Mat. Struct.*, 11, 423–434.
[20] Kalamkarov A.L., 1995. Torsion of a laminated composite shaft, *Proc. 15th Canadian Congress of Applied Mechanics (CANCAM'95)*, Victoria, BC, May/June 1995, Vol. 1, 230–231.
[21] Kalamkarov A.L., Kudryavtsev B.A., Bardzokas D., 1991. New generalized integral transforms in axisymmetric boundary-value problems in the mechanics of composites, *Mech. Comp. Mat.*, 27(6), 648–657.
[22] Allaire G., Amar M., 1999. Boundary layer tails in periodic homogenization, *ESAIM: Control Optim. Calc. Var.*, 4, 209–243.
[23] Mishuris G., Öchsner A., 2005. Edge effects connected with thin interfaces in composite materials, *Comp. Struct.*, 68, 409–417.
[24] Versieux H.M., Sarkis M., 2006. Numerical boundary corrector for elliptic equations with rapidly oscillating periodic coefficients, *Commun. Numer. Meth. Eng.*, 22, 577–589.

[25] Kalamkarov A.L., 1987. On the determination of effective characteristics of cellular plates and shells of periodic structure, *Mech. Solids*, 22, 175–179.

[26] Kalamkarov A.L., 1989. Thermoelastic problem for structurally non-uniform shells of regular structure, *J. Appl. Mech. Techn. Phys.*, 30, 981–988.

[27] Kalamkarov A.L., Kolpakov A.G., 1997. *Analysis, Design and Optimization of Composite Structures*, Wiley, Chichester, NY.

[28] Kalamkarov A.L., Kolpakov A.G., 1996. On the analysis and design of fiber reinforced composite shells, *Trans. ASME J. Appl. Mech.*, 63(4), 939–945.

[29] Andrianov I.V., Danishevs'kyy V.V., Kalamkarov A.L., 2006. Asymptotic justification of the three-phase composite model, *Comp. Struct.*, 77(3), 395–404.

[30] Christensen R.M., 2005. *Mechanics of Composite Materials*, Dover, New York.

[31] Kalamkarov A.L., Liu H.Q., 1998. A new model for the multiphase fiber-matrix composite materials, *Comp. B: Eng.*, 29B(5), 643–653.

[32] Kalamkarov A.L., Georgiades A., 2002. Modeling of smart composites on account of actuation, thermal conductivity and hygroscopic absorption, *Comp. B: Eng.*, 33 B(2), 141–152.

[33] Jiang C.P., Xu Y.L., Cheung Y.K., Lo S.H., 2004. A rigorous method for doubly periodic cylindrical inclusions under longitudinal shear and its applications, *Mech. Mater*, 36, 225–237.

[34] Rodríguez-Ramos R., Sabina F.J., Guinovart-Díaz R., Bravo-Castillero J., 2001. Closed-form expressions for the effective coefficients of fibre-reinforced composite with transversely isotropic constituents. I. Elastic and square symmetry, *Mech. Mater.*, 33, 223–235.

[35] Hassan E., Kalamkarov A.L., Georgiades A., Challagulla K., 2009. An asymptotic homogenization model for smart 3D grid-reinforced composite structures with generally orthotropic constituents, *Smart Mater. Struct.* 18(7), 075006 (16 pp).

[36] Reddy J.N., 1997. *Mechanics of Laminated Composite Plates: Theory and Analysis*, CRC Press, Boca Raton.

[37] Andrianov I.V., Danishevs'kyy V.V., Kalamkarov A.L., 2008. Micromechanical analysis of fiber-reinforced composites on account of influence of fiber coatings, *Comp. B: Eng.*, 39(5), 874–881.

[38] Hassan E., Georgiades A., Savi M., Kalamkarov A.L., 2011. Analytical and numerical analysis of 3D grid-reinforced orthotropic composite structures, *Int. J. Eng. Sci.*, 49(7), 589–605.

[39] Duvaut G., 1976. Analyse fonctionnelle et méchanique des milieux continues, in *Proc. 14th IUTAM Congr.*, North-Holland, Amsterdam, 119–132.

[40] Duvaut G., 1977. Comportement macroscopique d'une plaque perforée périodiquement, *Lect. Notes Math.*, 594, 131–145.

[41] Caillerie D., 1981. Equations de la diffusion stationnaire dans un domaine comportant une distribution périodique d'inclusions aplaties de grande conductivité, *C.R. Acad. Sci., Ser. 1*, 292(1), 115–118.

[42] Caillerie D., 1984. Thin elastic and periodic plates, *Math. Meth. Appl. Sci.*, 6, 159–191.

[43] Kohn R.V., Vogelius M., 1984. A new model for thin plates with rapidly varying thickness, *Int. J. Solids Struct.*, 20, 333–350.
[44] Kohn R.V., Vogelius M., 1985. A new model for thin plates with rapidly varying thickness, II: A convergence proof, *Quart. J. Appl. Math.*, 43, 1–22.
[45] Kohn R.V., Vogelius M., 1986. A new model for thin plates with rapidly varying thickness, III: Comparison of different scalings, *Quart. J. Appl. Math.*, 44, 35–48.
[46] Kalamkarov A.L., 1988. Geometrically nonlinear problem of a thin composite layer with undulating surfaces of periodic structure, *Mech. Solids*, 23(5), 40–45.
[47] Kalamkarov A.L., Duvaut G., Léné F., 2002. A new asymptotic model of flexible composite shells of a regular structure, *Int. J. Eng. Sci.*, 40, 333–343.
[48] Kalamkarov A.L., Kolpakov A.G., 2001. A new asymptotic model for a composite piezoelastic plate, *Int. J. Solids Struct.*, 38(34–35), 6027–6044.
[49] Kalamkarov A.L., Georgiades A., 2004. Asymptotic homogenization models for smart composite plates with rapidly varying thickness: Part I — Theory, *Int. J. Multiscale Comp. Eng.*, 2(1), 133–148.
[50] Georgiades A., Kalamkarov A.L., 2004. Asymptotic homogenization models for smart composite plates with rapidly varying thickness: Part II — Applications, *Int. J. Multiscale Comput. Eng.*, 2(1), 149–172.
[51] Kalamkarov A.L., Saha G., Georgiades A., 2007. General micromechanical modeling of smart composite shells with application to smart honeycomb sandwich structures, *Comp. Struct.*, 79(1), 18–33.
[52] Kalamkarov A.L., Georgiades A., Challagulla K., Saha G., 2006. Micromechanics of smart composite plates with periodically embedded actuators and rapidly varying thickness, *J. Thermoplastic Comp. Mat.*, 19(3), 251–276.
[53] Challagulla K., Georgiades A., Kalamkarov A.L., 2010. Asymptotic homogenization modeling of smart composite generally orthotropic grid-reinforced shells: Part I — Theory, *European J. Mech. — A/Solids*, 29(4), 530–540.
[54] Georgiades A., Challagulla K., Kalamkarov A.L., 2010. Asymptotic homogenization modeling of smart composite generally orthotropic grid-reinforced shells: Part II — Applications, *European J. Mech. — A/Solids*, 29(4), 541–556.
[55] Georgiades A., Kalamkarov A.L., Challagulla K., 2006. Asymptotic homogenization model for generally orthotropic reinforcing networks in smart composite plates, *Smart Mat. Struct.*, 15, 1197–1210.
[56] Hassan E., Kalamkarov A.L., Georgiades A., Challagulla K., 2009. Asymptotic homogenization model for smart 3D grid-reinforced composite structures with generally orthotropic constituents, *Smart Mat. Struct.*, 18(7), 075006.
[57] Hadjiloizi D., Georgiades A., Kalamkarov A.L., 2012. Dynamic modeling and determination of effective properties of smart composite plates with rapidly varying thickness, *Int. J. Eng. Sci.*, 56, 63–85.

[58] Challagulla K., Georgiades A., Kalamkarov A.L., 2007. Asymptotic homogenization modelling of thin composite network structures, *Comp. Struct.*, 79(3), 432–444.
[59] Challagulla K., Georgiades A., Kalamkarov A.L., 2007. Asymptotic homogenization model for three-dimensional network reinforced composite structures, *J. Mech. Mat. Struct.*, 2(4), 613–632.
[60] Challagulla K., Georgiades A., Saha G., Kalamkarov A.L., 2008. Micromechanical analysis of grid-reinforced thin composite generally orthotropic shells, *Comp. B: Eng.*, 39(4), 627–644.
[61] Kalamkarov A.L., Hassan E., Georgiades A., Savi M., 2009. Asymptotic homogenization model for 3D grid-reinforced composite structures with generally orthotropic reinforcements, *Comp. Struct.*, 89(2), 186–196.
[62] Saha G., Kalamkarov A.L., Georgiades A., 2007. Asymptotic homogenization modeling and analysis of effective properties of smart composite reinforced and sandwich shells, *Int. J. Mech. Sci.*, 49(2), 138–150.
[63] Saha G., Kalamkarov A.L., Georgiades A., 2007. Effective elastic characteristics of honeycomb sandwich composite shells made of generally orthotropic materials, *Comp. A: Appl. Sci. Manufact.*, 38(6), 1533–1546.
[64] Saha G., Kalamkarov A.L., 2009. Micromechanical thermoelastic model for sandwich composite shells made of generally orthotropic materials, *J. Sandwich Struct. Mat.*, 11(1), 27–56.
[65] Kalamkarov A.L., Veedu V., Ghasemi-Nejhad M., 2005. Mechanical properties modeling of carbon single-walled nanotubes: An asymptotic homogenization method, *J. Comp. Theor. Nanoscience*, 2(1), 124–131.
[66] Kalamkarov A.L., Georgiades A., Rokkam S., Veedu V., Ghasemi-Nejhad M., 2006. Analytical and numerical techniques to predict carbon nanotubes properties, *Int. J. Solids Struct.*, 43(22–23), 6832–6854.
[67] Kalamkarov A.L., Hassan E., Georgiades A., 2010. Micromechanical modeling of 3D grid-reinforced composite structures and nanocomposites, *J. Nanostructured Polymers Nanocomposites*, 6(1), 12–20.
[68] Yao N., Lordi V., 1998. Young's modulus of single-walled carbon nanotubes, *J. Appl. Phys.*, 84(4), 1939–1943.
[69] Cai J., Wang C.Y., Yu T., Yu S., 2009. Wall thickness of single-walled carbon nanotubes and its Young's modulus, *Physica Scripta*, 79, 025702 (5 pp).
[70] Vinson J.R., 1993. *The Behavior of Shells Composed of Isotropic and Composite Materials*, Kluwer, Dordrecht.
[71] Gibson R.F., 1994. *Principles of Composite Material Mechanics*, McGraw-Hill, New York.
[72] Christensen R.M., 2000. Mechanics of cellular and other low-density materials, *Int. J. Solids Struct.*, 37, 93–104.
[73] Georgiades A., Challagulla K., Kalamkarov A.L., 2006. Modeling of the thermopiezoelastic behavior of prismatic smart composite structures made of orthotropic materials, *Comp. Part B: Eng.*, 37(7–8), 569–582.
[74] Hadjiloizi D.A., Georgiades A., Kalamkarov A.L., Jothi S., 2013. Micromechanical model of piezo-magneto-thermo-elastic composite structures: Part I — Theory, *European J. Mech. A - Solids*, 39, 298–312.

[75] Hadjiloizi D.A., Georgiades A., Kalamkarov A.L., Jothi S., 2013. Micromechanical model of piezo-magneto-thermo-elastic composite structures: Part II — Applications, *European J. Mech. A - Solids*, 39, 313–327.

[76] Hadjiloizi D.A., Kalamkarov A.L., Metti Ch., Georgiades A., 2014. Analysis of smart piezo-magneto-thermo-elastic composite and reinforced plates: Part I — Model development, *Int. J. Curved Layered Struct.*, 1, 11–31.

[77] Hadjiloizi D.A., Kalamkarov A.L., Metti Ch., Georgiades A., 2014. Analysis of smart piezo-magneto-thermo-elastic composite and reinforced plates: Part II — Applications, *Int. J. Curved Layered Struct.*, 1, 32–58.

[78] Kalamkarov A.L., Andrianov I.V., Weichert D., 2012. Asymptotic analysis of perforated shallow shells, *Int. J. Eng. Sci.*, 53, 1–18.

[79] Andrianov I.V., Danishevs'kyy V.V., Kalamkarov A.L., 2012. Asymptotic analysis of perforated plates and membranes. Part 1. Static problems for small holes, *Int. J. Solids Struct.*, 49(2), 298–310.

[80] Kalamkarov A.L., Andrianov I.V., Starushenko G.A., 2014. Three-phase model for a composite material with cylindrical circular inclusions. Part I: Application of the Boundary shape perturbation method, *Int. J. Eng. Sci.*, 78, 154–177.

[81] Kalamkarov A.L., Andrianov, I.V. Starushenko G.A., 2014. Three-phase model for a composite material with cylindrical circular inclusions. Part II: Application of Padé approximants, *Int. J. Eng. Sci.*, 78, 178–191.

[82] Hadjiloizi D.A., Kalamkarov A.L., Georgiades A.V., 2018. Micromechanical analysis of piezo-magneto-thermo-elastic T-ribbed and P-ribbed plates, *Mech. Adv. Materials Struct.*, 25, 657–668.

[83] Hadjiloizi D.A., Kalamkarov A.L., Saha G.C., Christoforidis K., Georgiades A.V., 2017. Micromechanical modeling of thin composite and reinforced magnetoelectric plates — effective elastic, piezoelectric and piezomagnetic coefficients, *Comp. Struct.*, 172, 102–118.

[84] Hadjiloizi D.A., Kalamkarov A.L., Saha G.C., Christofi I., Georgiades A.V., 2017. Micromechanical modeling of thin composite and reinforced magnetoelectric plates — effective electrical, magnetic, thermal and product properties, *Comp. Part B: Eng.*, 113, 243–269.

[85] Hadjiloizi D.A., Kalamkarov A.L., Georgiades A.V., 2017. Plane stress analysis of magnetoelectric composite and reinforced plates: micromechanical modeling and application to laminated structures, *ZAMM (Zeit. Angew. Math. Mech.)*, 97, 761–785.

[86] Hadjiloizi D.A., Kalamkarov A.L., Georgiades A.V., 2017. Plane stress analysis of magnetoelectric composite and reinforced plates: micromechanical modeling and applications to wafer- and rib-reinforced plates and three-layered honeycomb shells, *ZAMM (Zeit. Angew. Math Mec.)* 97, 786–814.

[87] Christofi I., Hadjiloizi D.A., Kalamkarov A.L., Georgiades A.V., 2021. Micromechanical analysis of thermoelastic and magnetoelectric composite and reinforced shells, *Comp. Struct.* 259, 113426.

[88] Alam A., Saha G.C., Kalamkarov A.L., 2020. Micromechanical analysis of quantum dot-embedded smart nanocomposite materials, *Comp. C*, 3, 100062.

[89] Kalamkarov A.L., Andrianov A.V., Pacheco P.M.C.L., Savi M.A., Starushenko G.A., 2016. Asymptotic analysis of fiber-reinforced composites of hexagonal structure, *J. Multiscale Modelling* 7, 1650006.
[90] Kalamkarov A.L., Tornabene F., Pacheco P.M.C.L., Savi M.A., Saha G.C., 2017. Geometrically non-linear elastic model for a thin composite layer with wavy surfaces, *ZAMM (Zeits. Angew. Math. Mech.* 97, 1381–1392.
[91] Vignoli L.L., Savi M.A., Pacheco P.M.C.L., Kalamkarov A.L., 2019. Comparative analysis of micromechanical models for the elastic composite laminae, *Comp. B: Eng.,*, 174, 106961.
[92] Vignoli L.L., Savi M.A., Pacheco P.M.C.L., Kalamkarov A.L., 2020. Micromechanical analysis of strength of notched composite plates, *Comp. Struct.*, 253, 112827.
[93] Vignoli L.L., Savi M.A., Pacheco P.M.C.L., Kalamkarov A.L., 2020. Micromechanical analysis of longitudinal and shear strength of composite laminae, *J. Comp. Mater.*, 54, 4853–4873.
[94] Vignoli L.L., Savi M.A., Pacheco P.M.C.L., Kalamkarov A.L., 2020. Micromechanical analysis of transversal strength of composite laminae, *Comp. Struct.*, 250, 112546.
[95] Christofi I., Hadjiloizi D.A., Kalamkarov A.L., Georgiades A.V., 2021. Asymptotic homogenization of magnetoelectric reinforced shells: effective coefficients and influence of shell curvature, *Int. J. Solids Structures*, 228, 111105.

https://doi.org/10.1142/9781800611887_0002

Chapter 2

Scaling Functions in Spatially Random Composites

Martin Ostoja-Starzewski[*,‡] **and Shivakumar I. Ranganathan**[†,§]
[*]*Department of Mechanical Science and Engineering, Beckman Institute, and Institute for Condensed Matter Theory, University of Illinois at Urbana-Champaign, Urbana, IL 61801, USA*
[†]*International Consulting Associates, Inc., 3300 Fairfax Dr Suite 302, Arlington, VA 22201, USA*
[‡]*martinos@illinois.edu*
[§]*sranganathan@icaglobal.co*

Abstract

We review the key issues involved in scaling and homogenization of random composite materials. In the first place, this involves a Hill–Mandel condition in the setting of stochastic micromechanics. Within this framework, we introduce the concept of a *scaling function* that describes "finite-size scaling" of thermally conducting or elastic crystalline aggregates. While the finite size is represented by the *mesoscale*, the scaling function depends on an appropriate measure quantifying the single-crystal anisotropy. Based on the scaling function, we construct a *material scaling diagram*, from which one can assess the scaling trend from a statistical volume element (SVE) to a representative volume element (RVE) for many different materials. We demonstrate these concepts with the scaling of the fourth-rank elasticity and the second-rank thermal conductivity tensors. We also briefly discuss the trends in approaching the RVE for linear/nonlinear (thermo)elasticity, plasticity, and Darcy permeability.

Contents

2.1. Introduction

Determining the scale-dependent properties of spatially random materials given the properties of individual phases, volume fraction and the orientation distribution of phases is of importance in the design of composite materials with desired properties. To illustrate the effect of scale dependence, consider a simple tension test of a polycrystalline aggregate such as a bar of copper with randomly oriented grains. This is perhaps among the earliest concepts taught in an introductory solid mechanics or materials science course. The objective of conducting such a test would be to construct the stress–strain curve (true or engineering) and extract appropriate material properties (elastic modulus, yield stress, ultimate stress, toughness, etc). In extracting these properties, it is tacitly assumed that the sample size is "significantly larger" than the grain size which leads to the notion of a Representative Volume Element (RVE). According to Hill [1]:

> *A Representative Volume is a sample that (a) is structurally entirely typical of the whole mixture on average and (b) contains a sufficient number of inclusions for the apparent overall moduli*

to be effectively independent of the surface values of traction and displacement, so long as these values are macroscopically uniform.

The same idea was independently developed by Mandel and Dantu [2]. Strictly speaking, the apparent overall modulus becomes independent of the applied boundary conditions only at an infinite length scale, but for all practical purposes, one could approximate the size of the RVE at finite scales within a few percent error. Below the RVE, the response involves statistical scatter and such a volume element is therefore called the statistical volume element (SVE). Besides this situation of a domain containing a very large (mathematically infinite) set of microscale elements (e.g., grains or inhomogeneities), the RVE is very clearly defined for a unit in a periodic microstructure. We shall not pursue the latter situation here.

With reference to the example of the simple tension test on a copper bar, the above statement implies that the response of the specimen will be realization dependent if the sample size is not representative. It is of interest here to analyze: (i) the approach of the SVE towards the RVE, and (ii) establish scaling laws in a variety of random microstructures such as those depicted in Fig. 2.1. Indeed, these form but a small subset of a myriad of different microstructures (e.g., [3, 4]), but the methodology we outline on simple models of random composites can readily be applied to more complex material systems.

We employ the framework of stochastic (micro)mechanics consistent with the mathematical statement of the *Hill–Mandel condition* stated above. Such an approach is extremely versatile and can be used not only in linear elastic problems but also in bounding the response of hyperelastic materials, elasto-plastic materials, heat conduction, porous media; e.g., see [5–7]. In the following sections, we discuss the scale-dependent bounds and scaling laws on the various constitutive behaviors of random materials.

2.2. Conductivity of Random Polycrystals

Single crystals exhibit an anisotropic behavior in heat conduction. In general, thermal conductivity is a second-rank tensor that is completely defined by three independent components in the principal direction. For example, trigonal, hexagonal, and tetragonal single crystals demonstrate a uniaxial thermal character with two of the principal components of the conductivity tensor being equal. On the other hand, a randomly oriented polycrystal exhibits isotropic behavior at the RVE scale. Thus, the

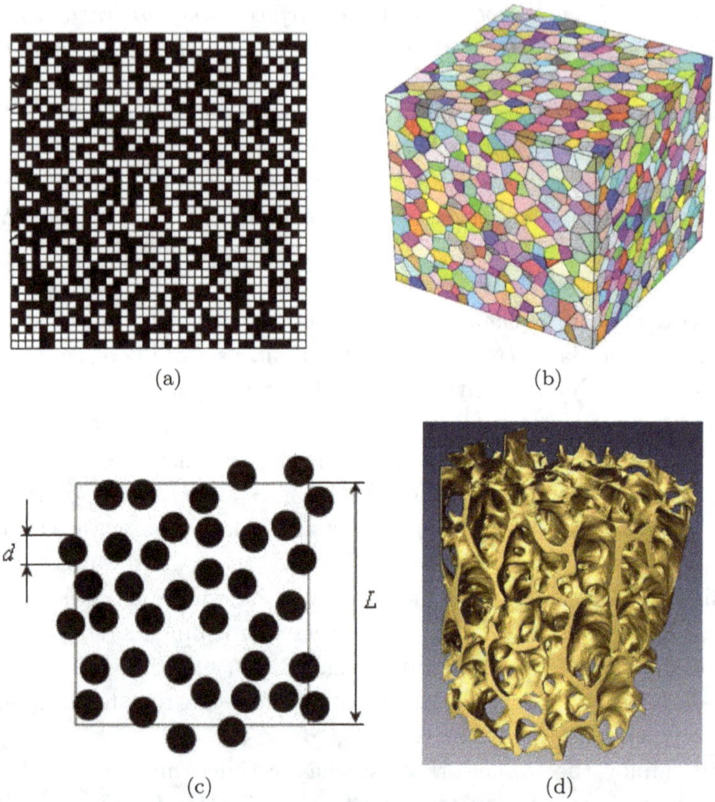

Fig. 2.1. (a) Random checkerboard in 2D. (b) Random polycrystal in 3D with 5000 grains. Color scale represents the different orientation (random) of each grain. (c) Circular-inclusion composite, showing a mesoscale window. (d) Microstructure of trabecular bone obtained by micro-CT imaging.

following are the key aspects at the intermediate scale of the SVE: (i) the sample response is in general anisotropic and realization dependent; (ii) the isotropic response may be recovered by averaging over a sufficient number of ensembles; (iii) the apparent conductivity is very much dependent on the applied boundary conditions; and (iv) with increasing length scales, the SVE approaches the RVE. It is apparent from these observations that in going from the level of a single crystal to that of a polycrystalline aggregate, the conductivity is scale dependent and isotropy is gradually approached indicating the reduction in the number of independent constants necessary to define the conductivity tensor completely. A related question is whether one can establish scaling laws unifying a class of materials. For instance, hematite and quartz possess a trigonal crystal structure at the single-crystal level and thereby one is tempted to seek similarities in their scale-dependent properties. In the subsequent discussion, we attempt to answer some of the above questions using stochastic (micro)mechanics as the tool and the Hill–Mandel condition as the basis.

2.2.1. The Hill–Mandel condition

Consider a specific realization $\mathbf{B}_\delta(\omega)$ of a random medium $\mathbf{B}_\delta$. Here, $\omega(\in \Omega)$ indicates a specific realization of the microstructure taken from the sample space Ω and the subscript δ, defined as

$$\delta = (N_G)^{\frac{1}{3}}, \tag{2.1}$$

is the dimensionless parameter specifying the *mesoscale* of the polycrystalline aggregate comprising N_G grains. In the following, δ will sometimes be referred to as window size.

Now, adapting the idea of the Hill–Mandel condition, one can establish the following relationship [6–9]:

$$\overline{\mathbf{q}\cdot\nabla T} = \overline{\mathbf{q}}\cdot\overline{\nabla T} \Leftrightarrow \int_{\partial B_\delta} (\mathbf{q}\cdot\mathbf{n} - \overline{\mathbf{q}}\cdot\mathbf{n})(T - \overline{\nabla T}\cdot\mathbf{x})dS = 0, \quad \forall x \in \partial B_\delta. \tag{2.2}$$

Here $\mathbf{q}$ is the heat flux, ∇T is the temperature gradient, and $\mathbf{x}$ is the position vector. The overbar operator $\overline{(\bullet)}$ indicates volume averaging. Equation (2.2) suggests three types of loadings:

(i) Essential Uniform Boundary Condition (EUBC): $T = \nabla T^0 \cdot \mathbf{x}$, (2.3a)

(ii) Natural Uniform Boundary Condition (NUBC): $\mathbf{q}\cdot\mathbf{n} = \mathbf{q}^0\cdot\mathbf{n}$, (2.3b)

(iii) Mixed-Orthogonal (MOBC): $(\mathbf{q}\cdot\mathbf{n} - \mathbf{q}^0\cdot\mathbf{n})(T - \nabla T^0\cdot\mathbf{x}) = 0$. (2.3c)

By increasing the mesoscale δ (effectively, the number of grains in $\mathbf{B}_\delta$) and by setting up stochastic boundary-value problems with the above boundary conditions and upon ensemble averaging, one obtains bounds on the constitutive response of the aggregate. Now, the condition (2.3a) results in a mesoscale (i.e., δ-dependent) conductivity tensor $\mathbf{C}^e_\delta$, (2.3b) is a mesoscale resistivity tensor $\mathbf{S}^n_\delta$, while (2.3c) yields a mesoscale conductivity or resistivity tensor (depending on the interpretation). The superscripts

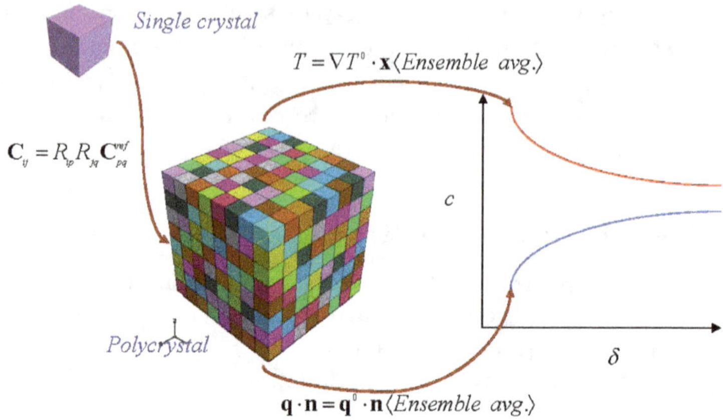

Fig. 2.2. Methodology for obtaining scale-dependent bounds in heat conduction.

e and n denote quantities obtained under essential and natural boundary conditions, respectively. The condition (2.3c) is understood with the stipulation that one must not simultaneously specify both the heat flux and the temperature gradient in any given direction on any portion of the boundary.

The methodology outlined here works, provided the hypotheses of spatial homogeneity and ergodicity hold for the random field $\boldsymbol{\Theta}(\mathbf{x}, \omega)$ of material parameters involved. In particular, we assume $\boldsymbol{\Theta}(\mathbf{x}, \omega)$ to be a wide-sense stationary (WSS) random field with a constant mean and finite-valued autocorrelation [6]:

$$\begin{aligned} \langle \boldsymbol{\Theta}(x_1) \rangle &= \mu, \\ \langle \boldsymbol{\Theta}(x_1)\boldsymbol{\Theta}(x_1 + h) \rangle &= R_{\Theta}(h) < \infty. \end{aligned} \tag{2.4a}$$

The operator $\langle \bullet \rangle$ indicates ensemble averaging. The random field $\boldsymbol{\Theta}(\mathbf{x}, \omega)$ is *mean ergodic* providing its spatial average equals the ensemble average:

$$\frac{1}{V}\int_V \boldsymbol{\Theta}(\mathbf{x}, \omega) dV = \overline{\boldsymbol{\Theta}(\mathbf{x})} = \langle \boldsymbol{\Theta}(\mathbf{x}) \rangle = \int_{\Omega} \boldsymbol{\Theta}(\mathbf{x}, \omega) dP. \tag{2.4b}$$

The proposed methodology is illustrated in Fig. 2.2. The single crystal has a reference conductivity tensor C^{ref}_{pq} with three independent constants in its principal directions c_1, c_2, and c_3. By using a set of uniformly generated rotation tensors (Fig. 2.3), the reference tensor is rotated to assign the material property for each individual crystal in the polycrystal. Using (2.3a) and (2.3b), the boundary-value problems are solved and mesoscale tensors established so that upon ensemble averaging one obtains bounds on the aggregate conductivity.

2.2.2. Bounds on the conductivity

At this point, we recall the ergodicity and WSS properties of the microstructure and obtain the hierarchy of scale-dependent bounds as follows:

$$\begin{aligned}\langle \mathbf{S}_1^n \rangle^{-1} &\leq \cdots \leq \langle \mathbf{S}_{\delta'}^n \rangle^{-1} \leq \langle \mathbf{S}_\delta^n \rangle^{-1} \leq \cdots \leq \mathbf{C}_\infty^{\text{eff}} \cdots \leq \langle \mathbf{C}_\delta^e \rangle \\ &\leq \langle \mathbf{C}_{\delta'}^e \rangle \cdots \leq \langle \mathbf{C}_1^e \rangle, \quad \forall\, \delta' \leq \delta. \end{aligned} \tag{2.5}$$

Such bounds date back to Ostoja-Starzewski and Schulte [8] and Jiang *et al.* [10]; see also [11]. Using (2.5) along with the definition of the isotropic conductivity tensor, we obtain the following hierarchy of bounds on the isotropic conductivity measure:

$$\begin{aligned} c^H &\leq \cdots \leq \langle c_{\delta'}^n \rangle \leq \langle c_\delta^n \rangle \leq \cdots \leq c_\infty^{\text{eff}} \\ &\leq \cdots \langle c_\delta^e \rangle \leq \langle c_{\delta'}^e \rangle \leq c^A, \quad \forall\, \delta' \leq \delta, \end{aligned} \tag{2.6}$$

where $1/c^H = (1/c_1 + 1/c_2 + 1/c_3)/3$ and $c^A = (c_1 + c_2 + c_3)/3$ are the harmonic (Reuss type) and the arithmetic mean (Voigt type) estimates of the conductivity. While the hierarchy type of behavior of mesoscale tensors appears evident, it has been found in Ostoja-Starzewski [12] that the coefficient of variation of the second invariant of $\mathbf{C}_\delta^e$ as well as $\mathbf{S}_\delta^e$ for several different planar random microstructures (all generated by homogeneous Poisson point fields) equals ~0.55.

2.2.3. Scaling function in heat conduction

For a given realization $\mathbf{B}_\delta(\omega)$ of a random medium $\mathbf{B}_\delta$ on some mesoscale δ, (2.3a) yields a mesoscale random conductivity tensor $\mathbf{C}_\delta^e(\omega)$ such that

$$\overline{\mathbf{q}_\delta}(\omega) = \mathbf{C}_\delta^e(\omega) \cdot \nabla T^0. \tag{2.7}$$

Similarly, (2.3b) yields a mesoscale random resistivity tensor $\mathbf{S}_\delta^n(\omega)$ such that

$$\overline{\nabla T_\delta}(\omega) = \mathbf{S}_\delta^n(\omega) \cdot \mathbf{q}^0. \tag{2.8}$$

In general, for any given realization $\omega(\in \Omega)$, $\mathbf{C}_\delta^e(\omega)$, and $\mathbf{S}_\delta^n(\omega)$ are anisotropic. We obtain an isotropic response only by assigning the crystal orientations uniformly (distributed uniformly on a unit sphere; also see [13]) and upon ensemble averaging over the realization space. Thus, the ensemble-averaged isotropic conductivity and resistivity tensors can be expressed as follows:

$$\langle \mathbf{C}_\delta^e \rangle = \langle c_\delta^e \rangle \mathbf{I}, \tag{2.9}$$

$$\langle \mathbf{S}_\delta^n \rangle = \frac{1}{\langle c_\delta^n \rangle} \mathbf{I}. \tag{2.10}$$

In the above, $\mathbf{I}$ represents the second-rank identity tensor, while $\langle c_\delta^e \rangle$ and $\langle c_\delta^n \rangle$ are ensemble-averaged isotropic conductivity measures under the essential and natural boundary conditions, respectively. By contracting (2.9) and (2.10), we obtain the following scalar equation:

$$\langle \mathbf{C}_\delta^e \rangle : \langle \mathbf{S}_\delta^n \rangle = 3 \frac{\langle c_\delta^e \rangle}{\langle c_\delta^n \rangle}. \tag{2.11}$$

In the limit $\delta \to \infty$ the conductivity tensor must be the exact inverse of the resistivity tensor, and so we obtain

$$\lim_{\delta \to \infty} \langle \mathbf{C}_\delta^e \rangle : \langle \mathbf{S}_\delta^n \rangle = 3. \tag{2.12}$$

Now, we postulate the following relationship between the left-hand side of (2.11) and (2.12), that is

$$\begin{aligned} \langle \mathbf{C}_\delta^e \rangle : \langle \mathbf{S}_\delta^n \rangle &= \lim_{\delta \to \infty} \langle \mathbf{C}_\delta^e \rangle : \langle \mathbf{S}_\delta^n \rangle + g(c_1, c_2, c_3, \delta) \\ &= \lim_{\delta \to \infty} \langle \mathbf{C}_\delta^e \rangle : \langle \mathbf{S}_\delta^n \rangle + g(k_1, k_2, c_3, \delta). \end{aligned} \tag{2.13}$$

In the above, $g(c_1, c_2, c_3, \delta)$ ["or" $g(k_1, k_2, c_3, \delta)$] defines the *scaling function* and $k_1 = c_1 c_3^{-1}$ and $k_2 = c_2 c_3^{-1}$ are two non-dimensional parameters. Substituting (2.13) and (2.12) into (2.11), we obtain

$$g(k_1, k_2, c_3, \delta) = 3 \left(\frac{\langle c_\delta^e \rangle}{\langle c_\delta^n \rangle} - 1 \right). \tag{2.14}$$

Notice that the right-hand side of (2.14) is dimensionless. Thus, the scaling function is dependent only on the non-dimensional parameters and takes the form $g(k_1, k_2, \delta)$ and can be determined numerically by the solution of boundary-value problems subject to the essential and natural uniform boundary conditions.

2.2.4. Some properties of and bounds on the scaling function

The scaling function $g(k_1, k_2, \delta)$ introduced in (2.14) has the following properties (see also [9, 14]):

$$g(k_1, k_2, \delta = \infty) = 0. \tag{2.15}$$

Again, the scaling function becomes null if the crystals are locally isotropic:

$$g(k_1 = k_2 = 1, \delta) = 0. \tag{2.16}$$

One can further establish the following bounds on the scaling function for aggregates made up of single crystals with uniaxial thermal character ($k_1 = k_2 = k$ for trigonal, hexagonal, and tetragonal single crystals. k is also a measure of a single crystal's anisotropy)

$$0 \leq g(k, \delta) \leq \frac{2}{3}\left(\sqrt{k} - \frac{1}{\sqrt{k}}\right)^2. \tag{2.17}$$

The lower bound in (2.17) is easily established using Eqs. (2.6) and (2.14). Also at any given scale the following inequality holds (using Eq. (2.6)):

$$\left(\frac{\langle c_\delta^e \rangle}{\langle c_\delta^n \rangle} - 1\right) \leq \left(\frac{c^A}{c^H} - 1\right). \tag{2.18}$$

Using the definitions of c^A and c^H in Eq. (2.6), we obtain the upper bound.

2.2.5. Numerical simulations

We proved several properties of the scaling function in Section 2.2.4. We now proceed to derive a suitable form for the scaling function based on numerical results. Since it is not practical here to perform numerical simulations on all known crystals, we restrict our study to the materials listed in Table 2.1.

We perform numerical simulations with the number of grains, $N_G = 2^0, 2^3, 2^6$, and 2^9 (i.e., $\delta = 1, 2, 4$, and 8). The following loading conditions have been imposed to run the stochastic boundary-value problems:

$$\text{Dirichlet problem: } \nabla T^0 = 4\vec{k}, \tag{2.19a}$$

$$\text{Neumann problem: } \mathbf{q}^0 = -18\vec{k}, \tag{2.19b}$$

where $\vec{k}$ is a unit vector along the z-direction. Based on the numerical results for these materials, one can postulate a suitable form for the scaling function and develop the material scaling diagram.

Table 2.1. Material parameters.

Crystal system	Thermal character	Thermal axes and orientation	Materials	Single crystal property (W/m.°C) $c_1 = c_2$	c_3	$k = c_1/c_3$	k^{-1}	Scaling function
Cubic	Isotropic	$c_1 = c_2 = c_3$	Aluminum	208	208	1	1	0
			Copper	410	410	1	1	0
Trigonal	Uniaxial	$c_1 = c_2 \neq c_3$	Calcite	4.18	4.98	0.84	1.19	$g_1(k, \delta)$
			Hematite	14.6	12.17	1.20	0.83	$\cong g_1(k, \delta)$
			Quartz	6.5	11.3	0.58	1.74	$g_2(k, \delta)$
Hexagonal	Uniaxial	$c_1 = c_2 \neq c_3$	Graphite	355	89	3.99	0.25	$g_3(k, \delta)$

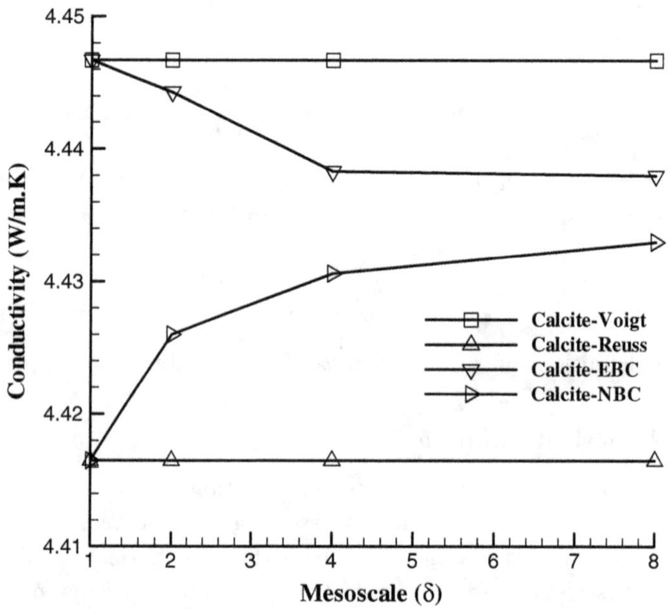

Fig. 2.3. Bounds on the aggregate response (calcite).

We begin the discussion by studying the scale-dependent response of calcite and hematite. As seen from Table 2.1, these materials have very different single-crystal conductivities. Notice that, although the anisotropy index of calcite is greater than 1 and that of hematite is less than 1, their product is almost equal to one. In other words, the anisotropy measure of calcite is almost the reciprocal of that of hematite. The scale-dependent bounds for calcite and hematite are shown in Figs. 2.3 and 2.4, respectively.

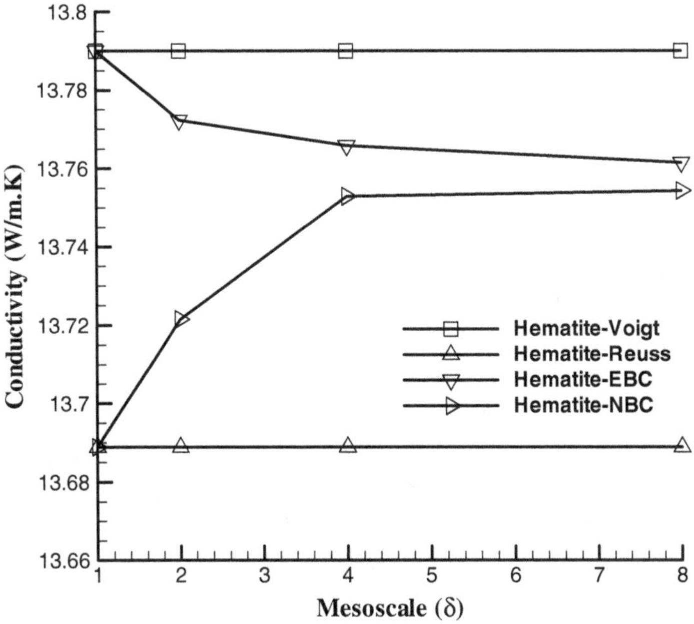

Fig. 2.4. Bounds on the aggregate response (hematite).

Notice that since the anisotropy index for these crystals is close to one, the Voigt and Reuss bounds are very close to one another. As observed in these plots, we obtain the upper bound on application of (2.19a) and the lower bound using (2.19b). Also, as we increase the number of grains (or as the mesoscale size δ increases), we obtain tighter bounds on the aggregate conductivity. The scaling functions for the calcite and hematite aggregates have similar trends and are quite close to one another (Fig. 2.5). We attribute the discrepancies to the finite number of realizations used to obtain the ensemble average and the slight differences in the anisotropy index. Based on these observations, we conclude that two materials with single-crystal anisotropy indices k and k^{-1} represent the same scaling function. Mathematically,

$$g(k,\delta) = g\left(\frac{1}{k},\delta\right). \tag{2.20}$$

Let us now consider the scale-dependent bounds on quartz and graphite as illustrated in Figs. 2.6 and 2.7. Owing to the higher single-crystal anisotropy index (or its reciprocal), the Voigt and Reuss bounds for quartz

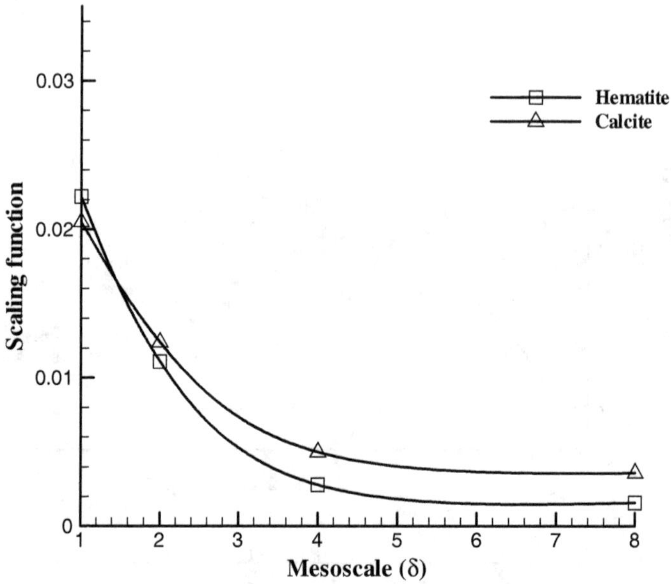

Fig. 2.5. Scaling function (calcite and hematite).

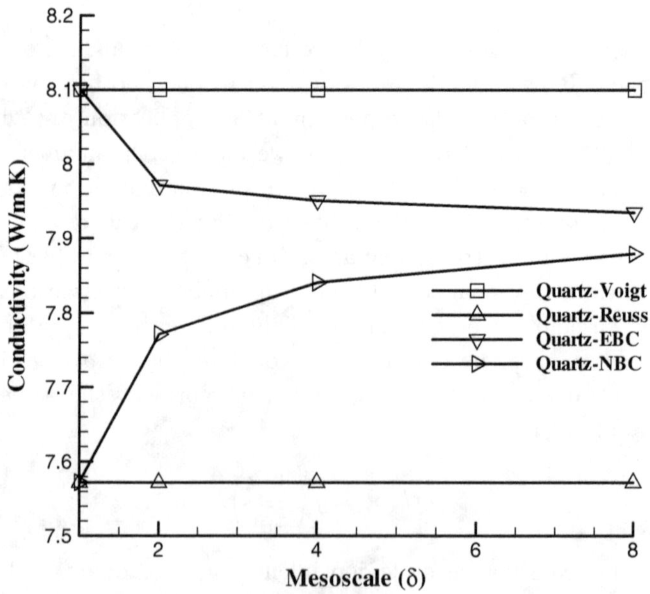

Fig. 2.6. Bounds on the aggregate response (quartz).

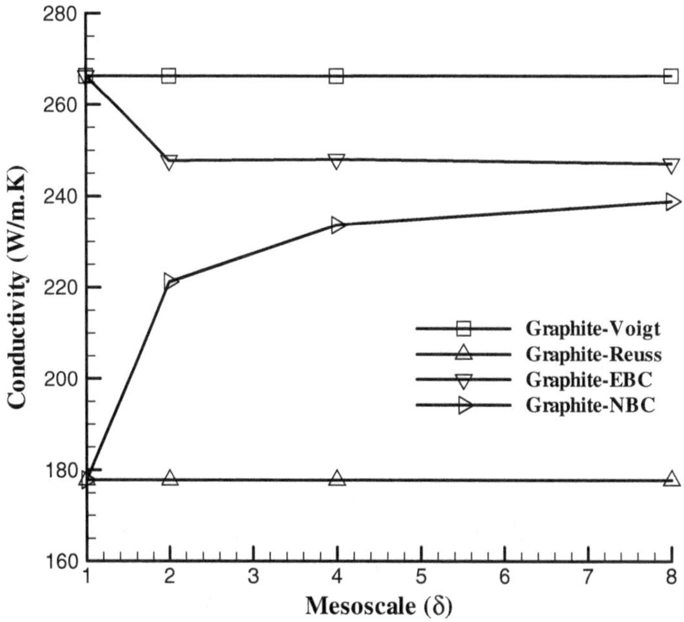

Fig. 2.7. Bounds on the aggregate response (graphite).

and graphite are much farther apart compared to calcite or hematite. Again, the application of (2.19a) and (2.19b) bounds the aggregate conductivity from above and below, respectively. The scaling functions for quartz and graphite are plotted in Figs. 2.8 and 2.9, respectively. We immediately notice that for any given scale, the scaling functions for graphite and quartz have much larger values than hematite or graphite. Since the anisotropy index (or its reciprocal) for graphite is much larger than that of quartz, its scaling function takes markedly higher values than that of quartz at all scales.

2.2.6. Constructing the scaling function

In Fig. 2.10, we compile the scaling function for all the different crystals considered in the previous section. It is useful to rewrite (2.17) as follows:

$$0 \leq \frac{3}{2\left(\sqrt{k} - \frac{1}{\sqrt{k}}\right)^2} g(k, \delta) \leq 1. \tag{2.21}$$

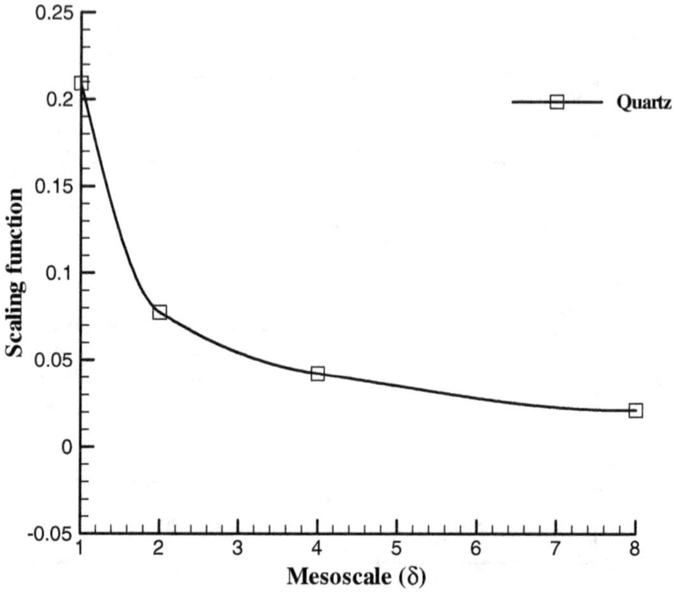

Fig. 2.8. Scaling function (quartz).

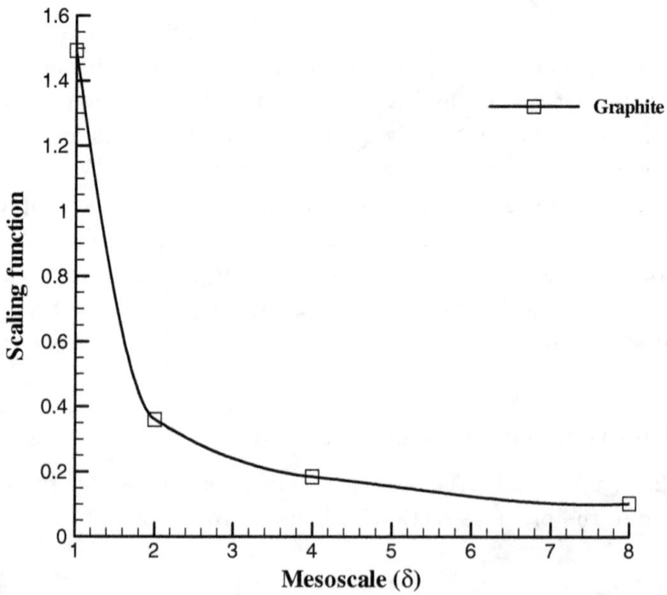

Fig. 2.9. Scaling function (graphite).

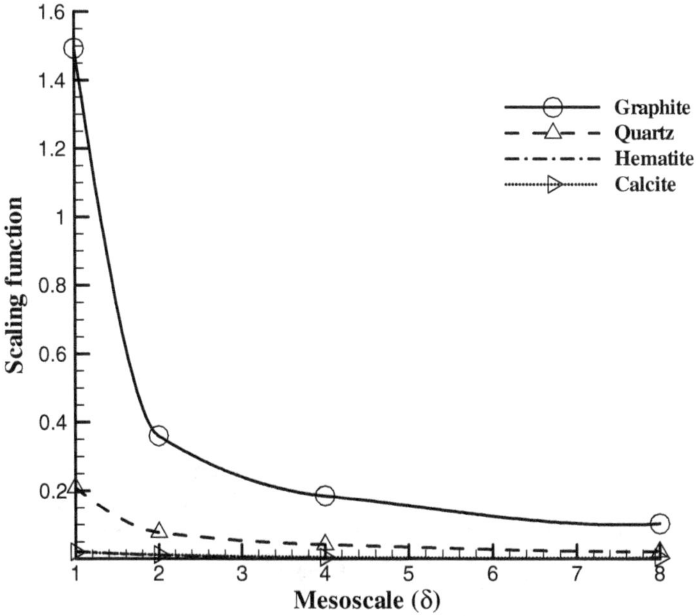

Fig. 2.10. Scaling function (compiled).

We now proceed to plot the rescaled scaling function defined in (2.21). Interestingly, from Fig. 2.11 it appears that rescaling bounds all the materials closely. We attribute the discrepancies (especially, for aggregates with few grains (small mesoscale δ)) to the finite number of realizations employed to construct the scaling function. The above essentially means that we could approximate the mean rescaled function $g^*(\delta)$ to be independent of the single-crystal anisotropy. The scaling function can now be redefined as follows:

$$g(k,\delta) = \frac{2}{3}\left(\sqrt{k} - \frac{1}{\sqrt{k}}\right)^2 g^*(\delta), \tag{2.22}$$

where $g^*(\delta)$ represents the material-independent rescaled function. Based on the mean values of $g^*(\delta)$ in Fig. 2.11, we construct the effective averaged rescaled function and fit it using an exponential function. The effective function and its fit are illustrated in Fig. 2.12. Based on this fit, $g^*(\delta)$ takes the following form:

$$g^*(\delta) = \exp[-0.9135(\delta - 1)^{0.5}]. \tag{2.23}$$

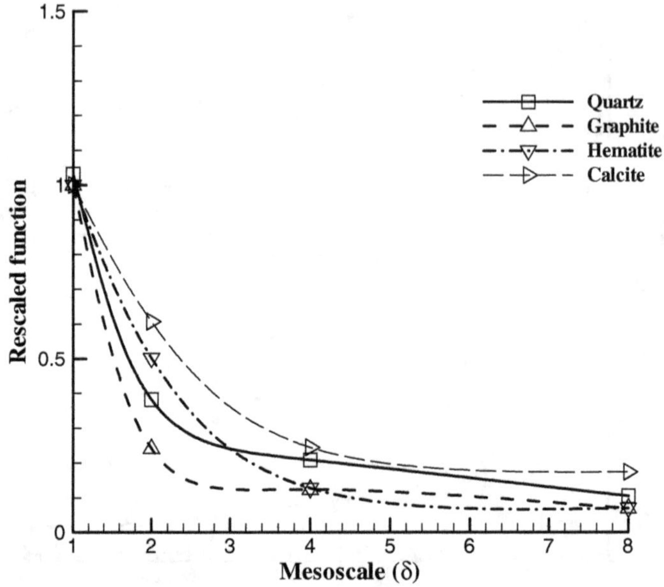

Fig. 2.11. Rescaled scaling function.

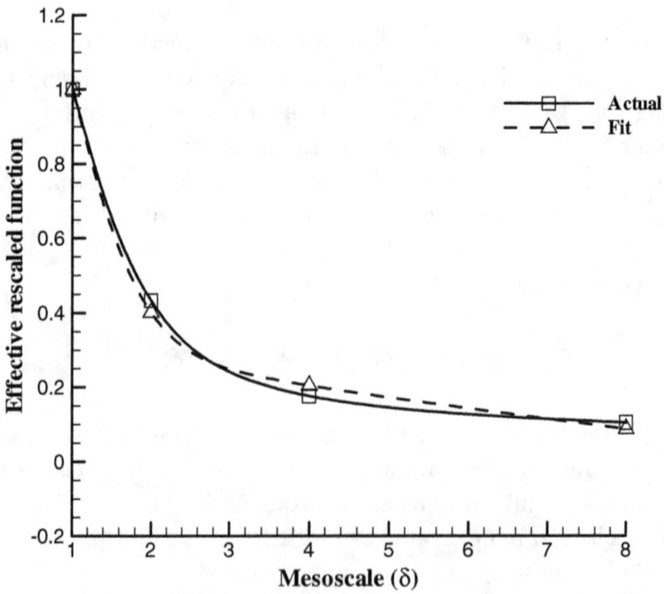

Fig. 2.12. Effective rescaled scaling function and fit.

In view of (2.22), the scaling function takes the following form:

$$g(k,\delta) = \frac{2}{3}\left(\sqrt{k} - \frac{1}{\sqrt{k}}\right)^2 \exp[-0.9135(\delta - 1)^{0.5}], \quad \delta = N_g^{\frac{1}{3}}. \tag{2.24}$$

This particular form of the scaling function satisfies all the properties defined in Section 2.2.4. Consistent with (2.20), it does not distinguish between k and k^{-1}. Equation (2.24) also suggests that the scaling function is exact in the single-crystal anisotropy index. The empirical form comes from its dependence on the window size. The form of (2.24) has been used to reconstruct the scaling function for the different polycrystals and is plotted in Fig. 2.13. It is evident from the plot that (2.24) captures the scaling function reasonably well.

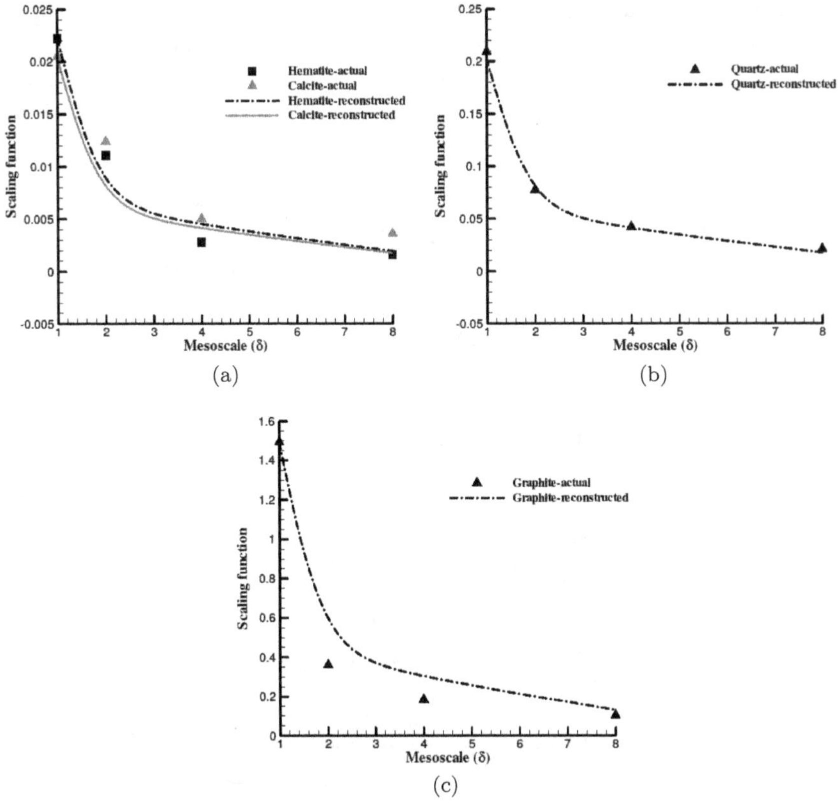

Fig. 2.13. Goodness of fit: (a) calcite and hematite, (b) quartz, (c) graphite.

It is now possible to construct the contours of the scaling function in the k–δ space based on Eq. (2.24) as illustrated in Fig. 2.14. Notice that, as the scaling function decreases, the curves shift towards a higher mesoscale and *vice versa*.

Theoretically the scaling function becomes zero only when the number of grains is infinite (i.e., $\delta \approx \infty$) or when the crystal is locally isotropic with $k = 1$ (such as cubic crystals). For all practical purposes, one defines the RVE based on some finite value of the scaling function. We chose a value of 0.01 for the scaling function to construct Fig. 2.15 since the discrepancy in the conductivity predictions for most crystals based on the essential and natural boundary conditions lie within 0.5%. The plots give an idea of the number of grains necessary to homogenize the aggregate response in the anisotropic heat conduction for hexagonal (Fig. 2.15(a)), trigonal (Fig. 2.15(b)), and tetragonal crystals (Fig. 2.15(c)). Since cubic crystals are locally isotropic, the number of grains necessary is trivial and equals one. We notice from these plots that two aggregates made up of single crystals with anisotropy indices k and k^{-1} scale identically. Also, we need

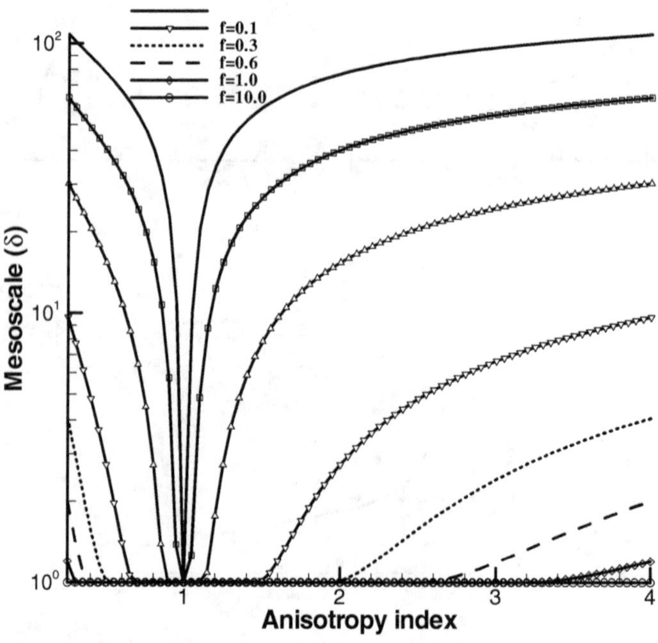

Fig. 2.14. Contours of scaling function (for $0.0001 \leq g \leq 10$).

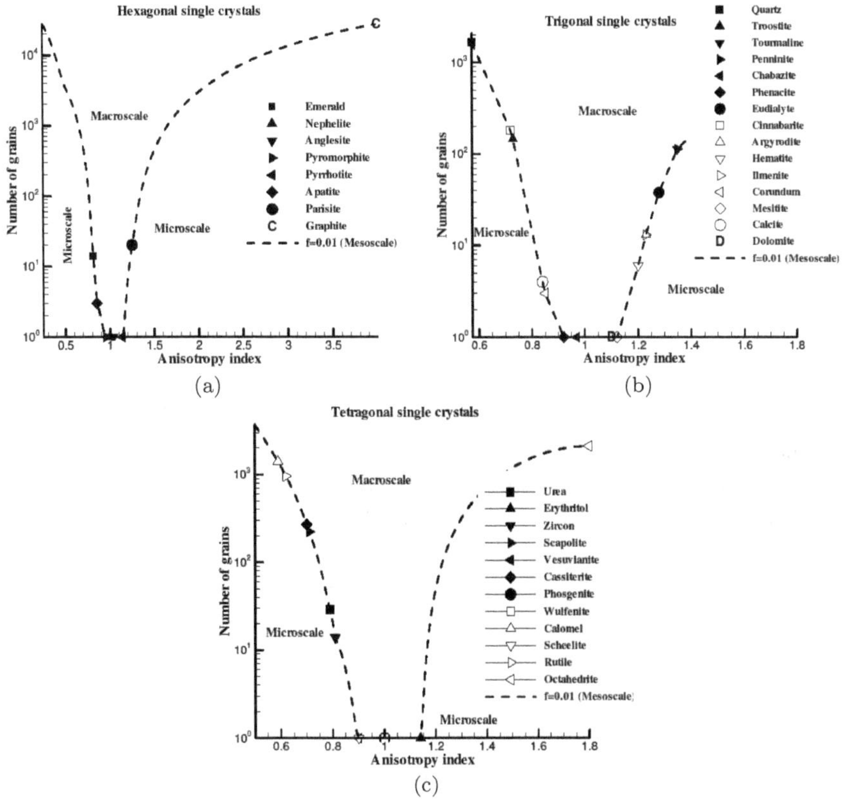

Fig. 2.15. Material scaling diagram at $g = 0.01$: (a) hexagonal single crystals, (b) trigonal single crystals, (c) tetragonal single crystals.

more grains (or crystals) to homogenize the aggregate response as k or k^{-1} are different from unity.

2.3. Conductivity of Planar Random Checkerboards

In this section, we focus on the scale dependence of conductivity and resistivity in two-phase planar random microstructures (see Fig. 2.1(a)) and, in particular, the resulting scaling function. The microstructures studied here are mixtures of two kinds of phases at nominal volume fractions 50 % at arbitrary contrasts. Although the effective, macroscale conductivity of such composites is well known (i.e., geometrical mean of the individual phases) at sufficiently large length scales, the conductivity at finite scales remains dependent on the boundary conditions. A wide range of checkerboards are analyzed under both essential and natural boundary conditions. Variety of material combinations was chosen to establish the form of the scaling function. It turns out that the scaling function for a random checkerboard is simply a function of the contrast and the dimensionless mesoscale, unlike the 3D polycrystal where it was a function of the single crystal anisotropy.

2.3.1. Governing equations

We study two-phase composite with geometries given by random checkerboards, such as shown in Fig. 2.1(a). Each phase ($i = 1, 2$) is characterized by an isotropic Fourier-type conductivity, with perfect bonding present at all the cell boundaries. Each square cell is sampled according to a strict-white-noise process, i.e., independent of states of all the other cells. Technically, the random checkerboard is a Bernoulli lattice process, generated at a nominal probability 1/2 of either phase (1 or 2). Clearly, this model is statistically homogeneous, isotropic, and ergodic. We are considering steady-state heat conduction without heat sinks or sources, so that the governing equation is

$$\nabla \cdot (c(\vec{x})\nabla T) = 0, \tag{2.25}$$

where, $c(\vec{x}) = \chi_1(\vec{x})\, c_1 + \chi_2(\vec{x})\, c_2$, $\vec{x}$ is a position vector, T is temperature, c_1 and c_2 are the conductivities of the individual phases and $\chi_i(\vec{x})$ is the indicator functions of the region occupied by phase i. The contrast ratio of material pairs is defined as

$$k = c_1/c_2. \tag{2.26}$$

It follows that the effective (macroscopic) conductivity should be isotropic and an explicit formula for it is well known [15]

$$c^{\text{eff}} = \sqrt{c_1 c_2}. \tag{2.27}$$

However, on finite scales and for any realization of the random checkerboard, the conductivity depends on the boundary conditions of the domain. Using Eqs. (2.3a) and (2.3b) on a variety of two phase checkerboards with varying contrast results in the following form for the scaling function [16]

$$g(\delta, k) = \frac{1}{2}\left(\sqrt{k} - \frac{1}{\sqrt{k}}\right)^2 \exp\left[-0.5\underline{3}(\delta - 1)^{0.69}\right]. \tag{2.28}$$

Subsequently, Dalaq and Ranganathan [17] proved that the variance of the trace of mesoscale conductivity tensor takes the following form:

$$\text{Variance} = \frac{(c_1 - c_2)^2}{\delta^2}, \tag{2.29}$$

a result that is consistent with Kanit *et al.* [18] and Matheron [19]. More recently, Raghavan *et al.* [20] generalized this study to two phase checkerboards with arbitrary volume fraction v_f in the context of electrical conductivity and obtained the following form for the scaling function:

$$g(v_f, k, \delta) = 2v_f\,(1 - v_f)\,\frac{1}{2}\left(\sqrt{k} - \frac{1}{\sqrt{k}}\right)^2 \exp\left[-0.7\underline{3}(\delta - 1)^{0.5}\right]. \tag{2.30}$$

2.4. Elastic Properties of Random Polycrystals

In this section, we consider the elastic properties of polycrystals at the microscale, mesoscale, and macroscale. Single crystals are typically anisotropic elastically and the extent of anisotropy can be quantified by using various measures of anisotropy [21–23]. At the microscale, a polycrystalline sample consists of relatively few grains exhibiting a realization-dependent anisotropic response. As the length scale increases and as the RVE is approached, the aggregate sample typically consists of many grains and the response becomes realization independent and isotropic. Much like our observations in the heat conduction problem, the elastic properties are indeed scale dependent and our attempt is to establish unifying scaling laws for a class of elastic crystals. For instance, copper and tantalum have a cubic crystal system at the single-crystal level and thereby one is tempted to seek similarities in their scale-dependent properties. With these objectives in mind, we re-state the Hill–Mandel condition for elastic properties and then employ stochastic (micro)mechanics to determine the scale-dependent elastic properties and to establish scaling laws.

2.4.1. The Hill–Mandel condition

The Hill–Mandel condition in elasticity stems from the consideration of the equivalence of energetic and mechanical interpretations of stored energy [2, 24–28]:

$$\begin{aligned} \overline{\sigma_{ij} : \varepsilon_{ij}} = \overline{\sigma_{ij}} : \overline{\varepsilon_{ij}} \Rightarrow \frac{1}{V}\int_{\partial V} \sigma'_{ij} : \varepsilon'_{ij}\,dV = 0 \\ \Leftrightarrow \int_{\partial B_\delta} (t_i - \overline{\sigma_{ij}} n_j)(u_i - \overline{\varepsilon_{ij}} x_j)\,dS = 0. \end{aligned} \tag{2.31}$$

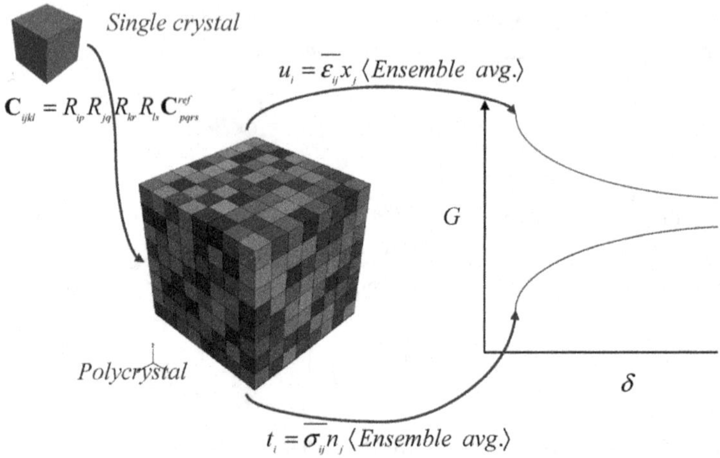

Fig. 2.16. Methodology for obtaining scale-dependent bounds for elastic polycrystals.

The three loadings that satisfy (2.31) are as follows:

(i) Kinematic Uniform Boundary Condition (KUBC): $u_i = \varepsilon^0_{ij}x_j$, (2.32a)

(ii) Static Uniform Boundary Condition (SUBC): $t_i = \sigma^0_{ij}n_j$, (2.32b)

(iii) Mixed-Orthogonal (MOBC): $(t_i - \sigma^0_{ij}n_j)(u_i - \varepsilon^0_{ij}x_j) = 0$. (2.32c)

With increasing mesoscale δ, we obtain scale-dependent bounds (Fig. 2.16) on the elastic response of the aggregate by setting up and solving boundary-value problems consistent with Eq. (2.32a). The condition (2.32c) is understood with the stipulation that one must not simultaneously specify both the traction and the displacement in any given direction on any portion of the boundary.

2.4.2. Bounds on the elastic response

At this point, we recall that the ergodicity and WSS properties of the microstructure, together with the variational principles of elasticity theory, imply a hierarchy of scale-dependent bounds on the elastic response

$$\begin{aligned} \langle \mathbf{S}^t_1 \rangle^{-1} \leq \cdots \leq \langle \mathbf{S}^t_{\delta'} \rangle^{-1} \leq \langle \mathbf{S}^t_{\delta} \rangle^{-1} \leq \cdots \leq \mathbf{C}^{\text{eff}}_{\infty} \cdots \leq \langle \mathbf{C}^d_{\delta} \rangle \\ \leq \langle \mathbf{C}^d_{\delta'} \rangle \cdots \leq \langle \mathbf{C}^d_1 \rangle, \quad \forall\, \delta' \leq \delta. \end{aligned} \tag{2.33}$$

The superscripts t and d denote quantities obtained under the application of traction and displacement boundary conditions, respectively. Such bounds date back to Huet [26] and Sab [29] and have been applied to microstructures with various geometries [6, 18, 30–34]. Using (2.33) along with the definition of the isotropic elasticity tensor, we obtain the following hierarchy of bounds on the shear and bulk moduli:

$$\begin{aligned} G^R &\leq \cdots \leq \langle G^t_{\delta'} \rangle \leq \langle G^t_{\delta} \rangle \leq \cdots \leq G^{\text{eff}}_{\infty} \\ &\leq \cdots \langle G^d_{\delta} \rangle \leq \langle G^d_{\delta'} \rangle \cdots \leq G^V, \end{aligned} \tag{2.34a}$$

$$\begin{aligned} K^R &\leq \cdots \leq \langle K^t_{\delta'} \rangle \leq \langle K^t_{\delta} \rangle \leq \cdots \leq K^{\text{eff}}_{\infty} \\ &\leq \cdots \langle K^d_{\delta} \rangle \leq \langle K^d_{\delta'} \rangle \cdots \leq K^V, \quad \forall\, \delta' \leq \delta, \end{aligned} \tag{2.34b}$$

where (G^R, K^R) and (G^V, K^V) represent, respectively, the Reuss and Voigt estimates of the shear and bulk moduli.

2.4.3. Elastic scaling function

For a given realization $\mathbf{B}_\delta(\omega)$ of a random medium $\mathbf{B}_\delta$ on some mesoscale δ, (2.32a) yields a mesoscale random stiffness tensor $\mathbf{C}^d_\delta(\omega)$ such that

$$\bar{\boldsymbol{\sigma}}_\delta(\omega) = \mathbf{C}^d_\delta(\omega) : \boldsymbol{\varepsilon}^0. \tag{2.35}$$

Similarly, (2.32b) yields a mesoscale random compliance tensor $\mathbf{S}^t_\delta(\omega)$ such that

$$\bar{\boldsymbol{\varepsilon}}_\delta(\omega) = \mathbf{S}^t_\delta(\omega) : \boldsymbol{\sigma}^0. \tag{2.36}$$

By uniformly distributing the crystal orientations and upon ensemble averaging, we recover the isotropic aggregate response. In such a case, the averaged stiffness and compliance tensors can be expressed in terms of the shear modulus G and the bulk modulus K as follows:

$$\langle \mathbf{C}^d_\delta \rangle = 2\langle G^d_\delta \rangle \mathbf{K} + 3\langle K^d_\delta \rangle \mathbf{J}, \tag{2.37}$$

$$\langle \mathbf{S}^t_\delta \rangle = \frac{1}{2\langle G^t_\delta \rangle}\mathbf{K} + \frac{1}{3\langle K^t_\delta \rangle}\mathbf{J}. \tag{2.38}$$

In the above, $\mathbf{J}$ and $\mathbf{K}$ represent the spherical and the deviatoric parts of the unit fourth-order tensor $\mathbf{I}$. By contracting (2.37) and (2.38), we obtain the following scalar equation:

$$\langle \mathbf{C}^d_\delta \rangle : \langle \mathbf{S}^t_\delta \rangle = 5\frac{\langle G^d_\delta \rangle}{\langle G^t_\delta \rangle} + \frac{\langle K^d_\delta \rangle}{\langle K^t_\delta \rangle}. \tag{2.39}$$

In the limit $\delta \to \infty$, the stiffness tensor must be the exact inverse of the compliance tensor, and so we obtain

$$\lim_{\delta\to\infty} \langle \mathbf{C}_\delta^d \rangle : \langle \mathbf{S}_\delta^t \rangle = 6. \tag{2.40}$$

Now, we postulate the following relationship between the left-hand side of (2.39) and (2.40), that is

$$\langle \mathbf{C}_\delta^d \rangle : \langle \mathbf{S}_\delta^t \rangle = \lim_{\delta\to\infty} \langle \mathbf{C}_\delta^d \rangle : \langle \mathbf{S}_\delta^t \rangle + f(C_{ij}, \delta), \tag{2.41}$$

where $f(C_{ij}, \delta)$ defines the *elastic scaling function*. The parameter C_{ij} represents all the independent single-crystal elastic constants depending on the crystal type. For aggregates made up of cubic single crystals, $C_{ij} \equiv (C_{11}, C_{12}, C_{44})$ and for triclinic systems, C_{ij} will include all the 21 independent single-crystal constants. Substituting (2.41) and (2.40) in (2.39), we obtain

$$f(C_{ij}, \delta) = 5\frac{\langle G_\delta^d \rangle}{\langle G_\delta^t \rangle} + \frac{\langle K_\delta^d \rangle}{\langle K_\delta^t \rangle} - 6. \tag{2.42}$$

For the specific case of cubic crystals, the bulk modulus is scale independent [35, 36] and (2.42) can be rewritten as

$$f(C_{11}, C_{12}, C_{44}, \delta) = 5\left(\frac{\langle G_\delta^d \rangle}{\langle G_\delta^t \rangle} - 1\right). \tag{2.43}$$

The scaling function $f(C_{ij}, \delta)$ introduced in (2.42) has the following properties:

$$f\left(C_{ij}, \delta = \infty\right) = 0. \tag{2.44}$$

Equation (2.44) states that the scaling function is identically zero at infinite mesoscale. Again, the scaling function becomes null if the crystals are locally isotropic, that is

$$f(\mathrm{iso}(C_{ij}), \delta) = 0. \tag{2.45}$$

The term $\mathrm{iso}(C_{ij})$ accounts for all the possible combinations of the single-crystal elastic constants that will ensure an isotropic single-crystal response. One can further establish the following bounds on the scaling function

$$f(C_{ij}, \infty) \leq f(C_{ij}, \delta) \leq f(C_{ij}, 1), \quad \forall 1 \leq \delta \leq \infty. \tag{2.46}$$

Using (2.44) and (2.42) in (2.46), we obtain

$$0 \le f(C_{ij}, \delta) \le A_4^U(1) = 5\frac{G^V}{G^R} + \frac{K^V}{K^R} - 6, \quad \forall 1 \le \delta \le \infty, \tag{2.47}$$

where the quantity $A_4^U(1) = 5G^V/G^R + K^V/K^R - 6$ represents the so-called *universal anisotropy index* quantifying the single-crystal anisotropy (see also [21]) of the fourth-rank elasticity tensor. That index is increasingly proving to be a new standard measure to many researchers in solid-state physics, materials science and geophysics; e.g., see [37] for the mapping of the entire Earth's surface in terms of $A_4^U(1) = 5G^V/G^R + K^V/K^R - 6$.

Based on (2.47) one can interpret the elastic scaling function in (2.42) as the evolution of the equivalent anisotropy in the mesoscale domain, thus

$$f(C_{ij}, \delta) = A_4^U(\delta). \tag{2.48}$$

The simplest form for (2.48) having a separable structure (see also [38]):

$$f(C_{ij}, \delta) = A_4^U(\delta) = A_4^U(1)h_4(\delta), \tag{2.49}$$

is a very good approximation for single-phase aggregates made up of single crystals of cubic type. The scaling function depends on the single-crystal anisotropy and the mesoscale, and takes the following form based on numerical simulations (see also [35]):

$$\begin{aligned} f(C_{ij}, \delta) &= A_4^U(\delta) = A_4^U(1)h_4(\delta) \\ &= \frac{6}{5}\left(\sqrt{A} - \frac{1}{\sqrt{A}}\right)^2 \exp[-0.767(\delta - 1)^{0.5}]. \end{aligned} \tag{2.50}$$

Again, as discussed previously in Section 2.2.3, one may determine the size of the RVE by choosing a convenient value for the scaling function. Based on (2.50), in Fig. 2.17 we plot the contours of the scaling function in the (A, δ) space. It is evident that for a fixed value of the scaling function, the mesoscale size increases with an increase in the single-crystal anisotropy. In other words, the higher is the single-crystal anisotropy, the greater is the number of grains necessary to homogenize the aggregate response. This property is again confirmed in Fig. 2.18 (plotted for $f = 0.23$) for a variety of aggregates made up of cubic single crystals. Again, notice the distinct regions (*microscale*, *limiting mesoscale*, and *macroscale*) in these plots.

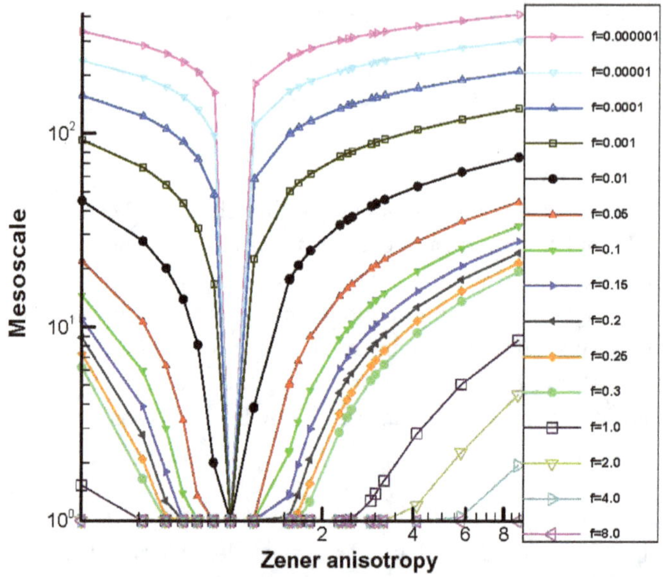

Fig. 2.17. Contours of the scaling function in the (A, δ) space.

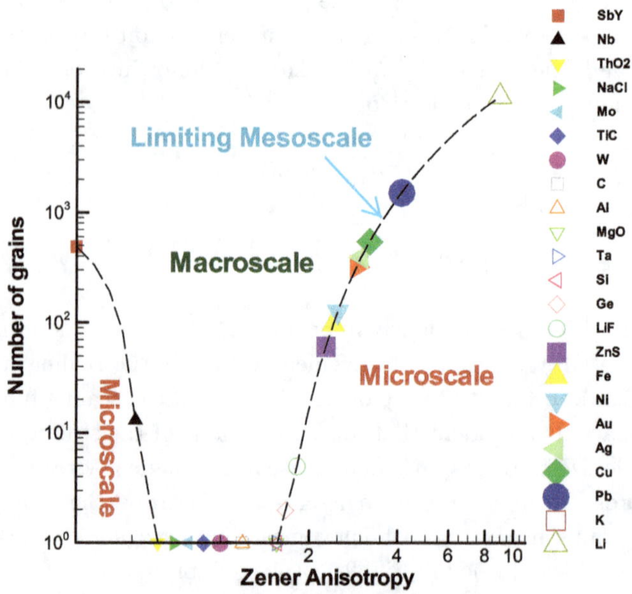

Fig. 2.18. Material scaling diagram for polycrystals made up of cubic single crystals ($f = 0.23$).

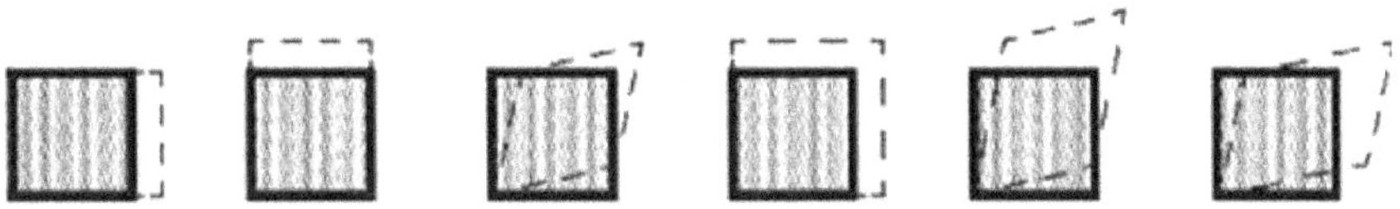

Fig. 2.19. Six tests (1 to 6 from left to right) used to determine the components of the in-plane stiffness and compliance tensors.

2.5. Elastic Properties of Planar Random Checkerboards

We now study the scaling of elastic properties in a two-phase checkerboard as shown in Fig. 2.1(a). For every microstructure, six numerical tests are run to determine the unknown components of the in-plane stiffness and compliance tensors as shown in Fig. 2.19. A total of 163,728 microstructural realizations were considered at varying contrasts, mesoscale and boundary conditions to establish the following form for the planar elastic scaling function (see [39])

$$f(K_1, K_2, \mu_1, \mu_2, \delta) = \left[2\frac{\mu^V}{\mu^R} + \frac{K^V}{K^R} - 3\right] \exp[-0.5\underline{8}(\delta - 1)^{0.44}], \qquad (2.51)$$

where K and μ represent the bulk and the shear modulus. The subscripts 1 and 2 represent the individual phases and the superscripts V and R represent the Voigt and Reuss bounds, respectively.

2.6. Scaling in Inelastic and Nonlinear Materials

2.6.1. Thermoelasticity

First recall that the thermal expansion phenomenon reflects a coupling between mechanical and thermal fields. It is grasped by the thermal strain coefficient α_{kl} (a second-rank tensor), a material property. Alternatively, one can also employ the thermal expansion stress coefficient Γ_{kl} given by the constitutive relation $\Gamma_{ij} = -C_{ijkl}\alpha_{kl}$, with C_{ijkl} being the stiffness tensor. For heterogeneous materials, that relation does not hold unless the dV element is a homogeneous domain within any constituent phase or the RVE for a homogenized composite. To study the RVE size in the latter case, one again uses the Hill–Mandel condition as the starting point, and follows the scaling concepts introduced elsewhere. Thus, besides Eqs. (2.33), (2.34a) and (2.34b), one also needs to set up bounds on the effective thermal expansion and the specific heat [40, 41]. In particular, depending on the relations between the material constants of a composite's constituents, for the thermal expansion coefficients we have two cases of hierarchies:

(i) $\alpha_1 > \alpha_2 \geq 0$ and $k_1 > k_2$:

$$\alpha^* \geq \cdots \geq \langle \alpha_\delta^n \rangle \geq \langle \alpha_{\delta'}^n \rangle \geq \cdots \geq \langle \alpha_1^n \rangle, \quad \forall \delta' \leq \delta; \tag{2.52}$$

(ii) $\alpha_1 > \alpha_2 \geq 0$ and $k_1 < k_2$:

$$\alpha^* \leq \cdots \leq \langle \alpha_\delta^n \rangle \leq \langle \alpha_{\delta'}^n \rangle \leq \cdots \leq \langle \alpha_1^n \rangle, \quad \forall \delta' \leq \delta. \tag{2.53}$$

Similarly, there are two cases of hierarchies for the thermal stress coefficients:

(i) $0 \geq \Gamma_1 > \Gamma_2$ and $k_1 > k_2$:

$$\Gamma^* \leq \cdots \leq \langle \Gamma_\delta^e \rangle \leq \langle \Gamma_{\delta'}^e \rangle \leq \cdots \leq \langle \Gamma_1^e \rangle, \quad \forall \delta' \leq \delta; \tag{2.54}$$

(ii) $0 \geq \Gamma_1 > \Gamma_2$ and $k_1 < k_2$:

$$\Gamma^* \geq \cdots \geq \langle \Gamma_\delta^e \rangle \geq \langle \Gamma_{\delta'}^e \rangle \geq \cdots \geq \langle \Gamma_1^e \rangle, \quad \forall \delta' \leq \delta. \tag{2.55}$$

2.6.2. Elasto-plasticity

Turning to a two-phase elastic-hardening plastic composite, the constitutive responses of both phases p $(=1,2)$ are taken in the following form [24]:

$$\begin{aligned}
d\varepsilon'_{ij} &= \frac{d\sigma'_{ij}}{2G_p} + h \cdot df_p \cdot \frac{\partial f_p}{\partial \sigma_{ij}} && \text{whenever } f_p = c_p \quad \text{and} \quad df_p \geq 0, \\
d\varepsilon'_{ij} &= d\sigma'_{ij}/2G_p && \text{whenever } f_p < c_p, \\
d\varepsilon &= d\sigma \cdot \frac{(1-2\nu_p)}{2G_p(1+\nu_p)} && \text{everywhere } \left(d\varepsilon = \frac{d\varepsilon_{ii}}{3},\ d\sigma = \frac{d\sigma_{ii}}{3} \right).
\end{aligned} \tag{2.56}$$

Here primes indicate deviatoric tensor components; G_p is a shear modulus, v_p is Poisson's ratio, f_p is a yield function, c_p is a material constant, and h is an isotropic hardening parameter.

Under monotonically increasing loading, elasto-plastic hardening composites can be treated as physically nonlinear elastic materials. Since the stiffness and compliance tensors are not constant any more, we consider the tangent stiffness and compliance moduli (C_δ^{Td} or S_δ^{Tt}), defined as

$$\overline{d\boldsymbol{\sigma}} = \mathbf{C}_\delta^{Td} : \overline{d\boldsymbol{\varepsilon}} = \mathbf{C}_\delta^{Td} : d\boldsymbol{\varepsilon}^0; \quad \overline{d\boldsymbol{\varepsilon}} = \mathbf{S}_\delta^{Tt} : \overline{d\boldsymbol{\sigma}} = \mathbf{S}_\delta^{Tt} : d\boldsymbol{\sigma}^0. \tag{2.57}$$

Here the superscript d (or t) indicates the response obtained under the displacement or traction boundary conditions. As noticed by Jiang *et al.* [42], there is a hierarchy of upper and lower bounds on the effective tangent modulus:

$$\begin{aligned}\langle \mathbf{S}_1^{TS}\rangle^{-1} &\equiv \langle \mathbf{S}_1^{Tt}\rangle^{-1} \le \cdots \le \langle \mathbf{S}_{\delta'}^{Tt}\rangle^{-1} \le \langle \mathbf{S}_\delta^{Tt}\rangle^{-1} \le \cdots \le (\mathbf{S}_\infty^{T})^{-1}\\ &\equiv \mathbf{C}_\infty^{T} \le \cdots \le \langle \mathbf{C}_\delta^{Td}\rangle \le \langle \mathbf{C}_{\delta'}^{Td}\rangle \le \cdots \le \langle \mathbf{C}_1^{Td}\rangle\\ &\equiv \langle \mathbf{C}_1^{TT}\rangle, \quad \forall\, \delta' \le \delta,\end{aligned} \tag{2.58}$$

where $\langle \mathbf{S}_1^{TS}\rangle^{-1}$ and $\langle \mathbf{C}_1^{TT}\rangle$ are recognized as the tangential Sachs and Taylor bounds, respectively. Sample results are shown in Figs. 2.20 and 2.21 [43]. Note that the flow rule on the mesoscale (under both kinds of boundary condition) for a composite made of phases with a normal flow rule is not necessarily normal.

An application of the above concepts to imperfect masonry structures is reported in [44, 45], while there is a computation of RVE size using the mixed-orthogonal boundary condition in [46]; see also [47]. A perspective on trends in multiscale plasticity is given in [48].

A recent experimental study investigating the RVE size is reported in [49]; see also [50–52]. A thermomechanics framework allowing an effective physical interpretation of the micromechanical internal variables and parameters in the RVE is developed in [53]; see also [54, 55]. The geodesic character of the strain field patterns, apparently almost self-evident in Fig. 2.19, has been investigated in [56].

2.6.3. Finite elasticity

The assumption inherent in the finite hyperelasticity theory is the existence of a strain energy function ψ per unit volume of an undeformed body, which depends on the deformation of the object and its material properties. The equation of state of the material in the reference configuration then takes the form $P_{ik} = \partial\psi/\partial F_{ik}$, where P_{ik} is the first Piola–Kirchhoff stress tensor and F_{ik} is the deformation gradient tensor.

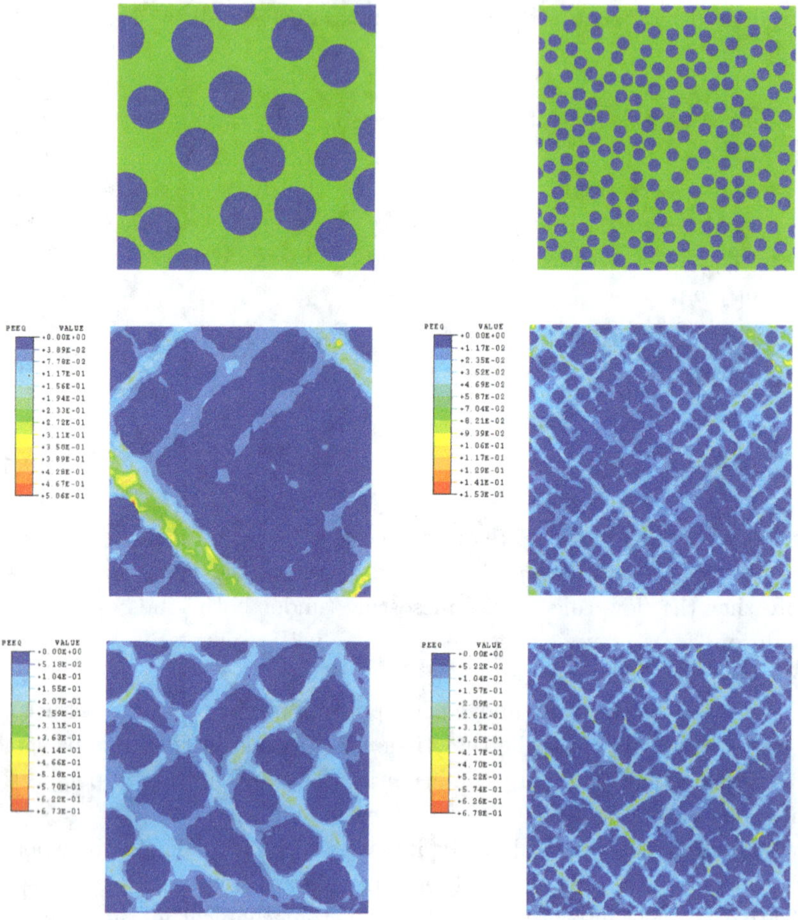

Fig. 2.20. Equivalent plastic strain patterns in matrix-inclusion composites (at mesoscale $\delta = 6$ and 20) under traction (middle row) and displacement (bottom row) boundary conditions [42].

These two tensors form a conjugate pair, which satisfies the Hill–Mandel condition:

$$\overline{P_{ij}F_{ij}} = \overline{P}_{ij}\overline{F}_{ij}. \tag{2.59}$$

Following the same procedure as in Section 2.2.2, we consider the functional:

$$P\{U_i\} = \int_{V_0} \psi(U_{i,k})dV - \int_{S_T} t_i^0 U_i dS, \tag{2.60}$$

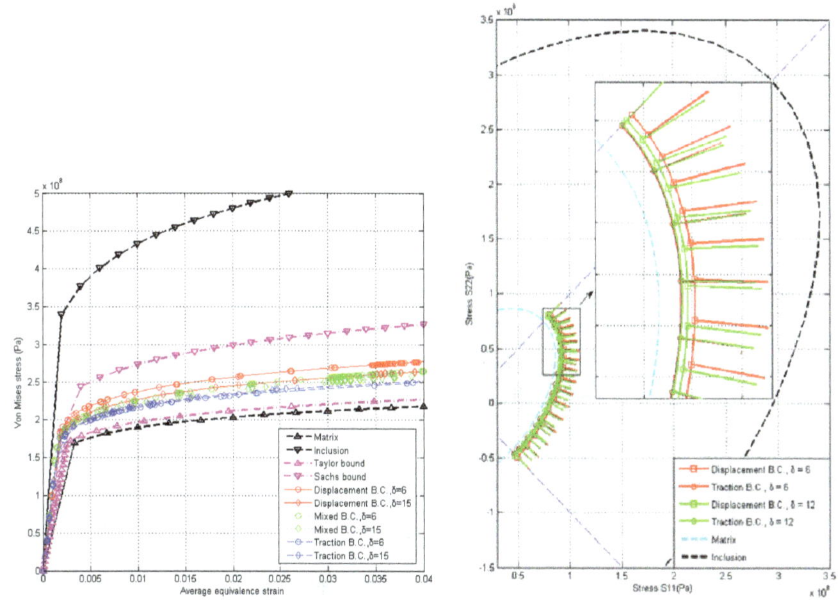

(a) (b)

Fig. 2.21. Ensemble-averaged elasto–plastic stress–strain curves (a) and yield surfaces (b), depending on the boundary conditions imposed.

where U_i is an admissible displacement field and t_i^0 is the specified boundary traction. The relation (2.60) represents the finite elasticity counterpart of the principle of the minimum potential energy for infinitesimal elastic deformation. The functional $P\{U_i\}$ assumes a local minimum for the actual solution u_i providing $\int_{V_0} \frac{\partial^2 \psi}{\partial u_{i,k} \partial u_{p,q}} d_{i,k} d_{p,q} dV > 0$, for all non-zero d_i such that $d_i = 0$ [57]. Under the uniform displacement boundary condition, which in nonlinear elasticity has the form $u_i = (F_{ik}^0 - \delta_{ik})x_k, \forall\, x_k \in S_0$, the upper bound on the effective response is [58, 59]:

$$\langle \Psi(\mathbf{F}^0, \Delta) \rangle \leq \langle \Psi(\mathbf{F}^0, \delta) \rangle \leq \langle \Psi(\mathbf{F}^0, \delta') \rangle \leq \langle \Psi(\mathbf{F}^0, 1) \rangle, \quad \text{for } \forall\, \delta' < \delta < \Delta, \tag{2.61}$$

where $\Psi(\omega, \mathbf{F}^0) = \int_{V_0} \psi(\omega, \mathbf{X}, \mathbf{F}) dV$, and Δ and 1 denote the RVE size and inhomogeneity size, respectively.

Considering the complementary energy principle [57], under the uniform traction boundary condition $t_i = P_{ik}^0 n_k, \forall\, x_j \in S_0$, we have another scale-dependent hierarchy of bounds on the effective properties:

$$\langle \Psi^*(\mathbf{P}^0)\rangle_\Delta \le \langle \Psi^*(\mathbf{P}^0)\rangle_\delta \le \langle \Psi^*(\mathbf{P}^0)\rangle_{\delta'} \le \langle \Psi^*(\mathbf{P}^0)\rangle_1, \quad \text{for } \forall\,\delta' < \delta < \Delta, \tag{2.62}$$

where $\Psi^*(\omega,\mathbf{P}^0) = \int_{V_0}\{\frac{\partial\psi}{\partial U_{ik}}U_{ik} - \psi\}dV$ and we assume $\int_{V_0}\frac{\partial^2\psi}{\partial u_{i,k}\partial u_{p,q}}d_{ik}$ $d_{pq}dV > 0$ for all non-zero d_{ik} satisfying $\frac{\partial}{\partial x_k}(\frac{\partial^2\psi}{\partial u_{i,k}\partial u_{p,q}}d_{pq}) = 0$ in V_0 and $\frac{\partial^2\psi}{\partial u_{i,k}\partial u_{p,q}}d_{pq}n_k = 0$ on S_T. $\Psi^*(\omega,\mathbf{P}^0)$ is a complementary strain energy function, which in nonlinear elasticity is generally unknown.

For the RVE-sized composite material, the application of different types of boundary conditions leads to a similar response and therefore $\Psi^*(\omega,\mathbf{P}_0) = \Psi^*(\omega,\mathbf{F}_0)$ and $\Psi(\omega,\mathbf{F}_0) = \Psi(\omega,\mathbf{P}_0)$. Then, the lower bound (2.62) can be written as

$$\begin{aligned}
&\langle \Psi(\mathbf{P}^0)\rangle_1 - V_0\langle \overline{\mathbf{P}:\mathbf{F}}\rangle_1 + V_0\langle \overline{\mathbf{P}:\mathbf{F}}\rangle_\Delta \\
&\quad\le \langle \Psi(\mathbf{P}^0)\rangle_{\delta'} - V_0\langle \overline{\mathbf{P}:\mathbf{F}}\rangle_{\delta'} + V_0\langle \overline{\mathbf{P}:\mathbf{F}}\rangle_\Delta \\
&\quad\le \langle \Psi(\mathbf{P}^0)\rangle_{\delta} - V_0\langle \overline{\mathbf{P}:\mathbf{F}}\rangle_{\delta} + V_0\langle \overline{\mathbf{P}:\mathbf{F}}\rangle_\Delta \\
&\quad\le \langle \Psi(\mathbf{P}^0)\rangle_\Delta, \quad \text{for } \forall\,\delta' < \delta < \Delta,
\end{aligned} \tag{2.63}$$

where we used the following relation between the complementary and strain energy functions:

$$\Psi^*(\omega,\mathbf{P}^0) = V_0(\overline{\mathbf{P}:\mathbf{F}}) - \Psi(\omega,\mathbf{P}^0). \tag{2.64}$$

Bounds (2.61) and (2.63) allow us to estimate the convergence rate and consequently the RVE size for any nonlinear composite satisfying the convexity requirement on the strain energy function. Sample results obtained by finite element analysis are shown in Figs. 2.22–2.24. A comparison of scaling in linear versus nonlinear thermoelastic materials

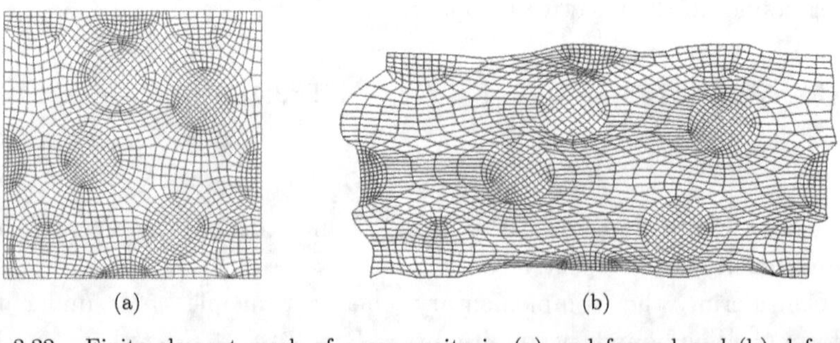

(a) (b)

Fig. 2.22. Finite element mesh of a composite in (a) undeformed and (b) deformed (under uniform traction boundary conditions) configurations.

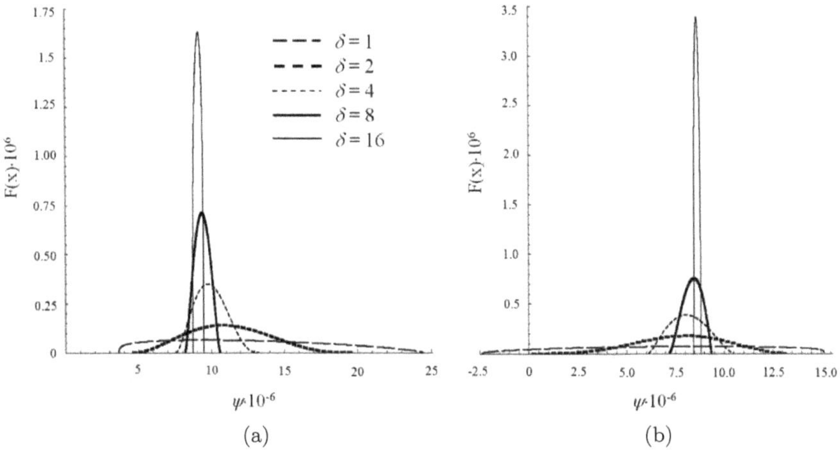

Fig. 2.23. Probability densities for the stored strain energy density of a nonlinear composite such as in Fig. 2.22 under (a) uniform displacement and (b) uniform traction boundary conditions.

is reported in [40]. See [60] for a related development in elastomers and Ghysels *et al.* [61] for an application of similar concepts in biological structures. A different approach, not involving the convexity assumption, has just been reported in Temizer and Wriggers [62].

2.6.4. Permeability of porous media

Darcy's law, which describes the permeability of a fluid-saturated material, has the form

$$\vec{U} = -\frac{\mathbf{K}}{\mu} \cdot \nabla p, \tag{2.65}$$

where $\vec{U}$ is the Darcy velocity, ∇p is the applied pressure gradient, μ is the fluid viscosity, and $\mathbf{K}$ is the permeability, is considered to be the governing partial differential equation for fluid flow in porous media at small Reynolds numbers. A flow distribution for one realization of particles is shown in Fig. 2.25.

Note that the relation (2.65) is a purely phenomenological equation (it is not derivable from the Navier–Stokes equation), and its validity needs assessment through a corresponding Hill–Mandel condition for flow in porous media. The latter is

$$\overline{p_{,i}U_i} = \overline{p_{,i}}\,\overline{U_i} \Leftrightarrow \int_{\partial B} (p - \overline{p_{,j}}x_j)(U_i n_i - \overline{U_i} n_i) dS = 0. \tag{2.66}$$

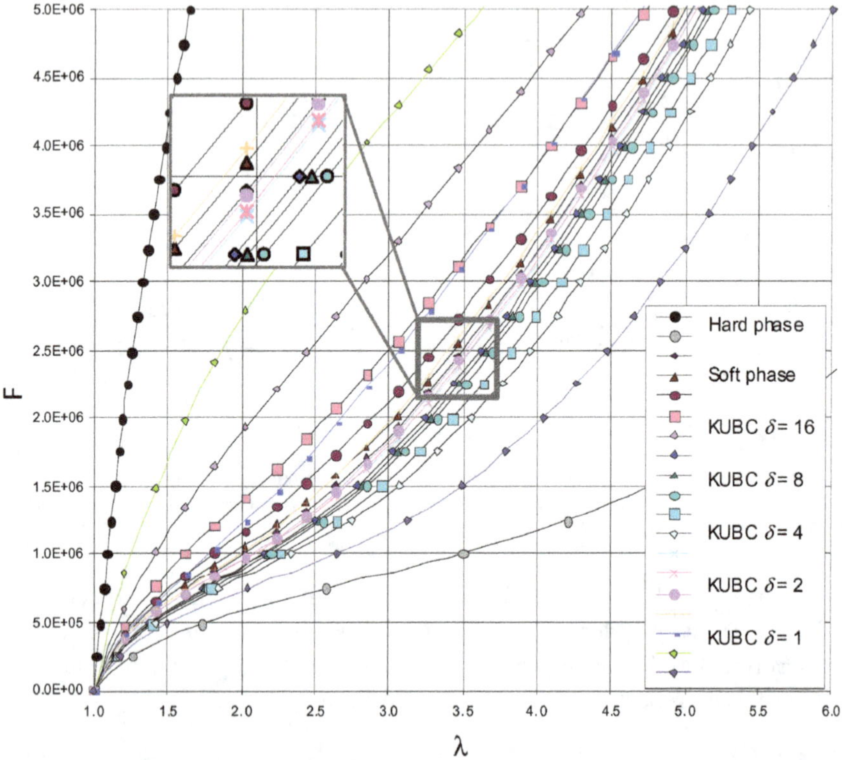

Fig. 2.24. Finite elasticity stress-strain responses of a random composite under uniaxial loading, KUBC (and SUBC) stand for kinematic (and static) uniform boundary conditions.

It follows that there are three possible boundary conditions: (i) $p = \overline{p_{,i}} x_i$, (ii) $U_i n_i = U_{0i} n_i$, and (iii) $(p - \overline{p_{,i}} x_i)(U_i n_i - U_{0i} n) = 0$ on ∂B satisfying (2.66). Following a derivation analogous to what was carried out in the above sections, one obtains the scale-dependent hierarchy of mesoscale bounds [63]:

$$\begin{aligned} \langle \mathbf{K}_\delta^{es} \rangle &\geq \langle \mathbf{K}_{2\delta}^{e} \rangle \geq \cdots \geq \mathbf{K}^{\text{eff}} \\ &= (\mathbf{R}^{\text{eff}})^{-1} \geq \cdots \geq \langle \mathbf{R}_{2\delta}^{n} \rangle^{-1} \geq \langle \mathbf{R}_{\delta}^{n} \rangle^{-1}, \quad \forall\, \delta' < \delta, \end{aligned} \tag{2.67}$$

where $\mathbf{R}$ is the resistance of porous media that satisfies $\mathbf{K} = \mathbf{R}^{-1}$ for homogenized media. Sample results are shown in Fig. 2.26.

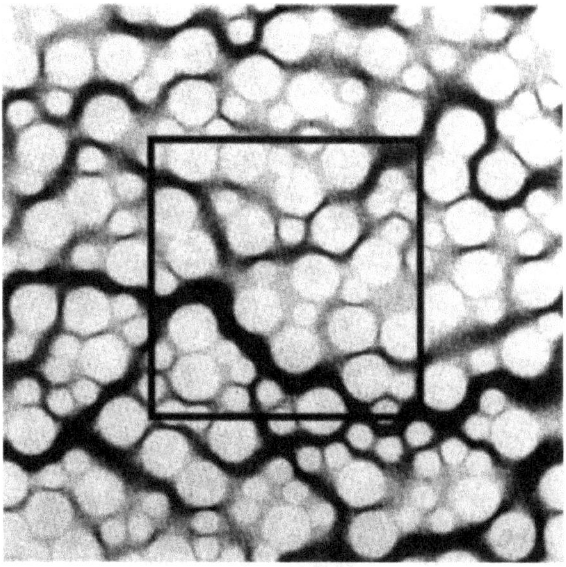

Fig. 2.25. Schematic of a flow distribution in one realization of particles in a random fluid-saturated material, leading to a question: On what scale is Darcy's law valid? A mesoscale window is also shown.

2.6.5. Comparative numerical results

In the preceding sections, we outlined basic theoretical results on scaling in linear elasticity, finite elasticity, elasto-plasticity, thermoelasticity, and permeability. In this section, we compare the relative convergence rates to the RVE of these different physical systems. For common reference, we consider a two-phase composite, consisting of circular inclusions randomly distributed in a matrix according to a Poisson point process with exclusion; see Fig. 2.1(c). Finite element analyses of such composites were conducted using ABAQUS 6.5. In order to compare the scaling from the SVE to the RVE between the physical systems, a so-called *discrepancy* is employed:

$$D = \frac{R_\delta^e - R_\delta^n}{(R_\delta^e + R_\delta^n)/2}. \tag{2.68}$$

Here R_δ^e is the response under the essential boundary conditions and R_δ^n is the response under the natural boundary conditions, both interpreted according to a given physical situation. For linear elasticity $R_\delta^e = tr\langle \mathbf{C}^e \rangle_\delta$,

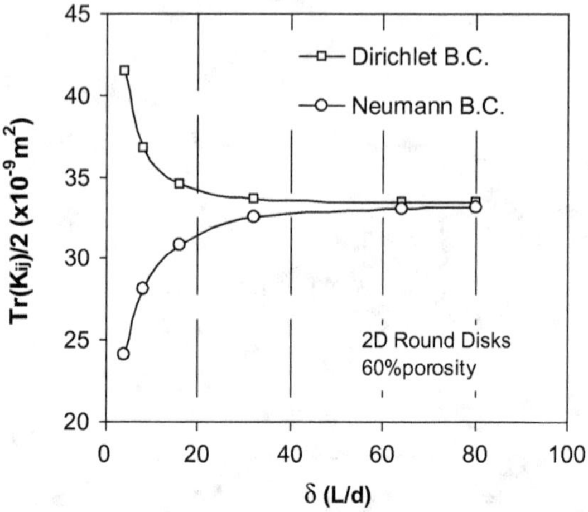

(a)

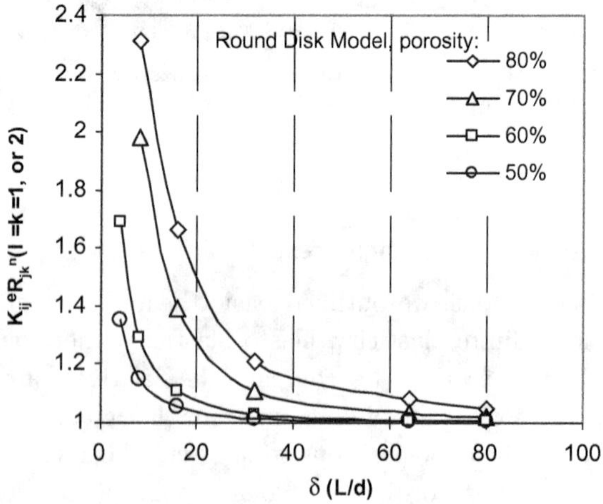

(b)

Fig. 2.26. (a) Effect of increasing window scale on the convergence of the hierarchy (2.67) of mesoscale bounds, obtained by computational fluid mechanics. (b) Effect of increasing window scale on the convergence of (2.67) as obtained by computational fluid mechanics.

$R^n_\delta = tr\langle \mathbf{S}^n\rangle^{-1}_\delta$, for linear thermoelasticity $R^e_\delta = \langle \Gamma^e_{ii}\rangle_\delta$, $R^n_\delta = -\langle C^n_{ijkl}\rangle_\delta \langle \alpha^n_{kl}\rangle_\delta$ (see Fig. 2.1), for plasticity $R^e_\delta = \langle h^e\rangle_\delta$, $R^n_\delta = \langle h^n\rangle_\delta$, and for nonlinear elasticity $R^e_\delta = \langle \Psi^e\rangle_\delta$, $R^n_\delta = \langle \mathbf{P}^n : \mathbf{F}^n - \Psi^n\rangle_\delta$.

The material parameters used for modeling and the computational results are summarized in Table 2.2. Here κ is the bulk modulus, μ is

Table 2.2. Mismatch and discrepancy values for the mesoscale $\delta = 16$.

	Mismatch	$D[\%]$
Linear elasticity	$\frac{\mu^{(i)}}{\mu^{(m)}} = 10, \quad \frac{\kappa^{(i)}}{\kappa^{(m)}} = 1$	2.28
Linear thermoelasticity	$\frac{\mu^{(i)}}{\mu^{(m)}} = 10, \quad \frac{\kappa^{(i)}}{\kappa^{(m)}} = 10, \quad \frac{\alpha^{(i)}}{\alpha^{(m)}} = \frac{1}{10}$	5.51
Plasticity	$\frac{h^{(i)}}{h^{(m)}} = 10, \quad \frac{E^{(i)}}{E^{(m)}} = 1$	2.29
Nonlinear elasticity	$\frac{\mu_0^{(i)}}{\mu_0^{(m)}} = 10, \quad \frac{\kappa_0^{(i)}}{\kappa_0^{(m)}} = 1$	5.86
Flow in porous media	$\frac{tr(\mathbf{K}^{(i)})}{tr(\mathbf{K}^{(m)})} = \infty$	27

the shear modulus, κ_0 is the initial bulk modulus, μ_0 is the initial shear modulus, and superscripts (i) and (m) denote inclusion and matrix accordingly. In order to compare physically different problems, linear-type constitutive relations were assumed, i.e., linear isotropic hardening in plasticity and neo-Hookean strain energy function in nonlinear elasticity. In linear thermoelasticity, we need to consider the mismatch not only of the thermal expansion coefficients but also of the elastic properties of both phases, otherwise α becomes a scale-dependent constant parameter.

As can be seen from the table, the discrepancy increases in the following order: linear elasticity, plasticity, linear thermoelasticity, and nonlinear elasticity. By nature, the mismatch between the solid and fluid phases in porous media is infinite, so the discrepancy value for permeability cannot be compared to the results in elasticity. The numerical results for this case show $D = 27\%$ for the sample size $\delta = 16$, which obviously cannot be considered homogeneous in the sense of Hill. While in linear elasticity the discrepancy D is not a function of strain, in nonlinear elasticity it strongly depends on deformation. In general, the value shown in Table 2.2 is the average over the considered deformation range.

2.7. Conclusions

Microstructural randomness is present in just about all solid materials. In this chapter, we reviewed the scaling from a statistical volume element (SVE) to a representative volume element (RVE). Using micromechanics, the RVE is approached in terms of two hierarchies of bounds stemming, respectively, from Dirichlet-type and Neumann-type boundary-value problems set up on the SVE. In [64] one can find a comprehensive introduction to basic homogenization theory, microstructural optimization, and the multifield analysis of heterogeneous elastic materials; see also [29, 65]. A separate area of research and application is the extensive methodology, based on mathematical morphology, used to model a wide range of random composite materials, typically employing a few parameters calibrated by the image analysis of real microstructures [66–69].

Besides the settings of conductivity, (thermo)elasticity, elasto-plasticity, and Darcy permeability discussed in this review as well as a more extensive one [7], the scaling from SVE to RVE has been (and continues to be) examined in various other fields of mechanics and physics, for instance: composite materials [70, 71], random plates [72], strength and damage [73], functionally graded materials [74], Fickian diffusion [75], solidification [76], thermodynamics and heat transfer [77, 78], granular media [79–81], and the morphogenesis of fractals [82]. Related efforts compare scale-dependent bounds, those resulting from periodic three-dimensional polycrystal models have been pursued in [83, 84], or determine the effects of the clustering of inclusions [85].

In the late 90s, the seminal paper by Forest and Sab [86] launched the effort to find a homogeneous (Cosserat) micropolar continuum smoothing a spatially heterogeneous Cauchy material. The objective of this line of research is to trade highly detailed information about a Cauchy-type microstructure for a less detailed description via an equivalent Cosserat continuum; see also [87–96] and, for the latest research on this subject, [97, 98]. A broad overview of multiscale models of microstructured materials — indeed, dating back to works by Navier, Cauchy, and Poisson — can be found in [99].

However, the presence of spatial randomness presents an extra challenge which has been undertaken in [100, 101] where a passage from a random microstructure of micropolar type of a linear elastic type in static setting has been made to a homogeneous micropolar medium.

We end with a note that the methodology outlined in this review forms a microstructure-mechanics basis for setting up mesoscale continuum random fields and stochastic finite element methods [102–105]. Such fields are essential for solving macroscopic boundary-value problems lacking a separation of the scales, i.e., where heterogeneous components are present but a finite element is not larger than the RVE. In that case, the finite element is analogous to the SVE. In view of the boundary conditions (2.32), the mesoscale random field is not only scale dependent, but also non-unique, its properties being anisotropic even if the RVE-level properties are isotropic. The stiffness matrix is dictated by the boundary condition (2.32a) while the flexibility matrix is dictated by (2.32b). These matrices enter, respectively, into minimum potential and complementary formulations of finite element methods.

While this chapter has been concerned with scaling and homogenization in random materials where a separation of scales can be attained, the subject matter meets even more formidable challenges in fractal media. A theoretical formulation in this direction, based on dimensional regularization, is under development [106–111].

Acknowledgments

This work was made possible by the NSF support under the grant CMMI-1030940. S.I.R. was partially supported by the start-up grant from Rowan University.

References

[1] Hill R., 1963. Elastic properties of reinforced solids: Some theoretical principles, *J. Mech. Phys. Solids*, 11, 357–372.

[2] Mandel J., Dantu P., 1963. Contribution à l'étude théorique et expérimentale du coefficient d'élasticité d'un milieu hétérogènes mais statisquement homogène, *Ann. Ponts Chaussées Paris*, 6, 115–145.

[3] Ohser J., Schladitz K., 2009. *3D Images of Materials Structures: Processing and Analysis*, Wiley-VCH, Weinheim.

[4] Kurzydłowski K.J., Ralph B., 1995. *The Quantitative Description of the Micro-structure of Materials*, CRC Press, Boca Raton.

[5] Ostoja-Starzewski M., 2006. Material spatial randomness: From statistical to representative volume element, *Probab. Eng. Mech.*, 20, 112–132.

[6] Ostoja-Starzewski M., 2008. *Microstructural Randomness and Scaling in Mechanics of Materials*, CRC Press, Boca Raton.

[7] Ostoja-Starzewski M., Kale S., Karimi P., Malyarenko A., Raghavan B., Ranganathan S.I., Zhang J., 2016. Scaling to RVE in random media, *Adv. Appl. Mech.*, 49, 111–211.

[8] Ostoja-Starzewski M., Schulte J., 1996. Bounding of effective thermal conductivities of multiscale materials by essential and natural boundary conditions, *Phys. Rev. B*, 54, 278–285.

[9] Ranganathan S.I., Ostoja-Starzewski M., 2008. Mesoscale conductivity and scaling function in aggregates of cubic, trigonal, hexagonal and tetragonal crystals, *Phys. Rev. B*, 77, 214308-1–214308-10.

[10] Jiang M., Jasiuk I., Ostoja-Starzewski M., 2002. Apparent thermal conductivity of periodic two-dimensional composites, *Comput. Mater. Sci.*, 25, 329–338.

[11] Özdemir I., Brekelmans W.A.M., Geers M.G.D., 2008. Computational homogenization for heat conduction in heterogeneous solids, *Int. J. Numer. Meth. Eng.*, 73, 185–204.

[12] Ostoja-Starzewski M., 2000. Universal material property in conductivity of planar random microstructures, *Phys. Rev. B*, 62(5) 2980–2983.

[13] Shoemake K., 1992. *Uniform Random Rotations*, *Graphics Gems III*, Academic Press, London.

[14] Ranganathan S.I., Ostoja-Starzewski M., 2009. Towards scaling laws in random polycrystals, *Int. J. Eng. Sci.* (*A.J.M. Spencer special issue*), 47, 1322–1330.

[15] Keller J.B., 1964. A theorem on the conductivity of a composite medium, *J. Math. Phys.*, 5, 548–549.

[16] Dalaq A.S., Ranganathan S.I., Ostoja-Starzewski M., 2013. Scaling function in conductivity of planar random checkerboards, *Comput. Mater. Sci.*, 79, 252–261.

[17] Dalaq A.S., Ranganathan S.I., 2015. Invariants of mesoscale thermal conductivity and resistivity tensors in random checkerboards, *Eng. Comput.*, 32, 1601–1618.

[18] Kanit T., Forest S., Galliet I., Monoury V., Jeulin D., 2003. Determination of the size of the representative volume element for random composites: Statistical and numerical approach, *Int. J. Solids Struct.*, 40, 3647–3679.

[19] Matheron G., 1971. The theory of regionalized variables and its applications. *Ecole Nationale Sup. des Mines*, 5, 211.

[20] Raghavan B.V., Ranganathan S.I., Ostoja-Starzewski M., 2015. Electrical properties of random checkerboards at finite scales, *AIP Adv.*, 5, 017131.

[21] Ranganathan S.I., Ostoja-Starzewski M., 2008. Universal elastic anisotropy index, *Phys. Rev. Lett.*, 101, 055504-1–4.

[22] Zener C., 1948. *Elasticity and Anelasticity of Metals*, University of Chicago Press, Chicago.

[23] Chung D.H., Buessem W.R., 1967. The elastic anisotropy of crystals, *J. Appl. Phys.* 38, 5.

[24] Hill R., 1950. *The Mathematical Theory of Plasticity*, Clarendon Press, Oxford.

[25] Suquet P.M., 1986. Elements of homogenization for solid mechanics, in *Homogenization Techniques for Composite Media*, Lecture Notes in Physics, *Vol. 272* (E. Sanchez-Palencia, A. Zaoui, eds.), Springer-Verlag, Berlin, pp. 193–278.

[26] Huet C., 1990. Application of variational concepts to size effects in elastic heterogeneous bodies, *J. Mech. Phys. Solids*, 38(6), 813–841.

[27] Huet C., 1999. Coupled size and boundary-condition effects in viscoelastic heterogeneous and composite bodies, *Mech. Mater.*, 31, 787–829.

[28] Hazanov S., Huet C., 1994. Order relationships for boundary conditions effect in the heterogeneous bodies smaller than the representative volume, *J. Mech. Phys. Solids*, 42, 1995–2011.

[29] Sab K., 1992. On the homogenization and the simulation of random materials, *Europ. J. Mech. A/Solids*, 11, 585–607.

[30] Jiang M., Alzebdeh K., Jasiuk I., Ostoja-Starzewski M., 2001. Scale and boundary conditions effects in elastic properties of random composites, *Acta Mech.*, 148(1–4), 63–78.

[31] Ostoja-Starzewski M., 1999. Scale effects in materials with random distributions of needles and cracks. *Mech. Mater.*, 31, 883–893.

[32] Ren Z.-Y., Zheng Q.-S., 2002. A quantitative study of minimum sizes of representative volume elements of cubic polycrystals — numerical experiments, *J. Mech. Phys. Solids*, 50, 881–893.

[33] Ren Z.-Y., Zheng Q.-S., 2004. A quantitative study of minimum sizes of representative volume elements of cubic polycrystals — numerical experiments, *Mech. Mater.*, 36, 1217–1229.

[34] Różanski A., Łydżba D., 2011. From digital image of microstructure to the size of representative volume element: B4C/Al composite, *Studia Geotech Mech.*, 33(1), 55–68.

[35] Ranganathan S.I., Ostoja-Starzewski M., 2008. Scaling function, anisotropy and the size of RVE in elastic random polycrystals, *J. Mech. Phys. Solids* 56, 2773–2791.

[36] Mendelson K.S., 1981. Bulk modulus of a polycrystal, *J. Phys. D: Appl. Phys.*, 14, 1307–1309.

[37] Walker A.M., Wookey J., 2012. MSAT — A new toolkit for the analysis of elastic and seismic anisotropy, *Comput. Geosci.*, 49, 81–90.

[38] Murshed M.R., Ranganathan S.I. 2017. Scaling laws in elastic polycrystals with individual grains belonging to any crystal class, *Acta Mechanica* (DOI: 10.1007/s00707-016-1774-3).

[39] Raghavan B.V., Ranganathan S.I., 2014. Bounds and scaling laws at finite scales in planar elasticity, *Acta Mechanica*, 225, 3007–3022.

[40] Khisaeva Z.F., Ostoja-Starzewski M., 2007. Scale effects in infinitesimal and finite thermoelasticity of random composites, *J. Thermal Stresses*, 30, 587–603.

[41] Du X., Ostoja-Starzewski M., 2006. On the scaling from statistical to representative volume element in thermoelasticity of random materials, *Netw. Heterogeneous Media*, 1, 259–274.

[42] Jiang M., Ostoja-Starzewski M., Jasiuk I., 2001. Scale-dependent bounds on effective elastoplastic response of random composites, *J. Mech. Phys. Solids*, 49, 655–673.

[43] Li W., Ostoja-Starzewski M., 2006. Yield of random elasto-plastic materials, *J. Mech. Mater. Struct.*, 1, 1055–1073.

[44] Zeman J., Sejnoha M., 2007. From random microstructures to representative volume elements, *Model. Simul. Mat. Sci. Eng.*, 15, S325–S335.

[45] Cavalagli N., Cluni F., Gusela V., 2011. Strength domain of non-periodic masonry by homogenization in generalized plane state, *Europ. J. Mech. A/Solids*, 30, 113–126.

[46] Salahouelhadj A., Haddadi H., 2010. Estimation of the size of the RVE for isotropic copper polycrystals by using elastic-plastic finite element homogenisation, *Comp. Mater. Sci.*, 48, 447–455.

[47] Markovic D., Ibrahimbegovic A., 2004. On micro-macro interface conditions for micro scale based FEM for inelastic behaviour of heterogeneous materials, *Comp. Meth. Appl. Mech. Eng.*, 193, 5503–5523.

[48] McDowell D.L., 2010. A perspective on trends in multiscale plasticity, *Int. J. Plast.*, 26, 1280–1309.

[49] Efstahiou C., Sehitoglu H., Lambros J., 2010. Multiscale strain measurements of plastic deforming polycrystalline titanium: Role of deformation heterogeneities, *Int. J. Plast.*, 26, 93–106.

[50] Rossoll A., Moser B., Mortensen A., 2005. Longitudinal deformation of fibre reinforced metals: Influence of fibre distribution on stiffness and flow stress, *Mech. Mater.*, 37, 1–17.

[51] Shen H., Brinson L.C., 2006. A numerical investigation of the effect of boundary conditions and representative volume element size for porous titanium, *JoMMS*, 1(7), 1179–1204.

[52] Yin X., Chen W., Ho T., McVeigh C., Liu W.K., 2008. Statistical volume element for predicting microstructure-constitutive property relations, *Comput. Meth. Appl. Mech. Eng.*, 197, 3516–2529.

[53] Einav I., Collins I.F., 2008. A thermomechanical framework of plasticity based on probabilistic micromechanics, *JoMMS*, 3(5), 867–892.

[54] Ostoja-Starzewski M., 2002. Microstructural randomness versus representative volume element in thermomechanics, *ASME J. Appl. Mech.*, 69, 25–35.

[55] Ostoja-Starzewski M., 2005. Scale effects in plasticity of random media: Status and challenges, *Int. J. Plast.*, 21, 1119–1160.

[56] Jeulin D., Li W., Ostoja-Starzewski M., 2008. On a geodesic property of strain field patterns in elasto-plastic composites, *Proc. Roy. Soc. Lond. A*, 464, 1217–1227.

[57] Lee S.J., Shield R.T., 1980. Variational principles in finite elastostatics, *ZAMP*, 31, 437–453.

[58] Khisaeva Z.F., Ostoja-Starzewski M., 2006. Mesoscale bounds in finite elasticity of random composites, *Proc. R. Soc. Lond. A*, 462, 1167–1180.

[59] Khisaeva Z.F., Ostoja-Starzewski M., 2006. On the size of RVE in finite elasticity of random composites, *J. Elast.*, 85, 153–173.

[60] Moraleda J., Segurado J., Llorca J., 2007. Finite deformation of porous elastomers: A computational micromechanics approach, *Phil. Mag.*, 87(35), 5607–5627.

[61] Ghysels P., Samaey G., Tijskens B., Van Liederkerke P., Ramon H., Roose D., 2009. Multi-scale simulation of plant tissue deformation using a model for individual cell mechanics, *Phys. Biol.*, 6, 1–14.

[62] Temizer I., Wriggers P., 2011. Homogenization in finite thermoelasticity, *J. Mech. Phys. Solids*, 59, 344–372.

[63] Du X., Ostoja-Starzewski M., 2006. On the size of representative volume element for Darcy law in random media, *Proc. R. Soc. Lond. A*, 462, 2949–296.

[64] Zohdi T.I., Wriggers P., 2008. *An Introduction to Computational Micromechanics*, Springer, Berlin.

[65] Böhm H.J., 2008. A short introduction to basic aspects of continuum micromechanics, ILSB Report 206, Vienna University of Technology, pp. 1–144.
[66] Jeulin D., 1997. *Advances in Theory and Application of Random Sets*, World Scientific, Singapore.
[67] Jeulin D., 1999. Morphological modeling of random composites, *Continuum Models and Discrete Systems*, 8, 199–206.
[68] Jeulin D., 2000. Random texture models for material structures, *Statist. Comput.*, 10(2), 121–132.
[69] Serra J., 1982. *Image Analysis and Mathematical Morphology*, Academic Press, London.
[70] Sriramula S., Chryssanthopoulos M.K., 2009. Quantification of uncertainty modelling in stochastic analysis of FRP composites, *Composites: Part A*, 40, 1673–1684.
[71] Xu X.F., Chen X., 2009. Stochastic homogenization of random elastic multiphase composites and size quantification of representative volume element, *Mech. Mater.*, 41, 174–186.
[72] Sab K., 2011. Homogenization of random plates, *Int. J. Multiscale Comp. Eng.*, 5, 503–513.
[73] Sab K., 2009. Overall ultimate yield strength of a quasi-periodic masonry, *C. R. Acad. Sci. Paris II*, 337, 603–609.
[74] Ostoja-Starzewski M., Jasiuk I., Wang W., Alzebdeh K., 1996. Composites with functionally graded interfaces: Meso-continuum concept and effective properties, *Acta Materialia*, 44(5), 2057–2066.
[75] Karim M.R., Krabbenhoft, K., 2009. Extraction of effective cement paste diffusivities from X-ray microtomography scans, *Transp. Porous Media*, 84, 371–388.
[76] Lee S., Sundararaghavan V., 2009. Multiscale modeling of moving interface problems with flux jumps: Application to solidification, *Comput. Mech.*, 44(3), 297–307.
[77] Ostoja-Starzewski M., 2002. Towards stochastic continuum thermodynamics, *J. Non-Equilib. Thermodyn.*, 27(4), 335–348.
[78] Ostoja-Starzewski M., Zubelewicz A., 2011. Powerless fluxes and forces, and change of scale in irreversible thermodynamics, *J. Phys. A: Math. Theor.*, 44, 335002.
[79] Ostoja-Starzewski M., Wang C., 1989. Linear elasticity of planar Delaunay networks: Random field characterization of effective moduli, *Acta Mech.*, 80, 61–80.
[80] Stroeven M., Askes, H., Sluys L.J., 2004. Numerical determination of representative volumes for granular materials, *Comput. Meth. Appl. Mech. Eng.*, 193, 3221–3238.
[81] Valavala P.K., Odegard G.M., Aifantis E.C., 2009. Influence of representative volume size on predicted elastic properties of polymer materials, *Modell. Simul. Mat. Sci. Eng.*, 17(4), 1–15.
[82] Li J., Ostoja-Starzewski M., 2010. Fractals in elastic-hardening plastic materials, *Proc. R. Soc. Lond. A*, 466, 603–621.

[83] Fritzen F., Böhlke T., Schnack E., 2009. Periodic three-dimensional mesh generation for crystalline aggregates based on Voronoi tessellations, *Comput. Mech.*, 43, 701–713.

[84] Sab K., Nedjar B., 2005. Periodization of random media and representative volume element size for linear composites, *C. R. Acad. Sci. Paris II*, 333, 187–195.

[85] Al-Ostaz A., Diwakar A., Alzebdeh K.I., 2007. Statistical model for characterizing random microstructure of inclusion composites, *J. Mater. Sci.*, 42, 7016–7030.

[86] Forest S., Sab K., 1998. Cosserat overall modeling of heterogeneous materials, *Mech. Res. Comm.*, 25, 449–454.

[87] Forest S., Barbe F., Cailleteaud G., 2000. Cosserat modelling of size effects in the mechanical behaviour of polycrystals and multi-phase materials, *Int. J. Solids Struct.*, 37, 7105–7126.

[88] Forest S., Pradel F., Sab K., 2001. Asymptotic analysis of Cosserat media, *Int. J. Solids Struct.*, 38, 4585–4608.

[89] Geers M.G.D., Kouznetsova V., Brekelmans W.A.M., 2001. Gradient-enhanced computational homogenization for the micro-macro scale transition, *J. Phys. IV*, 11, 145–152.

[90] Kouznetsova V., Brekelmans W.A.M., Baaijens F.P.T., 2001. An approach to micro-macromodeling of heterogeneous materials, *Comput. Mech.*, 27, 37–48.

[91] Kouznetsova V., Geers M.G.D., Brekelmans W.A.M., 2002. Multi-scale constitutive modelling of heterogeneous materials with a gradient-enhanced computational homogenization scheme, *Int. J. Numer. Meth. Engrg.*, 54, 1235–1260.

[92] Bouyge F., Jasiuk I., Boccara S., Ostoja-Starzewski M., 2002. A micromechanically based couple-stress model of an elastic orthotropic two-phase composite, *Europ. J. Mech./A: Solids*, 21(3), 465–481.

[93] Trovalusci P., Masiani R., 2003. Non-linear micropolar and classical continua for anisotropic discontinuous materials, *Int. J. Solids Struct.*, 40, 1281–1297.

[94] Trovalusci P., Masiani R., 2005. A multifield model for blocky materials based on multiscale description, *Int. J. Solids Struct.*, 42, 5778–5794.

[95] Chen H., Liu X., Hu G., Yuan H., 2009. Identification of material parameters of micropolar theory of composites by homogenization method, *Comp. Mater. Sci.*, 46, 733–737.

[96] Jänicke R., Deibels S., Sehlhorst H.-G., Düster A., 2009. Two-scale modelling of micromorphic continua: A numerical homogenization scheme, *Cont. Mech. Thermodyn.*, 21(4), 297–315.

[97] Addessi D., De Bellis M.L. and Sacco E., 2013. Micromechanical analysis of heterogeneous materials subjected to overall Cosserat strains, *Mech. Res. Commun.*, 54, 27–34.

[98] Addessi D., De Bellis M.L. and Sacco E., 2016. A micromechanical approach for the Cosserat modeling of composites, *Meccanica*, 51, 569–592.

[99] Trovalusci P., Capecchi D., Ruta G., 2009. Genesis of the multiscale approach for materials with microstructure, *Arch. Appl. Mech.*, 79(11), 981–007.

[100] Trovalusci P., De Bellis M.L., Ostoja-Starzewski M. and Murrali A., 2014. Particulate random composites homogenized as micropolar materials, *Meccanica*, 49, 2719–2727.

[101] Trovalusci P., Ostoja-Starzewski M., De Bellis M.L. and Murrali A., 2015. Scale-dependent homogenization of random composites as micropolar continua, *Europ. J. Mech./A:Solids*, 49, 396–407.

[102] Ostoja-Starzewski M., 1998. Random field models of heterogeneous materials, *Int. J. Solids Struct.*, 35(19), 2429–2455.

[103] Ostoja-Starzewski M., 1999. Microstructural disorder, mesoscale finite elements, and macroscopic response, *Proc. Roy. Soc. Lond. A*, 455, 3189–3199.

[104] Ostoja-Starzewski M., Wang X., 1999. Stochastic finite elements as a bridge between random material microstructure and global response, *Comp. Meth. Appl. Mech. Eng.*, 168(1–4), 35–49.

[105] Ostoja-Starzewski M., 2012. Stochastic finite elements: Where is the physics? *Theor. Appl. Mech.*, 38(4), 379–396.

[106] Tarasov V.E., 2010. *Fractional Dynamics: Applications of Fractional Calculus to Dynamics of Particles, Fields and Media*, Springer-Verlag, Berlin.

[107] Li J., Ostoja-Starzewski M. 2009. Fractal solids, product measures and fractional wave equations, *Proc. Roy. Soc. A*, 465, 2521–2536. Errata, 2010.

[108] Demmie P.N., Ostoja-Starzewski M., 2011. Waves in fractal media, *J. Elast.*, 104, 187–204.

[109] Li J., Ostoja-Starzewski M., 2011. Micropolar continuum mechanics of fractal media, *Int. J. Eng. Sci.*, 49, 1302–1310.

[110] Joumaa H., Ostoja-Starzewski M., Demmie P.N., 2012. Elastodynamics in micropolar fractal solids, *Math. Mech. Solids*, online (DOI: 10.1177/1081286512454557).

[111] Ostoja-Starzewski M., 2013. Electromagnetism on anisotropic fractal media, *ZAMP*, 64(2), 381–390.

https://doi.org/10.1142/9781800611887_0003

Chapter 3

Stroh-Like Formalism for General Thin-Laminated Plates and Its Applications

Chyanbin Hwu
Department of Aeronautics and Astronautics,
National Cheng Kung University,
Taiwan, R.O.C.
chwu@mail.ncku.edu.tw

Abstract

If laminates are unsymmetric, they will be stretched as well as bent even under pure in-plane forces or pure bending moments. The coupled stretching-bending theory of laminates was developed to study the mechanical behavior of thin-laminated plates. Since this theory considers the linear variation of displacements across the thickness direction, by separating the thickness dependence it is easy to get general solutions through the complex-variable approach. An elegant and powerful complex-variable method called Stroh formalism is well known for problems in two-dimensional linear anisotropic elasticity. In this chapter, its counterpart, generally called Stroh-like formalism, will be introduced to deal with the coupled stretching-bending theory of laminates. Moreover, its extension to hygrothermal problems and electro-elastic composite laminates, and its applications to the problems of holes and cracks in laminates will all be presented. Some representative numerical examples are then shown to illustrate the advantage and necessity of the analytical closed-form solutions obtained by the Stroh-like formalism.

Contents

3.1. Introduction

The virtually limitless combinations of ply materials, ply orientations, and ply stacking sequences offered by laminated plates considerably enhance the design flexibility inherent in composite structures. In practical applications, to take advantage of the designable characteristics of composite laminates, there is always the option of designing an unsymmetric laminated plate. In that case, stretching-bending coupling may occur no matter what kind of loading is applied on the laminated plates.

In most practical applications, the plates are usually designed with much smaller thickness than the other two dimensions; such configurations are commonly called *thin plates.* According to observations of the actual mechanical behavior of thin-laminated plates, *Kirchhoff's assumptions* are usually made in *classical lamination theory.* If the effects of transverse shear deformation are considered, *shear deformation theory* may be considered for relatively thick laminated plates. Both classical lamination theory and shear deformation theory are based upon the assumption of the linear variation of displacement fields in the thickness coordinate, and are usually called the *first-order plate theory.* Sometimes it is difficult to describe the deformation of thick laminates by first-order plate theory. By considering the warping of the cross-section that probably occurs in thick laminates, *high-order plate theory* was developed by assuming the displacement fields of higher-order variation in the thickness coordinate. In this chapter, the main concern is the applicability of the complex-variable method to general laminated plates; to avoid the complexity involved in the thickness direction only classical lamination theory for thin plates is considered.

It is well known that the complex-variable method is a powerful method for two-dimensional elasticity problems as well as plate bending problems. For two-dimensional linear anisotropic elasticity, there are two major complex-variable formalisms presented in the literature. One is *Lekhnitskii formalism* [1, 2], which starts with equilibrated stresses given in terms of stress functions followed by the compatibility equations, and the other is *Stroh formalism* [3–6], which starts with compatible displacements followed by the equilibrium equations. For plate bending analysis, *Lekhnitskii bending formalism* and *Stroh-like bending formalism* swap over. That is, Lekhnitskii bending formalism starts with compatible deflections whereas Stroh-like bending formalism starts with equilibrated bending moments. This interesting connection provides a hint to *mixed formalism* for the coupled stretching-bending analysis of general laminated plates [7], which is different from the *displacement formalism* initiated by Lu and Mahrenholtz [8], and improved by Cheng and Reddy [9] and Hwu [7]. Further discussion and interpretation of Stroh-like formalism were presented in [10–13]. In addition to the above-mentioned works, various Lekhnitskii-oriented complex-variable formulations and their applications have also been proposed in the literature, for example [14–19] with displacement formulation, [20–22] with mixed formulation, and [23–26] which considers all types of non-degenerate and degenerate anisotropic plates.

Through an appropriate combination of displacement formalism and mixed formalism, *Stroh-like formalism* was developed for the coupled stretching-bending analysis of general laminated plates [7]. Since there is much detail in the development of Stroh-like formalism, to save space in this chapter most of the results are presented without detailed derivation. The reader may refer to the book *Anisotropic Elastic Plates* [6] or the original papers cited in the following sections for the detailed derivation of Stroh-like formalism and its applications.

This chapter is divided into seven sections. Following the introduction in this section, Stroh-like formalism will be introduced in Section 3.2. The extension to hygrothermal problems and electro-elastic composite laminates will then be presented in Sections 3.3 and 3.4, respectively.

Since Stroh-like formalism has been purposely arranged into the form of Stroh formalism for two-dimensional linear anisotropic elasticity, almost all the mathematical techniques developed for two-dimensional problems can be transferred to the coupled stretching-bending problems. Thus, by simple analogy many unsolved lamination problems can be solved if their corresponding two-dimensional problems have been solved successfully. With this advantage, several applications of Stroh-like formalism can be found in the literature, for example [27–36]. In order to show the applications of Stroh-like formalism, problems with holes and cracks in laminates are considered in Section 3.5. These problems are: (i) holes in laminates subjected to uniform stretching and bending moments, (ii) holes in laminates subjected to uniform heat flow and moisture transfer, (iii) holes in electro-elastic laminates, and (iv) cracks in laminates. Some representative numerical examples are then shown in Section 3.6 to illustrate the advantage and necessity of the analytical explicit closed-form solutions presented in this chapter.

Recently, further applications have been made in boundary element analysis, such as [36–39]. This achievement was made mainly due to the Green's function derived based upon the Stroh-like formalism [32, 34, 35]. The Green's functions here are the solutions for an infinite laminate (symmetric or unsymmetric) with or without holes/cracks/inclusions subjected to concentrated forces and moments at an arbitrary point. The complete loading cases such as transverse loading, in-plane loading, out-of-plane bending moment and in-plane torsion are all considered. These solutions are important because analytically they can provide solutions for arbitrary loading through superposition, and numerically they can be employed as the fundamental solutions of boundary element method and as the kernel functions of integral equations to consider interactions between holes/inclusions and cracks.

Although the problems such as corners and singular integrals appear commonly for boundary element method and their solution techniques are well documented [40–43], most of them are restricted to two-dimensional or three-dimensional analysis with the fundamental solutions expressed in real form. Very few of them discussed the coupling between in-plane and plate bending problems with complex form fundamental solutions. To develop a boundary element for the coupled stretching-bending analysis of general laminated plates, most of the techniques available in the literature have been reconsidered and extended, such as auxiliary relationships for corners and analytical closed-form solutions for singular integrals [38, 39].

3.2. Stroh-Like Formalism

A commonly used orthotropic material in engineering applications is *unidirectional fiber-reinforced composite. Laminated composites* are made by laying various unidirectional fiber-reinforced composites. A single layer of a laminated composite is generally referred to as a *ply* or *lamina.* A single lamina is generally too thin to be directly used in engineering applications. Several laminae are bonded together to form a structure termed a *laminate.* The overall properties of laminates can be designed by changing the fiber orientation and the stacking sequence of the laminae. The most popular way to describe the overall properties and macromechanical behavior of a laminate is *classical lamination theory* [44]. According to observations of the actual mechanical behavior of laminates, *Kirchhoff's assumptions* are usually made in this theory: (a) The laminate consists of perfectly bonded laminae and the bonds are infinitesimally thin as well as non-shear-deformable. Thus, the displacements are continuous across lamina boundaries so that no lamina can slip relative to another. (b) A line originally straight and perpendicular to the middle surface of the laminate remains straight and perpendicular to the middle surface of the laminate when the laminate is deformed. In other words, transverse shear strains are ignored, i.e., $\gamma_{13} = \gamma_{23} = 0$. (c) The normals have constant length so that the strain perpendicular to the middle surface is ignored, i.e., $\varepsilon_3 = 0$.

Based upon Kirchhoff's assumptions, the displacement fields, the strain–displacement relations, the constitutive laws, and the equilibrium equations can be written as follows:

$$\begin{aligned} U_i &= u_i + x_3\beta_i, \quad \beta_1 = -w_{,1}, \quad \beta_2 = -w_{,2}, \\ \xi_{ij} &= \varepsilon_{ij} + x_3\kappa_{ij}, \quad \varepsilon_{ij} = (u_{i,j} + u_{j,i})/2, \quad \kappa_{ij} = (\beta_{i,j} + \beta_{j,i})/2, \\ N_{ij} &= A_{ijkl}\varepsilon_{kl} + B_{ijkl}\kappa_{kl}, \quad M_{ij} = B_{ijkl}\varepsilon_{kl} + D_{ijkl}\kappa_{kl}, \\ N_{ij,j} &= 0, \quad M_{ij,ij} + q = 0, \quad Q_i = M_{ij,j}, \quad i,j,k,l = 1,2, \end{aligned} \tag{3.1}$$

where U_1 and U_2 are the displacements in the x_1 and x_2 directions; u_1 and u_2 are their associated middle surface displacements; $\beta_i, i = 1,2$, are the negative slopes; w is the displacement in the x_3 direction; ξ_{ij}, ε_{ij}, and κ_{ij} are, respectively, the strains, the midplane strains, and the plate curvatures; N_{ij} and M_{ij} are the stress resultants and bending moments; A_{ijkl}, B_{ijkl}, and D_{ijkl} are, respectively, the tensors of *extensional, coupling,* and *bending stiffnesses*; Q_i is the transverse shear force; and q is the lateral distributed load applied on the laminates. The subscript comma stands for differentiation, e.g., $w_{,1} = \partial w/\partial x_1$, and repeated indices imply summation through 1 to 2.

In order to find a solution satisfying all the basic equations in (3.1) in the coupled stretching-bending analysis of general laminates, several different complex-variable methods have been proposed in the literature. The main differences between these methods are: (1) the choice of the primary and secondary functions, (2) the organization of the final expressions of the general solutions, (3) the establishment of a material eigen-relation, and (4) the consideration of degenerate anisotropic plates. These differences are discussed in the following.

(1) The choice of the primary and secondary functions starts from the consideration of the constitutive laws given in $(3.1)_3$, which can be rewritten in matrix form as

$$\begin{Bmatrix} \boldsymbol{N} \\ \boldsymbol{M} \end{Bmatrix} = \begin{bmatrix} \boldsymbol{A} & \boldsymbol{B} \\ \boldsymbol{B} & \boldsymbol{D} \end{bmatrix} \begin{Bmatrix} \boldsymbol{\varepsilon}_0 \\ \boldsymbol{\kappa} \end{Bmatrix}. \tag{3.2}$$

By inversion or semi-inversion, we have

$$\begin{Bmatrix} \boldsymbol{\varepsilon}_0 \\ \boldsymbol{\kappa} \end{Bmatrix} = \begin{bmatrix} \boldsymbol{A}^* & \boldsymbol{B}^* \\ \boldsymbol{B}^* & \boldsymbol{D}^* \end{bmatrix} \begin{Bmatrix} \boldsymbol{N} \\ \boldsymbol{M} \end{Bmatrix}, \quad \begin{Bmatrix} \boldsymbol{N} \\ \boldsymbol{\kappa} \end{Bmatrix} = \begin{bmatrix} \tilde{\boldsymbol{A}} & \tilde{\boldsymbol{B}} \\ -\tilde{\boldsymbol{B}}^T & \tilde{\boldsymbol{D}} \end{bmatrix} \begin{Bmatrix} \boldsymbol{\varepsilon}_0 \\ \boldsymbol{M} \end{Bmatrix}, \tag{3.3}$$

or

$$\begin{Bmatrix} \boldsymbol{\varepsilon}_0 \\ \boldsymbol{M} \end{Bmatrix} = \begin{bmatrix} \tilde{\boldsymbol{A}} & \tilde{\boldsymbol{B}} \\ -\tilde{\boldsymbol{B}}^T & \tilde{\boldsymbol{D}} \end{bmatrix}^{-1} \begin{Bmatrix} \boldsymbol{N} \\ \boldsymbol{\kappa} \end{Bmatrix} = \begin{bmatrix} \tilde{\boldsymbol{A}}^* & \tilde{\boldsymbol{B}}^* \\ -\tilde{\boldsymbol{B}}^{*T} & \tilde{\boldsymbol{D}}^* \end{bmatrix} \begin{Bmatrix} \boldsymbol{N} \\ \boldsymbol{\kappa} \end{Bmatrix}, \tag{3.4}$$

where

$$\begin{aligned} &\boldsymbol{A}^* = \boldsymbol{A}^{-1} + \boldsymbol{A}^{-1}\boldsymbol{B}\boldsymbol{D}^*\boldsymbol{B}\boldsymbol{A}^{-1}, \quad \boldsymbol{B}^* = -\boldsymbol{A}^{-1}\boldsymbol{B}\boldsymbol{D}^*, \\ &\boldsymbol{D}^* = (\boldsymbol{D} - \boldsymbol{B}\boldsymbol{A}^{-1}\boldsymbol{B})^{-1}, \\ &\quad \tilde{\boldsymbol{A}} = \boldsymbol{A} - \boldsymbol{B}\boldsymbol{D}^{-1}\boldsymbol{B}, \quad \tilde{\boldsymbol{B}} = \boldsymbol{B}\boldsymbol{D}^{-1}, \quad \tilde{\boldsymbol{D}} = \boldsymbol{D}^{-1}, \\ &\tilde{\boldsymbol{A}}^* = \boldsymbol{A}^{-1}, \quad \tilde{\boldsymbol{B}}^* = -\boldsymbol{A}^{-1}\boldsymbol{B}, \quad \tilde{\boldsymbol{D}}^* = \boldsymbol{D} - \boldsymbol{B}\boldsymbol{A}^{-1}\boldsymbol{B}. \end{aligned} \tag{3.5}$$

If the formulation takes the compatible displacements u_i, w as the primary functions, we have a single-valued midplane strain $\boldsymbol{\varepsilon}_0$ and curvature $\boldsymbol{\kappa}$. The stress resultants $\boldsymbol{N}$ and bending moments $\boldsymbol{M}$ can be obtained from (3.2). The equilibrium equations can then be expressed in terms of the displacements. The problems remaining are the system of partial differential equations of u_1, u_2, and w in which the plate properties are expressed by $\boldsymbol{A}$, $\boldsymbol{B}$, and $\boldsymbol{D}$. This is the usual step taken in the literature, for example, [7–9, 13, 14], and is generally called *displacement formalism.*

If the formulation takes the Airy stress function F and the plate deflection w as the primary functions, the stress resultants $\boldsymbol{N}$ are obtained directly from the Airy stress function, and the equilibrium equations for the in-plane problem will be satisfied automatically. A single-valued curvature $\boldsymbol{\kappa}$ is obtained through the second-order differentiation of the plate deflection w. With the equilibrated $\boldsymbol{N}$ and the compatible $\boldsymbol{\kappa}$, the midplane strain $\boldsymbol{\varepsilon}_0$ and bending moment $\boldsymbol{M}$ can be obtained from (3.4). The compatibility equation for the midplane strain $\boldsymbol{\varepsilon}_0$ and the equilibrium equations for plate bending moments $\boldsymbol{M}$ can then be expressed in terms of F and w. The problems remaining are the system of partial differential equations of F and w in which the plate properties are expressed by $\tilde{\boldsymbol{A}}^*$, $\tilde{\boldsymbol{B}}^*$, and $\tilde{\boldsymbol{D}}^*$. This is called *mixed formalism* and has been used in [20, 23].

An alternative choice for the above mixed primary functions is the compatible displacements u_1, u_2 and the equilibrated bending moment potentials ψ_1 and ψ_2. With this choice, we get the compatible midplane strain $\boldsymbol{\varepsilon}_0$ and the equilibrated bending moment $\boldsymbol{M}$. We get $\boldsymbol{N}$ and $\boldsymbol{\kappa}$ from the second equation of (3.3). The equilibrium equation of $\boldsymbol{N}$ and the compatibility equation of $\boldsymbol{\kappa}$ will then lead to the system of partial differential equations in terms of u_1, u_2, ψ_1, and ψ_2 in which the plate properties are expressed by $\tilde{\boldsymbol{A}}, \tilde{\boldsymbol{B}}$, and $\tilde{\boldsymbol{D}}$. This approach was taken by Hwu [7] and Lu [10].

(2) The organization of the final expressions of the general solutions can roughly be distinguished as being in *component form* or *matrix form.* The former expresses the general solutions for each component of the displacements, stress resultants, and bending moments, whereas the latter organizes the solutions in terms of vectors and matrices. Examples of component form expressions proposed in the literature are [14, 20] and examples of matrix form expressions are [7–9, 12, 23, 24].

(3) The establishment of a material eigen-relation is important in the complex-variable method since the material eigenvalues μ_α, which have been proved to be complex numbers, are the key parameters of the complex variables $z_\alpha(=x + \mu_\alpha y)$. In *Lekhnitskii formalism* this relation is represented by the characteristic polynomial equation, whereas in *Stroh formalism* it is represented by a standard eigen-relation $\mathbf{N}\boldsymbol{\xi} = \mu\boldsymbol{\xi}$ in which $\mathbf{N}$ is called the fundamental elasticity matrix and $\boldsymbol{\xi}$ is the material eigenvector. Using Lekhnitskii formalism, the analytical expressions of the material eigenvectors can be obtained explicitly in terms of the material eigenvalues.

Using the standard eigen-relation of Stroh formalism, many properties and identities that are useful for the derivation of explicit analytical solutions can be obtained for the material eigenvalues and eigenvectors, which is fundamental to the success of Stroh formalism in solving a large variety of problems in general anisotropic elasticity. In practical applications, these two different formalisms can benefit each other. One may refer to [14, 20, 23, 24] for Lekhnitskii formalism and [6–9, 12] for Stroh formalism.

(4) Most of the complex-variable methods proposed in the literature consider only *non-degenerate anisotropic plates*. That is, the formulation is based upon the assumption that the material eigenvalues are distinct. For *degenerate anisotropic plates*, some or all of the material eigenvalues are repeated and cannot yield a complete set of independent eigensolutions. Thus, the solutions constructed through the eigensolutions of distinct eigenvalues are not general enough and should be modified for degenerate anisotropic plates. To get a complete set of independent eigensolutions, one can add the generalized eigensolutions obtained by differentiating the eigensolutions with respect to the eigenvalue. Full consideration of degenerate anisotropic plates can be found in [23–26]. It should be noted that if a solution is obtained by the complex-variable method for non-degenerate anisotropic plates, it is applicable to degenerate plates analytically if its final expression does not contain the material eigenvector matrices, and it is also applicable to degenerate plates numerically if a small perturbation of the material eigenvalues is made [45].

As stated above, several different complex-variable methods have been proposed in the literature for the coupled stretching-bending analysis of general laminated plates. Their main differences have been described by the names: displacement formalism or mixed formalism, component form or matrix form, Lekhnitskii formalism or Stroh formalism, non-degenerate anisotropic plates or degenerate anisotropic plates. Since each method possesses its own merits, the best approach may be the combination of all their merits. From this viewpoint, the *Stroh-like formalism* proposed by Hwu [7] is the most appropriate one to be introduced in this chapter. In Hwu's Stroh-like formalism, the final expression of the general solution is written in matrix form and is derived from displacement formalism, whereas the associated material eigen-relation is derived from mixed formalism and the explicit expression of the material eigenvector is obtained by Lekhnitskii formalism. Although analytically it is valid only for non-degenerate anisotropic plates, numerically by perturbation it is applicable for all different types of anisotropic plates including degenerate anisotropic plates. Moreover, through the identities established using the material eigen-relation, many complex form solutions can be converted into real form. In this case, the final analytical solutions can still be applied to the degenerate anisotropic plates even if they have been derived for non-degenerate anisotropic plates.

By selecting u_1, u_2, β_1, and β_2 as the primary functions, general solutions satisfying all the basic equations in (3.1) with $q = 0$ have been obtained as [7]

$$\mathbf{u}_d = 2\,\mathrm{Re}\{\mathbf{A}\mathbf{f}(z)\}, \quad \boldsymbol{\phi}_d = 2\,\mathrm{Re}\{\mathbf{B}\mathbf{f}(z)\}, \tag{3.6a}$$

where

$$\mathbf{u}_d = \begin{Bmatrix} \mathbf{u} \\ \boldsymbol{\beta} \end{Bmatrix}, \quad \boldsymbol{\phi}_d = \begin{Bmatrix} \boldsymbol{\phi} \\ \boldsymbol{\psi} \end{Bmatrix},$$
$$\mathbf{A} = [\mathbf{a}_1\ \mathbf{a}_2\ \mathbf{a}_3\ \mathbf{a}_4], \quad \mathbf{B} = [\mathbf{b}_1\ \mathbf{b}_2\ \mathbf{b}_3\ \mathbf{b}_4],$$
$$\mathbf{f}(z) = \begin{Bmatrix} f_1(z_1) \\ f_2(z_2) \\ f_3(z_3) \\ f_4(z_4) \end{Bmatrix}, \quad z_k = x_1 + \mu_k x_2, \quad k = 1, 2, 3, 4, \tag{3.6b}$$

and

$$\mathbf{u} = \begin{Bmatrix} u_1 \\ u_2 \end{Bmatrix}, \quad \boldsymbol{\beta} = \begin{Bmatrix} \beta_1 \\ \beta_2 \end{Bmatrix} = \begin{Bmatrix} -w_{,1} \\ -w_{,2} \end{Bmatrix}, \quad \boldsymbol{\phi} = \begin{Bmatrix} \phi_1 \\ \phi_2 \end{Bmatrix}, \quad \boldsymbol{\psi} = \begin{Bmatrix} \psi_1 \\ \psi_2 \end{Bmatrix}. \tag{3.6c}$$

In Eq. (3.6c), $\phi_i, i = 1, 2$, are the stress functions related to the in-plane forces N_{ij}, and $\psi_i, i = 1, 2$, are the stress functions related to the bending moments M_{ij}, transverse shear forces Q_i, and effective transverse shear forces $V_i(= Q_i + M_{ij,j})$. Their relations are

$$\begin{aligned} &N_{11} = -\phi_{1,2}, \quad N_{22} = \phi_{2,1}, \quad N_{12} = \phi_{1,1} = -\phi_{2,2} = N_{21}, \\ &M_{11} = -\psi_{1,2}, \quad M_{22} = \psi_{2,1}, \quad M_{12} = \psi_{1,1} - \eta = -\psi_{2,2} + \eta = M_{21}, \\ &Q_1 = -\eta_{,2}, \quad Q_2 = \eta_{,1}, \\ &V_1 = -\psi_{2,22}, \quad V_2 = \psi_{1,11}, \end{aligned} \tag{3.7a}$$

where

$$\eta = \frac{1}{2}(\psi_{1,1} + \psi_{2,2}). \tag{3.7b}$$

The material eigenvalues μ_k and their associated eigenvectors $\mathbf{a}_k$ and $\mathbf{b}_k$ can be determined from the following eigen-relation

$$\mathbf{N}\boldsymbol{\xi} = \mu\boldsymbol{\xi}, \tag{3.8a}$$

where

$$\mathbf{N} = \mathbf{I}_t\mathbf{N}_m\mathbf{I}_t, \quad \boldsymbol{\xi} = \begin{Bmatrix} \mathbf{a} \\ \mathbf{b} \end{Bmatrix}, \tag{3.8b}$$

and

$$\mathbf{N} = \begin{bmatrix} \mathbf{N}_1 & \mathbf{N}_2 \\ \mathbf{N}_3 & \mathbf{N}_1^T \end{bmatrix}, \quad \mathbf{N}_m = \begin{bmatrix} (\mathbf{N}_m)_1 & (\mathbf{N}_m)_2 \\ (\mathbf{N}_m)_3 & (\mathbf{N}_m)_1^T \end{bmatrix}, \quad \mathbf{I}_t = \begin{bmatrix} \mathbf{I}_1 & \mathbf{I}_2 \\ \mathbf{I}_2 & \mathbf{I}_1 \end{bmatrix},$$
$$(\mathbf{N}_m)_1 = -\mathbf{T}_m^{-1}\mathbf{R}_m^T, \quad (\mathbf{N}_m)_2 = \mathbf{T}_m^{-1} = (\mathbf{N}_m)_2^T, \tag{3.8c}$$
$$(\mathbf{N}_m)_3 = \mathbf{R}_m\mathbf{T}_m^{-1}\mathbf{R}_m^T - \mathbf{Q}_m = (\mathbf{N}_m)_3^T, \quad \mathbf{I}_1 = \begin{bmatrix} \mathbf{I} & \mathbf{0} \\ \mathbf{0} & \mathbf{0} \end{bmatrix}, \quad \mathbf{I}_2 = \begin{bmatrix} \mathbf{0} & \mathbf{0} \\ \mathbf{0} & \mathbf{I} \end{bmatrix},$$

and $\mathbf{I}$ is a 2×2 unit matrix. Note that the material eigenvalues μ_k have been assumed to be distinct in the general solution (3.6). Moreover, the four pairs of material eigenvectors $(\mathbf{a}_k, \mathbf{b}_k)$, $k = 1, 2, 3, 4$, are assumed to be those corresponding to the eigenvalues with positive imaginary parts. For materials whose eigenvalues are repeated, a small perturbation in their value may be introduced to avoid degeneracy problems [45] or a modification of the general solution can be made [46].

In (3.8b), $\mathbf{N}_m$ is the 8×8 fundamental elasticity matrix of mixed formalism whose explicit expressions have been obtained in [47]. In (3.8c), the three 4×4 real matrices $\mathbf{Q}_m, \mathbf{R}_m$, and $\mathbf{T}_m$ are related to the matrices

$\tilde{\boldsymbol{A}}$, $\tilde{\boldsymbol{B}}$, and $\tilde{\boldsymbol{D}}$ by

$$\mathbf{Q}_m = \begin{bmatrix} \tilde{A}_{11} & \tilde{A}_{16} & \tilde{B}_{16}/2 & \tilde{B}_{12} \\ \tilde{A}_{16} & \tilde{A}_{66} & \tilde{B}_{66}/2 & \tilde{B}_{62} \\ \tilde{B}_{16}/2 & \tilde{B}_{66}/2 & -\tilde{D}_{66}/4 & -\tilde{D}_{26}/2 \\ \tilde{B}_{12} & \tilde{B}_{62} & -\tilde{D}_{26}/2 & -\tilde{D}_{22} \end{bmatrix}, \tag{3.9a}$$

$$\mathbf{R}_m = \begin{bmatrix} \tilde{A}_{16} & \tilde{A}_{12} & -\tilde{B}_{11} & -\tilde{B}_{16}/2 \\ \tilde{A}_{66} & \tilde{A}_{26} & -\tilde{B}_{61} & -\tilde{B}_{66}/2 \\ \tilde{B}_{66}/2 & \tilde{B}_{26}/2 & \tilde{D}_{16}/2 & \tilde{D}_{66}/4 \\ \tilde{B}_{62} & \tilde{B}_{22} & \tilde{D}_{12} & \tilde{D}_{26}/2 \end{bmatrix}, \tag{3.9b}$$

$$\mathbf{T}_m = \begin{bmatrix} \tilde{A}_{66} & \tilde{A}_{26} & -\tilde{B}_{61} & -\tilde{B}_{66}/2 \\ \tilde{A}_{26} & \tilde{A}_{22} & -\tilde{B}_{21} & -\tilde{B}_{26}/2 \\ -\tilde{B}_{61} & -\tilde{B}_{21} & -\tilde{D}_{11} & -\tilde{D}_{16}/2 \\ -\tilde{B}_{66}/2 & -\tilde{B}_{26}/2 & -\tilde{D}_{16}/2 & -\tilde{D}_{66}/4 \end{bmatrix}. \tag{3.9c}$$

Note that unlike the two-dimensional elastic case, $\mathbf{Q}_m$ and $\mathbf{T}_m$ defined in (3.9a) and (3.9c) are not positive definite. However, the existence of the inverse of $\mathbf{T}_m$ used in (3.8c) for the calculation of $(\mathbf{N}_m)_i$ is guaranteed, since it has been proved through the derivation of the explicit expressions of $(\mathbf{N}_m)_i$ [47] or the relation between displacement formalism and mixed formalism [7, 11].

With the general solutions presented in (3.6), the next important task is constructing an environment to help researchers find solutions satisfying the boundary conditions for a given problem. It is helpful if this environment contains the following features: (1) the relations for the physical quantities on the rotated coordinate; (2) the identities for the conversion of a complex form solution into a real form solution; and (3) the explicit expressions for the matrices used in the formulation.

(1) To be suitable for general problems, which may have arbitrary geometrical shape, it is always desirable to have the relations for the physical quantities expressed in other coordinates. If a quantity is identified as a tensor, the change to the components of the quantity can be made by

following the transformation laws of tensors. From this viewpoint, with the conventional approach one usually applies the transformation law of second-order tensors to obtain the bending moments (M_n, M_s, M_{ns}) and in-plane forces (N_n, N_s, N_{ns}), and applies the transformation laws of first-order tensors to obtain the transverse shear force (Q_n, Q_s), and then one uses the definition of effective transverse shear force to obtain (V_n, V_s). Here, the subscripts n and s denote, respectively, the normal and tangential directions of the normal-tangent $(n-s)$ coordinate system. In the Stroh-like formalism the secondary functions are the stress functions ϕ_1, ϕ_2, ψ_1, and ψ_2 instead of the stresses. Thus, it is more convenient if we express the relations calculating the stress resultants and bending moments in the $n-s$ coordinates using the stress functions instead of their quantities in the $x_1 - x_2$ coordinates. The relations obtained in [34] are as follows:

$$\begin{aligned}
&N_n = \mathbf{n}^T\mathbf{t}_n = \mathbf{n}^T\boldsymbol{\phi}_{,s}, && N_{ns} = \mathbf{s}^T\mathbf{t}_n = \mathbf{s}^T\boldsymbol{\phi}_{,s},\\
&N_s = \mathbf{s}^T\mathbf{t}_s = -\mathbf{s}^T\boldsymbol{\phi}_{,n}, && N_{sn} = \mathbf{n}^T\mathbf{t}_s = -\mathbf{n}^T\boldsymbol{\phi}_{,n} = N_{ns},\\
&M_n = \mathbf{n}^T\mathbf{m}_n = \mathbf{n}^T\boldsymbol{\psi}_{,s}, && M_{ns} = \mathbf{s}^T\mathbf{m}_n = \mathbf{s}^T\boldsymbol{\psi}_{,s} - \eta,\\
&M_s = \mathbf{s}^T\mathbf{m}_s = -\mathbf{s}^T\boldsymbol{\psi}_{,n}, && M_{sn} = \mathbf{n}^T\mathbf{m}_s = -\mathbf{n}^T\boldsymbol{\psi}_{,n} + \eta = M_{ns},\\
&Q_n = \eta_{,s}, && Q_s = -\eta_{,n}, \quad V_n = (\mathbf{s}^T\boldsymbol{\psi}_{,s})_{,s},\\
&V_s = -(\mathbf{n}^T\boldsymbol{\psi}_{,n})_{,n}, && \eta = (\mathbf{s}^T\boldsymbol{\psi}_{,s} + \mathbf{n}^T\boldsymbol{\psi}_{,n})/2,
\end{aligned} \tag{3.10}$$

where $\mathbf{t}_n$ and $\mathbf{m}_n$ are the surface traction and surface moment along the surface with normal $\mathbf{n}$, and $\mathbf{t}_s$ and $\mathbf{m}_s$ are the surface traction and surface moment along the surface with normal $\mathbf{s}$, which is perpendicular to the direction $\mathbf{n}$; $\mathbf{n}$ and $\mathbf{s}$ are the directions normal and tangential to the boundary and can be written as

$$\mathbf{n}^T = (-\sin\theta, \cos\theta), \quad \mathbf{s}^T = (\cos\theta, \sin\theta), \tag{3.11}$$

where θ denotes the angle from the positive x_1-axis to the direction s in a clockwise direction.

In addition to the relations shown in (3.10), another important relation that cannot be obtained directly through simple tensor transformation is the generalized material eigen-relation. This relation is established through a consideration of the dual coordinate system in which the displacements and stress functions are referred to the $x_1 - x_2$ coordinate system, whereas other terms such as the elastic constants are referred to a rotated coordinate system $x_1^*-x_2^*$. The generalized material eigen-relation presented in [48] is

$$\mathbf{N}(\omega)\boldsymbol{\xi}=\mu(\omega)\boldsymbol{\xi}, \tag{3.12a}$$

where

$$\mathbf{N}(\omega)=\mathbf{I}_t\mathbf{N}_m(\omega)\mathbf{I}_t,\quad \boldsymbol{\xi}=\begin{Bmatrix}\mathbf{a}\\ \mathbf{b}\end{Bmatrix}, \tag{3.12b}$$

$$\mathbf{N}(\omega)=\begin{bmatrix}\mathbf{N}_1(\omega) & \mathbf{N}_2(\omega)\\ \mathbf{N}_3(\omega) & \mathbf{N}_1^{\mathrm{T}}(\omega)\end{bmatrix},\quad \mathbf{N}_m(\omega)=\begin{bmatrix}(\mathbf{N}_m(\omega))_1 & (\mathbf{N}_m(\omega))_2\\ (\mathbf{N}_m(\omega))_3 & (\mathbf{N}_m(\omega))_1^T\end{bmatrix},$$

$$(\mathbf{N}_m(\omega))_1=-\mathbf{T}_m^{-1}(\omega)\mathbf{R}_m^T(\omega),\quad (\mathbf{N}_m(\omega))_2=\mathbf{T}_m^{-1}(\omega), \tag{3.12c}$$

$$(\mathbf{N}_m(\omega))_3=\mathbf{R}_m(\omega)\mathbf{T}_m^{-1}(\omega)\mathbf{R}_m^T(\omega)-\mathbf{Q}_m(\omega),$$

and

$$\mu(\omega)=\frac{-\sin\omega+\mu\cos\omega}{\cos\omega+\mu\sin\omega}. \tag{3.12d}$$

In (3.12), $\mathbf{Q}_m(\omega),\mathbf{R}_m(\omega)$, and $\mathbf{T}_m(\omega)$ are related to the matrices $\mathbf{Q}_m,\mathbf{R}_m$, and $\mathbf{T}_m$ defined in (3.9) by

$$\begin{aligned}\mathbf{Q}_m(\omega)&=\mathbf{Q}_m\cos^2\omega+(\mathbf{R}_m+\mathbf{R}_m^T)\sin\omega\cos\omega+\mathbf{T}_m\sin^2\omega,\\ \mathbf{R}_m(\omega)&=\mathbf{R}_m\cos^2\omega+(\mathbf{T}_m-\mathbf{Q}_m)\sin\omega\cos\omega-\mathbf{R}_m^T\sin^2\omega,\\ \mathbf{T}_m(\omega)&=\mathbf{T}_m\cos^2\omega-(\mathbf{R}_m+\mathbf{R}_m^T)\sin\omega\cos\omega+\mathbf{Q}_m\sin^2\omega,\end{aligned} \tag{3.12e}$$

in which ω denotes the angle between the transformed and original coordinates.

(2) Since all physical quantities are real, it is desirable to have real form solutions instead of complex form solutions. Moreover, if the solutions can be written in a real form that does not involve complex numbers — material eigenvalues μ_α and their associated material eigenvector matrices **A** and **B** — they will not have the degeneracy problems raised by the repetition of material eigenvalues. Therefore, if the solutions can be converted into real form, they are not only convenient for practical applications but are also applicable for all the different types of laminated plates including degenerate plates. Thus, the identities converting the complex form into real form are important for the analytical derivation using the Stroh-like formalism.

In the eigen-relation $\mathbf{N}\boldsymbol{\xi}=\mu\boldsymbol{\xi}$, $\mathbf{N}$ is real, whereas μ and $\boldsymbol{\xi}$ are complex. In other words, this eigen-relation is the foundation for converting the identities connecting the *complex* μ, **A**, and **B** to *real* **N**. Several identities deriving from this relation have been obtained for the Stroh formalism of two-dimensional anisotropic elasticity [5, 6]. Since the material eigen-relations (3.8) and (3.12) of the Stroh-like formalism have been purposely organized into the same forms as those of the Stroh formalism, all the identities derived from this eigen-relation can be applied to the Stroh-like formalism without further proof. Therefore, in this chapter no identities will be presented. For those interested in these identities, please refer to the books [5, 6] for anisotropic elasticity.

(3) The explicit expressions for the matrices used in the formulation are useful when one tries to find the analytical solutions for each component of the physical quantities such as displacements, strains, and stresses. The advantage of the matrix form is that the solution can be expressed in an elegant and easily programmed way. Moreover, through proper arrangement of the matrices, it is possible that one solution form will be suitable for several different kinds of problem. The typical example is the general solution shown in (3.6a), which is a matrix form solution suitable for two-dimensional elasticity problems, plate bending problems as well as coupled stretching-bending problems, and the materials considered can be any kinds of anisotropic or piezoelectric materials. In other words, this simple matrix form solution (3.6a) can be applied to a wide range of problems and materials, and hence, it is important to know the explicit expressions of the material eigenvector matrices $\mathbf{A}$ and $\mathbf{B}$ in this solution.

As stated in (2), a key feature of Stroh formalism is its eigen-relation relating the complex material eigenvectors to real material properties. Several real matrices connected through this eigen-relation, such as the fundamental elasticity matrix $\mathbf{N}$, the generalized fundamental elasticity matrix $\mathbf{N}(\theta)$, and the *Barnett–Lothe tensors* $\mathbf{L}$, $\mathbf{S}$, and $\mathbf{H}$, are crucial when the analytical solutions lead to real form expressions. Because these real matrices can be obtained directly from the material properties, they exist for all types of materials — degenerate or non-degenerate. In other words, if the final solution of a particular elasticity problem can be expressed in terms of these real matrices, even if they are derived using an assumption of non-degenerate materials, they are still valid for degenerate materials.

Therefore, no matter whether complex or real, getting the explicit expressions for $\mathbf{A}$, $\mathbf{B}$, $\mathbf{N}$, $\mathbf{N}(\theta)$, $\mathbf{L}$, $\mathbf{S}$, and $\mathbf{H}$ is an important stage in understanding the effects of the material properties in problems of anisotropic elasticity. For plane problems with anisotropic elastic materials, most of the explicit expressions have been well documented in [5, 6]. Their corresponding explicit expressions for general piezoelectric materials covering all the possible two-dimensional states such as generalized plane strain or plane stress and short circuit or open circuit, were provided in [49]. In my recent study [48], one can find the explicit expressions of the Stroh-like formalism for the coupled stretching-bending analysis.

3.3. Extended Stroh-Like Formalism — Hygrothermal Stresses

In the previous section, the constitutive relations pertain only to an environment with constant temperature and constant moisture, i.e., no temperature or moisture changes are allowed. However, in laminated structures hygrothermal changes occur frequently during fabrication and structural usage. When a homogeneous body is completely free to deform, the hygrothermal changes may produce only hygrothermal strains for which the body is free of stresses. However, for a composite laminate each individual lamina is not completely free to deform. Lamina stresses are therefore induced by the constraints placed on deformation by adjacent lamina. Thus, the study of hygrothermal stresses in laminates is important for practical engineering design. Like the extension of Stroh formalism to anisotropic thermoelasticity, in this section we extend the Stroh-like formalism to the hygrothermal stress analysis of laminates.

When hygrothermal effects are considered, the basic equations shown in (3.1) should be modified to

$$\begin{gathered}
U_i = u_i + x_3\beta_i, \quad T = T^0 + x_3T^*, \quad H = H^0 + x_3H^*, \\
\beta_1 = -w_{,1}, \quad \beta_2 = -w_{,2}, \\
\check{q}_i = -K^t_{ij}T^0_{,j} - K^{*t}_{ij}T^*_{,j} - K^t_{i3}T^*, \quad \check{m}_i = -K^h_{ij}H^0_{,j} - K^{*h}_{ij}H^*_{,j} - K^h_{i3}H^*, \\
\varepsilon_{ij} = \frac{1}{2}(u_{i,j} + u_{j,i}), \quad \kappa_{ij} = \frac{1}{2}(\beta_{i,j} + \beta_{j,i}), \\
N_{ij} = A_{ijkl}\varepsilon_{kl} + B_{ijkl}\kappa_{kl} - A^t_{ij}T^0 - A^h_{ij}H^0 - B^t_{ij}T^* - B^h_{ij}H^*, \\
M_{ij} = B_{ijkl}\varepsilon_{kl} + D_{ijkl}\kappa_{kl} - B^t_{ij}T^0 - B^h_{ij}H^0 - D^t_{ij}T^* - D^h_{ij}H^*, \\
N_{ij,j} = 0, \quad M_{ij,ij} + p = 0, Q_i = M_{ij,j}, \quad \check{q}_{i,i} + q = 0, \\
\check{m}_{i,i} + m = 0, \quad i,j,k,l = 1,2,
\end{gathered} \tag{3.13}$$

where U_i, T, and H are the displacements, temperature, and moisture content of the plates, u_i, T^0, and H^0 are the middle surface displacements, temperature, and moisture content, and β_i, T^*, and H^* are the negative slopes, and the rates of change of temperature and moisture content across the thickness. $\breve{q}_i$ and $\breve{m}_i$ denote the heat flux resultant and moisture transfer resultant; $A^t_{ij}, B^t_{ij}, D^t_{ij}$ and $A^h_{ij}, B^h_{ij}, D^h_{ij}$ are the corresponding tensors for the thermal and moisture expansion coefficients; K^t_{ij}, K^h_{ij} and K^{*t}_{ij}, K^{*h}_{ij} are the coefficients related to the heat conduction and moisture diffusion coefficients; p, q, and m are the lateral distributed load, heat flux, and moisture concentration transfer applied to the laminates.

Since the basic equations in (3.13) are quite general, it is not easy to find a solution satisfying all these basic equations. Here, two special cases that occur frequently in engineering applications are considered. One is the case when the temperature and moisture distributions depend on x_1 and x_2 only, i.e., $T^* = H^* = 0$, and the other is the case when the temperature and moisture distributions depend on x_3 only, i.e., $T = T^0 + x_3T^*$ and $H = H^0 + x_3H^*$ in which T^0, T^*, H^0, and H^* are constants independent of x_1 and x_2. The general solutions for these two special cases have been obtained in [28] and are called the *extended Stroh-like formalism* since they are the extensions of Stroh-like formalism.

Case 1: Temperature and moisture content depend on x_1 and x_2 only

In this case, the temperature and moisture content are assumed to vary in the x_1–x_2 plane and distribute uniformly in the thickness direction. The lateral distributed load, heat flux, and moisture concentration transfer applied on the laminates are neglected. With this consideration, $T^* = H^* = p = q = m = 0$, and the basic equations in (3.13) can be simplified. By following the steps for the Stroh formalism of two-dimensional thermoelasticity [50], the general solutions satisfying all the basic equations in (3.13) can be found as [28]

$$
\begin{aligned}
T &= 2Re\{g_t'(z_t)\}, \quad H = 2\mathrm{Re}\{g_h'(z_h)\},\\
\breve{q}_i &= -2\mathrm{Re}\{(K^t_{i1} + \tau_t K^t_{i2})g_t''(z_t)\},\\
\breve{m}_i &= -2\mathrm{Re}\{(K^h_{i1} + \tau_h K^h_{i2})g_h''(z_h)\},\\
\mathbf{u}_d &= 2\mathrm{Re}\{\mathbf{A}\mathbf{f}(z) + \mathbf{c}_t g_t(z_t) + \mathbf{c}_h g_h(z_h)\},\\
\boldsymbol{\phi}_d &= 2\mathrm{Re}\{\mathbf{B}\mathbf{f}(z) + \mathbf{d}_t g_t(z_t) + \mathbf{d}_h g_h(z_h)\},
\end{aligned}
\tag{3.14}
$$

in which $z_t = x_1 + \tau_t x_2, z_h = x_1 + \tau_h x_2$, and τ_t, τ_h and $(\mathbf{c}_t, \mathbf{d}_t)$, $(\mathbf{c}_h, \mathbf{d}_h)$ are thermal and moisture *eigenvalues* and *eigenvectors*, which can be determined by the following eigen-relations:

$$\begin{aligned} &K_{11}^t + 2\tau_t K_{12}^t + \tau_t^2 K_{22}^t = 0, \quad K_{11}^h + 2\tau_h K_{12}^h + \tau_h^2 K_{22}^h = 0, \\ &\mathbf{N}\boldsymbol{\eta}_t = \tau_t \boldsymbol{\eta}_t + \boldsymbol{\gamma}_t, \quad \mathbf{N}\boldsymbol{\eta}_h = \tau_h \boldsymbol{\eta}_h + \boldsymbol{\gamma}_h, \end{aligned} \tag{3.15a}$$

where

$$\mathbf{N} = \begin{bmatrix} \mathbf{N}_1 & \mathbf{N}_2 \\ \mathbf{N}_3 & \mathbf{N}_1^T \end{bmatrix}, \quad \boldsymbol{\eta}_t = \begin{Bmatrix} \mathbf{c}_t \\ \mathbf{d}_t \end{Bmatrix}, \quad \boldsymbol{\eta}_h = \begin{Bmatrix} \mathbf{c}_h \\ \mathbf{d}_h \end{Bmatrix}, \tag{3.15b}$$

and $\mathbf{N}$ is the fundamental elasticity matrix defined in (3.8c), $\boldsymbol{\gamma}_t$ and $\boldsymbol{\gamma}_h$ are two 8×1 complex vectors related to the elastic constants and the coefficients of thermal and moisture expansion [6].

Case 2: Temperature and moisture content depend on x_3 only

If the temperature and moisture content depend on x_3 only, the general solutions satisfying all the basic equations in (3.13) can be expressed as [28]

$$\begin{aligned} \mathbf{u}_d &= 2\mathrm{Re}\{\mathbf{Af}(z)\}, \\ \boldsymbol{\phi}_d &= 2\mathrm{Re}\{\mathbf{Bf}(z)\} - x_1\boldsymbol{\vartheta}_2 + x_2\boldsymbol{\vartheta}_1, \end{aligned} \tag{3.16a}$$

where

$$\boldsymbol{\vartheta}_i = \boldsymbol{\alpha}_i^t T^0 + \boldsymbol{\alpha}_i^h H^0 + \boldsymbol{\alpha}_i^{*t} T^* + \boldsymbol{\alpha}_i^{*h} H^*, \quad i = 1, 2. \tag{3.16b}$$

$\boldsymbol{\alpha}_i^t$ and $\boldsymbol{\alpha}_i^{*t}$ are defined by

$$\boldsymbol{\alpha}_1^t = \begin{Bmatrix} \boldsymbol{\alpha}_{A1}^t \\ \boldsymbol{\alpha}_{B1}^t \end{Bmatrix}, \quad \boldsymbol{\alpha}_2^t = \begin{Bmatrix} \boldsymbol{\alpha}_{A2}^t \\ \boldsymbol{\alpha}_{B2}^t \end{Bmatrix}, \quad \boldsymbol{\alpha}_{Ai}^t = \begin{Bmatrix} A_{1i}^t \\ A_{2i}^t \end{Bmatrix}, \quad \boldsymbol{\alpha}_{Bi}^t = \begin{Bmatrix} B_{1i}^t \\ B_{2i}^t \end{Bmatrix}, \tag{3.16c}$$

and

$$\boldsymbol{\alpha}_i^{*t} = \begin{Bmatrix} \boldsymbol{\alpha}_{Bi}^t \\ \boldsymbol{\alpha}_{Di}^t \end{Bmatrix}, \quad \boldsymbol{\alpha}_{Bi}^t = \begin{Bmatrix} B_{1i}^t \\ B_{2i}^t \end{Bmatrix}, \quad \boldsymbol{\alpha}_{Di}^t = \begin{Bmatrix} D_{1i}^t \\ D_{2i}^t \end{Bmatrix}, \quad i = 1, 2. \tag{3.16d}$$

The same expressions as (3.16c) and (3.16d) are defined for $\boldsymbol{\alpha}_i^h$ and $\boldsymbol{\alpha}_i^{*h}$ by replacing superscript t with h.

With the general solution shown in (3.14) or (3.16) for the generalized displacement and stress function vectors, the middle surface displacements u_i and slopes β_i can be obtained directly from the components of the generalized displacement vector $\mathbf{u}_d$. For the stress resultants N_{ij}, shear forces Q_i, effective shear forces V_i, and bending moments M_{ij}, we can utilize the relations shown in (3.7).

3.4. Expanded Stroh-Like Formalism — Electro-Elastic Laminates

Consider an electro-elastic laminate made of fiber-reinforced composites and piezoelectric materials. Although it may exhibit electric-elastic coupling effects that are more complicated than those of single-phase piezoelectric materials, a similar extension from pure elastic materials to piezoelectric materials for a two-dimensional analysis can still be applied to the coupled stretching-bending analysis of electro-elastic laminates. For piezoelectric anisotropic elasticity, to include the piezoelectric effects the constitutive laws, the strain–displacement relations, and the equilibrium equations can be written as follows [51]:

$$
\begin{cases} \sigma_{ij} = C^E_{ijkl}\varepsilon_{kl} - e_{kij}E_k, \\ D_j = e_{jkl}\varepsilon_{kl} + \omega^\varepsilon_{jk}E_k, \end{cases} \quad \varepsilon_{ij} = \frac{1}{2}(u_{i,j} + u_{j,i}),
$$

$$
\begin{cases} \sigma_{ij,j} = 0, \\ D_{i,i} = 0, \end{cases} \quad i, j, k, s = 1, 2, 3, \tag{3.17}
$$

in which $u_i, \sigma_{ij}, \varepsilon_{ij}, D_j$, and E_k are, respectively, the displacement, stress, strain, electric displacement (also called induction), and electric field; C^E_{ijkl}, e_{kij}, and ω^ε_{jk} are, respectively, the elastic stiffness tensor at constant electric field, piezoelectric stress tensor, and dielectric permittivity tensor at constant strain. By letting

$$
\begin{aligned}
&D_j = \sigma_{4j}, \quad -E_j = u_{4,j} = \varepsilon_{4j}, \quad j = 1, 2, 3, \\
&C_{ijkl} = C^E_{ijkl}, \quad i, j, k, l = 1, 2, 3, \\
&C_{ij4l} = e_{lij}, \quad i, j, l = 1, 2, 3, \\
&C_{4jkl} = e_{jkl}, \quad j, k, l = 1, 2, 3, \\
&C_{4j4l} = -\omega^\varepsilon_{jl}, \quad j, l = 1, 2, 3,
\end{aligned} \tag{3.18}
$$

the basic equations in (3.17) can be rewritten in an *expanded notation* as

$$
\begin{gathered}
\sigma_{Ij} = C_{IjKl}\varepsilon_{Kl}, \quad \varepsilon_{Ij} = \frac{1}{2}(u_{I,j} + u_{j,I}), \quad \sigma_{Ij,j} = 0, \\
I, K = 1, 2, 3, 4, \quad j, l = 1, 2, 3,
\end{gathered} \tag{3.19}
$$

where $u_{j,4} = 0$. Through the use of the expanded notation for piezoelectric materials, the basic equations for the coupled mechanical-electrical analysis of electro-elastic composite laminates can be rewritten in the form of (3.1) by expanding the range of certain subscripts from 2 to 4 [29]. And, hence the general solutions satisfying the basic equations can also be written in the same matrix form as shown in (3.6) for the Stroh-like formalism [6]. The only difference is the dimension and content of the matrices. Thus, this formalism is called *expanded Stroh-like formalism* due to the expansion of the matrix dimension.

3.5. Holes and Cracks

Due to the stress concentration induced by the existence of holes and cracks, anisotropic plates containing holes or cracks have been studied extensively in two-dimensional problems. Owing to the mathematical complexity, not many analytical results have been presented for coupled stretching-bending problems of holes and cracks in general composite laminates. In this section we will show some results for laminates containing holes or cracks presented in the literature using the Stroh-like formalism.

3.5.1. Holes in laminates under uniform stretching and bending moments

Consider an unbounded composite laminate with an elliptical hole subjected to in-plane forces $N_{11} = N_{11}^{\infty}$, $N_{22} = N_{22}^{\infty}$, $N_{12} = N_{12}^{\infty}$, and out-of-plane bending moments $M_{11} = M_{11}^{\infty}$, $M_{22} = M_{22}^{\infty}$, $M_{12} = M_{12}^{\infty}$, at infinity (Fig. 3.1). The contour of the elliptical hole is represented by

$$x_1 = a\cos\varphi, \quad x_2 = b\sin\varphi, \tag{3.20}$$

where $2a$ and $2b$ are the major and minor axes of the ellipse and φ is a real parameter. There is no load around the edge of the elliptical hole, i.e.,

$$N_n = N_{ns} = 0, \quad M_n = V_n = 0, \tag{3.21}$$

along the hole boundary.

A solution satisfying all the basic equations in (3.1) and the boundary condition (3.21) has been found as [27]

$$\begin{aligned} \mathbf{u}_d &= x_1\mathbf{d}_1^{\infty} + x_2\mathbf{d}_2^{\infty} - \mathrm{Re}\{\mathbf{A}\langle\zeta_\alpha^{-1}\rangle\mathbf{B}^{-1}(a\mathbf{m}_2^{\infty} - ib\mathbf{m}_1^{\infty})\}, \\ \boldsymbol{\phi}_d &= x_1\mathbf{m}_2^{\infty} - x_2\mathbf{m}_1^{\infty} - \mathrm{Re}\{\mathbf{B}\langle\zeta_\alpha^{-1}\rangle\mathbf{B}^{-1}(a\mathbf{m}_2^{\infty} - ib\mathbf{m}_1^{\infty})\}, \end{aligned} \tag{3.22}$$

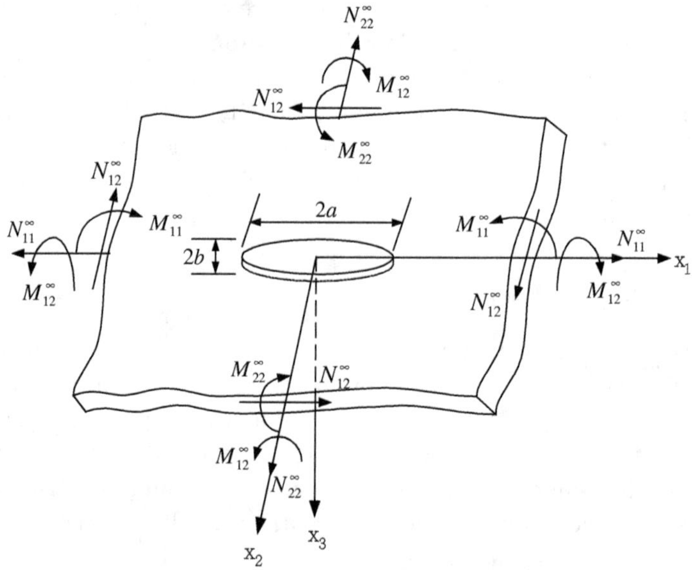

Fig. 3.1. A composite laminate weakened by an elliptical hole subjected to in-plane forces and out-of-plane bending moments.

in which the angular bracket $\langle\,\rangle$ stands for a diagonal matrix in which each component is varied according to the subscript α, and

$$\zeta_\alpha = \frac{z_\alpha + \sqrt{z_\alpha^2 - a^2 - \mu_\alpha^2 b^2}}{a - i\mu_\alpha b}. \tag{3.23}$$

$\mathbf{m}_1^\infty = (N_{11}^\infty, N_{12}^\infty, M_{11}^\infty, M_{12}^\infty)^T, \mathbf{m}_2^\infty = (N_{12}^\infty, N_{22}^\infty, M_{12}^\infty, M_{22}^\infty)^T$, and $\mathbf{d}_1^\infty$ and $\mathbf{d}_2^\infty$ are the vectors containing the quantities for the shear strains and curvatures at infinity, which are related to $\mathbf{m}_1^\infty$ and $\mathbf{m}_2^\infty$ by

$$\begin{Bmatrix} \mathbf{d}_1^\infty \\ \mathbf{d}_2^\infty \end{Bmatrix} = \begin{bmatrix} \mathbf{Q}^* & \mathbf{R}^* \\ \mathbf{R}^{*T} & \mathbf{T}^* \end{bmatrix} \begin{Bmatrix} \mathbf{m}_1^\infty \\ \mathbf{m}_2^\infty \end{Bmatrix}, \tag{3.24a}$$

where

$$\mathbf{Q}^* = \begin{bmatrix} A_{11}^* & A_{16}^*/2 & B_{11}^* & B_{16}^*/2 \\ A_{16}^*/2 & A_{66}^*/4 & B_{61}^*/2 & B_{66}^*/4 \\ B_{11}^* & B_{61}^*/2 & D_{11}^* & D_{16}^*/2 \\ B_{16}^*/2 & B_{66}^*/4 & D_{16}^*/2 & D_{66}^*/4 \end{bmatrix}, \tag{3.24b}$$

$$\mathbf{R}^* = \begin{bmatrix} A_{16}^*/2 & A_{12}^* & B_{16}^*/2 & B_{12}^* \\ A_{66}^*/4 & A_{26}^*/2 & B_{66}^*/4 & B_{62}^*/2 \\ B_{61}^*/2 & B_{21}^* & D_{16}^*/2 & D_{12}^* \\ B_{66}^*/4 & B_{26}^*/2 & D_{66}^*/4 & D_{26}^*/2 \end{bmatrix}, \tag{3.24c}$$

$$\mathbf{T}^* = \begin{bmatrix} A_{66}^*/4 & A_{26}^*/2 & B_{66}^*/4 & B_{62}^*/2 \\ A_{26}^*/2 & A_{22}^* & B_{26}^*/2 & B_{22}^* \\ B_{66}^*/4 & B_{26}^*/2 & D_{66}^*/4 & D_{26}^*/2 \\ B_{62}^*/2 & B_{22}^* & D_{26}^*/2 & D_{22}^* \end{bmatrix}, \tag{3.24d}$$

and $A_{ij}^*, B_{ij}^*, D_{ij}^*$ are the components of $\boldsymbol{A}^*, \boldsymbol{B}^*, \boldsymbol{D}^*$ defined in (3.5).

According to the relations given in (3.10) we know that the calculation of the stress resultants and bending moments relies upon the calculation of the differentials $\boldsymbol{\phi}_{d,s}$ and $\boldsymbol{\phi}_{d,n}$. Along the hole boundary, the real form expressions have been obtained as $\boldsymbol{\phi}_{d,s} = \mathbf{0}$ and

$$\begin{aligned} \boldsymbol{\phi}_{d,n} = \sin\theta \left\{ \mathbf{G}_1(\theta)\mathbf{m}_1^\infty - \left[\mathbf{I} + \frac{a}{b}\mathbf{G}_3(\theta)\right]\mathbf{m}_2^\infty \right\} \\ - \cos\theta \left\{ \left[\mathbf{I} + \frac{b}{a}\mathbf{G}_3(\theta)\right]\mathbf{m}_1^\infty + \mathbf{G}_1(\theta)\mathbf{m}_2^\infty \right\}, \end{aligned} \tag{3.25a}$$

where

$$\mathbf{G}_1(\theta) = \mathbf{N}_1^T(\theta) - \mathbf{N}_3(\theta)\mathbf{S}\mathbf{L}^{-1}, \quad \mathbf{G}_3(\theta) = -\mathbf{N}_3(\theta)\mathbf{L}^{-1}. \tag{3.25b}$$

$\mathbf{L}$, $\mathbf{S}$, and $\mathbf{H}$ are three real matrices which are called the Barnett–Lothe tensors defined by

$$\mathbf{H} = 2i\mathbf{A}\mathbf{A}^T, \quad \mathbf{L} = -2i\mathbf{B}\mathbf{B}^T, \quad \mathbf{S} = i(2\mathbf{A}\mathbf{B}^T - \mathbf{I}). \tag{3.26}$$

Note that, as it has been shown by Wu[52], in some cases to satisfy the requirement of single-valued displacement an additional term should be added in the solutions of (3.22) and (3.25a).

3.5.2. Holes in laminates under uniform heat flow and moisture transfer

Using the extended Stroh-like formalism presented in Section 3.3, the hygrothermal stresses in composite laminates disturbed by an elliptical hole subjected to uniform heat flow and moisture transfer in the $x_1 - x_2$ plane or x_3 direction were solved in [28]. To save space only the solutions for the problem under uniform heat flow and moisture transfer in the x_3 direction are presented in this section.

If the temperature and moisture content are assumed to vary linearly in the thickness direction, and $T = T_u$ and $H = H_u$ on the top surface and $T = T_l$ and $H = H_l$ on the bottom surface, we have

$$T^0 = \frac{T_l + T_u}{2}, \quad H^0 = \frac{H_l + H_u}{2}, \quad T^* = \frac{T_l - T_u}{h}, \quad H^* = \frac{H_l - H_u}{h}, \tag{3.27}$$

where h is the thickness of the laminated plate.

The field solution of this problem has been obtained as

$$\begin{aligned} \mathbf{u}_d &= \mathrm{Re}\{\mathbf{A}\langle\zeta_\alpha^{-1}\rangle\mathbf{B}^{-1}(a\boldsymbol{\vartheta}_2 - ib\boldsymbol{\vartheta}_1)\}, \\ \boldsymbol{\phi}_d &= \mathrm{Re}\{\mathbf{B}\langle\zeta_\alpha^{-1}\rangle\mathbf{B}^{-1}(a\boldsymbol{\vartheta}_2 - ib\boldsymbol{\vartheta}_1)\} - x_1\boldsymbol{\vartheta}_2 + x_2\boldsymbol{\vartheta}_1, \end{aligned} \tag{3.28}$$

where $\boldsymbol{\vartheta}_1$ and $\boldsymbol{\vartheta}_2$ are related to T^0, H^0, T^*, and H^* by the equation shown in (3.16b). Along the hole boundary, the real form expressions of $\boldsymbol{\phi}_{d,n}$ have been obtained as

$$\begin{aligned} \boldsymbol{\phi}_{d,n} = \cos\theta &\left[\boldsymbol{\vartheta}_1 + \mathbf{G}_1(\theta)\boldsymbol{\vartheta}_2 + \frac{b}{a}\mathbf{G}_3(\theta)\boldsymbol{\vartheta}_1\right] \\ &+ \sin\theta\left[\boldsymbol{\vartheta}_2 - \mathbf{G}_1(\theta)\boldsymbol{\vartheta}_1 + \frac{a}{b}\mathbf{G}_3(\theta)\boldsymbol{\vartheta}_2\right]. \end{aligned} \tag{3.29}$$

3.5.3. Holes in electro-elastic laminates under uniform loads and charges

Consider an unbounded electro-elastic composite laminate with an elliptical hole subjected to the uniform generalized forces $N_{11}^\infty, N_{22}^\infty, N_{12}^\infty, N_{41}^\infty, N_{42}^\infty$, and uniform generalized moments $M_{11}^\infty, M_{22}^\infty, M_{12}^\infty$, $M_{41}^\infty, M_{42}^\infty$ at infinity. The generalized forces and moments N_{4i} and $M_{4i}, i = 1, 2$, are related to the electric displacement [29]. The contour of the elliptical hole is represented by (3.20). If the hole edge and the upper and lower surfaces of the laminate are free of traction and electric charge, the boundary conditions of this problem can be expressed by the same matrix form equation as that for the laminates discussed in Section 3.5.1. The only difference is that the dimension of the vectors $\boldsymbol{\phi}_d$, $\mathbf{m}_1^\infty$, and $\mathbf{m}_2^\infty$ is now 6×1 instead of 4×1. Due to the equivalence of the mathematical formulation, the solutions of this problem can also be expressed by the matrix form expressions shown in (3.22)–(3.25).

3.5.4. Cracks in laminates

Consider an unbounded composite laminate containing a through-thickness crack loaded at infinity and the crack is assumed to lie on the x_1-axis with its center located at the origin. Since an elliptical opening can be considered to be a crack of length $2a$ by letting the minor axis b be zero, the field solutions for cracks can therefore be obtained from the solutions for problems for elliptical holes with $b = 0$. After deriving the field solutions, the stress intensity factors $\mathbf{k}$ can be calculated directly through the following definition:

$$\mathbf{k} = \begin{Bmatrix} K_{II} \\ K_I \\ K_{IIB} \\ K_{IB} \end{Bmatrix} = \lim_{r\to 0} \sqrt{2\pi r} \begin{Bmatrix} N_{12} \\ N_{22} \\ M_{12} \\ M_{22} \end{Bmatrix} = \lim_{r\to 0} \sqrt{2\pi r}\mathbf{m}_2, \tag{3.30}$$

where r is the distance ahead of the crack tip; $K_{II}, K_I, K_{IIB}, K_{IB}$ are, respectively, the stress intensity factors of the shearing mode, opening mode, twisting mode, and bending mode. From the relations given in (3.7) and the definition (3.30), we get

$$\mathbf{k} = \lim_{r\to 0} \sqrt{2\pi r}(\boldsymbol{\phi}_{d,1} - \eta \mathbf{i}_3), \tag{3.31a}$$

where

$$\mathbf{i}_3^T = (0\ 0\ 1\ 0). \tag{3.31b}$$

Using the relation (3.31) and the solutions for the problems of elliptical holes $(3.22)_2$ with $b = 0$, the explicit solutions for the stress intensity factors can be obtained as [53]

$$\mathbf{k} = \sqrt{\pi a}\{\mathbf{m}_2^\infty - \eta_0 \mathbf{i}_3\}, \tag{3.32a}$$

where

$$\eta_0 = \{M_{12}^\infty + (\mathbf{G}_1\mathbf{m}_2^\infty)_4\}/2, \tag{3.32b}$$

and $\mathbf{G}_1 = \mathbf{G}_1(0)$ which is defined in (3.25b).

3.6. Numerical Examples

For the analytical closed-form solutions presented in the previous section, several numerical examples are illustrated in this section. All the examples consider an unbounded laminate composed of different combinations of graphite/epoxy fiber-reinforced composite laminae. Each lamina thickness is 1 mm, and the material properties of the graphite/epoxy are

$$\begin{gathered} E_1 = 181\,\text{GPa}, \quad E_2 = 10.3\,\text{GPa}, \quad G_{12} = 7.17\,\text{GPa}, \quad \nu_{12} = 0.28, \\ \alpha_{11}^t = 0.02 \times 10^{-6}/^\circ\text{C}, \quad \alpha_{22}^t = 22.5 \times 10^{-6}/^\circ\text{C}, \\ k_{11}^t = 1.5\,\text{W/m}^\circ\text{C}, \quad k_{22}^t = 0.5\,\text{W/m}^\circ\text{C}, \end{gathered} \tag{3.33}$$

where E_1 and E_2 are the Young's moduli in the x_1 and x_2 directions, respectively; G_{12} is the shear modulus in the x_1–x_2 plane; ν_{12} is the major Poisson's ratio and is related by $\nu_{21}E_1 = \nu_{12}E_2$ to the minor Poisson's ratio ν_{21}; α_{11}^t and α_{22}^t are the coefficients of thermal expansion in the fiber and its transverse directions, respectively; and k_{11}^t and k_{22}^t are the coefficients of heat conduction. All the other values of $\alpha_{ks}^t, k_{ij}^t, i \neq j$, are zero. Note that the properties given in (3.33) are for the plane-stress condition of orthotropic materials, whose properties in the thickness direction are not required. With the material properties (3.33) and layup sequence, the field deformations and stresses as well as the resultant forces and moments around the hole boundary can be calculated by following the procedure shown in the flowchart Fig. 3.2.

3.6.1. Holes

Consider a $[+45/0/+45/-45]$ unsymmetric laminate containing a circular hole. Figure 3.3 shows the resultant forces and bending moments around the hole boundary under different loading conditions. From this figure, we see that if an unsymmetric composite laminate is subjected to in-plane forces only or out-of-plane bending moments only, both bending moments and in-plane forces will be induced around the hole boundary, which is reasonable and expected due to the existence of the coupling stiffnesses.

To show the effect of hole shape, five different ratios, $b/a = 5$, 3, 1, 1/3, 1/5, were considered. Figure 3.4 shows that the maximum values of N_s increase as the ratio b/a increases. Moreover, the maximum values of N_s are located at $\varphi = 90^\circ$ and 270° when the ratio $b/a \geq 1$. On the other hand, when the ratio $b/a < 1$ the location of the maximum values has a tendency, although not quite clear, to shift to near $\varphi = 180^\circ$ and $\varphi = 360^\circ$. A similar situation occurs for the variation of M_s and M_{ns} [27].

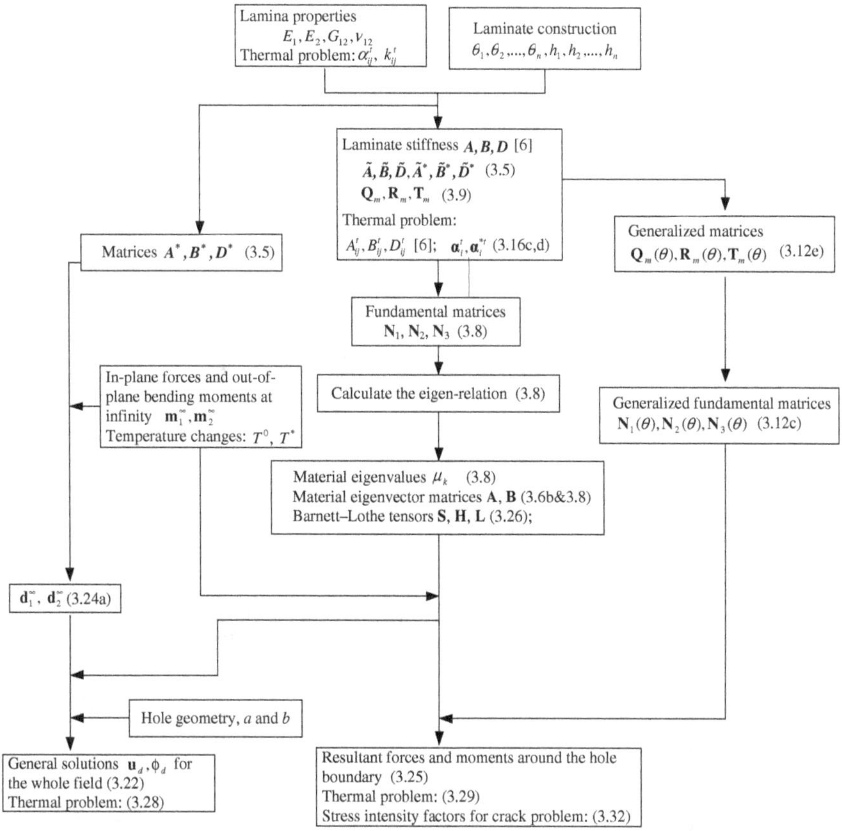

Fig. 3.2. Flowchart for the calculation of deformations and stresses, and resultant forces and moments, around the hole boundary, and the stress intensity factors for crack problems.

3.6.2. Thermal environment

To show the necessity of the analytical solutions, here we consider the simplest case where only in-plane stretching occurs under uniform temperature changes in a unidirectional laminate. Consider a $[0]_4$ unidirectional laminate containing a through-thickness circular hole subjected to a temperature change from 0°C to 100°C over the entire laminate, i.e., $T^0 = 100$°C and $T^* = 0$ in (3.13). Figure 3.5 compares the resultant hoop stress N_s around the hole boundary obtained by the solution (3.14) with three different element types, Plane42, Shell99, and Solid45, from the commercial finite element software ANSYS. In the ANSYS

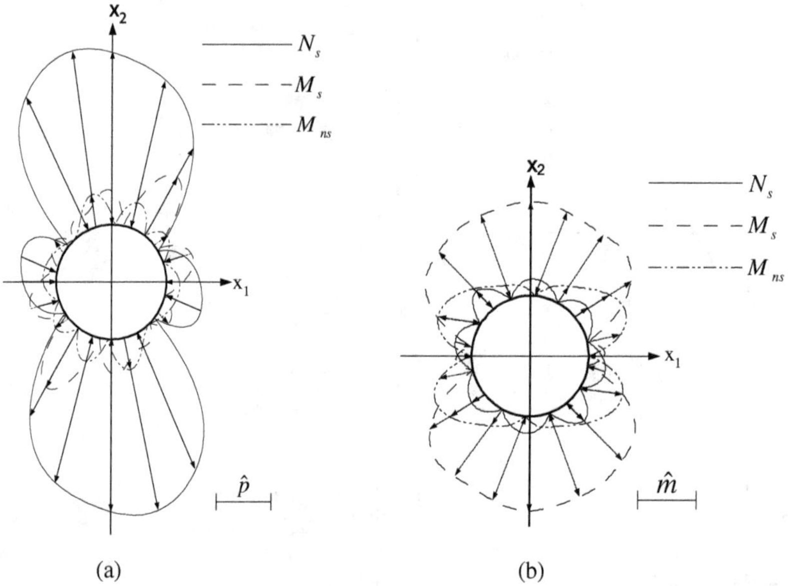

Fig. 3.3. Forces and moments around the circular hole in an unsymmetric composite laminate under different loading conditions. (a) $N_{11}^{\infty} = \hat{p}$; (b) $M_{11}^{\infty} = \hat{m}$ [27].

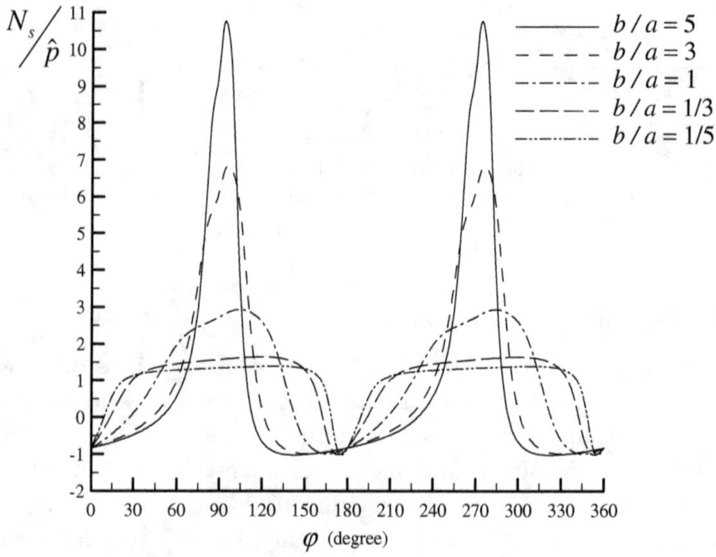

Fig. 3.4. Force around an elliptical hole in an unsymmetric laminate subjected to $N_{11}^{\infty} = \hat{p}, M_{11}^{\infty} = \hat{m}$ ($\hat{m} = \hat{p} \times 1$) [27].

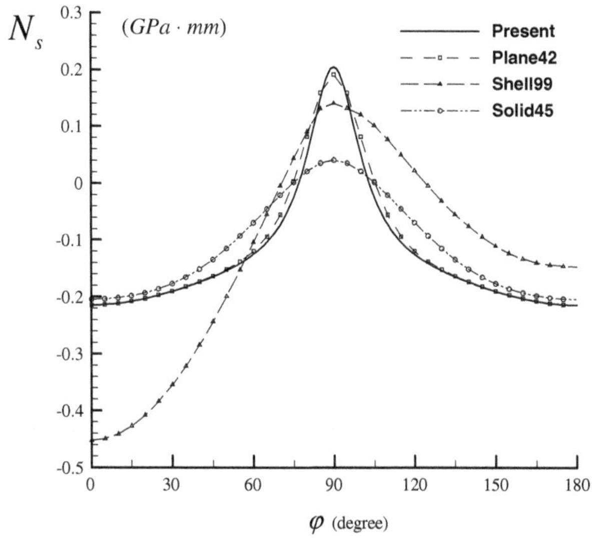

Fig. 3.5. Resultant force N_s around the circular hole boundary when the laminate is subjected to a uniform temperature change [28].

simulation, 72 nodes around the hole boundary and a 1:100 hole/plate ratio were used to approximate the unbounded laminate. Figure 3.5 shows that the results for Plane42 agree well with the present solution while a larger discrepancy occurs for the other element types. To illustrate the reason for this, numerical values of the resultant tractions N_n/N_s and N_{ns}/N_s around the hole boundary, which should vanish for a traction-free hole, are plotted in Figs. 3.6(a) and 3.6(b). These two figures provide strong evidence for why a larger discrepancy occurs for Shell99 and Solid45 since they did not satisfy the traction-free hole boundary condition, which is satisfied by Plane42 approximately and by the present solution exactly. Because the element type Plane42 can only be used for in-plane problems, for general stretching-bending coupling problems one must choose a shell or solid element. It is then expected that a large discrepancy between the analytical solutions and those of ANSYS will occur for general unsymmetric laminates. In other words, due to the approximate nature of any finite element software, to avoid the calculation of inaccurate solutions it is important to have a good reference such as the exact solutions for unbounded plates shown in this chapter. Several numerical examples for unsymmetric laminates under a uniform heat flow in the $x_1 - x_2$ plane or in the x_3 direction can be found in [28].

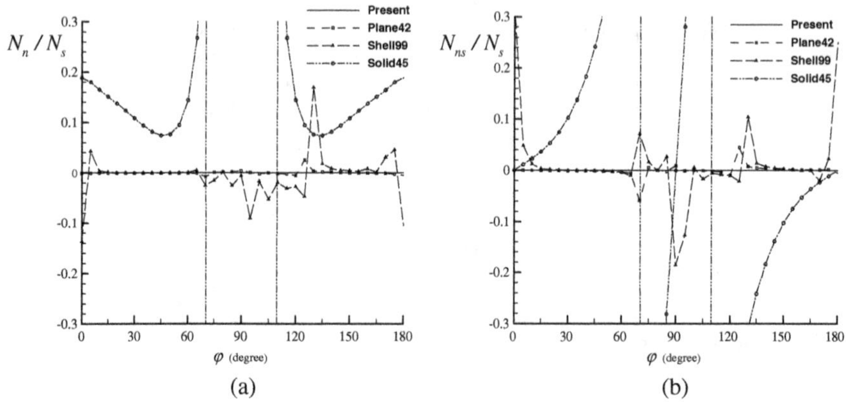

Fig. 3.6. Resultant forces around the hole boundary under uniform temperature changes. (a) N_n/N_s; (b) N_{ns}/N_s [28].

3.6.3. Electro-elastic coupling

Due to the complexity of electro-elastic coupling, not many numerical examples can be found in the literature. To learn how to calculate the physical responses for electro-elastic laminates subjected to in-plane forces, out-of-plane bending moments, or electric displacements, one may refer to [29] for [E/0/45/E] composed of two layers of a graphite/epoxy fiber-reinforced composite in the middle and two piezoelectric layers of left-hand quartz at the top and bottom. From the discussions therein, we see that to avoid numerical ill-conditioning before employing the explicit solutions shown in this chapter a proper dimensional adjustment is necessary, since the elastic stiffness constants are usually of the order above 9, whereas the dielectric permittivity is usually of the order below −12, if SI units are used. For example, the constitutive relation shown in (3.17) can be rewritten as

$$\begin{aligned} \sigma_{ij}/E_0 &= (C^E_{ijkl}/E_0)\varepsilon_{kl} - e_{kij}(E_k/E_0), \\ D_j &= e_{jkl}\varepsilon_{kl} + (E_0\omega^{\varepsilon}_{jk})(E_k/E_0), \end{aligned} \tag{3.34}$$

in which E_0 is a reference number, such as $10^9\,\mathrm{N/m^2}$, used for scale adjustment. In numerical calculations, Eq. (3.34) allows the replacement of the material properties C^E_{ijkl} and $\omega^{\varepsilon}_{jk}$ by C^E_{ijkl}/E_0 and $E_0\omega^{\varepsilon}_{jk}$. With this replacement the output values of the stresses and electric fields are σ_{ij}/E_0 and E_k/E_0, which should be multiplied by E_0 to return them to their original units.

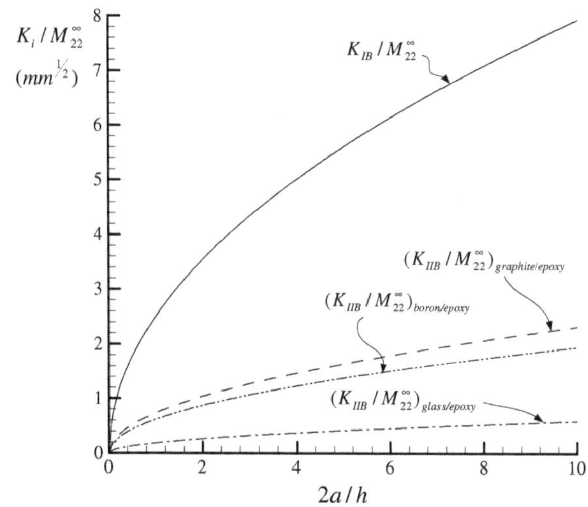

Fig. 3.7. Stress intensity factors of composite laminates.

3.6.4. Cracks

From the solution shown in (3.32) we see that the stress intensity factor K_{IIB} depends on the material properties, and all the other stress intensity factors are independent of the material properties. To illustrate the effect of the material, two additional materials were considered in crack problems besides the graphite/epoxy laminates. One was glass/epoxy and the other was boron/epoxy. Their material properties are

Glass/epoxy:

$$E_1 = 38.6\,\text{GPa}, \quad E_2 = 8.27\,\text{GPa}, \quad G_{12} = 4.14\,\text{GPa}, \quad \nu_{12} = 0.26.$$

Boron/epoxy:

$$E_1 = 209\,\text{GPa}, \quad E_2 = 19\,\text{GPa}, \quad G_{12} = 6.6\,\text{GPa}, \quad \nu_{12} = 0.21.$$

Figure 3.7 shows that the larger the crack length $2a$ the higher the stress intensity factors K_{IB} and K_{IIB}, and the Mode II stress intensity factor K_{IIB} depends on both the crack geometry and the material properties.

3.7. Conclusions

From the presentation in this chapter, it can be seen that most of the features of the Stroh formalism for two-dimensional anisotropic elasticity have been preserved in the Stroh-like formalism for coupled stretching-bending analysis. Using this advantage, most of the problems in coupled stretching-bending analysis, which could not be solved previously have now been solved by referring to the solutions obtained for the two-dimensional elasticity problems.

Acknowledgments

The author would like to thank the National Science Council, Taiwan, R.O.C. for support through Grant NSC 96-2221-E-006-174.

References

[1] Lekhnitskii S.G., 1963. *Theory of Elasticity of an Anisotropic Body*, MIR, Moscow.
[2] Lekhnitskii S.G., 1968. *Anisotropic Plates*, Gordon and Breach Science Publishers, New York.
[3] Stroh A.N., 1958. Dislocations and cracks in anisotropic elasticity, *Phil. Mag.*, 7, 625–646.
[4] Stroh A.N., 1962. Steady state problems in anisotropic elasticity, *J. Math. Phys.*, 41, 77–103.
[5] Ting T.C.T., 1996. *Anisotropic Elasticity: Theory and Applications*, Oxford Science Publications, New York.
[6] Hwu C., 2010. *Anisotropic Elastic Plates*, Springer, New York.
[7] Hwu C., 2003. Stroh-like formalism for the coupled stretching-bending analysis of composite laminates, *Int. J. Solids Struct.*, 40(13–14), 3681–3705.
[8] Lu P., Mahrenholtz O., 1994. Extension of the Stroh formalism to an analysis of bending of anisotropic elastic plates, *J. Mech. Phys. Solids*, 42(11), 1725–1741.
[9] Cheng Z.Q., Reddy J.N., 2002. Octet formalism for Kirchhoff anisotropic plates, *Proc. Royal Soc. London, Ser. A*, 458, 1499–1517.
[10] Lu P., 2004. A Stroh-type formalism for anisotropic thin plates with bending-extension coupling, *Arch. Appl. Mech.*, 73, 690–710.
[11] Cheng Z.Q., Reddy J.N., 2005. Structure and properties of the fundamental elastic plate matrix, *ZAMM Z. Angew. Math. Mech.*, 85(10), 721–739.
[12] Fu Y.B., Brookes D.W., 2006. Edge waves in asymmetrically laminated plates, *J. Mech. Phys. Solids*, 54, 1–21.
[13] Fu Y.B., 2007. Hamiltonian interpretation of the Stroh formalism in anisotropic elasticity, *Proc. Royal Soc. London, Ser. A*, 463, 3073–3087.
[14] Becker W., 1991. A complex potential method for plate problems with bending extension coupling, *Arch. Appl. Mech.*, 61, 318–326.
[15] Becker W., 1992. Closed-form analytical solutions for a Griffith crack in a non-symmetric laminate plate, *Compos. Struct.*, 21, 49–55.

[16] Becker W., 1993. Complex method for the elliptical hole in an unsymmetric laminate, *Arch. Appl. Mech.*, 63, 159–169.
[17] Becker W., 1995. Concentrated forces and moments on laminates with bending extension coupling, *Compos. Struct.*, 30, 1–11.
[18] Zakharov D.D., Becker W., 2000. Boundary value problems for unsymmetric laminates, occupying a region with elliptic contour, *Compos. Struct.*, 49, 275–284.
[19] Zakharov D.D., Becker W., 2000. Singular potentials and double-force solutions for anisotropic laminates with coupled bending and stretching, *Arch. Appl. Mech.*, 70, 659–669.
[20] Chen P., Shen Z., 2001. Extension of Lekhnitskii's complex potential approach to unsymmetric composite laminates, *Mech. Res. Commun.*, 28(4), 423–428.
[21] Chen P., Shen Z., 2001. Green's functions for an unsymmetric laminate with an elliptic hole, *Mech. Res. Commun.*, 28(5), 519–524.
[22] Chen P., Nie H., 2004. Green's function for bending problem of an unsymmetrical laminated plate with bending-extension coupling containing an elliptic hole, *Arch. Appl. Mech.*, 73, 846–856.
[23] Yin W.L., 2003. General solutions of anisotropic laminated plates, *ASME J. Appl. Mech.*, 70(4), 496–504.
[24] Yin W.L., 2003. Structures and properties of the solution space of general anisotropic laminates, *Int. J. Solids Struct.*, 40, 1825–1852.
[25] Yin W.L., 2005. Green's function of anisotropic plates with unrestricted coupling and degeneracy, Part 1: The infinite plate, *Compos. Struct.*, 69, 360–375.
[26] Yin W.L., 2005. Green's function of anisotropic plates with unrestricted coupling and degeneracy, Part 2: Other domains and special laminates, *Compos. Struct.*, 69, 376–386.
[27] Hsieh M.C., Hwu C., 2003. Explicit solutions for the coupled stretching-bending problems of holes in composite laminates, *Int. J. Solids Struct.*, 40(15), 3913–3933.
[28] Hsieh M.C., Hwu C., 2006. Hygrothermal stresses in unsymmetric laminates disturbed by elliptical holes, *ASME J. Appl. Mech.*, 73, 228–239.
[29] Hwu C., Hsieh M.C., 2005. Extended Stroh-like formalism for the electro-elastic composite laminates and its applications to hole problems, *Smart Mater. Struct.*, 14, 56–68.
[30] Cheng Z.Q., Reddy J.N., 2003. Green's functions for infinite and semi-infinite anisotropic thin plates, *J. Appl. Mech.*, 70, 260–267.
[31] Cheng Z.Q., Reddy J.N., 2004. Laminated anisotropic thin plate with an elliptic inhomogeneity, *Mech. Mater.*, 36, 647–657.
[32] Hwu C., Tan C.Z., 2006. In-plane/out-of-plane concentrated forces and moments on composite laminates with elliptical elastic inclusions, *Int. J. Solids Struct.*, 44, 6584–6606.

[33] Hwu C., 2003. Stroh-Like complex variable formalism for bending theory of anisotropic plates, *ASME J. Appl. Mech.*, 70, 696–707.
[34] Hwu C., 2004. Green's function for the composite laminates with bending extension coupling, *Compos. Struct.*, 63, 283–292.
[35] Hwu C., 2005. Green's functions for holes/cracks in laminates with stretching bending coupling, *ASME J. Appl. Mech.*, 72, 282–289.
[36] Hwu C., 2010. Boundary integral equations for general laminated plates with coupled stretching-bending deformation, *Proc. Royal Soc. London, Ser. A*, 466, 1027–1054.
[37] Hwu C., 2012. Boundary element formulation for the coupled stretching-bending analysis of thin laminated plates, *Engineering Anal. Boundary Elements*, 36, 1027–1039.
[38] Hwu C., Chang H.W., 2015. Coupled stretching-bending analysis of laminated plates with corners via boundary elements, *Compos. Struct.*, 120, 300–314.
[39] Hwu C., Chang H.W., 2015. Singular integrals in boundary elements for coupled stretching-bending analysis of unsymmetric laminates, *Compos. Struct.*, 132, 933–943.
[40] Brebbia C.A., 1980. *The Boundary Element Method for Engineers*, 2nd edn., Pentech Press, London.
[41] Banerjee P.K., Butterfield R., 1981. *Boundary Element Methods in Engineering Science*, McGraw-Hill, New York.
[42] Gaul L., Kögl M., Wagner M., 2003. *Boundary Element Methods for Engineers and Scientists*, Springer-Verlag, Berlin.
[43] Ameen M., 2001. *Boundary Element Analysis: Theory and Programming*, Narosa Publishing House, New Delhi.
[44] Jones R.M., 1974. *Mechanics of Composite Materials*, Scripta, Washington, DC.
[45] Hwu C., Yen W.J., 1991. Green's functions of two-dimensional anisotropic plates containing an elliptic hole, *Int. J. Solids Struct.*, 27(13), 1705–1719.
[46] Ting T.C.T., Hwu C., 1988. Sextic formalism in anisotropic elasticity for almost non-semisimple matrix N, *Int. J. Solids Struct.*, 24(1), 65–76.
[47] Hsieh M.C., Hwu C., 2002. Explicit expressions of the fundamental elasticity matrices of Stroh-like formalism for symmetric/unsymmetric laminates, *Chin. J. Mech.-Ser. A*, 18(3), 109–118.
[48] Hwu C., 2010. Some explicit expressions of Stroh-like formalism for coupled stretching-bending analysis, *Int. J. Solids Struct.*, 47, 526–536.
[49] Hwu C., 2008. Some explicit expressions of extended Stroh formalism for two-dimensional piezoelectric anisotropic elasticity, *Int. J. Solids Struct.*, 45, 4460–4473.
[50] Hwu C., 1990. Thermal stresses in an anisotropic plate disturbed by an insulated elliptic hole or crack, *ASME J. Appl. Mech.*, 57(4), 916–922.
[51] Rogacheva N.N., 1994. *The Theory of Piezoelectric Shells and Plates*, CRC Press, London.

[52] Wu K.C., Hsiao P.S., 2015. An exact solution for an anisotropic plate with an elliptic hole under arbitrary remote uniform moments, *Compos. Part B*, 75, 281–287.

[53] Hsieh M.C., 2005. *Various coupling analyses of composite laminates with holes/cracks*, PhD thesis, Institute of Aeronautics and Astronautics, National Cheng Kung University.

https://doi.org/10.1142/9781800611887_0004

Chapter 4

Classical, Refined, Zig-Zag, Layer-Wise Models and Best Theory Diagrams for Laminated Structures

Erasmo Carrera[*,‡], **Maria Cinefra**[†,§], **and Marco Petrolo**[*,¶]
[*]*MUL*^{2} *Lab, Department of Mechanical and Aerospace Engineering, Politecnico di Torino, Corso Duca degli Abruzzi 24, 10129 Torino, Italy*
[†]*Department of Mechanics, Mathematics and Management, Politecnico di Bari, Via Orabona 4, 70125 Bari, Italy*
[‡]*erasmo.carrera@polito.it*
[§]*maria.cinefra@poliba.it*
[¶]*marco.petrolo@polito.it*

Abstract

This chapter overviews classical and advanced theories for laminated plates and shell structures. Findings from existing historical reviews are used to confirm that the advanced theories can be grouped and referred to as Lekhnitskii multilayered theories, Ambartsumian multilayered theories, and Reissner multilayered theories. The unified formulation proposed by the first author, which is known as CUF (Carrera Unified Formulation), is used to make numerical assessments of various laminated plate/shell theories. The chapter provides details of a recent reliable finite element formulation for laminated shell analysis. It is embedded in the CUF framework, and it leads to the classical, zig-zag, and layer-wise models as particular cases. Numerical mechanisms such as shear and membrane locking are contrasted by developing an appropriate choice of shape functions and mixed assumed shear strain techniques. Furthermore, the Best Theory Diagram (BTD) is introduced as a tool to evaluate the accuracy of any structural model against a reference solution. The BTD is obtained through an axiomatic/asymptotic method (AAM) developed by the authors and using genetic algorithms. BTDs for plate and shell Equivalent Single Layer models (ESL) are presented, and guidelines and recommendations are provided for the proper development of refined structural theories.

Contents

4.1. Introduction

Two-dimensional (2D) modeling of multilayered plates and shells requires appropriate theories. The discontinuity of physical/mechanical properties in the thickness direction makes theories that were originally developed for one-layered structures inadequate, such as the Cauchy–Poisson–Kirchhoff–Love thin plate/shell theories [1–4], or the Reissner and Mindlin [5, 6] first-order shear deformation theory (FSDT) as well as higher-order models such as that by Hildebrand, Reissner, and Thomas [7]. These theories are in fact not able to reproduce piecewise continuous displacement and transverse stress fields in the thickness direction, which are usually experienced by multilayered structures. These two fields are often described in the literature as *zig-zag effects* and *interlaminar continuity,* respectively (see also the three-dimensional solutions reported by Pagano [8]). In [9] these two effects have been summarized using the acronym C_z^0-requirements, that is displacements and transverse stresses must be

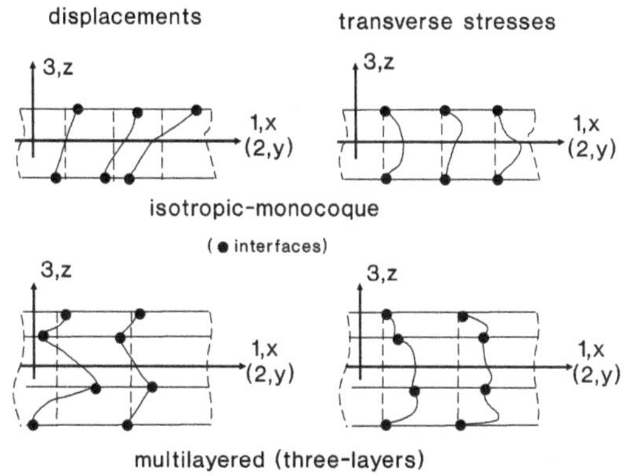

Fig. 4.1. C_z^0-requirements. Single-layered and three-layered structures.

C^0-continuous functions in the z-thickness directions. A qualitative comparison of displacement and stress fields in a one-layered and a multilayered structure is depicted in Fig. 4.1. This picture clearly shows that theories designed for one-layer structures are inappropriate to analyze multilayered ones.

Some refinements of classical models as well as theories designed for multilayered structures have been proposed in the literature over the last four decades. Due to the form of displacement fields (see Fig. 4.1), the latter are often referred to as "zig-zag" theories. For a complete review of this topic, readers who are interested can refer to the many available survey articles on beams, plates, and shells. Among these, excellent reviews are quoted in the articles by Ambartsumian [10], Librescu and Reddy [11], Grigolyuk and Kulikov [12], Kapania and Raciti [13], Kapania [14], Noor and co-authors [15–17], Reddy and Robbins [18], Carrera [19], as well as in the books by Librescu [20] and Reddy [21]. These articles review theories that deal with layer-wise models (LWMs) and equivalent single-layer models (ESLMs). Following Reddy [21], it is intended that the number of displacement variables is kept independent of the number of constitutive layers in the ESLM, while the same variables are independent in each layer for LWM cases.

Although these review works are excellent, in the authors' opinion, there still exists the need for a historical review with the aim of giving clear answers to the following questions:

(1) Who first presented a zig-zag theory for a multilayered structure?

(2) How many different and independent ESL zig-zag theories have been proposed in open literature?
(3) Who first proposed the theories for question 2?
(4) Are the original works well recognized and mentioned correctly in subsequent articles?
(5) What are the main differences among the available approaches to multilayered structures?

The answers to these five points could be extremely useful to analyze of layered structures. Furthermore, it will give an insight into early and, equally very interesting, ideas and methods such as those by Lekhnitskii [22], which could be extended and applied to further problems.

This chapter is, therefore, a historical review of "zig-zag" theories, which can describe what have previously been called C_z^0-requirements, given questions (1)–(5). These topics have already been documented in the historical note by the first author [23]. The findings in that paper are reconsidered in the first part of this chapter.

The present chapter considers mostly ESLMs. For the sake of completeness, a few comments on layer-wise cases are given in a separate section. A further limitation of the present chapter is that it is restricted to axiomatic-type approaches, although a novel axiomatic/asymptotic approach is presented in the last part of this work. The three multilayered theories discussed introduce initial assumptions: stress function forms were assumed by Lekhnitskii, transverse shear stress fields were assumed by Ambartsumian, while both displacements and transverse shear stresses were assumed in the framework of the mixed theorem proposed by Reissner. Therefore, those works which are based on asymptotic expansion such as those in [24–26] have not been discussed in the present chapter.

The second part of this chapter considers the development of a refined shell finite element formulation, which is based on the Carrera Unified Formulation (CUF) [27, 28].

The most common mathematical models used to describe shell structures may be classified into two classes according to their different physical assumptions. The Koiter model [29] is based on the Kirchhoff hypothesis. The Naghdi model [30] is based on the Reissner–Mindlin assumptions, which take into account transverse shear deformation. It is known that when a finite element method is used to discretize a physical model, numerical *locking* may arise from hidden constraints that are not well represented in the finite element approximation. In the Naghdi model both transverse shear and membrane constraints appear as the shell thickness becomes very small, thus locking may arise. The most common approaches proposed to overcome the locking phenomenon are the standard displacement formulation with higher-order elements [31, 32] or techniques of reduced-selective integration [33, 34]. But these introduce other numerical problems.

Concerning works by Bathe and others [35–37], the present authors have employed the mixed interpolation of tensorial components (MITC) method, coupled to CUF, to overcome the locking phenomenon. This method has been applied to both ESL and LW variable kinematic models contained in CUF to analyze multilayered structures. Nine-node cylindrical shell elements have been considered. The performance of the new element has been tested by solving benchmark problems involving very thin shells as well as multilayered shells. The results show that the element has good convergence and robustness when the thicknesses become very small. In particular, the study of multilayered structures demonstrates that the zig-zag and LW models provide more accurate solutions than the simple ESL models.

The present work represents an extension of [38]. In particular, the axiomatic/asymptotic method (AAM) and the Best Theory Diagram (BTD) were added. The AAM has been recently introduced [39–42] as a method to evaluate the influence of each primary variable of a structural model for the solution of a given problem. Starting from an axiomatic choice of a model, asymptotic-like results can be obtained by evaluating the effectiveness of each variable on characteristic parameters, such as the thickness, orthotropic ratio or stacking sequence. Via the AMM reduced advanced models can be built by retaining only those variables that influence the solution. The AMM has been recently applied to various structural problems involving beam [40, 43], plate [44–49], and shell [50, 51] models. The AAM can be used to build the BTD that is a curve in which, for a given problem, the number of unknown variables of a structural model is plotted against the accuracy of the model [52–55]. In particular, all those models that provide the best accuracy with the minimum number of variables belong to the BTD. Such a curve can be obtained by direct use of the AAM, or by coupling the AAM with a genetic algorithm. The former approach is computationally less expensive. In this chapter, some BTDs are presented, with particular attention paid to ESL models for composite plates and shells.

4.2. Who First Proposed a Zig-Zag Theory?

To the best of the authors' knowledge, Lekhnitskii should be considered as the first contributor to the theory for multilayered structures. In [22], in fact, Lekhnitskii proposed a splendid method able to describe zig-zag effects (for both in-plane and through-the-thickness displacements) and interlaminar continuous transverse stresses. This is proved by Fig. 4.2, taken from Lekhnitskii's pioneering work [22], which shows an interlaminar continuous transverse shear stress field (τ^1 and τ^2 are shear stresses in the layers 1 and 2, respectively) with discontinuous derivatives at the layer interface (the first author thanks Prof Shifrin, who provided the original article in Russian, and D. Carrera for providing an Italian translation of the same article). In other words, the C_z^0–requirements of Fig. 4.1 were entirely accounted for by Lekhnitskii [22].

The authors believe it would be of relevant interest to quote the original derivations made by Lekhnitskii. It is, in fact, difficult to obtain the original article by Lekhnitskii, which has no English translation. Furthermore, the theory proposed by Lekhnitskii is very interesting and the method used could be a starting point for future developments. The following detailed

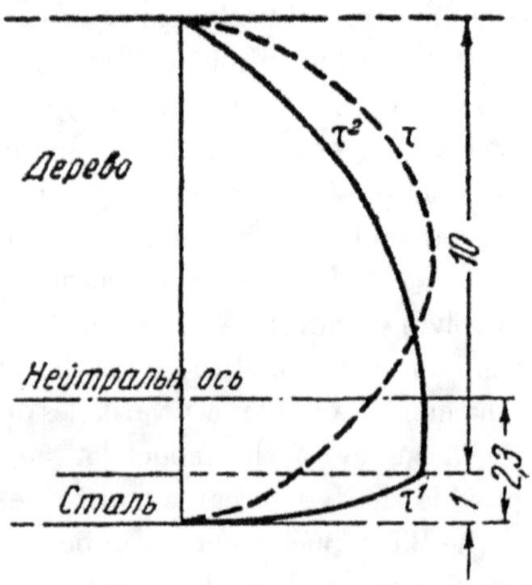

Fig. 4.2. C_z^0-form of a transverse shear stress in a two-layered structure.

derivation is therefore taken directly from Lekhnitskii's original paper, written in Russian. A few changes in notation are made. A briefer treatment can be found in the English translation of the book [56, Chapter 3, Section 18, p. 74].

This section closes with a few remarks on the theory proposed by Lekhnitskii:

(1) Lekhnitskii's theory describes the zig-zag form of both longitudinal and through-the-thickness displacements, in particular:
   (a) the longitudinal displacements u^k have a cubic order in the z–thickness direction;
   (b) the through-the-thickness displacement w^k varies according to a parabolic order in z.
(2) Lekhnitskii's theory gives the interlaminar continuous transverse stresses σ_{zz} and σ_{xz}.
(3) The stresses obtained by Lekhnitskii fulfill the 3D indefinite equilibrium equations (this fundamental property is intrinsic in the used stress function formulation).
(4) Stresses and displacements were obtained by employing:
   (a) compatibility conditions for stress functions;
   (b) strain–displacement relations;
   (c) compatibility conditions for displacements at the interface:

$$u^{k-1} = u^k, \quad w^{k-1} = w^k, \quad k = 2, N_l; \tag{4.1}$$

   (d) homogeneous conditions at the bottom and top surfaces for the transverse stresses:

$$\sigma^1_{zz} = \sigma^{N_l}_{zz} = 0, \quad \sigma^1_{xz} = \sigma^{N_l}_{xz} = 0, \quad \text{for } z = 0, h; \tag{4.2}$$

   (e) interlaminar equilibrium for the transverse stresses:

$$\sigma^{k-1}_{zz} = \sigma^k_{zz}, \quad \sigma^{k-1}_{xz} = \sigma^k_{xz}, \quad k = 2, N_l. \tag{4.3}$$

(5) No post-processing is used to recover transverse stresses.
(6) The thickness stresses σ_{zz} are neglected. Nevertheless, the Poisson effects on the thickness displacement w^k are fully retained.
(7) Full retention of Koiter's recommendation would require a different assumption for the stress functions (the authors do not know any work that does so).

Although Lekhnitskii's theory was published in the mid-1930s and reported in a short paragraph of the English edition of his book [56], it has been systematically forgotten in the recent literature. An exception is the work by Ren [57–59], which is documented in the next paragraph.

4.3. The Lekhnitskii–Ren Theory

This is the first of the three discussed theories. It is named after the author of the original work, Lekhnitskii and the author who first extended the work to plates, Ren. Due to the original stress function formulation, the present approach could also be referred to as a "stress approach".

To the best of the authors' knowledge, Ren is the only scientist who has used Lekhnitskii's work as described in the previous section. In two papers [57, 58], Ren has, in fact, extended Lekhnitskii's theory to orthotropic and anisotropic plates. Further applications to vibration and buckling were made in a third paper written in collaboration with Owen [59]. These three papers are the unique contributions known to the authors that have been made under the framework of Lekhnitskii's theory. As these three papers have been published in journals that are easily available worldwide, a full description of Ren's extension of Lekhnitskii's theory to plates has been, therefore, omitted. Nevertheless, it is of interest to make a few comments on Ren's work in order to make explicit the stress and displacement fields that were introduced by Ren to analyze the response of anisotropic plates.

On the basis of the form of τ_{xz}^k obtained by Lekhnitskii, it appeared reasonable to Ren, see [57], to assume the following distribution of transverse shear stresses in a laminated plate, composed of N_l orthotropic layers (x, y, and z are the coordinates of the reference system depicted in Fig. 4.3):

$$\begin{aligned}\sigma_{xz}^k(x,y,z) &= \xi_x(x,y)a^k(z) + \eta_x(x,y)c^k(z),\\ \sigma_{yz}^k(x,y,z) &= \xi_y(x,y)b^k(z) + \eta_y(x,y)g^k(z).\end{aligned} \tag{4.4}$$

Four independent function of x, y were introduced to describe the transverse shear stresses. The layer constants are parabolic functions of the thickness coordinate z. As in Lekhnitskii, the displacement fields are obtained by integrating the strain–displacement relations.

In contrast to the work by Lekhnitskii, it is underlined that the transverse strain ϵ_{zz} was discarded by Ren. This assumption contrasts with Koiter's recommendation already mentioned. The constants of

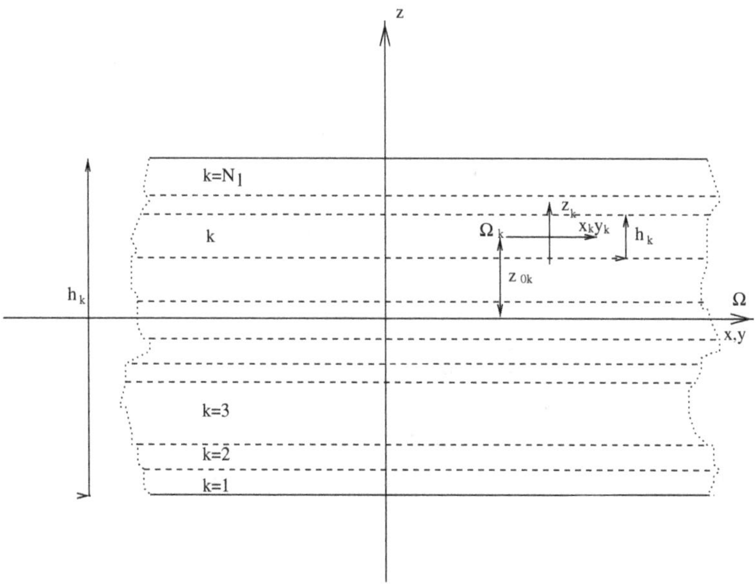

Fig. 4.3. Multilayered plate.

integration are determined by imposing compatibility conditions for the displacements at the interface. The displacement field assumes the following form:

$$\begin{aligned}
u^k(x, y, z) &= u_0(x, y) - w_{,x} + \xi_x(x, y)A^k(z) + \eta_x(x, y)C^k(z), \\
v^k(x, y, z) &= v_0(x, y) - w_{,y} + \xi_y(x, y)B^k(z) + \eta_y(x, y)G^k(z), \\
w(x, y, z) &= w_0(x, y),
\end{aligned} \tag{4.5}$$

where $A^k(z)$, $B^k(z)$, $C^k(z)$, and $G^k(z)$ are obtained by integrating the corresponding $a^k(z)$, $b^k(z)$, $c^k(z)$, and $g^k(z)$. That is, Eqs. (4.5) represent a piecewise continuous displacement field in the thickness direction z, which is cubic in each layer. An extension to generally anisotropic layers has been provided by the same author in the article already mentioned [58].

The displacement model of Eqs. (4.5) can be used in the framework of known variational statements, such as the principle of virtual displacements (PVD) to formulate the governing equations for anisotropic plates as well as finite element models. This was done in [57–59]. No shell applications of the Lekhnitskii–Ren theory are known to the authors.

4.4. The Ambartsumian–Whitney–Rath–Das Theory

This is the second of the three discussed theories. Ambartsumian was the author of the original work [60–63]; Whitney [64] extended the theory to anisotropic plates and introduced the theory to the scientific community in the West; Rath and Das [65], extended Whitney's work to shell geometries.

The Ambartsumian–Whitney–Rath–Das (AWRD) approach has the peculiarity of having the same number of unknown variables as first-order shear deformation theory, i.e., three displacements and two rotations (or shear strains). It was originated by Ambartsumian [60, 61] who restricted the formulation to orthotropic layers. Here attention will focus on the work by Whitney [64] who first applied and extended it to generally anisotropic and symmetrical and asymmetrical plates. For simplicity, only symmetrical laminated plates are outlined. Details can be read in the above-mentioned articles and books. The transverse shear stresses are assumed to be:

$$\begin{aligned} \sigma_{xz}^k(x,y,z) &= [Q_{55}^k f(z) + a_{55}^k]\phi_x(x,y) + [Q_{45}^k f(z) + a_{45}^k]\phi_y(x,y), \\ \sigma_{yz}^k(x,y,z) &= [Q_{45}^k f(z) + a_{55}^k]\phi_x(x,y) + [Q_{44}^k f(z) + a_{44}^k]\phi_y(x,y). \end{aligned} \tag{4.6}$$

The Ambartsumian case can be obtained by putting $Q_{45}^k = a_{45}^k = 0$. $f(z)$ is a function of the thickness coordinate which is assumed to be different in the symmetrical and unsymmetrical laminate cases. A parabolic form for $f(z)$ has mostly been considered (an explicit formula for unsymmetrical cases was given by Whitney). The layer constants a_{44}^k, a_{45}^k, and a_{55}^k are determined by imposing the continuity conditions of transverse shear stresses at the interface while top–bottom homogeneous conditions are used to determine the form of $f(z)$. Notice that the top–bottom inhomogeneous conditions for transverse shear stresses were addressed by Ambartsumian [60, 61], along with a method to compute transverse normal stresses. These two facts have not been addressed in subsequent work.

The transverse shear strains related to the assumed transverse shear stress fields are as follows:

$$\begin{aligned} \gamma_{xz}^k(x,y,z) &= \left[f(z) + S_{55}^k a_{55}^k + S_{45}^k a_{45}^k\right]\phi_x(x,y) \\ &\quad + \left[S_{55}^k a_{45}^k + S_{45}^k a_{44}^k\right]\phi_y(x,y), \\ \gamma_{yz}^k(x,y,z) &= \left[S_{44}^k a_{44}^k + S_{45}^k a_{45}^k\right]\phi_x(x,y) \\ &\quad + \left[f(z) + S_{44}^k a_{44}^k + S_{45}^k a_{55}^k\right]\phi_y(x,y), \end{aligned} \tag{4.7}$$

in which the following compliances have been introduced:

$$S_{55}^k = \frac{Q_{55}^k}{D}, \quad S_{45}^k = -\frac{Q_{45}^k}{D}, \quad S_{44}^k = \frac{Q_{44}^k}{D}, \quad D = Q_{44}^k Q_{55}^k - (Q_{45}^k)^2.$$

By assuming the transverse displacement is constant in the thickness direction, i.e., $\epsilon_{zz} = 0$, on integrating the shear strains, the displacement field has the following form:

$$\begin{aligned} u^k(x,y,z) &= -zw_{,x} + [J(z) + g_1^k(z)]\phi_x(x,y) + g_2^k(z)\phi_y(x,y), \\ v^k(x,y,z) &= -zw_{,y} + [J(z) + g_3^k(z)]\phi_y(x,y) + g_4^k(z)\phi_x(x,y), \\ w^k(x,y,z) &= w_0(x,y,z), \end{aligned} \tag{4.8}$$

where

$$\begin{aligned} J(z) &= \textstyle\int f(z)dz, \\ g_1^k(z) &= [S_{55}^k a_{55}^k + S_{45}^k a_{45}^k]z + d_1^k, \\ g_2^k(z) &= [S_{55}^k a_{55}^k + S_{45}^k a_{45}^k]z + d_2^k, \\ g_3^k(z) &= [S_{55}^k a_{55}^k + S_{45}^k a_{45}^k]z + d_3^k, \\ g_4^k(z) &= [S_{55}^k a_{55}^k + S_{45}^k a_{45}^k]z + d_4^k. \end{aligned} \tag{4.9}$$

d_1^k, d_2^k, d_3^k, and d_4^k are calculated by imposing the compatibility of the in-plane displacement at each interface. Equation (4.8) are the starting point for any analytical or computational study of multilayered plates.

An extension to doubly curved shells and a dynamic case of Whitney's work was made by Rath and Das [65].

Dozens of papers have been presented over recent decades that deal with zig-zag effects and interlaminar continuous transverse shear stresses, and which have stated that new theories were being proposed. The authors believe that these articles should be considered as simplified cases of the AWRD theory or the AWRD theory itself. Unfortunately, the original work and authors (Ambartsumian, Whitney, Rath, and Das) are not mentioned, or rarely cited, in the literature lists of this large number of articles. This historical unfairness has been corrected in [23].

4.5. The Reissner–Murakami–Carrera Theory

A third approach to laminated structures originated in two papers by Reissner [66, 67] in which a mixed variational equation, namely the Reissner mixed variational theorem (RMVT) was proposed. The displacement and transverse stress variables are independently assumed in RMVT. This third approach is the only one that was entirely developed in the West. Reissner [66] proposed a mixed theorem and traced the manner in which it could be developed; Murakami [68, 69], a student under Prof Reissner in San Diego, was the first to develop a plate theory on the basis of RMVT and introduced fundamental ideas on the application of RMVT in the framework of ESLM; Carrera [9, 70] presented a systematic way to use RMVT to develop plate and shell theories and introduced a weak form of Hooke's law (WFHL), which reduces mixed theories to classical models with only displacement variables.

RMVT fulfills completely and *a priori* the C_z^0-requirements by assuming two independent fields for displacements $\boldsymbol{u} = \{u, v, w\}$, and transverse stresses $\boldsymbol{\sigma}_n = \{\sigma_{xz}, \sigma_{yz}, \sigma_{zz}\}$ (bold letters denote arrays). Briefly, RMVT puts 3D indefinite equilibrium equations (and related equilibrium conditions at the boundary surfaces, which for brevity are not written here) and compatibility equations for transverse strains in a variational form. The 3D equilibrium equations in the dynamic case are as follows:

$$\sigma_{ij,j} - \rho\, \ddot{u}_i = p_i \quad i, j = 1, 2, 3, \tag{4.10}$$

where ρ is the mass density and double dots denote acceleration while $(p_1, p_2, p_3) = \boldsymbol{p}$ are volume loadings. The compatibility conditions for transverse stresses can be written by evaluating transverse strains in two ways: using Hooke's law $\boldsymbol{\epsilon}_{nH} = \{\epsilon_{xz_H}, \epsilon_{yz_H}, \epsilon_{zz_H}\}$ and using a geometrical relation $\boldsymbol{\epsilon}_{nG} = \{\epsilon_{xz_G}, \epsilon_{yz_G}, \epsilon_{zz_G}\}$; the subscript n denotes transverse/normal components. Hence

$$\boldsymbol{\epsilon}_{nH} - \boldsymbol{\epsilon}_{nG} = 0. \tag{4.11}$$

RMVT therefore states:

$$\int_V (\delta \boldsymbol{\epsilon}_{pG}^T \boldsymbol{\sigma}_{pH} + \delta \boldsymbol{\epsilon}_{nG}^T \boldsymbol{\sigma}_{nM} + \delta \boldsymbol{\sigma}_{nM}^T (\boldsymbol{\epsilon}_{nG} - \boldsymbol{\epsilon}_{nH}))dV = \int_V \rho \delta \boldsymbol{u}\, \ddot{\boldsymbol{u}}\, dV + \delta L_e. \tag{4.12}$$

The superscript T signifies an array transposition and V denotes the 3D multilayered body volume while the subscript p denotes in-plane components, respectively. Therefore, $\boldsymbol{\sigma}_p = \{\sigma_{xx}, \sigma_{yy}, \sigma_{xy}\}$ and $\boldsymbol{\epsilon}_p = \{\epsilon_{xx}, \epsilon_{yy}, \epsilon_{xy}\}$. The subscript H underlines that stresses are computed via Hooke's law. The variation of the internal work has been split into in-plane and out-of-plane parts and involves the stress from Hooke's law and the strain from the geometrical relations (subscript G). δL_e is the virtual variation of the work done by the external layer-force $\boldsymbol{p}$. Subscript M underlines that transverse stresses are those of the assumed model.

The first application of RMVT was due to Murakami [68, 69], who developed a refinement of Reissner–Mindlin type theories. First a zig-zag form of the displacement field was introduced by means of two "zig-zag" functions (D_x, D_y):

$$\begin{aligned} u^k(x,y,z) &= u_0(x,y) + z\phi_x(x,y) + \xi_k(-1)^k D_x(x,y), \\ v^k(x,y,z) &= v_0(x,y) + z\phi_y(x,y) + \xi_k(-1)^k D_y(x,y), \\ w(x,y,z) &= w_0(x,y). \end{aligned} \tag{4.13}$$

$\xi_k{=}2z_k/h_k$ is a dimensionless layer coordinate (z_k is the physical coordinate of the kth layer whose thickness is h_k). The exponent k changes the sign of the zig-zag term in each layer. This trick reproduces the discontinuity of the first derivative of the displacement variables in the z-direction. The geometrical meaning of the zig-zag function is explained in Figs. 4.4 and 4.5.

The transverse shear stresses fields were assumed to be parabolic by Murakami [68] in each layer and interlaminar continuous according to the following formula:

$$\begin{aligned} \sigma^k_{xz}(x,y,z) &= \sigma^{kt}_{xz}(x,y)F_0(z_k) + F_1(z_k)^k_x(x,y) + \sigma^{kb}_{xz}(x,y)F_2(z_k), \\ \sigma^k_{yz}(x,y,z) &= \sigma^{kt}_{yz}(x,y)F_0(z_k) + F_1(z_k)R^k_y(x,y) + \sigma^{kb}_{yz}(x,y)F_2(z_k), \end{aligned} \tag{4.14}$$

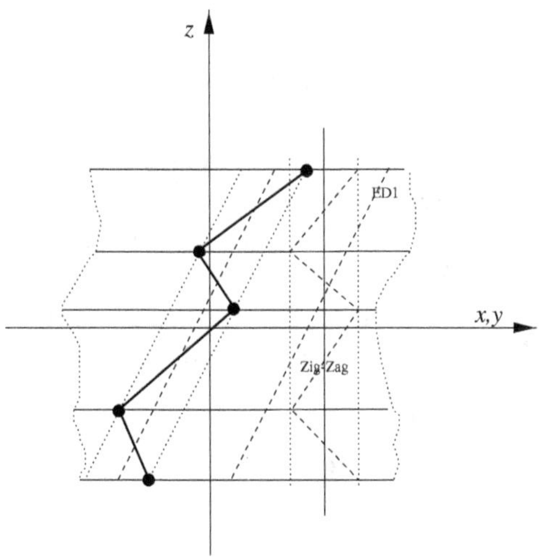

Fig. 4.4. Geometrical meaning of Murakami's zig-zag function. Linear case.

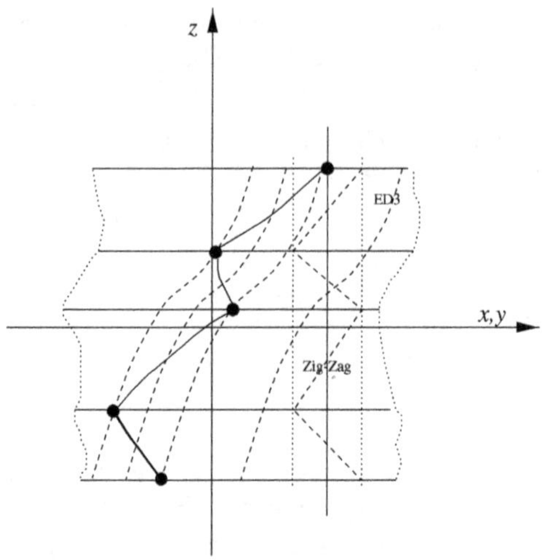

Fig. 4.5. Geometrical meaning of Murakami's zig-zag function. Higher-degree case.

where $\sigma_{xz}^{kt}(x,y)$, $\sigma_{yz}^{kt}(x,y)$, $\sigma_{xz}^{kb}(x,y)$, and $\sigma_{yz}^{kb}(x,y)$ are the top and bottom values of the transverse shear stresses, while $R_x^k(x,y)$, and $R_y^k(x,y)$ are the layer stress resultants. The introduced layer thickness coordinate polynomials hold:

$$F_0 = -1/4 + \xi_k + 3\xi_k^2,\ F_1 = \frac{3-12\xi_k^2}{2h_k},\ F_2 = -1/4 - \xi_k + 3\xi_k^2.$$

The homogeneous and inhomogeneous boundary conditions at the top–bottom plate surfaces can be linked to the introduced stress field.

Toledano and Murakami [71] introduced transverse normal strain and stress effects by using a third-order displacement field for both in-plane and out-of-plane components and a fourth-order transverse stress field for both shear and normal components. This paper is the first paper in the ESLM framework in which Koiter's recommendation is retained.

A generalization of RMVT to plate/shell theories has been provided by Carrera [28, 72–78]. The displacements and transverse stress components were assumed as follows:

$$\begin{aligned} \boldsymbol{u}^k &= F_t\boldsymbol{u}_t^k + F_b\boldsymbol{u}_b^k + F_r\boldsymbol{u}_r^k = F_\tau\boldsymbol{u}_\tau^k, && \tau = t,b,r,\\ & && r = 2,3,\ldots,N, \quad (4.15)\\ \boldsymbol{\sigma}_{nM}^k &= F_t\boldsymbol{\sigma}_{nt}^k + F_b\boldsymbol{\sigma}_{nb}^k + F_r\boldsymbol{\sigma}_{nr}^k = F_\tau\boldsymbol{\sigma}_{n\tau}^k, && k = 1,2,\ldots,N_l. \end{aligned}$$

The subscripts t and b denote values for the top and bottom surface layer, respectively. These two terms consist of the linear part of the expansion. The thickness functions $F_\tau(\xi_k)$ can now be defined at the kth layer level:

$$F_t = \frac{P_0 + P_1}{2}, \quad F_b = \frac{P_0 - P_1}{2}, \quad F_r = P_r - P_{r-2},$$
$$r = 2, 3, \ldots, N, \tag{4.16}$$

in which $P_j = P_j(\xi_k)$ is the jth order Legendre polynomial defined in the ξ_k domain: $-1 \le \xi_k \le 1$. For instance, the first five Legendre polynomials are as follows:

$$P_0 = 1, \quad P_1 = \xi_k, \quad P_2 = (3\xi_k^2 - 1)/2, \quad P_3 = \frac{5\xi_k^3}{2} - \frac{3\xi_k}{2},$$
$$P_4 = \frac{35\xi_k^4}{8} - \frac{15\xi_k^2}{4} + \frac{3}{8}.$$

The chosen functions have the following properties:

$$\xi_k = \begin{cases} 1 & \text{when } F_t = 1;\ F_b = 0;\ F_r = 0, \\ -1 & \text{when } F_t = 0;\ F_b = 1;\ F_r = 0. \end{cases} \tag{4.17}$$

The top and bottom values have been used as unknown variables. Such a choice makes the model particularly suitable, in view of the fulfillment of the C_z^0-requirements. The interlaminar transverse shear and normal stress continuity can therefore be linked by simply writing:

$$\boldsymbol{\sigma}_{nt}^k = \boldsymbol{\sigma}_{nb}^{k+1}, \quad k = 1, N_l - 1. \tag{4.18}$$

In those cases in which the top/bottom plate/shell stress values are prescribed (zero or imposed values), the following additional equilibrium conditions must be accounted for:

$$\boldsymbol{\sigma}_{nb}^1 = \bar{\boldsymbol{\sigma}}_{nb}, \quad \boldsymbol{\sigma}_{nt}^{N_l} = \bar{\boldsymbol{\sigma}}_{nt}, \tag{4.19}$$

where the over-bar denotes the imposed values for the plate boundary surfaces.

Examples of the application of RMVT to laminated plates in the equivalent single-layer model were presented in the already mentioned articles [68, 69, 71]. The results obtained for the cylindrical bending of cross-ply symmetrically laminated plates showed an improvement in describing the in-plane response with respect to the first-order shear deformation theory [69]. Applications to unsymmetrically laminated plates were presented in [71]. Shell applications based on [69] were developed by Bhaskar and Varadan [79] and Jing and Tzeng [80]. Bhaskar and Varadan [79] underlined the severe limitation of the transverse shear stress *a priori* evaluated by the assumed model. Finite element applications of this model have been developed. The linear analysis of thick plates was discussed by Rao and Meyer-Piening [81]. Linear and geometrically nonlinear static and dynamic analyses were considered by Carrera [72, 82] and co-authors [83]. Partial implementations to shell elements were proposed by Bhaskar and Varadan [84]. A full shell implementation has recently been given by Brank and Carrera [85].

The limitations of the equivalent single-layer analysis were known to Toledano and Murakami [71] who applied RMVT in a multilayered model. A linear in-plane displacement expansion was expressed in terms of the interface values in each layer while the transverse shear stresses were assumed parabolic. It was shown that the accuracy of the resulting theories was independent of layout. Transverse normal stress and related effects were discarded and the analysis showed severe limitations when analyzing thick plates. A more comprehensive evaluation of layer-wise theories for linear and parabolic expansions was made by the first author in [73] where applications to the static analysis of plates were presented. Subsequent work extended the analysis to dynamic cases [28, 74, 77, 78] and shell geometry [70, 75, 76]. A more exhaustive overview work of based on the Reissner theorem has been provided in [19].

4.6. Remarks on the Theories

In the authors' opinion the work by Lekhnitskii is the most relevant contribution to multilayered structure modeling:

- L1. This is the *first* work to account for the C_z^0-requirements.
- L2. Even though Lekhnitskii restricted his analysis to a cantilevered multilayered beam, he quoted explicit formulas for transverse stresses and displacement fields (Eqs. (4.4) and (4.5)), which are valid at all points of the considered beam. This could be extremely useful in assessing new analytical and numerical models.
- L3. The work by Lekhnitskii shows how multilayered structures problems can be handled. For instance, it is clear in [22] that the inclusion of a transverse normal stress would require a different choice of stress functions.

- L4. The stress function formulation leads to in-plane and transverse stress fields which fulfill "by definition" the 3D equilibrium equations. Stresses were calculated by Lekhnitskii by solving a boundary-value problem for the compatibility equations written in terms of a stress function. In particular, the evaluation of transverse stresses does not require any post-processing procedure such as Hooke's law or integration of 3D equilibrium equations.
- L5. Although transverse normal stresses are neglected, the transverse displacement varies in the beam thickness according to a piecewise-parabolic distribution. A direct attempt to include the transverse normal stress effect would require an appropriate choice for the stress function.

Concerning Lekhnitskii–Ren plate theory observe that:

- LR1. The transverse shear stresses are continuous at the interfaces and parabolic in each layer. Furthermore, homogeneous conditions are fulfilled at the top–bottom plate surfaces.
- LR2. Four independent functions defined on Ω are used to express transverse shear stresses. Layer constants, which are parabolic in each layer, are used to describe the transverse shear stresses.
- LR3. Expressions for the layer constants were given by Ren. In other words, their calculation does not require any imposition of transverse shear stresses.
- LR4. The in-plane displacements are continuous at each interface and are cubic in each layer.
- LR5. Seven independent variables defined on Ω were used to describe the displacement and stress fields in the laminated plates. Four are used for the transverse shear stresses and three for the displacements corresponding to the chosen reference surface Ω.
- LR6. According to Lekhnitskii, Ren neglects the transverse normal stress σ_{zz}. In contrast to Lekhnitskii, the transverse normal strain ϵ_{zz} is discarded by Ren.
- LR7. The transverse shear stresses are calculated by Ren directly using Eqs. (4.4). Hooke's law is not used and integration of the 3D equilibrium equations is not required.

Regarding the Ambartsumian–Whitney–Rath–Das theory notice that:

- AWRD1. As LR1.
- AWRD2. Two independent functions defined on Ω are used to express transverse shear stresses (Eqs. (4.6)).

- AWRD3. Layer constants, parabolic in each layer, must be computed by imposing transverse shear stress continuity at each interface while the form of the $f(z)$ function is found by imposing top–bottom layer homogeneous conditions.
- AWRD4. As LR4.
- AWRD5. Five independent variables defined on Ω are used to describe the displacement and stress fields in a laminated plate/shell, which is two less than LR.
- AWRD6. As LR6.
- AWRD7. The literature shows that much better evaluations for transverse shear stresses are obtained via integration of the 3D equilibrium equations, with respect to Eqs. (4.6).
- AWRD8. The extension to a shell requires a reformulation of the displacement models and related layer constants.

For the Reissner–Murakami–Carrera theory observe that:

- RMC1. As LR1. In this case, homogeneous as well as inhomogeneous conditions for transverse stresses can be included.
- RMC2. At least $2N_l + 1$ independent variables must be used for each transverse stress component. However, these variables can be expressed in terms of the displacement variables using a weak form of Hooke's law.
- RMC3. The in-plane displacements are continuous at each interface and can be chosen linear or of higher order in each layer.
- RMC4. The number of independent variables can be chosen arbitrarily according to RMC3.
- RMC5. Interlaminar continuous transverse normal stresses/strains can be easily described by the RMC theory. These effects were, in fact, included in the early development of the RMC theory, fulfilling the fundamental Koiter's recommendation.
- RMC6. As for the AWRD theory much better evaluations for transverse stresses are obtained via integration of the 3D equilibrium equations, with respect to assumed forms, e.g. Eqs. (4.14).
- RMC7. The extension to a shell does not require any changes in either displacement or stress fields.

4.7. A Brief Discussion on Layer-Wise Theories

The previous discussion has been restricted to ESLM. In this class of theories the number of unknown variables does not depend on the number of layers (it is intended that for the RMC theory this restriction is only for displacement variables). The use of independent variables in each layer, as in the layer-wise description, increases computational costs. On the other hand, such a choice permits one to include "naturally" the zig-zag form of displacements in the thickness direction and in general can significantly improve for the response of very thick structures. In this respect, the authors' experience suggests that the layer-wise description is mandatory for thick plate/shell analyses and in any other problems in which the response is essentially a layer response. In particular, in [78] the first author showed that the use of a sufficiently high order for the displacement fields in the layers could lead to a description with acceptable accuracy of the transverse stresses directly computed by Hooke's law. Many layer-wise theories have been proposed. So-called global/local approaches have also been proposed, see [21]. Excellent overviews can be found in the review articles and books mentioned in the Introduction.

To the best of the authors' knowledge, there is no layer-wise theory based on the LR approach. Works with a layer-wise description in the framework of the AWRD theories have recently been discussed by Cho and Averill [86]. Studies on the use of the RMC theory have been made for plates by Toledano and Murakami [71] and extended to higher-order cases (including normal stress effects), dynamics, and shells in Carrera's articles [73, 78].

4.8. Best Theory Diagrams via the Axiomatic/Asymptotic Method

The structural models discussed in this chapter are axiomatic. In fact, some hypotheses on the mechanical behavior of structures are formulated and then applied to the mathematical model. An alternative approach is the asymptotic method which makes use of expansions of characteristic parameters of the structures (e.g., the length-to-thickness ratio) to build an asymptotic series. Those terms that exhibit the same order of magnitude as the parameter when it vanishes are retained. The axiomatic approach is simpler but is affected by the lack of information about the accuracy of the approximated theory on the exact 3D solution. In other words, it is not usually possible to apriori evaluate the accuracy of an axiomatic theory. The asymptotic method is mathematically more cumbersome and may require the use of many characteristic parameters, but can be seen as a step towards the development of approximated theories with known accuracy with respect to the 3D exact solution which, in the beam case, is a method that can approximate the 3D energy through 1D terms with known accuracy.

The AAM method can be seen as a methodology that, starting from axiomatic hypotheses, leads to asymptotic-like results. The AAM evaluates the influence of each expansion term, or variable, and eliminate all those terms that do not contribute to the solution for a given set of characteristic parameters. The AAM can, therefore, reduce the computational cost of refined models without affecting their accuracies by developing reduced refined models with fewer degrees of freedom but as accurate as the full models.

4.8.1. The axiomatic/asymptotic method

The AAM is a tool to build reduced refined models, and it consists of the following steps:

(1) Parameters such as the geometry, BCs, materials and layer layouts are fixed.
(2) A set of output parameters is chosen, such as displacement or stress components.
(3) A starting theory is fixed (axiomatic part); that is, the displacement variables to be analyzed are defined; usually, a theory which provides 3D-like solutions is chosen; a reference solution is defined (in the present work the LW4 was adopted, since this fourth-order model offers an excellent agreement with the three-dimensional solutions [27]).
(4) The CUF is used to generate the governing equations for the theories considered.
(5) The effectiveness of each term of the adopted expansion is evaluated by evaluating the error due to its deactivation; a term is considered as non-effective if the error is negligible; the deactivation of a term is obtained using a penalty technique.
(6) The most suitable structural model for a given structural problem is then obtained by discarding all the non-effective variables.

A graphical notation was introduced to show the results. This consists of a table of three lines, and as many columns as the number of the displacement variables used in the expansion. The displacement field of a fourth-order

Table 4.1. Symbolic representation of the reduced kinematic model with v_2 discarded.

▲	▲	▲	▲	▲
▲	▲	△	▲	▲
▲	▲	▲	▲	▲

ESL model (ESL4) can be defined as

$$\begin{aligned} u &= u_0 + zu_1 + z^2u_2 + z^3u_3 + z^4u_4, \\ v &= v_0 + zv_1 + z^2v_2 + z^3v_3 + z^4v_4, \\ w &= w_0 + zw_1 + z^2w_2 + z^3w_3 + z^4w_4. \end{aligned} \tag{4.20}$$

Table 4.1 shows a reduced model in which v_2 is deactivated; that is,

$$\begin{aligned} u &= u_0 + zu_1 + z^2u_2 + z^3u_3 + z^4u_4, \\ v &= v_0 + zv_1 + \qquad\quad + z^3v_3 + z^4v_4, \\ w &= w_0 + zw_1 + z^2w_2 + z^3w_3 + z^4w_4. \end{aligned} \tag{4.21}$$

4.8.2. The Best Theory Diagram

It is possible to associate to each reduced refined model the number of the active terms and its error computed on a reference solution as reported in Fig. 4.6. Each black dot represents a reduced refined model and its position on the Cartesian plane is defined considering its error and the number of the active terms. Furthermore, the graphical representation of the active/non-active terms is reported for some reduced models. It is possible to note that some of them present the lowest error for a given number of active terms. These models are labeled in Fig. 4.6 as 1, 2, 3, 4, 5 and they represent the Pareto front for the considered problem. This Pareto front is defined as the Best Theory Diagram (BTD). This curve can be constructed for several problems, for example by considering several types of materials, geometries and boundary conditions. Moreover, the information reported in a BTD makes it possible to evaluate the minimum number of terms that have to be used to achieve the desired accuracy.

The number of all possible combinations of active/not-active terms for a given refined model is equal to 2^M where M is the number of the terms of a model. In the case of an ESL4 model, M is 15. As the expansion order increases, the number of the combinations to consider also increases. A genetic algorithm was used to construct a BTD with a small

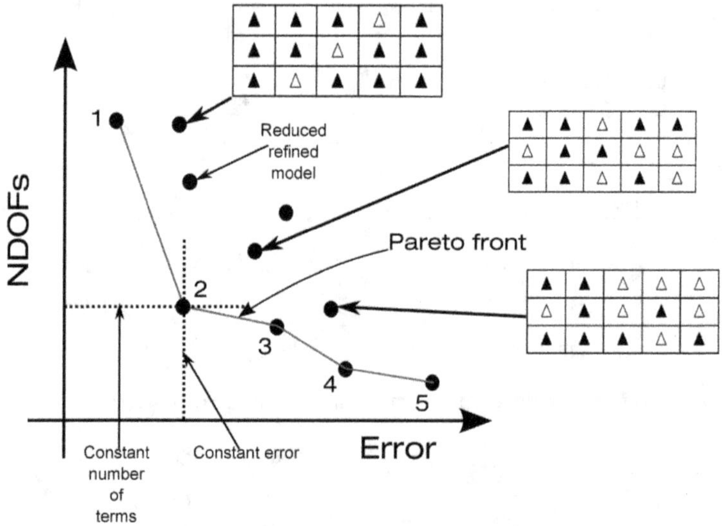

Fig. 4.6. The Best Theory Diagram.

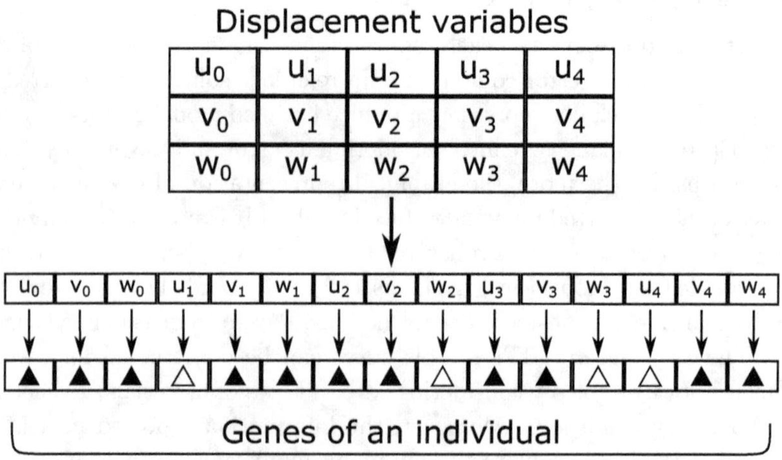

Fig. 4.7. Displacement variables of a refined model and genes of an individual.

computational effort. In fact, each model is considered as an individual. The genes are the terms of the expansion, and each gene can be active or not active, as shown in Fig. 4.7. Each individual is, therefore, described by the number of active terms and its error computed with respect to a reference solution. Through these two parameters, it is possible to apply the dominance rule to evaluate the individual fitness. The generation of new refined theories starting from a generic population is inspired to the reproduction of bacteria; for each individual (i.e., for each structural model) a number of copies are created according to its dominance and then, some mutations are applied to vary the set of new individuals. The purpose of this analysis is to find the individuals that belong to the Pareto front, that is, the subset of individuals that are dominated by no other individuals. In all cases, the number of generations, i.e., iterations, needed is equal to 10, and the number of the initial population is equal to 400. The error of the reduced models with respect to a reference solution is evaluated through the following formula:

$$e = 100\,\frac{\sum_{i=1}^{N_p} |Q^i - Q^i_{\text{ref}}|}{\max Q_{\text{ref}}} \cdot \frac{1}{N_p}, \tag{4.22}$$

where Q can be a stress or displacement component, and N_p is the number of points along the thickness on which Q is evaluated.

4.9. CUF Shell Finite Elements

The efficient load-carrying capabilities of shell structures make them very useful in a variety of engineering applications. The continuous development of new structural materials leads to ever more complex structural designs that require careful analysis. Although analytical techniques are very important, the use of numerical methods to solve mathematical shell models of complex structures has become an essential ingredient in the design process. The finite element method (FEM) has been the fundamental numerical procedure in the analysis of shells.

In this section, a new shell finite element approach based on variable kinematic models within the Carrera Unified Formulation [27, 28] is presented. Elements with nine nodes and cylindrical geometry are considered. Referring to Bathe and others [35–37], the MITC method is used to overcome the locking phenomenon. The governing equations are derived in the framework of the CUF in order to apply FEM. Some numerical results are provided to show the efficiency of the new element.

4.9.1. Geometry of cylindrical shells

Let us consider a cylindrical shell. In a system of Cartesian coordinates (O, x, y, z), the region occupied by the mid-surface of the shell is as follows:

$$S = \{(x, y, z) \in R^3 : \; -L/2 < x < L/2, \; y^2 + z^2 = R^2\}, \tag{4.23}$$

where L and R are the length and the radius of the shell, respectively. Let us consider a curvilinear coordinate system (α, β, z) placed at the center of the upper part of the mid-surface. The 3D medium corresponding to the shell is defined by the 3D chart given by

$$\boldsymbol{\Phi}(\alpha, \beta, z) = \boldsymbol{\phi}(\alpha, \beta) + z\boldsymbol{a}_3(\alpha, \beta), \tag{4.24}$$

where $\boldsymbol{a}_3$ is the unit vector normal to the tangent plane. Then, the mid-surface S of the cylindrical shell is described by the following 2D chart:

$$\begin{cases} \phi_1(\alpha, \beta) = \alpha, \\ \phi_2(\alpha, \beta) = R\sin(\beta/R), \\ \phi_3(\alpha, \beta) = R\cos(\beta/R). \end{cases} \tag{4.25}$$

With this choice, the region $\Omega \subset R^2$ corresponding to the mid-surface S is the rectangle:

$$\Omega = \{(\alpha, \beta) : \; -L/2 < \alpha < L/2, \; -R\pi < \beta < R\pi\}. \tag{4.26}$$

Using these geometrical assumptions, the strain–displacement relations can be obtained by considering the linear part of the *3D Green–Lagrange strain tensor*. Remembering that in the unified formulation the unknowns are the components of the displacement $u_\tau(\alpha, \beta)$, $v_\tau(\alpha, \beta)$, and $w_\tau(\alpha, \beta)$, for $\tau = 0, 1, \ldots, N$, the *geometrical relations* for the kth layer of a multilayer cylindrical shell can be written as follows:

$$\begin{aligned}
\varepsilon^k_{\alpha\alpha} &= F_\tau u^k_{\tau,\alpha}, \\
\varepsilon^k_{\beta\beta} &= F_\tau \left[\left(1 + \frac{z_k}{R_k}\right) \frac{w^k_\tau}{R_k} + \left(1 + \frac{z_k}{R_k}\right) v^k_{\tau,\beta} \right], \\
\varepsilon^k_{\alpha\beta} &= F_\tau \left[u^k_{\tau,\beta} + \left(1 + \frac{z_k}{R_k}\right) v^k_{\tau,\alpha} \right], \\
\varepsilon^k_{\alpha z} &= w^k_{\tau,\alpha} F_\tau + u^k_\tau F_{\tau,z}, \\
\varepsilon^k_{\beta z} &= F_\tau \left[w^k_{\tau,\beta} - \frac{v^k_\tau}{R_k} \right] + F_{\tau,z} \left[\left(1 + \frac{z_k}{R_k}\right) v^k_\tau \right], \\
\varepsilon^k_{zz} &= w^k_\tau F_{\tau,z},
\end{aligned} \tag{4.27}$$

where R_k is the radius of the mid-surface of the layer k. The thickness functions F_τ are Taylor functions $(1, z, z^2, \ldots)$ if the approach used is ESL or combinations of Legendre polynomials if the approach is LW (see Eqs. (4.16)). For more details of the geometrical description and the procedure to obtain the strain–displacement relations, the reader can refer to [87].

The previous geometrical relations can be expressed in matrix form as follows:

$$\begin{aligned} \boldsymbol{\varepsilon}_p^k &= (\boldsymbol{D}_p^k + \boldsymbol{A}_p^k)\boldsymbol{u}^k, \\ \boldsymbol{\varepsilon}_n^k &= (\boldsymbol{D}_{n\Omega}^k + \boldsymbol{D}_{nz}^k - \boldsymbol{A}_n^k)\boldsymbol{u}^k, \end{aligned} \tag{4.28}$$

where subscripts (p) and (n) indicate in-plane and normal components, respectively, and the differential operators are defined as follows:

$$\boldsymbol{D}_p^k = \begin{bmatrix} \partial_\alpha & 0 & 0 \\ 0 & H_k\partial_\beta & 0 \\ \partial_\beta & H_k\partial_\alpha & 0 \end{bmatrix}, \quad \boldsymbol{D}_{n\Omega}^k = \begin{bmatrix} 0 & 0 & \partial_\alpha \\ 0 & 0 & \partial_\beta \\ 0 & 0 & 0 \end{bmatrix},$$

$$\boldsymbol{D}_{nz}^k = \partial_z \cdot \boldsymbol{A}_{nz}^k = \partial_z \begin{bmatrix} 1 & 0 & 0 \\ 0 & H_k & 0 \\ 0 & 0 & 1 \end{bmatrix}, \tag{4.29}$$

$$\boldsymbol{A}_p^k = \begin{bmatrix} 0 & 0 & 0 \\ 0 & 0 & \frac{1}{R_k}H_k \\ 0 & 0 & 0 \end{bmatrix}, \quad \boldsymbol{A}_n^k = \begin{bmatrix} 0 & 0 & 0 \\ 0 & \frac{1}{R_k} & 0 \\ 0 & 0 & 0 \end{bmatrix}, \tag{4.30}$$

where $H_k = (1 + z_k/R_k)$.

4.9.2. MITC method

Considering the components of the strain tensor in the local coordinate system (ξ, η, z), the MITC shell elements are formulated using — instead of the strain components directly computed from the displacements — an interpolation of these strain components within each element using a specific interpolation strategy for each component. The corresponding interpolation points — called the *tying points* — are shown in Fig. 4.8 for a nine-node shell element (MITC9 shell element). For more details see [87].

The interpolating functions are arranged in the following arrays:

$$\begin{aligned} N_{m1} &= [N_{A1}, N_{B1}, N_{C1}, N_{D1}, N_{E1}, N_{F1}], \\ N_{m2} &= [N_{A2}, N_{B2}, N_{C2}, N_{D2}, N_{E2}, N_{F2}], \\ N_{m3} &= [N_P, N_Q, N_R, N_S]. \end{aligned} \tag{4.31}$$

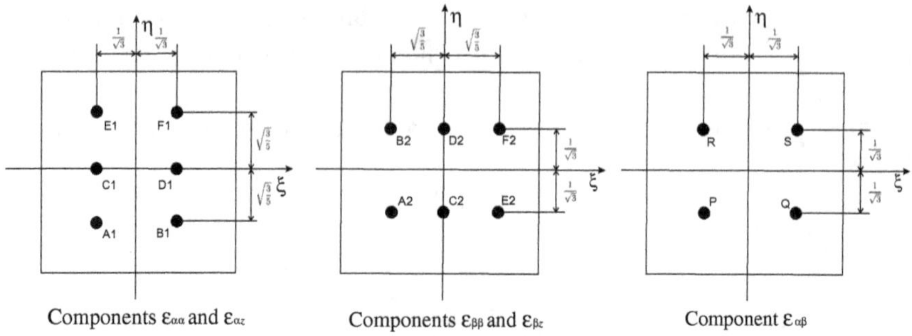

Fig. 4.8. Tying points for MITC9 shell finite element.

For convenience, we will indicate with the subscripts $m1$, $m2$, and $m3$ the quantities calculated for the points $(A1, B1, C1, D1, E1, F1)$, $(A2, B2, C2, D2, E2, F2)$, and (P, Q, R, S), respectively.

According to the MITC method, the strain components are interpolated on the tying points as follows:

$$\boldsymbol{\varepsilon}_p^k = \begin{bmatrix} \varepsilon_{11}^k \\ \varepsilon_{22}^k \\ \varepsilon_{12}^k \end{bmatrix} = \begin{bmatrix} N_{m1} & 0 & 0 \\ 0 & N_{m2} & 0 \\ 0 & 0 & N_{m3} \end{bmatrix} \begin{bmatrix} \varepsilon_{11_{m1}}^k \\ \varepsilon_{22_{m2}}^k \\ \varepsilon_{12_{m3}}^k \end{bmatrix} = [\boldsymbol{N1}] \begin{bmatrix} \varepsilon_{11_{m1}}^k \\ \varepsilon_{22_{m2}}^k \\ \varepsilon_{12_{m3}}^k \end{bmatrix}$$

$$\boldsymbol{\varepsilon}_n^k = \begin{bmatrix} \varepsilon_{13}^k \\ \varepsilon_{23}^k \\ \varepsilon_{33}^k \end{bmatrix} = \begin{bmatrix} N_{m1} & 0 & 0 \\ 0 & N_{m2} & 0 \\ 0 & 0 & 1 \end{bmatrix} \begin{bmatrix} \varepsilon_{13_{m1}}^k \\ \varepsilon_{23_{m2}}^k \\ \varepsilon_{33}^k \end{bmatrix} = [\boldsymbol{N2}] \begin{bmatrix} \varepsilon_{13_{m1}}^k \\ \varepsilon_{23_{m2}}^k \\ \varepsilon_{33}^k \end{bmatrix}, \tag{4.32}$$

which define the matrixes $\boldsymbol{N1}$ and $\boldsymbol{N2}$.

Applying the finite element method, the unknown displacements are interpolated on the nodes of the element by means of the Lagrangian shape functions N_i (for $i = 1, \ldots, 9$):

$$\boldsymbol{u}^k = F_\tau N_i \boldsymbol{q}_{\tau_i}^k, \tag{4.33}$$

where $\boldsymbol{q}_{\tau_i}^k$ are the nodal displacements and the unified formulation is applied. Substituting in Eqs. (4.28) the geometrical relations become:

$$\begin{aligned} \boldsymbol{\varepsilon}_p^k &= F_\tau(\boldsymbol{D}_p^k + \boldsymbol{A}_p^k)(N_i \boldsymbol{I})\boldsymbol{q}_{\tau_i}^k \\ \boldsymbol{\varepsilon}_n^k &= F_\tau(\boldsymbol{D}_{n\Omega}^k - \boldsymbol{A}_n^k)(N_i \boldsymbol{I})\boldsymbol{q}_{\tau_i}^k + F_{\tau,z}\boldsymbol{A}_{nz}^k(N_i \boldsymbol{I})\boldsymbol{q}_{\tau_i}^k, \end{aligned} \tag{4.34}$$

where $\boldsymbol{I}$ is a 3×3 identity matrix.

If the MITC technique is applied, the geometrical relations are rewritten as follows:

$$\begin{aligned}\varepsilon^{k\tau}_{p_{im}} &= F_\tau[\boldsymbol{C}^k_{3_{im}}]\boldsymbol{q}^k_{\tau_i},\\ \varepsilon^{k\tau}_{n_{im}} &= F_\tau[\boldsymbol{C}^k_{1_{im}}]\boldsymbol{q}^k_{\tau_i} + F_{\tau,z}[\boldsymbol{C}^k_{2_{im}}]\boldsymbol{q}^k_{\tau_i},\end{aligned} \tag{4.35}$$

where the introduced matrixes are as follows:

$$\begin{aligned}[\boldsymbol{C}^k_{1_{im}}] &= [\boldsymbol{N2}]\begin{bmatrix}[(\boldsymbol{D}^k_{n\Omega} - \boldsymbol{A}^k_n)(N_i\boldsymbol{I})]_{m1}(1,:)\\ [(\boldsymbol{D}^k_{n\Omega} - \boldsymbol{A}^k_n)(N_i\boldsymbol{I})]_{m2}(2,:)\\ [(\boldsymbol{D}^k_{n\Omega} - \boldsymbol{A}^k_n)(N_i\boldsymbol{I})](3,:)\end{bmatrix},\\ [\boldsymbol{C}^k_{2_{im}}] &= [\boldsymbol{N2}]\begin{bmatrix}[\boldsymbol{A}^k_{nz}(N_i\boldsymbol{I})]_{m1}(1,:)\\ [\boldsymbol{A}^k_{nz}(N_i\boldsymbol{I})]_{m2}(2,:)\\ [\boldsymbol{A}^k_{nz}(N_i\boldsymbol{I})](3,:)\end{bmatrix},\\ [\boldsymbol{C}^k_{3_{im}}] &= [\boldsymbol{N1}]\begin{bmatrix}[(\boldsymbol{D}^k_p + \boldsymbol{A}^k_p)(N_i\boldsymbol{I})]_{m1}(1,:)\\ [(\boldsymbol{D}^k_p + \boldsymbol{A}^k_p)(N_i\boldsymbol{I})]_{m2}(2,:)\\ [(\boldsymbol{D}^k_p + \boldsymbol{A}^k_p)(N_i\boldsymbol{I})]_{m3}(3,:)\end{bmatrix}.\end{aligned} \tag{4.36}$$

(1, :), (2, :), and (3, :), respectively, indicate that the first, second, or third line of the relevant matrix is considered.

4.9.3. Governing equations

This section presents the derivation of the governing equations based on the *principle of virtual displacements* (PVD) for a multilayered shell subjected to mechanical loads. CUF can be used to obtain the so-called *fundamental nuclei*, which are simple matrices representing the basic elements from which the stiffness matrix of the whole structure can be computed.

The PVD for a multilayered shell with N_l layers is

$$\sum_{k=1}^{N_l}\int_{\Omega_k}\int_{A_k}\left\{\delta\boldsymbol{\epsilon}^{k}_{pG}{}^{T}\boldsymbol{\sigma}^k_{pC} + \delta\boldsymbol{\epsilon}^{k}_{nG}{}^{T}\boldsymbol{\sigma}^k_{nC}\right\}d\Omega_k dz_k = \sum_{k=1}^{N_l}\delta L^k_e, \tag{4.37}$$

where Ω_k and A_k are the integration domains in the plane (α,β) and the z-direction, respectively, and T indicates the transpose of a vector. The first member of the equation represents the variation of internal work δL^k_{int} and δL^k_e is the external work. G means geometrical relations and C constitutive relations. The first step in deriving the fundamental nuclei is the substitution of the *constitutive equations* (C) in the variational statement of PVD, which are as follows:

$$\begin{aligned}\boldsymbol{\sigma}_{pC}^{k} &= \boldsymbol{\sigma}_{p_{jn}}^{ks} = \boldsymbol{C}_{pp}^{k}\,\boldsymbol{\varepsilon}_{p_{jn}}^{ks} + \boldsymbol{C}_{pn}^{k}\,\boldsymbol{\varepsilon}_{n_{jn}}^{ks},\\ \boldsymbol{\sigma}_{nC}^{k} &= \boldsymbol{\sigma}_{n_{jn}}^{ks} = \boldsymbol{C}_{np}^{k}\,\boldsymbol{\varepsilon}_{p_{jn}}^{ks} + \boldsymbol{C}_{nn}^{k}\,\boldsymbol{\varepsilon}_{n_{jn}}^{ks}\end{aligned} \tag{4.38}$$

with

$$\begin{aligned}\boldsymbol{C}_{pp}^{k} &= \begin{bmatrix} C_{11}^{k} & C_{12}^{k} & C_{16}^{k}\\ C_{12}^{k} & C_{22}^{k} & C_{26}^{k}\\ C_{16}^{k} & C_{26}^{k} & C_{66}^{k}\end{bmatrix} \quad \boldsymbol{C}_{pn}^{k} = \begin{bmatrix} 0 & 0 & C_{13}^{k}\\ 0 & 0 & C_{23}^{k}\\ 0 & 0 & C_{36}^{k}\end{bmatrix},\\ \boldsymbol{C}_{np}^{k} &= \begin{bmatrix} 0 & 0 & 0\\ 0 & 0 & 0\\ C_{13}^{k} & C_{23}^{k} & C_{36}^{k}\end{bmatrix} \quad \boldsymbol{C}_{nn}^{k} = \begin{bmatrix} C_{55}^{k} & C_{45}^{k} & 0\\ C_{45}^{k} & C_{44}^{k} & 0\\ 0 & 0 & C_{33}^{k}\end{bmatrix},\end{aligned} \tag{4.39}$$

and C are the material coefficients.

Then, one substitutes the geometrical relations (4.35) and the constitutive equations (4.38) into the variational statement (4.37) to obtain the governing equation system:

$$\delta \boldsymbol{u}_{\tau i}^{k\ T} : \boldsymbol{K}_{uu}^{k\tau sij}\boldsymbol{u}_{sj}^{k} = \boldsymbol{P}_{u\tau i}^{k}, \tag{4.40}$$

where $\boldsymbol{K}_{uu}^{k\tau sij}$ is the fundamental nucleus of the stiffness array, which is expanded according to the indexes τ, s and i, j in order to obtain the matrix for the whole shell. $\boldsymbol{P}_{u\tau}^{k}$ is the fundamental nucleus of the external mechanical load. The explicit form of the stiffness fundamental nucleus is the following:

$$\begin{aligned}K_{11}^{k\tau sij} &= C_{55}^{k} N_{i_{m1}} \lhd N_{m1}N_{n1} \rhd_{\Omega_k} N_{j_{n1}} \lhd F_{\tau,z}F_{s,z} \rhd_{A_k}\\ &\quad + C_{11}^{k} N_{i,\alpha_{m1}} \lhd N_{m1}N_{n1} \rhd_{\Omega_k} N_{j,\alpha_{n1}} \lhd F_{\tau}F_{s} \rhd_{A_k}\\ &\quad + C_{16}^{k} N_{i,\beta_{m3}} \lhd N_{m3}N_{n1} \rhd_{\Omega_k} N_{j,\alpha_{n1}} \lhd F_{\tau}F_{s} \rhd_{A_k}\\ &\quad + C_{16}^{k} N_{i,\alpha_{m1}} \lhd N_{m1}N_{n3} \rhd_{\Omega_k} N_{j,\beta_{n3}} \lhd F_{\tau}F_{s} \rhd_{A_k}\\ &\quad + C_{66}^{k} N_{i,\beta_{m3}} \lhd N_{m3}N_{n3} \rhd_{\Omega_k} N_{j,\beta_{n3}} \lhd F_{\tau}F_{s} \rhd_{A_k},\\ K_{12}^{k\tau sij} &= -C_{45}^{k}\frac{1}{R_k} N_{i_{m1}} \lhd N_{m1}N_{n2} \rhd_{\Omega_k} N_{j_{n2}} \lhd F_{\tau,z}F_{s} \rhd_{A_k}\\ &\quad + C_{45}^{k} N_{i_{m1}} \lhd N_{m1}N_{n2} \rhd_{\Omega_k} N_{j_{n2}} \lhd H_k F_{\tau,z}F_{s,z} \rhd_{A_k}\end{aligned}$$

$$
\begin{aligned}
&+C_{12}^{k} N_{i,\alpha_{m1}} \lhd N_{m1} N_{n2} \rhd_{\Omega_k} N_{j,\beta_{n2}} \lhd H_k F_\tau F_s \rhd_{A_k}\\
&+C_{16}^{k} N_{i,\alpha_{m1}} \lhd N_{m1} N_{n3} \rhd_{\Omega_k} N_{j,\alpha_{n3}} \lhd H_k F_\tau F_s \rhd_{A_k}\\
&+C_{26}^{k} N_{i,\beta_{m3}} \lhd N_{m3} N_{n2} \rhd_{\Omega_k} N_{j,\beta_{n2}} \lhd H_k F_\tau F_s \rhd_{A_k}\\
&+C_{66}^{k} N_{i,\beta_{m3}} \lhd N_{m3} N_{n3} \rhd_{\Omega_k} N_{j,\alpha_{n3}} \lhd H_k F_\tau F_s \rhd_{A_k},
\end{aligned} \tag{4.41}
$$

$$
\begin{aligned}
K_{13}^{k\tau sij} =\; & C_{13}^{k} N_{i,\alpha_{m1}} \lhd N_{m1} N_j \rhd_{\Omega_k} \lhd F_\tau F_{s,z} \rhd_{A_k}\\
&+C_{36}^{k} N_{i,\beta_{m3}} \lhd N_{m3} N_j \rhd_{\Omega_k} \lhd F_\tau F_{s,z} \rhd_{A_k}\\
&+C_{12}^{k} \frac{1}{R_k} N_{i,\alpha_{m1}} \lhd N_{m1} N_{n2} \rhd_{\Omega_k} N_{j_{n2}} \lhd H_k F_\tau F_s \rhd_{A_k}\\
&+C_{26}^{k} \frac{1}{R_k} N_{i,\beta_{m3}} \lhd N_{m3} N_{n2} \rhd_{\Omega_k} N_{j_{n2}} \lhd H_k F_\tau F_s \rhd_{A_k}\\
&+C_{55}^{k} N_{i_{m1}} \lhd N_{m1} N_{n1} \rhd_{\Omega_k} N_{j,\alpha_{n1}} \lhd F_{\tau,z} F_s \rhd_{A_k}\\
&+C_{45}^{k} N_{i_{m1}} \lhd N_{m1} N_{n2} \rhd_{\Omega_k} N_{j,\beta_{n2}} \lhd F_{\tau,z} F_s \rhd_{A_k},
\end{aligned}
$$

$$
\begin{aligned}
K_{21}^{k\tau sij} =\; & -C_{45}^{k} \frac{1}{R_k} N_{i_{m2}} \lhd N_{m2} N_{n1} \rhd_{\Omega_k} N_{j_{n1}} \lhd F_\tau F_{s,z} \rhd_{A_k}\\
&+C_{45}^{k} N_{i_{m2}} \lhd N_{m2} N_{n1} \rhd_{\Omega_k} N_{j_{n1}} \lhd H_k F_{\tau,z} F_{s,z} \rhd_{A_k}\\
&+C_{12}^{k} N_{i,\beta_{m2}} \lhd N_{m2} N_{n1} \rhd_{\Omega_k} N_{j,\alpha_{n1}} \lhd H_k F_\tau F_s \rhd_{A_k}\\
&+C_{16}^{k} N_{i,\alpha_{m3}} \lhd N_{m3} N_{n1} \rhd_{\Omega_k} N_{j,\alpha_{n1}} \lhd H_k F_\tau F_s \rhd_{A_k}\\
&+C_{26}^{k} N_{i,\beta_{m2}} \lhd N_{m2} N_{n3} \rhd_{\Omega_k} N_{j,\beta_{n3}} \lhd H_k F_\tau F_s \rhd_{A_k}\\
&+C_{66}^{k} N_{i,\alpha_{m3}} \lhd N_{m3} N_{n3} \rhd_{\Omega_k} N_{j,\beta_{n3}} \lhd H_k F_\tau F_s \rhd_{A_k},
\end{aligned}
$$

$$
\begin{aligned}
K_{22}^{k\tau sij} =\; & C_{22}^{k} N_{i,\beta_{m2}} \lhd N_{m2} N_{n2} \rhd_{\Omega_k} N_{j,\beta_{n2}} \lhd H_k^2 F_\tau F_s \rhd_{A_k}\\
&+C_{26}^{k} N_{i,\beta_{m2}} \lhd N_{m2} N_{n3} \rhd_{\Omega_k} N_{j,\alpha_{n3}} \lhd H_k^2 F_\tau F_s \rhd_{A_k}\\
&+C_{26}^{k} N_{i,\alpha_{m3}} \lhd N_{m3} N_{n2} \rhd_{\Omega_k} N_{j,\beta_{n2}} \lhd H_k^2 F_\tau F_s \rhd_{A_k}\\
&+C_{66}^{k} N_{i,\alpha_{m3}} \lhd N_{m3} N_{n3} \rhd_{\Omega_k} N_{j,\alpha_{n3}} \lhd H_k^2 F_\tau F_s \rhd_{A_k}\\
&+C_{44}^{k} \frac{1}{R_k^2} N_{i_{m2}} \lhd N_{m2} N_{n2} \rhd_{\Omega_k} N_{j_{n2}} \lhd F_\tau F_s \rhd_{A_k}\\
&-C_{44}^{k} \frac{1}{R_k} N_{i_{m2}} \lhd N_{m2} N_{n2} \rhd_{\Omega_k} N_{j_{n2}} \lhd H_k F_\tau F_{s,z} \rhd_{A_k}\\
&-C_{44}^{k} \frac{1}{R_k} N_{i_{m2}} \lhd N_{m2} N_{n2} \rhd_{\Omega_k} N_{j_{n2}} \lhd H_k F_{\tau,z} F_s \rhd_{A_k}\\
&+C_{44}^{k} N_{i_{m2}} \lhd N_{m2} N_{n2} \rhd_{\Omega_k} N_{j_{n2}} \lhd H_k^2 F_{\tau,z} F_{s,z} \rhd_{A_k},
\end{aligned} \tag{4.42}
$$

$$
\begin{aligned}
K_{23}^{k\tau sij} &= C_{22}^k \frac{1}{R_k} N_{i,\beta_{m2}} \lhd N_{m2} N_{n2} \rhd_{\Omega_k} N_{j_{n2}} \lhd H_k^2 F_\tau F_s \rhd_{A_k} \\
&+ C_{23}^k N_{i,\beta_{m2}} \lhd N_{m2} N_j \rhd_{\Omega_k} \lhd H_k F_\tau F_{s,_z} \rhd_{A_k} \\
&+ C_{26}^k \frac{1}{R_k} N_{i,\alpha_{m3}} \lhd N_{m3} N_{n2} \rhd_{\Omega_k} N_{j_{n2}} \lhd H_k^2 F_\tau F_s \rhd_{A_k} \\
&+ C_{36}^k N_{i,\alpha_{m3}} \lhd N_{m3} N_j \rhd_{\Omega_k} \lhd H_k F_\tau F_{s,_z} \rhd_{A_k} \\
&- C_{45}^k \frac{1}{R_k} N_{i_{m2}} \lhd N_{m2} N_{n1} \rhd_{\Omega_k} N_{j,\alpha_{n1}} \lhd F_\tau F_s \rhd_{A_k} \\
&- C_{44}^k \frac{1}{R_k} N_{i_{m2}} \lhd N_{m2} N_{n2} \rhd_{\Omega_k} N_{j,\beta_{n2}} \lhd F_\tau F_s \rhd_{A_k} \\
&+ C_{45}^k N_{i_{m2}} \lhd N_{m2} N_{n1} \rhd_{\Omega_k} N_{j,\alpha_{n1}} \lhd H_k F_{\tau,_z} F_s \rhd_{A_k} \\
&+ C_{44}^k N_{i_{m2}} \lhd N_{m2} N_{n2} \rhd_{\Omega_k} N_{j,\beta_{n2}} \lhd H_k F_{\tau,_z} F_s \rhd_{A_k}, \\
K_{31}^{k\tau sij} &= C_{55}^k N_{i,\alpha_{m1}} \lhd N_{m1} N_{n1} \rhd_{\Omega_k} N_{j_{n1}} \lhd F_\tau F_{s,_z} \rhd_{A_k} \\
&+ C_{45}^k N_{i,\beta_{m2}} \lhd N_{m2} N_{n1} \rhd_{\Omega_k} N_{j_{n1}} \lhd F_\tau F_{s,_z} \rhd_{A_k} \\
&+ C_{12}^k \frac{1}{R_k} N_{i_{m2}} \lhd N_{m2} N_{n1} \rhd_{\Omega_k} N_{j,\alpha_{n1}} \lhd H_k F_\tau F_s \rhd_{A_k} \\
&+ C_{13}^k \lhd N_i N_{n1} \rhd_{\Omega_k} N_{j,\alpha_{n1}} \lhd F_{\tau,_z} F_s \rhd_{A_k} \\
&+ C_{26}^k \frac{1}{R_k} N_{i_{m2}} \lhd N_{m2} N_{n3} \rhd_{\Omega_k} N_{j,\beta_{n3}} \lhd H_k F_\tau F_s \rhd_{A_k} \\
&+ C_{36}^k \lhd N_i N_{n3} \rhd_{\Omega_k} N_{j,\beta_{n3}} \lhd F_{\tau,_z} F_s \rhd_{A_k}, \\
K_{32}^{k\tau sij} &= C_{22}^k \frac{1}{R_k} N_{i_{m2}} \lhd N_{m2} N_{n2} \rhd_{\Omega_k} N_{j,\beta_{n2}} \lhd H_k^2 F_\tau F_s \rhd_{A_k} \\
&+ C_{23}^k \lhd N_i N_{n2} \rhd_{\Omega_k} N_{j,\beta_{n2}} \lhd H_k F_{\tau,_z} F_s \rhd_{A_k} \\
&+ C_{26}^k \frac{1}{R_k} N_{i_{m2}} \lhd N_{m2} N_{n3} \rhd_{\Omega_k} N_{j,\alpha_{n3}} \lhd H_k^2 F_\tau F_s \rhd_{A_k} \\
&+ C_{36}^k \lhd N_i N_{n3} \rhd_{\Omega_k} N_{j,\alpha_{n3}} \lhd H_k F_{\tau,_z} F_s \rhd_{A_k} \\
&- C_{45}^k \frac{1}{R_k} N_{i,\alpha_{m1}} \lhd N_{m1} N_{n2} \rhd_{\Omega_k} N_{j_{n2}} \lhd F_\tau F_s \rhd_{A_k} \qquad (4.43) \\
&- C_{44}^k \frac{1}{R_k} N_{i,\beta_{m2}} \lhd N_{m2} N_{n2} \rhd_{\Omega_k} N_{j_{n2}} \lhd F_\tau F_s \rhd_{A_k} \\
&+ C_{45}^k N_{i,\alpha_{m1}} \lhd N_{m1} N_{n2} \rhd_{\Omega_k} N_{j_{n2}} \lhd H_k F_\tau F_{s,_z} \rhd_{A_k} \\
&+ C_{44}^k N_{i,\beta_{m2}} \lhd N_{m2} N_{n2} \rhd_{\Omega_k} N_{j_{n2}} \lhd H_k F_\tau F_s \rhd_{A_k},
\end{aligned}
$$

$$
\begin{aligned}
K_{33}^{k\tau sij} &= C_{22}^{k}\frac{1}{R_k^2}N_{i_{m2}} \lhd N_{m2}N_{n2} \rhd_{\Omega_k} N_{j_{n2}} \lhd H_k^2 F_\tau F_s \rhd_{A_k} \\
&+ C_{23}^{k}\frac{1}{R_k}N_{i_{m2}} \lhd N_{m2}N_j \rhd_{\Omega_k} \lhd H_k F_\tau F_{s,z} \rhd_{A_k} \\
&+ C_{23}^{k}\frac{1}{R_k} \lhd N_i N_{n2} \rhd_{\Omega_k} N_{j_{n2}} \lhd H_k F_{\tau,z} F_s \rhd_{A_k} \\
&+ C_{33}^{k} \lhd N_i N_j \rhd_{\Omega_k} \lhd F_{\tau,z} F_{s,z} \rhd_{A_k} \\
&+ C_{55}^{k} N_{i,\alpha_{m1}} \lhd N_{m1}N_{n1} \rhd_{\Omega_k} N_{j,\alpha_{n1}} \lhd F_\tau F_s \rhd_{A_k} \\
&+ C_{45}^{k} N_{i,\beta_{m2}} \lhd N_{m2}N_{n1} \rhd_{\Omega_k} N_{j,\alpha_{n1}} \lhd F_\tau F_s \rhd_{A_k} \\
&+ C_{45}^{k} N_{i,\alpha_{m1}} \lhd N_{m1}N_{n2} \rhd_{\Omega_k} N_{j,\beta_{n2}} \lhd F_\tau F_s \rhd_{A_k} \\
&+ C_{44}^{k} N_{i,\beta_{m2}} \lhd N_{m2}N_{n2} \rhd_{\Omega_k} N_{j,\beta_{n2}} \lhd F_\tau F_s \rhd_{A_k},
\end{aligned}
$$

where $\lhd \cdots \rhd_{\Omega_k}$ indicates $\int_{\Omega_k} \ldots d\Omega_k$ and $\lhd \cdots \rhd_{A^k}$ indicates $\int_{A_k} \ldots dz_k$.

4.10. Numerical Examples

The model introduced does not involve an approximation of the geometry of the shell and it accurately describes the curvature of the shell. However, the locking phenomenon is still present. In this work, the model is combined with a simple displacement formulation. The CUF, coupled with the MITC method, allows us to increase the degree of approximation by increasing the order of expansion of the displacements in the thickness direction and the number of elements used. Firstly, the reliability of the model is analyzed. Two classical discriminating test problems are considered: the pinched cylinder with a diaphragm [80], which is the most severe test for both inextensional bending modes and complex membrane states; and the Scordelis–Lo problem [81], which is extremely useful for determining the ability of a finite element to accurately solve complex states of a membrane strain.

The pinched shell has been analyzed in [32] and the essential shape is shown in Fig. 4.9. It is simply supported at each end by a rigid diaphragm and singularly loaded by two opposing forces acting at the midpoint of the shell. Due to the symmetry of the structure the computations have been performed, using a uniform decomposition, on an octave of the shell. The physical data given in Table 4.2 have been assumed. The following

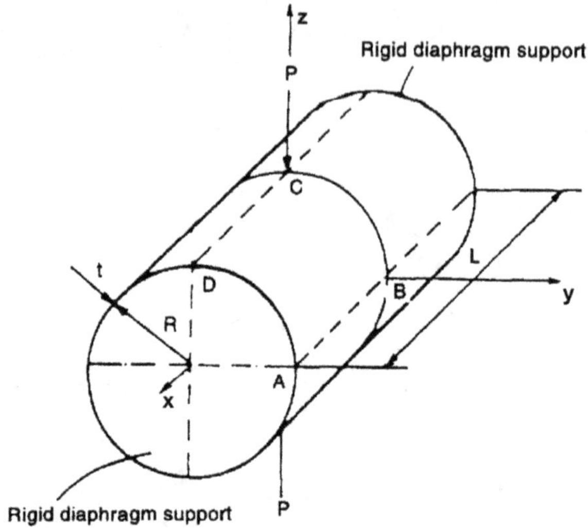

Fig. 4.9. Pinched shell.

Table 4.2. Physical data for pinched shell.

Young's modulus	E	3×10^6 psi $= 20.684 \times 10^9$ N/m^2
Poisson's ratio	ν	0.3
Load	P	1 lb $= 0.154$ Kg
Length	L	600 in $= 15.24$ m
Radius	R	300 in $= 7.62$ m
Thickness	h	3 in $= 0.0762$ m

symmetry conditions and boundary conditions are applied:

$$\begin{aligned} v_s(\alpha, 0) &= 0, \\ u_s(0, \beta) &= 0, \\ v_s(\alpha, R\pi/2) &= 0, \\ v_s(L/2, \beta) &= w_s(L/2, \beta) = 0 \end{aligned} \tag{4.44}$$

with $s = 0, 1, \ldots, N$.

In Table 4.3, the transversal displacement at the loaded point C is presented for several decompositions $[n \times n]$ and different theories. The high-order equivalent single-layer theories in the CUF are indicated with the acronym ESLN, where N is the order of expansion. The exact solution is given by Flügge in Ref. [88] 1.8248×10^{-5} in. The table shows that the

Table 4.3. Pinched shell. Transversal displacement w in $\times 10^5$.

Theory	[4 × 4]	[10 × 10]	[13 × 13]
Koiter	1.7891	1.8231	1.8253
Naghdi	1.7984	1.8364	1.8398
ESL1	1.9212	1.9583	1.9617
ESL2	1.7805	1.8361	1.8408
ESL3	1.7818	1.8380	1.8428
ESL4	1.7818	1.8380	1.8428

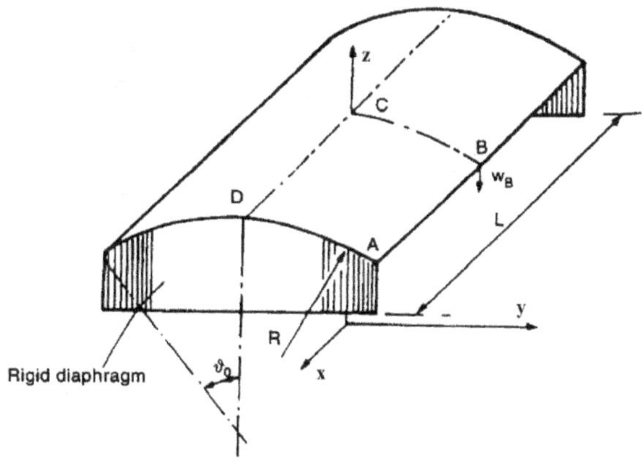

Fig. 4.10. Scordelis–Lo roof.

MITC9 element has good convergence and robustness on increasing the mesh size. According to Reddy [21], the results obtained with high-order theories are greater than the reference value because Flügge uses a classical shell theory. Indeed, the solution calculated with the Koiter model for mesh [13 × 13] is very close to the exact solution, while the Naghdi model, which takes into account the shear energy, gives a higher value, as one would expect. The ESL theory with linear expansion (ESL1) produces such a high value because a correction for Poisson locking has been applied (for details of Poisson locking one can refer to [90]), but for cylindrical shell structures this correction causes problems. The remaining theories give almost the same results and they converge to the same value (1.842×10^{-5} in) by increasing the order of expansion and the number of elements used.

The second problem (the Scordelis–Lo problem [89]) concerns a cylindrical shell known in the literature as a barrel vault, see Fig. 4.10. The shell is simply supported on diaphragms, is free on its other sides, and

Table 4.4. Physical data for barrel vault.

Young's modulus	E	4.32×10^6 lb/h^2 = 20.684×10^9 N/m^2
Poisson's ratio	ν	0.0
Load	P	90 lb/ft^2 = 4309.224 N/m^2
Length	L	50 ft = 15.24 m
Radius	R	25 ft = 7.62 m
Thickness	h	0.25 ft = 0.0762 m
Angle	θ_0	$2\pi/9$ rad

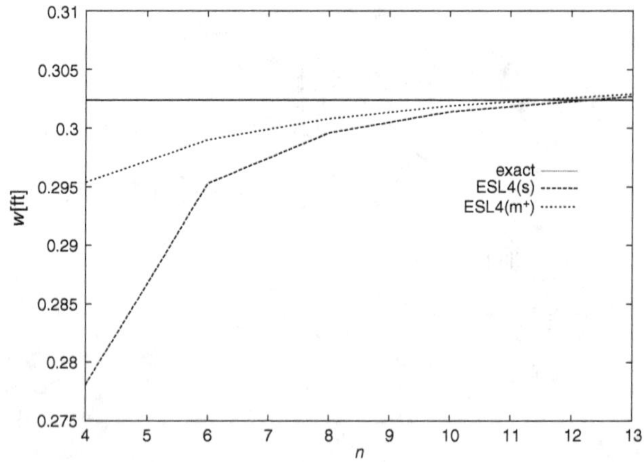

Fig. 4.11. Scordelis–Lo problem. Transversal displacement w [ft] at the point B of the mid-surface S.

is loaded by its own weight P. The physical data given in Table 4.4 have been assumed. The computations were performed only on a quarter of the shell, using a uniform decomposition.

The exact solution for this problem is given by McNeal and Harder in [91] in terms of the transversal displacement at the point B: 0.3024 ft. In Fig. 4.11, the solution is given for several decompositions $[n \times n]$. The performance of the MITC9 element in which a correction for both shear and membrane locking has been applied (m^+) is compared with an element in which only shear locking has been corrected (s). The figure confirms the conclusions for the pinched shell: the results converge to the exact solution on increasing the number of elements used. Moreover, the figure shows that for thin shells $(h/R = 0.01)$ the correction for membrane locking is essential because for coarse meshes the solution (m^+) is much higher than the (s) solution. One can conclude that the MITC9 element is completely locking free. The theory used for this analysis is ESL4 but the behavior is the same as for the other models.

Finally, a multilayered shell was analyzed in order to show the superiority of LW and the zig-zag models compared to ESL models. The orthotropic cylindrical shell studied by Varadan and Bhaskar in [92] is considered. The ends are simply supported. The loading is an internal sinusoidal pressure, applied normal to the shell surface, given by:

$$p_z^+ = \hat{p}_z^+ \sin\left(\frac{m\pi\alpha}{a}\right) \sin\left(\frac{n\pi\beta}{b}\right), \tag{4.45}$$

where the wave numbers are $m = 1$ and $n = 8$. The amplitude of the load $\hat{p}_z^+$ is assumed to be 1. a is the length and b the circumference of the cylinder.

The cylinder is made up of three equal layers with lamination $(90°/0°/90°)$. Each layer is assumed to be made of a square symmetric unidirectional fibrous composite material with the following properties:

$$\begin{aligned} E_L/E_T &= 25, \\ G_{LT}/E_T &= 0.5, \\ G_{TT}/E_T &= 0.2, \\ \nu_{LT} = \nu_{TT} &= 0.25, \end{aligned} \tag{4.46}$$

where L is the direction parallel to the fibers and T is the transverse direction. The length a of the cylinder is assumed to be $4R_\beta$, and the radius R_β, referred to the mid-surface of the whole shell, is 10. Since the cylinder is a symmetric structure and it is symmetrically loaded, the computations were performed only on an octave of the shell, using a uniform decomposition.

The solution is given in terms of the transversal displacement w for different values of the thickness ratio R_β/h, where h is the global thickness of the cylinder. According to [92], the following dimensionless parameter is used:

$$\bar{w} = w\frac{10E_L h^3}{\hat{p}_z^+ R_\beta^4}. \tag{4.47}$$

The results are presented in Table 4.5 and are compared with the 3D-elasticity solution given by Varadan and Bhaskar in [92]. The transversal displacement is calculated on the mid-surface of the multilayered shell $(z = 0)$, at the max-loading point. An $[8 \times 8]$ mesh was used, which is

Table 4.5. Varadan and Bhaskar. Dimensionless transversal displacement at the max-loading point in $z = 0$.

R_β/h	2	4	10	50	100	500
3D	10.1	4.009	1.223	0.5495	0.4715	0.1027
ESL4	9.682	3.782	1.1438	0.5456	0.4707	0.1029
ESL3	9.664	3.785	1.1439	0.5456	0.4707	0.1029
ESL2	8.280	2.971	0.9540	0.5378	0.4692	0.1029
ESL1	8.925	3.015	0.9559	0.5380	0.4696	0.1034
Naghdi	8.421	2.872	0.9382	0.5370	0.4688	0.1029
Koiter	0.4094	0.4796	0.5205	0.5209	0.4656	0.1029
ESLZ3	9.791	3.987	1.224	0.5493	0.4715	0.1029
ESLZ2	9.596	3.866	1.191	0.5479	0.4712	0.1029
ESLZ1	10.228	3.901	1.191	0.5457	0.4694	0.1028
LW4	10.267	4.032	1.225	0.5493	0.4715	0.1029
LW3	10.256	4.031	1.225	0.5493	0.4715	0.1029
LW2	9.789	3.971	1.223	0.5493	0.4715	0.1029
LW1	9.689	3.874	1.191	0.5477	0.4710	0.1029

sufficient to ensure numerical convergence. Equivalent single-layer (ESLN), zig-zag (ESLZN), and layer-wise (LWN) theories in the CUF are employed for the analysis. The classical Koiter's and Naghdi's models were also used for comparison. One can note that the solution obtained with the classical models is completely wrong, while the ESL theories give a more accurate solution by increasing the order of expansion N, especially for high-thickness ratios. If one also takes into account the zig-zag effects in the displacements using Murakami's zig-zag function (ESLZ models), the results improve again and the ESLZ3 theory provides approximately the 3D solution even for very thick shells. Finally, the table shows that the LW theories give the best results even when the order of expansion is not high ($N = 2, 3$), according to the assertions made in the introduction of this chapter about C^0_z-requirements. This behavior is particularly visible for thick shells ($R_\beta/h = 2, 4$). For very thin shells ($R_\beta/h = 500$) all the theories converge to the 3D solution and this fact demonstrates once again the numerical efficiency of the new approach. Note that the LW3 and LW4 models give a solution slightly higher than the 3D solution for very thick shells. This is due to a curvature approximation along the thickness, which can be easily eliminated by considering the shell to be composed of thinner fictitious layers with the same properties.

4.11. Best Theory Diagrams

4.11.1. Plates

Two plates were considered with symmetric and asymmetric laminations. The material properties are the following: $E_L = 40 \times 10^9$ Pa, $E_T = E_z = 1 \times 10^9$ Pa, $G_{LT} = 0.5 \times 10^9$ Pa, $G_z = 0.6 \times 10^9$ Pa, $\nu = 0.25$. The ply sequences considered are 0°/90°/0° for the symmetric plate and 0°/90° for the asymmetric plate; each layer has the same thickness. A transverse pressure was applied to the top surface of the plate,

$$p_z = p_z^0 \sin\left(\frac{m\,\pi}{a}\,x\right) \sin\left(\frac{n\,\pi}{b}\,y\right), \tag{4.48}$$

where $m = n = 1$. All the reduced models are developed for stress σ_{xx}, which is computed at [a/2, b/2, z] with $-\frac{h}{2} \leq z \leq \frac{h}{2}$, where h is the total thickness of the plate.

First, BTDs were obtained using the ESL4 model baseline and LW4 as the reference to evaluate the accuracy of the models. Figure 4.12 shows the assessment of the genetic algorithm to compute the BTD. In fact, all the 2^{15} models given by the combinations of the 15 terms of the ESL4 were evaluated. The BTD evaluated by the genetic algorithm perfectly matches the lower boundary of the cloud of points representing all the possible plate

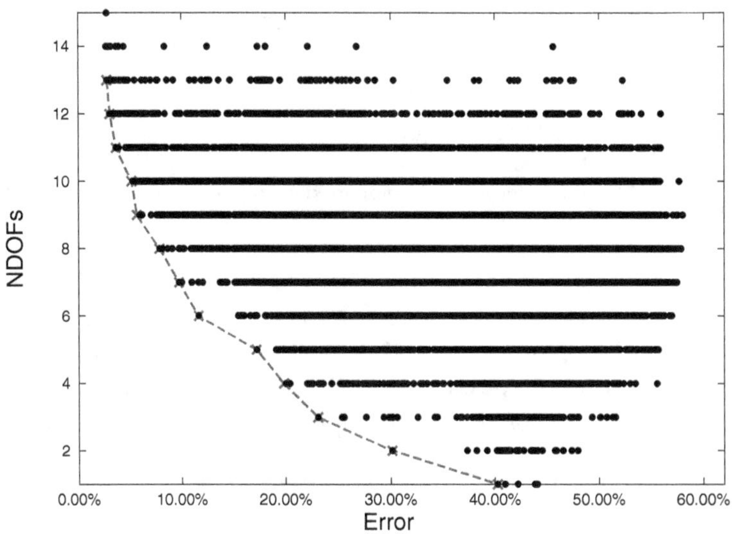

Fig. 4.12. BTD for the asymmetric plate, $a/h = 2.5$.

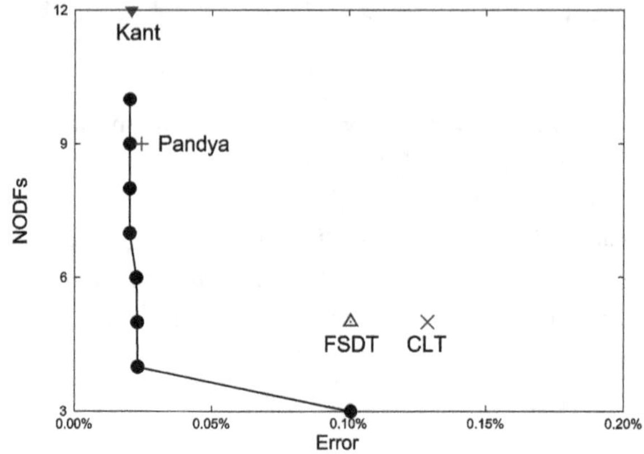

(a) $a/h = 50$

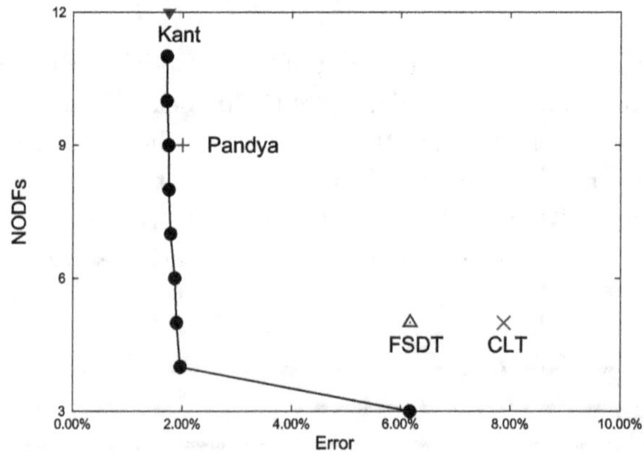

(b) $a/h = 5$

Fig. 4.13. BTDs for a simply-supported symmetric laminated plate.

models given by the ESL4 baseline. The ESL4 model BTDs for different plate geometries are reported in Figs. 4.13 and 4.14, for a symmetric and an asymmetric plate, respectively. Four theories from the literature are also reported in the same graphs, that is, the CLT and FSDT theories and the theories of Pandya [93] and Kant [94]. Pandya's model is given by

$$\begin{aligned} u &= u_0 + z\, u_1 + z^2\, u_2 + z^3\, u_3, \\ v &= v_0 + z\, v_1 + z^2\, v_2 + z^3\, v_3, \\ w &= w_0, \end{aligned} \tag{4.49}$$

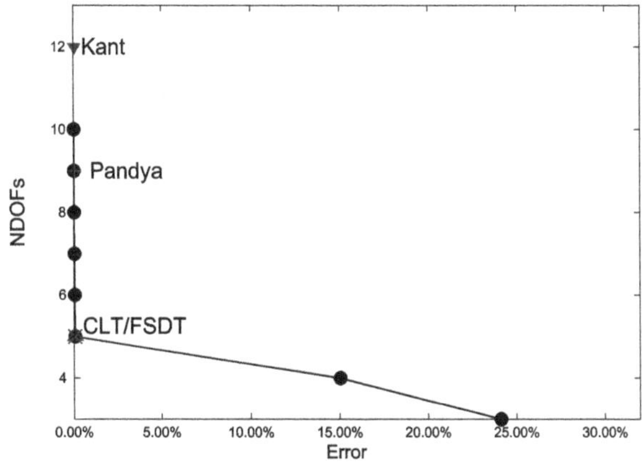

(a) $a/h = 50$

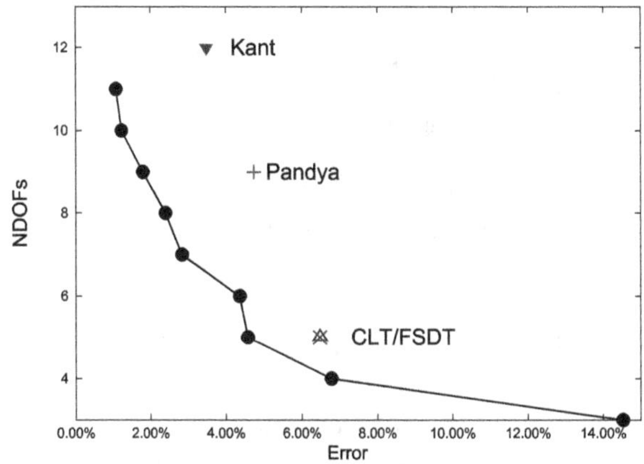

(b) $a/h = 5$

Fig. 4.14. BTDs for a simply-supported asymmetric laminated plate.

whereas, Kant's one is

$$\begin{aligned} u &= u_0 + z\,u_1 + z^2\,u_2 + z^3\,u_3, \\ v &= v_0 + z\,v_1 + z^2\,v_2 + z^3\,v_3, \\ w &= w_0 + z\,w_1 + z^2\,w_2 + z^3\,w_3. \end{aligned} \tag{4.50}$$

Table 4.6 shows the best plate models with seven degrees of freedom and their accuracy. For instance, the best model for the thin, symmetric plate is the following:

Table 4.6. Best plate models with seven unknown variables for the simply-supported plate.

	0°/90°/0°					0°/90°				
					$a/h = 50$					
	▲	▲	△	▲	△	▲	▲	▲	▲	△
	△	▲	△	△	△	▲	▲	△	△	△
	▲	▲	▲	△	△	▲	△	△	△	△
Error	2.0339×10^{-2} %					4.0673×10^{-2} %				
					$a/h = 5$					
	▲	▲	▲	▲	△	▲	▲	▲	△	▲
	△	▲	△	△	△	▲	▲	△	△	△
	▲	▲	△	△	△	▲	△	△	△	△
Error	1.7773 %					2.8403 %				

$$
\begin{aligned}
u &= u_0 + z\, u_1 + z^3\, u_3, \\
v &= z\, v_1, \\
w &= w_0 + z\, w_1 + z^2\, w_2.
\end{aligned}
\tag{4.51}
$$

The analysis of the results suggests that:

- In most cases, many variables of the ESL4 models can be discarded without accuracy penalties. In particular, some 10–12 out of 15 ESL4 variables are generally enough to obtain the best accuracy the ESL4 can provide.
- In the case of thick plates, classical models and those from literature do not belong to the BTD. In other words, their accuracy can be obtained with fewer degrees of freedom; or, conversely, there exist models based on the same number of degrees of freedom that are more accurate.
- For thin shells, ESL4 can be as accurate as LW4 for σ_{xx}. On the other hand, in the case of thick shells, the best accuracy of ESL4 is 98–99% of the LW4.
- The use of a genetic algorithm is an efficient and reliable approach to obtaining BTDs with reduced computational costs.

More results on the BTD for composite plates can be found in [53].

4.11.2. Shells

BTDs for laminated shells are presented in this section. The material properties are $E_L/E_T = 25$, $\nu = 0.25$, $G_{LT}/E_T = G_{TT}/E_T = 0.5$, $G_{Lz}/E_T = 0.2$ and the dimensions of the shell are $a = 4\, R_\beta$ and $b = 2\,\pi\, R_\beta$. The transverse pressure load (see Eq. (4.48)) is applied internally, m and

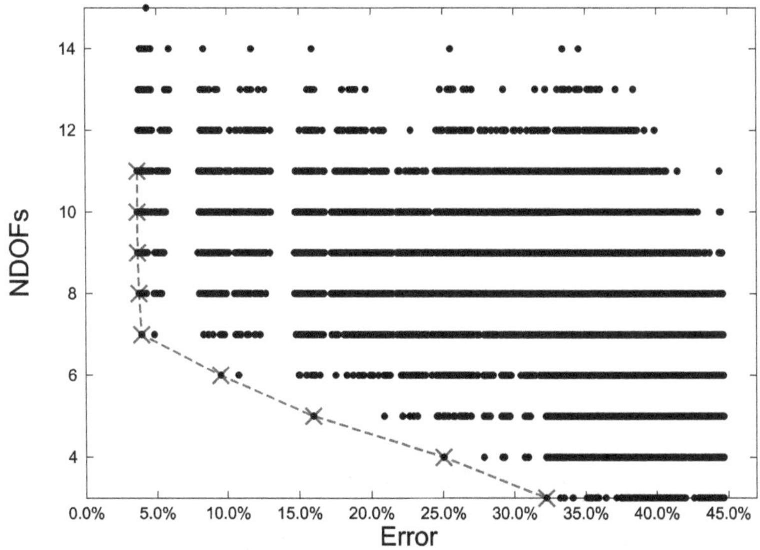

Fig. 4.15. BTD based on $\sigma_{\alpha\alpha}$ for the symmetric composite shell, $R_\beta/h = 4$.

n are equal to 1 and 4, respectively. BTDs were obtained using the ESL4 baseline, whereas the LW4 was used as reference solution.

A symmetric lamination was considered first, $90°/0°/90°$. Figure 4.15 shows the BTD for the $\sigma_{\alpha\alpha}$ in the thick shell case. The BTD was obtained by means of the genetic algorithm, and it perfectly matches the boundaries of the best accuracy models that were computed by considering all the combinations (2^{15}) of the ESL4. Figure 4.16 shows the BTDs that were obtained by considering u_z and $\sigma_{\alpha\alpha}$, thin and thick shells were investigated. The accuracy of CLT, FSDT, and Pandya models are compared with BTD in Fig. 4.17. Table 4.7 shows the accuracy and computational cost of the classical models and of the BTD models that have the same amount of DOFs.

A $0°/90°$ stacking sequence was considered as the second assessment for the composite shell case, and, again, LW4 was used as the reference solution. Figure 4.18 shows BTD for $\sigma_{\alpha\alpha}$.

Figure 4.19 present a comparison between the BTD for $\sigma_{\alpha\alpha}$ and the one for u_z. Table 4.8 shows the accuracy and computational cost of the classical models and of the BTD models that have the same amount of DOFs. The results suggest the following:

- Geometry, stacking sequence, and the displacement/stress variable influence the BTD to a great extent.

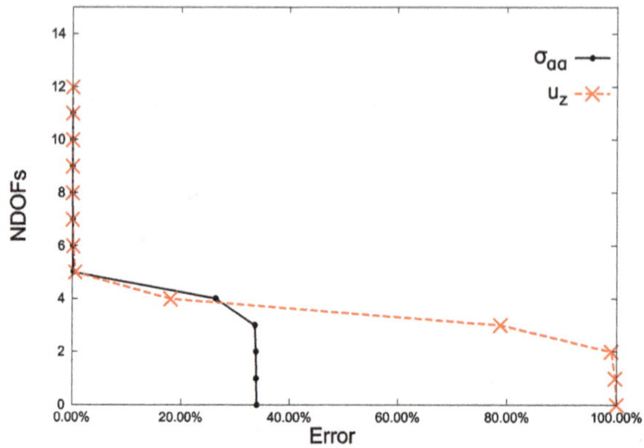

(a) $R_\beta/h = 100$

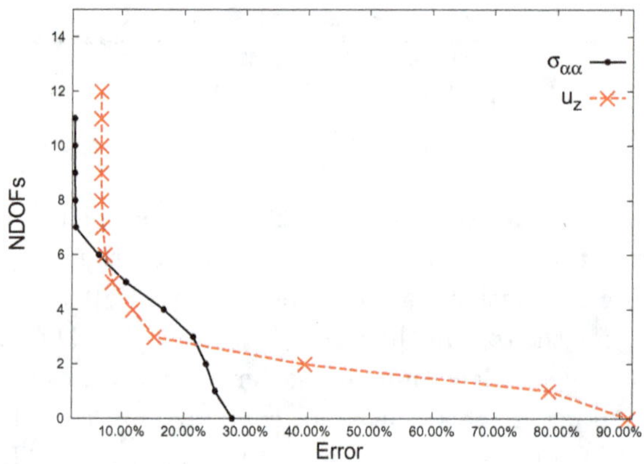

(b) $R_\beta/h = 4$

Fig. 4.16. BTDs for u_z and $\sigma_{\alpha\alpha}$, symmetric composite shell.

- Classical models are often quite distant from the BTD curve. This means that they can be improved both in accuracy and computational cost standpoint.
- The BTD can provide insightful guidelines for the development of refined models. For instance, the accuracy of the FSDT model in detecting $\sigma_{\alpha\alpha}$ for a thin, symmetric shell can be improved by considering a third-order term instead of a first-order term (i.e., $u_{\beta 4}$ instead of $u_{\beta 2}$).

More results on the BTD for composite shells can be found in [54].

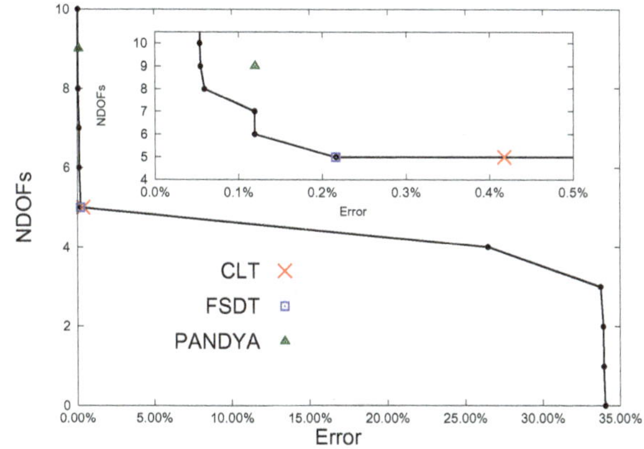

(a) $R_\beta/h = 100$

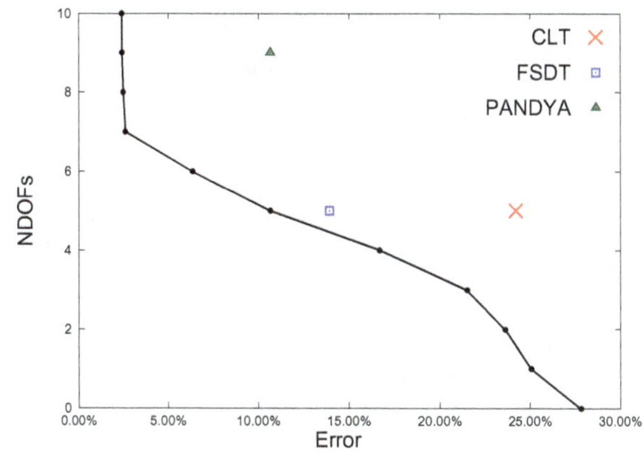

(b) $R_\beta/h = 4$

Fig. 4.17. Comparison among BTD and other theories for the symmetric composite shell, $\sigma_{\alpha\alpha}$.

4.12. Conclusions

In this chapter, it has been shown that there are three independent ways of introducing "zig-zag" theories for the analysis of multilayered plates and shells. In particular, it has been established that:

- Lekhnitskii [22] was the first to propose a theory for multilayered structures that describes the zig-zag form of a displacement field in the thickness direction and the interlaminar equilibrium of transverse stresses.

Table 4.7. Accuracy and computational cost of classical and BTD ESL models for the symmetric composite shell.

	$\sigma_{\alpha\alpha}$		u_z	
	$R_\beta/h = 100$, $M_e/M = 9/15$			
	Pandya	BTD	Pandya	BTD
	▲ ▲ ▲ ▲ △	▲ ▲ △ △ △	▲ ▲ ▲ ▲ △	▲ ▲ ▲ △ ▲
	▲ ▲ ▲ ▲ △	▲ ▲ △ ▲ △	▲ ▲ ▲ ▲ △	▲ ▲ △ ▲ △
	▲ △ △ △ △	▲ ▲ ▲ ▲ △	▲ △ △ △ △	▲ ▲ △ △ △
Error	0.1201 %	0.0547 %	0.2485 %	0.0771 %
	$R_\beta/h = 4$, $M_e/M = 9/15$			
	Pandya	BTD	Pandya	BTD
	▲ ▲ ▲ ▲ △	▲ ▲ △ △ △	▲ ▲ ▲ ▲ △	▲ ▲ △ △ △
	▲ ▲ ▲ ▲ △	▲ ▲ △ ▲ ▲	▲ ▲ ▲ ▲ △	▲ ▲ ▲ ▲ ▲
	▲ △ △ △ △	▲ ▲ ▲ △ △	▲ △ △ △ △	▲ ▲ △ △ △
Error	10.6406 %	2.4431 %	9.6671 %	6.7721 %
	$M_e/M = 5/15$			
	FSDT	BTD	FSDT	BTD
	▲ ▲ △ △ △	▲ ▲ △ △ △	▲ ▲ △ △ △	△ ▲ △ △ △
	▲ ▲ △ △ △	▲ △ △ ▲ △	▲ ▲ △ △ △	▲ △ △ ▲ △
	▲ △ △ △ △	▲ △ △ △ △	▲ △ △ △ △	▲ ▲ △ △ △
Error	13.9300 %	10.6558 %	27.6149 %	8.5433 %

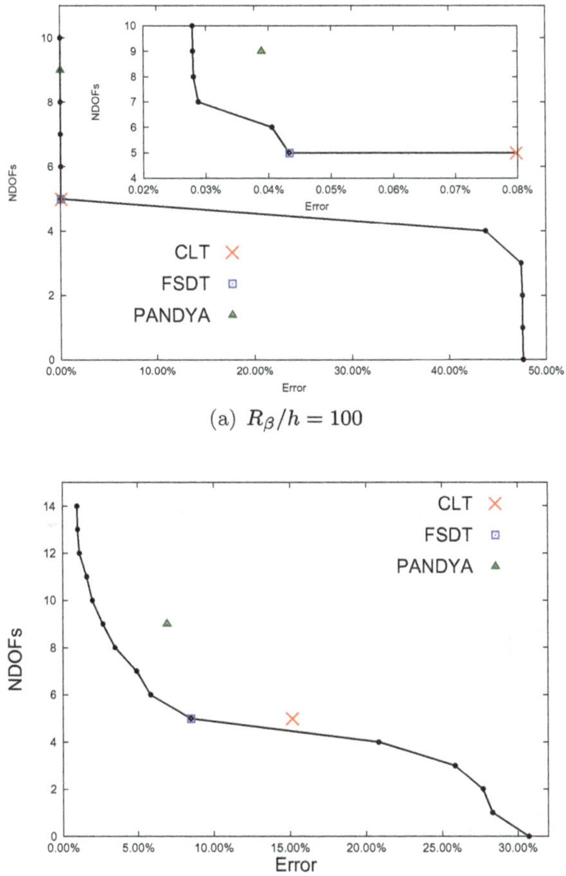

(a) $R_\beta/h = 100$

(b) $R_\beta/h = 4$

Fig. 4.18. BTD for the asymmetric composite shell, $\sigma_{\alpha\alpha}$.

- Three different and independent theories are proposed in the literature. Apart from the one by Lekhnitskii [22], the other two approaches were based on work by Ambartsumian [62, 73] and Reissner [66], respectively.
- Based on the authors' historical considerations, which are documented in this chapter, it is suggested that these three approaches are called the Lekhnitskii–Ren, Ambartsumian–Whitney–Rath–Das, and Reissner–Murakami–Carrera theories, respectively.
- As far as the Ambartsumian–Whitney–Rath–Das theory is concerned, it should be underlined that other developments, even though derived independently by other authors (such as those originated by Yu [95],

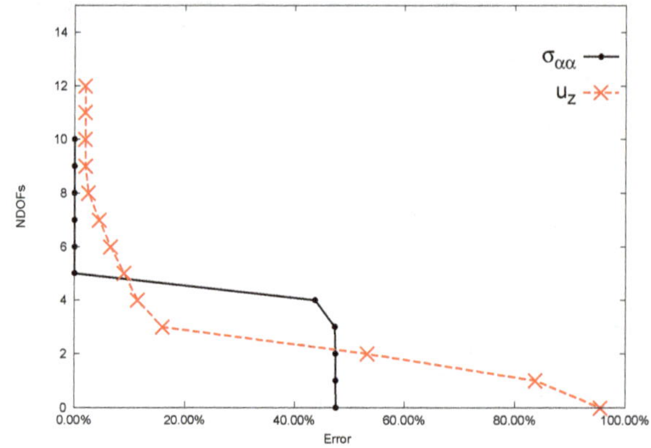

(a) $R_\beta/h = 100$

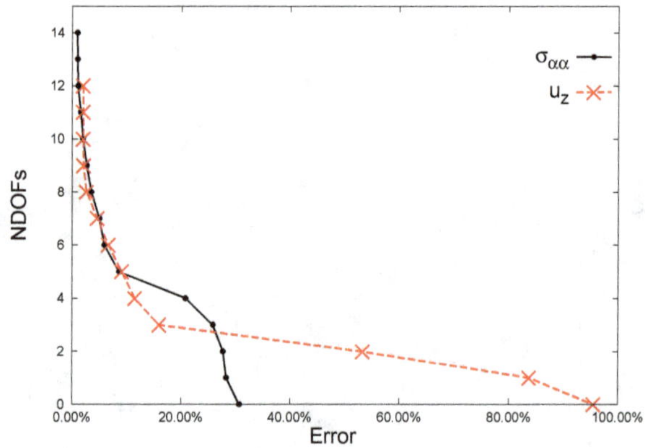

(b) $R_\beta/h = 4$

Fig. 4.19. BTD for the asymmetric composite shells, $\sigma_{\alpha\alpha}$ and u_z.

Chou and Carleone [96], Disciuva [97], Bhaskar and Varadan [98], Cho and Parmerter [99], among others), are applications of the AWRD theory.

Even though most of this discussion has been about the so-called ESLMs, these being more relevant for the subject of this chapter, a brief outline of LWMs was given in Section 4.7.

The author would encourage scientists who are working on the analysis of multilayer structures to return to the fundamental work by Lekhnitskii [22], Ambartsumian [62, 63], and Reissner [66]. There is, in fact, a significant amount to learn from these works and probably more

Table 4.8. Accuracy and computational cost of classical and BTD models for the asymmetric composite shell, $\sigma_{\alpha\alpha}$.

	$R_\beta/h = 100$									
	$M_e/M = 9/15$									
	Pandya					BTD				
	▲	▲	▲	▲	△	▲	▲	▲	△	▲
	▲	▲	▲	▲	△	▲	▲	△	△	△
	▲	△	△	△	△	▲	▲	▲	△	△
Error	0.0389 %					0.0279 %				
	$R_\beta/h = 4$									
	$M_e/M = 9/15$									
	Pandya					BTD				
	▲	▲	▲	▲	△	▲	▲	△	△	△
	▲	▲	▲	▲	△	▲	▲	▲	▲	△
	▲	△	△	△	△	▲	▲	▲	△	△
Error	6.9251 %					2.6671 %				
	$M_e/M = 5/15$									
	FSDT					BTD				
	▲	▲	△	△	△	▲	▲	△	△	△
	▲	▲	△	△	△	▲	▲	△	△	△
	▲	△	△	△	△	▲	△	△	△	△
Error	8.4968 %					8.4968 %				

could be done, on the basis of these fundamental works, to obtain a better understanding of the mechanics of multilayered structures. In particular, future developments could be to extend Lekhnitskii's theory as well as the Reissner theorem. This latter, in the authors' opinion, is the natural tool for the analysis of multilayered structures.

As a final remark, the authors are clearly aware that this historical review may be not complete. The authors are aware that other significant articles and papers could exist on this subject that have not been considered. However, what has been quoted in this chapter will help to assign the right credit concerning the contributions and contributors to multilayered theory.

The second part of this chapter discussed the development of a refined shell finite element approach based on the Carrera Unified Formulation. The CUF has been coupled to the MITC method to overcome the locking phenomenon that affects finite element analysis. The reliability of the approach has been tested by considering classical discriminating problems, such as the pinched cylinder studied in [88] and the Scordelis–Lo problem analyzed in [89], and the approach has shown good convergence and robustness on growing the mesh size. Moreover, the accuracy of

Table 4.9. Available theories for laminated structures.

Theory	σ_{nM}	ZZ	IC	ϵ_{zz}	σ_{zz}
Classical [1–6]					
CUF-ESL [9, 27]				•	•
L [22]		•	•	•	
LR [56–58]		•	•		
AWRD [59–64]		•	•		
RMC [65–71]	•	•	•	•	•
CUF-LW-D [78]		•		•	•
CUF-LW-M [72–78]	•	•	•	•	•

the solution has been demonstrated to improve by increasing the order of expansion of the displacements in the thickness direction. Finally, the orthotropic multilayered cylinder studied by Varadan and Bhaskar in [92] was considered. From this analysis one can conclude that for the study of multilayered structures it is mandatory to consider zig-zag effects in the displacements in order to obtain the 3D solution. This is possible by introducing Murakami's zig-zag function in the ESL models or by using the LW models briefly discussed in this chapter, which allow us to use independent variables in each layer. This gives the best results.

For clarity, Table 4.9 summarizes the features of the theories cited in this chapter for the analysis of laminated structures:

- Classical = classical models such as Kirchoff–Love, Reissner–Mindlin, and so on;
- CUF-ESL = equivalent single-layer theories contained in the CUF, in which a high order of expansion in the thickness direction is used for both the in-plane and ransversal displacements;
- L = Lekhnitskii theory;
- LR = Lekhnitskii–Ren theory;
- AWRD = Ambartsumian–Whitney–Rath–Das theory;
- RMC = Reissner–Murakami–Carrera theory, based on ESL approach for displacement variables;
- CUF-LW-D = layer-wise models contained in the CUF, based on the principle of virtual displacements (PVD);
- CUF-LW-M = layer-wise models contained in the CUF, based on the Reissner mixed variational theorem (RMVT).

The features, considered in the table, are as follows:

- σ_{nM} = the transverse shear and normal stresses are unknown variables with the displacements;
- ZZ = zig-zag effects are considered in the displacements;
- IC = interlaminar continuity of the transverse stresses is fulfilled;
- ϵ_{zz} = thickness stretching effects are considered, $\epsilon_{zz} \neq 0$;
- σ_{zz} = Koiter's recommendation is fulfilled, $\sigma_{zz} \neq 0$.

The symbol • indicates that the theory satisfies the corresponding feature. The last part of this chapter has presented some results related to the Axiomatic/Asymptotic Method (AAM). The AAM evaluates the influence of each unknown variable of a structural theory on the solution of a given problem. In particular, the AAM can be used to build the Best Theory Diagram. A structural model belonging to the BTD is the one that provides the best accuracy for a given number of unknown variables. Results on ESL plates and shells have been presented and proved that

- In most cases, the computational cost of refined models can be reduced significantly without undermining their accuracy.
- Often, classical models and refined models from literature do not belong to the BTD. In other words, there exist models that can provide better accuracies with the same computational costs and models as accurate but computationally less expensive.
- The construction of BTD is problem dependent. In fact, many characteristic parameters, such as the thickness or the stacking sequence, can modify the BTD significantly. Furthermore, the output control variable adopted to build the BTD influences it as well.
- The BTD can be easily built by coupling the CUF, the AAM, and genetic algorithms.
- The systematic use of BTD can be seen as a tool to evaluate the accuracy of any given structural model. In other words, the BTD can be a powerful method to develop structural models that, for a given number of variables, can provide the best accuracy.

References

[1] Cauchy A.L., 1828. Sur l'équilibre et le mouvement d'une plaque solide, *Exercises Mat.*, 3, 328–355.

[2] Poisson S.D., 1829. Mémoire sur l'équilibre et le mouvement des corps élastique, *Mem. Acad. Sci.*, 8, 357.

[3] Kirchhoff G., 1850. Über das Gleichgewicht und die Bewegung einer elastishen Scheibe, *J. Angew. Math.*, 40, 51–88.

[4] Love A.E.H., 1927. *The Mathematical Theory of Elasticity*, 4th edn., Cambridge University Press, Cambridge.

[5] Reissner E., 1945. The effect of transverse shear deformation on the bending of elastic plates, *J. Appl. Mech.*, 12, 69–76.

[6] Mindlin R.D., 1951. Influence of rotatory inertia and shear in flexural motions of isotropic elastic plates, *J. Appl. Mech.*, 18, 1031–1036.

[7] Hildebrand F.B., Reissner E., Thomas G.B., 1938. Notes on the foundations of the theory of small displacements of orthotropic shells. NACA TN-1833, Washington, DC.

[8] Pagano N.J., 1969. Exact solutions for composite laminates in cylindrical bending, *J. Compos. Mater.*, 3, 398–411.

[9] Carrera E., 1995. A class of two dimensional theories for multilayered plates analysis, *Atti Accademia delle Scienze di Torino, Mem Sci. Fis.*, 19–20, 49–87.

[10] Ambartsumian S.A., 1962. Contributions to the theory of anisotropic layered shells, *Appl. Mech. Rev.*, 15, 245–249.

[11] Librescu L., Reddy J.N., 1986. A critical review and generalization of transverse shear deformable anisotropic plates, in *Euromech Colloquium 219, Refined Dynamical Theories of Beams, Plates and Shells and Their Applications*, Kassel, September, pp. 32–43.

[12] Grigolyuk E.I., Kulikov G.M., 1988. General directions of the development of theory of shells, *Mekhanica Kompozitnykh Materialov*, 24, 287–298.

[13] Kapania R.K., Raciti S., 1989. Recent advances in analysis of laminated beams and plates, *AIAA J.*, 27, 923–946.

[14] Kapania R.K., 1989. A review on the analysis of laminated shells, *J. Press. Vess. Tech.*, 111, 88–96.

[15] Noor A.K., Burton W.S., 1989. Assessment of shear deformation theories for multilayered composite plates, *Appl. Mech. Rev.*, 41, 1–18.

[16] Noor A.K., Burton W.S., 1990. Assessment of computational models for multilayered composite shells, *Appl. Mech. Rev.*, 43, 67–97.

[17] Noor A.K., Burton W.S., Bert C.W., 1996. Computational model for sandwich panels and shells, *Appl. Mech. Rev.*, 49, 155–199.

[18] Reddy J.N., Robbins D.H., 1994. Theories and computational models for composite laminates, *Appl. Mech. Rev.*, 47, 147–165.

[19] Carrera E., 2001. Developments, ideas and evaluations based upon the Reissner's Mixed Theorem in the modeling of multilayered plates and shells, *Appl. Mech. Rev.*, 54(4), 301–329.

[20] Librescu L., 1975. *Elasto-Statics and Kinetics of Anisotropic and Heterogeneous Shell-type Structures*, Noordhoff Int., Leyden, Netherlands.

[21] Reddy J.N., 1997. *Mechanics of Laminated Composite Plates, Theory and Analysis*, CRC Press, Boca Raton.

[22] Lekhnitskii S.G., 1935. Strength calculation of composite beams, *Vestn. Inzh. Tekh.*, 9.

[23] Carrera E., 2003. Historical review of zig-zag theories for multilayered plates and shells, *Appl. Mech. Rev.*, 56(3), 287–308.
[24] Fettahlioglu O.A., Steele C.R., 1974. Asymptotic solutions for orthotropic non-homogeneous shells of revolution, *J. Appl. Mech.*, 41(3), 753–758.
[25] Hodges D.H., Lee B.W., Atilgan A.R., 1993. Application of the variational-asymptotic method to laminated composite plates, *AIAA J.*, 31, 1674–1983.
[26] Satyrin V.G., Hodges D.H., 1996. On asymptotically correct linear laminated plate theory, *Int. J. Solids Struct.*, 33, 3649–3671.
[27] Carrera E., 2003. Theories and finite elements for multilayered plates and shells: a unified compact formulation with numerical assessment and benchmarking, *Arch. Comput. Meth. Eng.*, 97, 10–215.
[28] Carrera E., 1999. Transverse normal stress effects in multilayered plates, *J. Appl. Mech.*, 66, 1004–1012.
[29] Koiter W.T., 1970. On the foundations of the linear theory of thin elastic shell, *Proc. Kon. Nederl. Akad. Wetensch.*, 73, 169–195.
[30] Naghdi P.M., 1972. The theory of shells and plates, *Handbuch Phys.*, 6, 425–640.
[31] Hakula H., Leino Y., Pitkäranta J., 1996. Scale resolution, locking, and higher-order finite element modelling of shells, *Comp. Meth. Appl. Mech. Eng.*, 133, 155–182.
[32] Chinosi C., Della Croce L., Scapolla T., 1998. Hierarchic finite elements for thin Naghdi shell model, *Int. J. Solids Struct.*, 35, 1863–1880.
[33] Zienkiewicz O.C., Taylor R.L., Too J.M., 1971. Reduced integration techniques in general analysis of plates and shells, *Int. J. Num. Meth. Eng.*, 3, 275–290.
[34] Stolarski H., Belytschko T., 1981. Reduced integration for shallow-shell facet elements, in *New Concepts in Finite Element Analysis* (T.J.R. Hughes, *et al.*, eds), ASME, New York, pp. 179–194.
[35] Bathe K.-J., Dvorkin E., 1986. A formulation of general shell elements the use of mixed interpolation of tensorial components, *Int. J. Num. Meth. Eng.*, 22, 697–722.
[36] Chinosi C., Della Croce L., 1998. Mixed-interpolated elements for thin shell, *Comm. Num. Meth. Eng.*, 14, 1155–1170.
[37] Huang H.-C., 1987. Membrane locking and assumed strain shell elements, *Comp. Struct.*, 27(5), 671–677.
[38] Carrera E., Cinefra M., 2014. Classical, refined, zig-zag and layer-wise models for laminated structures, in *Computational and Experimental Methods in Structures: Volume 5 — Mathematical Methods and Models in Composites* (V. Mantič, ed.), Imperial College Press, pp. 135–172.
[39] Carrera E., Petrolo M., 2010. Guidelines and recommendation to contruct theories for metallic and composite plates, *AIAA J.*, 48(12), 2852–2866.
[40] Carrera E., Petrolo M., 2011. On the effectiveness of higher-order terms in refined beam theories, *J. Appl. Mech.*, 78.
[41] Carrera E., Giunta G., Petrolo M., 2011. *Beam Structures: Classical and Advanced Theories*, John Wiley & Sons.

[42] Carrera E., Cinefra M., Petrolo M., Zappino E., 2014. *Finite Element Analysis of Structures through Unified Formulation*, John Wiley & Sons.

[43] Carrera E., Miglioretti F., Petrolo M., 2011. Computations and evaluations of higher-order theories for free vibration analysis of beams, *J. Sound Vib.*, 331, 4269–4284.

[44] Carrera E., Miglioretti F., Petrolo M., 2012. Accuracy of refined finite elements for laminated plate analysis, *Comp. Struct.*, 93(5), 1311–1327.

[45] Carrera E., Miglioretti F., Petrolo M., 2011. Guidelines and recommendations on the use of higher order finite elements for bending analysis of plates, *Int. J. Comput. Meth. Eng. Sci. Mech.*, 12(6), 303–324.

[46] Mashat D.S., Carrera E., Zenkour A.M., Al Khateeb S.A., 2013. Axiomatic/asymptotic evaluation of multilayered plate theories by using single and multi-points error criteria, *Comp. Struct.*, 106, 393—406.

[47] Petrolo M., Cinefra M., Lamberti A., Carrera E., 2015. Evaluation of mixed theories for laminated plates through the axiomatic/asymptotic method, *Comp. B: Eng.*, 76, 260–272.

[48] Cinefra M., Lamberti A., Zenkour A.M., Carrera E., 2015. Axiomatic/ asymptotic technique applied to refined theories for piezoelectric plates, *Mech. Adv. Mat. Struct.*, 22(1–2), 107–124.

[49] Petrolo M., Lamberti A., 2016. Asymptotic/axiomatic analysis of refined layer-wise theories for composite and sandwich plates, *Mech. Adv. Mat. Struct.*, 23(1), 28–42.

[50] Mashat D.S., Carrera E., Zenkour A.M., Al Khateeb S.A., 2014. Use of axiomatic/asymptotic approach to evaluate various refined theories for sandwich shells, *Comp. Struct.*, 109, 139–149.

[51] Mashat D.S., Carrera E., Zenkour A.M., Al Khateeb S.A., Lanberti A., 2014. Evaluation of refined theories for multilayered shells via axiomatic/ asymptotic method, *J. Mech. Sci. Technol.*, 28(11), 4663–4672.

[52] Carrera E., Miglioretti F., 2012. Selection of appropriate multilayered plate theories by using a genetic like algorithm, *Comp. Struct.*, 94(3), 1175–1186.

[53] Carrera E., Cinefra M., Petrolo M., 2011. A best theory diagram for metallic and laminated shells, in *Shell-like Structures* (H. Altenbach, V.A. Eremeyev, eds.), Springer, Berlin, pp. 681–698.

[54] Carrera E., Cinefra M., Lamberti A., Petrolo M., 2015. Results on best theories for metallic and laminated shells including layer-wise models, *Comp. Struct.*, 126, 285–298.

[55] Petrolo M., Lamberti A., Miglioretti F., 2016. Best theory diagram for metallic and laminated composite plates, *Mech. Adv. Mater. Struct.*, 23(9), 1114–1130.

[56] Lekhnitskii S.G., 1968. *Anisotropic Plates*, translated from the 2nd Russian edn by S.W. Tsai and T. Cheron, Bordon and Breach, Science Publishers, New York.

[57] Ren J.G., 1986. A new theory of laminated plates, *Compos. Sci. Tech.*, 26, 225–239.

[58] Ren J.G., 1986. Bending theory of laminated plates, *Compos. Sci. Tech.*, 27, 225–248.

[59] Ren J.G., D.R. Owen J., 1989. Vibration and buckling of laminated plates, *Int. J. Solids Struct.*, 25, 95–106.

[60] Ambartsumian S.A., 1958. On a theory of bending of anisotropic plates, *Investiia Akad. Nauk. SSSR, Ot. Tekh. Nauk.*, No. 4.

[61] Ambartsumian S.A., 1958. On a general theory of anisotropic shells, *Prikl. Mat. Mekh.*, 22(2), 226–237.

[62] Ambartsumian S.A., 1961. Theory of anisotropic shells, *Fizmatgiz. Moskwa*, translated from Russian, NASA TTF–118.

[63] Ambartsumian S.A., 1969. *Theory of Anisotropic Plates*, translated from Russian by T. Cheron and edited by J.E. Ashton, Technomic Pub. Co.

[64] Whitney J.M., 1969. The effects of transverse shear deformation on the bending of laminated plates, *J. Compos. Mater.*, 3, 534–547.

[65] Rath B.K., Das Y.C., 1973. Vibration of layered shells, *J. Sound Vib.*, 28, 737–757.

[66] Reissner E., 1984. On a certain mixed variational theory and a proposed application, *Int. J. Num. Meth. Eng.*, 20, 1366–1368.

[67] Reissner E., 1986. On a mixed variational theorem and on a shear deformable plate theory, *Int. J. Num. Meth. Eng.*, 23, 193–198.

[68] Murakami H., 1985. Laminated composite plate theory with improved in-plane responses, *ASME Proc. PVP Conf.*, New Orleans, June 24–26, 98(2), 257–263.

[69] Murakami H., 1986. Laminated composite plate theory with improved in-plane response, *J. Appl. Mech.*, 53, 661–666.

[70] Carrera E., 1999. A study of transverse normal stress effects on vibration of multilayered plates and shells, *J. Sound Vib.*, 225, 803–829.

[71] Toledano A., Murakami H., 1987. A high-order laminated plate theory with improved in-plane responses, *Int. J. Sol. Struct.*, 23, 111–131.

[72] Carrera E., 1997. An improved Reissner–Mindlin-Type model for the electromechanical analysis of multilayered plates including piezo-layers, *J. Intel. Mat. Syst. Struct.*, 8, 232–248.

[73] Carrera E., 1998. Evaluation of layer-wise mixed theories for laminated plates analysis, *AIAA J.*, 36, 830–839.

[74] Carrera E., 1998. Layer-wise mixed models for accurate vibration analysis of multilayered plates, *J. Appl. Mech.*, 65, 820–828.

[75] Carrera E., 1999. A Reissner's mixed variational theorem applied to vibrational analysis of multilayered shell, *J. Appl. Mech.*, 66(1), 69–78.

[76] Carrera E., 1999. Transverse normal stress effects in multilayered plates, *J. Appl. Mech.*, 66, 1004–1012.

[77] Carrera E., 2000. Single-layer vs multi-layers plate modelings on the basis of Reissner's mixed theorem, *AIAA J.*, 38, 342–343.

[78] Carrera E., 2000. *A priori* vs. *a posteriori* evaluation of transverse stresses in multilayered orthotropic plates, *Compos. Struct.*, 48, 245–260.

[79] Bhaskar K., Varadan T.K., 1992. Reissner's new mixed variational principle applied to laminated cylindrical shells, *J. Press. Vessel Tech.*, 114, 115–119.

[80] Jing H., Tzeng K.G., 1993. Refined shear deformation theory of laminated shells, *AIAA J.*, 31, 765–773.

[81] Rao K.M., Meyer-Piening H.R., 1990. Analysis of thick laminated anisotropic composites plates by the finite element method, *Comp. Struct.*, 15, 185–213.
[82] Carrera E., 1996. C^0 Reissner–Mindlin multilayered plate elements including zigzag and interlaminar stresses continuity, *Int. J. Num. Meth. Eng.*, 39, 1797–1820.
[83] Carrera E., Kröplin B., 1997. Zig-zag and interlaminar equilibria effects in large deflection and postbuckling analysis of multilayered plates, *Mech. Comp. Mat. Struct.*, 4, 69–94.
[84] Bhaskar K., Varadan T.K., 1991. A higher-order theory for bending analysis of laminated shells of revolution, *Comp. Struct.*, 40, 815–819.
[85] Brank B., Carrera E., 2000. Multilayered shell finite element with interlaminar continuous shear stresses: a refinement of the Reissner–Mindlin formulation, *Int. J. Num. Meth. Eng.*, 48, 843–874.
[86] Cho Y.B., Averill R.C., 2000. First order zig-zag sub-laminate plate theory and finite element model for laminated composite and sandwich panels, *Comp. Struct.*, 50, 1–15.
[87] Chapelle D., Bathe K.-J., 2003. *The Finite Element Analysis of Shells. Fundamentals*, Springer, Berlin.
[88] Flügge W., 1960. *Stresses in Shells*, 2nd edn., Springer, Berlin.
[89] Scordelis A., Lo K.S., 1964. Computer analysis in cylindrical shells, *J. Amer. Concr. Inst.*, 61, 561–593.
[90] Carrera E., Brischetto S., 2008. Analysis of thickness locking in classical, refined and mixed theories for layered shells, *Comp. Struct.*, 85(1), 83–90.
[91] McNeal R.H., Harder R.L., 1985. A proposed standard set of problems to test finite element accuracy, *Fin. Elem. Anal. Des.*, 1, 3–20.
[92] Varadan T.K., Bhaskar K., 1991. Bending of laminated orthotropic cylindrical shells — an elasticity approach, *Comp. Struct.*, 17, 141–156.
[93] Pandya B., Kant T., 1988. Finite element analysis of laminated composite plates using high-order displacement model, *Comp. Sci. Technol*, 32, 137–55.
[94] Kant T., Manjunatha B., 1988. An unsymmetric FRC laminate C^0 finite element model with 12 degrees of freedom per node, *Eng. Comput.*, 5(3), 292–308.
[95] Yu Y.Y., 1959. A new theory of elastic sandwich plates one dimensional case, *J. Appl. Mech.*, 37, 1031–1036.
[96] Chou P.C., Carleone J., 1973. Transverse shear in laminated plates theories, *AIAA J.*, 11, 1333–1336.
[97] Disciuva M., 1984. A refinement of the transverse shear deformation theory for multilayered plates, *Aerotecnica Missili e Spazio*, 63, 84–92.
[98] Bhaskar B., Varadan T.K., 1989. Refinement of higher-order laminated plate theories, *AIAA J.*, 27, 1830–1831.
[99] Cho M., Parmerter R.R., 1993. Efficient higher order composite plate theory for general lamination configurations, *AIAA J.*, 31, 1299–1305.

https://doi.org/10.1142/9781800611887_0005

Chapter 5

A Modeling Framework for the Analysis of Instabilities and Delamination in Composite Shells

José Reinoso[*,‡], **Marco Paggi**[†,§], **and Antonio Blázquez**[*,¶]
[*]*Escuela Técnica Superior de Ingeniería, Universidad de Sevilla, Camino de los Descubrimientos s/n, 41092, Sevilla, Spain*
[†]*IMT School for Advanced Studies Lucca, Piazza San Francesco 19, 55100 Lucca, Italy*
[‡]*jreinoso@us.es*
[§]*marco.paggi@imtlucca.it*
[¶]*abg@us.es*

Abstract

The onset and development of instabilities and failure in thin-walled composite structures can limit their incorporation in different engineering systems. In this chapter, a robust modeling framework to account for such events is presented. On the shell theory side, a thorough revision with regard to different continuum-based shell parametrizations, and the corresponding extension for composite structures is performed. With reference to delamination modeling, a nonlinear cohesive interface element which incorporated geometric and material nonlinear effects is herein considered. A consistent variational formulation and the corresponding finite element discretization are accordingly derived. Both numerical strategies are combined for the nonlinear analysis of different composite structures. The results obtained are compared with experimental data and previous semi-analytical investigations, demonstrating the reliability of the proposed methodology for the analysis of complex thin-walled composite applications involving instabilities.

Contents

5.1. Introduction

5.1.1. Review of shell formulations

Thin-walled structures (shells) are abundantly found in nature and in a wide range of engineering systems. From the practical standpoint, they have been used for the production of fuselage components in aircraft and wing designs, car body constructions, bioengineering systems, recent stretchable electronic devices, among many other applications.

The thorough analysis of shells frequently presents several challenging tasks. The main difficulties arise from their curved and slender character. The curvature effect is utilized to carry transverse external actions in an optimal manner by in-plane actions. Due to their intrinsic peculiarities, the development of theories and models for shells has attracted the attention of many authors over the last two centuries aiming at exploiting their excellent load-deflection performances. In this context, names like Euler, Bernoulli, Kirchhoff, Timoshenko, Love, Reissner, Mindlin, Naghdi are worth to be mentioned due to their notable contributions in the topic. In the last decades, one of the driving forces promoting the advance in the structural analysis of shells is the development of computer-based discretization methods, where the finite element method (FEM) is certainly predominant.

The basic ingredient for the development of shell theories regards the introduction of mechanical assumptions in order to simplify the complex three-dimensional behavior to condensed states. This procedure is usually denominated as *dimensional reduction concept* [1, 2], where the static (stress) and kinematic-derived (strains) quantities are usually referred to the midsurface of the body. Inspired by this idea, the so-called classical shell models relying on the Kirchhoff–Love and Reissner–Mindlin theories were proposed.

However, in a general sense, shells models can be classified following different criteria [2]. One possible categorization can be established according to the number of independent kinematic parameters involved, which are directly associated with the mechanical assumption across the thickness. In this context, the so-called Kirchhoff–Love shell theory only considers two in-plane displacements and the transverse deflection as independent kinematic parameters, and is referred as the 3-parameter (3-p) model. Based on its kinematic hypothesis, the normal direction is coupled with the deformation of the shell midsurface, and therefore the director vector remaining straight and normal to the shell midsurface along the deformation process. The consistency of the 3-p model is preserved through the modification of the constitutive law equations assuming a plane stress state. Note also that the neglecting the transverse shear strains limits the use of this formulation to thin shells. Moreover, one of the main difficulties for their computational implementation into FE codes is the requirement of C^1-continuity order of the shape functions for the interpolation of the geometry and the kinematic field. This latter condition can be overcome using recent higher order continuity formulations based on isogeometric procedures [3], see in particular the contributions by Kiendl *et al.* [4] and Bischoff and coworkers [5, 6].

The consideration of the transverse shear effects leads to the Reissner–Mindlin shell models, where five independent kinematic parameters are defined, i.e., three displacements and two rotations. This formulation is sometimes denominated as First-Order Shear Deformation Theory (FOSDT) in the related literature, which attains independent rotation of the shell director vector with respect to the midsurface. In comparison to the 3-p models, 5-p approaches reduce the error with respect to three-dimensional body, becoming well suited for the analysis of moderately thick shells and dynamic problems.

Standard formulations relying on the previous 3-p and 5-p models and their corresponding FE formulations have been extensively used in applications involving small strains. Nevertheless, the inextensibility of the shell director vector in both theories precludes again the use of unmodified three-dimensional constitutive formulations.

Focusing the attention on laminated shells, a straightforward manner to present the different models is the *Carrera's unified formulation* (CUF) [7–9], which considers a basic kernel for the development of different models. Assuming the previously referred hypothesis concerning the kinematics (3-p or 5-p models) through the thickness, classical so-called *equivalent single layer* (ESL) approaches can be developed, see Chapter 5 in [10] and [11] and the references therein given. ESL approaches, whereby a weighted average of the physical properties of each ply over the thickness direction is recalled [11, 12], are especially suitable for laminates with similar mechanical properties for adjacent layers. This is the case of many Carbon Fibre Reinforced Polymeric (CFRP) composite laminates, for which such formulations provide sufficient accurate global results at laminate level for thin and moderately thick laminates.

Holistic ESL models (which are based on the assumed C^n_z-continuity, i.e., C^n-continuity along the thickness direction) do not generally fulfill equilibrium conditions at the interfaces between layers, also referred as *interlaminar continuity of transverse stresses* (IC) [7]. More reliable estimations of such stress components can be obtained via *a posterior* equilibrium considerations, though notable deviations can arise for thick laminates [10]. IC of the transverse stress components together with the material discontinuity along the interfaces originate the well-known zig-zag (ZZ) effect. In this concern, a possible method to approximate such effect into ESL models is the *Murakami's zig-zag function* (MZZF) [13], which has been comprehensively assessed in [8, 9, 14].

However, the aforementioned 3-p and 5-p models turned out to be rather inaccurate in cases where finite strains might lead to considerable thickness variation are involved. In these situations, the effects of transverse shear and normal strains are not negligible any longer, playing a pivotal role in composite structures, where interlaminar stresses undoubtedly contribute to the initiation and growth of delamination events. The inaccuracy of such formulations have been revealed in estimating the 3D stress field using

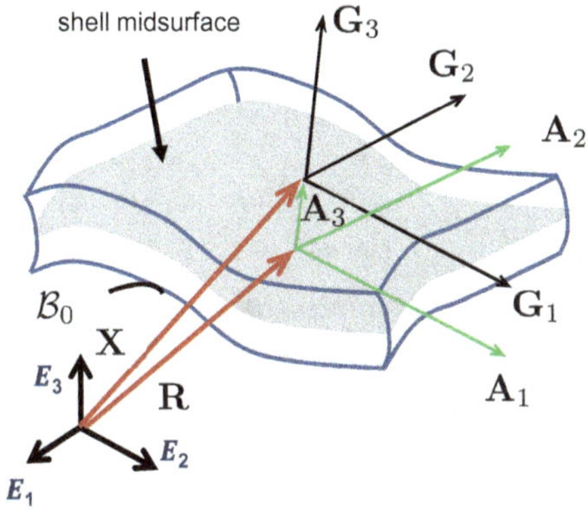

Fig. 5.1. Parametrization of the shell body in the reference configuration.

Classical Laminate Theory (CLT) adopting the 3-p kinematics, FOSDT, and Higher Order Shear Deformation Theories (HOSDT) [11].

A potential alternative to overcome these limitations with a mathematical root is the interpretation of the shell theory as a certain approximation of the three-dimensional equations. This approach is based on the seminal works performed by Ericksen and Truesdell [15] and Nagdhi [16] and can be interpreted as a step back into the 3D continuum, allowing the use of constitutive laws without any additional manipulation. The point of departure concerns with the parametrization of the three-dimensional shell body by the curvilinear coordinates $\boldsymbol{\xi} = \{\xi^1, \xi^2, \xi^3\}$, where $A(\xi^1, \xi^2, \xi^3) = A(\boldsymbol{\xi}) = M \times F \subset \mathbb{R}^3$ is the manifold of the body: $M(\xi^1, \xi^2) \subset \mathbb{R}^2$ denoting the parameter space of the shell reference surface and $F(\xi^3) \subset \mathbb{R}$ standing for the thickness space (Fig. 5.1). The previous parametrization is defined within the bi-unit cube $\square = [-1, 1] \times [-1, 1] \times [-1, 1]$. In this setting, the representation of any arbitrary point in the reference, $\mathbf{X}$, and current, $\mathbf{x}$, configurations[a] can be expressed as an infinite sum of functions of the in-plane coordinates expanded in thickness direction:

$$\mathbf{X}(\boldsymbol{\xi}) = \sum_{N=0}^{\infty} \left(\xi^3\right)^N \mathbf{R}^{(N)}(\xi^1, \xi^2), \tag{5.1}$$

$$\mathbf{x}(\boldsymbol{\xi}) = \sum_{N=0}^{\infty} \left(\xi^3\right)^N \mathbf{r}^{(N)}(\xi^1, \xi^2). \tag{5.2}$$

[a]In the subsequent developments, the capital letters refer to the reference configuration of the shell, whereas the small letters are associated with the current configuration. The Latin indices range from 1 to 3, whereas the Greek indices vary from 1 to 2.

In Eqs. (5.1) and (5.2), the terms $\mathbf{R}^{(0)}$, $\mathbf{r}^{(0)}$, corresponding to index $N = 0$ and respectively denoted as $\mathbf{R}$ and $\mathbf{r}$ in the sequel, identify the position vectors of the shell midsurface in the reference and current configurations; $\mathbf{R}^{(N)}$, $\mathbf{r}^{(N)}$ (index $N \geq 1$) denote the directors vectors into the three-dimensional shell space for the reference and current configurations, respectively; and N in $\left(\xi^3\right)^N$ indicates an exponent, whereas in $\mathbf{r}^{(N)}$ and $\mathbf{R}^{(N)}$ stands for a superscript.

The displacement field for a particular shell model, whose definition reads $\mathbf{u}(\boldsymbol{\xi}) := \mathbf{x}(\boldsymbol{\xi}) - \mathbf{X}(\boldsymbol{\xi})$, can be expressed in terms of the thickness coordinate as

$$\mathbf{u}(\boldsymbol{\xi}) = \sum_{N=0}^{\infty} \left(\xi^3\right)^N \mathbf{v}^{(N)}(\xi^1, \xi^2). \tag{5.3}$$

The approximation of the displacement field given in Eq. (5.3) enables choosing a different interpolation order for the in-plane and transverse displacements. Moreover, the use of additional kinematic restrictions with regard to the inextensibility of the shell direction vector, $\mathbf{R}^{(1)}$ (also identified by $\mathbf{D}$) can lead to Kirchhoff–Love and Reissner–Mindlin-type models [2].

Restricting the series expansion up to the first-order term in ξ^3, the 6-parameter formulation renders a linearly varying displacement field along the thickness direction:

$$\mathbf{u}(\boldsymbol{\xi}) = \mathbf{v}(\xi^1, \xi^2) + \frac{H}{2}\xi^3 \mathbf{w}(\xi^1, \xi^2), \tag{5.4}$$

where $\mathbf{v}$ and $\mathbf{w}$ respectively identify $\mathbf{v}^{(0)}$ and $\mathbf{v}^{(1)}$ according to the series expansion given in Eq. (5.3). In Eq. (5.4), $\mathbf{v}$ and $\mathbf{w}$ stand for the displacement field of the shell midsurface and the difference vector between the reference and deformed shell director vectors, respectively, and H denotes the reference shell thickness. As was amply discussed in the related literature, see [2, 17, 18] where the authors developed the so-called 7-p model, 6-p formulations suffer from the so-called Poisson thickness locking pathology due to the presence of spurious transverse normal stresses. This deficiency has been usually tackled using two methodologies:

- Through the assumption of a quadratic interpolation of the transverse displacement, see [19]:

$$\mathbf{u}(\boldsymbol{\xi}) = \mathbf{v}(\xi^1, \xi^2) + \frac{H}{2}\xi^3 \mathbf{w}(\xi^1, \xi^2) + \frac{H}{2}\left(\xi^3\right)^2 \bar{\mathbf{w}}(\xi^1, \xi^2), \tag{5.5}$$

being $\bar{\mathbf{w}} = \mathbf{v}^{(2)}$ the one component vector which accounts for the thickness stretching in line with Eq. (5.3).

- Using a hybrid-mixed variational formulation that includes the necessary seventh parameter as an internal degree of freedom at the element level. This latter approach constitutes the basis of the shell model introduced by Bischoff, Ramm and coauthors [12, 17, 20, 21], whose formulation will be used in the sequel using two different parametrizations, see Section 5.2.

Further elaborations of the kinematic scheme given in Eq. (5.3) can lead to refined formulations, usually called as multi-director approaches (relying on the concept of p–refinement). The general displacement field for p-refinement formulations reads

$$\mathbf{u}\left(\boldsymbol{\xi}\right) = \mathbf{v}\left(\xi^1,\xi^2\right) + \sum_{N=1}^{\infty} \left(\xi^3\right)^N \mathbf{w}^{(N)}\left(\xi^1,\xi^2\right). \tag{5.6}$$

The truncation of the Taylor series in terms of the coordinate ξ^3 up to cubic order leads to the recently developed 12-parameter formulation [22], whose displacement field is given by

$$\mathbf{u}(\boldsymbol{\xi}) = \mathbf{v}(\xi^1,\xi^2) + \frac{H}{2}\xi^3\mathbf{w}(\xi^1,\xi^2) + \frac{H}{2}\left(\xi^3\right)^2\bar{\mathbf{w}}(\xi^1,\xi^2) + \frac{H}{2}\left(\xi^3\right)^3\hat{\mathbf{w}}(\xi^1,\xi^2), \tag{5.7}$$

with $\hat{\mathbf{w}}$ corresponding to vector $\mathbf{v}^{(3)}$ of the series expansion introduced in Eq. (5.3). Note that in Eq. (5.7), $\bar{\mathbf{w}}$ and $\hat{\mathbf{w}}$ are vectors with three components, which contribute to the alleviation of Poisson thickness locking. This form of the 12-p model allows the use of C^0-continuity order of the shape functions for the corresponding FE discretization, making it amenable from the numerical point of view.

However, it is clearly perceivable the dramatic increase of the computational cost of p-refinement formulations at the element level in conjuction with the difficulties to provide a clear mechanical interpretation of the model higher order parameters.

Alternatively to the previous refinement scheme, h-refinement models can be considered as potential modeling methods, which generally correspond to the Layer–Wise (LW) approaches, see Chapter 12 in [10] and [11] among many others. These models encompass the subdivision of the shell body into several kinematic layers over the thickness and assume separate displacement field expansions for each of them. Therefore, the resulting displacement field exhibits a C^0_z-continuity (i.e., C^0-continuity along the thickness direction). As was amply discussed in the related literature, LW formulations are especially suitable for modeling laminates with significant stiffness difference between the adjoining layers as is the case of sandwich structures, see [12] and the references therein given. According to this approach, the interpolation of the displacement field presents the following scheme:

$$\mathbf{u}(\boldsymbol{\xi}) = \sum_{N=1}^{n_{\mathrm{KL}}} \Phi_N(\xi^3)\, \mathbf{w}^{(N)}(\xi^1, \xi^2), \tag{5.8}$$

where n_{KL} is the number of *kinematic* layers (several kinematic layers can be defined in each material layer) through the shell thickness, Φ_N are known functions in the thickness coordinate (ξ^3). The resulting model significantly increases the number of unknowns of the system and additionally requires the establishment of the displacement continuity and equilibrium conditions across the interfaces between layers via a set of constraint equations. It is also worth noting that each of the composing layers has its own kinematics model (5-p, 6-p, etc.), therefore the zig-zag effect over the thickness can be properly represented. Based on their intrinsic nature, LW models incur into higher numerical costs comparing with ESL models, this aspect becomes of notable matter in the case of nonlinear FE analysis and for the generation of the meshes. Nevertheless, the applicability and reliability of LW formulations have been proved through their application to different laminate configurations using standard FE [23–26] and isogeometric parametrizations [27].

5.1.2. Finite element formulations for shells

From the numerical perspective, several shell finite elements have been proposed in the last decades. Focusing the attention on those allowing the use of 3D constitutive relations, two basic shell formulations can be distinguished: (i) three-dimensional shell elements (surface-based FE meshes), which model a reference surface of the body [19, 20, 28–30], and (ii) solid shell elements (solid-based FE meshes) which accounts for the parametrization of the top and bottom surfaces of the body and completely avoid the use of rotational degrees of freedom for updating the shell normal along the deformation process, see [31–37] to quote a few of them.

The improvement of the accuracy of the previous elements based on low-order kinematic interpolations can be accomplished through hybrid-mixed formulations. One popular approach to alleviate several locking deficiencies is the use of the Enhanced Assumed Strain (EAS) method. This methodology was introduced by Simo and Rifai [38] relying on the multi-field Hu–Washizu variational principle, and subsequently exploited in [17, 35, 39–42] among many others. In particular, the three-field functional, where the displacement, the enhanced strain and the stress fields are primary variables, is reduced to a two-field functional by recalling the orthogonality condition between the incompatible strains and the stress field [40]. This leads to the elimination of the stress field from the discrete FE equations.

Stress interpolations which are not removed at the element level leads to hybrid stress elements [43], which are based on the two-field Hellinger–Reissner variational form. Alternative mixed-element formulations on the basis of the Hu–Washizu principle consider the approximation of the displacements, the strains and the stresses as independent primary fields, see [44–48]. In this latter element topology, the independent stress and the enhanced strain interpolations are not assumed to be orthogonal (Hybrid-EAS formulations), and therefore the stress field is not removed from the resulting FE equations. This methodology has been exploited by Klinkel and coauthors [47] through a solid shell approach for linear and nonlinear problems showing a superior in-plane bending performance. Recently, a different Hybrid-EAS model solid shell formulation was proposed by Vu-Quoc and Tan [48] for the analysis of thick laminated structures, which exhibits an excellent level of accuracy with regard to the estimation of the interlaminar stress field.

The previous EAS- and hybrid-based formulations have been also combined with alternative strategies in order to remove additional locking effects. This is the case of the Assumed Natural Strain (ANS) [49, 50] method to alleviate transverse shear and trapezoidal locking deficiencies, and Reduced Integration (RI) [31, 37, 51] techniques, among other different alternatives.

5.1.3. Instabilities in thin-walled composite engineering systems

Structural behavior of shells may be extremely sensitive to different parameters such as supporting conditions, material defects and orientations in composites, geometrical imperfections, among other effects. These aspects can greatly affect their mechanical responses and might becoming undesired and unexpected or even chaotic [2]. Nonlinear effects in shells involving finite strains, material nonlinearities, buckling and postbuckling including features with regard to Ruga mechanics, which analyzes wrinkling, folding and creasing phenomena among other effects, have been extensively analyzed in the last years.

With special interest in the aerospace and aeronautical industries, the use of CFRP composites in the construction of stiffened panels has received a notable attention in the last two decades, see [52–55] and the references therein given. In these applications, due to the high slenderness of such components, the applied external loadings can promote the appearance and development of structural instabilities (postbuckling evolution) once the local buckling load of the specimen has been overcome. However, these panels exhibit an additional significant load carrying capacity prior to the corresponding collapsing point. With the aim of the fully exploitation of this reserve of strength, several numerical studies reproduced the overall postbuckling response using different shell formulations [53, 54]. Focusing on the investigation of the collapsing mechanisms associated with damage, inter-laminar failure events have been triggered using the so-called Virtual Crack Closure Technique (VCCT) [56] and Cohesive Zone Models (CZMs) [57] strategies, whereas the simulation of intra-laminar damage has been mostly performed using continuum damage-based techniques [58].

An additional application for thin-walled structures concerns with Ruga mechanics in thin layer–substrate systems. One of the configurations most used in practice comprises a flexible thick substrate and a thin coating, usually with a functionality such a protecting layer against impact, including an electronic device, among others. There exists a wide range of applicability of thin layer–substrate systems such as printable solar cells, flexible tilt sensors [59], thermal barrier coatings [60, 61], among many others. As stated above, failure of such systems generally involves the formation of creasing, folding, wrinkling, which can originates the initiation and evolution of delamination events at the thin layer–substrate interface due to the notable mismatch between the corresponding mechanical properties [62–64]. These complex scenarios have been observed in experimental investigations, see [63, 65]. The analysis of the mechanical performance of these systems has been carried out using semi-analytical procedures based on perturbation analysis in order to estimate the critical membrane force and wrinkles wave-length that provokes the delamination of the layer. Subsequent investigations have also incorporated the use of FE-based simulations in order to estimate the postbuckling evolution along with delamination events predominantly using CZMs as a consequence of their versatility [60, 66, 67].

5.1.4. Overview

The chapter is organized as follows. Section 5.2 outlines the principal aspects of the 7-parameter shell model. The composite material models considered in this research are given in Section 5.3. The interface cohesive model is described in Section 5.4. The variational basis and specific details of the corresponding FE formulation of the proposed modeling framework are addressed in Section 5.5. The assessment of the proposed methodology by means of several applications is presented in Section 5.6. Finally, the main conclusions are summarized in Section 5.7.

5.2. Shell Formulation: 7-Parameter Model

This section outlines the basic aspects of the structural model herewith considered for the simulation of thin-walled composite structures. Two different kinematic parametrization are considered: (i) the three-dimensional shell model (surface-based FE meshes) [18, 20], which is presented in Section 5.2.2, and (ii) the solid shell model [33], whose fundamental aspects are given in Section 5.2.3.

5.2.1. Differential geometry and fundamental equations

For the derivation of three-dimensional seven-parameter shell model, let us consider a shell body that occupies a region in the Euclidean space $\mathcal{B}_0 \in \mathbb{R}^3$ in the reference configuration and $\mathcal{B}_t \in \mathbb{R}^3$ in the current configuration. The deformation of the shell is considered through a nonlinear mapping $\varphi(\mathbf{X}, t)$, which transforms any material point $\mathbf{X} \in \mathcal{B}_0$ onto spatial points $\mathbf{x} \in \mathcal{B}_t$: $\varphi(\mathbf{X}, t) : \mathcal{B}_0 \times [0, t] \to \mathbb{R}^3$, where $[0, t]$ is the time step interval, see Fig. 5.2. The displacement derived deformation gradient is defined as follows: $\mathbf{F}^u := \partial_{\mathbf{X}}\varphi(\mathbf{X}, t)$, where $J = \det[\mathbf{F}^u]$ is the Jacobian of the transformation, and $\partial_{\mathbf{X}}$ identifies the partial derivative with respect to the Lagrangian frame.

The curvilinear basis vectors $\mathbf{G}_i$ and $\mathbf{g}_i$ in the reference and current configurations, respectively, are given by:

$$\mathbf{G}_i = \frac{\partial \mathbf{X}(\boldsymbol{\xi})}{\partial \xi^i}; \quad \mathbf{g}_i = \frac{\partial \mathbf{x}(\boldsymbol{\xi})}{\partial \xi^i} = \mathbf{G}_i + \frac{\partial \mathbf{u}(\boldsymbol{\xi})}{\partial \xi^i}; \quad i = 1, 2, 3. \tag{5.9}$$

The contravariant basis vectors are defined using the following standard relationships: $\mathbf{G}_i \cdot \mathbf{G}^j = \delta_i^j$ and $\mathbf{g}_i \cdot \mathbf{g}^j = \delta_i^j$, where δ_i^j is the Kronecker delta. The metric coefficients are given by: $g_{ij} = \mathbf{g}_i \cdot \mathbf{g}_j$ and $G_{ij} = \mathbf{G}_i \cdot \mathbf{G}_j$.

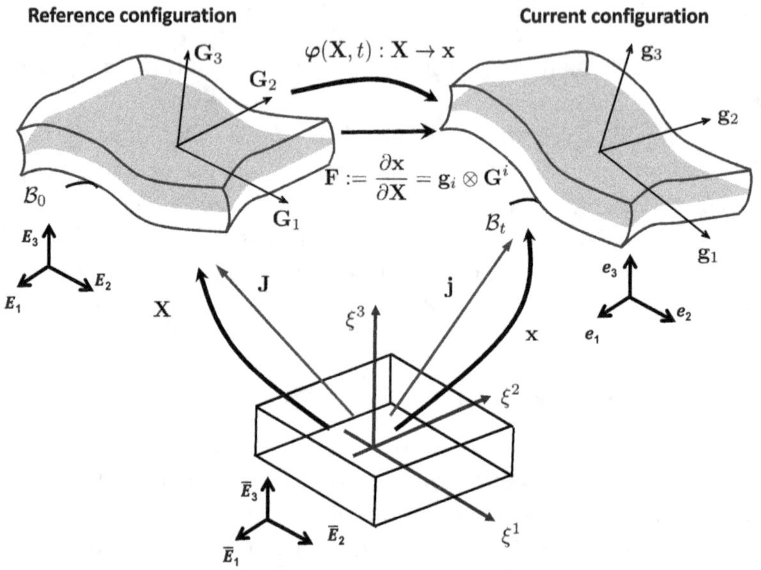

Fig. 5.2. Description of the shell body in the curvilinear setting. The coordinate systems for the reference $\mathcal{B}_0$ and current $\mathcal{B}_t$ configurations are denoted by $\{\mathbf{E}_i\}_{i=1,3}$ and $\{\mathbf{e}_i\}_{i=1,3}$, respectively; the points in the reference and current configurations are denoted by $\mathbf{X}$ and $\mathbf{x}$ respectively, whereas the corresponding Jacobi matrices are given by $\mathbf{J}$ and $\mathbf{j}$.

The Jacobi matrices referred to the transformations between the parametric space in the reference, $\mathbf{J}(\boldsymbol{\xi})$, and in the current, $\mathbf{j}(\boldsymbol{\xi})$, configurations read:

$$\mathbf{J}(\boldsymbol{\xi}) = [\mathbf{G}_1, \mathbf{G}_2, \mathbf{G}_3]^{\mathrm{T}}, \quad \mathbf{j}(\boldsymbol{\xi}) = [\mathbf{g}_1, \mathbf{g}_2, \mathbf{g}_3]^{\mathrm{T}}. \tag{5.10}$$

The respective covariant basis vectors on the shell midsurface ($\xi^3 = 0$) are obtained by particularizing Eq. (5.9):

$$\mathbf{A}_\alpha = \frac{\partial \mathbf{R}(\xi^1, \xi^2)}{\partial \xi^\alpha} = \mathbf{R}_{,\alpha}\,; \quad \mathbf{a}_\alpha = \frac{\partial \mathbf{r}(\xi^1, \xi^2)}{\partial \xi^\alpha} = \mathbf{r}_{,\alpha}\,; \quad \alpha = 1, 2, \tag{5.11}$$

where $\mathbf{R}(\xi^1, \xi^2)$ and $\mathbf{r}(\xi^1, \xi^2)$ denote the mid-surface shell position vectors in the reference and in the current configurations, respectively. The shell director vector in the reference configuration $\mathbf{A}_3$ is defined perpendicular to the covariant in-plane vectors $\mathbf{A}_1$ and $\mathbf{A}_2$:

$$\mathbf{A}_3(\xi^1, \xi^2) = \frac{H}{2} \frac{\mathbf{A}_1 \times \mathbf{A}_2}{|\mathbf{A}_1 \times \mathbf{A}_2|}, \tag{5.12}$$

where H is the initial shell thickness that is used for normalizing $\mathbf{A}_3$.

Through the previous definitions, the position vector of an arbitrary point in the reference and in the current configurations are accordingly given by

$$\mathbf{X}(\boldsymbol{\xi}) = \mathbf{R}(\xi^1, \xi^2) + \xi^3 \frac{H}{2} \mathbf{A}_3(\xi^1, \xi^2), \tag{5.13}$$

$$\mathbf{x}(\boldsymbol{\xi}) = \mathbf{r}(\xi^1, \xi^2) + \xi^3 \frac{h}{2} \mathbf{A}_3(\xi^1, \xi^2), \tag{5.14}$$

where h is the updated shell thickness. The kinematic field, $\mathbf{u}(\boldsymbol{\xi})$, can be expressed as follows:

$$\mathbf{u}(\boldsymbol{\xi}) = \left[\mathbf{r}(\xi^1, \xi^2) - \mathbf{R}(\xi^1, \xi^2)\right] + \xi^3 \left[\mathbf{a}_3(\xi^1, \xi^2) - \mathbf{A}_3(\xi^1, \xi^2)\right], \tag{5.15}$$

decomposing $\mathbf{u} = \mathbf{v} + \xi^3 \mathbf{w}$ into

$$\mathbf{v}(\xi^1, \xi^2) = \mathbf{r}(\xi^1, \xi^2) - \mathbf{R}(\xi^1, \xi^2); \quad \mathbf{w}(\xi^1, \xi^2) = \mathbf{a}_3(\xi^1, \xi^2) - \mathbf{A}_3(\xi^1, \xi^2). \tag{5.16}$$

In the previous definitions, $\mathbf{v}(\xi^1, \xi^2)$ is the displacement of the shell midsurface, whereas the vector $\mathbf{w}(\xi^1, \xi^2)$ (difference vector) accounts for the update the shell director along the deformation process, whose computations depend on the particular parametrization.

The definition of the displacement derived Green–Lagrange strain tensor reads:

$$\mathbf{E}^u := \frac{1}{2}\left[(\mathbf{F}^u)^{\mathrm{T}} \mathbf{F}^u - \mathbb{I}_2\right] = \frac{1}{2}\left[g_{ij} - G_{ij}\right] \mathbf{G}^i \otimes \mathbf{G}^j, \tag{5.17}$$

where $\mathbb{I}_2$ is the material covariant metric. The compatible Green–Lagrange strain tensor admits the following decomposition:

$$E^u_{ij} = p_{ij} + \frac{H}{2}\xi^3 q_{ij} + \frac{H^2}{4}(\xi^3)^2 s_{ij}, \quad \text{with} \quad i, j = 1, 2, 3, \tag{5.18}$$

where p_{ij}, q_{ij} and s_{ij} are the constant, linear and quadratic strain components in the thickness coordinate ξ^3. It is noted that the contributions of the components s_{ij} are usually neglected in thin-walled applications due to their minor role regarding the mechanical performance [20] in small strain conditions.

The constant strain components in Eq. (5.18) are defined as follows:

$$p_{\alpha\beta} := \frac{1}{2}\left[\mathbf{a}_\alpha \cdot \mathbf{a}_\beta - \mathbf{A}_\alpha \cdot \mathbf{A}_\beta\right], \tag{5.19}$$

$$p_{\alpha 3} := \frac{1}{2}\left[\mathbf{a}_\alpha \cdot \mathbf{A}_3 - \mathbf{A}_\alpha \cdot \mathbf{A}_3\right], \tag{5.20}$$

$$p_{33} := \frac{1}{2}\left[\mathbf{a}_3 \cdot \mathbf{a}_3 - \mathbf{A}_3 \cdot \mathbf{A}_3\right], \tag{5.21}$$

where $p_{\alpha\beta}$ denote the membrane terms of the Kirchhoff–Love model, $p_{\alpha 3}$ stem from the Reissner–Mindlin shear forces hypothesis, and p_{33} is the component associated with normal strains along the thickness direction. Accordingly, the linear strain components take the form:

$$q_{\alpha\beta} := \frac{1}{H}\left[\mathbf{a}_\alpha \cdot \mathbf{a}_{3,\beta} + \mathbf{a}_\beta \cdot \mathbf{a}_{3,\alpha} - \mathbf{A}_\alpha \cdot \mathbf{A}_{3,\beta} - \mathbf{A}_\beta \cdot \mathbf{A}_{3,\alpha}\right], \tag{5.22}$$

$$q_{\alpha 3} := \frac{1}{H}\left[\mathbf{a}_{3,\alpha} \cdot \mathbf{a}_3 - \mathbf{A}_{3,\alpha} \cdot \mathbf{A}_3\right], \tag{5.23}$$

$$q_{33} := 0, \tag{5.24}$$

where $q_{\alpha\beta}$ stand for the curvature changes (q_{11} and q_{22}) associated with the Kirchhoff–Love model, q_{12} is the twisting strain component of Kirchhoff–Love formulation and $q_{\alpha 3}$ identify the transverse shear curvatures.

Recalling [12, 20, 21], the previous parametrization (6-p model) yields to constant transverse normal strain components for pure bending applications since $p_{33} \neq 0$ and $q_{33} = 0$. However, under such conditions and according to the Theory of Elasticity, the transverse normal strain is linearly varying over the thickness. Consequently the 6-p formulation suffers from the well-known Poisson thickness locking, incurring in an error of the order ν^2 (being ν the Poisson ratio of the material) when unmodified 3D constitutive laws are employed. To alleviate this deficiency, two main models can be adopted: (i) the use of a quadratic interpolation for the displacement field over the shell thickness as proposed in [19, 68], and (ii) the advocation of the enhanced assumed strain concept [12, 20, 21, 35], which mathematically can be expressed as follow:

$$E_{33} = E_{33}^u + \xi^3 \tilde{\beta}_{33}(\xi^1, \xi^2), \tag{5.25}$$

where $\xi^3 \tilde{\beta}_{33}(\xi^1, \xi^2)$ is the linear strain term in the transverse direction.

From the operative standpoint, this latter option is particularly interesting due to several reasons: (1) the incompatible modes can be condensed out at the element level, so that the computational efficiency is preserved, (2) it can be directly implemented into commercial and research FE codes, (3) it allows the use of standard meshing techniques without any additional feature for considering the intermediate node along the thickness direction that the formulation developed in [19, 68] required. Consequently, highly efficient and locking free shell elements can be formulated.

The corresponding second Piola–Kirchhoff static quantities n^{ij} and m^{ij} energetically conjugated to p_{ij} and q_{ij}, respectively, are defined as follows:

$$n^{ij} := \int_{-1}^{+1} S^{ij}\hat{\mu}\, \mathrm{d}\xi^3; \quad m^{ij} := \int_{-1}^{+1} S^{ij}\xi^3\hat{\mu}\, \mathrm{d}\xi^3. \tag{5.26}$$

The terms n^{ij} and m^{ij} denote the constant (forces) and linear (moments) stress resultants per unit length. The constitutive tensor of the shell, $\mathbb{C}$ (with components $\mathbb{C}^{ijkl}$), is also referred to the shell mid-surface via the numerical integration across the thickness:

$$D_k^{ijkl} = \int_{-1}^{+1} (\xi^3)^k \mathbb{C}^{ijkl} \left(\frac{H}{2}\right)^{k+1} \hat{\mu}\, \mathrm{d}\xi^3, \quad k = 0, 1, 2, \tag{5.27}$$

where $\hat{\mu}$ regards the transformation of the volume integrals into an integration over the shell mid-surface through the shifter tensor $\mathbf{Z} := \mathbf{G}_i \otimes \mathbf{A}^i$ [20]. The constitutive operator can be obtained performing the consistent linearization of the stress tensor via the directional derivative concept [69, 70]:

$$\Delta\mathbf{S} = \frac{\partial\mathbf{S}}{\partial\mathbf{E}} : \Delta\mathbf{E} = \mathbb{C} : \Delta\mathbf{E}. \tag{5.28}$$

The constitutive tensor $\mathbb{C}$ for composite structures can be expressed as a function of the thickness coordinate ξ^3, whose integration along this direction, Eq. (5.27), is generally carried out using numerical methods. The constitutive relation referred to the shell midsurface takes the form:

$$\begin{bmatrix} n^{ij} \\ m^{ij} \end{bmatrix} = \begin{bmatrix} D_0^{ijkl} & D_0^{ijkl} \\ D_1^{ijkl} & D_2^{ijkl} \end{bmatrix} \begin{bmatrix} p_{ij} \\ q_{ij} \end{bmatrix}, \tag{5.29}$$

where the stress and strain components of the model are arranged in the following vectors:

$$\left[n^{ij}, m^{ij}\right] = \left[n^{11}, n^{12}, n^{13}, n^{22}, n^{23}, n^{33}, m^{11}, m^{12}, m^{13}, m^{22}, m^{23}, m^{33}\right]^{\mathrm{T}}, \tag{5.30a}$$

$$\left[p_{ij}, q_{ij}\right] = \left[p_{11}, p_{12}, p_{13}, p_{22}, p_{23}, p_{33}, q_{11}, q_{12}, q_{13}, q_{22}, q_{23}, q_{33}\right]^{\mathrm{T}}. \tag{5.30b}$$

Note that differing from alternative shell formulations, the current approach embodies a complete description of the Green–Lagrange strain tensor, $\mathbf{E}^u$, and its energetically conjugated second Piola–Kirchhoff stress tensor $\mathbf{S}$. This aspect yields to the use of unmodified of three-dimensional material laws.

The mechanical significances of the static $[n^{ij}, m^{ij}]$ and kinematic quantities $[p_{ij}, q_{ij}]$ of the current 7-p model have been comprehensively described in [18], providing a clear and engineering interpretation to such components.

5.2.2. Three-dimensional shell parametrization

The first possible shell parametrization of the 6-p model herein considered is outlined in Fig. 5.3. The deformation of the reference shell surface is accounted for using six local degrees of freedom, which are arranged by a displacement vector of the shell surface $\mathbf{v}(\xi^1, \xi^2)$ and a difference vector $\mathbf{w}(\xi^1, \xi^2)$ to update the shell director. This kinematic representation induces the following relationships:

$$\mathbf{r}(\xi^1, \xi^2) = \mathbf{R}(\xi^1, \xi^2) + \mathbf{v}(\xi^1, \xi^2); \quad \mathbf{a}_3(\xi^1, \xi^2) = \mathbf{A}_3(\xi^1, \xi^2) + \mathbf{w}(\xi^1, \xi^2). \tag{5.31}$$

The current parametrization offers several appealing aspects, which as summarized as follows: (i) the simple linear structure of the kinematic field, Eq. (5.31), (ii) the avoidance of rotational degrees of freedom for arbitrary large displacements differing from alternative parametrizations [30], and (iii) the stretching of the shell director vector. However, note that a careful attention should be paid to the fact that the boundary conditions associated with the shell director vector should be accordingly defined, see [18, 55] for a comprehensive discussion.

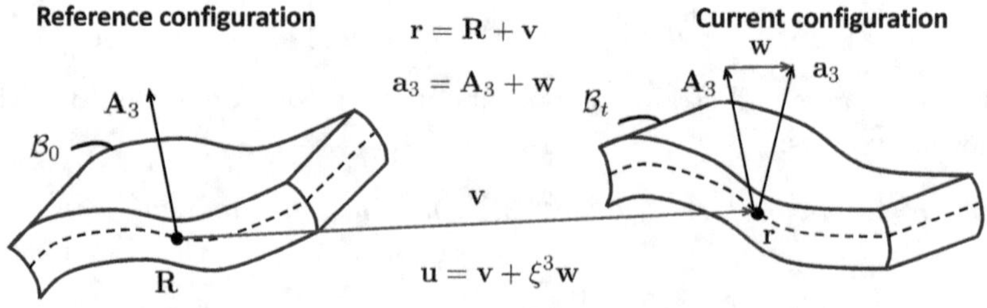

Fig. 5.3. Three-dimensional shell parametrization: linear parametrization of the shell midsurface and the director vector.

5.2.3. Solid shell parametrization

The second parametrization complies with the so-called solid shell concept [33], which considers two displacements fields $\mathbf{v}(\xi^1, \xi^2)$ and $\mathbf{w}(\xi^1, \xi^2)$ for the description shell kinematics, relating a pair of material points on the top and bottom surfaces of the shell, see Fig. 5.4. Complying with this parametrization, the position vector of any reference material point is given by

$$\mathbf{X}(\boldsymbol{\xi}) = \frac{1}{2}\left[1 + \xi^3\right]\mathbf{X}_t(\xi^1, \xi^2) + \frac{1}{2}\left[1 - \xi^3\right]\mathbf{X}_b(\xi^1, \xi^2), \tag{5.32}$$

where the position vectors $\mathbf{X}_t$ and $\mathbf{X}_b$ correspond to the top and bottom surfaces of the shell in the reference configuration, respectively. This expression, Eq. (5.32), can be rearranged as follows:

$$\mathbf{X}(\boldsymbol{\xi}) = \frac{1}{2}\left[\mathbf{X}_t(\xi^1, \xi^2) + \mathbf{X}_b(\xi^1, \xi^2)\right] + \frac{1}{2}\xi^3\left[\mathbf{X}_t(\xi^1, \xi^2) - \mathbf{X}_b(\xi^1, \xi^2)\right], \tag{5.33}$$

with

$$\mathbf{R}(\xi^1, \xi^2) = \frac{1}{2}\left[\mathbf{X}_t(\xi^1, \xi^2) + \mathbf{X}_b(\xi^1, \xi^2)\right]. \tag{5.34}$$

$$\mathbf{A}_3(\xi^1, \xi^2) = \frac{1}{2}\xi^3\left[\mathbf{X}_t(\xi^1, \xi^2) - \mathbf{X}_b(\xi^1, \xi^2)\right]. \tag{5.35}$$

A similar parametrization can be also performed for the position vector in the current configuration:

$$\mathbf{x}(\boldsymbol{\xi}) = \mathbf{r}(\xi^1, \xi^2) + \xi^3\mathbf{a}_3(\xi^1, \xi^2). \tag{5.36}$$

Therefore, the kinematic field is again expressed according to the following relationship:

$$\mathbf{u}(\boldsymbol{\xi}) = \mathbf{X}(\boldsymbol{\xi}) - \mathbf{x}(\boldsymbol{\xi}) = \mathbf{v}(\xi^1, \xi^2) + \xi^3\mathbf{w}(\xi^1, \xi^2), \tag{5.37}$$

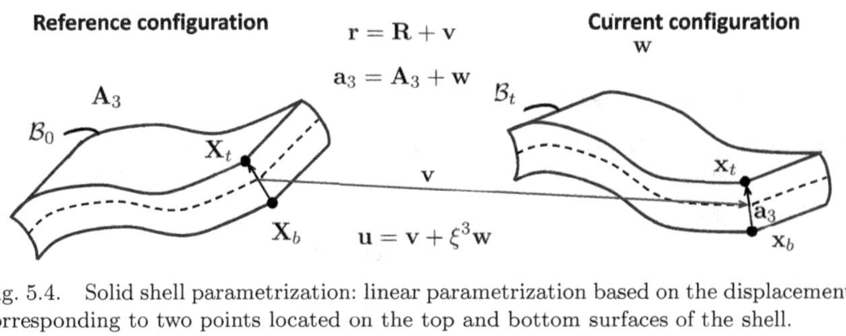

Fig. 5.4. Solid shell parametrization: linear parametrization based on the displacements corresponding to two points located on the top and bottom surfaces of the shell.

where $\mathbf{v}$ and $\mathbf{w}$ denote the displacement vectors of the shell midsurface and the director vector, respectively. These vectors take the following particular form:

$$\mathbf{v}(\xi^1, \xi^2) = \frac{1}{2}\left[\mathbf{u}_t(\xi^1, \xi^2) + \mathbf{u}_b(\xi^1, \xi^2)\right], \tag{5.38}$$

$$\mathbf{w}(\xi^1, \xi^2) = \frac{1}{2}\left[\mathbf{u}_t(\xi^1, \xi^2) - \mathbf{u}_b(\xi^1, \xi^2)\right], \tag{5.39}$$

where $\mathbf{u}_t$ and $\mathbf{u}_b$ are the displacement vectors of the top and bottom surfaces of the shell, respectively.

From the operative standpoint, the current solid shell model offers some advantages with respect to the surface-based parametrization discussed in Section 5.2.2. The most interesting aspects are as follows: (i) the prevention of material overlapping in highly complex structures, especially in corner locations where surface-based shell parametrizations can present severe difficulties, (ii) the simpler adaption of FE models from geometrical data using CAD packages, (iii) the suitable use in applications concerning double-sided contact, (iv) the improvement with regard to the connection between thin and thick regions in the model, among many others. Based on the previous considerations, the solid shell parametrization has been one of the most used approaches for the development of new shell elements in the last two decades [31, 35–37].

5.3. Constitutive Formulations for the Shell

In this investigation, two basic constitutive models for composite materials complying with the Kichhoff–Saint–Venant formulation (with a linear relationship $\mathbf{S} = \mathbb{C} : \mathbf{E}$) are considered. The first material type is used to model layered composite shells (Section 5.3.1), whereas the second material formulation concerns with power-based functionally graded materials (Section 5.3.2).

5.3.1. Layered composite shells

The first type of composite structures under consideration concerns laminated CFRP composites advocating the ESL model. In particular, for such applications, an orthotropic material law for each lamina and perfectly bonded behavior are assumed, i.e., no delamination events are taken into consideration.

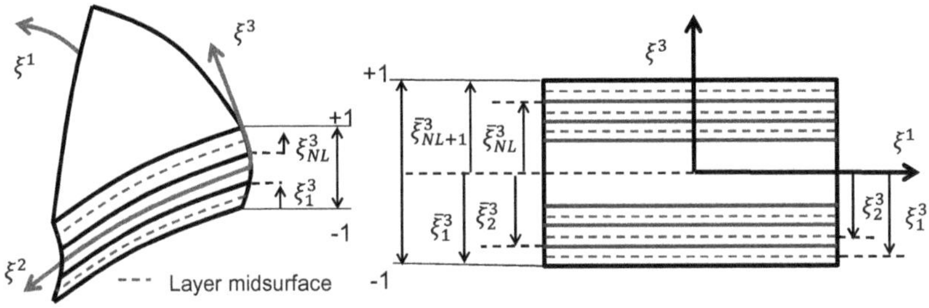

Fig. 5.5. Sketch of the laminated shell structure: local layer setting and definition of auxiliary natural coordinates over the shell thickness $\bar{\xi}^3_i$.

In line with the ESL approach, the computation of the resulting constitutive tensor for laminates can be carried out by considering the dependency of the laminate disposal with respect to the thickness coordinate ξ^3 (Fig. 5.5) in the form:

$$\mathbb{C}(\xi^3)\begin{cases}\mathbb{C}_{N_L} & \bar{\xi}^3_{N_L} \leq \xi^3 \leq \bar{\xi}^3_{N_L+1} = +1\\ \mathbb{C}_{N_L-1} & \bar{\xi}^3_{N_L-1} \leq \xi^3 \leq \bar{\xi}^3_{N_L}\\ \ldots\ldots & \ldots\ldots\\ \mathbb{C}_2 & \bar{\xi}^3_2 \leq \xi^3 \leq \bar{\xi}^3_3\\ \mathbb{C}_1 & -1 = \bar{\xi}^3_1 \leq \xi^3 \leq \bar{\xi}^3_2.\end{cases} \tag{5.40}$$

The thickness coordinate in Eq. (5.40) varies from $\xi^3 \in [-1, +1]$ after being scaled by the total laminate thickness H, which is given by $H = \sum_{i=1}^{N_L} H_i$, identifying N_L the total number of laminas and H_i the individual thickness of each ply. The bottom and top coordinates surfaces of each lamina i are denoted by $\bar{\xi}^3_i$ and $\bar{\xi}^3_{i+1}$, respectively, with $-1 \leq \bar{\xi}^3_i \leq +1$ at the lamina level. Accordingly, the definition of $\bar{\xi}^3_i$ reads:

$$\bar{\xi}^3_i = -1 + \frac{2}{H}\sum_{j=1}^{i-1} H_j \quad i = 2, \ldots, N_L. \tag{5.41}$$

The shell coordinate midsurface of each layer is denoted as ξ^3_i, which reads

$$\xi^3_i = -1 + \frac{H_i}{H} + \frac{2}{H}\sum_{j=1}^{i-1} H_j \quad i = 1, \ldots, N_L. \tag{5.42}$$

The modified version of Eq. (5.27) for laminates can be expressed as follows:

$$D_k^{mnop} = \sum_{i=1}^{N_L} \frac{H_i}{H^{k+1}} \int_{-1}^{1} \hat{\mu}_{\zeta_i} \left[-H - H_i(1-\zeta_i) + 2\sum_{j=1}^{i} H_j \right] \mathbb{C}_i^{mnop} \, \mathrm{d}\zeta_L, \tag{5.43}$$

with $k = 0, 1, 2$ and where i stands for the current lamina, and ξ^3 renders:

$$\xi^3 = -1 + \frac{1}{H}\left[- H_L(1-\zeta_L) + 2\sum_{i=1}^{L} H_i \right]. \tag{5.44}$$

5.3.2. Functionally graded isotropic shells

Metal-ceramic functionally graded (FG) shells have been extensively used as thermal barrier coatings (TBCs), nuclear fusion reactors, among many applications. In these materials, the volume fractions of two or more constituents vary continuously and smoothly within the structure, usually along the thickness direction. From the modeling standpoint, in agreement with [66, 71], for two-constituent FG shells, a variation of the material properties over the shell thickness is assumed to obey the following power-based relationship:

$$\iota(\xi^3) = \iota_m f_m + \iota_c f_c, \tag{5.45}$$

where the subscripts m and c identify the metallic and the ceramic components, respectively, f is the volume fraction of the corresponding phase, and ι is a generic material property. The volume fraction of each constituent can be determined as follows:

$$f_c = \left[\frac{\xi^3}{H} + \frac{1}{2} \right]^n, \tag{5.46}$$

$$f_m = 1 - f_c, \tag{5.47}$$

where n represents a volume fraction exponent. Invoking the standard Coleman–Noll approach [72], the second Piola–Kirchhoff stress tensor, $\mathbf{S}(\xi^3)$, and the constitutive tensor, $\mathbb{C}(\xi^3)$, as a function of ξ^3 read

$$\mathbf{S}(\xi^3) = \partial_{\mathbf{E}} \Psi(\xi^3) = S^{ij}(\xi^3) \mathbf{G}_i \otimes \mathbf{G}_j, \tag{5.48}$$

$$\mathbb{C}(\xi^3) = \partial_{\mathbf{EE}} \Psi(\xi^3) = C^{ijkl}(\xi^3) \mathbf{G}_i \otimes \mathbf{G}_j \otimes \mathbf{G}_k \otimes \mathbf{G}_l. \tag{5.49}$$

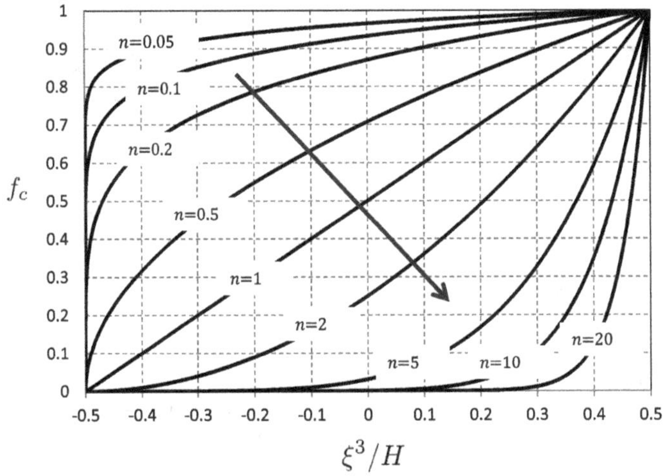

Fig. 5.6. Functionally graded composites based on power-based variation of the volume fraction of ceramic material f_c through the thickness.

Based on the variation law given in Eqs. (5.46)–(5.47), the constitutive operator (assuming a constant Poisson ratio over the shell thickness) can be expressed as

$$\mathbb{C}^{ijkl}(\xi^3) = \mathbb{C}_c^{ijkl} f_c + \mathbb{C}_m^{ijkl} f_m. \tag{5.50}$$

In the current investigation, for metal–ceramic FG shells, the case in which the exponent n is equal to zero identifies a fully ceramic structure, whereas when n tends to infinity a fully metallic body is obtained. Figure 5.6 depicts the evolution of the ceramic constituent volume fraction within the shell body for different values of n.

5.4. Cohesive Interface for Large Deformation Analysis

This section briefly revisits the fundamental aspects of the large deformation finite interface formulation henceforth employed. The development of interface elements adopting different nonlinear cohesive traction separation laws (TSL) between the two flanks of a crack has been an intensively covered topic in the last years, see [57, 73, 74] to quote a few of them. Recently, further contributions have also incorporated the role of geometrically nonlinear effects concerning the description of the interface performance, see [75–77] and the references therein given.

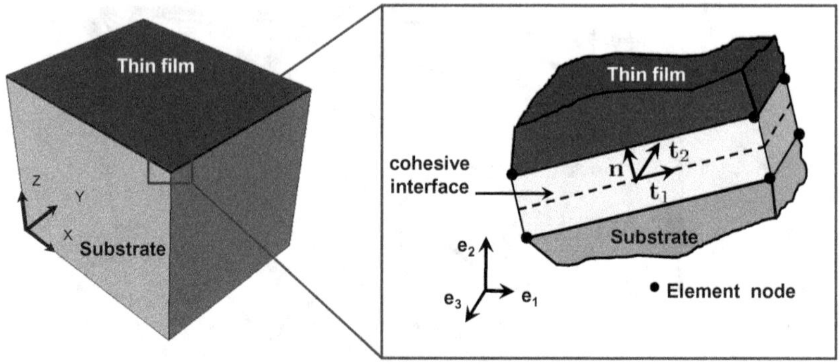

Fig. 5.7. Description of the thin layer–substrate systems: solid shell parametrization is used for the thin layer, whereas a large deformation cohesive zone model is included at the interface layer–substrate for triggering delamination events.

The current interface formulation is included into the applications with regard to postbuckling delamination of thin film substrate systems and relies on the formulation developed in [77, 78]. It is noting that the proposed interface model is fully compatible to the solid shell parametrization addressed in Section 5.2.3, and consequently no further considerations associated with the kinematic coupling between both element topologies are required. This fact contracts with alternative models, which need the implementation of additional features as is the use of first-order shear deformation theory to compute the gaps at the interface elements [79].

The starting point of the interface derivation concerns with the assumption of the existence of an interface between the two primary bodies in the reference configuration $\mathcal{B}_0^{(1)}$, $\mathcal{B}_0^{(2)} \subset \mathbb{R}^3$, which is identified between the thin film and the substrate in Fig. 5.7. The assumed cohesive interface occupies the surface $S_0 \subset \mathbb{R}^2$ in the reference configuration, and is mapped along the deformation process onto the current surface $s_0 \subset \mathbb{R}^2$.

The relation between the second Piola–Kirchhoff traction vector $\mathbf{T} = (\tau_1, \tau_2, \sigma)$ and the interface gaps $\mathbf{g}_{loc} = (g_{\text{loc},t1}, g_{\text{loc},t2}, g_{\text{loc},n})$ in the local reference system, where the corresponding TSL is defined, follows the cohesive interface model proposed in [80]. In particular, τ_1 and τ_2 identify the in-plane tractions associated with shear fracture events, respectively, whereas σ denotes traction component associated with fracture Mode I. Their respective energetically conjugated variables are the displacement gaps between the two flanks of the interface, which are denoted by $g_{\text{loc},t1}$, $g_{\text{loc},t2}$, and $g_{\text{loc},n}$.

The postulation of the cohesive TSL is given by the following relationship:

$$\tau_1 = \tau_{\max,1}\frac{g_{\mathrm{loc},t1}}{l_{t1c}}P(\lambda);\ \tau_2 = \tau_{\max,2}\frac{g_{\mathrm{loc},t2}}{l_{t2c}}P(\lambda);\ \sigma = \sigma_{\max}\frac{g_{\mathrm{loc},n}}{l_{nc}}P(\lambda), \tag{5.51}$$

where $\sigma_{\max}$, $\tau_{\max,1}$ and $\tau_{\max,2}$ are the critical cohesive tractions corresponding to the fracture Mode I, and shear, respectively; l_{nc}, l_{t1c}, and l_{t2c} identify the respective critical relative displacements. The dimensionless separation parameter λ allows mixed mode fracture to be modeled. This factor and the parameter $P(\lambda)$ are defined as follows:

$$\lambda = \sqrt{\left(\frac{g_{\mathrm{loc},n}}{l_{nc}}\right)^2 + \left(\frac{g_{\mathrm{loc},t1}}{l_{t1c}}\right)^2 + \left(\frac{g_{\mathrm{loc},t2}}{l_{t2c}}\right)^2}, \tag{5.52}$$

$$P(\lambda) = \begin{cases} \frac{27}{4}\left(1 - 2\lambda + \lambda^2\right), & \text{for } 0 \leq \lambda \leq 1, \\ 0, & \text{otherwise.} \end{cases} \tag{5.53}$$

A schematic representation of this TSL for mixed mode in-plane conditions is shown in Fig. 5.8.

In order to incorporate the current interface model within the context of implicit nonlinear finite elements the tangent operator is derived according

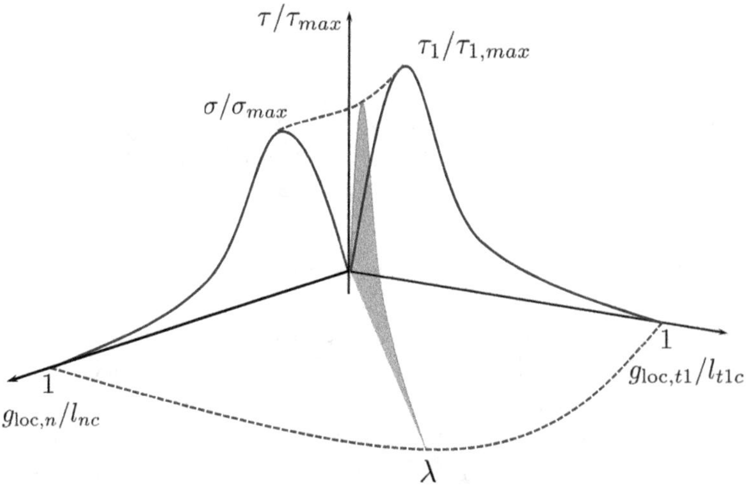

Fig. 5.8. Mixed-mode TSL for the polynomial-based law according to [80], $\tau/\tau_{\max}$ identifying an equivalent critical traction that accounts for mixed fracture conditions.

to standard arguments. The particular form of this operator reads

$$\mathbb{C}_c = \begin{bmatrix} \dfrac{\partial \tau_1}{\partial g_{\text{loc,t1}}} & \dfrac{\partial \tau_1}{\partial g_{\text{loc,t2}}} & \dfrac{\partial \tau_1}{\partial g_{\text{loc,n}}} \\ \dfrac{\partial \tau_2}{\partial g_{\text{loc,t1}}} & \dfrac{\partial \tau_2}{\partial g_{\text{loc,t2}}} & \dfrac{\partial \tau_2}{\partial g_{\text{loc,n}}} \\ \dfrac{\partial \sigma}{\partial g_{\text{loc,t1}}} & \dfrac{\partial \sigma}{\partial g_{\text{loc,t2}}} & \dfrac{\partial \sigma}{\partial g_{\text{loc,n}}} \end{bmatrix}. \tag{5.54}$$

A detailed derivation of the terms arising in Eq. (5.54) are reported in [77], which are omitted here for the sake of brevity.

5.5. Computational Framework and Finite Element Formulation

This section is devoted to present the variational basis of the current modeling framework. For the shells under consideration, the proposed formulation uses the multi-field Hu–Washizu principle, which is the fundamental basis of the EAS method. This variational principle holds for both of shell parametrizations herein considered, namely the three-dimensional shell (Section 5.2.2) and the solid shell (Section 5.2.3) formulation. The corresponding FE discretizations of both models are covered in Section 5.5.2 based on first-order shell elements.

5.5.1. Variational basis

The central result for the construction of the FE formulation of the proposed modeling framework concerns with the weak form of the balance of linear momentum of the complete system by means of the standard Galerkin procedure:

$$\delta\Pi_{\text{int}} - \delta\Pi_{\text{ext}} = \delta\Pi_{\text{int,b}} + \delta\Pi_{\text{int,c}} + \delta\Pi_{\text{ext}} = \mathcal{R}_{\text{int}} - \mathcal{R}_{\text{ext}} = 0, \ \forall \delta\mathbf{u} \in \mathcal{V}, \tag{5.55}$$

where $\delta\Pi_{\text{int}}$ stands for the internal contribution, which includes the terms associated with the bulk bodies $\delta\Pi_{\text{int,b}}$, whereas $\delta\Pi_{\text{int,c}}$ identifies the contribution of the cohesive interface; the term $\delta\Pi_{\text{ext}}$ denotes the contribution of the external actions applied to the system (body and surface actions), and $\mathcal{V} = \left\{\delta\mathbf{u} \in [H^1(\mathcal{B}_0)] : \delta\mathbf{u} = \mathbf{0} \text{ on } \partial\mathcal{B}_{0,u}\right\}$ is any kinematic admissible virtual displacements, which satisfy the essential boundary conditions.

The Lagrangian version of the internal contributions of the continuum shell body and the cohesive interface can be expressed as:

$$\mathcal{R}_{\text{int,b}} = \delta\Pi_{\text{int,b}}(\mathbf{u}, \delta\mathbf{u}) = \int_{\mathcal{B}_0} \mathbf{S} : \delta\mathbf{E}\ \mathrm{d}\Omega, \tag{5.56}$$

$$\mathcal{R}_{\text{int,c}} = \delta\Pi_{\text{int,c}}(\mathbf{u}, \delta\mathbf{u}) = \int_{S_0} \delta\mathbf{g}_{\text{loc}}^{\text{T}}\,\mathbf{T}\mathrm{d}S. \tag{5.57}$$

Through the directional derivative concept, the standard linearization of Eqs. (5.56)–(5.57) for the case of pure-displacement formulation is given by

$$\begin{aligned}\Delta\mathcal{R}_{\text{int,b}} &= \Delta\delta\Pi_{\text{int,b}}(\mathbf{u}, \delta\mathbf{u}, \Delta\mathbf{u})\\ &= \int_{\mathcal{B}_0} \Delta\mathbf{S} : \delta\mathbf{E}\ \mathrm{d}\Omega + \int_{\mathcal{B}_0} \mathbf{S} : \Delta\delta\mathbf{E}\ \mathrm{d}\Omega, \end{aligned}\tag{5.58}$$

$$\begin{aligned}\Delta\mathcal{R}_{\text{int,c}} &= \Delta\delta\Pi_{\text{int,c}}(\mathbf{u}, \delta\mathbf{u}, \Delta\mathbf{u})\\ &= \int_{S_0} \Delta\delta\mathbf{g}_{\text{loc}}^{\text{T}}\,\mathbf{T}\mathrm{d}\partial\Omega + \int_{S_0} \delta\mathbf{g}_{\text{loc}}^{\text{T}}\,\Delta\mathbf{T}\mathrm{d}\partial\Omega. \end{aligned}\tag{5.59}$$

Recalling the Hu–Washizu variational principle, we adopt the additive decomposition of $\mathbf{E}$ into displacement derived $\mathbf{E}^u$, and an incompatible $\tilde{\mathbf{E}}$, counterparts [20, 35]: $\mathbf{E} = \mathbf{E}^u + \tilde{\mathbf{E}}$. Therefore, for the shell body, the following internal contribution to the weak form, with $\mathcal{R}_{\text{int,b}} = \mathcal{R}_{\text{int,s}}$, can be recast:

$$\begin{aligned}\mathcal{R}_{\text{int,s}} &= \delta\Pi_{\text{int,s}}(\mathbf{u}, \delta\mathbf{u}, \tilde{\mathbf{E}}, \delta\tilde{\mathbf{E}})\\ &= \int_{\mathcal{B}_0} \mathbf{S} : \delta\mathbf{E}^u\ \mathrm{d}\Omega + \int_{\mathcal{B}_0} \mathbf{S} : \delta\tilde{\mathbf{E}}\ \mathrm{d}\Omega,\ \ \forall\delta\mathbf{u} \in \mathcal{V},\ \delta\tilde{\mathbf{E}} \in \mathcal{V}^{\tilde{E}}, \end{aligned}\tag{5.60}$$

where $\mathcal{V}^{\tilde{E}} = [L_2(\mathcal{B}_0)]$ is the admissible space corresponding to the incompatible strains. Note that the stress field is eliminated from Eq. (5.60) by imposing the so-called orthogonality condition between the interpolation spaces associated with the stress and the enhanced strain fields [38, 40]. Accordingly, assuming a field of incremental deformation in terms of the displacement field and the enhanced strains, the linearization of Eq. (5.60) results in:

$$\begin{aligned}&\Delta\delta\Pi_{\text{int,s}}(\mathbf{u}, \delta\mathbf{u}, \Delta\mathbf{u}, \tilde{\mathbf{E}}, \delta\tilde{\mathbf{E}}, \Delta\tilde{\mathbf{E}})\\ &\quad= \int_{\mathcal{B}_0} \Delta\mathbf{S} : \delta\mathbf{E}^u\ \mathrm{d}\Omega + \int_{\mathcal{B}_0} \mathbf{S} : \Delta\delta\mathbf{E}^u\ \mathrm{d}\Omega + \int_{\mathcal{B}_0} \Delta\mathbf{S} : \delta\tilde{\mathbf{E}}\ \mathrm{d}\Omega, \end{aligned}\tag{5.61}$$

where the linearization of the stress field including the incompatible strains takes the form:

$$\Delta \mathbf{S} = \frac{\partial \mathbf{S}}{\partial \mathbf{E}^u} : \Delta \mathbf{E}^u + \frac{\partial \mathbf{S}}{\partial \tilde{\mathbf{E}}} : \Delta \tilde{\mathbf{E}}. \tag{5.62}$$

The linearized virtual Green–Lagrange strain tensor that is derived from the displacement field reads:

$$\Delta \delta \mathbf{E}^u = \frac{1}{2} \left[\delta \mathbf{g}_i \cdot \Delta \mathbf{g}_j + \Delta \mathbf{g}_i \cdot \delta \mathbf{g}_j \right] \mathbf{G}^i \otimes \mathbf{G}^j. \tag{5.63}$$

5.5.2. Shell finite element discretization

In this section, the FE discretization schemes of the shell models outlined in Sections 5.2.2 and 5.2.3 are derived. The discretization is constructed in the sense of n_e non-overlapping elements defined in the Lagrangian setting, such that $\mathcal{B}_0 \approx \bigcup_{e=1}^{n_e} \mathcal{B}_0^{(e)}$.

5.5.2.1. *Displacement formulation supplemented by EAS*

The standard shape functions defined on the shell midsurface for linear interpolation strategy read

$$N^A = \frac{1}{4} \left(1 + \xi_A^1 \xi^1\right) \left(1 + \xi_A^2 \xi^2\right), \quad \text{with } \xi_A^1, \xi_A^2 = \pm 1. \tag{5.64}$$

Recalling the isoparametric concept, both the *three-dimensional shell* and the *solid shell* parametrizations adopt the following scheme for the discretization of the geometry in the reference and current configurations:

$$\mathbf{X} = \mathbf{R} + \xi^3 \mathbf{A}_3 \approx \sum_{I=1}^{4} N^I \mathbf{R}^I + \xi^3 N^I \mathbf{A}_3^I = \mathbf{N}\mathbf{R}^e + \xi^3 \mathbf{N}\mathbf{A}_3^e, \tag{5.65}$$

$$\mathbf{x} = \mathbf{r} + \xi^3 \mathbf{a}_3 \approx \sum_{I=1}^{4} N^I \mathbf{r}^I + \xi^3 N^I \mathbf{a}_3^I = \mathbf{N}\mathbf{r}^e + \xi^3 \mathbf{N}\mathbf{a}_3^e. \tag{5.66}$$

In Eqs. (5.65)–(5.66), the discrete midsurface position vectors in the reference and in the current configurations are identified by $\mathbf{R}^I$ and $\mathbf{r}^I$, respectively, which are collected by their corresponding operators at element level $\mathbf{R}^e$ and $\mathbf{r}^e$; $\mathbf{A}_3^I$ and $\mathbf{a}_3^I$ stand for the nodal director vectors in the reference and current configurations, respectively, which are arranged into the operators at element level $\mathbf{A}_3^e$ and $\mathbf{a}_3^e$; finally, the operator $\mathbf{N}$ contains the shape functions at the element level.

In this concern, it is worth mentioning that, whereas the *three-dimensional shell* formulation leads to a 4-node shell element, the *solid shell* parametrization results in a 8-node shell element. Note also that in contrast to standard 3D continuum elements, the solid shell parametrization assumes a particular orientation, where ξ^3 defines the thickness direction.

The interpolation of the displacement field $\mathbf{u}$, its variation $\delta\mathbf{u}$ and its increment $\Delta\mathbf{u}$ read:

$$\mathbf{u} \approx \tilde{\mathbf{N}}\mathbf{d}; \quad \delta\mathbf{u} \approx \tilde{\mathbf{N}}\delta\mathbf{d}; \quad \Delta\mathbf{u} \approx \tilde{\mathbf{N}}\Delta\mathbf{d}, \tag{5.67}$$

with

$$\tilde{\mathbf{N}}\mathbf{d} = \mathbf{N}\mathbf{v} + \xi^3\mathbf{N}\mathbf{w}. \tag{5.68}$$

The interpolation scheme given in Eq. (5.68) also holds for the variation and increment of the displacement field. In the previous expressions, $\mathbf{d}$ represents the nodal displacement vector at the element level, whilst the operator $\tilde{\mathbf{N}}$ also depends upon the particular shell parametrization, see [55, 81] for further details. The displacement derived strain field ($\mathbf{E}^u$), its variation ($\delta\mathbf{E}^u$) and its increment ($\Delta\mathbf{E}^u$) are interpolated through a suitable *compatibility* operator ($\mathbf{B}$) as follows:

$$\mathbf{E}^u \approx \mathbf{B}(\mathbf{d})\mathbf{d}, \quad \delta\mathbf{E}^u \approx \mathbf{B}(\mathbf{d})\delta\mathbf{d}, \quad \Delta\mathbf{E}^u \approx \mathbf{B}(\mathbf{d})\Delta\mathbf{d}. \tag{5.69}$$

The operator $\mathbf{B}$ contains the derivatives of the shape functions with respect to the global coordinate setting, and again its particular form depends on the chosen shell parametrization.

The interpolation of the incompatible strain field is expressed in terms of the operator $\mathbf{M}$ that contains the trial functions to alleviate the following pathologies: (i) membrane and in-plane shear locking, (ii) volumetric locking and (iii) Poisson thickness locking. The interpolation of the enhanced strain field $\tilde{\mathbf{E}}$, its variation $\delta\tilde{\mathbf{E}}$ and its increment $\Delta\tilde{\mathbf{E}}$ renders

$$\tilde{\mathbf{E}} \approx \mathbf{M}(\boldsymbol{\xi})\varsigma, \quad \delta\tilde{\mathbf{E}} \approx \mathbf{M}(\boldsymbol{\xi})\delta\varsigma, \quad \Delta\tilde{\mathbf{E}} \approx \mathbf{M}(\boldsymbol{\xi})\Delta\varsigma. \tag{5.70}$$

Note that since the operator $\mathbf{M}$ is defined in the parametric space, a posterior transformation into the global setting is required, see Section 5.5.2.2.

The insertion of the previous discretization scheme into the linearized system outlined in Eq. (5.61) leads to the following coupled algebraic system:

$$\begin{bmatrix} \mathbf{k}_{dd} & \mathbf{k}_{d\varsigma} \\ \mathbf{k}_{\varsigma d} & \mathbf{k}_{\varsigma\varsigma} \end{bmatrix} \begin{bmatrix} \Delta\mathbf{d} \\ \Delta\varsigma \end{bmatrix} = \begin{bmatrix} \mathbf{f}_{\text{ext}} \\ 0 \end{bmatrix} - \begin{bmatrix} \mathbf{f}_{\text{int}} \\ \mathbf{f}_{\text{EAS}} \end{bmatrix}, \tag{5.71}$$

where $\mathbf{f}_{\text{ext}}$ identifies the external force vector, whereas $\mathbf{f}_{\text{int}}$ and $\mathbf{f}_{\text{EAS}}$ stand for the internal force vectors associated with the kinematic and enhanced strain fields, respectively:

$$\mathbf{f}_{\text{int}} = \int_{\mathcal{B}_0} \mathbf{B}^{\text{T}}\mathbf{S}\ \text{d}\Omega; \ \ \mathbf{f}_{\text{EAS}} = \int_{\mathcal{B}_0} \mathbf{M}^{\text{T}}\mathbf{S}\ \text{d}\Omega. \tag{5.72}$$

The resulting matrices take the form:

$$\mathbf{k}_{dd} = \int_{\mathcal{B}_0} \left(\mathbf{B}^{\text{T}}\mathbb{C}\mathbf{B} + \left[\frac{\partial \mathbf{B}}{\partial \mathbf{d}}\right]^{\text{T}} \mathbf{S} \right) \text{d}\Omega; \quad \mathbf{k}_{d\varsigma} = \int_{\mathcal{B}_0} \mathbf{B}^{\text{T}}\mathbb{C}\mathbf{M}\ \text{d}\Omega; \tag{5.73}$$

$$\mathbf{k}_{\varsigma d} = \int_{\mathcal{B}_0} \mathbf{M}^{\text{T}}\mathbb{C}\mathbf{B}\ \text{d}\Omega; \ \ \mathbf{k}_{\varsigma\varsigma} = \int_{\mathcal{B}_0} \mathbf{M}^{\text{T}}\mathbb{C}\mathbf{M}\ \text{d}\Omega. \tag{5.74}$$

Eliminating the enhanced strains due to their discontinuity across the the element boundaries [20], the final element stiffness matrix and internal residual vector have the following form:

$$\tilde{\mathbf{k}}_{dd} = \mathbf{k}_{dd} - \mathbf{k}_{d\varsigma} \left[\mathbf{k}_{\varsigma\varsigma}\right]^{-1} \mathbf{k}_{\varsigma d}, \quad \tilde{\mathbf{f}}_{\text{int}} = \mathbf{f}_{\text{int}} - \mathbf{k}_{d\varsigma} \left[\mathbf{k}_{\varsigma\varsigma}\right]^{-1} \mathbf{f}_{\text{EAS}}. \tag{5.75}$$

The numerical implementation of the current EAS strategy in the commercial FE package `ABAQUS` has been amply discussed by the authors in [55], therefore particular details are omitted here for the sake of brevity.

5.5.2.2. *Interpolation of the incompatible strains*

As a consequence of its versatility, different designs with regard to the interpolation of the incompatible strain field has been proposed in the last years, see [34, 41, 55, 82] and the references therein given.

The enhanced part of the strain field is expressed in different basis as

$$\tilde{\mathbf{E}} = \tilde{E}_{ij}(\mathbf{G}^i \otimes \mathbf{G}^j) = \tilde{\epsilon}^0_{kl}(\mathbf{A}^k_0 \otimes \mathbf{A}^l_0), \tag{5.76}$$

where $\tilde{\epsilon}^0_{kl}$ represent these components at the element center with the internal basis $\mathbf{A}_0$ at this location. Operating in equation Eq. (5.76), one obtains:

$$\tilde{E}_{ij} = \tilde{\epsilon}^0_{kl}(\mathbf{A}^k_0 \cdot \mathbf{G}_i)(\mathbf{A}^l_0 \cdot \mathbf{G}_j) \rightarrow \tilde{\mathbf{E}} = (\mathbf{T}_0)^{-\text{T}}\tilde{\boldsymbol{\epsilon}}^0, \tag{5.77}$$

denoting $\mathbf{T}_0$ a transformation matrix between the basis at the element centre and the contravariant basis system at the integration points.

The strain field at the element centre is interpolated as

$$\tilde{\epsilon}^0 = \left[\frac{\det\mathbf{J}_0}{\det\mathbf{J}}\right]\tilde{\mathbf{M}}\varsigma, \tag{5.78}$$

where $\tilde{\mathbf{M}}$ is the interpolation matrix for the additional strain fields which are defined in the isoparametric space of the element; ς is the vector of the internal strain parameters at the element level, whereas $\mathbf{J}$ and $\mathbf{J}_0$ identify the Jacobian and its evaluation at the element center, respectively. After some manipulations, the enhanced strain tensor takes the form

$$\tilde{\boldsymbol{E}} = (\mathbf{T}_0)^{-\mathrm{T}}\left[\frac{\det\mathbf{J}_0}{\det\mathbf{J}}\right]\tilde{\mathbf{M}}\varsigma = \mathbf{M}\varsigma. \tag{5.79}$$

In the current research, a particular version of the matrix $\mathbf{M}$ through the consideration of 22 enhanced strains is chosen. According to the strain arrangement given in Eq. (5.30), This matrix adopts the following representation in a condensed format:

$$\tilde{\mathbf{M}} = \begin{bmatrix}\tilde{\mathbf{M}}_p \\ \tilde{\mathbf{M}}_q\end{bmatrix} = \begin{bmatrix}\tilde{\mathbf{M}}_7 & \mathbf{0}_7 & \mathbf{0}_4 & \mathbf{0}_4 \\ \mathbf{0}_7 & \tilde{\mathbf{M}}_7 & \tilde{\mathbf{M}}_4^{q_{33}} & \tilde{\mathbf{M}}_4\end{bmatrix}, \tag{5.80}$$

with

$$\tilde{\mathbf{M}}_7 = \begin{bmatrix} \xi^1 & 0 & 0 & 0 & \xi^1\xi^2 & 0 & 0 \\ 0 & 0 & \xi^1 & \xi^2 & 0 & 0 & \xi^1\xi^2 \\ 0 & 0 & 0 & 0 & 0 & 0 & 0 \\ 0 & \xi^2 & 0 & 0 & 0 & \xi^1\xi^2 & 0 \\ 0 & 0 & 0 & 0 & 0 & 0 & 0 \\ 0 & 0 & 0 & 0 & 0 & 0 & 0 \end{bmatrix}, \tag{5.81}$$

$$\tilde{\mathbf{M}}_4 = \begin{bmatrix} 0 & 0 & 0 & 0 \\ 0 & 0 & 0 & 0 \\ \xi^1 & \xi^1\xi^2 & 0 & 0 \\ 0 & 0 & 0 & 0 \\ 0 & 0 & \xi^2 & \xi^1\xi^2 \\ 0 & 0 & 0 & 0 \end{bmatrix} \quad \tilde{\mathbf{M}}_4^{q_{33}} = \begin{bmatrix} 0 & 0 & 0 & 0 \\ 0 & 0 & 0 & 0 \\ 0 & 0 & 0 & 0 \\ 0 & 0 & 0 & 0 \\ 0 & 0 & 0 & 0 \\ 1 & \xi^1 & \xi^2 & \xi^1\xi^2 \end{bmatrix}, \tag{5.82}$$

being $\mathbf{0}_n$ null matrix of order $6 \times n$. Note that the matrix $\tilde{\mathbf{M}}_4^{q_{33}}$ only affects the strain component q_{33}, as is described in [18].

It is noting that the implementation of the EAS method requires the computation of $[\mathbf{k}_{\varsigma\varsigma}]^{-1}$ for each iteration at each time step along the solution process. This fact significantly increases the time cost associated with the current computational methodology to alleviate locking pathologies [55]. Based on these previous considerations concerning the computational costs, it is interesting to optimize the number of EAS parameters depending on the application under analysis. In this concern, according to [41], an alternative option for the strain components $p_{\alpha\beta}$ and $q_{\alpha\beta}$ for structured meshes is reducing the 7-in-plane EAS parameters to 4. This leads to the substitution of the matrix $\tilde{\mathbf{M}}_7$ in Eq. (5.80) by $\tilde{\mathbf{M}}_4^*$. The form of $\tilde{\mathbf{M}}_4^*$ is given by

$$\tilde{\mathbf{M}}_4^* = \begin{bmatrix} \xi^1 & 0 & 0 & 0 \\ 0 & 0 & \xi^1 & \xi^2 \\ 0 & 0 & 0 & 0 \\ 0 & \xi^2 & 0 & 0 \\ 0 & 0 & 0 & 0 \\ 0 & 0 & 0 & 0 \end{bmatrix}. \tag{5.83}$$

For structured meshes, the bilinear terms can be also eliminated from $\tilde{\mathbf{M}}_4$ and $\tilde{\mathbf{M}}_4^{q_{33}}$ leading to $\tilde{\mathbf{M}}_2^*$ and $\tilde{\mathbf{M}}_3^{q_{33}}$, respectively:

$$\tilde{\mathbf{M}}_2^* = \begin{bmatrix} 0 & 0 \\ 0 & 0 \\ \xi^1 & 0 \\ 0 & 0 \\ 0 & \xi^2 \\ 0 & 0 \end{bmatrix} \quad \tilde{\mathbf{M}}_3^{q_{33}} = \begin{bmatrix} 0 & 0 & 0 \\ 0 & 0 & 0 \\ 0 & 0 & 0 \\ 0 & 0 & 0 \\ 0 & 0 & 0 \\ 1 & \xi^1 & \xi^2 \end{bmatrix}. \tag{5.84}$$

Through the insertion of $\tilde{\mathbf{M}}_4^*$, $\tilde{\mathbf{M}}_2^*$ and $\tilde{\mathbf{M}}_3^{q_{33}}$ replacing $\tilde{\mathbf{M}}_7$, $\tilde{\mathbf{M}}_4$ and $\tilde{\mathbf{M}}_4^{q_{33}}$ respectively, the size of $\tilde{\mathbf{M}}$ can be reduced to 13:

$$\tilde{\mathbf{M}} = \begin{bmatrix} \tilde{\mathbf{M}}_p \\ \tilde{\mathbf{M}}_q \end{bmatrix} = \begin{bmatrix} \tilde{\mathbf{M}}_4^* & \mathbf{0}_4 & \mathbf{0}_3 & \mathbf{0}_2 \\ \mathbf{0}_4 & \tilde{\mathbf{M}}_4^* & \tilde{\mathbf{M}}_3^{q_{33}} & \tilde{\mathbf{M}}_2^* \end{bmatrix}. \tag{5.85}$$

This option employs 13 EAS parameters and it has been also examined in the current research. The corresponding results to this latter choice, which are omitted here for brevity reasons, were very satisfactory since meshes created were very uniform and had good in-plane aspect element ratio.

In addition, alternative EAS scheme proposed by Rah *et al.* [83] to alleviate volumetric locking and for highly distorted meshes have been also numerically implemented and assessed via standard benchmark problems. However, for the applications under analysis in this research, the enhancement scheme given in [83] did not offer significant improvements of the numerical results since compressible materials and regular FE meshes were considered.

Finally, it is worth noting that complete polynomials regarding the strain interpolations can be achieved following the scheme proposed by Andelfinger and Ramm [82], though this option notably increases the numerical demands.

5.5.2.3. *The ANS method*

This methodology alleviates typical locking effects in brick elements in the thin limit, which arise from the poor displacement interpolation over the thickness.

According to the Dvorkin–Bathe procedure [49], the ANS interpolation of the transverse shear components considers four collocation points, which are defined in the natural space as (Fig. 5.9): $\boldsymbol{\xi}_A = (0, -1, 0)$, $\boldsymbol{\xi}_B = (1, 0, 0)$, $\boldsymbol{\xi}_C = (0, 1, 0)$ and $\boldsymbol{\xi}_D = (-1, 0, 0)$. Consequently, the modified interpolation reads:

$$\begin{bmatrix} p_{13}^{ANS} \\ p_{23}^{ANS} \end{bmatrix} = \begin{bmatrix} (1-\xi^2)p_{13}(\boldsymbol{\xi}_A) + (1+\xi^2)p_{13}(\boldsymbol{\xi}_C) \\ (1+\xi^1)p_{23}(\boldsymbol{\xi}_B) + (1-\xi^1)p_{13}(\boldsymbol{\xi}_D) \end{bmatrix}. \tag{5.86}$$

The assumed strain interpolations proposed in [50] for the treatment of the thickness strain component q_{33} account for the definition of four

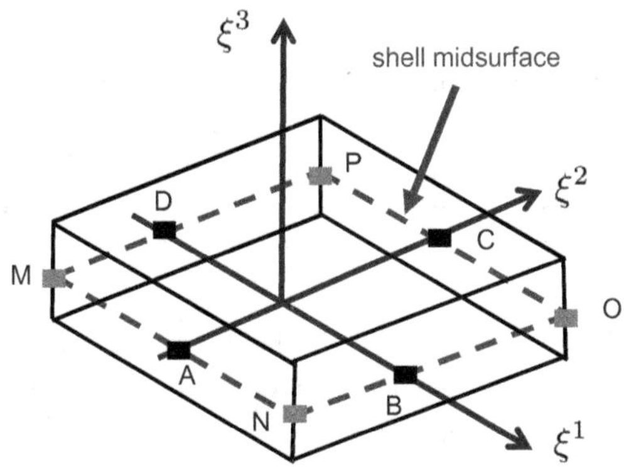

Fig. 5.9. Application of ANS method: identification of collocation points for the alleviation of transverse shear ($\boldsymbol{\xi}_A = (0, -1, 0)$, $\boldsymbol{\xi}_B = (1, 0, 0)$, $\boldsymbol{\xi}_C = (0, 1, 0)$ and $\boldsymbol{\xi}_D = (-1, 0, 0)$), and trapezoidal locking ($\boldsymbol{\xi}_M = (-1, -1, 0)$, $\boldsymbol{\xi}_N = (1, -1, 0)$, $\boldsymbol{\xi}_O = (1, 1, 0)$ and $\boldsymbol{\xi}_P = (-1, 1, 0)$) pathologies.

collocation points located on the corners of the shell midsurface (Fig. 5.9): $\boldsymbol{\xi}_M = (-1,-1,0)$, $\boldsymbol{\xi}_N = (1,-1,0)$, $\boldsymbol{\xi}_O = (1,1,0)$ and $\boldsymbol{\xi}_P = (-1,1,0)$. The modified interpolation scheme associated with this component is given by

$$p_{33}^{ANS} = \sum_{m=1}^{4} N^m(\xi^1,\xi^2)p_{33}, \qquad m = M,N,O,P$$
$$N^m(\xi^1,\xi^2) = \frac{1}{4}\left(1+\xi_m^1\xi^1\right)\left(1+\xi_m^2\xi^2\right), \quad \text{with } \xi_m^1,\xi_m^2 = \pm 1. \tag{5.87}$$

Following [20], the ANS method is combined with the EAS method in the current formulation in order to modify the interpolation associated with the transverse shear strains components p_{13} and p_{23}, and the transverse normal component p_{33} in both shell parametrizations under consideration.

5.5.3. Interface finite element discretization

In this section, the FE formulation of the cohesive interface model previously introduced (Section 5.4) is derived. As customary in interface formulations, the definition of a reference surface (usually the interface midsurface) is required [78]. The reference $\overline{\mathbf{X}}_c$ and current $\overline{\mathbf{x}}_c$ position vectors along the midsurface of the interface are computed according to the following interpolation scheme:

$$\overline{\mathbf{X}}_c \approx \overline{\mathbf{X}}_c^e = \mathbf{N}\mathbf{M}_c\mathbf{X}_c^n; \quad \overline{\mathbf{x}}_c \approx \overline{\mathbf{x}}_c^e = \mathbf{N}\mathbf{M}_c\mathbf{x}_c^n, \tag{5.88}$$

where $\mathbf{M}_c$ stands for a suitable averaging operator; $\overline{\mathbf{X}}_c^e$ and $\overline{\mathbf{x}}_c^e$ denote the reference and current discrete nodal positions at the element level, respectively, whilst the nodal positions are given by $\mathbf{X}_c^n$ and $\mathbf{x}_c^n$; $\mathbf{N}$ is the operator collecting isoparametric shape functions in the natural space $\{\xi^1, \xi^2\} \in [-1,1]\times[-1,1]$, where ξ^1, ξ^2 identify the natural coordinates of the interface.

The displacement field of the interface, $\overline{\mathbf{u}}_c$, is interpolated as follows:

$$\overline{\mathbf{u}}_c \approx \overline{\mathbf{u}}_c^e = \mathbf{N}\mathbf{M}_c\mathbf{d}_c, \tag{5.89}$$

where $\overline{\mathbf{u}}_c^e$ identifies the discrete nodal displacements at the element level, and $\mathbf{d}_c$ denotes the nodal displacement vector of the interface.

The global gap vector $\mathbf{g}_{\text{glob}}$ reads:

$$\mathbf{g}_{\text{glob}} \approx \mathbf{g}_{\text{glob}}^e = \mathbf{N}\mathbf{L}_c\mathbf{d}_c = \mathbf{B}_c\mathbf{d}_c, \tag{5.90}$$

where $\mathbf{L}_c$ is the so-called difference operator to compute the relative displacements at the interface between the two flanks, and $\mathbf{B}_c$ denotes the strain operator of the interface. The global gap vector is then referred to a local setting in order to evaluate the cohesive law by means of the rotation operator $\mathbf{R}_c$ as follows:

$$\mathbf{g}^e_{\text{loc}} = \mathbf{R}_c(\mathbf{d}_c)\mathbf{g}^e_{\text{glob}}. \tag{5.91}$$

The virtual local discrete gaps, $\delta\mathbf{g}^e_{\text{loc}}$, and the increment of this variation, $\Delta\delta\mathbf{g}^e_{\text{loc}}$, can be expressed as follows:

$$\delta\mathbf{g}^e_{\text{loc}} = \left[\mathbf{R}_c\mathbf{B}_c + \frac{\partial \mathbf{R}_c}{\partial \mathbf{d}_c}\mathbf{B}_c\mathbf{d}_c\right]\delta\mathbf{d}_c; \quad \Delta\delta\mathbf{g}^e_{\text{loc}} = \left[2\frac{\partial \mathbf{R}_c}{\partial \mathbf{d}_c}\mathbf{B}_c\delta\mathbf{d}_c\right]\Delta\mathbf{d}_c. \tag{5.92}$$

The increment of the cohesive tractions, which is required for the computation of the element stiffness matrix, reads:

$$\Delta\mathbf{T}^e = \mathbb{C}_c\left[\mathbf{R}_c\mathbf{B}_c + \frac{\partial \mathbf{R}_c}{\partial \mathbf{d}_c}\mathbf{B}_c\mathbf{d}_c\right]\Delta\mathbf{d}_c, \tag{5.93}$$

where $\mathbb{C}_c$ stands for the constitutive tangent tensor, see Eq. (5.54).

With the previous definitions at hand and after some additional operations, the interface element formulation proposed in [77] for large deformation analysis can be obtained.

5.6. Representative Applications

The modeling framework described above has been implemented into the FE package `ABAQUS` by means of the user-defined capability `UEL`. The objective of this section is to assess the performance of the proposed formulation involving geometric instabilities and buckling delamination events in different engineering systems. We focus the study on the following applications: (i) the postbuckling response of a CFRP composite stiffened panel up to collapse (Section 5.6.1), which is modeled using the three-dimensional shell parametrization complying with the ESL approach, (ii) the wrinkling-delamination analysis of thin layer–substrate systems, which are discretized using the solid shell element and the large deformation cohesive element previously outlined (Section 5.6.2).

5.6.1. Postbuckling analysis of composite stiffened panel

The first application is concerned with the analysis of the postbuckling evolution of a CFRP cylindrical composite stiffened panel under uniform pressure loading conditions up to failure [53, 55, 84]. This panel is part of an actual fairing of an aeronautical engine, satisfying the demanded tolerances

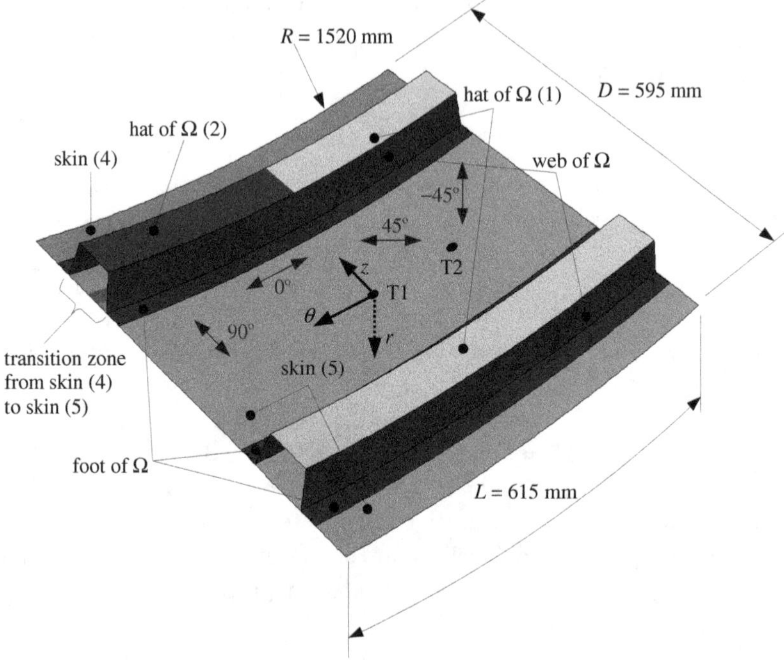

Fig. 5.10. Geometric definition of composite stiffened panel, where different laminate disposals and material orientations are identified.

Table 5.1. Mechanical properties of tape and fabric materials.

Material	E_1 (GPa)	E_2 (GPa)	G_{12}, G_{13} (GPa)	G_{23} (GPa)	ν_{12}
Tape	131	9.75	4.65	3	0.3
Fabric	61	61	4.9	3	0.05

in terms of geometry, material properties, flaws, among other aspects. The experimental program is conducted in the test facilities of the Grupo de Elasticidad y Resistencia de Materiales, Universidad de Sevilla, Spain.

Figure 5.10 portrays a sketch of the panel, whose main dimensions are: radius 1520 mm, arc length 615 mm, and width 594 mm. Two circumferential stringers with Ω-transverse section are symmetrically disposed (height 55 mm, with hat and flange widths of 60 and 25 mm, respectively), with a separation of 235 mm between their closer flanges. Table 5.1 reports the mechanical properties of the unidirectional and fabric plies carbon-epoxy laminates, whose nominal lamina thickness are equal

Table 5.2. Lay-ups of each zone of the panel, where the subscript \$ denotes perfect symmetric laminate except for the last layer (one less ply).

Ω hat(1)	Tape($\underline{0}$), Fabric (0,45)	$[45/\underline{0}/\underline{0}/0/\underline{0}/\underline{0}/45]$
Ω hat(2)	Tape($\underline{0}$), Fabric(0,45)	$[45/\underline{0}/\underline{0}/\underline{0}/0/\underline{0}/\underline{0}/\underline{0}/45]$
Ω web/flange	Fabric	[45/0/45]
Skin (4)	Tape	[45/−45/0/90/90/−45/45/0/0]\$
Skin (5)	Tape	[45/−45/0/90/0/−45/45]

to 0.184 and 0.28 mm, respectively. The circumferential direction identifies the reference material orientation, see Fig. 5.10. The stringer-skin pairs were co-bonded using and adhesive layer of 0.184 mm in thickness, with mechanical properties: $E = 3.216$ GPa, $\nu = 0.3$.

The laminate disposition at each zone of the panel is reported in Table 5.2. The stringers are manufactured from fabric material with a stacking sequence [45/0/45], being reinforced at the top hats. The stacking sequence of the skin corresponds to [45/−45/0/90/0/−45/45] in the region skin (5), being the number of plies is increased up to 17 at the region skin (4). Therefore, there exists a transition zone between both regions with steps of two laminas.

The external pressure is applied by means of a combination of gas and liquid using an external hermetic box. The panel is fastened to the hermetic box, which is specifically designed for this purpose. The mechanical response of the specimen during the test is recorded using 28 strain gages and two displacement transducers LVDT (locations T1 and T2 in Fig. 5.10) [84].

The experimental program is conducted in several stages with a final test up to the collapsing point of the specimen [84], which occurs at around 0.41 bars. Figure 5.11 shows the evolution of the displacement transducer at location T1 against the applied external loading. In this graph, a clear nonlinear evolution is observed once the buckling load of the specimen is overcome (which is estimated around 0.1 bar).

The numerical analysis of this specimen is performed using the layered version of the 7-p shell model based on the three-dimensional parametrization. The corresponding numerical model of the panel consists of a monolithic approach for the skin-stringers pairs with the following mesh characteristics: 3600 elements for each of the stringers and 6600 elements for the skin of the panel. The numerical analysis comprises: (i) linear buckling simulations, and (ii) subsequent general nonlinear static analyses. Fully clamped boundary conditions are set to the external boundaries of the model with the aim of reproducing the actual gripping system.

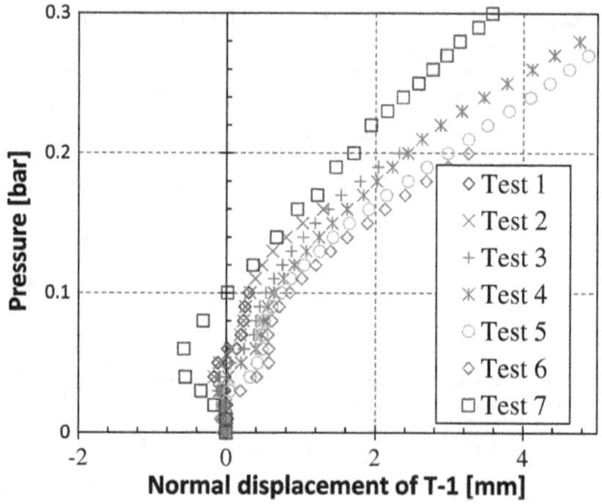

Fig. 5.11. Experimental evolution of the displacement transducer at the location T-1.

The geometric imperfections of the specimen are incorporated into the simulations through considering a perturbed geometry, which is based on the adjustment of the actual geometry of the panel by means of a certain combination of the previously estimated buckling modes [53]. With this information at hand the least square method (LSM) is employed to perform the best adjustment. The general scheme of this methodology can be expressed according to the following expression:

$$X_i(n) = X_i^o(n) + \sum_{k=1}^{k=M} a_k u_i^k(n), \tag{5.94}$$

where the perturbed position of node n, $X_i(n)$ can be obtained through combining the lowest M buckling modes; $X_i^o(n)$, $i = 1, 2, 3$ denote the nominal coordinates of node n; $u_i^k(n)$ identify the components of the displacement of the node n corresponding to the kth numerical buckling mode.

In [53], from 2 to 30 buckling mode shapes were used for the application of the proposed methodology. The coefficients a_i in Eq. (5.94) when 2 (label P02) and 5 (label P05) modes are combined are given in Table 5.3. Examining the reported data, it can be seen that the initial maximum geometric imperfections are about 50% of the nominal skin thickness.

Figure 5.12 shows a zoom of the experimental-numerical correlation at the location T-1 [84]. In this graph, it can be seen the strong dependency of the final deformed shape of the panel upon the number of buckling

Table 5.3. Adjusted combination using 2 (P02) and 5 (P05) buckling mode shapes.

Label	a_1	a_2	a_3	a_4	a_5
P02	–4.02E-4	4.09E-4			
P05	–4.59E-4	2.52E-4	–6.59E-4	–1.55E-4	2.14E-4

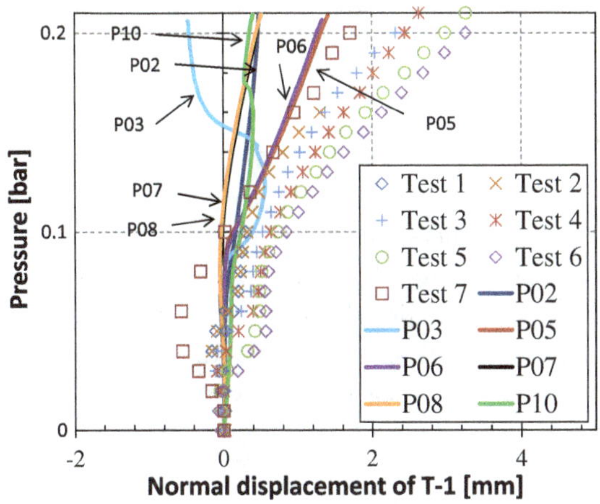

Fig. 5.12. Evolution for the transducer T-1: experimental-numerical correlation.

modes selected to perturb the nominal geometry of the panel. Note that a satisfactory adjustment to the experimental data up to around the first buckling load predicted ($P_{cr} = 0.1$ bar) is obtained for all the combinations herein shown. Nevertheless, beyond this load level, the best agreement between the numerical and the experimental data is achieved when the first 5 (P05) or 6 (P06) buckling modes are combined in order to perturb the nominal geometry. These data are in accordance with respect to those reported in [53]. These differences (Fig. 5.12) are very similar to those obtained using the built-in ABAQUS shell formulations S4R and SC8R, and they are mainly attributed to lack of precise information regarding the actual surface of the panel via experimental techniques. A possible manner to reduce such discrepancies would concern the use of scanning systems or alternative procedures to measure the complete surface of the specimen. Future research actions are planned in order to shed light to the role of the geometrical imperfections in the postbuckling evolution of slender composites structures.

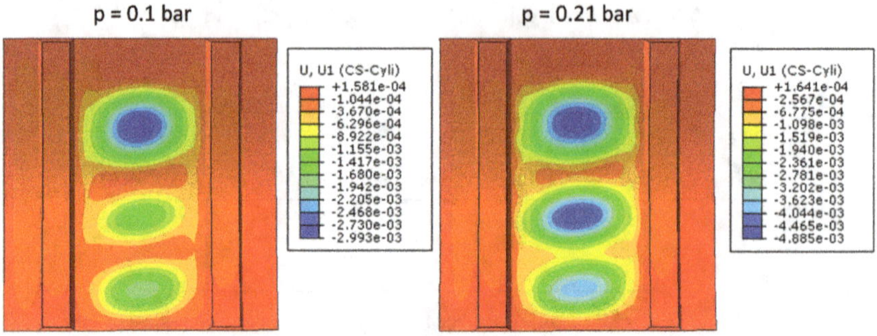

Fig. 5.13. Postbuckling radial displacements estimations using the ESL version of the 7-p three-dimensional shell element.

Finally, Fig. 5.13 depicts the estimated radial displacement along the postbuckling evolution using the proposed layered 7-p formulation. The final deformed shape is predicted to be characterized by two main half-waves, where one of them is subsequently subdivided into two sub-waves. These deformation patterns satisfactorily agree with the experimental observations, showing the reliability of the proposed approach [55, 84].

5.6.2. Wrinkling–delamination analysis of composite systems

In what follows, the analysis of wrinkling–delamination in composite systems is carried out using the proposed computational framework. In particular, the numerical investigation comprises the following stages combining the solid shell element and the cohesive interface for large deformation analyses: (i) the estimation of the wrinkling loadings by means of linear buckling analysis; (ii) postbuckling analyses assuming perfect interfaces for the thin layer–substrate, i.e., no delamination failure is considered; and (iii) postbuckling simulations including the potential development of delamination events.

Wrinkling and buckling-induced delamination of FG TBCs bonded to soft substrates has been investigated using perturbation-based semi-analytical procedures, whose results are used in order to calibrate subsequent FE simulations [60].

The system herein analyzed is composed by a substrate and a thin layer, whose respective thicknesses are denoted by $\bar{H}$ and h, whereas L and B identify their span and width, respectively, see Fig. 5.14. The following boundary conditions (Fig. 5.15) are defined in order to reproduce the

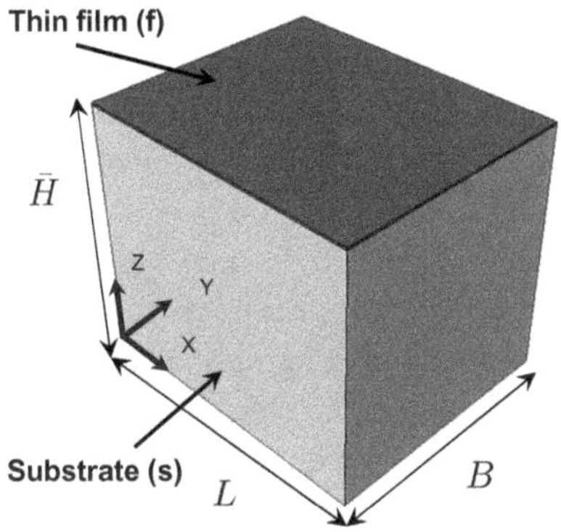

Fig. 5.14. Mechanical system FG thin layer over soft substrate under uniform compression loading: geometric definitions.

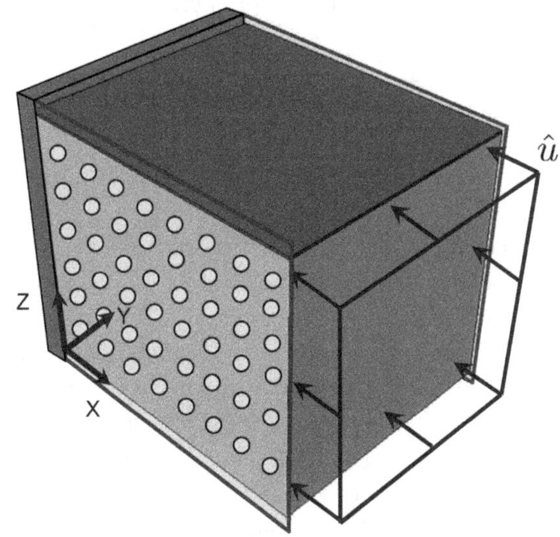

Fig. 5.15. Mechanical system FG thin layer over soft substrate under uniform compression loading: supporting conditions.

conditions in [60]: (i) the vertical displacement is restrained at the surface $z = 0$; (ii) the horizontal displacement at the surface $x = 0$ in x-direction is constrained; (iii) the displacements in y-direction are restrained at the surfaces $y = 0, B$; (iv) the applied external displacement in x-axis ($\hat{u}$) is imposed on the plane $x = L$, leading to uniform strain conditions strain $\varepsilon = \hat{u}/L$ within the domain.

In the current investigation, the analysis of FG thin layer–substrate systems is performed considering a significant mismatch between the mechanical properties. In particular, we set the following ratio between the Young's moduli of the layer, E_f, and the substrate, E_s: $E_f/E_s \geq 1000$. Kirchhoff–Saint–Venant material law is assigned to the thin layer and substrate, whilst the nonlinear CZM discussed in Section 5.4 is employed for modeling delamination failure at the interface between both entities. The system is discretized using conforming meshes in terms of mesh density and order of displacements interpolation. Regarding the FE-topology, solid shell finite elements correspond to the thin layer, whereas cohesive elements and standard continuum elements are used for the interface and the substrate, respectively.

The first stage of the simulation regards the computation of linear buckling analysis. Derived from analytical techniques [60], the estimation critical strain ε_c, at which wrinkling is predicted to take place, reads

$$\varepsilon_c = \frac{1}{4}\left[\frac{3\overline{E}_s}{\overline{E}_f}\right]^{2/3}\left[1-\frac{1}{4}\left(\frac{1-2\nu_s}{1-\nu_s}\right)^2\right]^{-2/3}, \tag{5.95}$$

where $\overline{E}_s$ and $\overline{E}_f$ are the plane strain Young's moduli of the substrate and of the film, respectively.

Figure 5.16 shows the evolution of the numerically estimated critical strain ε_c with respect to the thickness ratio $\bar{H}/h$ in comparison with the analytical values [60]. In this graph, a satisfactory agreement between both the current FE framework and the semi-analytical approaches can be observed. This linear buckling analysis is further employed for FG coating, whose stiffness ratios between the film $E_{f,x}$, x meaning the grading ratio, and the substrate E_s are reported in Table 5.4. Figure 5.17 depicts the comparison between semi-analytical critical strain (using the equivalent Young's moduli averaged over the FG thickness) and the lowest numerically estimated eigenvalue for an infinitely thick substrate $\bar{H}/h > 20$. Again, a good adjustment between both methods with a maximum deviation of 1.4% between the corresponding data is achieved.

The postbuckling and the postbuckling-delamination analyses of the current system complete the current application. In particular, the postbuckling evolution beyond the critical strain is conducted assuming perfectly bonded layer–substrate interface (without damage capabilities). The nominal geometry is perturbed using the lowest numerically computed eigenmode according to the scheme given in Eq. (5.94). For the particular

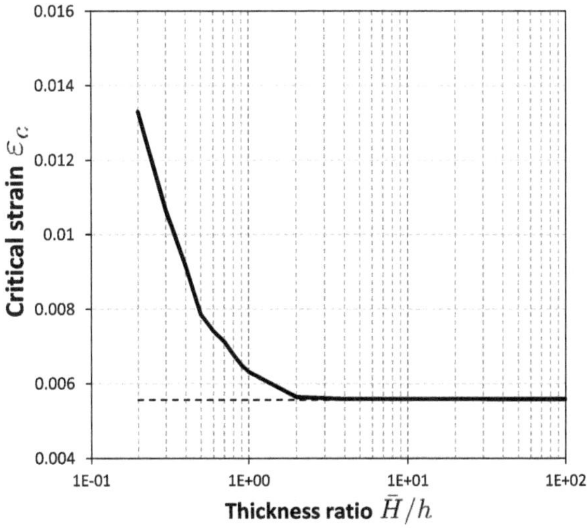

Fig. 5.16. Estimation of the critical strain: Effect of the H/h ratio on the FE analysis of wrinkling for $E_f/E_s = 1000$, see Eq. (5.46).

Table 5.4. Elastic property ratios using an equivalent Young's modulus over the shell thickness for FG coatings. $E_{f,m}$ and $E_{f,c}$ denote the Young's moduli for pure metallic and ceramic fractions, respectively; $E_{f,x}$ denotes the equivalent value after the integration of the variable Young's modulus over the thickness, where x denotes the grading exponent, see Eq. (5.46).

$E_{f,m}/E_s$	$E_{f,5}/E_s$	$E_{f,2}/E_s$	$E_{f,1}/E_s$	$E_{f,0.5}/E_s$	$E_{f,0.2}/E_s$	$E_{f,0.1}/E_s$	$E_{f,c}/E_s$
1000	1193	1386	1579	1771	1964	2052	2154

case of $n = 0.1$, Eq. (5.46), Fig. 5.18 shows the wrinkle amplitude for different imperfection magnitudes as a function of the nominal strain in comparison with the semi-analytical procedure presented in [60], where $*h$ identifies the perturbation magnitude being the coefficient $*$ the scaling value with respect to the thin layer. The semi-analytical value for wrinkle amplitude obtained by Mei *et al.* [60] beyond the critical strain reads

$$A_w(\lambda) = \frac{\lambda}{\pi}\sqrt{\varepsilon - \varepsilon_c}, \tag{5.96}$$

where λ denotes the wrinkle wavelength.

Analyzing the data shown in this graph and setting $\lambda_c/h = 50$ and $\varepsilon_c = 0.00336213$, it can be seen that, as expected, the FE results depend on

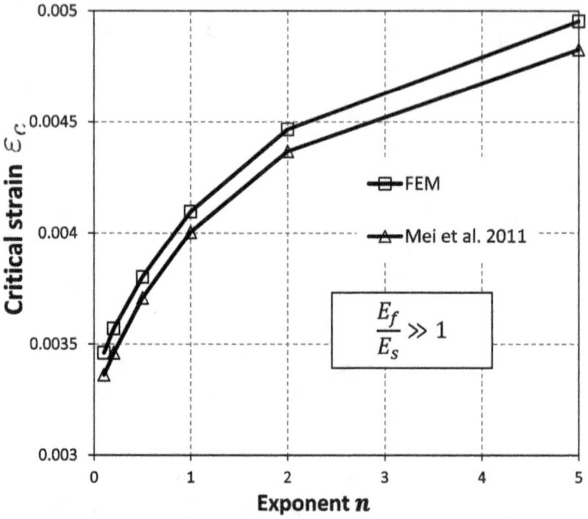

Fig. 5.17. Comparison between the FE and semi-analytical critical strain predictions for several grading exponent values, see Eq. (5.46).

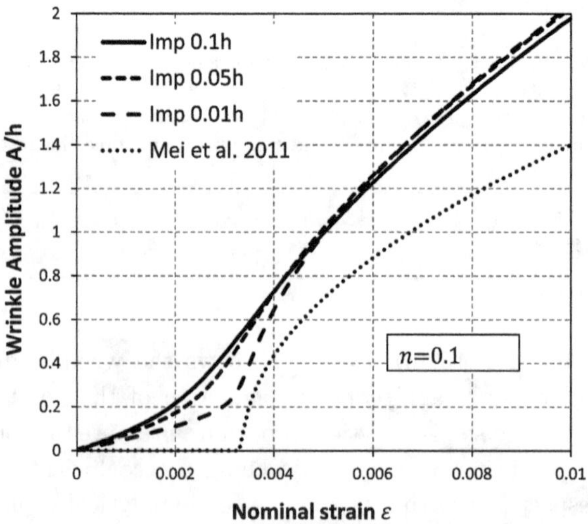

Fig. 5.18. Comparison between the FE and semi-analytical results for wrinkle amplitude-nominal strain evolution with $n = 0.1$ using different magnitudes of the geometric imperfection.

the magnitude of the initial imperfection. Thus, the lower the scaling value is set, the closer agreement with the semi-analytical results is achieved. Nevertheless, it is noting that the FE data slightly overestimate the value of semi-analytical data at mid postbuckling stages.

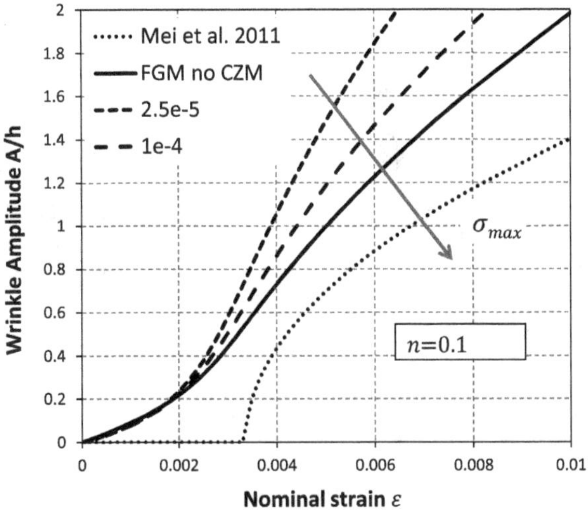

Fig. 5.19. Wrinkle amplitude-nominal strain evolution for $n = 0.1$ for different value of the peak interface strength $\sigma_{\max}$ using an initial imperfection of about 10% of the film thickness.

Finally, postbuckling-delamination computations using the proposed CZ formulation is performed using the previous configuration. In this concern, we analyze the role of the maximum interfacial normal traction ($\sigma_{\max}$), which can originate delamination events. Mei *et al.* [60] estimated a closed-form for the maximum normal traction which provokes the initiation of the wrinkle induced delamination as

$$\sigma_{\max} = \frac{4(1-\nu_s)^2 \bar{E}_s}{3 - 4\nu_s} \sqrt{\varepsilon - \varepsilon_c}. \tag{5.97}$$

The previous expression is taken as the peak Mode I CZM traction for the interface model and particularizes for the corresponding material properties of the system. Figure 5.19 depicts the wrinkle amplitude evolution as a function of the nominal strain for different values of $\sigma_{\max}$ and using an initial imperfection magnitude of about 10% of the film thickness. In this evolution, it can be observed that the delamination is predicted to be initiated immediately prior to the occurrence of wrinkling. This effect is a direct consequence of the notable elastic mismatch between both entities of the system, namely the thin layer and the substrate. This trend can be more clearly observed for lower values of the interface strength, leading to higher wrinkle amplitudes. Moreover, note that beyond the nominal strain of delamination initiation, both nonlinear phenomena (delamination and wrinkling) concomitantly evolve [60]. In order to illustrate this aspect,

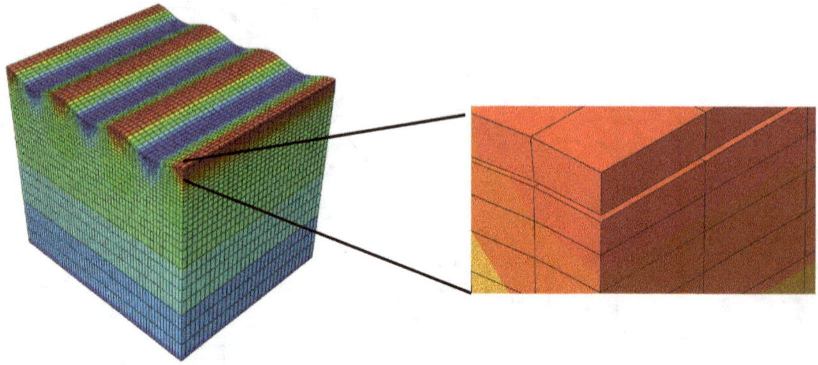

Fig. 5.20. Postbuckling deformation pattern and layer–substrate delamination detail.

Fig. 5.20 shows the postbuckling deformation pattern at an intermediate stage and a layer–substrate delamination detail at one of the corners of the system.

5.7. Concluding Remarks

In this chapter, a modeling framework for instabilities and delamination in composite shells has been presented. On the side of shell theory, several possible continuum-based parametrizations were comprehensively revisited. In particular, two shell models have been considered and extended to deal with composite layers: (i) the three-dimensional shell parametrization complying with the ESL for layered CFRP structures, and (ii) the solid shell parametrization, which has been used for the instability and delamination analysis of thin layer–substrate systems. Specific aspects with regard to the variational basis and the corresponding FE formulation have been addressed.

The performance of the present methods has been assessed by means of two applications. Nonlinear finite element simulations comprising linear buckling, postbuckling and postbuckling-delamination analyses have been successfully carried out. The corresponding results for the applications under consideration have demonstrated the applicability and reliability of the proposed approach in comparison with alternative semi-analytical procedures and experimental data.

Further research activities are currently being considered for the analysis of similar phenomena in complex multi-layered systems. In particular the use of structured interfaces to optimized the inter-laminar behavior, 3D roughness induced imperfections and the combination of intra-laminar and inter-laminar failure events are under investigation. These developments feature a new frontier regarding modeling instabilities and failure in composite structures.

Acknowledgments

JR and AB are grateful to the support of the Spanish Ministry of Economy and Competitiveness (Projects DPI2012-37187, MAT2015-71036-P and MAT2015-71309-P) and the Andalusian Government (Projects of Excellence P11-TEP-7093 and P12-TEP-1050). MP is thankful to the European Research Council for supporting the ERC Starting Grant "Multi-field and multi-scale Computational Approach to Design and Durability of PhotoVoltaic Modules" — CA2PVM, under the European Union's Seventh Framework Programme (FP/2007-2013)/ERC Grant Agreement No. 306622. J. Reinoso acknowledges financial support from the Ministry of Science, Innovation and Universities (Project PGC2018-099197-B-I00), the Consejería de Economía y Conocimiento, Junta de Andalucía, and the European Regional Development Fund (Projects US-1265577 and P20-00595).

The authors would like to dedicate this chapter to Prof. Ekkehard Ramm. His thorough knowledge of the topic has been of crucial importance along the progress of the current research by means of very fruitful discussions and suggestions. Especially, JR would like to acknowledge Prof. Ekkehard Ramm for becoming a reference person from the scientific and (more importantly) personal points of view.

References

[1] Ahmad S., Irons B.M., Zienkiewicz O.C., 1970. Analysis of thick shell and thin shell structures by curved finite elements, *Int. J. Numer. Meth. Eng.*, 2, 419–459.

[2] Bischoff M., Wall, W.A., Bletzinger K.-U., Ramm E., 2004. Models and finite elements for thin-walled structures, in: E. Stein, R. de Borst, T.J.R. Hughes, eds., *Encyclopedia of Computational Mechanics, Volume 2: Solids and Structures*, Wiley, Chapter 3.

[3] Cottrell J.A., Hughes T.J.R., Bazilevs Y., 2009. *Isogeometric Analysis: Toward Integration of CAD and FEA*, Wiley.

[4] Kiendl J., Bletzinger K.U., Linhard J., Wuechner R., 2009. Isogeometric shell analysis with Kirchhoff–Love elements, *Comput. Method. Appl. Mech. Eng.*, 198(49), 3902–3914.

[5] Echter R., Oesterle B., Bischoff M., 2013. A hierarchic family of isogeometric shell finite elements, *Comput. Method. Appl. Mech. Eng.*, 254, 170–180.
[6] Oesterle B., Ramm E., Bischoff M., 2016. A shear deformable, rotation-free isogeometric shell formulation, *Comput. Method Appl. Mech. Eng.*, 307, 235–255.
[7] Carrera E., 2001. Developments, ideas, and evaluations based upon Reissner's mixed variational theorem in the modeling of multilayered plates and shells, *Appl. Mech. Rev.*, 54, 301–328.
[8] Carrera E., Cinefra M., Lamberti A., Petrolo M., 2015. Results on best theories for metallic and laminated shells including layer–wise models, *Compos. Struct.*, 126, 285–298.
[9] Rodrigues J.D., Roque C.C., Ferreira A.J.M., Carrera E., Cinefra M., 2011. RBF-FD collocation and a unified formulation for bending, vibration and buckling analysis of laminated plates, according to Murakami's zig-zag theory, *Compos. Struct.*, 93, 1613–1620.
[10] Reddy J.N., 1997. *Mechanics of Laminated Composite Plates, Theory and Analysis*, CRC Press.
[11] Carrera E., 2002. Theories and finite elements for multilayered anisotropic, composite plates and shells. *Arch. Comput. Method Eng.*, 9, 87–140.
[12] Braun M., Bischoff M., Ramm E., 1994. Nonlinear shell formulations for complete three-dimensional constitutive laws including composites and laminates, *Comput. Mech.*, 15, 1–18.
[13] Murakami H., 1986. Laminated composite plate theory with improved in-plane response, *J. Appl. Mech.*, 53, 661–666.
[14] Carrera E., 2004. On the use of Murakami's zig-zag function in the modeling of layered plates and shells, *Comput. Struct.*, 82, 541–554.
[15] Ericksen J.L., Truesdell C., 1958. Exact theory of stress and strain in rods and shells, *Arch. Ration. Mech. Anal.*, 1:295–323.
[16] Naghdi P.M., 1972. *The Theory of Shells*, Handbuch der Physik, Vol. VI/2, Springer, Berlin.
[17] Büchter N., Ramm E., 1992. Shell theory versus degeneartion- a comparison in large rotation finite element analysis, *Int. J. Numer. Meth. Eng.*, 34, 39–59.
[18] Bischoff M., Ramm E. 2000. On the physical significance of higher order kinematic and static variables in a three-dimensional shell formulation, *Int. J. Solids Struct.*, 37, 6933–6960.
[19] Parisch H., 1995. A continuum-based shell theory for non-linear applications, *Int. J. Numer. Meth. Eng.*, 38, 1855–1883.
[20] Bischoff M., Ramm E. 1997. Shear deformable shell elements for large strains and rotations, *Int. J. Numer. Meth. Eng.*, 40, 4427–4449.
[21] Büchter N., Ramm E., Roehl D., 1994. Three-dimensional extension of nonlinear shell formulation based on the enhanced assumed strain concept, *Int. J. Numer. Meth. Eng.*, 37, 2551–2568.
[22] Rivera M.G., Reddy J.N., Amabili M., 2016. A new twelve-parameter spectral/hp shell finite element for large deformation analysis of composite shells, *Compos. Struct.*, 151, 183–196.

[23] Ferreira A.J.M., Araújo A.L., Neves A.M.A., Rodrigues J.D., Carrera E., Cinefra M., Mota Soares C.M., 2013. A finite element model using a unified formulation for the analysis of viscoelastic sandwich laminates, *Compos. B Eng.*, 203(45), 1258–1264.

[24] Dehkordi M.B., Khalili S.M.R., Carrera E., 2016. Non-linear transient dynamic analysis of sandwich plate with composite face-sheets embedded with shape memory alloy wires and flexible core-based on the mixed LW (layer-wise)/ESL (equivalent single layer) models, *Compos. B Eng.*, 87, 59–74.

[25] Robbins Jr D.H., Reddy J.N., 1993. Modeling of thick composites using a layer-wise theory, *Int. J. Numer. Meth. Eng.*, 36, 655–677.

[26] Tornabene F., 2016. General higher-order layer-wise theory for free vibrations of doubly-curved laminated composite shells and panels, *Mech. Adv. Mater. Struct.*, 23, 1046–1067.

[27] Thai C.H., Ferreira A.J.M., Carrera E., Nguyen-Xuan H., 2013. Isogeometric analysis of laminated composite and sandwich plates using a layerwise deformation theory, *Compos. Struct.*, 104, 196–214.

[28] Arciniega R.A., Reddy J.N., 2007. Tensor–based finite element formulation for geometrically nonlinear analysis of shell structures, *Comput. Method Appl. Mech. Eng.*, 196, 1048–1073.

[29] Sansour C., 1995. A theory and finite element formulation of shells at finite deformations including thickness change: circumventing the use of a rotation tensor, *Arch. Appl. Mech.*, 10:194–216.

[30] Simo J.C., Rifai M.S., Fox D.D., 1990. On a stress resultant geometrically exact shell model. Part IV: Variable thickness shells with through-the-thickness stretching, *Comput. Method Appl. Mech. Eng.*, 81, 91–126.

[31] Alves de Sousa R.J.A., Cardoso R.P.R., Fontes Valente R.A., Yoon J.W., Grácio J.J., Natal Jorge R.M., 2005. A new one-point quadrature enhanced assumed strain (EAS) solid-shell element with multiple integration points along thickness: Part I — geometrically linear applications, *Int. J. Numer. Meth. Eng.*, 62(7), 952–977.

[32] Başar Y., Itskov M., Eckstein A., 2000. Composite laminates: nonlinear interlaminar stress analysis by multi-layer shell elements, *Comput. Method. Appl. Mech. Eng.*, 185, 367–397.

[33] Hartmann S., Brunssen S., Ramm E., Wohlmuth B., 2007. Unilateral non-linear dynamic contact of thin-walled structures using a primal–dual active set energy, *Int. J. Numer. Meth. Eng.*, 70, 883–912.

[34] Hauptmann R., Schweizerhof K., Doll S., 2000. Extension of the solid-shell concept for application to large elastic and large elastoplastic deformations, *Int. J. Numer. Meth. Eng.*, 49, 1121–1141.

[35] Klinkel S., Wagner W., 1997. A geometrical nonlinear brick element based on the EAS method, *Int. J. Numer. Meth. Eng.*, 40, 4529–4545.

[36] Miehe C., 1998. A theoretical and computational model for isotropic elastoplastic stress analysis in shells at large strains, *Comput. Method. Appl. Mech. Eng.*, 155, 193–233.

[37] Reese S., 2007. A large deformation solid-shell concept based on reduced integration with hourglass stabilization, *Int. J. Numer. Meth. Eng.*, 69, 1671–1716.
[38] Simo J.C., Rifai S., 1990. A class of mixed assumed strain methods and the method of incompatible modes, *Int. J. Numer. Meth. Eng.*, 29, 1595–1638.
[39] Glaser S., Armero F., 1997. On the formulation of enhanced strain finite elements in finite deformations. *Eng. Comput.*, 14, 759–791.
[40] Simo J.C., Armero F., 1992. Geometrically nonlinear enhanced strain mixed methods and the method of incompatible modes, *Int. J. Numer. Meth. Eng.*, 33, 1413–1449.
[41] Vu-Quoc L., Tan X.G., 2003. Optimal solid shells for non-linear analyses of multilayer composites. I. Statics, *Comput. Method Appl. Mech. Eng.*, 192, 975–1016.
[42] Wriggers P., Reese S., 1996. A note on enhanced strain methods for large deformations, *Comput. Method Appl. Mech. Eng.*, 135, 201–209.
[43] Pian T., Sumihara K., 1984. Rotational approach for assumed stress finite elements, *Int. J. Numer. Meth. Eng.*, 20, 1685–1695.
[44] Gruttmann F., Wagner W., 2006. Structural analysis of composite laminates using a mixed hybrid shell element. *Comput. Mech.*, 37, 479–497.
[45] Kasper E., Taylor R.L., 2000. A mixed-enhanced strain method. Part I: geometrically linear problems, *Comput. Struct.*, 75, 237–250.
[46] Kasper E., Taylor R.L., 2000. A mixed-enhanced strain method. Part II: geometrically nonlinear problems, *Comput. Struct.*, 75, 251–260.
[47] Klinkel S., Gruttmann F., Wagner W., 2006. A robust non-linear solid shell element based on a mixed variational formulation, *Comput. Method Appl. Mech. Eng.*, 195, 179–201.
[48] Vu-Quoc L., Tan X., 2013. Efficient hybrid-EAS solid element for accurate stress prediction in thick laminated beams, plates, and shells, *Comput. Method Appl. Mech. Eng.*, 253, 337–355.
[49] Bathe K.J., Dvorkin E.N., 1985. A four-node plate bending element based on Mindlin/Reissner plate theory and a mixed interpolation, *Int. J. Numer. Meth. Eng.*, 21, 367–383.
[50] Betsch P., Gruttmann F., Stein E., 1996. A 4-node finite shell element for the implementation of general hyperelastic 3D-elasticity at finite strains, *Comput. Method. Appl. Mech. Eng.*, 130, 57–79.
[51] Schwarze M., Reese S., 2009. A reduced integration solid-shell element based on the EAS and the ANS concept — geometrically linear problems, *Int. J. Numer. Meth. Eng.*, 80, 1322–1355.
[52] Bisagni C., Cordisco P., 2006. Post–buckling and collapse experiments of stiffened composite cylindrical shells subjected to axial loading and torque, *Compos. Struct.*, 73, 138–149.
[53] Blázquez A., Reinoso J., París F., Cañas J., 2012. Analysis in the postbuckling regime of a pressurized stiffened panel. Part II: Numerical analysis and effect of the geometric imperfections, *Compos. Struct.*, 94(5), 1544–1554.

[54] Falzon B.G., Stevens K.A., Davies G.O., 2000. Postbuckling behaviour of a blade–stiffened composite panel loaded in uniaxial compression, *Compos. A-Appl. Sci. Manuf.*, 31(5), 459–468.
[55] Reinoso J., Blázquez A., 2016. Application and finite element implementation of 7-parameter shell element for geometrically nonlinear analysis of layered CFRP composites, *Compos. Struct.*, 139, 263–276.
[56] Falzon B.G., Davies G.A.O., Greenhalgh E., 2001. Failure of thick-skinned stiffener runout sections loaded in uniaxial compression, *Compos. Struct.*, 53, 223–233.
[57] Turón A., Camanho P.P., Costa J., Dávila C.G., 2006. A damage model for the simulation of delamination in advanced composites under variable-mode loading, *Mech. Mater.*, 38, 1072–1089.
[58] Maimí P., Camanho P.P., Mayugo J.A., Dávila C.G., 2007. A continuum damage model for composite laminates: part I — constitutive model, *Mech. Mater.*, 39, 897–908.
[59] Rogers J.A., Someya T., Huang Y., 2010. Materials and mechanics for stretchable electronics, *Science*, 327, 1603–1607.
[60] Mei H., Landis C.M., Huang R., 2011. Concomitant wrinkling and buckle-delamination of elastic thin films on compliant substrates, *Mech. Mater.*, 43, 627–642.
[61] Huang Z.Y., Hong W., Suo Z., 2005. Nonlinear analysis of wrinkling in a film bonded to a compliant substrate, *J. Mech. Phys. Solids*, 53, 2101–2118.
[62] Hutchinson J.W., 2013. The role of nonlinear substrate elasticity in the wrinkling of thin films, *Phil. Trans. R. Soc. A.*, 371, 20120422.
[63] Roccabianca S., Bigoni D., Gei M., 2011. Long wavelength bifurcations and multiple neutral axes of elastic layered structures subject to finite bending, *J. Mech. Mater. Struct.*, 6, 511–527.
[64] Zang J., Zhao X., Cao Y., Hutchinson J.W., 2012. Localized ridge wrinkling of stiff films on compliant substrates, *J. Mech. Phys. Solids*, 60, 1265–1279.
[65] Jiang H., Khang D.Y., Fei H., Kim H., Huang Y., Xiao J., Rogers J.A., 2008. Finite width effect of thin-films buckling on compliant substrate: Experimental and theoretical studies, *J. Mech. Phys. Solids*, 56, 2585–2598.
[66] Reinoso J., Paggi M., Areias P., 2016. A finite element framework for the interplay between delamination and buckling of rubber-like bi-material systems and stretchable electronics, *J. Eur. Ceram. Soc.*, 36(9), 2371–2382.
[67] Reinoso J., Paggi M., Rolfes R., 2016. A computational framework for the interplay between delamination and wrinkling in functionally graded thermal barrier coatings, *Comp. Mater. Sci.*, 116, 82–95.
[68] Hosseini S., Remmers J.J.C., Verhoosel C.V., Borst R., 2013. An isogeometric solid-like shell element for nonlinear analysis, *Int. J. Numer. Meth. Eng.*, 95(3), 238–256.
[69] Bonet J., Wood R.D., 1997. *Nonlinear Continuum Mechanics for Finite Element Analysis*, Cambridge University Press.
[70] Holzapfel G., 2000. *Nonlinear Solid Mechanics*, Wiley.

[71] Arciniega R.A., Reddy J.N., 2007. Large deformation analysis of functionally graded shells, *Int. J. Solids Struct.*, 44, 2036–2052.

[72] Coleman B.D., Noll W., 1963. The thermodynamics of elastic materials with heat conduction and viscosity, *Arch. Ration. Mech. Anal.*, 13, 167–178.

[73] Paggi M., Wriggers P., 2011. A nonlocal cohesive zone model for finite thickness interfaces — Part I: Mathematical formulation and validation with molecular dynamics, *Comp. Mater. Sci.*, 50(5), 1625–1633.

[74] Paggi M., Wriggers P., 2011. A nonlocal cohesive zone model for finite thickness interfaces — Part II: FE implementation and application to polycrystalline materials, *Comp. Mater. Sci.*, 50(5), 1634–1643.

[75] van den Bosch M.J., Schreurs P.J.G., Geers M.G.D., 2008. On the development of a 3D cohesive zone element in the presence of large deformations, *Comput. Mech.*, 42, 171–180.

[76] Qiu Y., Crisfield M.A., Alfano G., 2001. An interface element formulation for the simulation of delamination with buckling, *Eng. Frac. Mech.*, 68, 1755–1776.

[77] Reinoso J., Paggi M., 2014. A consistent interface element formulation for geometrical and material nonlinearities, *Comput. Mech.*, 54, 1569–1581.

[78] Paggi M., Reinoso J., 2015. An anisotropic large displacement cohesive zone model for fibrillar and crazing interfaces, *Int. J. Solids Struct.*, 69–70, 106–120.

[79] Dávila C.G., Camanho P.P., Turón A., 2007. *Cohesive elements for shells*, Technical Report, NASA/Technical Publication, TP-2007-214869.

[80] Tvergaard V., 1990. Effect of fiber debonding in a whisker-reinforced metal, *Mat. Sci. Eng. A*, 107, 23–40.

[81] Reinoso J., Blázquez A., 2016. Geometrically nonlinear analysis of functionally graded power-based and carbon nanotubes reinforced composites using a fully integrated solid shell element, *Compos. Struct.*, 152, 277–294.

[82] Andelfinger U., Ramm E., 1993. EAS — Elements for two-dimensional, three-dimensional, plate and shells and their equivalence to HR-elements. *Int. J. Numer. Meth. Eng.*, 36, 1413–1449.

[83] Rah K., Van Paepegem W., Habraken A.M., Degrieck J., de Sousa R.A., Valente R.A.F., 2013. Optimal low-order fully integrated solid-shell elements, *Comput. Mech.*, 51(3), 309–326.

[84] Reinoso J., Blázquez A., París F., Cañas J., Meléndez J.C., 2012. Postbuckling behaviour of a pressurized stiffened composite panel. Part I: Experimental analysis, *Compos. Struct.*, 94(5), 1533–1543.

https://doi.org/10.1142/9781800611887_0006

Chapter 6

Bifurcation of Elastic Multilayers

Davide Bigoni[*,§], **Massimiliano Gei**[†,¶], **and Sara Roccabianca**[‡,||]
[*]Department of Civil, Environmental and Mechanical Engineering, University of Trento, Via Mesiano 77, I-38123 Trento, Italy
[†]Department of Engineering and Architecture, University of Trieste, Via A. Valerio 6/1, I-34127 Trieste, Italy
[‡]College of Engineering, Michigan State University, East Lansing MI 48824, USA
[§]*davide.bigoni@unitn.it*
[¶]*massimiliano.gei@dia.units.it*
[||]*roccabis@egr.msu.edu*

Abstract

The occurrence of a bifurcation in a multilayer structure during loading sets a limit on its deformability, and therefore represents an important factor in the design of composites. Since bifurcation is strongly influenced by the contact conditions at the interfaces between the layers, the mechanical modeling of these conditions is crucial. The theory of incremental bifurcation is reviewed for elastic multilayers, when these are subject to a finite strain before bifurcation, corresponding to uniform tension/compression and finite bending. The interlaminar contact is described by introducing linear imperfect interfaces. Results are critically discussed in view of applications and available experiments.

Contents

6.1. Introduction

Natural geological formations, biological materials and human-made structures, such as sandwich panels, submarine coatings, microelectronic devices, ceramic capacitors, are often made of layers of different materials bonded together, to form so-called *multilayers*. Large strains are imposed on these structures (i) as an industrial need (for instance when forming metallic multilayers [1], 'wrapping' engineered tissues around tubular supports to create artificial blood vessels [2] or bending of multilayer flexible solar cells [3]), (ii) under working conditions (for instance when multilayer films are employed for flexible packaging) or (iii) as a natural process (for instance during morphogenesis of arteries or geological formations or when the leaf of a plant bends to trap an insect, to disperse seeds or to resist dehydration). In all these cases, the occurrence of various forms of bifurcation sets limits to deformation performance. For instance compressive strain is limited by buckling and subsequent folding (see the example on the left-hand side of Fig. 6.1), uniform tensile strain may terminate with shear band formation and growth, while uniform flexure may

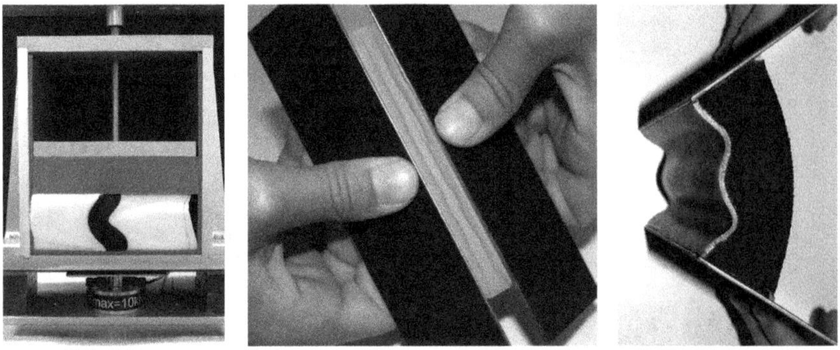

Fig. 6.1. Left: A stiff (30 mm thick, neoprene) layer bonded by two compliant (100 mm thick foam) layers in a rigid-wall, confined compression apparatus (note that separation between sample and wall has occurred on the right upper edge of the sample). Center: Creases on the compressive side of a rubber strip, coated on the tensile side with a 0.4 mm thick polyester transparent film, subject to flexure. Right: Bifurcation of a two-layer rubber block under finite bending, evidencing long-wavelength bifurcation modes (the stiff layer, made of natural rubber, is on the compressive side of a neoprene block).

lead to the formation of bifurcation modes such as creases and undulations (see the example in the center and on the right-hand side of Fig. 6.1). Bifurcation is therefore an important factor in the design of multilayered materials, and so it has been the focus of a thorough research effort, initiated by Maurice A. Biot [4] and continued by many others. In particular, elastic layered structures deformed in plane strain and subject to a uniform state of stress have been analyzed by Dorris and Nemat-Nasser [5], Steif [6–9], Papamichos, Vardoulakis and Muhlhaus [10], Dowaikh and Ogden [11], Benallal, Billardon and Geymonat [12], Triantafyllidis and Lehner [13], Triantafyllidis and Leroy [14], Shield, Kim and Shield [15], Ogden and Sotiropoulos [16] and Steigmann and Ogden [17] as a bifurcation problem of an isolated layer subject to uniform tension or compression [18–20]. A perturbative approach for a layered structure idealized as a constrained Cosserat material near failure of ellipticity has been introduced by Bigoni and Gourgiotis [21]. Layered structures subject to finite bending have been considered by Roccabianca, Bigoni and Gei [22, 23], who found solutions both for the non-uniform state of stress that develops during flexure[a] and for the related incremental bifurcation problem. These findings relied on a

[a]The solution of finite flexure of an elastic multilayered structure is interesting from different points of view, since the stress state induced by bending is complex (it may involve for instance the presence of more than one neutral axis) and strongly influences bifurcation.

Fig. 6.2. Bifurcation through compression of a finely layered metamorphic rock has induced severe folding. This is an example of a so-called *accommodation structure* (Trearddur Bay, Holyhead, N. Wales, UK; the coin in the photo is a pound).

generalization of previous results for plane-strain bending of an elastic block given by Rivlin [24] and on analyses of incremental bifurcations [25–31].

The bifurcation loads and modes are strongly sensitive to the bonding conditions between the layers, which may be perfect (as in the case of the rock shown in Fig. 6.2), but often may involve the possibility of slip and detachments, resulting in so-called *delaminations* (as in the cases shown in Fig. 6.3). A simple way to account for this crucial behavior is to introduce interfacial laws at the contact between layers. The simplest model of these laws is linear and can be formulated by assuming that the interface has null [32–35] or finite [36–38] thickness. We will limit our attention to zero-thickness linear interfaces, across which the nominal traction increment remains continuous, but linearly related to the jump in incremental displacement, which is unrestricted. For simplicity, the materials forming the multilayer are assumed hyperelastic and incompressible, according to the general framework laid out by Biot [4], in which Mooney–Rivlin and Ogden materials [39], as well as the J_2-deformation theory of plasticity materials, are particular cases. Therefore, the constitutive laws are broad enough to embrace the behavior of rubber, plastics, and geological materials, but also of ductile metals subject to proportional loading, as they can be represented in terms of the J_2-deformation theory.

After the introduction of the constitutive laws for the material and the interfaces (Section 6.2), we start with the problem of an elastic incompressible structure made of straight layers connected through linear

Fig. 6.3. Bifurcation through compression with detachment of layers. Upper part, left: A stiff (1 mm thick) plastic coating has detached from the foam substrate to which it was initially glued. Upper part, right: Three layers of foam subject to compression show folding with detachment, clearly visible near the edges of the sample. Lower part: A severely bent layer in a folded rock formation near Millook Haven (UK) where detachment has occurred.

interfaces and deformed in a state of uniform biaxial stress, for which incremental bifurcations are sought (Section 6.3). We conclude with the case of finite bending of a layered elastic block, deformed under plane strain (Section 6.4).

6.2. Notations and Governing Equations

The notations employed in this chapter and the main equations governing equilibrium in finite and incremental elasticity are now briefly reviewed (see [40]). Let $\boldsymbol{x}^0$ denote the position of a material point in some stress-free reference configuration B_0 of an elastic body. A deformation $\boldsymbol{\xi}$ is applied, mapping points of B_0 to those of the current configuration B indicated by $\boldsymbol{x} = \boldsymbol{\xi}(\boldsymbol{x}^0)$. We identify its deformation gradient by $\boldsymbol{F}$, i.e., $\boldsymbol{F} = \mathsf{grad}\, \boldsymbol{\xi}$, and we define the right $\boldsymbol{C}$ and the left $\boldsymbol{B}$ Cauchy–Green tensors as $\boldsymbol{C} = \boldsymbol{F}^T\boldsymbol{F}$ and $\boldsymbol{B} = \boldsymbol{F}\boldsymbol{F}^T$.

For isotropic incompressible elasticity, the constitutive equations can be written as a relationship between the Cauchy stress $\boldsymbol{T}$ and $\boldsymbol{B}$ as follows:

$$\boldsymbol{T} = -\pi\boldsymbol{I} + \alpha_1\boldsymbol{B} + \alpha_{-1}\boldsymbol{B}^{-1}, \quad \det\boldsymbol{B} = 1, \tag{6.1}$$

where π is an arbitrary Lagrange multiplier representing hydrostatic pressure, and α_1 and α_{-1} are coefficients (such that $\alpha_1 > 0$ and $\alpha_{-1} \leq 0$), which may depend on the deformation.

Alternatively, the principal stresses T_i ($i = 1, 2, 3$), which are aligned with the Eulerian principal axes, can be obtained in terms of a strain-energy function W, which can be viewed as a function of the principal stretches λ_i ($i = 1, 2, 3$). For an incompressible material, these relationships take the form (index i not summed)

$$T_i = -\pi + \lambda_i \frac{\partial W(\lambda_1, \lambda_2, \lambda_3)}{\partial \lambda_i}, \quad \lambda_1\lambda_2\lambda_3 = 1. \tag{6.2}$$

Equations (6.1) and (6.2) are linked through the following equations [34]:

$$\begin{aligned}\alpha_1 &= \frac{1}{\lambda_1^2 - \lambda_2^2}\left[\frac{(T_1 - T_3)\lambda_1^2}{\lambda_1^2 - \lambda_3^2} - \frac{(T_2 - T_3)\lambda_2^2}{\lambda_2^2 - \lambda_3^2}\right],\\ \alpha_{-1} &= \frac{1}{\lambda_1^2 - \lambda_2^2}\left[\frac{T_1 - T_3}{\lambda_1^2 - \lambda_3^2} - \frac{T_2 - T_3}{\lambda_2^2 - \lambda_3^2}\right],\end{aligned} \tag{6.3}$$

which express the coefficients α_1 and α_{-1} in terms of the strain-energy function of the material.

In the absence of body forces, equilibrium is expressed in terms of the first Piola–Kirchhoff stress tensor $\boldsymbol{S} = \boldsymbol{T}\boldsymbol{F}^{-T}$ (note that for incompressible materials $\det\boldsymbol{F} = 1$) as $\mathsf{div}\,\boldsymbol{S} = \boldsymbol{0}$, an equation defined on B_0.

The loss of uniqueness of plane-strain incremental boundary-value problems is investigated, so that the incremental displacements are given by

$$\boldsymbol{u}(\boldsymbol{x}) = \dot{\boldsymbol{\xi}}(\boldsymbol{x}^0), \tag{6.4}$$

where, as in the following, a superposed dot is used to denote a first-order increment and an updated Lagrangian formulation (where the governing equations are defined in the current configuration B) is adopted.

The incremental counterpart of equilibrium is expressed by div $\boldsymbol{\Sigma} = \mathbf{0}$, where the updated incremental first Piola–Kirchhoff stress is given by

$$\boldsymbol{\Sigma} = \dot{\boldsymbol{S}}\boldsymbol{F}^T, \quad \dot{\boldsymbol{S}} = \dot{\boldsymbol{T}}\boldsymbol{F}^{-T} - \boldsymbol{T}\boldsymbol{L}^T\boldsymbol{F}^{-T}. \tag{6.5}$$

The linearized constitutive equation is

$$\boldsymbol{\Sigma} = \mathbb{C}\boldsymbol{L} - \dot{\pi}\boldsymbol{I}, \tag{6.6}$$

where $\boldsymbol{L} = \operatorname{grad} \boldsymbol{u}$ and $\mathbb{C}$ is the fourth-order tensor of instantaneous elastic moduli (possessing the major symmetries). Incompressibility requires that $\operatorname{tr} \boldsymbol{L} = 0$. Since $\boldsymbol{\Sigma} = \dot{\boldsymbol{T}} - \boldsymbol{T}\boldsymbol{L}^T$ (see Eq. (6.5)), the balance of rotational momentum yields $\Sigma_{12} - \Sigma_{21} = T_2L_{12} - T_1L_{21}$, and a comparison with Eq. (6.6) shows that (no sum on indices i and j)

$$C_{ijji} + T_i = C_{jiji} \quad (i \neq j). \tag{6.7}$$

For a hyperelastic material, the components of $\mathbb{C}$ can be defined in terms of the strain-energy function W.

For the plane problem addressed here, they depend on two incremental moduli [4], namely

$$\mu = \frac{\lambda}{2}\left(\frac{\lambda^4+1}{\lambda^4-1}\frac{d\hat{W}}{d\lambda}\right), \quad \mu_* = \frac{\lambda}{4}\left(\frac{d\hat{W}}{d\lambda} + \lambda\frac{d^2\hat{W}}{d\lambda^2}\right), \tag{6.8}$$

where $\hat{W} = W(\lambda, 1/\lambda, 1)$, due to incompressibility. In the following, examples are given for two specific materials both of which are initially isotropic elastic solids. One is the Mooney–Rivlin material, for which

$$W = \frac{\mu_0}{2}(\lambda_1^2 + \lambda_2^2 - 2), \tag{6.9}$$

where λ_1 and λ_2 are the principal in-plane stretches and μ_0 is the shear modulus in the undeformed configuration. Due to incompressibility $\lambda = \lambda_1$ and $\lambda_2 = 1/\lambda$, so that

$$T_1 = \mu_0(\lambda^2 - \lambda^{-2}) \quad \text{and} \quad \mu = \mu^* = \frac{\mu_0}{2}(\lambda^2 + \lambda^{-2}), \tag{6.10}$$

where the former is the uniaxial tension law (along axis x_1). Notice that the ratio between T_1 and μ is

$$\frac{T_1}{\mu} = \frac{2(\lambda^2 - \lambda^{-2})}{\lambda^2 + \lambda^{-2}}, \tag{6.11}$$

and its value always ranges between -2 (infinite compression) and 2 (infinite tension). The other material analyzed in this section is the J_2-deformation theory solid introduced by Hutchinson and Neale [41], for which

$$W = \frac{K}{N+1}\varepsilon^{N+1}, \quad \mu = \frac{K\varepsilon^N \coth(2\varepsilon)}{2}, \quad \mu^* = \frac{KN\varepsilon^{N-1}}{4}, \tag{6.12}$$

where K is a material parameter, $N \in \,]0,1]$ is the hardening exponent and ε is the maximum principal logarithmic strain ($\varepsilon = \ln\lambda$). The uniaxial stress–strain law turns out to be $T_1 = K\varepsilon^N$. For this material, the governing equilibrium equations become hyperbolic when [42]

$$\varepsilon_{sb} = \sqrt{N[2\varepsilon_{sb}\coth(2\varepsilon_{sb}) - N]}, \tag{6.13}$$

a threshold which corresponds to the emergence of shear bands in the deformed solid.

At the interfaces between layers we employ the compliant interface model of Suo, Ortiz and Needleman [32] and Bigoni, Ortiz and Needleman [33] for which the jump between incremental stress and incremental displacement can be written, in components (in a reference system with the axis 1 orthogonal to the interface), as

$$\Sigma_{11} = \mathcal{S}_{1m}(u_m^+ - u_m^-), \quad \Sigma_{21} = \mathcal{S}_{2m}(u_m^+ - u_m^-); \tag{6.14}$$

here $\mathcal{S}_{ij}$, the instantaneous stiffness of the interface, is a 2×2 constant matrix which components have dimension [stress/length]. It is important to notice that the model depends on the situation, as in the present case, in which the stress vector at the interface is null for the fundamental path. The limiting cases of a traction-free and perfectly bonded interface correspond to $\mathcal{S}_{ij} \equiv 0$ and to $\mathcal{S}_{ij} \to \infty$, respectively. $\mathcal{S}_{11}$ represents the normal stiffness and $\mathcal{S}_{22}$ the shear stiffness of the interface. $\mathcal{S}_{12}$ and $\mathcal{S}_{21}$ are the coupling between the normal and shear responses and, in the applications, will be chosen equal to zero. In (6.14), the terms $()^+$ and $()^-$ indicate quantities for the two sides of the interface. In addition to (6.14), continuity of traction across the interface has to be imposed, namely

$$\boldsymbol{\Sigma}^+\boldsymbol{n} = \boldsymbol{\Sigma}^-\boldsymbol{n}. \tag{6.15}$$

6.3. Uniaxial Tension/Compression of an Elastic Multilayer

In this section, bifurcation is analysed for a multilayered elastic structure with straight interfaces separating orthotropic, incompressible layers deformed in plane-strain tension and compression. The fundamental path is characterized by finite, uniform deformations, and the loss of uniqueness in the form of waves of vanishing velocity is considered. The materials in the layers obey a general hyperelastic incompressible constitutive law and specific results are presented for Mooney–Rivlin and J_2-deformation theory materials. Different boundary conditions are imposed at the external surfaces of the multilayered structure, namely, traction free, and bonding to an elastic or undeformable substrate. The possibility of shear-band instability, due to the loss of ellipticity as seen in the equilibrium equations, is also analyzed.

6.3.1. Equations for a layer

A laminated structure composed of n-layers is considered, subject to homogeneous large deformation in the fundamental path, so that equilibrium and compatibility are trivially satisfied. Plane-strain conditions are assumed with the principal directions of deformation aligned normal and parallel to the layers (Fig. 6.4), with the additional assumption that each layer, along the fundamental path, is subjected to a uniaxial stress along direction x_2. The possibility of bifurcation from the homogeneous state is investigated by adopting an updated Lagrangian formulation of the field equations where the current configuration is taken as a reference.

The material is a nonlinear, orthotropic, incompressible elastic solid and obeys the incremental constitutive equation (6.6). In the absence of body forces, incremental equilibrium requires div $\boldsymbol{\Sigma} = \mathbf{0}$. In each layer,

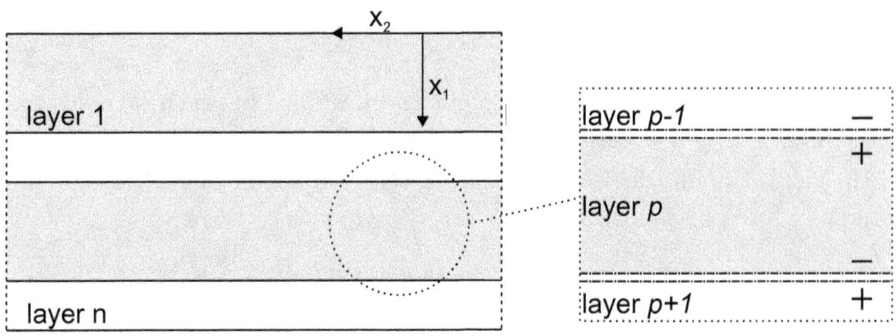

Fig. 6.4. Sketch of the laminated structure. Note that a linear interface is present at each junction between layers.

non-homogeneous incremental solutions are considered in the form

$$u_j = w_j(x_1)e^{ikx_2}(j = 1, 2), \quad \dot{p} = q(x_1)e^{ikx_2}. \tag{6.16}$$

The functions $w_j(x_1)$ and $q(x_1)$ will, in general, differ from layer to layer, but the wave number k is taken to be the same for all layers. A chain substitution of Eq. (6.16) into the constitutive law (6.6) and, finally, into the incremental equilibrium equations yields a system of three constant-coefficient ordinary differential equations for the three unknown functions $w_j(x_1)$ and $q(x_1)$. The solution is

$$\begin{aligned} w_1(x_1) &= b_1 e^{\tau_1 x_1} + b_2 e^{\tau_2 x_1} + b_3 e^{\tau_3 x_1} + b_4 e^{\tau_4 x_1}, \\ w_2(x_1) &= \frac{i}{k}[\tau_1 b_1 e^{\tau_1 x_1} + \tau_2 b_2 e^{\tau_2 x_1} + \tau_3 b_3 e^{\tau_3 x_1} + \tau_4 b_4 e^{\tau_4 x_1}], \\ q(x_1) &= \frac{1}{2}[(C_{2222} - C_{1111} + M)(\tau_1 b_1 e^{\tau_1 x_1} + \tau_2 b_2 e^{\tau_2 x_1}) \\ &\quad + (C_{2222} - C_{1111} - M)(\tau_3 b_3 e^{\tau_3 x_1} + \tau_4 b_4 e^{\tau_4 x_1})], \end{aligned} \tag{6.17}$$

in which $M = \sqrt{L^2 - 4C_{1212}C_{2121}}$ and $L = 2C_{1221} + 2C_{1122} - C_{1111} - C_{2222}$. Coefficients τ_s ($s = 1, \ldots, 4$, $\tau_2 = -\tau_1$, $\tau_4 = -\tau_3$) are the eigenvalues of the equilibrium equations and depend on k, μ, μ^* and T_2. Using the standard classification of regimes, coefficients τ_s may be: (i) real numbers in the elliptic imaginary regime, (ii) two complex conjugate pairs in the elliptic complex regime, (iii) purely imaginary numbers in the hyperbolic regime and (iv) two purely imaginary and two real numbers in the parabolic regime. Departure from the elliptic range corresponds to the occurrence of shear bands. In the following, examples are given for two previously introduced materials, namely, the Mooney–Rivlin and the J_2-deformation theory constitutive models.

Focusing now on the conditions at the interface between layers p and $p+1$ in Fig. 6.4, a substitution of $w_j(x_1)$ and $q(x_1)$ into Eqs. (6.14) and (6.15) yields the interfacial conditions in terms of coefficients b_s^p and $b_s^{(p+1)}$. In matrix form these are

$$\boldsymbol{H}^{p^-}\boldsymbol{b}^p = \boldsymbol{H}^{(p+1)^+}\boldsymbol{b}^{(p+1)}, \tag{6.18}$$

where vectors $\boldsymbol{b}^p$ and $\boldsymbol{b}^{(p+1)}$ collect coefficients b_s for the two layers sharing the interface, while $\boldsymbol{H}^{p^-}$ and $\boldsymbol{H}^{(p+1)^+}$ are the interfacial matrices for layer

− and +, respectively

$$\boldsymbol{H}^{p^-} = \begin{bmatrix} (e^{\tau x_1}\tau\Gamma)_1^{p^-} & (e^{\tau x_1}\tau\Gamma)_2^{p^-} & (e^{\tau x_1}\tau\Gamma)_3^{p^-} & (e^{\tau x_1}\tau\Gamma)_4^{p^-} \\ (e^{\tau x_1}\Delta)_1^{p^-} & (e^{\tau x_1}\Delta)_2^{p^-} & (e^{\tau x_1}\Delta)_3^{p^-} & (e^{\tau x_1}\Delta)_4^{p^-} \\ (e^{\tau x_1}[\tau\Gamma+\Theta])_1^{p^-} & (e^{\tau x_1}[\tau\Gamma+\Theta])_2^{p^-} & (e^{\tau x_1}[\tau\Gamma+\Theta])_3^{p^-} & (e^{\tau x_1}[\tau\Gamma+\Theta])_4^{p^-} \\ (e^{\tau x_1}[ik\Delta+\Xi])_1^{p^-} & (e^{\tau x_1}[ik\Delta+\Xi])_2^{p^-} & (e^{\tau x_1}[ik\Delta+\Xi])_3^{p^-} & (e^{\tau x_1}[ik\Delta+\Xi])_4^{p^-} \end{bmatrix},$$

$$\boldsymbol{H}^{(p+1)^+} = \begin{bmatrix} (e^{\tau x_1}\tau\Gamma)_1^{(p+1)^+} & (e^{\tau x_1}\tau\Gamma)_2^{(p+1)^+} & (e^{\tau x_1}\tau\Gamma)_3^{(p+1)^+} & (e^{\tau x_1}\tau\Gamma)_4^{(p+1)^+} \\ (e^{\tau x_1}\Delta)_1^{(p+1)^+} & (e^{\tau x_1}\Delta)_2^{(p+1)^+} & (e^{\tau x_1}\Delta)_3^{(p+1)^+} & (e^{\tau x_1}\Delta)_4^{(p+1)^+} \\ (e^{\tau x_1}\Theta)_1^{(p+1)^+} & (e^{\tau x_1}\Theta)_2^{(p+1)^+} & (e^{\tau x_1}\Theta)_3^{(p+1)^+} & (e^{\tau x_1}\Theta)_4^{(p+1)^+} \\ (e^{\tau x_1}\Xi)_1^{(p+1)^+} & (e^{\tau x_1}\Xi)_2^{(p+1)^+} & (e^{\tau x_1}\Xi)_3^{(p+1)^+} & (e^{\tau x_1}\Xi)_4^{(p+1)^+} \end{bmatrix}, \tag{6.19}$$

where the entries in the matrices are

$$\begin{gathered} (e^{\tau x_1}\tau\Gamma)_s^{p^-} = e^{\tau_s^p x_1^{p^-}}\,\tau_s^p\Gamma_s^p, \quad (e^{\tau x_1}\Delta)_s^{p^-} = e^{\tau_s^p x_1^{p^-}}\,\Delta_s^p, \\ (e^{\tau x_1}[\tau\Gamma+\Theta])_s^{p^-} = e^{\tau_s^p x_1^{p^-}}\,[\tau_s^p\Gamma_s^p + \Theta_s^{p^-}], \\ (e^{\tau x_1}[ik\Delta+\Xi])_s^{p^-} = e^{\tau_s^p x_1^{p^-}}\,[ik\Delta_s^p + \Xi_s^{p^-}], \end{gathered} \tag{6.20}$$

and the expressions for Γ_s^p, Δ_s^p, $\Theta_s^{p^-}$ and $\Xi_s^{p^-}$ are, respectively,

$$\begin{gathered} \Gamma_1^p = \Gamma_2^p = \left[\frac{C_{1111}^p}{2} + \frac{C_{2222}^p}{2} - C_{1122}^p + \frac{M^p}{2}\right], \\ \Gamma_3^p = \Gamma_4^p = \left[\frac{C_{1111}^p}{2} + \frac{C_{2222}^p}{2} - C_{1122}^p - \frac{M^p}{2}\right], \\ \Delta_s^p = \left[C_{1221}^p + C_{1212}^p\left(\frac{\tau_s^p}{k}\right)^2\right], \\ \Theta_s^{p^-} = \mathcal{S}_{11}^{p^-} + i\mathcal{S}_{12}^{p^-}\frac{\tau_s^p}{k}, \quad \Xi_s^{p^-} = \mathcal{S}_{21}^{p^-} + i\mathcal{S}_{22}^{p^-}\frac{\tau_s^p}{k}. \end{gathered} \tag{6.21}$$

Relation (6.18) holds at every interface. To complete the analysis, the boundary conditions at the external surfaces 1^+ and n^- need to be set.

6.3.1.1. *Traction free at the external surface of the multilayer*

With reference to the external surface 1^+ of the multilayer, vanishing of the nominal tractions requires

$$\Sigma_{11}^{1^+} = 0, \quad \Sigma_{21}^{1^+} = 0, \tag{6.22}$$

which can be written in matrix form as

$$\boldsymbol{C}^{1^+} \boldsymbol{b}^1 = \boldsymbol{0},$$
$$\boldsymbol{C}^{1^+} = \begin{bmatrix} (e^{\tau x_1}\tau\Gamma)_1^{1^+} & (e^{\tau x_1}\tau\Gamma)_2^{1^+} & (e^{\tau x_1}\tau\Gamma)_3^{1^+} & (e^{\tau x_1}\tau\Gamma)_4^{1^+} \\ (e^{\tau x_1}\Delta)_1^{1^+} & (e^{\tau x_1}\Delta)_2^{1^+} & (e^{\tau x_1}\Delta)_3^{1^+} & (e^{\tau x_1}\Delta)_4^{1^+} \end{bmatrix}. \tag{6.23}$$

A similar result can be obtained for the free boundary n^-.

6.3.1.2. *Bonding to an elastic half-space at the external surface of the multilayer*

When an elastic half-space is coated with a multilayer, the elastic solution has to decay within it with depth $x_1 \to +\infty$ (or $x_1 \to -\infty$), a condition implying vanishing of the two coefficients b_s corresponding to the eigenvalues τ_s with positive (or negative) real part. Therefore, the interfacial matrices for half-spaces at the upper (label 1) and lower (label n) external surfaces of the multilayer are

$$\boldsymbol{H}^{1^-} = \begin{bmatrix} (e^{\tau x_1}\tau\Gamma)_1^{1^-} & (e^{\tau x_1}\tau\Gamma)_3^{1^-} \\ (e^{\tau x_1}\Delta)_1^{1^-} & (e^{\tau x_1}\Delta)_3^{1^-} \\ (e^{\tau x_1}[\tau\Gamma+\Theta])_1^{1^-} & (e^{\tau x_1}[\tau\Gamma+\Theta])_3^{1^-} \\ (e^{\tau x_1}[ik\Delta+\Xi])_1^{1^-} & (e^{\tau x_1}[ik\Delta+\Xi])_3^{1^-} \end{bmatrix},$$
$$\boldsymbol{H}^{(n+1)^+} = \begin{bmatrix} (e^{\tau x_1}\tau\Gamma)_2^{(n+1)^+} & (e^{\tau x_1}\tau\Gamma)_4^{(n+1)^+} \\ (e^{\tau x_1}\Delta)_2^{(n+1)^+} & (e^{\tau x_1}\Delta)_4^{(n+1)^+} \\ (e^{\tau x_1}\Theta)_2^{(n+1)^+} & (e^{\tau x_1}\Theta)_4^{(n+1)^+} \\ (e^{\tau x_1}\Xi)_2^{(n+1)^+} & (e^{\tau x_1}\Xi)_4^{(n+1)^+} \end{bmatrix}. \tag{6.24}$$

6.3.1.3. *Bonding to an undeformable substrate at the external surface of the multilayer*

In the case when the external surface of the multilayer is jointed to a smooth undeformable constraint, the normal component of the velocity and the tangential nominal traction have to vanish. With reference to the surface 1^+ these conditions are

$$v_1^{1^+} = 0, \quad \Sigma_{21}^{1^+} = 0, \tag{6.25}$$

which in matrix form become

$$\boldsymbol{C}^{1^+}\boldsymbol{b}^1 = \boldsymbol{0},$$
$$\boldsymbol{C}^{1^+} = \begin{bmatrix} e^{\tau_1^1 x_1^{1^+}} & e^{\tau_2^1 x_1^{1^+}} & e^{\tau_3^1 x_1^{1^+}} & e^{\tau_4^1 x_1^{1^+}} \\ (e^{\tau x_1}\Delta)_1^{1^+} & (e^{\tau x_1}\Delta)_2^{1^+} & (e^{\tau x_1}\Delta)_3^{1^+} & (e^{\tau x_1}\Delta)_4^{1^+} \end{bmatrix}. \tag{6.26}$$

6.3.1.4. *Bonding to an undeformable substrate with a compliant interface at the external surface of the multilayer*

In this case, the interfacial constitutive law, Eq. (6.14), is used between the external elastic layer of the multilayer and an undeforming substrate, which behaves as a rigid constraint. At the external surface 1^+, this boundary condition is

$$\boldsymbol{C}^{1^+}\boldsymbol{b}^1 = \boldsymbol{0},$$
$$\boldsymbol{C}^{1^+} =$$
$$\begin{bmatrix} (e^{\tau x_1}[\tau\Gamma - \Theta])_1^{1^+} & (e^{\tau x_1}[\tau\Gamma - \Theta])_2^{1^+} & (e^{\tau x_1}[\tau\Gamma - \Theta])_3^{1^+} & (e^{\tau x_1}[\tau\Gamma - \Theta])_4^{1^+} \\ (e^{\tau x_1}[ik\Delta - \Xi])_1^{1^+} & (e^{\tau x_1}[ik\Delta - \Xi])_2^{1^+} & (e^{\tau x_1}[ik\Delta - \Xi])_3^{1^+} & (e^{\tau x_1}[ik\Delta - \Xi])_4^{1^+} \end{bmatrix}, \tag{6.27}$$

whilst when imposed at the external surface $n-$, it becomes

$$\boldsymbol{C}^{n^-}\boldsymbol{b}^n = \boldsymbol{0},$$
$$\boldsymbol{C}^{n^-} =$$
$$\begin{bmatrix} (e^{\tau x_1}[\tau\Gamma + \Theta])_1^{n^-} & (e^{\tau x_1}[\tau\Gamma + \Theta])_2^{n^-} & (e^{\tau x_1}[\tau\Gamma + \Theta])_3^{n^-} & (e^{\tau x_1}[\tau\Gamma + \Theta])_4^{n^-} \\ (e^{\tau x_1}[ik\Delta + \Xi])_1^{n^-} & (e^{\tau x_1}[ik\Delta + \Xi])_2^{n^-} & (e^{\tau x_1}[ik\Delta + \Xi])_3^{n^-} & (e^{\tau x_1}[ik\Delta + \Xi])_4^{n^-} \end{bmatrix}. \tag{6.28}$$

6.3.2. Bifurcation criterion

The set of equations for interfacial and boundary conditions forms a linear system, where the coefficients b_s of all layers are the unknowns. When elastic half-spaces are not present, the dimension of the linear system is $4n \times 4n$. When one elastic half-space or two half-spaces are considered as external boundary conditions, the order of the linear system becomes $2(2n-1)$ or $4(n-1)$, respectively.

A critical bifurcation condition is attained when a non-trivial solution is possible. This occurs when the system is singular. In terms of interfacial matrices, the system can be written as follows:

$$\begin{bmatrix} \boldsymbol{C}^{1^+} & & \vdots & & 0 \\ \boldsymbol{H}^{1^-} & -\boldsymbol{H}^{2^+} & \vdots & & \\ \cdots & \cdots & \cdots & \cdots & \cdots \\ & & \vdots & \boldsymbol{H}^{(n-1)^-} & -\boldsymbol{H}^{n^+} \\ 0 & & \vdots & & -\boldsymbol{C}^{n^-} \end{bmatrix} \begin{bmatrix} \boldsymbol{b}^1 \\ \boldsymbol{b}^2 \\ \vdots \\ \boldsymbol{b}^{n-1} \\ \boldsymbol{b}^n \end{bmatrix} = \boldsymbol{0} \Rightarrow \boldsymbol{Y}\boldsymbol{b} = \boldsymbol{0}, \quad (6.29)$$

and the bifurcation criterion becomes $\det(\boldsymbol{Y}) = \boldsymbol{0}$. The system can be reduced using the transfer matrix method [43, 44]. In particular, using the interfacial condition (6.18), the vector $\boldsymbol{b}^p$ can be expressed in terms of $\boldsymbol{b}^{p+1}$ as

$$\boldsymbol{b}^p = (\boldsymbol{H}^{p^-})^{-1}\boldsymbol{H}^{(p+1)^+}\boldsymbol{b}^{p+1}, \quad (6.30)$$

so that, as a consequence, $\boldsymbol{b}^1$ can be given as a function of $\boldsymbol{b}^n$

$$\boldsymbol{b}^1 = \boldsymbol{\Omega}\boldsymbol{b}^n, \quad \boldsymbol{\Omega} = (\boldsymbol{H}^{1^-})^{-1}\boldsymbol{H}^{2^+}(\boldsymbol{H}^{2^-})^{-1}\boldsymbol{H}^{3^+}\ldots(\boldsymbol{H}^{(n-1)^-})^{-1}\boldsymbol{H}^{n^+}, \quad (6.31)$$

where $\boldsymbol{\Omega}$ is the transfer matrix. The linear system is therefore reduced to four equations in four unknowns:

$$\begin{matrix} \boldsymbol{C}^{1^+}\boldsymbol{b}^1 = \boldsymbol{0} \\ \boldsymbol{C}^{n^-}\boldsymbol{b}^n = \boldsymbol{0} \end{matrix} \Rightarrow \begin{matrix} \boldsymbol{C}^{1^+}\boldsymbol{\Omega}\boldsymbol{b}^n = \boldsymbol{0} \\ \boldsymbol{C}^{n^-}\boldsymbol{b}^n = \boldsymbol{0} \end{matrix} \Rightarrow \begin{bmatrix} \boldsymbol{C}^{1^+}\boldsymbol{\Omega} \\ \boldsymbol{C}^{n^-} \end{bmatrix} \boldsymbol{b}^n = \boldsymbol{0} \Rightarrow \boldsymbol{X}\boldsymbol{b}^n = \boldsymbol{0}, \quad (6.32)$$

and the bifurcation criterion becomes $\det(\boldsymbol{X}) = \boldsymbol{0}$. Finally notice that, when the number of the layers increases, numerical difficulties due to ill-conditioning may be encountered using the transfer matrix method. An investigation of these problems, well known in the case of infinitesimal elasticity [45–47], would be interesting, but falls beyond the scope of this chapter.

6.3.3. Results and discussion

The above-described general formulation is applied now to a few simple bifurcation problems. As already remarked, $\mathcal{S}_{12} = \mathcal{S}_{21} = 0$ is assumed. Moreover, the analysis is limited for simplicity to $\mathcal{S}_{11} = \mathcal{S}_{22}$. Results for different interfacial compliances are calculated in terms of the ratio c/h, where h is the thickness of a representative layer and c is given by

$$c = \frac{\mu_1^*}{\mathcal{S}_{11}}, \tag{6.33}$$

where μ_1^* is for layer 1. Parameter c/h is zero for perfect bonding and infinite when the interface becomes a separation surface between two disjointed layers.

In Figs. 6.5 and 6.8, the logarithmic strain ε versus $\bar{k}h$ (where $\bar{k} = k/2\pi$ is the inverse of the wavelength of the bifurcation mode) is shown for a J_2-deformation theory material. Cauchy stress replaces ε in Fig. 6.6 for the Mooney–Rivlin material. A null transversal stress $T_1 = 0$ has been imposed in the fundamental path for all analyzed cases. For the J_2-material, loss of

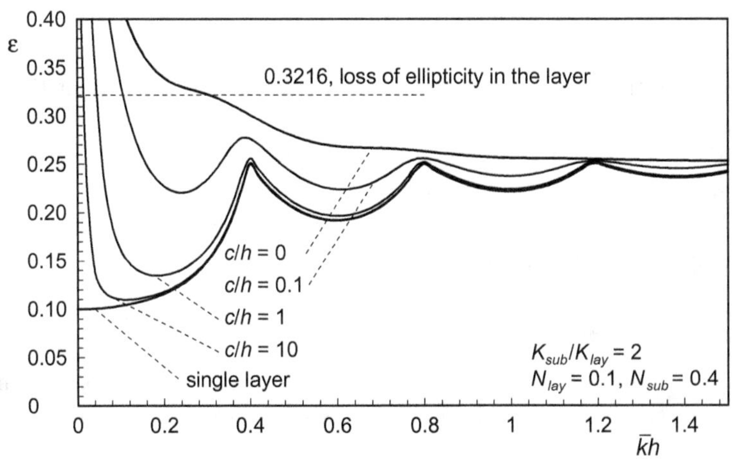

Fig. 6.5. Bifurcation logarithmic strain for a layer bonded to a half-space and loaded under plane-strain uniaxial tension for a J_2-deformation theory material.

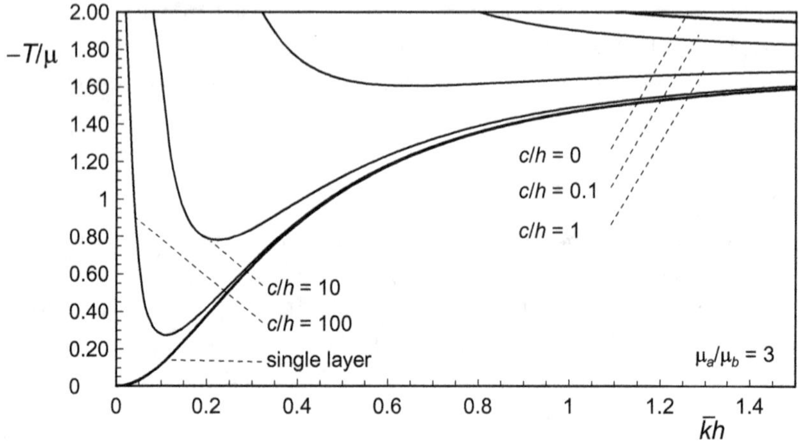

Fig. 6.6. Bifurcation stress for a periodic multilayer in which the representative cell is made of three layers jointed through an imperfect interface and externally bonded to a smooth undeformable substrate. The structure, made of Mooney–Rivlin material, is loaded under plane-strain uniaxial compression.

ellipticity may occur before bifurcation into a diffuse mode. In particular, condition (6.13) gives $\varepsilon_{sb} = 0.3216$ for $N = 0.1$.

6.3.3.1. *Layer bonded to a half-space*

The compression case was analyzed by Bigoni, Ortiz and Needleman [33] and therefore only the behavior under tension is investigated here, for a J_2-deformation theory material (Fig. 6.5). The substrate is stiffer than the layer: $K_{\text{sub}}/K_{\text{lay}} = 2$, $N_{\text{sub}} = 0.4$, $N_{\text{lay}} = 0.1$. The effect of the interfacial compliance gives a strong reduction in the bifurcation critical strain. In the short wavelength limit ($\bar{k}h \to \infty$), all curves tend to the surface instability value for the layer ($\varepsilon = 0.2524$). For sufficiently large wavelength modes shear bands occur in the layer.

6.3.3.2. *Periodic multilayered structures*

Following [7, 13], a periodic multilayered structure can be analyzed under certain restrictions as a bifurcation problem of a representative cell (Fig. 6.7), subject to the boundary conditions of contact with smooth undeformable substrates. We consider the layers joined with the imperfect interface defined by Eq. (6.14).

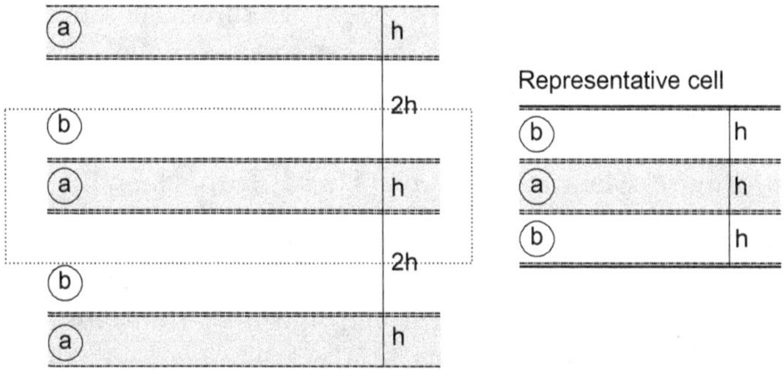

Fig. 6.7. Periodic multilayer structure and representative cell.

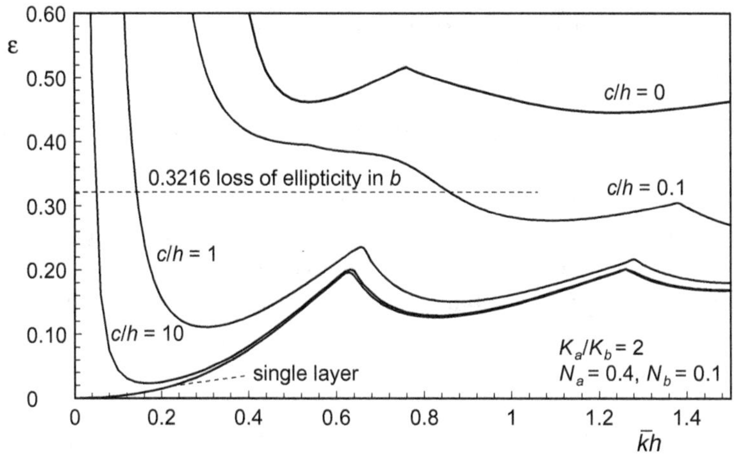

Fig. 6.8. Bifurcation logarithmic strain for a periodic multilayer in which the representative cell is made of three layers jointed through an imperfect interface and externally bonded to a smooth undeformable substrate. The representative cell analyzed to model a periodic multilayer is made of J_2-deformation theory material and is loaded under plane-strain uniaxial compression.

In the example, a ratio $\mu_a/\mu_b = 3$ is assumed for the Mooney–Rivlin material (Fig. 6.6, where T replaces the uniaxial stress T_2). In the case of perfect bonding, bifurcation is impossible when $\bar{k}h < 1.1$ and the interface instability ($T/\mu = -1.9216$) is approached when $\bar{k}h \to \infty$. When the interfacial compliance increases, the bifurcation load reduces, and the single layer solution [18] is recovered in the limit case of complete separation. The case of compression for J_2-deformation theory material is analyzed for the values $K_a/K_b = 2$, $N_a = 0.4$, $N_b = 0.1$ (Fig. 6.8). In all cases, a portion of the curves falls beyond the loss of ellipticity threshold. This portion, in which homogeneous deformation is terminated by strain localization in the weaker layer, becomes larger as the interface becomes stiffer.

6.4. Bending of Elastic Multilayers with Imperfect Interfaces

In this section, we consider elastic multilayers subject to finite flexure, in which the different layers are jointed with imperfect interfaces allowing for full transmission of normal traction and imperfect transmission of shear traction, which are linearly related to a possible jump in the tangential incremental displacement. These conditions are again given by Eqs. (6.14), written now in a cylindrical coordinate system, but with normal stiffness $\mathcal{S}_r \to \infty$. Note that such an interface is not 'activated' during finite bending of a multilayer (since shear tractions are not present at the interfaces between different layers), so that the solution for finite flexure is identical both for perfect and imperfect bonding when $\mathcal{S}_r \to \infty$, but the bifurcation thresholds are strongly affected by the tangential stiffness of the interface $\mathcal{S}_\theta$.

The solution for pure bending of an elastic layered thick plate (of initial 'global' dimensions $l_0 \times h_0$, see Fig. 6.9) made up of N layers jointed through interfaces, which allow complete transmission of normal tractions, follows from an 'appropriate assembling' of solutions relative to the bending of all layers taken separately, a problem analyzed by Rivlin [24]. This solution

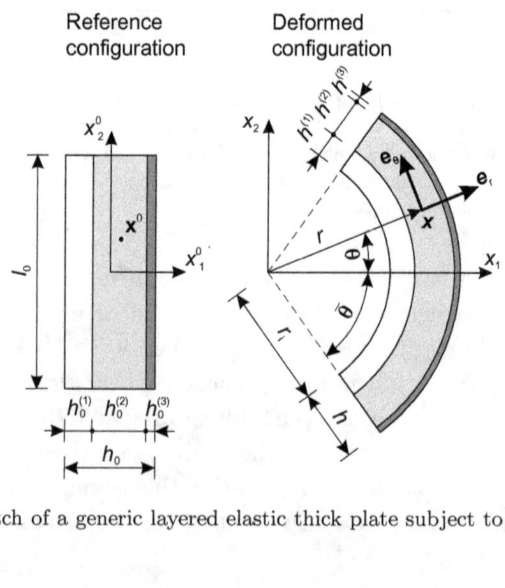

Fig. 6.9. Sketch of a generic layered elastic thick plate subject to finite bending.

is now briefly explained, with reference to a generic layer (the sth) of the considered multilayer (see [22] for more details).

6.4.1. Kinematics

With reference to Fig. 6.9, the generic layer, denoted by the superscript '(s)' ($s = 1, \ldots, N$), is considered in the reference stress-free configuration of a Cartesian coordinate system $O_0^{(s)} x_1^{0(s)} x_2^{0(s)} x_3^{0(s)}$, centered at its centroid, with basis vectors $\boldsymbol{e}_i^0$ ($i = 1, 2, 3$), $x_1^{0(s)} \in [-h_0^{(s)}/2, h_0^{(s)}/2]$, $x_2^{0(s)} \in [-l_0/2, l_0/2]$, and with $x_3^{0(s)}$ denoting the out-of-plane coordinate.

The deformed configuration of each layer is a sector of a cylindrical tube of semi-angle $\bar{\theta}$, which can be referred to as a cylindrical coordinate system $O^{(s)} r^{(s)} \theta^{(s)} z^{(s)}$, with basis vectors $\boldsymbol{e}_r$, $\boldsymbol{e}_\theta$ and $\boldsymbol{e}_z$, $r^{(s)} \in [r_i^{(s)}, r_i^{(s)} + h^{(s)}]$, $\theta^{(s)} \in [-\bar{\theta}, +\bar{\theta}]$, and with out-of-plane coordinate $z^{(s)}$ (Fig. 6.9).

The deformation is prescribed so that a line at constant $x_1^{0(s)}$ transforms to a circular arc at constant $r^{(s)}$, while a line at constant $x_2^{0(s)}$ remains straight but inclined at constant $\theta^{(s)}$. The out-of-plane deformation is null, so that $x_3^{0(s)} = z^{(s)}$. The incompressibility constraint means that

$$r_i^{(s)} = \frac{l_0 h_0^{(s)}}{2\bar{\theta} h^{(s)}} - \frac{h^{(s)}}{2}, \tag{6.34}$$

where $h^{(s)}$ is the current thickness of the circular sector, to be determined.

The deformation is described by the functions

$$r^{(s)} = r^{(s)}(x_1^{0(s)}), \quad \theta^{(s)} = \theta^{(s)}(x_2^{0(s)}), \quad z^{(s)} = x_3^{0(s)}, \tag{6.35}$$

so that the deformation gradient takes the form

$$\boldsymbol{F}^{(s)} = \frac{dr^{(s)}}{dx_1^{0(s)}} \boldsymbol{e}_r \otimes \boldsymbol{e}_1^0 + r^{(s)} \frac{d\theta^{(s)}}{dx_2^{0(s)}} \boldsymbol{e}_\theta \otimes \boldsymbol{e}_2^0 + \boldsymbol{e}_z \otimes \boldsymbol{e}_3^0, \tag{6.36}$$

and we can therefore identify the principal stretches as

$$\lambda_r^{(s)} = \frac{dr^{(s)}}{dx_1^{0(s)}}, \quad \lambda_\theta^{(s)} = r^{(s)} \frac{d\theta^{(s)}}{dx_2^{0(s)}} \quad \text{and} \quad \lambda_z^{(s)} = 1. \tag{6.37}$$

Imposition of the incompressibility constraint with Eq. (6.35) yields

$$r^{(s)} = \sqrt{\frac{2}{\alpha^{(s)}} x_1^{0(s)} + \beta^{(s)}}, \quad \theta^{(s)} = \alpha^{(s)} x_2^{0(s)}, \tag{6.38}$$

so that, using Eq. (6.37), the principal stretches can be evaluated as

$$\lambda_r^{(s)} = \frac{1}{\alpha^{(s)} r^{(s)}}, \quad \lambda_\theta^{(s)} = \alpha^{(s)} r^{(s)} \quad \text{and} \quad \lambda_z^{(s)} = 1, \tag{6.39}$$

where $\alpha^{(s)}$ and $\beta^{(s)}$ (Eq. (6.38)) are constants which can be determined by imposing the boundary conditions, which for the sth layer, are the following:

- At $x_2^{0(s)} = \pm l_0/2$, $\theta^{(s)} = \pm\bar{\theta}$, which from Eq. $(6.38)_2$, $\theta^{(s)} = \pm\alpha^{(s)} l_0/2$, yield

$$\alpha^{(s)} = \frac{2\bar{\theta}}{l_0}, \tag{6.40}$$

  where it is worth noting that $\alpha^{(s)}$ is now independent of the index s;
- At $x_1^{0(s)} = -h_0^{(s)}/2$, $r^{(s)} = r_i^{(s)}$, which from Eqs. (6.34) and $(6.38)_1$, $r_i^{(s)} = r^{(s)}(-h_0^{(s)}/2)$, yield

$$\beta^{(s)} = r_i^{(s)^2} + \frac{l_0 h_0^{(s)}}{2\bar{\theta}}. \tag{6.41}$$

The N layers are assumed to be imperfectly bonded to each other as previously explained, so that continuity of the radial displacements is preserved, and therefore the interfaces do not affect the bending solution. Therefore, we have

$$r_i^{(s)} = r_i^{(s-1)} + h^{(s-1)} \quad (s = 2, \ldots, N), \tag{6.42}$$

with $r_i^{(1)}$ given by $r_i^{(1)} = l_0 h_0^{(1)}/(2\bar{\theta} h^{(1)}) - h^{(1)}/2$ (see Eq. (6.34)). Repeated use of Eqs. (6.34) and (6.42) can be employed to express all thicknesses $h^{(s)}$ $(s = 2, \ldots, N)$ in terms of the thickness of the first layer $h^{(1)}$, which remains the sole kinematical unknown of the problem. In particular, since Eq. (6.42) is imposed at each of the $N - 1$ interfaces between layers, all radial coordinates $r^{(s)}$ share the same origin O of a new cylindrical coordinate system $Or\theta z$, common to all deformed layers (Fig. 6.9 on the right); therefore, the index s on the local current coordinates will be omitted in the following, so that the deformed configuration of the multilayer will be described in terms of the global system $Or\theta z$. From the kinematic analysis, all the stretches are obtained in the multilayer and represented as

$$\lambda_r = \frac{l_0}{2\bar{\theta} r}, \quad \lambda_\theta = \frac{2\bar{\theta} r}{l_0} \quad \text{and} \quad \lambda_z = 1; \tag{6.43}$$

moreover, the current thickness of the sth layer $h^{(s)}$ becomes a function of $h^{(s-1)}$, namely

$$h^{(s)} = -\frac{l_0 h_0^{(s-1)}}{2\bar{\theta} h^{(s-1)}} - \frac{h^{(s-1)}}{2} + \sqrt{\left(\frac{l_0 h_0^{(s-1)}}{2\bar{\theta} h^{(s-1)}} + \frac{h^{(s-1)}}{2}\right)^2 + \frac{l_0 h_0^{(s)}}{\bar{\theta}}}$$
$$(s = 2, \ldots, N). \quad (6.44)$$

We may conclude that all current thicknesses are known once the thickness of the first layer $h^{(1)}$ is known (and this will be determined from the solution of the boundary-value problem described in the following section).

6.4.2. Stress

Let us analyze now the stress state within the multilayer and consider that the Cauchy stress tensor in a generic layer s can be written as

$$\boldsymbol{T}^{(s)} = T_r^{(s)} \boldsymbol{e}_r \otimes \boldsymbol{e}_r + T_\theta^{(s)} \boldsymbol{e}_\theta \otimes \boldsymbol{e}_\theta + T_z^{(s)} \boldsymbol{e}_z \otimes \boldsymbol{e}_z, \quad (6.45)$$

where, from the constitutive equations (6.2),

$$T_r^{(s)} = -\pi^{(s)} + \lambda_r \frac{\partial W^{(s)}}{\partial \lambda_r}, \quad T_\theta^{(s)} = -\pi^{(s)} + \lambda_\theta \frac{\partial W^{(s)}}{\partial \lambda_\theta}, \quad (6.46)$$
$$T_z^{(s)} = -\pi^{(s)} + \left.\frac{\partial W^{(s)}}{\partial \lambda_z}\right|_{\lambda_z=1}.$$

Since the stretches only depend on r, the chain rule of differentiation

$$\frac{d\,\cdot}{dr} = \frac{\partial\,\cdot}{\partial \lambda_r}\frac{d\lambda_r}{dr} + \frac{\partial\,\cdot}{\partial \lambda_\theta}\frac{d\lambda_\theta}{dr}, \quad (6.47)$$

together with Eqs. (6.46) and the derivatives of stretches with respect to r (calculated from Eq. (6.39)), can be used in the equilibrium equations

$$\frac{\partial T_r^{(s)}}{\partial r} + \frac{T_r^{(s)} - T_\theta^{(s)}}{r} = 0, \quad \frac{\partial T_\theta^{(s)}}{\partial \theta} = 0, \quad (6.48)$$

to obtain the identities

$$\frac{dW^{(s)}}{dr} = -\frac{T_r^{(s)} - T_\theta^{(s)}}{r} = \frac{dT_r^{(s)}}{dr}. \quad (6.49)$$

Therefore, identifying λ_θ with λ, for a Mooney–Rivlin material (Eq. (6.9)), we arrive at the expressions

$$T_r^{(s)} = \hat{W}^{(s)} + \gamma^{(s)} = \frac{\mu_0^{(s)}}{2}\left(\lambda^2 + \frac{1}{\lambda^2}\right) + \gamma^{(s)},$$
$$T_\theta^{(s)} = (\lambda \hat{W}^{(s)})' + \gamma^{(s)} = \frac{\mu_0^{(s)}}{2}\left(3\lambda^2 - \frac{1}{\lambda^2}\right) + \gamma^{(s)}, \tag{6.50}$$

where $\hat{W}^{(s)}(\lambda) = W^{(s)}(1/\lambda, \lambda, 1)$, $\gamma^{(s)}$ is an integration constant and $()'$ denotes differentiation with respect to the stretch λ. The component $T_z^{(s)}$ can be inferred from Eq. (6.46).

Constants $\gamma^{(s)}$ $(s = 1, \ldots, N)$ and thickness $h^{(1)}$ can be calculated by imposing (i) continuity of tractions at interfaces between layers ($N-1$ equations), and (ii) traction-free boundary conditions at the external boundaries of the multilayer (two equations). Considering N layers, the traction continuity at the interfaces is

$$T_r^{(s-1)}(r_i^{(s-1)} + h^{(s-1)}) = T_r^{(s)}(r_i^{(s)}) \quad (s = 2, \ldots, N), \tag{6.51}$$

while null traction at the external surfaces of the multilayer yields

$$T_r^{(1)}(r_i^{(1)}) = 0, \quad T_r^{(N)}(r_i^{(N)} + h^{(N)}) = 0. \tag{6.52}$$

Therefore, $\gamma^{(N)}$ can be calculated from Eq. $(6.52)_2$ and specified for a Mooney–Rivlin strain-energy function as

$$\gamma^{(N)} = -\frac{\mu_0^{(N)}}{2}\left[(\alpha r_e^{(N)})^2 + \frac{1}{(\alpha r_e^{(N)})^2}\right], \tag{6.53}$$

while employing Eq. (6.51), the following recursive formulas are obtained

$$\gamma^{(s-1)} = \frac{\mu_0^{(s)} - \mu_0^{(s-1)}}{2}\left[(\alpha r_e^{(s-1)})^2 + \frac{1}{(\alpha r_e^{(s-1)})^2}\right] + \gamma^{(s)} \quad (s = 2, \ldots, N), \tag{6.54}$$

where $r_i^{(s)} = r_e^{(s-1)}$ (see Eq. (5.42)).

Considering now Eq. $(6.52)_1$ and evaluating $\gamma^{(1)}$ from Eq. (6.54) written for $s = 2$, we obtain an implicit expression to be solved for $h^{(1)}$

$$\frac{\mu_0^{(1)}}{2}\left[(\alpha r_i^{(1)})^2 + \frac{1}{(\alpha r_i^{(1)})^2}\right] + \frac{\mu_0^{(2)} - \mu_0^{(1)}}{2}\left[(\alpha r_e^{(1)})^2 + \frac{1}{(\alpha r_e^{(1)})^2}\right] + \gamma^{(2)} = 0, \tag{6.55}$$

in which $r_i^{(1)}$, $r_e^{(1)}$ and $\gamma^{(2)}$ are all functions of $h^{(1)}$, through Eqs. (6.44) and (6.54).

The obtained solution allows determination of the complex stress and strain fields within a thick, multilayered plate, when subject to finite bending. For instance, we show in Fig. 6.10 the deformed geometries for a four-layer structure (with $l_0/h_0 = 1$, thickness ratios: $h_0^{(b)}/h_0^{(a)} = 2$, $h_0^{(c)}/h_0^{(a)} = 3$ and $h_0^{(d)}/h_0^{(a)} = 4$ and stiffness ratios: $\mu^{(a)}/\mu^{(d)} = 27$, $\mu^{(b)}/\mu^{(d)} = 9$ and $\mu^{(c)}/\mu^{(d)} = 3$), together with graphs of the dimensionless Cauchy principal stresses $T_r(r)/\mu^{(a)}$ (the transverse component) and $T_\theta(r)/\mu^{(a)}$ (the circumferential component).

Note that the transverse stress is always compressive, while the distribution of $T_\theta(r)$ strongly depends on the stiffness of the layer under consideration and always has a null resultant, so that it is equivalent to the bending moment loading the plate. For all cases, the neutral axis (the line corresponding to vanishing circumferential stress) is drawn. Note that in

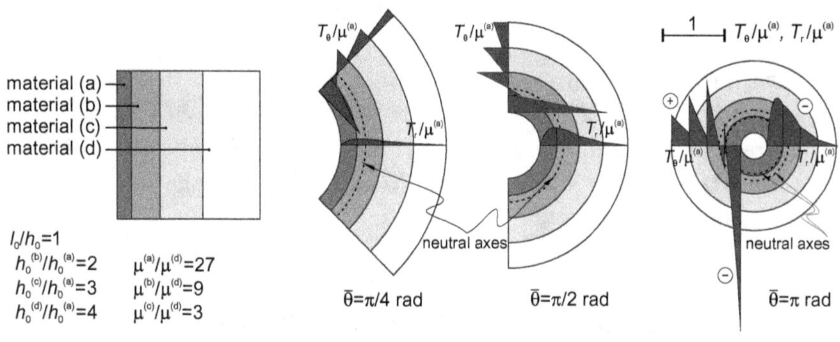

Fig. 6.10. Undeformed and deformed shapes and internal stress states for finite bending of a Mooney–Rivlin four-layer structure with $l_0/h_0 = 1$, thickness ratios: $h_0^{(b)}/h_0^{(a)} = 2$, $h_0^{(c)}/h_0^{(a)} = 3$ and $h_0^{(d)}/h_0^{(a)} = 4$ and stiffness ratios: $\mu^{(a)}/\mu^{(d)} = 27$, $\mu^{(b)}/\mu^{(d)} = 9$ and $\mu^{(c)}/\mu^{(d)} = 3$. Dashed lines represent the neutral axes. Note that two neutral axes are visible in the figure on the right.

the sketch on the right *two neutral axes* are visible, an interesting feature which may occur, depending on the geometry and on the properties of layers, for a multilayered plate under finite bending (see [23] for details).

6.4.3. Incremental bifurcations superimposed on finite bending of an elastic multilayered structure

We address in this section the plane-strain incremental bifurcation problem of the multilayered thick plate subject to the previously solved (Section 6.4) finite bending deformation. For simplicity we consider the problem of a bilayered structure made of Mooney–Rivlin material, but consideration of additional layers or different constitutive equations is straightforward. The incremental equilibrium is again expressed in terms of the updated incremental first Piola–Kirchhoff stress $\boldsymbol{\Sigma}$ by

$$\mathsf{div}\boldsymbol{\Sigma} = \mathbf{0}, \tag{6.56}$$

where $\boldsymbol{\Sigma}$ is given by Eq. (6.5) in terms of the gradient of incremental displacements $\boldsymbol{L}$, which in cylindrical components can be written as

$$\boldsymbol{L} = u_{r,r}\boldsymbol{e}_r \otimes \boldsymbol{e}_r + \frac{u_{r,\theta} - u_\theta}{r}\boldsymbol{e}_r \otimes \boldsymbol{e}_\theta + u_{\theta,r}\boldsymbol{e}_\theta \otimes \boldsymbol{e}_r + \frac{u_r + u_{\theta,\theta}}{r}\boldsymbol{e}_\theta \otimes \boldsymbol{e}_\theta, \tag{6.57}$$

and is subject to the constraint $\mathsf{tr}\ \boldsymbol{L} = 0$ (incremental incompressibility), namely,

$$ru_{r,r} + u_r + u_{\theta,\theta} = 0. \tag{6.58}$$

The linearized constitutive equation is given by Eq. (6.6), and for a Mooney–Rivlin material, the components of $\mathbb{C}$ can be written as functions of two incremental moduli, denoted by μ and μ_*, Eq. (6.10), and depending on the value of the current strain. In cylindrical coordinates, the non-vanishing components of $\mathbb{C}$ are [18, 48]

$$\begin{aligned} &C_{rrrr} = C_{\theta\theta\theta\theta} = 2\mu_* + p, \quad &&C_{\theta r\theta r} = \mu - \Gamma, \\ &C_{r\theta r\theta} = \mu + \Gamma, &&C_{r\theta\theta r} = C_{\theta rr\theta} = \mu + p, \end{aligned} \tag{6.59}$$

where Γ and p are given by

$$\Gamma = \frac{T_\theta - T_r}{2}, \quad \text{and} \quad p = -\frac{T_\theta + T_r}{2}, \tag{6.60}$$

describing the state of prestress. Therefore, the incremental constitutive equations (6.6) take, for each layer, the explicit form

$$
\begin{aligned}
\Sigma_{rr} &= -\dot{\pi} + (2\mu_* + p)u_{r,r},\\
\Sigma_{\theta\theta} &= -\dot{\pi} + (2\mu_* + p)\frac{u_r + u_{\theta,\theta}}{r},\\
\Sigma_{r\theta} &= (\mu + \Gamma)\frac{u_{r,\theta} - u_\theta}{r} + (\mu + p)u_{\theta,r},\\
\Sigma_{\theta r} &= (\mu + p)\frac{u_{r,\theta} - u_\theta}{r} + (\mu - \Gamma)u_{\theta,r}.
\end{aligned}
\tag{6.61}
$$

We seek bifurcations represented by an incremental displacement field in the form

$$
\begin{cases}
u_r(r,\theta) = f(r)\cos n\theta,\\
u_\theta(r,\theta) = g(r)\sin n\theta,\\
\dot{\pi}(r,\theta) = k(r)\cos n\theta,
\end{cases}
\tag{6.62}
$$

so that Eq. (6.58) can be reformulated as

$$
g = -\frac{(f + rf')}{n},
\tag{6.63}
$$

and the incremental equilibrium equations as

$$
\begin{aligned}
k' &= Df'' + \left(C_{,r} + D_{,r} + \frac{C + 2D}{r}\right)f' + \frac{E(1 - n^2)}{r^2}f,\\
k &= \frac{r^2 C}{n^2}f''' + \frac{F + 3C}{n^2}rf'' + \left(\frac{F}{n^2} - D\right)f' - \frac{1 - n^2}{n^2}\frac{F}{r}f,
\end{aligned}
\tag{6.64}
$$

where coefficients C, D, E and F can be expressed (for a Mooney–Rivlin material) as

$$
\begin{aligned}
&C = \mu - \Gamma = \frac{\mu_0}{\lambda^2}, \qquad D = 2\mu_* - \mu = \frac{\mu_0}{2}\frac{\lambda^4 + 1}{\lambda^2},\\
&E = \mu + \Gamma = \mu_0\lambda^2, \quad F = rC_{,r} + C = -\frac{\mu_0}{\lambda^2}.
\end{aligned}
\tag{6.65}
$$

By differentiating Eq. $(6.64)_2$ with respect to r and substituting the result into Eq. $(6.64)_1$, a single differential equation in terms of $f(r)$ is obtained

$$
\begin{aligned}
&r^4 f'''' + 2r^3 f''' - (3 + n^2(\lambda^4 + 1))r^2 f''\\
&\quad + (3 + n^2(1 - 3\lambda^4))rf' + (n^2 - 1)(3 + n^2\lambda^4)f = 0,
\end{aligned}
\tag{6.66}
$$

defining the function $f(r)$ within a generic layer. Once $f(r)$ is known for each layer, the other functions, $g(r)$ and $k(r)$, can be calculated by employing Eqs. (6.63) and (6.64)$_2$, respectively, so that function $f(r)$ becomes the primary unknown.

The differential equation (6.66) for the functions $f^{(s)}(r)$ $(s = 1, \ldots, N)$ is complemented by the following boundary conditions:

- Continuity of incremental tractions at interfaces:

$$\Sigma_{rr}^{(s)}\big|_{r=r_e^{(s)}} = \Sigma_{rr}^{(s+1)}\big|_{r=r_i^{(s+1)}}, \quad \Sigma_{\theta r}^{(s)}\big|_{r=r_e^{(s)}} = \Sigma_{\theta r}^{(s+1)}\big|_{r=r_i^{(s+1)}}; \qquad (6.67)$$

- Continuity of the radial component of the incremental displacement at the interfaces:

$$u_r^{(s)}\big|_{r=r_e^{(s)}} = u_r^{(s+1)}\big|_{r=r_i^{(s+1)}}; \qquad (6.68)$$

- Imperfect 'shear-type' interface (obtained from Eq. (6.14) taking $\mathcal{S}_r \to \infty$)

$$\Sigma_{\theta r}^{(s)}\big|_{r=r_e^{(s)}} = \mathcal{S}_\theta\big(u_\theta^{(s+1)^+} - u_\theta^{(s)^-}\big), \qquad (6.69)$$

  where $\mathcal{S}_\theta$ is a positive shear stiffness coefficient, so that perfect bonding is recovered in the limit $\mathcal{S}_\theta \to \infty$;
- For dead-load tractions on the external surfaces, the boundary conditions at $r = r_i^{(1)}$ and $r = r_e^{(N)}$ are

$$\Sigma_{rr}^{(1),(N)}\big|_{r=r_i^{(1)},r_e^{(N)}} = 0, \quad \Sigma_{\theta r}^{(1),(N)}\big|_{r=r_i^{(1)},r_e^{(N)}} = 0. \qquad (6.70)$$

On the boundaries $\theta = \pm\bar{\theta}$ we require that shear stresses and incremental normal displacements vanish

$$\Sigma_{r\theta}^{(s)}\big|_{\theta=\pm\bar{\theta}} = 0, \quad u_\theta^{(s)}\big|_{\theta=\pm\bar{\theta}} = 0, \qquad (6.71)$$

a condition which is achieved if $\sin n\bar{\theta} = 0$ (see Eq. (6.62)) or equivalently using Eq. (6.40), if

$$n = \frac{2m\pi}{\alpha l_0} \quad (m \in \mathbb{N}). \qquad (6.72)$$

6.4.4. An example: Bifurcation of a bilayer

The critical angle $\bar{\theta}_{cr}$ and the critical stretch λ_{cr} (on the compressive side of the specimen) for a bilayer at bifurcation are shown in Fig. 6.11 as functions of the aspect ratio l_0/h_0 (the unloaded height of the specimen is

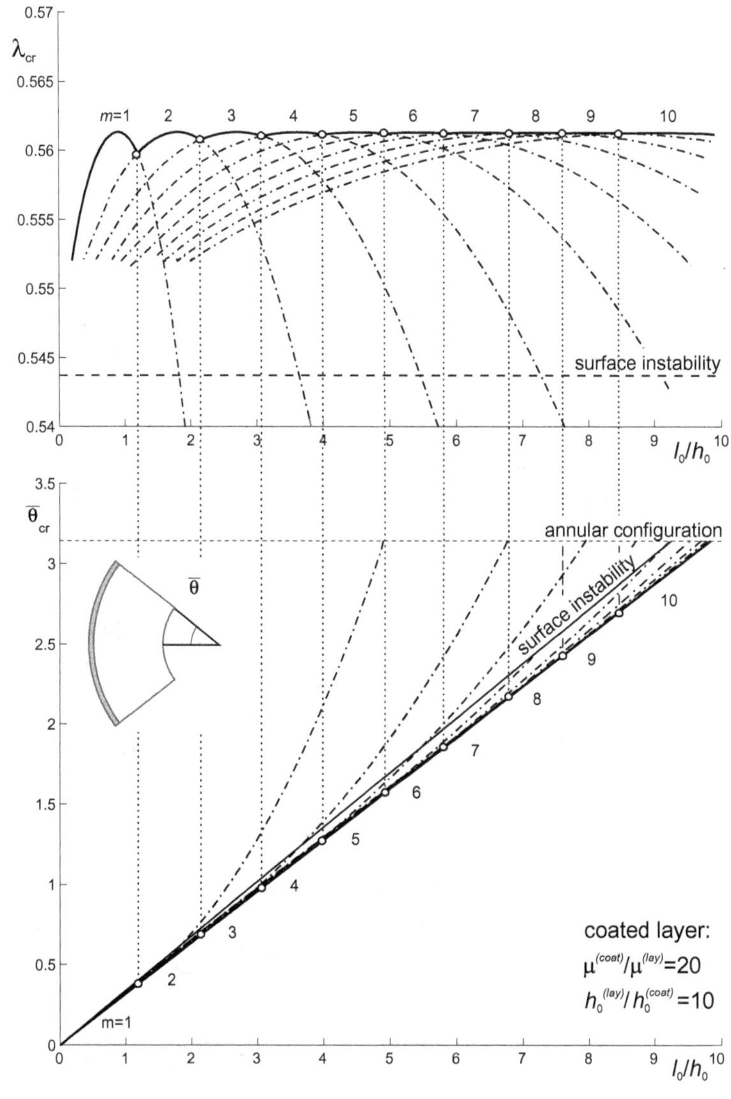

Fig. 6.11. Critical angle $\bar{\theta}_{cr}$ and critical stretch λ_{cr} (evaluated at the internal boundary, $r = r_i^{(1)}$) versus aspect ratio l_0/h_0 of a Mooney–Rivlin coated bilayer subject to bending with $h_0^{(\text{lay})}/h_0^{(\text{coat})} = 10$ and $\mu^{(\text{coat})}/\mu^{(\text{lay})} = 20$. The coating is located on the tensile side of the structure. In both plots, a small circle denotes a transition between two different integer values of m (the parameter which sets the circumferential wave number).

l_0 and global thickness is h_0, see Fig. 6.9), for the thickness and stiffness ratios $h_0^{(\text{lay})}/h_0^{(\text{coat})} = 10$ and $\mu^{(\text{coat})}/\mu^{(\text{lay})} = 20$, respectively. In the figure, bifurcation curves are shown for different values of the integer parameter m which, through Eq. (6.72), defines the circumferential wave number n. Obviously, for a given value of l_0/h_0 the bifurcation threshold is set by the value of m providing the minimum (or maximum) value of the critical angle (or stretch).

In the same figure, the threshold for surface instability of the 'soft' layer material ($\lambda_{\text{surf}} \approx 0.545$ [4]) is also shown. It can be deduced from the figure that a diffuse mode, which set the bifurcation thresholds, always exists before surface instability for each aspect ratio l_0/h_0. It is important to observe that the occurrence of the critical diffuse mode is very close to the surface instability when the coating is located on the tensile side of the specimen (Fig. 6.11). The critical angle at bifurcation is given in Fig. 6.12 as a function of the aspect ratio l_0/h_0, for two values of the

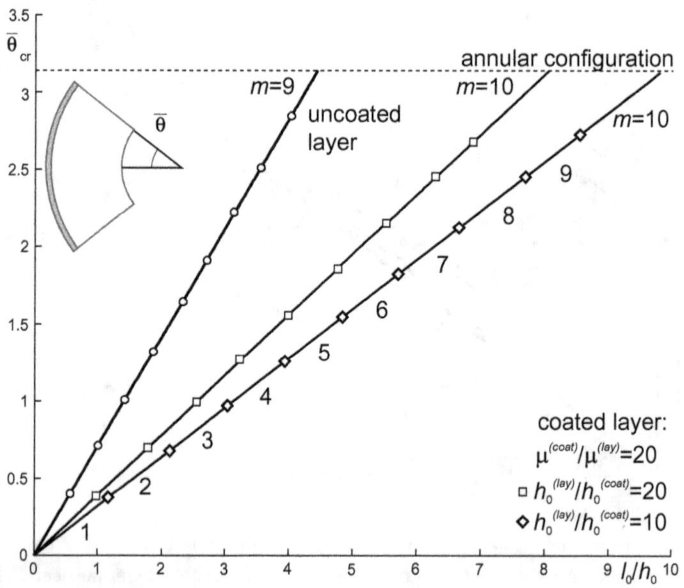

Fig. 6.12. Comparison between the critical angle $\bar{\theta}_{cr}$ at bifurcation versus aspect ratio l_0/h_0 of two Mooney–Rivlin coated bilayers subject to bending with the coating on the tensile side of the structure, with $\mu^{(\text{coat})}/\mu^{(\text{lay})} = 20$ and $h_0^{(\text{lay})}/h_0^{(\text{coat})} = 10$ and 20, respectively. On each curve, a small symbol denotes a transition between two different integer values of m (the parameter which sets the circumferential wave number). Bifurcation angles for a single, uncoated layer are also shown.

coating thickness, $h_0^{(\mathrm{lay})}/h_0^{(\mathrm{coat})} = \{10,\ 20\}$, and when the coating layer is on the tensile side. In the same figure, results for the uncoated layer are also shown for comparison.

It is evident from the figures that the bifurcation solution for a single layer is approximated by a straight line, so that we can define an approximate solution

$$\bar{\theta}_{cr} = 0.712\, l_0/h_0, \tag{6.73}$$

which is very useful for applications. We may also notice that a linear relation between $\bar{\theta}_{cr}$ and l_0/h_0 is found for the bilayer (Figs. 6.11, 6.12 and 6.16); however, the inclination of such lines depends on the elastic and thickness contrasts between the layer and coating, so that it is difficult to obtain a simple formula like Eq. (6.73) in this case.

The effects of an imperfect interface on bifurcations of a layered block under bending have never been analyzed, so we limit the discussion to a simple situation, while a more detailed presentation will be the subject of future research. The simple example analyzed in Figs. 6.13 and 6.14 pertains to a uniform elastic block divided into two identical layers through an imperfect interface of stiffness $\mathcal{S}_\theta$ and $\mathcal{S}_r \to \infty$. Results presented in Figs. 6.13 and 6.14 are in terms of the critical bending angle for bifurcation $\bar{\theta}_{cr}$ versus the initial 'global' aspect ratio l_0/h_0, as a function of the dimensionless interfacial stiffness parameter $\mathcal{S}_\theta h_0/\mu_0$. Results for several values of this parameter (ranging between 0 and 1000) are shown in Fig. 6.13, while those results for only two (namely, 0 and 10) are shown in Fig. 6.14.

Only the smallest circumferential number $m = 1$ was considered for Fig. 6.13, so that $\bar{\theta}$ is not always 'critical', since for low values of the aspect ratio l_0/h_0 the onset of instability is associated with higher values of m.

A general conclusion that can be drawn from the results shown in Figs. 6.13 and 6.14 is that the bifurcation threshold strongly depends on the dimensionless parameter $\mathcal{S}_\theta h_0/\mu_0$, which yields an important decrease in the bifurcation angles with respect to the perfectly bonded case, approached when $\mathcal{S}_\theta h_0/\mu_0 \to \infty$.

6.4.5. Experiments on coated and uncoated rubber blocks under bending

To substantiate the theoretical results for the bifurcation of layered structures subject to finite bending, Roccabianca, Bigoni and Gei [22, 23]

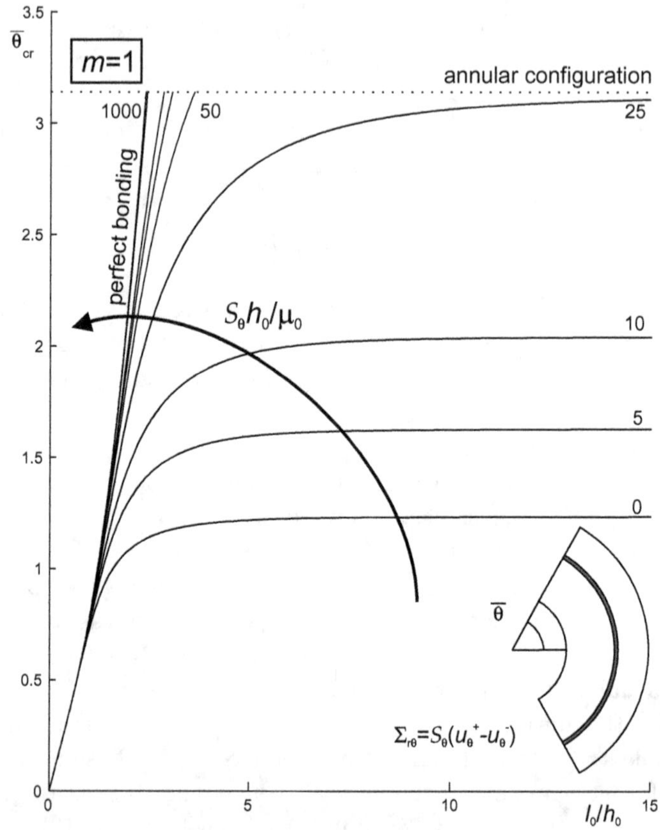

Fig. 6.13. Critical angle $\bar{\theta}_{cr}$ at bifurcation ($m = 1$) versus the 'global' aspect ratio l_0/h_0 for two Mooney–Rivlin identical layers subjected to flexure and jointed through a 'shear-type' imperfect interface of dimensionless stiffness $\mathcal{S}_\theta h_0/\mu_0$. Perfect bonding corresponds to $\mathcal{S}_\theta h_0/\mu_0 \to \infty$.

designed and performed experiments, similar to those initiated by Gent and Cho [49, 50]. In these experiments, a finite flexure was imposed on uncoated and coated elastic blocks (made of natural rubber), glued to two metallic platelets, which were forced to bend by a simple screw loading device (Fig. 6.15 left; see also [22]).

Different coatings and blocks were tested. Bending results for three uncoated rubber strips (made of natural rubber with a ground-state shear modulus $\mu^{(\mathrm{lay})} \cong 1\,\mathrm{N/mm^2}$) and ten coated strips of the same dimensions with two types of coating (both made of a polyester transparent film having $\mu^{(\mathrm{coat})} \cong 500\,\mathrm{N/mm^2}$ but with different thicknesses), all situated

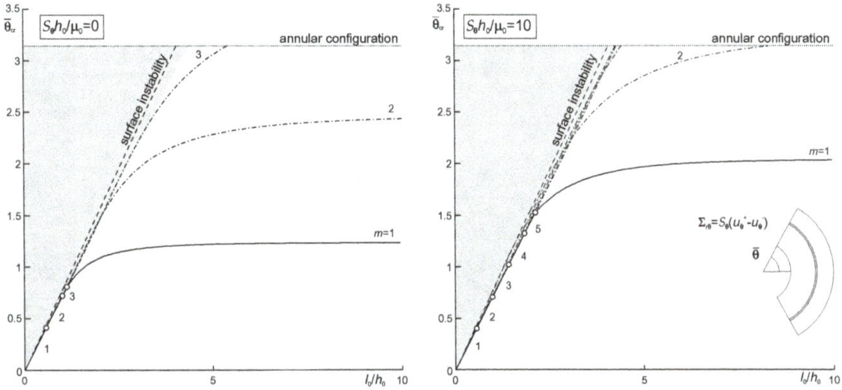

Fig. 6.14. Bifurcation angles $\bar{\theta}_{cr}$ versus the 'global' aspect ratio l_0/h_0 for two Mooney–Rivlin identical layers subjected to flexure and jointed through a 'shear-type' imperfect interface as in Fig. 6.13. Left: $\mathcal{S}_\theta h_0/\mu_0 = 0$; right: $\mathcal{S}_\theta h_0/\mu_0 = 10$. The small numbers near a curve denote the value of the circumferential number m. The lower boundary of the grey region is the bifurcation threshold for perfect bonding.

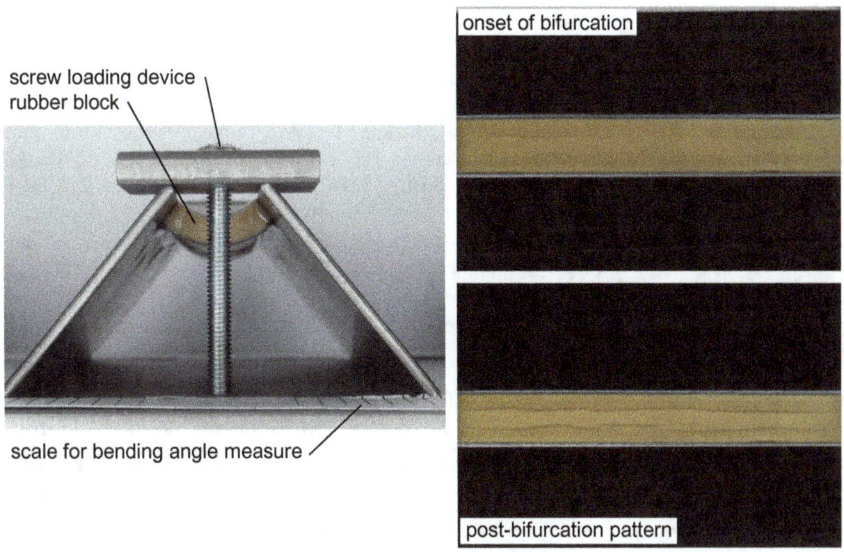

Fig. 6.15. Left: Device used to impose a finite bending (of semi-angle $\bar{\theta}$ equal to 35° in the photo). Right: Bifurcation of a $20 \times 4 \times 100\,\text{mm}^3$ rubber block, coated with two polyester 0.2 mm thick films on the tensile side. Top: Onset of bifurcation ($\bar{\theta} = 40°$, creases become visible). Lower: Post-bifurcation pattern ($\bar{\theta} = 50°$, creases invade the whole specimen).

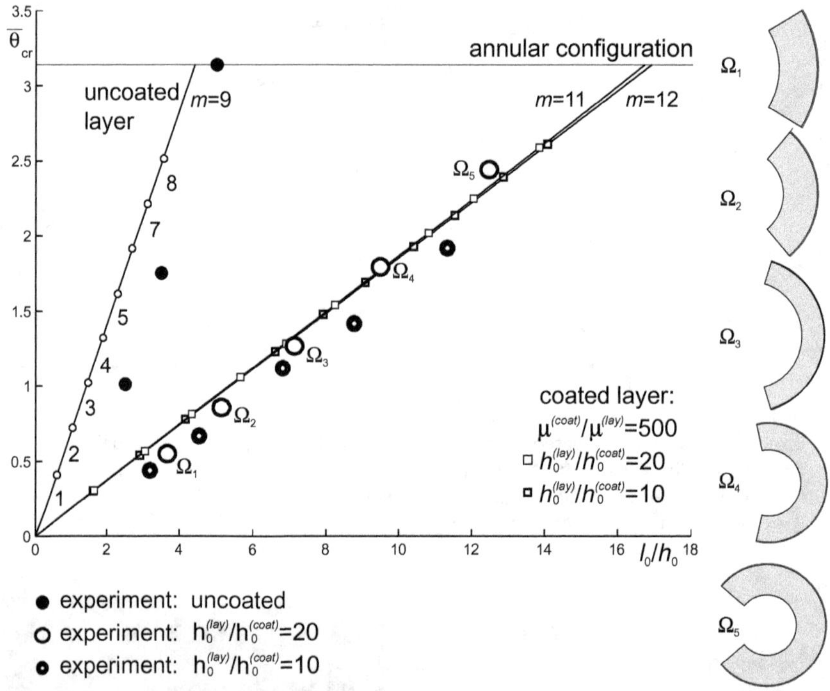

Fig. 6.16. Experimental results versus theoretical predictions for the bifurcation opening semi-angle $\bar{\theta}_{cr}$ of uncoated and coated rubber strips subject to finite bending, versus the aspect ratio l_0/h_0 of the undeformed configuration. The shear moduli ratio $\mu^{(\mathrm{coat})}/\mu^{(\mathrm{lay})}$ of the coated layers has been taken equal to 500, while two thickness ratios $h_0^{(\mathrm{lay})}/h_0^{(\mathrm{coat})}$ equal to 20 and 10 were considered. The critical theoretical configurations (for $h_0^{(\mathrm{lay})}/h_0^{(\mathrm{coat})} = 20$) corresponding to bifurcation points Ω_i $(i = 1, \ldots, 5)$ are sketched the right on the figure.

on the tensile side of the structure, are shown in Fig. 6.16. At a certain stage of finite bending, namely at a certain bending semi-angle $\bar{\theta}_{cr}$, creases can be detected on the surface of the sample, as in Fig. 6.15 on the right (and in Fig. 6.1 in the center). This has been identified with the appearance of small wavelength bifurcations and compared with theoretical predictions for uncoated layers and for a layer with a stiff coating on the tensile side of the specimen, in terms of the critical bending semi-angle $(\bar{\theta}_{cr})$ at bifurcation versus the aspect ratio of the samples, Fig. 6.16. Experiments demonstrate that the trend predicted by the theory is qualitatively very well followed, while quantitatively experimental values for bifurcation angles are often a bit lower than the theoretical predictions, a result consistent with observations by Gent and Cho [49]. The fact that experimental results substantiate theoretical predictions allows us to conclude that bifurcation theory can be successfully employed to predict the deformational capabilities of a composite plate subject to finite bending.

In the case when the coating is applied to the compressed side, long wavelengths become visible in the experiment, as qualitatively demonstrated in Fig. 6.1 on the right (see also [23]), while quantitative evaluation still requires further investigation.

6.5. Conclusions

The load-carrying capacity of laminated structures is often limited by the occurrence of various instabilities at different structural levels. Among these, delamination is the best known. Accordingly, there is a large literature where bifurcations and instabilities of multilayers are analysed from a variety of perspectives.

We have shown that the theory of incremental bifurcation of prestressed elastic solids, in which each layer is treated as an elastic nonlinear continuum and plate-like approximations are not introduced, can be effectively used to find threshold loads for delamination involving complex bifurcation modes. The presented framework is broad enough to include several constitutive laws modeling the mechanical response of (i) the interfaces (for instance spring-like or shear-type junctions) and (ii) the layers (for instance Mooney–Rivlin and J_2-deformation theory of plasticity materials).

The bifurcation analysis, carried out for different deformation paths including finite tension/compression of straight layers and finite bending, reveals a number of different instabilities that may occur in a multilayer, including Euler buckling, necking, surface instability and various wave-like modes. The occurrence of one or another form of instability is strongly related to the interfacial conditions between the layers.

Acknowledgments

D.B. and M.G. gratefully acknowledge the support from the European Union Seventh Framework Programme under contract no. ERC-AdG-340561-INSTABILITIES. M.G. acknowledges support from University of Trieste through grant FRA 2021 'NEO-PHONON'.

References

[1] Yanagimoto J., Oya T., Kawanishi S., Tiesler N., Koseki T., 2010. Enhancement of bending formability of brittle sheet metal in multilayer metallic sheets, *CIRP Annals Manufacturing Tech.*, 59, 287–290.

[2] L'Heureux N., *et al.*, 2006. Human tissue-engineered blood vessel for adult arterial revascularization, *Nature Med.*, 12, 361–365.

[3] Fan Z., *et al.*, 2009. Three-dimensional nanopillar-array photovoltaics on low-cost and flexible substrates, *Nature Mater.*, 8, 648–653.

[4] Biot M.A., 1965. *Mechanics of Incremental Deformations*, Wiley, New York.

[5] Dorris J.F., Nemat-Nasser S., 1980. Instability of a layer on a half space, *J. Appl. Mech.*, 47, 304–312.

[6] Steif P.S., 1986. Bimaterial interface instabilities in plastic solids, *Int. J. Solids Struct.*, 22, 195–207.

[7] Steif P.S., 1986. Periodic necking instabilities in layered plastic solids, *Int. J. Solids Struct.*, 22, 1571–1578.

[8] Steif P.S., 1987. An exact two-dimensional approach to fiber microbuckling, *Int. J. Solids Struct.*, 23, 1235–1246.

[9] Steif P.S., 1990. Interfacial instabilities in an unbounded layered solid, *Int. J. Solids Struct.*, 26, 915–925.

[10] Papamichos E., Vardoulakis I., Muhlhaus H.B., 1990. Buckling of layered elastic media: A Cosserat-continuum approach and its validation, *Int. J. Numer. Anal. Meth. Geomech.*, 14, 473–498.

[11] Dowaikh M.A., Ogden R.W., 1991. Interfacial waves and deformations in pre-stressed elastic media, *Proc. R. Soc. Lond. A*, 443, 313–328.

[12] Benallal A., Billardon R., Geymonat G., 1993. Bifurcation and localization in rate-independent materials: Some general considerations, in *CISM Lecture Notes No. 327*, Springer, Berlin, pp. 1–47.

[13] Triantafyllidis N., Lehner F.K., 1993. Interfacial instability of density-stratified 2–layer systems under initial stress, *J. Mech. Phys. Solids*, 41, 117–142.

[14] Triantafyllidis N., Leroy Y.M., 1994. Stability of a frictional material layer resting on a viscous half-space, *J. Mech. Phys. Solids*, 42, 51–110.

[15] Shield T.W., Kim K.S., Shield R.T., 1994. The buckling of an elastic layer bonded to an elastic substrate in plane strain, *J. Appl. Mech.*, 61, 231–235.

[16] Ogden R.W., Sotiropoulos D.A., 1995. On interfacial waves in prestressed layered incompressible elastic solid, *Proc. R. Soc. Lond. A*, 450, 319–341.

[17] Steigmann D.J., Ogden R.W., 1997. Plane deformations of elastic solids with intrinsic boundary elasticity, *Proc. R. Soc. Lond. A*, 453, 853–877.

[18] Hill R., Hutchinson J.W., 1975. Bifurcation phenomena in the plane tension test, *J. Mech. Phys. Solids*, 23, 239–264.

[19] Young N.J.B., 1976. Bifurcation phenomena in the plane compression test, *J. Mech. Phys. Solids*, 24, 77–91.

[20] Needleman A., 1979. Non-normality and bifurcation in plane strain tension or compression, *J. Mech. Phys. Solids*, 27, 231–254.

[21] Bigoni D., Gourgiotis P.A., 2016. Folding and faulting of an elastic continuum, *Proc. Roy. Soc. Lond. A*, 472, 20160018.

[22] Roccabianca S., Gei M., Bigoni D., 2010. Plane strain bifurcations of elastic layered structures subject to finite bending: Theory versus experiments, *IMA J. Appl. Math.*, 75, 525–548.

[23] Roccabianca S., Bigoni D., Gei M., 2011. Long wavelength bifurcations and multiple neutral axes of elastic layered structures subject to finite bending, *J. Mech. Mat. Struct.*, 6, 511–527.

[24] Rivlin R.S., 1949. Large elastic deformations of isotropic materials. V. The problem of flexure, *Proc. R. Soc. Lond. A*, 195, 463–473.

[25] Triantafyllidis N., 1980. Bifurcation phenomena in pure bending, *J. Mech. Phys. Solids*, 28, 221–245.

[26] Aron M., Wang Y., 1995. Remarks concerning the flexure of a compressible nonlinearly elastic rectangular block, *J. Elasticity*, 40, 99–106.

[27] Aron M., Wang Y., 1995. On deformations with constant modified stretches describing the bending of rectangular blocks, *Quart. J. Mech. Appl. Math.*, 48, 375–387.

[28] Dryburgh G., Ogden R.W., 1999. Bifurcation of an elastic surface-coated incompressible isotropic elastic block subject to bending, *Z. Angew. Math. Phys.*, 50, 822–838.

[29] Bruhns O., Xiao H., Meyers A., 2002. Finite bending of a rectangular block of an elastic Hencky material, *J. Elasticity*, 66, 237–256.

[30] Bruhns O., Gupta N.K., Meyers A.T.M., Xiao H., 2003. Bending of an elastoplastic strip with isotropic and kinematic hardening, *Arch. Appl. Mech.*, 72, 759–778.

[31] Coman C.D., Destrade M., 2008. Asymptotic results for bifurcation in pure bending of rubber blocks, *Quart. J. Mech. Appl. Math.*, 61, 395–414.

[32] Suo Z., Ortiz M., Needleman A., 1992. Stability of solids with interfaces, *J. Mech. Phys. Solids*, 40, 613–640.

[33] Bigoni D., Ortiz M., Needleman A., 1997. Effect of interfacial compliance on bifurcation of a layer bonded to a substrate, *Int. J. Solids Struct.*, 34, 4305–4326.

[34] Bigoni D., Gei M., 2001. Bifurcations of a coated, elastic cylinder, *Int. J. Solids Struct.*, 38, 5117–5148.

[35] Gei M., Genna F., Bigoni D., 2002. An interface model for the periodontal ligament. *ASME J. Biomech. Eng.*, 124, 538-546.

[36] Bigoni D., Movchan A.B., 2002. Statics and dynamics of structural interfaces in elasticity, *Int. J. Solids Struct.*, 39, 4843–4865.

[37] Bertoldi K., Bigoni D., Drugan W.J., 2007. Structural interfaces in linear elasticity. Part I: Nonlocality and gradient approximations, *J. Mech. Phys. Solids*, 55, 1–34.

[38] Brun M., Guenneau S., Movchan A.B., Bigoni D., 2010. Dynamics of structural interfaces: Filtering and focussing effects for elastic waves, *J. Mech. Phys. Solids*, 58, 1212–1224.

[39] Ogden R.W., 1982. Elastic deformations of rubberlike solids, in *Mechanics of Solids, The Rodney Hill, 60th Anniversary Volume* (H.G. Hopkins, M.J. Sewell, eds), Pergamon Press, Oxford, pp. 499–537.

[40] Bigoni D., 2012. *Nonlinear solid mechanics: bifurcation theory and material instability*, Cambridge University Press, New York.

[41] Hutchinson J.W., Neale K.W., 1979. Finite strain J_2-deformation theory, *Proc. IUTAM Symp. on Finite Elasticity* (D.E. Carlson, R.T. Shield, eds), Martinus Nijhoff, The Hague, Boston, London, pp. 237–247.

[42] Hutchinson J.W., Tvergaard V., 1981. Shear band formation in plane strain, *Int. J. Solids Struct.*, 17, 451–470.

[43] Thomson W.T., 1950. Transmission of elastic waves through a stratified solid medium, *J. Appl. Mech.*, 21, 89–93.

[44] Bufler H., 1965. Die Druckstabilitt rechteckiger Verbundplatten, *Ingenieur Archiv*, 31, 109–128.

[45] Pilkey W.D., 1993. *Formulas for Stress, Strain, and Structural Matrices*, John Wiley and Sons, New York.

[46] Su X.Y., Tian J.Y., Pao Y-H., 2002. Application of the reverberation-ray matrix to the propagation of elastic waves in a layered solid, *Int. J. Solids Struct.*, 39, 5447–5463.

[47] Tian J.Y., Yang W.X., Su X.Y., 2006. Transient elastic waves in a transversely isotropic laminate impacted by axisymmetric load, *J. Sound Vibr.*, 289, 94–108.

[48] Gei M., Ogden R.W., 2002. Vibration of a surface-coated elastic block subject to bending, *Math. Mech. Solids*, 7, 607–628.

[49] Gent A.N., Cho I.S., 1999. Surface instabilities in compressed or bent rubber blocks, *Rubber Chem. Technol.*, 72, 253–262.

[50] Gent A.N., 2005. Elastic instabilities in rubber, *Int. J. Non-linear Mech.*, 40, 165–175.

https://doi.org/10.1142/9781800611887_0007

Chapter 7

Instabilities Associated with Loss of Ellipticity in Fiber-Reinforced Nonlinearly Elastic Solids

José Merodio[*,‡] **and Ray W. Ogden**[†,§]
[*]*Departamento de Matemática Aplicada a las Tecnologías de la Información y las Comunicaciones,*
Escuela Técnica Superior de Ingeniería de Sistemas Informáticos,
Universidad Politécnica de Madrid, 28031, Madrid, Spain
[†]*School of Mathematics and Statistics,*
University of Glasgow, Glasgow G12 8QQ, UK
[‡]*jose.merodio@upm.es*
[§]*raymond.ogden@glasgow.ac.uk*

Abstract

In fiber-reinforced materials several different instability mechanisms can be operative. These include fiber kinking, fiber splitting, fiber debonding and matrix failure, *inter alia*. These phenomena are analyzed in this chapter on the basis of the notion of loss of ellipticity. For this purpose, the material is considered as elastic and is modeled by a constitutive law consisting of an isotropic contribution associated with a matrix material, in which the reinforcing fibers are embedded, and a transversely isotropic contribution (the reinforcing model) associated with the fiber direction. Loss of ellipticity is associated with surfaces of discontinuity in the material, and is assessed for two different reinforcing models, both for compressible and incompressible materials in two and three dimensions. Depending on the choice of reinforcing model and the prevailing state of deformation, loss of ellipticity may be associated with a surface of discontinuity at different orientations relative to the fiber direction. In the first part of the chapter, weak surfaces of discontinuity (associated with discontinuity of the second derivative of the deformation field) are examined, while the second part is concerned with strong surfaces of discontinuity (associated with discontinuity in the first derivative of the deformation field) and the transition between weak and strong surfaces.

Contents

7.1. Introduction

Fibrous composite materials are susceptible to failure from several different mechanisms, including fiber kinking [1–8], fiber splitting [9–11], fiber debonding [12, 13] and matrix failure [14, 15]. For an extensive review of a wide range of compressive failure mechanisms we refer to [16], and for further discussion of experimental results, analysis of kink band propagation and failure to [17–22].

These failure mechanisms can be modeled on the basis of the continuum mechanics of fiber-reinforced materials. In particular, in the setting of the nonlinear theory of elasticity, certain types of failure have been characterized in terms of loss of ellipticity of the differential equations that govern the deformation of the material, as exemplified in the papers [23–32].

Loss of ellipticity is associated with the emergence of a surface (or surfaces) of discontinuity in the material. Such a surface is referred to as a *weak* (*discontinuity*) *surface* if the second gradient of the deformation field is discontinuous, and a *strong* (*discontinuity*) *surface* if the deformation gradient itself is discontinuous. For a given elastic constitutive law, the equation which defines the discontinuity surface determines both the

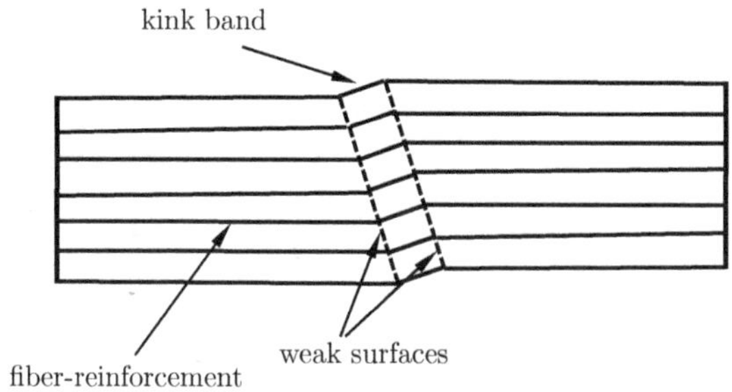

Fig. 7.1. Parallel reinforcing fibers under compression leading to the formation of a kink band with weak surfaces whose normals are close to parallel to the fiber direction exterior to the kink band.

deformation at the initiation of the discontinuity surface and the direction of the normal to the surface. A strong (discontinuity) surface is also weak, but a weak (discontinuity) surface is not in general strong, although a weak surface can develop into a strong one, and a strong surface can degenerate into a weak one. For a general analysis of discontinuity relations for functions with particular applications in solid mechanics, we refer to the article [33].

Different failure mechanisms may be associated with different relative orientations of the normal to the weak surface and the fiber direction. This is illustrated, for example, by the initiation of fiber kinking when the fibers are compressed, in which case the weak surface normal is oriented close to the fiber direction [1]. A kink band is depicted in Fig. 7.1, which shows parallel fibers under compression and the associated weak surfaces.

Fiber splitting is another failure mechanism and occurs, for example, in glass fiber composites and corresponds to weak surfaces that are almost parallel to the fibers. Fiber splitting and fiber kinking can arise simultaneously [9, 10], as depicted in Fig. 7.2.

Another possible failure mechanism is debonding between a fiber and the matrix when the fibers are under tension [13], as modeled in [12], which may also be interpreted in terms of a weak surface that is close to parallel to the fiber direction. This scenario is depicted in Fig. 7.3, which shows just a single fiber embedded in a matrix.

When the fibers are extended rather than compressed matrix failure can occur [14, 15] and is associated with weak surfaces normal to the fiber-reinforcement, as depicted in Fig. 7.4.

To provide a theoretical description of these phenomena we focus on the analysis of constitutive equations for nonlinearly elastic solids and their ellipticity and loss of ellipticity, following in the first instance the approach

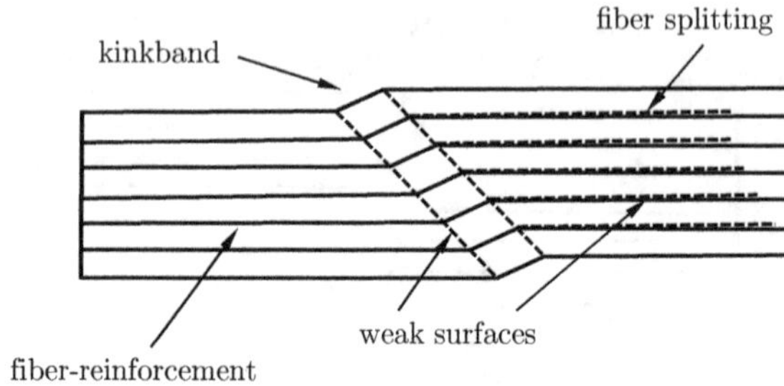

Fig. 7.2. Parallel reinforcing fibers under compression leading to the formation of a kink band, with weak surfaces whose normals are close to parallel to the fiber direction exterior to the kink band, combined with fiber splitting for which the weak surfaces are close to parallel to the fiber-reinforcement direction.

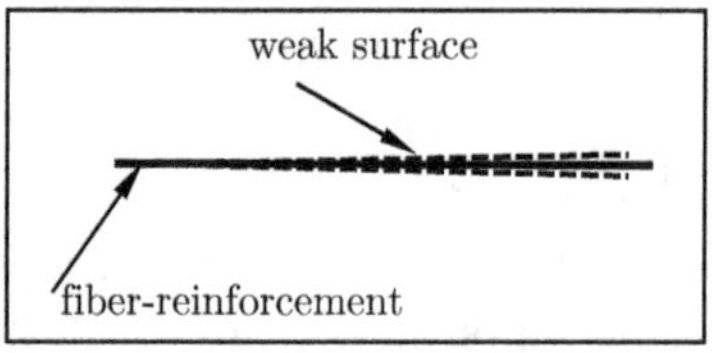

Fig. 7.3. Fiber debonding depicted for a single fiber under tension embedded within a matrix. This may be associated with a weak surface that lies close to parallel to the fiber direction.

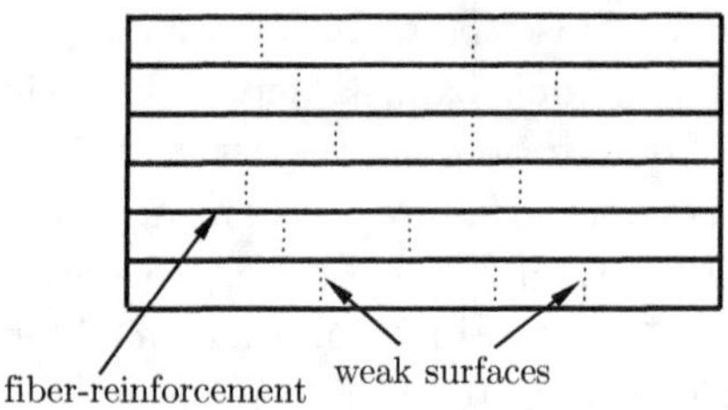

Fig. 7.4. Depiction of parallel fibers under extension embedded within a matrix leading to possible matrix failure with weak surfaces normal to the direction of fiber-reinforcement.

in [24] for compressible materials and [23] for incompressible materials. Initially, we consider the equations governing weak surfaces and then go on to analyze strong surfaces. Constitutive equations that admit loss of ellipticity have been the subject of attention in a number of different contexts, exemplified in [3, 6, 7, 23, 24, 34–40]. The analysis herein is based on considering the elastic strain-energy function to consist of an isotropic model that describes the matrix within which the fibers are embedded and a reinforcing model that characterizes the properties of the fibers.

The material model is described in Section 7.2 following a summary of the required aspects of the kinematics of deformation. The notions of ellipticity, strong ellipticity and loss of ellipticity of the equations governing equilibrium of the material are then described for compressible materials in three dimensions, and the form of strain-energy function is highlighted for different reinforcing models that are used subsequently.

Specialization to plane strain is then discussed in Section 7.3, in which the ellipticity status of the different reinforcing models is established. Depending on the reinforcing model and its convexity status, loss of ellipticity can occur when the fibers are compressed or extended and the different weak surfaces discussed above can be predicted.

In Section 7.4 a parallel, but briefer, analysis of loss of ellipticity is provided for incompressible materials, both for three dimensions and the plane strain specialization. For incompressible materials, attention is then turned in Section 7.5 to the analysis of strong discontinuities for which purpose the plane strain specialization is adopted. An expression for the jump in the deformation gradient across a strong surface is obtained and related to the normal to the discontinuity surface and the strength of the discontinuity. The energetics of a strong discontinuity are then examined in terms of a so-called discontinuity driving traction. Two simple examples are then examined, one involving a kink surface orthogonal to the initial fiber direction and one that shows how, as the strength of a strong discontinuity tends to zero, the strong discontinuity degenerates into a weak discontinuity.

Finally, Section 7.6 contains a brief concluding discussion.

7.2. Compressible Materials in Three Dimensions

For comprehensive introductions to continuum mechanics, in particular to applications in nonlinear elasticity theory, we refer to the standard texts [41, 42]. In this section, we focus on the main aspects of the theory that are needed in this chapter.

7.2.1. Kinematics

We first consider a material body occupying a *reference configuration* when unloaded and stress free, and we denote this configuration by $\mathcal{B}_\mathrm{r}$, within which material points are labeled by the position vector $\mathbf{X}$ relative to a suitably chosen origin. The Cartesian components of $\mathbf{X}$ are denoted X_α, $\alpha \in \{1,2,3\}$. Once subject to loads the material is deformed and occupies the *deformed configuration*, denoted $\mathcal{B}$, in which the material point $\mathbf{X}$ has position vector $\mathbf{x}$, with Cartesian components x_i, $i \in \{1,2,3\}$. Note, in particular, that here and henceforth we associate Greek indices with $\mathcal{B}_\mathrm{r}$ and Roman indices with $\mathcal{B}$.

Of special importance is the *deformation gradient tensor*, which is denoted $\mathbf{F}$ and has Cartesian components $F_{i\alpha} = \partial x_i/\partial X_\alpha$. From this are formed two symmetric tensors, the left, $\mathbf{B}$, and right, $\mathbf{C}$, Cauchy–Green deformation tensors, which are defined by

$$\mathbf{B} = \mathbf{F}\mathbf{F}^\mathrm{T}, \quad \mathbf{C} = \mathbf{F}^\mathrm{T}\mathbf{F}. \tag{7.1}$$

The principal invariants of these two tensors, sometimes referred to as isotropic invariants, are the same and defined, through $\mathbf{C}$, by

$$I_1 = \mathrm{tr}\,\mathbf{C}, \quad I_2 = \frac{1}{2}[I_1^2 - \mathrm{tr}\,(\mathbf{C}^2)] = I_3\mathrm{tr}\,(\mathbf{C}^{-1}), \quad I_3 = \det\mathbf{C}. \tag{7.2}$$

Reinforcing fibers are modeled as a continuous distribution with their direction locally defined by the unit vector $\mathbf{M}$ (the so-called preferred direction) in $\mathcal{B}_\mathrm{r}$, and since fiber directions may vary $\mathbf{M}$ will in general depend on $\mathbf{X}$. This allows us to introduce two additional (in general independent) invariants which depend on both $\mathbf{C}$ and $\mathbf{M}$. These are typically denoted I_4 and I_5 and defined by

$$I_4 = \mathbf{M}\cdot(\mathbf{C}\mathbf{M}) = \mathbf{m}\cdot\mathbf{m}, \quad I_5 = \mathbf{M}\cdot(\mathbf{C}^2\mathbf{M}) = \mathbf{m}\cdot(\mathbf{B}\mathbf{m}), \tag{7.3}$$

wherein we have introduced the notation $\mathbf{m}$ for the push forward of $\mathbf{M}$ from $\mathcal{B}_\mathrm{r}$ to $\mathcal{B}$ defined by

$$\mathbf{m} = \mathbf{F}\mathbf{M}. \tag{7.4}$$

The eigenvalues of $\mathbf{B}$ and $\mathbf{C}$ are the squares of the principal stretches $\{\lambda_1, \lambda_2, \lambda_3\}$, in terms of which the invariants are expressed as

$$I_1 = \lambda_1^2 + \lambda_2^2 + \lambda_3^2, \quad I_2 = I_3(\lambda_1^{-2} + \lambda_2^{-2} + \lambda_3^{-2}), \quad I_3 = \lambda_1^2\lambda_2^2\lambda_3^2, \tag{7.5}$$

$$I_4 = \lambda_1^2 M_1^2 + \lambda_2^2 M_2^2 + \lambda_3^2 M_3^2 = m_1^2 + m_2^2 + m_3^2, \tag{7.6}$$

$$I_5 = \lambda_1^4 M_1^2 + \lambda_2^4 M_2^2 + \lambda_3^4 M_3^2 = \lambda_1^2 m_1^2 + \lambda_2^2 m_2^2 + \lambda_3^2 m_3^2, \tag{7.7}$$

where (M_1, M_2, M_3) and (m_1, m_2, m_3) are the components of $\mathbf{M}$ and $\mathbf{m}$ referred to the principal axes of $\mathbf{C}$ and $\mathbf{B}$, respectively.

The invariant I_4 measures changes in the length of the fiber that was in the direction $\mathbf{M}$ in $\mathcal{B}_r$ and represents the square of the stretch of such a fiber. The invariant I_5 does not in general have such a clear interpretation and we refer to [23, 24] for a discussion of this point.

7.2.2. Elasticity

The material properties are here considered to be elastic, and characterized in terms of a strain-energy function. In this section we do not apply any internal constraints to the material, but in Section 7.4 we adopt the constraint of incompressibility. The material is assumed to be transversely isotropic with the transverse isotropy defined locally in terms the fiber direction $\mathbf{M}$. Then the strain energy is an isotropic function of the two tensors $\mathbf{C}$ and $\mathbf{M} \otimes \mathbf{M}$, which means that in its most general form it can be expressed as a function of the five invariants I_1, I_2, I_3, I_4, I_5. Thus, the strain-energy function, denoted W and defined per unit volume in $\mathcal{B}_r$, is written

$$W = W(I_1, I_2, I_3, I_4, I_5). \tag{7.8}$$

For background on the mathematical theory of fiber-reinforced material, we refer to [43].

The nominal stress tensor (the transpose of the first Piola–Kirchhoff stress tensor) is denoted by $\mathbf{S}$ and in general given by

$$\mathbf{S} = \frac{\partial W}{\partial \mathbf{F}}, \tag{7.9}$$

and, for the considered material, use of the formulas

$$\frac{\partial I_1}{\partial \mathbf{F}} = 2\mathbf{F}^{\mathrm{T}}, \quad \frac{\partial I_2}{\partial \mathbf{F}} = 2I_1\mathbf{F}^{\mathrm{T}} - 2\mathbf{F}^{\mathrm{T}}\mathbf{F}\mathbf{F}^{\mathrm{T}}, \quad \frac{\partial I_3}{\partial \mathbf{F}} = 2I_3\mathbf{F}^{-1}, \tag{7.10}$$

$$\frac{\partial I_4}{\partial \mathbf{F}} = 2\mathbf{M} \otimes \mathbf{F}\mathbf{M}, \quad \frac{\partial I_5}{\partial \mathbf{F}} = 2(\mathbf{M} \otimes \mathbf{F}\mathbf{C}\mathbf{M} + \mathbf{C}\mathbf{M} \otimes \mathbf{F}\mathbf{M}), \tag{7.11}$$

enables $\mathbf{S}$ to be expanded out in the form

$$\begin{aligned} \mathbf{S} = {} & 2W_1\mathbf{F}^{\mathrm{T}} + 2W_2(I_1\mathbf{I} - \mathbf{C})\mathbf{F}^{\mathrm{T}} + 2I_3W_3\mathbf{F}^{-1} \\ & + 2W_4\mathbf{M} \otimes \mathbf{F}\mathbf{M} + 2W_5(\mathbf{M} \otimes \mathbf{F}\mathbf{C}\mathbf{M} + \mathbf{C}\mathbf{M} \otimes \mathbf{F}\mathbf{M}), \end{aligned} \tag{7.12}$$

where we have adopted the shorthand notation $W_i = \partial W/\partial I_i$ for $i = 1, 2, 3, 4, 5$, and $\mathbf{I}$ is the identity tensor in $\mathcal{B}_r$.

The Cauchy stress tensor $\boldsymbol{\sigma}$ is related to $\mathbf{S}$ via the connection $J\boldsymbol{\sigma} = \mathbf{FS}$, where $J = \det \mathbf{F} = I_3^{1/2}$, and hence

$$J\boldsymbol{\sigma} = 2W_1\mathbf{B} + 2W_2(I_1\mathbf{B} - \mathbf{B}^2) + 2I_3W_3\mathbf{I} \\ + 2W_4\mathbf{m} \otimes \mathbf{m} + 2W_5(\mathbf{m} \otimes \mathbf{Bm} + \mathbf{Bm} \otimes \mathbf{m}), \qquad (7.13)$$

$\mathbf{I}$ now being the identity tensor in $\mathcal{B}$ (here we use the same notation for the two, in general different, identity tensors).

We have assumed that the configuration $\mathcal{B}_\mathrm{r}$ is stress free. Thus, $\boldsymbol{\sigma}$ (and hence $\mathbf{S}$) must vanish when $\mathbf{F} = \mathbf{I}$, in which case $I_1 = I_2 = 3$ and $I_3 = I_4 = I_5 = 1$. Thus, from (7.13), it follows that

$$W_1(3,3,1,1,1) + 2W_2(3,3,1,1,1) + W_3(3,3,1,1,1) = 0, \qquad (7.14)$$

$$W_4(3,3,1,1,1) + 2W_5(3,3,1,1,1) = 0. \qquad (7.15)$$

We also assume that W itself vanishes in $\mathcal{B}_\mathrm{r}$, so that

$$W(3,3,1,1,1) = 0. \qquad (7.16)$$

Connections between the first and second derivatives of W evaluated in $\mathcal{B}_\mathrm{r}$ and the classical elastic constants of linear transversely isotropic elasticity that ensure consistency with the classical theory were given in [24] in terms of the Voigt notation for the linear elastic constants given in, for example, [44, p. 160].

7.2.3. Equilibrium and ellipticity considerations

Here we do not consider the effect of body forces, so that in terms of the nominal stress tensor the equation of equilibrium may be written Div $\mathbf{S} = \mathbf{0}$. In Cartesian components, we then have

$$\mathcal{A}_{\alpha i\beta j}x_{j,\alpha\beta} = 0, \qquad (7.17)$$

where the subscripts following a comma indicate differentiation with respect to the relevant coordinate, the usual summation convention for repeated indices is adopted and $\mathcal{A}_{\alpha i\beta j}$ are the components of the elasticity tensor $\boldsymbol{\mathcal{A}}$ given by

$$\mathcal{A}_{\alpha i\beta j} = \frac{\partial^2 W}{\partial F_{i\alpha}\partial F_{j\beta}}. \qquad (7.18)$$

Note that this has the pairwise symmetry $\mathcal{A}_{\alpha i\beta j} = \mathcal{A}_{\beta j\alpha i}$.

Now we consider $\mathbf{F}$ to be continuous throughout $\mathcal{B}_r$. We also assume that its gradient Grad $\mathbf{F}$ is continuous in $\mathcal{B}_r$ except that it may be discontinuous across a surface, denoted $\mathcal{S}_r$, in $\mathcal{B}_r$. Then $\mathcal{S}_r$ is called a *weak (discontinuity) surface.* Let $\mathbf{N}$ denote a unit normal vector to $\mathcal{S}_r$, and let N_α, $\alpha = 1, 2, 3$, be its components. The components of Grad $\mathbf{F}$ are $x_{i,\alpha\beta}$ and the jump (discontinuity) in $x_{i,\alpha\beta}$ across $\mathcal{S}_r$ is given by

$$[\![x_{i,\alpha\beta}]\!] = a_i N_\alpha N_\beta, \tag{7.19}$$

where $[\![\bullet]\!]$ denotes the difference in the enclosed quantity evaluated on the two sides of $\mathcal{S}_r$ and a_i, $i = 1, 2, 3$, are the components of a vector $\mathbf{a}$, which at this point is unknown.

Since $\mathbf{F}$, and hence $\boldsymbol{\mathcal{A}}$, is continuous, application of (7.19) to the equilibrium equation (7.17) on $\mathcal{S}_r$ gives

$$\mathcal{A}_{\alpha i\beta j} a_j N_\alpha N_\beta = 0. \tag{7.20}$$

Let the vector $\mathbf{n}$ be defined through the equation

$$\mathbf{N} = \mathbf{F}^{\mathrm{T}}\mathbf{n}. \tag{7.21}$$

Then (7.20) can be expressed in the form

$$\mathcal{A}_{0piqj} a_j n_p n_q = 0, \tag{7.22}$$

which is the Eulerian counterpart of (7.20) and where $\mathcal{A}_{0piqj}$ (the components of the fourth-order updated elasticity tensor $\boldsymbol{\mathcal{A}}_0$) are defined as the push forwards from $\mathcal{B}_r$ to $\mathcal{B}$ of the components $\mathcal{A}_{\alpha i\beta j}$ by

$$J\mathcal{A}_{0piqj} = F_{p\alpha} F_{q\beta} \mathcal{A}_{\alpha i\beta j} \tag{7.23}$$

(see, e.g., [41] for detailed discussion of this and similar connections). The pairwise symmetry $\mathcal{A}_{0piqj} = \mathcal{A}_{0qjpi}$ is inherited from that of $\mathcal{A}_{\alpha i\beta j}$.

According to Nanson's formula, $\mathbf{n}\mathrm{d}a = J\mathbf{F}^{-\mathrm{T}}\mathbf{N}\mathrm{d}A$, where $\mathrm{d}A$ and $\mathrm{d}a$ are area elements in $\mathcal{B}_r$ and $\mathcal{B}$, respectively, $\mathbf{N}$ and $\mathbf{n}$ being the associated normals (unit vectors in this case, as distinct from in (7.21) where $\mathbf{n}$ is not in general a unit vector). Let $\mathcal{S}$ be the image of $\mathcal{S}_r$ in $\mathcal{B}$. It follows that $\mathbf{n}$, defined either by (7.21) or by Nanson's formula, is normal to $\mathcal{S}$. We note that, strictly, it is $\mathcal{S}$ which is the actual discontinuity surface, while $\mathcal{S}_r$ is its preimage representation in $\mathcal{B}_r$.

Next, we write (7.22) in vector form as

$$\mathbf{Q}(\mathbf{n})\mathbf{a} = \mathbf{0}, \tag{7.24}$$

where we have introduced the *acoustic tensor* $\mathbf{Q}(\mathbf{n})$, with components defined by

$$Q_{ij} = \mathcal{A}_{0piqj} n_p n_q \tag{7.25}$$

enjoying the symmetry $Q_{ji} = Q_{ij}$.

If Eq. (7.24) is satisfied for some (non-zero) $\mathbf{n}$ then $\mathbf{Q}(\mathbf{n})$ is singular and the equation

$$\det \mathbf{Q}(\mathbf{n}) = 0 \tag{7.26}$$

determines such vectors $\mathbf{n}$, which are normal to surfaces of weak discontinuity. For an $\mathbf{n}$ satisfying (7.26) the corresponding eigenvector $\mathbf{a}$ of $\mathbf{Q}$ is obtained from (7.24). It follows from (7.24) that, for any deformation for which $\mathbf{n}$ and $\mathbf{a}$ are so determined,

$$[\mathbf{Q}(\mathbf{n})\mathbf{a}] \cdot \mathbf{a} \equiv \mathcal{A}_{0piqj} n_p n_q a_i a_j = 0, \tag{7.27}$$

and for a weak surface this condition is therefore necessarily satisfied.

If Eq. (7.27) does not hold for *any* pair of non-zero vectors $\mathbf{n}$ and $\mathbf{a}$ at some deformation, then the system of equations (7.17) is said to be *elliptic*, in which case no weak surface can exist at that deformation for the considered material model. The general ellipticity requirement is then

$$[\mathbf{Q}(\mathbf{n})\mathbf{a}] \cdot \mathbf{a} \equiv \mathcal{A}_{0piqj} n_p n_q a_i a_j \neq 0 \tag{7.28}$$

for *all* non-zero vectors $\mathbf{a}$ and $\mathbf{n}$. Of particular importance is the special case of this for which

$$[\mathbf{Q}(\mathbf{n})\mathbf{a}] \cdot \mathbf{a} \equiv \mathcal{A}_{0piqj} n_p n_q a_i a_j > 0 \quad \text{for all} \qquad \mathbf{a} \neq \mathbf{0},\ \mathbf{n} \neq \mathbf{0}. \tag{7.29}$$

This is known as the *strong ellipticity condition*. When strong ellipticity holds $\mathbf{Q}(\mathbf{n})$ is positive definite for all non-zero vectors $\mathbf{n}$.

For any given strain-energy function W evaluation of $\mathcal{A}_{0piqj} n_p n_q a_i a_j$ enables the ellipticity status of W to be determined. In particular, if the deformation gradient $\mathbf{F}$ satisfies (7.28) for that W for every pair of non-zero vectors $\mathbf{a}$ and $\mathbf{n}$ it is said to be an *elliptic deformation*. Moreover, if all possible deformations for a particular W are elliptic then W itself is said to be an *elliptic strain-energy function*. An example of an elliptic strain-energy function is the incompressible isotropic neo-Hookean material model (see, e.g., [23]). In contrast, for a given W, if the deformation gradient $\mathbf{F}$ satisfies Eq. (7.27) for some pair of non-zero vectors $\mathbf{a}$ and $\mathbf{n}$, then $\mathbf{F}$ is said to be a non-elliptic deformation for that W. Then, $\mathbf{n}$ is normal to a weak surface $\mathcal{S}$ in the deformed configuration $\mathcal{B}$, and the corresponding $\mathbf{N}$ is normal to its preimage $\mathcal{S}_\mathrm{r}$ in $\mathcal{B}_\mathrm{r}$. Here we have focused on weak surfaces, across which Grad $\mathbf{F}$ is discontinuous but $\mathbf{F}$ is continuous. In Section 7.5 we shall turn our attention to surfaces of strong discontinuity, i.e., surfaces across which $\mathbf{F}$ is discontinuous. These surfaces are also weak surfaces since a discontinuity in $\mathbf{F}$ entails a discontinuity in Grad $\mathbf{F}$ and they are also signaled by the loss of ellipticity condition (7.27).

7.2.4. Reinforcing models

In this section, we specialize the energy function $W = W(I_1, I_2, I_3, I_4, I_5)$ so that it contains information about the isotropic matrix within which reinforcing fibers are embedded and the fibers themselves. For this purpose we consider the isotropic matrix (or *isotropic base material*) to be characterized by the strain-energy function denoted $W^{\mathrm{iso}}(I_1, I_2, I_3)$, to which is added a contribution $W^{\mathrm{fib}}(I_4, I_5)$ associated with the fibers through their direction $\mathbf{M}$ in the reference configuration, this also being known as the *reinforcing model.* The superscripts 'iso' and 'fib' stand for 'isotropic' and 'fiber', respectively. Thus, we write

$$W = W(I_1, I_2, I_3, I_4, I_5) = W^{\mathrm{iso}}(I_1, I_2, I_3) + W^{\mathrm{fib}}(I_4, I_5). \tag{7.30}$$

The condition (7.14) now applies to $W^{\mathrm{iso}}(I_1, I_2, I_3)$ alone, and (7.15) to $W^{\mathrm{fib}}(I_4, I_5)$. Thus,

$$W_1^{\mathrm{iso}}(3,3,1) + 2W_2^{\mathrm{iso}}(3,3,1) + W_3^{\mathrm{iso}}(3,3,1) = 0, \tag{7.31}$$

$$W_4^{\mathrm{fib}}(1,1) + 2W_5^{\mathrm{fib}}(1,1) = 0, \tag{7.32}$$

and we may also set $W^{\mathrm{iso}}(3,3,1) = W^{\mathrm{fib}}(1,1) = 0$.

We now specialize further by considering the effect of the fiber-reinforcement through the invariants I_4 and I_5 separately. First, we specialize $W^{\mathrm{fib}}(I_4, I_5)$ to a function of I_4 alone: $W^{\mathrm{fib}}(I_4, I_5) = F(I_4)$. The term W_4 in the expression (7.13) for the Cauchy stress then yields a component $2I_4F'(I_4)$ in the direction $\mathbf{m}$. We require this to be positive (negative) when fibers in this direction are extended (contracted), so that

$$F'(I_4) > 0\ (< 0) \quad \text{for} \qquad I_4 > 1\ (< 1), \quad F'(1) = 0. \tag{7.33}$$

Next, we consider a reinforcing model of the form $W^{\text{fib}}(I_4, I_5) = G(I_5)$. The counterpart of (7.33) is then

$$G'(I_5) > 0\ (< 0) \quad \text{for} \qquad I_5 > 1\ (< 1), \quad G'(1) = 0. \tag{7.34}$$

In general, however, $I_5 > 1$ does not correspond to fiber extension, although it is easy to show that $I_4 > 1$ implies $I_5 > 1$. On the other hand $I_5 < 1$ implies $I_4 < 1$. In the following sections we shall assume that the inequalities (7.33) and (7.34) hold when $F(I_4)$ and $G(I_5)$ are used. In general $F(I_4)$ and $G(I_5)$ need not be convex functions.

7.3. Specialization to Plane Strain Deformations

In this section, in order to facilitate illustrations, we specialize by considering just plane strain deformations. In particular, we evaluate the extent of ellipticity of the specific examples of material models identified in Section 7.2.4 and we obtain conditions on the reinforcing models that elucidate the ellipticity status of the energy function (7.30). Attention is now restricted to plane deformations in the (X_1, X_2) plane with the fiber-reinforcement direction $\mathbf{M}$ within this plane. For plane strain we then have $x_3 = X_3$ and x_1 and x_2 depend only on X_1 and X_2. It follows that the only non-zero (in general) components of the deformation gradient are F_{11}, F_{12}, F_{21}, F_{22}, with $F_{33} = 1$, so that the out-of-plane principal stretch is $\lambda_3 = 1$. The corresponding components of $\mathbf{C}$ are C_{11}, $C_{12} = C_{21}$, C_{22} and $C_{33} = 1$. The isotropic invariants (7.5) now specialize to

$$I_1 = \lambda_1^2 + \lambda_2^2 + 1, \quad I_2 = \lambda_1^2 + \lambda_2^2 + \lambda_1^2\lambda_2^2, \quad I_3 = \lambda_1^2\lambda_2^2. \tag{7.35}$$

Let $\mathbf{M}$ have components M_1 and M_2 in the (X_1, X_2) plane ($M_3 = 0$). Then (7.6) and (7.7) specialize to

$$I_4 = \lambda_1^2 M_1^2 + \lambda_2^2 M_2^2, \quad I_5 = \lambda_1^4 M_1^2 + \lambda_2^4 M_2^2, \tag{7.36}$$

respectively. It may then be deduced that

$$I_2 = I_1 + I_3 - 1, \quad I_5 = (I_1 - 1)I_4 - I_3. \tag{7.37}$$

This means, in particular, that under the plane strain specialization only three of the original five invariants are independent. Thus, the strain-energy function can be specialized accordingly. We choose (I_1, I_3, I_4) as the independent invariants and denote the specialization of the strain-energy function by $\hat{W}$, which is defined by

$$\hat{W}(I_1, I_3, I_4) = W(I_1, I_1 + I_3 - 1, I_3, I_4, (I_1 - 1)I_4 - I_3). \tag{7.38}$$

Let $\mathbf{F}$ now denote the in-plane specialization of the deformation gradient and $\mathbf{M}$ the corresponding in-plane fiber direction. Then, $(7.10)_{1,3}$ and $(7.11)_1$ specialize to

$$\frac{\partial I_1}{\partial \mathbf{F}} = 2\mathbf{F}^{\mathrm{T}}, \quad \frac{\partial I_3}{\partial \mathbf{F}} = 2I_3\mathbf{F}^{-1}, \quad \frac{\partial I_4}{\partial \mathbf{F}} = 2\mathbf{M} \otimes \mathbf{FM}. \tag{7.39}$$

Let $\mathbf{S}$ similarly denote the plane strain specialization of the nominal stress tensor (7.12), which then specializes to

$$\mathbf{S} = 2\hat{W}_1\mathbf{F}^{\mathrm{T}} + 2I_3\hat{W}_3\mathbf{F}^{-1} + 2\hat{W}_4\mathbf{M} \otimes \mathbf{m}, \tag{7.40}$$

where $\mathbf{m} = \mathbf{FM}$ is the in-plane version of (7.4). We emphasize that (7.40) includes only the in-plane part of the three-dimensional nominal stress. The only non-zero out-of-plane component S_{33} is required (in general) to maintain the plane strain deformation and can be obtained from (7.12) if needed. The associated in-plane specialization of the Cauchy stress $\boldsymbol{\sigma}$ is given by $J\boldsymbol{\sigma} = \mathbf{FS}$, with $J = \det \mathbf{F}$. Hence the in-plane specialization of (7.13) is

$$J\boldsymbol{\sigma} = 2\hat{W}_1\mathbf{B} + 2\hat{W}_3\mathbf{I} + 2\hat{W}_4\mathbf{m} \otimes \mathbf{m}, \tag{7.41}$$

with $\mathbf{B} = \mathbf{FF}^{\mathrm{T}}$ the in-plane specialization of the left Cauchy–Green deformation tensor and $\mathbf{I}$ the in-plane identity tensor.

The planar strain-energy function $\hat{W}$ inherits the restrictions (7.14) and (7.15) in the (stress free) reference configuration identified for W in the form

$$\hat{W}_1(3, 1, 1) + \hat{W}_3(3, 1, 1) = 0, \quad \hat{W}_4(3, 1, 1) = 0, \tag{7.42}$$

together with $\hat{W}(3, 1, 1) = 0$.

Next, we note that in Section 7.2.3 general expressions for the components of $\boldsymbol{\mathcal{A}}_0$ were given, but their expanded forms in terms of invariants were not provided since they are rather lengthy. In the plane strain specialization, however, they are manageable but still quite lengthy. Explicitly they are given by

$$\begin{aligned} J\mathcal{A}_{0piqj} &= 4\hat{W}_{11}B_{pi}B_{qj} + 4I_3\hat{W}_{13}(B_{pi}\delta_{qj} + B_{qj}\delta_{pi}) \\ &\quad + 4I_3^2\hat{W}_{33}\delta_{pi}\delta_{qj} + 4I_3\hat{W}_{34}(\delta_{pi}m_q m_j + \delta_{qj}m_p m_i) \\ &\quad + 4\hat{W}_{14}(B_{pi}m_q m_j + B_{qj}m_p m_i) + 4\hat{W}_{44}m_p m_i m_q m_j \\ &\quad + 2\hat{W}_1\delta_{ij}B_{pq} + 2I_3\hat{W}_3(2\delta_{pi}\delta_{qj} - \delta_{pj}\delta_{qi}) + 2\hat{W}_4\delta_{ij}m_p m_q, \end{aligned} \tag{7.43}$$

in which the indices run over values 1 and 2 only and δ_{ij} denotes the Kronecker delta. In obtaining (7.43) use has been made of the second derivatives of the invariants I_1, I_3 and I_4 with respect to $\mathbf{F}$, which, in component form, are given by

$$F_{p\alpha}F_{q\beta}\frac{\partial^2 I_1}{\partial F_{i\alpha}\partial F_{j\beta}} = 2\delta_{ij}B_{pq}, \quad F_{p\alpha}F_{q\beta}\frac{\partial^2 I_4}{\partial F_{i\alpha}\partial F_{j\beta}} = 2\delta_{ij}m_p m_q, \tag{7.44}$$

$$F_{p\alpha}F_{q\beta}\frac{\partial^2 I_3}{\partial F_{i\alpha}\partial F_{j\beta}} = 4I_3\delta_{ip}\delta_{jq} - 2I_3\delta_{iq}\delta_{jp}. \tag{7.45}$$

These may be obtained from more general formulas such as those in the appendix of [45] or by direct differentiation of the component forms of the appropriate expressions for the first derivatives in (7.10) and (7.11).

Referred to the principal axes of $\mathbf{B}$ the (in-plane) components of $\mathbf{Q}$, obtained from (7.43) and the definition (7.25), are given by

$$\begin{aligned} JQ_{ij} &= 4\hat{W}_{11}\lambda_i^2\lambda_j^2 n_i n_j + 4I_3\hat{W}_{13}(\lambda_i^2 + \lambda_j^2)n_i n_j \\ &\quad + 4I_3^2\hat{W}_{33}n_i n_j + 4I_3\hat{W}_{34}(\mathbf{n}\cdot\mathbf{m})(n_i m_j + n_j m_i) \\ &\quad + 4\hat{W}_{14}(\mathbf{n}\cdot\mathbf{m})(\lambda_i^2 n_i m_j + \lambda_j^2 n_j m_i) + 4\hat{W}_{44}(\mathbf{n}\cdot\mathbf{m})^2 m_i m_j \\ &\quad + 2\hat{W}_1\delta_{ij}(\lambda_1^2 n_1^2 + \lambda_2^2 n_2^2) + 2I_3\hat{W}_3 n_i n_j + 2\hat{W}_4\delta_{ij}(\mathbf{n}\cdot\mathbf{m})^2, \end{aligned} \tag{7.46}$$

in which the indices i and j take values 1 and 2 only. The plane strain specialization of the strong ellipticity condition (7.29) then takes the form

$$\begin{aligned} &2\hat{W}_{11}[\mathbf{a}\cdot(\mathbf{Bn})]^2 + 4I_3\hat{W}_{13}[\mathbf{a}\cdot(\mathbf{Bn})](\mathbf{a}\cdot\mathbf{n}) + 2I_3^2\hat{W}_{33}(\mathbf{a}\cdot\mathbf{n})^2 \\ &\quad + [4I_3\hat{W}_{34}(\mathbf{a}\cdot\mathbf{n}) + 4\hat{W}_{14}[\mathbf{a}\cdot(\mathbf{Bn})] + 2\hat{W}_{44}(\mathbf{a}\cdot\mathbf{m})(\mathbf{n}\cdot\mathbf{m})](\mathbf{a}\cdot\mathbf{m})(\mathbf{n}\cdot\mathbf{m}) \\ &\quad + \hat{W}_1\mathbf{n}\cdot(\mathbf{Bn}) + I_3\hat{W}_3(\mathbf{a}\cdot\mathbf{n})^2 + \hat{W}_4(\mathbf{n}\cdot\mathbf{m})^2 > 0 \end{aligned} \tag{7.47}$$

for all (in-plane) unit vectors $\mathbf{a}$ and $\mathbf{n}$. It should be emphasized, however, that $\mathbf{m} = \mathbf{FM}$ is not in general a unit vector.

It is worth mentioning that the inequality (7.47) may also be expressed as positive definiteness of Q_{ij}, i.e.,

$$Q_{11}(\mathbf{n}) > 0, \quad Q_{11}(\mathbf{n})Q_{22}(\mathbf{n}) - [Q_{12}(\mathbf{n})]^2 > 0 \tag{7.48}$$

for all (in-plane) unit vectors $\mathbf{n}$, the components $Q_{ij}(\mathbf{n})$ being given by (7.46).

For a brief discussion of the specialization of the above inequalities to isotropic materials and to linear transversely isotropic materials, in which cases necessary and sufficient conditions for strong ellipticity to hold are obtained, we refer to [24, 46], while prior analysis for isotropy can be found in, for example, [47, 48].

To proceed with the discussion of loss of ellipticity we now focus on strain-energy functions of the decoupled form (7.30), but specialized to plane strain. In particular, we first consider the reinforcing model to depend only on I_4. Thus

$$\hat{W}(I_1, I_3, I_4) = \hat{W}^{\text{iso}}(I_1, I_3) + \hat{W}^{\text{fib}}(I_4). \tag{7.49}$$

A basic assumption is that the strong ellipticity condition (7.47) holds in the undeformed reference configuration. We also assume here and henceforth that $\hat{W}^{\text{iso}}(I_1, I_3)$ on its own satisfies the strong ellipticity condition so that, by continuity, incipient failure of ellipticity is dependent on $\hat{W}^{\text{fib}}(I_4)$ losing ellipticity as the deformation proceeds on some path from the reference configuration. In respect of (7.49) the strong ellipticity condition (7.47) takes the form

$$\text{SE}^{\text{iso}} > 0, \quad \text{SE}^{\text{iso}} + (\mathbf{m}\cdot\mathbf{n})^2[F' + 2(\mathbf{a}\cdot\mathbf{m})^2F''] > 0 \tag{7.50}$$

for all (in-plane) unit vectors $\mathbf{n}$, where we have adopted the notations

$$\begin{aligned}\text{SE}^{\text{iso}} &= 2\hat{W}^{\text{iso}}_{11}[\mathbf{a}\cdot(\mathbf{Bn})]^2 + 4I_3\hat{W}^{\text{iso}}_{13}[\mathbf{a}\cdot(\mathbf{Bn})](\mathbf{a}\cdot\mathbf{n})\\ &\quad + 2I_3^2\hat{W}^{\text{iso}}_{33}(\mathbf{a}\cdot\mathbf{n})^2 + \hat{W}^{\text{iso}}_{1}\mathbf{n}\cdot(\mathbf{Bn}) + I_3\hat{W}^{\text{iso}}_{3}(\mathbf{a}\cdot\mathbf{n})^2\end{aligned} \tag{7.51}$$

and

$$F(I_4) = \hat{W}^{\text{fib}}(I_4) \tag{7.52}$$

for subsequent convenience and a prime signifies differentiation with respect to I_4. In the following, we examine the contribution to (7.50) of the reinforcing model $F(I_4)$.

7.3.1. Ellipticity of the reinforcing model $F(I_4)$

Only the latter two terms in $(7.50)_2$ depend on $\mathbf{m} = \mathbf{FM}$ and $F(I_4)$. Since $\mathbf{n}$ can be chosen freely, it follows that the ellipticity status of the model (7.49) is strongly dependent on the sign of

$$F'(I_4) + 2(\mathbf{a} \cdot \mathbf{m})^2 F''(I_4). \tag{7.53}$$

Since $\mathbf{a}$ may be chosen independently of $\mathbf{n}$ so that $\mathbf{a} \cdot \mathbf{m} = \mathbf{0}$ then for (7.53) to be non-negative it is necessary that $F'(I_4) \geq 0$.

If $F''(I_4) < 0$ then since, from $(7.3)_1$, $\mathbf{m} \cdot \mathbf{m} = I_4$ we have

$$F'(I_4) + 2(\mathbf{a} \cdot \mathbf{m})^2 F''(I_4) \geq F'(I_4) + 2I_4 F''(I_4).$$

It follows that (7.53) is non-negative for all $\mathbf{m}$ if and only if

$$F'(I_4) \geq 0, \quad F'(I_4) + 2I_4 F''(I_4) \geq 0. \tag{7.54}$$

Thus, given that we have assumed that $\hat{W}^{\text{iso}}(I_1, I_3)$ is strongly elliptic, i.e., $\text{SE}^{\text{iso}} > 0$, Eq. (7.54) is sufficient for the strong ellipticity condition (7.50) to hold.

We recall from (7.33) that $F'(I_4) \gtrless 0$ according as $I_4 \gtrless 1$, and it follows that $F''(1) \geq 0$. Since $\mathbf{n}$ may be chosen so that $\mathbf{n} \cdot \mathbf{m} = 0$ the reinforcing term in (7.50) loses ellipticity for any deformation, but does not affect the overall strong ellipticity in this case. Moreover, if $\mathbf{n} \cdot \mathbf{m} \neq 0$ the contribution of the reinforcing term to (7.50) is strictly positive if and only if

$$F'(I_4) > 0, \quad F'(I_4) + 2I_4 F''(I_4) > 0. \tag{7.55}$$

7.3.1.1. *Failure of ellipticity*

We have already assumed that the isotropic matrix material is strongly elliptic, so that $\text{SE}^{\text{iso}} > 0$. We are now concerned with assessing the deformations for which the strong ellipticity condition $(7.50)_2$ just fails, which requires that

$$\text{SE}^{\text{iso}} + (\mathbf{m} \cdot \mathbf{n})^2 [F' + 2(\mathbf{a} \cdot \mathbf{m})^2 F''] \geq 0, \tag{7.56}$$

for all unit vectors $\mathbf{a}$ and $\mathbf{n}$, with equality holding for one or more pairs of unit vectors $(\mathbf{a}, \mathbf{n})$. Let the region of deformation gradient space (which includes the reference configuration $\mathbf{F} = \mathbf{I}$) for which the strong ellipticity condition (7.50) holds be denoted by E. Its boundary, denoted $\partial\mathsf{E}$, is then defined by the equality in (7.56).

In view of the strong ellipticity $\mathrm{SE}^{\mathrm{iso}} > 0$ the failure of ellipticity (7.56) can only occur once the reinforcing model $F(I_4)$ has lost ellipticity, i.e., when either $F'(I_4) < 0$ or $F'(I_4) + 2I_4F''(I_4) < 0$ prior to the deformation within E reaching the boundary $\partial\mathsf{E}$.

To take the analysis further we now decompose the acoustic tensor $\mathbf{Q}$ for the material model (7.49) with the notation (7.52) into its isotropic and reinforcing fiber contributions by writing $\mathbf{Q} = \mathbf{Q}^{\mathrm{iso}} + \mathbf{Q}^{\mathrm{fib}}$, where $\mathbf{Q}^{\mathrm{iso}}$ is the acoustic tensor associated with $\hat{W}^{\mathrm{iso}}(I_1, I_3)$ and $\mathbf{Q}^{\mathrm{fib}}$ that with $F(I_4)$. Then, we obtain

$$\det \mathbf{Q} = \det \mathbf{Q}^{\mathrm{iso}} + Q_{11}^{\mathrm{iso}}Q_{22}^{\mathrm{fib}} + Q_{22}^{\mathrm{iso}}Q_{11}^{\mathrm{fib}} - 2Q_{12}^{\mathrm{iso}}Q_{12}^{\mathrm{fib}} + \det \mathbf{Q}^{\mathrm{fib}}, \tag{7.57}$$

where, explicitly,

$$J^2 \det \mathbf{Q}^{\mathrm{fib}} = 4(\mathbf{n}\cdot\mathbf{m})^4 F'(F' + 2I_4F''), \tag{7.58}$$

and, because of the strong ellipticity of $\hat{W}^{\mathrm{iso}}(I_1, I_3)$, we have

$$\det \mathbf{Q}^{\mathrm{iso}} \equiv Q_{11}^{\mathrm{iso}}Q_{22}^{\mathrm{iso}} - Q_{12}^{\mathrm{iso}\,2} > 0, \quad Q_{11}^{\mathrm{iso}} > 0, \quad Q_{22}^{\mathrm{iso}} > 0. \tag{7.59}$$

Also, bearing in mind (7.48), we have

$$JQ_{11}^{\mathrm{fib}} = 2(\mathbf{n}\cdot\mathbf{m})^2(F' + 2F''m_1^2), \quad JQ_{22}^{\mathrm{fib}} = 2(\mathbf{n}\cdot\mathbf{m})^2(F' + 2F''m_2^2). \tag{7.60}$$

It is straightforward to deduce from (7.58) and (7.60) that neither Q_{11}^{fib} nor Q_{22}^{fib} can vanish before $\det \mathbf{Q}^{\mathrm{fib}}$ within E, and it is therefore appropriate to examine loss of ellipticity of $F(I_4)$ based on $\det \mathbf{Q}^{\mathrm{fib}}$.

Suppose that any material parameter associated with the matrix is significantly smaller than any associated with the reinforcement, and let ε (the ratio of a reinforcing to a matrix material constant) be a (large in general) dimensionless parameter that measures the strength of the reinforcement compared with the strength of the matrix material. Then, we write $F(I_4) = \varepsilon\tilde{F}(I_4)$, where $\tilde{F}(I_4)$ has the same order of magnitude as $\hat{W}^{\mathrm{iso}}(I_1, I_3)$, so that (7.57) can be expressed as a quadratic in ε, namely

$$a + b\varepsilon + c\varepsilon^2, \tag{7.61}$$

where

$$a = \det \mathbf{Q}^{\mathrm{iso}}, \quad b = Q_{11}^{\mathrm{iso}}\tilde{Q}_{22}^{\mathrm{fib}} + Q_{22}^{\mathrm{iso}}\tilde{Q}_{11}^{\mathrm{fib}} - 2Q_{12}^{\mathrm{iso}}\tilde{Q}_{12}^{\mathrm{fib}}, \quad c = \det \tilde{\mathbf{Q}}^{\mathrm{fib}}, \tag{7.62}$$

and $\tilde{\mathbf{Q}}^{\mathrm{fib}}$ is defined as $\mathbf{Q}^{\mathrm{fib}}/\varepsilon$, a, b and c all having the same order of magnitude.

A necessary condition for loss of ellipticity is therefore $\det \mathbf{Q}^{\text{fib}} < 0$, and this requires that $\mathbf{n} \cdot \mathbf{m} \neq 0$. When this holds loss of ellipticity will certainly occur if ε is sufficiently large. Given the fiber direction, the point at which loss of ellipticity occurs depends on the deformation, the vectors $\mathbf{a}$ and $\mathbf{n}$, and the strength of the reinforcement. We now illustrate the type of weak surface that can be predicted for the cases in which $F''(I_4) \geq 0$ and $F''(I_4) < 0$ separately.

It is clear from (7.50) that ellipticity cannot fail if $\mathbf{n} \cdot \mathbf{m} = 0$, so that a weak surface cannot be parallel to the fiber-reinforcement direction. A particular example leading to this conclusion was provided in [6], and a parallel result for an incompressible material was given in [23].

Case (a): Fiber kinking. First we consider $F''(I_4) \geq 0$. Then, for loss of ellipticity it is necessary that $F'(I_4) < 0$, which means that fibers are subject to compression. For a given $\mathbf{n}$ the reinforcement term

$$(\mathbf{m} \cdot \mathbf{n})^2[F' + 2(\mathbf{a} \cdot \mathbf{m})^2 F''] \tag{7.63}$$

in (7.50) takes on its least value when $\mathbf{a} \cdot \mathbf{m} = 0$ and its minimum value is $I_4 F'(I_4)$, which corresponds to $\mathbf{n}$ being parallel to $\mathbf{m}$, i.e., $\mathbf{n} = \hat{\mathbf{m}}$, where $\hat{\mathbf{m}} = \mathbf{m}/|\mathbf{m}|$. The magnitude of this term increases as I_4 decreases from unity and loss of ellipticity is initiated when the negative value of $I_4 F'(I_4)$ balances the positive value of SE^{iso} in (7.50) for $\mathbf{n} = \hat{\mathbf{m}}$.

As a simple example consider the situation in which $\hat{\mathbf{m}}$ is an eigenvector of $\mathbf{B}$ corresponding to the stretch λ_1. Then for $\mathbf{a} \cdot \mathbf{m} = 0$ the left-hand side of $(7.50)_2$ reduces to $\lambda_1^2 \hat{W}_1^{\text{iso}}(I_1, I_3) + I_4 F'(I_4)$. But, from $(7.50)_1$ and (7.51), it follows that $\hat{W}_1^{\text{iso}}(I_1, I_3) > 0$. Thus, $\lambda_1^2 \hat{W}_1^{\text{iso}}(I_1, I_3) + I_4 F'(I_4)$ will vanish for some $I_4 < 1$ since the reinforcing term is dominant except for relatively weak reinforcement, and for strong reinforcement it will vanish for I_4 close to unity.

Since $\mathbf{n} = \hat{\mathbf{m}}$ and $\mathbf{a} \cdot \hat{\mathbf{m}} = 0$ the weak surface is normal to the fiber direction and the failure mechanism associated with this situation is interpreted as the initiation of *fiber kinking*, which occurs under fiber compression. Loss of ellipticity is associated not only with a weak discontinuity but also with strong discontinuities, and, as we have already noted, a strong discontinuity is also a weak discontinuity (but not the converse in general). However, fiber kinking is actually a strong discontinuity and the weak surface is also a strong surface in this case. Here we have considered the situation in which $\mathbf{a} \cdot \hat{\mathbf{m}} = 0$, but it is possible that, for $\mathbf{a} \cdot \hat{\mathbf{m}} \neq 0$, $[\mathbf{Q}^{\text{iso}}(\mathbf{n})\mathbf{a}] \cdot \mathbf{a}$ has a smaller value than when $\mathbf{a} \cdot \hat{\mathbf{m}} = 0$, in which case ellipticity could be lost on a deformation path prior to the point at which a kink is initiated.

Case (b): Fiber debonding. Second, we focus on the case $F''(I_4) < 0$. If $F'(I_4) + 2I_4F''(I_4) \geq 0$ then overall ellipticity cannot be lost, but if $F'(I_4) + 2I_4F''(I_4) < 0$ then the minimum value of (7.63) is

$$I_4F' + 2I_4^2F'' \tag{7.64}$$

for either $F' > 0$ or $F' < 0$, and this corresponds to $\mathbf{a} = \mathbf{n} = \hat{\mathbf{m}}$.

Since $F''(1) > 0$ then, by continuity, $F''(I_4) > 0$ when I_4 is near 1. In fiber compression, therefore, loss of ellipticity is most likely to arise in the mode discussed in (a). In contrast, in fiber extension, since then $F'(I_4) > 0$, loss of ellipticity can only occur if $F'(I_4) + 2I_4F''(I_4)$ becomes negative, which entails loss of convexity of F.

Again we consider the example for which $\hat{\mathbf{m}}$ is an eigenvector of $\mathbf{B}$. Then in fiber *extension* ellipticity can be lost when $\mathbf{n} \cdot \mathbf{m}$ is small since the negative contribution to (7.63) then balances the positive contribution due to $\hat{W}^{\text{iso}}(I_1, I_3)$ provided the reinforcement is sufficiently strong. In this case the weak surface is close to parallel to the fiber direction and the relevant failure mechanism can be interpreted as *debonding*.

7.3.2. Ellipticity of a hybrid reinforcing model

For the model (7.49) considered in the foregoing section, the isotropic $\hat{W}_1^{\text{iso}}(I_1, I_3)$ and reinforcing $\hat{W}^{\text{fib}}(I_4)$ contributions are decoupled. In this section, we consider a reinforcing model that depends not only on I_4 but also on the isotropy. First, note that as well as I_4 the invariant I_5 reflects the anisotropy. Thus, a possible alternative model to (7.49) that might be considered is

$$\hat{W}(I_1, I_3, I_4) = \hat{W}^{\text{iso}}(I_1, I_3) + \hat{W}^{\text{fib}}(I_5). \tag{7.65}$$

For plane strain, from $(7.37)_2$, I_5 is given by $I_5 = (I_1 - 1)I_4 - I_3$, in which case the reinforcing model depends on all three invariants I_1, I_3, I_4. To simplify the resulting analysis, but without changing the substance of the results, we adopt a slightly modified version of I_5, denoted I_5^* and defined by

$$I_5^* = (I_1 - 1)I_4 - 1, \tag{7.66}$$

noting that for incompressible materials $I_3 = 1$ and $I_5^* = I_5$. Then (7.65) is replaced by

$$\hat{W}(I_1, I_3, I_4) = \hat{W}^{\mathrm{iso}}(I_1, I_3) + \hat{W}^{\mathrm{fib}}(I_5^*), \tag{7.67}$$

and for simplicity we also use the notation $G(I_5^*) = W^{\mathrm{fib}}(I_5^*)$.

From (7.41) the contribution of the term G to the normal stress in the deformed fiber direction is easily seen to be $2J^{-1}I_4[\hat{\mathbf{m}}\cdot(\mathbf{B}\hat{\mathbf{m}})+I_1-1]G'(I_5^*)$. Since the coefficient of $G'(I_5^*)$ is positive this stress is positive (negative) when $G'(I_5^*) > (<)\, 0$ according as $I_5^* > (<)\, 1$, i.e., the conditions (7.34) with I_5 replaced by I_5^*.

By specializing the general plane strain strong ellipticity condition (7.47) for the model (7.67) we obtain

$$\mathrm{SE}^{\mathrm{iso}} + J(\mathbf{Q}^{\mathrm{fib}}\mathbf{a})\cdot\mathbf{a} > 0, \tag{7.68}$$

where $\mathrm{SE}^{\mathrm{iso}}$ is defined in (7.51) while $\mathbf{Q}^{\mathrm{fib}}$ is here obtained from G to give

$$\begin{aligned} J(\mathbf{Q}^{\mathrm{fib}}\mathbf{a})\cdot\mathbf{a} = {} & 2G''\left[I_4\mathbf{a}\cdot(\mathbf{B}\mathbf{n}) + (I_1-1)(\mathbf{n}\cdot\mathbf{m})(\mathbf{a}\cdot\mathbf{m})\right]^2 \\ & + G'\left[I_4\mathbf{n}\cdot(\mathbf{B}\mathbf{n}) + (I_1-1)(\mathbf{n}\cdot\mathbf{m})^2\right]. \end{aligned} \tag{7.69}$$

The coefficient of G' in (7.69) is strictly positive while that of G'' is non-negative. Thus, the sign of this expression depends on the signs of G' and G''. For ellipticity to be lost $(\mathbf{Q}^{\mathrm{fib}}\mathbf{a})\cdot\mathbf{a}$ must be negative, and we now examine two examples for which ellipticity does fail for this model.

Case (c): Fiber kinking and fiber splitting. We recall from Section 7.3.1.1 that strong ellipticity holds when $\mathbf{m}\cdot\mathbf{n} = 0$ since in this case the reinforcing model involving F does not contribute to $(7.50)_2$. This is not the case for (7.69) since for $\mathbf{m}\cdot\mathbf{n} = 0$ it becomes

$$2I_4^2G''[\mathbf{a}\cdot(\mathbf{B}\mathbf{n})]^2 + I_4G'\mathbf{n}\cdot(\mathbf{B}\mathbf{n}). \tag{7.70}$$

With $\mathbf{m}\cdot\mathbf{n} = 0$, for a first illustration we consider the situation in which $\mathbf{n}$ coincides with the principal axis of $\mathbf{B}$ corresponding to the stretch λ_1, and we set $\mathbf{a} = \hat{\mathbf{m}}$. Then the left-hand side of (7.68) simplifies to $\lambda_1^2[\hat{W}_1^{\mathrm{iso}}(I_1, I_3) + I_4G'(I_5^*)]$. Since $\hat{W}_1^{\mathrm{iso}}(I_1, I_3) > 0$ and $G'(1) = 0$ this expression is positive in the reference configuration but can vanish as I_5^* decreases from unity since $G'(I_5^*) < 0$ for $I_5^* < 1$. Thus ellipticity can be lost for $I_5^* < 1$ and, since $\mathbf{m}\cdot\mathbf{n} = 0$, the associated weak surface is aligned with the fiber direction; we interpret the corresponding failure mechanism as fiber splitting [9, 10]. The reinforcing model $F(I_4)$ is not able to predict this mechanism. Since $\mathbf{n}$ is an eigenvector of $\mathbf{B}$ and $\mathbf{m}\cdot\mathbf{n} = 0$ then $\mathbf{m}$ is also an eigenvector of $\mathbf{B}$ and corresponds to the stretch λ_2, and hence $I_5^* < 1$ implies $(\lambda_1^2 + \lambda_2^2)\lambda_2^2 < 2$. For an isochoric deformation $\lambda_1\lambda_2 = 1$, in which case $\lambda_2 < 1$, so that the fiber is subject to compression.

For the second example we consider $\mathbf{n} = \hat{\mathbf{m}}$ with $\mathbf{a}\cdot\mathbf{n} = 0$ and again take $\mathbf{n}$ to be an eigenvector of $\mathbf{B}$ corresponding to the stretch λ_1, so that $I_4 = \lambda_1^2$. The left-hand side of (7.68) becomes $\lambda_1^2[\hat{W}_1^{\text{iso}}(I_1, I_3) + (2\lambda_1^2 + \lambda_2^2)G'(I_5^*)]$. Ellipticity can be lost as I_5^* decreases from 1 at a value nearer to 1 than in the first example, and in this case the weak surface is normal to the fiber direction. In particular, if $\lambda_1 < 1$ the fiber direction is compressed and the loss of ellipticity is associated with fiber kinking, similarly to the situation for the $F(I_4)$ reinforcement.

These two mechanisms may arise simultaneously with weak surfaces along and normal to the fiber direction, as illustrated in the experimental work [9, 10].

More generally, eventual loss of ellipticity is guaranteed as I_5^* decreases from 1 since the terms in G dominate the left-hand side of (7.68) for strong reinforcement.

Case (d): Fiber debonding and matrix failure. Case (c) was associated with $I_5^* < 1$ and $G'(I_5^*) < 0$. We now consider what may happen when $I_5^* > 1$ and $G'(I_5^*) > 0$. Given that $\text{SE}^{\text{iso}} > 0$ it is clear from (7.69) that loss of ellipticity requires $G''(I_5^*) < 0$, and we highlight two possible occurrences of weak surfaces. First, if $\mathbf{n}\cdot\mathbf{m} = 0$ the weak surface is parallel to the fiber direction. If, for example, $\mathbf{m}$ is an eigenvector of $\mathbf{B}$ corresponding to the stretch λ_1 and $\mathbf{a} = \mathbf{n}$ then $\mathbf{n}$ is an eigenvector corresponding to the stretch λ_2, $I_4 = \lambda_1^2$, $I_5^* = (\lambda_1^2 + \lambda_2^2)\lambda_1^2 - 1$ and (7.70) becomes

$$\lambda_1^2\lambda_2^2[G'(I_5^*) + 2I_4G''(I_5^*)], \tag{7.71}$$

which can become negative for suitable functions $G(I_5^*)$, and it is a simple matter to construct such functions. Note that for an isochoric deformation $I_5^* = \lambda_1^4$ and since $I_5^* > 1$ the fiber is extended, and the relevant failure mechanism in this case is debonding.

The second example is when $\mathbf{a} = \mathbf{n} = \hat{\mathbf{m}}$ is an eigenvector of $\mathbf{B}$ corresponding to the stretch λ_1, so the weak surface is normal to the fiber direction. Then $I_4 = \lambda_1^2$, I_5^* is as in the first example, and (7.70) takes the form

$$\lambda_1^2(2\lambda_1^2 + \lambda_2^2)[G'(I_5^*) + 2(2\lambda_1^2 + \lambda_2^2)G''(I_5^*)], \tag{7.72}$$

and again suitable functions $G(I_5^*)$ will allow this to become negative. The loss of ellipticity in this case is then interpreted as matrix failure.

In summary, the loss of ellipticity outlined in cases (a)–(d) above can be thought of as predicting the features highlighted in Figs. 7.1–7.4. We emphasize that in the foregoing analysis we have focused on compressible materials. A parallel treatment for incompressible materials is contained in [23], and in the remainder of this chapter, indeed, we restrict attention to incompressible materials as we turn attention to strong discontinuities. As a prelude to the discussion of strong discontinuities we summarize in the following section the equations appropriate for incompressible elastic materials.

7.4. Incompressible Materials

For incompressible materials the deformation gradient satisfies the constraint $\det \mathbf{F} = 1$ and hence $I_3 = 1$. In terms of the principal stretches this is expressed as

$$\lambda_1\lambda_2\lambda_3 = 1. \tag{7.73}$$

Then, with the omission of I_3 the strain-energy function (7.8) specializes to

$$W(I_1, I_2, I_4, I_5) \tag{7.74}$$

and the expression (7.9) for the nominal stress tensor is adjusted accordingly as

$$\mathbf{S} = \frac{\partial W}{\partial \mathbf{F}} - p\mathbf{F}^{-1}, \tag{7.75}$$

where p is a Lagrange multiplier associated with the incompressibility constraint. When expanded in terms of the invariants, analogously to (7.12), it becomes

$$\begin{aligned} \mathbf{S} = {} & 2W_1\mathbf{F}^{\mathrm{T}} + 2W_2(I_1\mathbf{I} - \mathbf{C})\mathbf{F}^{\mathrm{T}} - p\mathbf{F}^{-1} + 2W_4\mathbf{M}\otimes\mathbf{FA} \\ & + 2W_5(\mathbf{M}\otimes\mathbf{FCA} + \mathbf{CA}\otimes\mathbf{FA}), \end{aligned} \tag{7.76}$$

and the corresponding Cauchy stress tensor is

$$\begin{aligned} \boldsymbol{\sigma} = {} & 2W_1\mathbf{B} + 2W_2(I_1\mathbf{B} - \mathbf{B}^2) - p\mathbf{I} + 2W_4\mathbf{m}\otimes\mathbf{m} \\ & + 2W_5(\mathbf{m}\otimes\mathbf{Ba} + \mathbf{Ba}\otimes\mathbf{m}). \end{aligned} \tag{7.77}$$

The restrictions (7.14), (7.15) and (7.16) in the reference configuration are replaced by

$$2W_1(3,3,1,1) + 4W_2(3,3,1,1) - p_0 = 0, \tag{7.78}$$

$$W_4(3,3,1,1) + 2W_5(3,3,1,1) = 0, \tag{7.79}$$

and $W(3,3,1,1) = 0$, respectively, where p_0 is the value of p in the reference configuration.

The equilibrium equation (7.17) is replaced by

$$\mathcal{A}_{\alpha i\beta j}x_{j,\alpha\beta} - p_{,i} = 0, \tag{7.80}$$

where $\mathcal{A}_{\alpha i\beta j}$ is defined as in (7.18) with the incompressibility constraint applied after the differentiation of W with respect to the components of $\mathbf{F}$. The counterpart of (7.20) is

$$\mathcal{A}_{\alpha i\beta j}a_i N_\alpha N_\beta - qn_i = 0, \tag{7.81}$$

which may be rewritten as

$$\mathbf{Q}(\mathbf{n})\mathbf{a} - q\mathbf{n} = \mathbf{0}, \tag{7.82}$$

where $Q_{ij} = \mathcal{A}_{0piqj}n_p n_q$. This is formally the same as in the compressible case, but now the incompressibility constraint is in force, and this requires $\mathbf{a}\cdot\mathbf{n} = 0$.

By taking the dot product of (7.82) with $\mathbf{n}$ we obtain

$$q = [\mathbf{Q}(\mathbf{n})\mathbf{a}]\cdot\mathbf{n}, \tag{7.83}$$

and hence (7.82) may be rewritten as

$$[\mathbf{Q}(\mathbf{n}) - \mathbf{n}\otimes\mathbf{Q}(\mathbf{n})\mathbf{n}]\mathbf{a} = \mathbf{0}. \tag{7.84}$$

The strong ellipticity condition is in this case

$$[\mathbf{Q}(\mathbf{n})\mathbf{a}]\cdot\mathbf{a} \equiv \mathcal{A}_{0piqj}n_p n_q a_i a_j > 0 \tag{7.85}$$

as in (7.29) but now subject to $\mathbf{a}\cdot\mathbf{n} = 0$.

Note that, following the method of [49], the eigenvalue problem (7.84) can be recast as the two-dimensional problem

$$\hat{\mathbf{Q}}(\mathbf{n})\mathbf{a} = \mathbf{0}, \tag{7.86}$$

where $\hat{\mathbf{Q}}(\mathbf{n}) = \hat{\mathbf{I}}\mathbf{Q}(\mathbf{n})\hat{\mathbf{I}}$ is the projection of $\hat{\mathbf{Q}}(\mathbf{n})$ on to the space normal to $\mathbf{n}$ and $\hat{\mathbf{I}} = \mathbf{I} - \mathbf{n}\otimes\mathbf{n}$.

For the considered incompressible material the model (7.30) now takes the form

$$W = W^{\mathrm{iso}}(I_1, I_2) + W^{\mathrm{fib}}(I_4, I_5) \tag{7.87}$$

and may be specialized by replacing the latter term with $F(I_4)$ or $G(I_5^*)$ as in Section 7.3.1 or Section 7.3.2, respectively.

7.4.1. Plane strain

For plane strain the connection $(7.37)_2$ specializes to

$$I_5 = (I_1 - 1)I_4 - 1 \tag{7.88}$$

and the strain-energy function can be written in terms of the two invariants I_1 and I_4 according to

$$\hat{W}(I_1, I_4) = W(I_1, I_1, I_4, (I_1 - 1)I_4 - 1), \tag{7.89}$$

so that the planar nominal stress is

$$\mathbf{S} = 2\hat{W}_1\mathbf{F}^{\mathrm{T}} + 2\hat{W}_4\mathbf{M} \otimes \mathbf{FM} - \hat{p}\mathbf{F}^{-1}, \tag{7.90}$$

where $\hat{p}$ is the planar counterpart of the Lagrange multiplier p. The reference configuration restrictions (7.42) become

$$\hat{W}_1(3, 1) - \hat{p}_0 = 0, \quad \hat{W}_4(3, 1) = 0, \tag{7.91}$$

along with $\hat{W}(3, 1) = 0$, $\hat{p}_0$ being the reference configuration specialization of $\hat{p}$.

The (plane) components of $\boldsymbol{\mathcal{A}}_0$ are

$$\begin{aligned} \mathcal{A}_{0piqj} &= 4\hat{W}_{11}B_{pi}B_{qj} + 2\hat{W}_1 B_{pq}\delta_{ij} + 4\hat{W}_{14}(B_{pi}m_q m_j + B_{qj}m_p m_i) \\ &\quad + 4\hat{W}_{44}m_p m_q m_i m_j + 2\hat{W}_4 m_p m_q \delta_{ij}. \end{aligned} \tag{7.92}$$

Referred to the principal axes of $\mathbf{B}$ the corresponding strong ellipticity condition takes the form

$$\begin{aligned} &2\hat{W}_{11}(\lambda_1^2 - \lambda_2^2)^2 n_1^2 n_2^2 + \hat{W}_1(\lambda_1^2 n_1^2 + \lambda_2^2 n_2^2) \\ &\quad + 4\hat{W}_{14}(\lambda_1^2 - \lambda_2^2)n_1 n_2(n_1 m_1 + n_2 m_2)(n_2 m_1 - n_1 m_2) \\ &\quad + 2\hat{W}_{44}(n_1 m_1 + n_2 m_2)^2(n_2 m_1 - n_1 m_2)^2 + \hat{W}_4(n_1 m_1 + n_2 m_2)^2 > 0, \end{aligned} \tag{7.93}$$

and we note, in particular, that in the reference configuration the necessary and sufficient conditions

$$\hat{W}_1(3,1) > 0, \quad 2\hat{W}_1(3,1) + \hat{W}_{44}(3,1) > 0 \tag{7.94}$$

for (7.94) to hold may be deduced on use of $(7.91)_2$.

7.5. Strong Discontinuities

7.5.1. Piecewise homogeneous deformation gradients

In this section, we consider strong discontinuities, i.e., discontinuities in the deformation gradient, for an incompressible elastic material. Thus, as with weak discontinuities, solutions of the equilibrium equations lose their global smoothness across the discontinuity surface. Suppose then that the system of equations has a solution $\mathbf{x}(\mathbf{X})$ for which $\mathbf{F}$, $\mathbf{S}$ and p are continuous in some region $\mathcal{D}_\mathrm{r}$ in the reference configuration except on some smooth surface $\mathcal{S}_\mathrm{r}$ across which $\mathbf{x}(\mathbf{X})$ is continuous but $\mathbf{F}$, $\mathbf{S}$ and p are discontinuous. The surface $\mathcal{S}_\mathrm{r}$ separates the material into two connected regions with different deformation gradients and stresses, thus corresponding to two different phases of the same material.

For illustration it suffices to focus on homogeneous deformations so that $\mathbf{F}$ is uniform on either side of $\mathcal{S}_\mathrm{r}$, $\mathcal{S}_\mathrm{r}$ is a plane and the fibers are parallel in the reference configuration. Since we are considering plane strain deformations $\mathcal{S}_\mathrm{r}$ is orthogonal to the considered plane and its intersection with the plane corresponds to a straight line.

Let $\mathcal{D}$ denote the image of $\mathcal{D}_\mathrm{r}$ under the deformation and $\mathcal{S}$ that of $\mathcal{S}_\mathrm{r}$. Then we refer to the discontinuity surface $\mathcal{S}$ as a *kink surface* in $\mathcal{D}$ and its preimage $\mathcal{S}_\mathrm{r}$ as a kink surface in $\mathcal{D}_\mathrm{r}$. We also denote by Π_r^+ and Π_r^- the two regions (half-spaces) separated by $\mathcal{S}_\mathrm{r}$ in $\mathcal{D}_\mathrm{r}$ and by Π^+ and Π^- their images in $\mathcal{D}$, as depicted in Fig. 7.5, in which are also shown the unit vectors $\mathbf{N}$ and $\mathbf{n}$, which are normal to $\mathcal{S}_\mathrm{r}$ and $\mathcal{S}$, respectively, and $\mathbf{L}$ and $\mathbf{l}$, which are aligned with the discontinuity surfaces within the considered plane and make angles φ and α, respectively, with the horizontal axis. Also shown in Fig. 7.5 is a typical fiber parallel to the $\mathbf{E}_1$ direction in $\mathcal{D}_\mathrm{r}$ which becomes kinked in $\mathcal{D}$, with the so-called *kinking angle* ϕ, which indicates the direction change of that part of the fiber in Π^- compared with its initial direction. For a general analysis of stress induced two-phase piecewise homogeneous deformations and an analysis of their stability, we refer to [50].

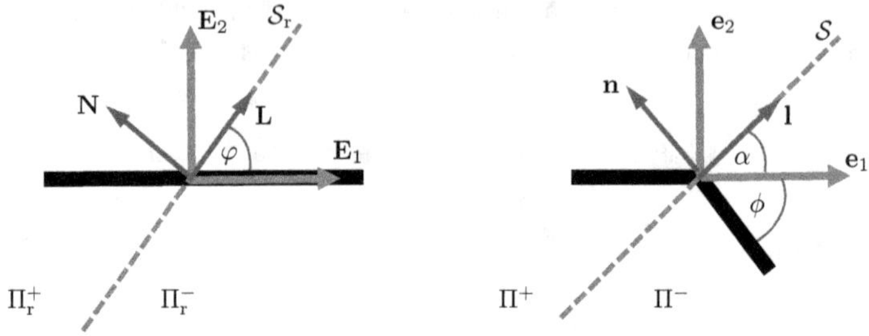

Fig. 7.5. Schematic of the intersection of the discontinuity plane $\mathcal{S}_\mathrm{r}$ in the reference region $\mathcal{D}_\mathrm{r}$ in the $\mathbf{E}_1, \mathbf{E}_2$ plane — the dashed line in left-hand figure in the direction $\mathbf{L}$ with normal $\mathbf{N}$, and their images in the deformed region $\mathcal{D}$ in the $\mathbf{e}_1, \mathbf{e}_2$ plane in the right-hand figure. The region $\mathcal{D}_\mathrm{r}$ is separated by $\mathcal{S}_\mathrm{r}$ into the two parts Π_r^+ and Π_r^-, and similarly $\mathcal{S}$ separates $\mathcal{D}$ into the two parts Π^+ and Π^-. See the text for more details.

Now suppose that the deformation from Π_r^+ to Π^+ is a pure homogeneous strain with deformation gradient $\mathbf{F}^+$ given by

$$\mathbf{F}^+ = \lambda\,\mathbf{e}_1 \otimes \mathbf{E}_1 + \lambda^{-1}\,\mathbf{e}_2 \otimes \mathbf{E}_2 + \mathbf{e}_3 \otimes \mathbf{E}_3, \tag{7.95}$$

where, by incompressibility and the plane strain assumption, the principal stretches may be expressed as $\lambda_1 = \lambda$, $\lambda_2 = \lambda^{-1}$, $\lambda_3 = 1$. The associated right and left Cauchy–Green deformation tensors, $\mathbf{C}^+ = (\mathbf{F}^+)^\mathrm{T}\mathbf{F}^+$ and $\mathbf{B}^+ = \mathbf{F}^+(\mathbf{F}^+)^\mathrm{T}$, respectively, are

$$\mathbf{C}^+ = \lambda^2\,\mathbf{E}_1 \otimes \mathbf{E}_1 + \lambda^{-2}\,\mathbf{E}_2 \otimes \mathbf{E}_2 + \mathbf{E}_3 \otimes \mathbf{E}_3, \tag{7.96}$$

$$\mathbf{B}^+ = \lambda^2\,\mathbf{e}_1 \otimes \mathbf{e}_1 + \lambda^{-2}\,\mathbf{e}_2 \otimes \mathbf{e}_2 + \mathbf{e}_3 \otimes \mathbf{e}_3. \tag{7.97}$$

The fiber-reinforcement maintains its initial direction under the deformation $\mathbf{F}^+$, but, since the deformation gradient suffers a jump across the discontinuity surface let the deformation gradient from Π_r^- to Π^- be denoted $\mathbf{F}^-$. As indicated above, this results in the fiber direction having an abrupt change across the kink surface $\mathcal{S}$, thus forming a kink in the fiber, which is characterized by the kinking angle ϕ shown in Fig. 7.5.

Let $\mathbf{L}$ and $\mathbf{N}$ be the tangent and the normal unit vectors to $\mathcal{S}_\mathrm{r}$ shown in Fig. 7.5. Then

$$\mathbf{L} = L_1\,\mathbf{E}_1 + L_2\,\mathbf{E}_2 = \cos\varphi\,\mathbf{E}_1 + \sin\varphi\,\mathbf{E}_2, \tag{7.98}$$

$$\mathbf{N} = N_1\,\mathbf{E}_1 + N_2\,\mathbf{E}_2 = -\sin\varphi\,\mathbf{E}_1 + \cos\varphi\,\mathbf{E}_2, \tag{7.99}$$

where L_α and N_α, $\alpha = 1, 2$, are the components of $\mathbf{L}$ and $\mathbf{N}$. Similarly, for $\mathcal{S}$, again with reference to Fig. 7.5,

$$\mathbf{l} = l_1\,\mathbf{e}_1 + l_2\,\mathbf{e}_2 = \cos\alpha\,\mathbf{e}_1 + \sin\alpha\,\mathbf{e}_2, \tag{7.100}$$

$$\mathbf{n} = n_1\,\mathbf{e}_1 + n_2\,\mathbf{e}_2 = -\sin\alpha\,\mathbf{e}_1 + \cos\alpha\,\mathbf{e}_2, \tag{7.101}$$

where l_i and n_i, $i = 1, 2$, are the components of $\mathbf{l}$ and $\mathbf{n}$.

Next, we note that, since the deformation is homogeneous, we have $\mathbf{x} = \mathbf{F}^+\mathbf{X}$ in Π_r^+ and $\mathbf{x} = \mathbf{F}^-\mathbf{X}$ in Π_r^-, and $\mathbf{x}$ is continuous. Hence

$$\mathbf{F}^+\mathbf{X} = \mathbf{F}^-\mathbf{X} \quad \text{on} \quad \mathcal{S}_\mathrm{r}. \tag{7.102}$$

It follows that $\mathbf{F}^+\mathbf{L} = \mathbf{F}^-\mathbf{L}$, and hence $\mathbf{F}^{-\mathrm{T}}\mathbf{N}$ is continuous across $\mathcal{S}_\mathrm{r}$ and $|\mathbf{FL}| = |\mathbf{F}^{-\mathrm{T}}\mathbf{N}|$. This continuity can also be deduced from Nanson's formula $\mathbf{n}\mathrm{d}a = \mathbf{F}^{-\mathrm{T}}\mathbf{N}\mathrm{d}A$ for an incompressible material connecting area elements $\mathbf{n}\mathrm{d}a$ and $\mathbf{N}\mathrm{d}A$ in the deformed and reference configurations, which applies not just in two dimensions. Thus $\mathbf{F}^\mathrm{T}\mathbf{n}$ is continuous, $\mathrm{d}a = |\mathbf{FL}|\mathrm{d}A$, and hence, in two dimensions, Nanson's formula can be recast as

$$\mathbf{N} = |\mathbf{FL}|\,\mathbf{F}^\mathrm{T}\mathbf{n}. \tag{7.103}$$

Expressed in terms of the components of $\mathbf{F}^+$ (or $\mathbf{F}^-$) this provides a connection between the angles α and φ, namely

$$\tan\alpha = \frac{F_{21} + F_{22}\tan\varphi}{F_{11} + F_{12}\tan\varphi}, \tag{7.104}$$

within which either $\mathbf{F} = \mathbf{F}^+$ or $\mathbf{F} = \mathbf{F}^-$ can be used. In particular, if evaluated for $\mathbf{F}^+$ from (7.95) it gives

$$\tan\alpha = \lambda^{-2}\tan\varphi. \tag{7.105}$$

In view of the continuity conditions discussed above we deduce that the deformation gradient $\mathbf{F}^-$ is given by

$$\mathbf{F}^- = \mathbf{F}^+ + k\mathbf{l}\otimes\mathbf{F}^{+\mathrm{T}}\mathbf{n}, \tag{7.106}$$

where the parameter k is a measure of the discontinuity in the deformation gradient and is referred to as the *discontinuity strength*.

It follows that the non-zero components of $\mathbf{F}^-$ are given by

$$F_{11}^- = \lambda(1 + kn_1n_2), \quad F_{12}^- = \lambda^{-1}kn_2^2, \tag{7.107}$$

$$F_{21}^- = -\lambda kn_1^2, \quad F_{22}^- = \lambda^{-1}(1 - kn_1n_2), \tag{7.108}$$

and $F_{33}^- = 1$. The corresponding components of the right and left Cauchy–Green deformation tensors $\mathbf{C}^-$ and $\mathbf{B}^-$ are

$$\begin{aligned} C_{11}^- &= (1 + 2kn_1n_2 + k^2n_1^2)\lambda^2, \\ C_{22}^- &= (1 - 2kn_1n_2 + k^2n_2^2)\lambda^{-2}, \\ C_{12}^- &= k(n_2^2 - n_1^2) + k^2n_1n_2, \end{aligned} \tag{7.109}$$

with $C_{33}^- = 1$, and

$$\begin{aligned} B_{11}^- &= (1 + kn_1n_2)^2\lambda^2 + k^2n_2^4\lambda^{-2}, \\ B_{22}^- &= (1 - kn_1n_2)^2\lambda^{-2} + k^2n_1^4\lambda^2, \\ B_{12}^- &= k(n_2^2\lambda^{-2} - n_1^2\lambda^2) - k^2(n_2^2\lambda^{-2} + n_1^2\lambda^2)n_1n_2, \end{aligned} \tag{7.110}$$

with $B_{33}^- = 1$.

On use of (7.108), an expression for the *kinking angle* ϕ shown in Fig. 7.5 can be obtained in terms of k and the angle α via

$$\tan\phi = -\frac{F_{21}^-}{F_{11}^-} = \frac{kn_1^2}{(1 + kn_1n_2)} = \frac{k\tan^2\alpha}{1 + \tan^2\alpha - k\tan\alpha}. \tag{7.111}$$

As well as $\mathbf{F}$, the nominal stress $\mathbf{S}$ and Lagrange multiplier p have discontinuities across $\mathcal{S}_\mathrm{r}$, and we denote by $\mathbf{S}^+$, p^+ and $\mathbf{S}^-$, p^- their values in Π_r^+ and Π_r^-, respectively. For these two regions the constitutive equation (7.75) then gives

$$\begin{cases} \mathbf{S}^+ = \dfrac{\partial W}{\partial \mathbf{F}}(\mathbf{F}^+) - p^+(\mathbf{F}^+)^{-1} \quad \text{in} \quad \Pi_\mathrm{r}^+, \\ \mathbf{S}^- = \dfrac{\partial W}{\partial \mathbf{F}}(\mathbf{F}^-) - p^-(\mathbf{F}^-)^{-1} \quad \text{in} \quad \Pi_\mathrm{r}^-. \end{cases} \tag{7.112}$$

The equilibrium equation $\mathrm{Div}\,\mathbf{S}^+ = \mathbf{0}$ is satisfied in $\mathcal{D}_\mathrm{r}^+$ and $\mathrm{Div}\,\mathbf{S}^- = \mathbf{0}$ in $\mathcal{D}_\mathrm{r}^-$ excluding $\mathcal{S}_\mathrm{r}$, while on $\mathcal{S}_\mathrm{r}$ the traction must be continuous, i.e.,

$$[\![\mathbf{S}^\mathrm{T}]\!]_-^+\mathbf{N} = \mathbf{0} \quad \text{on } \mathcal{S}_\mathrm{r}, \tag{7.113}$$

where $[\![\bullet]\!]_-^+ = \bullet^+ - \bullet^-$, indicating evaluation of the two sides of $\mathcal{S}_\mathrm{r}$. Note that since we are restricting attention to plane strain $S_{3\alpha} = 0$, $\alpha = 1, 2$, and hence the out-of-plane component of (7.113) is automatically satisfied. From (7.112), Eq. (7.113) becomes

$$\left[\!\left[\left(\frac{\partial W}{\partial \mathbf{F}}\right)^{\mathrm{T}} - p\mathbf{F}^{-\mathrm{T}} \right]\!\right]_{-}^{+} \mathbf{N} = \mathbf{0}. \tag{7.114}$$

7.5.2. Energy considerations

We are considering strong discontinuities in the context of elastostatics, otherwise referred to as *elastostatic discontinuities*, and we note that here we do not use the terminology 'elastostatic shocks' adopted in [51]. Specifically, we examine quasi-static motion of an elastic material body that includes an elastostatic discontinuity so that the equilibrium equations are maintained (no inertia terms). Such a motion will in general be dissipative and the existence of an elastostatic discontinuity therefore modifies the mechanical energy balance of the body.

We begin with the three-dimensional equations of compressible elasticity, but the formulas that follow also apply to incompressible materials although they involve minor modifications in their derivation (as shown in [51]). We use 'time' t as a parameter that tracks the evolution of the deformation. Let $\mathcal{D} \subset \mathcal{B}$, with boundary $\partial\mathcal{D}$, denote a sub-domain of the deformed configuration $\mathcal{B}$ and $\mathcal{D}_{\mathrm{r}}$, with boundary $\partial\mathcal{D}_{\mathrm{r}}$, its preimage in the reference configuration $\mathcal{B}_{\mathrm{r}}$. Then, for each such domain the inequality

$$\int_{\partial\mathcal{D}} \mathbf{t}\cdot\mathbf{v}\,\mathrm{d}a - \frac{\mathrm{d}}{\mathrm{d}t}\int_{\mathcal{D}_{\mathrm{r}}} W(\mathbf{F})\mathrm{d}V \geq 0 \tag{7.115}$$

must be satisfied, where $\mathbf{t} = \boldsymbol{\sigma}\mathbf{n}$ is the traction vector on $\partial\mathcal{D}$ expressed in terms of the Cauchy stress tensor and the unit outward normal $\mathbf{n}$ to $\partial\mathcal{D}$, $\mathbf{v}$ is the (quasi-static) particle velocity on $\partial\mathcal{D}$, W is the strain energy per unit volume in $\mathcal{D}_{\mathrm{r}}$ and $\mathrm{d}V$ is the volume element therein. The first integral in (7.115) may also be written in Lagrangian form (as an integral over $\partial\mathcal{D}_{\mathrm{r}}$) by noting the connection $\mathbf{t}\mathrm{d}a = \mathbf{T}\mathrm{d}A$, where $\mathbf{T} = \mathbf{S}^{\mathrm{T}}\mathbf{N}$ is the nominal traction vector and $\mathrm{d}A$ is the area element on $\partial\mathcal{D}_{\mathrm{r}}$, which is the preimage of the area element $\mathrm{d}a$. The first term in this inequality represents the work done by the external traction acting on $\partial\mathcal{D}$, while the second term represents the rate of storage of elastic energy.

In general the left-hand side of (7.115) is non-zero and equal to the work needed to move the elastostatic discontinuity $\mathcal{S}$ through the material in $\mathcal{D}$. As shown in [51] this work can be expressed as an integral over the discontinuity surface $\mathcal{S}_r$ (the preimage of $\mathcal{S}$), namely

$$\int_{\mathcal{S}_r} T\mathbf{V}\cdot\mathbf{N}\,\mathrm{d}A, \tag{7.116}$$

where the scalar T is the so-called *discontinuity driving traction* and $\mathbf{V}\cdot\mathbf{N}$ is the normal component of the velocity of $\mathcal{S}_r$. Note that $\mathbf{V}\cdot\mathbf{N}\,\mathrm{d}A = \mathbf{v}_{\mathcal{S}}\cdot\mathbf{n}\,\mathrm{d}a$, where, on use of Nanson's formula, the actual velocity $\mathbf{v}_{\mathcal{S}}$ of $\mathcal{S}$ in $\mathcal{D}$ is related to $\mathbf{V}$ by $\mathbf{V} = J\mathbf{F}^{-1}\mathbf{v}_{\mathcal{S}}$.

In Lagrangian form (7.115) can now be recast as

$$\int_{\partial\mathcal{D}_r} \mathbf{T}\cdot\mathbf{v}\,\mathrm{d}A - \frac{\mathrm{d}}{\mathrm{d}t}\int_{\mathcal{D}_r} W(\mathbf{F})\mathrm{d}V = \int_{\mathcal{S}_r} T\mathbf{V}\cdot\mathbf{N}\,\mathrm{d}A \geq 0. \tag{7.117}$$

As discussed in [51], the discontinuity driving traction T can be thought of as the magnitude of a 'fictitious' nominal traction vector acting on the elastostatic discontinuity by the surrounding material, and it was shown in [51] that T is expressible in the form

$$T = [\![W(\mathbf{F}) - (\mathbf{S}^{\mathrm{T}}\mathbf{N})\cdot(\mathbf{F}\mathbf{N})]\!]_-^+. \tag{7.118}$$

Since $\mathbf{S}^{\mathrm{T}}\mathbf{N} = \mathbf{T}$ is continuous this may also be written

$$T = [\![W(\mathbf{F})]\!]_-^+ - \mathbf{T}\cdot([\![\mathbf{F}]\!]_-^+\mathbf{N}). \tag{7.119}$$

We emphasize that $\mathbf{S}$ here is the nominal stress tensor, whereas $\mathbf{S}^{\mathrm{T}}$ is the first Piola–Kirchhoff stress tensor, which was employed in [51].

If, during the quasi-static motion, $T = 0$, the motion is said to be dissipation free, and an equilibrium state for which $T = 0$ is said to satisfy the *Maxwell condition*, which is also known as a condition for *neutral stability*, in which case neither Π_r^+ nor Π_r^- is energetically favored.

If the elastostatic discontinuity is advancing into Π_r^+ then $\mathbf{V}\cdot\mathbf{N} > 0$, while advancement into Π_r^- requires $\mathbf{V}\cdot\mathbf{N} < 0$. Then, based on the energy balance in (7.117), the existence of an energetically admissible elastostatic discontinuity requires

- $T \leq 0$ when the discontinuity moves so that Π_r^- is being converted into the favored Π_r^+ ($\mathbf{V}\cdot\mathbf{N} < 0$).
- $T \geq 0$ when the discontinuity moves so that Π_r^+ is being converted into the favored Π_r^- ($\mathbf{V}\cdot\mathbf{N} > 0$).

If we restrict attention to two dimensions and an incompressible material then, by the connections $\mathbf{T}\mathrm{d}A = \mathbf{t}\mathrm{d}a$, $|\mathbf{FL}|\mathrm{d}A = \mathrm{d}a$, (7.103) and (7.106) it follows that (7.119) specializes to

$$T = [\![W(\mathbf{F})]\!]_-^+ + k\,\mathbf{t}\cdot\mathbf{l}, \tag{7.120}$$

where $\mathbf{t}\cdot\mathbf{l}$ is the Cauchy shear traction on $\mathcal{S}$ and we recall that k is the strength of the elastostatic discontinuity.

7.5.3. Illustrations

For definiteness we now consider a simple (incompressible) material model, namely the standard reinforcing model for which the strain-energy function has the form

$$W = \frac{1}{2}\mu(I_1 - 3) + \frac{1}{2}\mu\gamma(I_4 - 1)^2, \tag{7.121}$$

where the constant $\mu > 0$ is the reference configuration shear modulus of the matrix material governed by the neo-Hookean model and the constant $\gamma > 0$ is a measure of the strength of the reinforcement in the reinforcing model $F(I_4)$ defined by the second term in (7.121). From (7.77) the Cauchy stress is then obtained as

$$\boldsymbol{\sigma} = \mu\mathbf{B} - p\mathbf{I} + 2\mu\gamma(I_4 - 1)\mathbf{m}\otimes\mathbf{m}. \tag{7.122}$$

For the plane strain specialization considered in the previous section it follows that in respect of (7.121) a kink can be characterized in terms of the parameters $\mu, \gamma, \lambda, \alpha, k, p^+, p^-$. The angle φ is then determined from (7.105), the deformation gradient $\mathbf{F}^-$ from (7.106), and the kinking angle ϕ from (7.111).

We now apply the traction continuity condition (7.113) expressed equivalently as $[\![\boldsymbol{\sigma}]\!]_-^+\mathbf{n} = \mathbf{0}$ on $\mathcal{S}$. As already noted, the out-of-plane component is satisfied identically, while the two in-plane components yield

$$\begin{aligned} &2\mu\gamma k\lambda^2[(3\lambda^2 - 1)n_2 + k\lambda^2(1 + 2n_2^2)n_1 + k^2\lambda^2 n_1^2 n_2]n_1^2 \\ &\quad + \mu k(\lambda^2 n_1^2 + \lambda^{-2}n_2^2)n_2 + (p^+ - p^-)n_1 = 0, \end{aligned} \tag{7.123}$$

$$\begin{aligned} &2\mu\gamma k\lambda^2[(\lambda^2 - 1)n_1 + 2k\lambda^2 n_1^2 n_2 + k^2\lambda^2 n_1^3]n_1^2 \\ &\quad + \mu k(\lambda^2 n_1^2 + \lambda^{-2}n_2^2)n_1 - (p^+ - p^-)n_2 = 0. \end{aligned} \tag{7.124}$$

Note that if $n_1 = 0$ then these equations are independent of γ and cannot both be satisfied, thus recovering the well-known result that the neo-Hookean material does not admit kinks.

By multiplying (7.123) by n_1 and (7.124) by n_2 and subtracting the two resulting equations we obtain

$$p^- - p^+ = 2\mu\gamma k\lambda^4 n_1^3(kn_1 + 2n_2), \tag{7.125}$$

and by elimination of $p^+ - p^-$ from the equations we obtain

$$P(\lambda, n_1, n_2, k, \gamma) \equiv 2\gamma\lambda^2[\lambda^2 - 1 + 2\lambda^2 n_2^2 + k\lambda^2(kn_1 + 3n_2)n_1]n_1^2 \\ +\lambda^2 n_1^2 + \lambda^{-2}n_2^2 = 0, \tag{7.126}$$

wherein the notation $P(\lambda, n_1, n_2, k, \gamma)$ is defined.

By evaluating T from (7.120) in respect of the strain-energy function (7.121) and then using (7.126) we obtain

$$T = \frac{1}{2}\mu\gamma k^3\lambda^4 n_1^3(kn_1 + 2n_2) = \frac{1}{2}\mu\gamma k^3\lambda^4 \sin^3\alpha(k\sin\alpha - 2\cos\alpha), \tag{7.127}$$

and hence, from (7.125),

$$T = \frac{1}{4}k^2(p^- - p^+). \tag{7.128}$$

Note that for a non-trivial solution for $T = 0$ the discontinuity strength k and the angle α in $\mathcal{S}$ are related by $\tan\alpha = 2/k$, and then $p^+ = p^-$. Moreover, (7.126) then gives

$$\tan^2\alpha = \lambda^4[2\gamma(1 - \lambda^2) - 1], \tag{7.129}$$

which yields a real α in the range $(0, \pi)$ if and only if $\gamma > 1/2$ and $\lambda^2 < 1 - 1/(2\gamma) < 1$ and also provides an expression for the dependence of k on λ for any given $\gamma > 1/2$.

We now assume that $\mathbf{F}^+$ and p^+ are prescribed. Bearing in mind that $\mathbf{F}^-$ is given by (7.106), we require to find p^-, α, k that satisfy (7.125) and (7.126), the latter providing a connection between k and α and the former an expression for p^-. Let $\mathbf{F}^+$ be given in terms of the stretch λ by (7.95). Then, for any given value of γ, a mechanically equilibrated kink can exist if and only if there are real values of k and α such that (7.126) is satisfied.

Note that since $P(\lambda, -n_1, n_2, -k, \gamma) = P(\lambda, n_1, n_2, k, \gamma)$ follows from (7.126), $(\pi - \alpha, -k)$ is a solution of $P(\lambda, n_1, n_2, k, \gamma) = 0$ whenever (α, k) is, which allows us to restrict attention to values of α in $(0, \pi)$.

We now exemplify the above theory by considering two simple problems.

7.5.3.1. *Example 1: An orthogonal kink*

First, we consider an orthogonal kink $\mathcal{S}$, so that $\alpha = \pi/2$, and hence $n_1 = -1$, $n_2 = 0$. From (7.126) we then obtain

$$k = \pm\sqrt{\frac{1}{\lambda^2}\left(1 - \frac{1}{2\gamma}\right) - 1}, \tag{7.130}$$

allowing k to be either positive or negative. For a real solution k it is necessary that $\gamma > 1/2$ and

$$0 < \lambda \leq \hat{\lambda}(\gamma) < 1, \tag{7.131}$$

where

$$\hat{\lambda}(\gamma) = \sqrt{1 - \frac{1}{2\gamma}}. \tag{7.132}$$

In this case the kink driving traction in (7.127) is given by $T = \mu\gamma k^4\lambda^4/2$.

The value of I_4^- is $(m_1^-)^2 + (m_2^-)^2 = (F_{11}^-)^2 + (F_{21}^-)^2 = \lambda^2(1 + k^2) = \hat{\lambda}(\gamma)^2$, and hence the fiber stretch in the region Π^- is $\hat{\lambda}(\gamma)$. Thus, from (7.131) the fiber is under contraction on both sides of an orthogonal kink. Equation (7.111) gives $\phi = \tan^{-1}(k)$ so that, with reference to Fig. 7.5, $k > 0$ implies $0 < \phi < \pi/2$, whereas for $k < 0$ the fiber direction is reflected in the $\mathbf{e}_1$-axis.

Suppose λ is decreased continuously from $\lambda = 1$. An orthogonal kink first becomes possible when λ reaches the value $\hat{\lambda}(\gamma)$, at which point $k = 0$. This represents a weak kink with $\phi = 0$. As λ then decreases from $\hat{\lambda}(\gamma)$ an orthogonal strong kink can develop, with k given by (7.130). Whether the value of k is positive or negative the fiber stretch remains constant in Π^- as λ decreases, and the kinking angle ϕ increases in magnitude with the magnitude of k via $\tan\phi = k$. The dependence of k on λ is illustrated in Fig. 7.6 for three values of γ. Note that $\lambda = \hat{\lambda}(\gamma)$ for $k = 0$, with $\hat{\lambda}(\gamma) \to 1$ as $\gamma \to \infty$.

7.5.3.2. *Example 2: A weak kink*

In this second example we consider a weak kink obtained by letting k tend to zero in the expression (7.126), which then yields, after replacing n_2^2 by $1 - n_1^2$,

$$4\gamma\lambda^4 n_1^4 - (6\gamma\lambda^4 - 2\gamma\lambda^2 + \lambda^2 - \lambda^{-2})n_1^2 - \lambda^{-2} = 0. \tag{7.133}$$

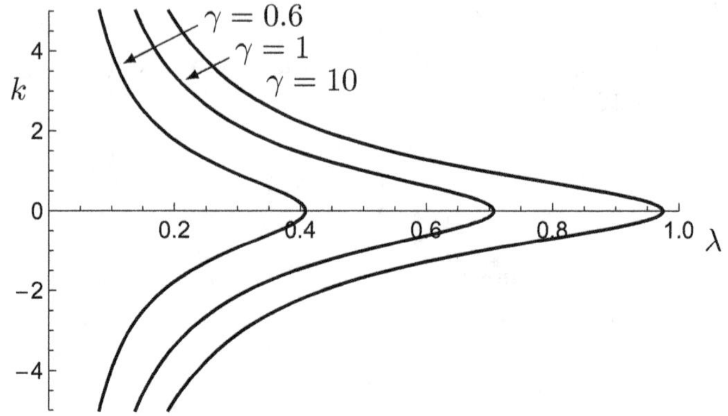

Fig. 7.6. Plots of k versus λ for $\gamma = 0.6, 1, 10$ based on Eq. (7.130). The curves cut the $k = 0$ axis at $\lambda = \hat{\lambda}(\gamma)$ given by Eq. (7.132).

This equation has a unique positive solution n_1^2 in the range $0 < n_1^2 \leq 1$ with positive λ provided $\gamma > 1/2$, in which case Eqs. (7.131) and (7.132) again apply. Indeed, it is easy to show that Eqs. (7.131) and (7.132) are required for loss of the strong ellipticity condition (7.94) applied to the model (7.121) since the left-hand side of (7.94) reduces to precisely minus half of the left-hand side of (7.133), i.e., continuity of traction is equivalent to loss of ellipticity in this case.

Thus, provided $\gamma > 1/2$, two possible weak kinks can emerge, corresponding to $n_2 = \pm\sqrt{1 - n_1^2}$ and symmetrically disposed with respect to the axes $\mathbf{e}_1$ and $\mathbf{e}_2$, with the kink orientation given in terms of the angle α: if $n_2 > 0\,(< 0)$, $0 < \alpha < \pi/2$ $(\pi/2 < \alpha < \pi)$. In the limiting case when $\alpha = \pi/2$ the two weak kinks coincide with $\lambda = \tilde{\lambda}(\gamma)$, but if $\lambda < \tilde{\lambda}(\gamma)$ then the weak kinks are necessarily separated with $0 < n_1^2 < 1$ and distinct from the orthogonal kink considered in Example 1.

Again, provided $\gamma > 1/2$, for each λ satisfying $0 < \lambda \leq \tilde{\lambda}(\gamma)$ both an orthogonal kink solution ($\alpha = \pi/2$) and weak kink solutions ($k \to 0$) of (7.126) can occur (note that $\gamma = 1/2$ requires $\lambda \to 0$). Under a compressive loading programme with λ decreasing from 1 the various solutions first emerge when $\lambda = \tilde{\lambda}(\gamma)$ after which, as λ decreases further, the associated orthogonal weak kink solution develops into four kink solutions: two symmetric orthogonal strong kinks with k given by (7.130) ($\alpha = \pi/2$) and two symmetric weak kinks with $k = 0$ and $\alpha = \pi/2 \pm \sin^{-1}\sqrt{1 - n_1^2}$, where n_1^2 is the unique solution of (7.133) satisfying $0 < n_1^2 < 1$.

7.6. Concluding Remarks

In this chapter, we have reviewed some of the recent research concerned with the theoretical prediction of certain types of failure in fiber-reinforced nonlinearly elastic materials on the basis of loss of ellipticity of the governing equations of equilibrium. Attention has been focused on materials that consist of an isotropic matrix and embedded parallel fibers modeled as a continuous distribution of a preferred direction. The associated mathematical model combines additively the strain-energy function of the isotropic matrix and that of a transversely isotropic strain-energy function, the latter being referred to as a reinforcing model.

We have shown how loss of ellipticity could be related to different failure mechanisms, including fiber kinking and splitting in compression and fiber debonding or matrix failure in tension, and we have considered only homogeneous deformations, so that boundary conditions are not an issue. However, the analysis of instabilities such as fiber buckling has not been included, and such instabilities may be initiated prior to the loss of ellipticity.

The second law of thermodynamics, exemplified by Eq. (7.117), must be satisfied for solutions for which the deformation field is continuous but the deformation gradient is discontinuous. It is consistent with either a zero discontinuity driving traction (so that the discontinuity does not propagate), with a positive driving traction (the discontinuity surface $\mathcal{S}$ moves into the region Π^+) or a negative driving traction (the discontinuity surface $\mathcal{S}$ moves into the region Π^-).

The list of possible failure mechanisms mentioned above is not exhaustive. Other examples might include loss of ellipticity associated with an incipient shear band. However, the development of a shear band (or a kink band) requires a material model that involves a length scale that is not available in pure elasticity theory, and consideration of such models is beyond the scope of the present work.

References

[1] Budiansky B., Fleck N.A., 1993. Compressive failure of fiber composites, *J. Mech. Phys. Solids*, 41, 183–211.

[2] Vogler T.J., Kyriakides S., 1997. Initiation and axial propagation of kink bands in fiber composites, *Acta Materialia*, 45, 2443–2454.

[3] Jensen H.M., Christoffersen J., 1997. Kink band formation in fiber reinforced materials, *J. Mech. Phys. Solids*, 45, 1121–1136.

[4] Budiansky B., Fleck N.A., Amazigo J.C., 1998. On kink-band propagation in fiber composites, *J. Mech. Phys. Solids*, 46, 1637–1653.

[5] Moran P.M., Shih C.F., 1998. Kink band formation and band broadening in ductile matrix fiber composites: experiments and analysis, *Int. J. Solids Struct.*, 35, 1709–1722.

[6] Merodio J., Pence T.J., 2001. Kink surfaces in a directionally reinforced neo-Hookean material under plane deformation: I. Mechanical equilibrium, *J. Elasticity*, 62, 119–144.

[7] Merodio J., Pence T.J., 2001. Kink surfaces in a directionally reinforced neo-Hookean material under plane deformation: II. Kink band stability and maximally dissipative band broadening, *J. Elasticity*, 62, 145–170.

[8] Wadee M.A., Völlmecke C., Haley J.F., Yiatros S., 2012. Geometric modelling of kink banding in laminated structures, *Philos. Trans. R. Soc. Lond. A*, 370, 1827–1849.

[9] Lee S.H., Waas A.M., 1999. Compressive response and failure of fiber reinforced unidirectional composites, *Int. J. Fracture*, 100, 275–306.

[10] Lee S.H., Yerramalli C.S., Waas A.M., 2000. Compressive splitting response of glass-fiber reinforced unidirectional composites, *Compos. Sci. Technol.*, 60, 2957–2966.

[11] Prabhakar P., Waas A.M., 2013. Interaction between kinking and splitting in the compressive failure of unidirectional fiber reinforced laminated composites, *Compos. Struct.*, 98, 85–92.

[12] Hutchinson J.W., Jensen H.M., 1990. Models of fiber debonding and pullout in brittle composites with friction, *Mech. Mater.*, 9, 139–163.

[13] Piggott M.R., 1997. Why interface testing by single-fiber methods can be misleading, *Compos. Sci. Technol.*, 57, 965–974.

[14] Okabe T., Komotori J., Shimizu M., Takeda N., 1999. Mechanical behavior of SiC fiber reinforced brittle-matrix composites, *J. Mater. Sci.*, 34, 3405–3412.

[15] Liao K., Reifsnider K.L., 2000. A tensile model for unidirectional fiber-reinforced brittle matrix composite, *Int. J. Fracture*, 106, 95–115.

[16] Fleck N.A., 1997. Compressive failure of fiber composites, *Adv. Appl. Mech.*, 33, 43–117.

[17] Hsu S.-Y., Vogler T.J., Kyriakides S., 1999). On the axial propagation of kink bands in fiber composites: Part II analysis, *Int. J. Solids Struct.*, 36, 575–595.

[18] Vogler T.J., Kyriakides S. (1999). On the axial propagation of kink bands in fiber composites: Part I experiments, *Int. J. Solids Struct.*, 36, 557–574.

[19] Pimenta S., Gutkin R., Pinho S.T., Robinson P., 2009. A micromechanical model for kink–band formation: Part I — experimental study and numerical modelling, *Composites Sci. Technol.*, 69, 948–955.

[20] Pimenta S., Gutkin R., Pinho S.T., Robinson P., 2009. A micromechanical model for kink–band formation: Part II — analytical modelling, *Composites Sci. Technol.*, 69, 956–964.

[21] Davidson P., Waas A.M., 2016. Mechanics of kinking in fiber-reinforced composites under compressive loading, *Math. Mech. Solids*, 21, 667–684.

[22] Zidek R.A.E., Völlmecke C., 2014. On the influence of material non-linearities in geometric modeling of kink band instabilities in unidirectional fiber composites, *Int. J. Non-Lin. Mech.*, 62, 23–32.

[23] Merodio J., Ogden R.W., 2002. Material instabilities in fiber-reinforced nonlinearly elastic solids under plane deformation, *Arch. Mech.*, 54, 525–552.

[24] Merodio J., Ogden,, R.W., 2003. Instabilities and loss of ellipticity in fiber-reinforced compressible non-linearly elastic solids under plane deformation, *Int. J. Solids Struct.*, 40, pp. 4707–4727.

[25] Merodio J., Ogden R.W., 2005. On tensile instabilities and ellipticity loss in fiber-reinforced incompressible nonlinearly elastic solids, *Mech. Res. Comm.*, 32, 290–299.

[26] Merodio J., Ogden R.W., 2005. Mechanical response of fiber-reinforced incompressible nonlinearly elastic solids, *Int. J. Non-Lin. Mech.*, 40, 213–227.

[27] Merodio J., Ogden R.W., 2005. Tensile instabilities and ellipticity in fiber-reinforced compressible nonlinearly elastic solids, *Int. J. Eng. Sci.*, 43, 697–706.

[28] Merodio J., Ogden R.W., 2005. Remarks on tensile instabilities and ellipticity for a compressible reinforced nonlinearly elastic solids, *Q. Appl. Math.*, 63, 325–333.

[29] Merodio J., Ogden R.W., 2006. The influence of the invariant I_8 on the stress-deformation and ellipticity characteristics of doubly fiber-reinforced nonlinearly elastic solids, *Int. J. Non-Lin. Mech.*, 41, 556–563.

[30] Kassianidis K., Merodio J., Ogden R.W., Pence T.J., 2008. Azimuthal shear of a transversely isotropic elastic solid, *Math. Mech. Solids*, 13, 690–724.

[31] Dorfmann A., Merodio J., Ogden R.W., 2010. Non-smooth solutions in the azimuthal shear of an anisotropic nonlinearly elastic material, *J. Eng. Math.*, 68, 27–36.

[32] El Hamdaoui M., Merodio J., Ogden R.W., 2015. Loss of ellipticity in the combined helical, axial and radial elastic deformations of a fiber-reinforced circular cylindrical tube, *Int. J. Solids Struct.*, 63, 99–108.

[33] Hill R., 1961. Discontinuity relations in mechanics of solids, in eds. Sneddon, I.N. and Hill, R., *Progress in Solid Mechanics*, Vol. 2, pp. 245–276. North-Holland, Amsterdam.

[34] Knowles J.K., Sternberg E., 1978. On the failure of ellipticity and the emergence of discontinuous deformation gradients in plane finite elastostatics, *J. Elasticity*, 8, 329–379.

[35] Abeyaratne R.C., 1980. Discontinuous deformation gradients in plane finite elastostatics of incompressible materials, *J. Elasticity*, 10, 255–293.

[36] Triantafyllidis N., Abeyaratne R.C., 1983. Instability of a finitely deformed fiber-reinforced elastic material, *J. Appl. Mech.*, 50, 149–156.

[37] Zee L., Sternberg E., 1983. Ordinary and strong ellipticity in the equilibrium theory of incompressible hyperelastic solids, *Arch. Ration. Mech. Anal.*, 83, 50–90.

[38] Rosakis P., 1990. Ellipticity and deformations with discontinuous gradients in plane elastostatics of compressible solids, *Arch. Ration. Mech. Anal.*, 109, 1–37.

[39] Horgan C.O., 1996. Remarks on ellipticity for the generalized Blatz-Ko constitutive model for a compressible nonlinearly elastic solid, *J. Elasticity*, 42, 165–176.

[40] Qiu G.Y., Pence T.J., 1997. Loss of ellipticity in plane deformation of a simple directionally reinforced incompressible nonlinearly elastic solid, *J. Elasticity*, 49, 31–63.

[41] Ogden R.W., 1984. *Non-linear Elastic Deformations*, Ellis Horwood, Chichester.

[42] Holzapfel G.A., 2000. *Nonlinear Solid Mechanics. A Continuum Approach for Engineering*, John Wiley & Sons, Chichester.

[43] Spencer A.J.M., 1972. *Deformations of Fiber-reinforced Materials*, Oxford University Press.

[44] Love A.E.H., 1944. *A Treatise on the Mathematical Theory of Elasticity*, Dover Publications, New York.

[45] Shams M., Destrade M., Ogden R.W., 2011. Initial stresses in elastic solids: constitutive laws and acoustoelasticity, *Wave Motion*, 48, 552–567.

[46] Merodio J., Ogden R.W., 2003. A note on strong ellipticity for transversely isotropic linearly elastic solids, *Q. J. Mech. Appl. Math.*, 56, 589–591.

[47] Knowles J.K., Sternberg E., 1977. On the failure of ellipticity of the equations for elastostatic plane strain, *Arch. Ration. Mech. Anal.*, 63, 321–336.

[48] Dowaikh M.A., Ogden R.W., 1991. On surface waves and deformations in a compressible elastic half-space, *Stab. Appl. Anal. Cont. Media*, 1, 27–45.

[49] Scott N.H., Hayes M., 1985. A note on wave propagation in internally constrained hyperelastic materials, *Wave Motion*, 7, 601–605.

[50] Fu Y.B., Freidin A.B., 2004. Characterization and stability of twophase piecewise-homogeneous deformations, *Proc. R. Soc. Lond. A*, 460, 3065–3094.

[51] Knowles J.K., 1979. On the dissipation associated with equilibrium shocks in finite elasticity, *J. Elasticity*, 9, 131–158.

https://doi.org/10.1142/9781800611887_0008

Chapter 8

Propagation of Rayleigh Waves in Anisotropic Media and an Inverse Problem in the Characterization of Initial Stress

Kazumi Tanuma[*,‡] **and Chi-Sing Man**[†,§]
[*]*Department of Mathematics, Faculty of Science and Technology, Gunma University, Kiryu 376-8515, Japan*
[†]*Department of Mathematics, University of Kentucky, Lexington, KY 40506-0027, USA*
[‡]*tanuma@gunma-u.ac.jp*
[§]*cman@uky.edu*

Abstract

Composite materials behave in the long wavelength limit as if they are homogeneous but, more often than not, carry strong anisotropy. We review the Stroh formalism for dynamic elasticity and apply this formalism to derive perturbation formulas for the phase velocity and polarization ratio of Rayleigh waves that propagate along the free surface of a prestressed half-space where, when unstressed, the principal part of the elasticity tensor is orthotropic or transversely isotropic. An objective of the present study is to examine the possibility of and the problems and issues regarding the determination of the prestress in an otherwise transversely isotropic composite material by boundary measurements of angular dependence of Rayleigh-wave phase velocity or polarization ratio.

Contents

8.1. Introduction

In the long wavelength limit composite materials behave as if they are homogeneous but, more often than not, carry strong anisotropy (see, e.g., Chapter17 of [1],[2], Chapter11 of [3], and [4]).[a] The Stroh formalism is a powerful and elegant mathematical method developed for the analysis of problems of two-dimensional deformations in anisotropic elasticity[6–12]. It reveals simple structures hidden in the equations of motion (or equilibrium) for such problems and provides a systematic approach for tackling these equations. In particular the Stroh formalism is an effective tool in the study of Rayleigh waves, long since a topic of utmost importance in non-destructive evaluation, seismology, and materials science.

In this chapter, we will use the Stroh formalism to derive a perturbation formula for the phase velocities of Rayleigh waves that propagate along various directions on the free surface of a prestressed half-space where, when unstressed, the principal part of the elasticity tensor is orthotropic or transversely isotropic. The effects of the initial stress are grouped under the perturbative part of the constitutive equation. In our previous study [13], we derived a perturbation formula for the phase velocity of Rayleigh waves that

[a]Here we cite one concrete example. Consider wave propagation in a composite that consists of an elastic homogeneous matrix reinforced by a random distribution of aligned continuous cylindrical elastic fibers. It is well established that such a composite can be modeled as a transversely isotropic homogeneous elastic medium when the wavelength of the propagating wave is much longer than the fiber diameter. See, for example, the theoretical and experimental study of Datta and Ledbetter[5], where they calculated and measured the macroscopic elastic constants of a composite that consisted of an aluminum 6061 alloy matrix reinforced by 0.14-mm-diameter boron fibers. They used two methods in their measurements of the elastic constants. In the resonance method they used oscillations of frequencies that ranged from 30 to 50 kHz. In the ultrasonic-velocity method they used waves at frequencies near 10 MHz.

propagate along the free surface of a prestressed half-space; that formula expresses the shift of the phase velocity of the Rayleigh waves from its value for the comparative unstressed and isotropic medium. Unlike that work, here we allow the base material to be orthotropic or transversely isotropic (and the anisotropy may be strong), which is typical of directional fiber-reinforced composite materials (see e.g., [1, 2, 4, 14], Chapter 15 of [15], and [16]). An objective of the present study is to examine the possibility of and the problems and issues regarding the determination of the prestress in an otherwise transversely isotropic composite material by boundary measurements of angular dependence of Rayleigh-wave phase velocity or polarization ratio.

The present chapter reviews the essence of the Stroh formalism for dynamic elasticity, summarizes an approach based on this formalism to derive a first-order perturbation formula for the phase velocity of Rayleigh waves, and outlines what information on the prestress of a body of a composite material can, in principle, be delivered by boundary measurements of Rayleigh-wave phase velocities. One objective of this chapter is to introduce the reader to the full paper [17].

The polarization ratio r_R of Rayleigh waves is defined as the ratio of the maximum longitudinal component to the maximum normal component of the displacements at the surface on which the Rayleigh waves propagate. Junge *et al.* [18] studied the relative polarization of Rayleigh waves for several materials (mild steel, four aluminum alloys, polystyrene, and brass) under uniaxial stress; here by relative polarization is meant the ratio of r_R and its value for the comparative isotropic medium where the (applied) prestress is zero. They observed that "the relative polarization is *more sensitive* to applied stress than the relative wave speeds ..., while being less sensitive to uncertainties in the values of the TOE-constants [i.e., third-order elastic constants]". These properties of the polarization ratio represent possible advantages of using it in non-destructive evaluation of stress. The observation of Junge *et al.* was largely corroborated by our earlier study [19] where we derived a first-order perturbation formula for the polarization ratio of Rayleigh waves when the base material is isotropic. Here taking advantage of the opportunity to revise this chapter for the second edition, we present in a new final section a first-order perturbation formula for the polarization ratio of Rayleigh waves that propagate along the free surface of a prestressed half-space where the base material is orthotropic or transversely isotropic. The possibility of using this formula for the determination of the prestress in an otherwise transversely isotropic composite material by boundary measurements of the polarization ratios of Rayleigh waves will be briefly discussed.

8.2. Basic Elasticity in Anisotropic Materials with Initial Stress

Here we consider a macroscopically homogeneous, anisotropic, prestressed elastic medium and small elastic motions superimposed on it. Let $\mathbb{R}$ denote the set of real numbers. Suppose that the medium occupies a region B in $\mathbb{R}^3$ with a smooth boundary. We take the given initial configuration of B as the reference configuration for the description of the elastic motions.

Let $\boldsymbol{E} = \boldsymbol{E}(\boldsymbol{u}) = (\varepsilon_{ij})_{i,j=1,2,3}$ be the infinitesimal strain tensor

$$\varepsilon_{ij} = \frac{1}{2}\left(\frac{\partial u_i}{\partial x_j} + \frac{\partial u_j}{\partial x_i}\right), \tag{8.1}$$

where $\boldsymbol{u} = \boldsymbol{u}(\mathbf{x}) = (u_1, u_2, u_3)$ is the displacement at the place $\mathbf{x}$ pertaining to the superimposed small elastic motion and (x_1, x_2, x_3) are the Cartesian coordinates of $\mathbf{x}$.

In the theoretical context of linear elasticity with initial stress (see [20] and the references therein), the constitutive equation can be written as

$$\boldsymbol{S} = \overset{\circ}{\boldsymbol{T}} + \boldsymbol{H}\overset{\circ}{\boldsymbol{T}} + \mathbb{L}[\boldsymbol{E}], \tag{8.2}$$

where $\boldsymbol{S} = (S_{ij})_{i,j=1,2,3}$ is the first Piola–Kirchhoff stress, $\overset{\circ}{\boldsymbol{T}} = (\overset{\circ}{T}_{ij})_{i,j=1,2,3}$ the initial stress, $\boldsymbol{H} = (\partial u_i/\partial x_j)_{i,j=1,2,3}$ the displacement gradient, and $\mathbb{L}$ the incremental elasticity tensor. The initial stress $\overset{\circ}{\boldsymbol{T}}$ is symmetric

$$\overset{\circ}{T}_{ij} = \overset{\circ}{T}_{ji}, \quad i,j = 1,2,3 \tag{8.3}$$

and the incremental elasticity tensor $\mathbb{L}$, regarded as a fourth-order tensor that maps symmetric tensors $\boldsymbol{E}$ onto symmetric tensors, has major and minor symmetries

$$L_{ijkl} = L_{klij} = L_{jikl}, \quad i,j,k,l = 1,2,3. \tag{8.4}$$

Then we can rewrite (8.2) componentwise as

$$S_{ij} = \overset{\circ}{T}_{ij} + \sum_{k,l=1}^{3} (\overset{\circ}{T}_{jl}\delta_{ik} + L_{ijkl})\frac{\partial u_k}{\partial x_l}, \tag{8.5}$$

where δ_{ik} is Kronecker's delta. Both $\mathbb{L}$ and $\overset{\circ}{\boldsymbol{T}}$ depend implicitly on the formative history of the material body in the given initial configuration, but we will keep this dependence implicit and simply take $\mathbb{L}$ and $\overset{\circ}{\boldsymbol{T}}$ as given fourth-order and second-order tensors, respectively.

Note that when $\overset{\circ}{\boldsymbol{T}} = \mathbf{0}$ the constitutive equation (8.2) is reduced to the generalized Hooke's law

$$\boldsymbol{S} = \mathbb{C}[\boldsymbol{E}] \quad \text{or} \quad S_{ij} = \sum_{k,l=1}^{3} C_{ijkl}\varepsilon_{kl},$$

where $\mathbb{L}$ now collapses to the classical elasticity tensor $\mathbb{C}$. Furthermore, we say that the elastic material is isotropic if the elasticity tensor $\mathbb{C} = \left(C_{ijkl}\right)_{1\le i,j,k,\ell\le 3}$ satisfies

$$C_{ijkl} = \sum_{p,q,r,s=1}^{3} Q_{ip}Q_{jq}Q_{kr}Q_{ls}C_{pqrs}, \quad i,j,k,l = 1,2,3 \tag{8.6}$$

for any orthogonal tensor $\boldsymbol{Q} = (Q_{ij})_{i,j=1,2,3}$. For isotropic materials, the components $C_{ijkl}(i,j,k,l = 1,2,3)$ can be written as

$$C_{ijkl} = \lambda\,\delta_{ij}\delta_{kl} + \mu(\delta_{ik}\delta_{jl} + \delta_{il}\delta_{kj})$$

with the Lamé constants λ and μ. We say that the elastic material is anisotropic if it is not isotropic.

We will study dynamic deformations of the elastic medium in B. Let t denote the time and ρ the uniform mass density of the material in the given initial (prestressed) configuration. We assume that the initial stress $\overset{\circ}{\boldsymbol{T}}$ satisfies the equation of equilibrium with the body force (per unit mass) $\boldsymbol{b} = \boldsymbol{b}(\mathbf{x}) = (b_1, b_2, b_3)$:

$$\sum_{j=1}^{3} \frac{\partial}{\partial x_j}\overset{\circ}{T}_{ij} + \rho b_i = 0, \quad i = 1,2,3.$$

Substitution of (8.5) into the equation of motion with the body force $\boldsymbol{b}$, namely

$$\sum_{j=1}^{3} \frac{\partial}{\partial x_j} S_{ij} + \rho b_i = \rho\frac{\partial^2}{\partial t^2}u_i, \quad i = 1,2,3,$$

leads to the elastic wave equation written in terms of the displacement

$$\sum_{j=1}^{3} \frac{\partial}{\partial x_j}\left(B_{ijkl}\frac{\partial u_k}{\partial x_l}\right) = \rho\frac{\partial^2}{\partial t^2}u_i, \quad i = 1,2,3, \tag{8.7}$$

where

$$B_{ijkl} = \delta_{ik}\overset{\circ}{T}_{jl} + L_{ijkl} \tag{8.8}$$

are the effective elastic coefficients which have the major symmetry

$$B_{ijkl} = B_{klij}, \quad i, j, k, l = 1, 2, 3 \tag{8.9}$$

but do not have the minor symmetries because of the term $\delta_{ik}\overset{\circ}{T}_{jl}$.

Hereafter in this chapter we assume the strong ellipticity condition:

$$\text{The matrix } \Big(\sum_{j,l=1}^{3} B_{ijkl}\xi_j\xi_l\Big)_{i\downarrow k\to 1,2,3} \text{ is positive definite for}$$
$$\text{any non-zero vector } \boldsymbol{\xi} = (\xi_1, \xi_2, \xi_3) \in \mathbb{R}^3.\text{[b]} \tag{8.10}$$

This condition is guaranteed if the incremental elasticity tensor $\mathbb{L}$ satisfies the strong convexity condition:

$$\sum_{i,j,k,l=1}^{3} L_{ijkl}\varepsilon_{ij}\varepsilon_{kl} > 0 \text{ for any non-zero } 3\times 3 \text{ real symmetric matrix } (\varepsilon_{ij})$$

and if the initial stress $\overset{\circ}{\boldsymbol{T}}$ is sufficiently small.

We have treated $\overset{\circ}{\boldsymbol{T}}$ and $\mathbb{L}$ as if they are independent. In fact they are generally related through their mutual dependence on the formative history of the material body in question. In particular $\mathbb{L}$ generally depends on $\overset{\circ}{\boldsymbol{T}}$. There are models for the dependency of $\mathbb{L}$ on $\overset{\circ}{\boldsymbol{T}}$. The simplest case of which is found when $\mathbb{L}$ is an isotropic bilinear function of $\overset{\circ}{\boldsymbol{T}}$ and $\boldsymbol{E}$ relative to the initial configuration. In this case $\mathbb{L}$ is given by

$$\begin{aligned}\mathbb{L}(\overset{\circ}{\boldsymbol{T}})[\boldsymbol{E}] &= \lambda(\operatorname{tr}\boldsymbol{E})\boldsymbol{I} + 2\mu\boldsymbol{E} + \beta_1(\operatorname{tr}\boldsymbol{E})(\operatorname{tr}\overset{\circ}{\boldsymbol{T}})\boldsymbol{I} + \beta_2(\operatorname{tr}\overset{\circ}{\boldsymbol{T}})\boldsymbol{E}\\ &\quad + \beta_3\Big((\operatorname{tr}\boldsymbol{E})\overset{\circ}{\boldsymbol{T}} + (\operatorname{tr}\boldsymbol{E}\overset{\circ}{\boldsymbol{T}})\boldsymbol{I}\Big) + \beta_4(\boldsymbol{E}\overset{\circ}{\boldsymbol{T}} + \overset{\circ}{\boldsymbol{T}}\boldsymbol{E}),\end{aligned} \tag{8.11}$$

where λ and μ are the Lamé constants, β_i $(i = 1, \ldots, 4)$ are material constants and $\mathbf{I}$ denotes the 3×3 identity matrix (see [21]). For instance classical acoustoelastic theory, where the initial stress is caused by a deformation of an isotropic elastic material from a stress-free natural configuration, in effect uses (see [19, Sec. 9]) the constitutive equation (8.2)

[b] The subscript $i \downarrow k \to 1, 2, 3$ means that i and k are the row and column number, respectively, and both numbers run from 1 to 3.

with $\mathbb{L}$ given by (8.11) and the four β_i coefficients expressed in terms of three third-order elastic constants and the Lamé constants of the isotropic elastic material.

We take as the sole reference configuration the initial configuration, which already has the initial stress and material anisotropy, and we make no presumption of their origin. Thus our study begins with the constitutive relation (8.2) and our theory is applicable so long as infinitesimal motions superimposed on this initial configuration are elastic. We assume the symmetries (8.3) for $\overset{\circ}{\boldsymbol{T}}$ and (8.4) for $\mathbb{L}$ but do not assume any other material symmetry.

In Section 8.3, we give a brief review of the Stroh formalism for elastic wave equations. This formalism, which reveals simple structures hidden in the equations of anisotropic elasticity, is powerful and elegant. One of its applications is a systematic study of Rayleigh waves in anisotropic materials presented in Sections 8.4, 8.5 and 8.7.

8.3. The Stroh Formalism for Dynamic Elasticity

In this section, we review the Stroh formalism for elastic wave equations in a manner as self-contained, readable, and brief as possible. Rayleigh waves propagate along free surfaces, i.e., these waves produce no traction at the boundary. However, the formalism in this section applies to surface waves that do not necessarily satisfy a zero-traction condition at the boundary. In the subsequent sections, we will restrict our attention to Rayleigh waves.

All the theorems, lemmas, etc. in Sections 8.3 and 8.4 can be found in [8–12]. In the following, except for several important results, we shall refrain from referring to the literature at each point.

We assume that the elastic medium and the initial stress are homogeneous, and hence the effective elastic coefficients $\left(B_{ijkl}\right)_{i,j,k,l=1,2,3}$ are independent of the position vector $\mathbf{x}$. In this case (8.7) becomes

$$\sum_{j,k,l=1}^{3} B_{ijkl}\frac{\partial^2 u_k}{\partial x_j \partial x_l} = \rho\frac{\partial^2}{\partial t^2}u_i, \quad i = 1, 2, 3. \tag{8.12}$$

The only assumptions we make for $(B_{ijkl})_{i,j,k,l=1,2,3}$ are their major symmetry (8.9) and the strong ellipticity condition (8.10).

Let $\mathbf{x} = (x_1, x_2, x_3)$ be the position vector and let $\mathbf{m} = (m_1, m_2, m_3)$ and $\mathbf{n} = (n_1, n_2, n_3)$ be orthogonal unit vectors in $\mathbb{R}^3$. Let $\mathbb{C}$ denote the set of complex numbers.[c] We consider the motion of a homogeneous elastic medium that occupies the half-space $\mathbf{n}\cdot\mathbf{x} = n_1x_1 + n_2x_2 + n_3x_3 \le 0$, and seek solutions to (8.12) of the form

$$\boldsymbol{u} = (u_1, u_2, u_3) = \mathbf{a}\, e^{-\sqrt{-1}\,k(\mathbf{m}\cdot\mathbf{x}+p\mathbf{n}\cdot\mathbf{x}-vt)} \in \mathbb{C}^3 \tag{8.13}$$

in $\mathbf{n}\cdot\mathbf{x} \le 0$ (cf., e.g., [10, Sec. 2]).

When $\mathrm{Im}\, p$, the imaginary part of $p \in \mathbb{C}$, is positive, a solution of the preceding form describes a surface wave in $\mathbf{n}\cdot\mathbf{x} \le 0$; it propagates along the surface $\mathbf{n}\cdot\mathbf{x} = 0$ in the direction of $\mathbf{m}$ with the wave number k and the phase velocity $v > 0$, has the polarization defined by a constant vector $\mathbf{a}$, and decays exponentially as $\mathbf{n}\cdot\mathbf{x} \longrightarrow -\infty$. When $\mathrm{Im}\, p < 0$, the solution blows up as $\mathbf{n}\cdot\mathbf{x} \longrightarrow -\infty$. We exclude such a solution.

A surface wave described by the form (8.13) depends on the projection of $\mathbf{x}$ on the plane spanned by the two orthogonal unit vectors $\mathbf{m}$ and $\mathbf{n}$.

Let us determine the equations that $\mathbf{a} \in \mathbb{C}^3$ and $p \in \mathbb{C}$ in (8.13) must satisfy. Substituting (8.13) into (8.12) and noting that

$$\frac{\partial \boldsymbol{u}}{\partial x_j} = -\sqrt{-1}\,k(m_j + pn_j)\mathbf{a}\, e^{-\sqrt{-1}\,k(\mathbf{m}\cdot\mathbf{x}+p\mathbf{n}\cdot\mathbf{x}-vt)}$$

and

$$\frac{\partial \boldsymbol{u}}{\partial t} = \sqrt{-1}\,kv\mathbf{a}\, e^{-\sqrt{-1}\,k(\mathbf{m}\cdot\mathbf{x}+p\,\mathbf{n}\cdot\mathbf{x}-vt)},$$

we get

$$\begin{aligned}
&\left(\sum_{j,l=1}^{3} B_{ijkl}(m_j + pn_j)(m_l + pn_l) - \rho v^2 \delta_{ik}\right)_{i\downarrow k\rightarrow 1,2,3} \mathbf{a} \\
&= \left(\sum_{j,l=1}^{3} B_{ijkl}\, m_j m_l - \rho v^2 \delta_{ik} + p\left(\sum_{j,l=1}^{3} B_{ijkl}\, m_j n_l + \sum_{j,l=1}^{3} B_{ijkl}\, n_j m_l\right)\right. \\
&\quad \left. + p^2 \sum_{j,l=1}^{3} B_{ijkl}\, n_j n_l\right)_{i\downarrow k\rightarrow 1,2,3} \mathbf{a} = \mathbf{0},
\end{aligned} \tag{8.14}$$

[c] We also use the symbol $\mathbb{C}$ to denote the classical elasticity tensor. It should be clear from the context what the symbol means when it appears.

where δ_{ik} is the Kronecker delta. We introduce the 3×3 real matrices

$$\mathbf{Q} = \left(\sum_{j,l=1}^{3} B_{ijkl}\, m_j m_l - \rho v^2 \delta_{ik} \right)_{i\downarrow k \rightarrow 1,2,3},$$

$$\mathbf{R} = \left(\sum_{j,l=1}^{3} B_{ijkl}\, m_j n_l \right)_{i\downarrow k \rightarrow 1,2,3}, \quad \mathbf{T} = \left(\sum_{j,l=1}^{3} B_{ijkl}\, n_j n_l \right)_{i\downarrow k \rightarrow 1,2,3}. \tag{8.15}$$

Taking the major symmetry (8.9) into account, we rewrite equation (8.14) as

$$[\mathbf{Q} + p(\mathbf{R} + \mathbf{R}^T) + p^2 \mathbf{T}]\mathbf{a} = \mathbf{0}, \tag{8.16}$$

where the superscript T denotes transposition. For the existence of a non-trivial vector $\mathbf{a} \neq \mathbf{0}$, we observe that p satisfies the sextic equation

$$\det[\mathbf{Q} + p(\mathbf{R} + \mathbf{R}^T) + p^2 \mathbf{T}] = 0. \tag{8.17}$$

Lemma 8.1.

(1) *Matrices* $\mathbf{Q}$ *and* $\mathbf{T}$ *are symmetric. Moreover, matrix* $\mathbf{T}$ *is positive definite.*
(2) *When* $v = 0$, *the characteristic roots* p_α $(1 \leq \alpha \leq 6)$, *i.e., the solutions to the sextic equation* (8.17), *are not real and they occur in complex conjugate pairs.*

Proof. The symmetries of $\mathbf{Q}$ and $\mathbf{T}$ follow immediately from the major symmetry (8.9) and the positive definiteness of $\mathbf{T}$ from the strong ellipticity condition (8.10). Assertion (2) also follows easily from (8.10). For details, refer, for example, to (2) of Lemma 1.1 of [12]. □

Next we examine the traction on the surface $\mathbf{n} \cdot \mathbf{x} = 0$ produced by the surface-wave solution (8.13). Since the outward unit normal of this surface is the vector $\mathbf{n}$, the traction $\boldsymbol{t}$ on $\mathbf{n} \cdot \mathbf{x} = 0$ is given by

$$\boldsymbol{t} = \left(\sum_{j=1}^{3} S_{ij} n_j \right)_{i\downarrow 1,2,3} \Bigg|_{\mathbf{n}\cdot\mathbf{x}=0}$$

which, by (8.5), (8.8), (8.13) and (8.15), becomes[d]

$$\begin{aligned} \boldsymbol{t} &= \Big(\sum_{j,k,l=1}^{3} B_{ijkl}\, \frac{\partial u_k}{\partial x_l}\, n_j \Big)_{i\downarrow 1,2,3}\bigg|_{\mathbf{n}\cdot\mathbf{x}=0} \\ &= -\sqrt{-1}\, k \Big(\sum_{j,l=1}^{3} B_{ijkl}\, (m_l + p n_l)\, n_j \Big)_{i\downarrow k\to 1,2,3} \mathbf{a}\, e^{-\sqrt{-1}\, k(\mathbf{m}\cdot\mathbf{x} - vt)} \\ &= -\sqrt{-1}\, k\, [\mathbf{R}^T + p\mathbf{T}]\mathbf{a}\, e^{-\sqrt{-1}\, k(\mathbf{m}\cdot\mathbf{x} - vt)}. \end{aligned}$$

Hence we define a vector $\mathbf{l} \in \mathbb{C}^3$ as

$$\mathbf{l} = [\mathbf{R}^T + p\mathbf{T}]\,\mathbf{a}. \tag{8.18}$$

Then

$$\boldsymbol{t} = -\sqrt{-1}\, k\, \mathbf{l}\, e^{-\sqrt{-1}\, k(\mathbf{m}\cdot\mathbf{x} - vt)} \tag{8.19}$$

is the traction on the surface $\mathbf{n}\cdot\mathbf{x} = 0$ produced by (8.13).

By (1) of Lemma 8.1, $\mathbf{T}^{-1}$ exists. Hence from (8.18) we get

$$p\,\mathbf{a} = -\mathbf{T}^{-1}\mathbf{R}^T\,\mathbf{a} + \mathbf{T}^{-1}\mathbf{l}, \quad p\,\mathbf{l} = [p\,\mathbf{R}^T + p^2\mathbf{T}]\mathbf{a}. \tag{8.20}$$

The last equation becomes, by (8.16) and the first equation of (8.20),

$$\begin{aligned} p\,\mathbf{l} &= -[\mathbf{Q} + p\mathbf{R}]\mathbf{a} = -\mathbf{Q}\mathbf{a} - \mathbf{R}(-\mathbf{T}^{-1}\mathbf{R}^T\mathbf{a} + \mathbf{T}^{-1}\mathbf{l}) \\ &= [-\mathbf{Q} + \mathbf{R}\mathbf{T}^{-1}\mathbf{R}^T]\mathbf{a} - \mathbf{R}\mathbf{T}^{-1}\mathbf{l}. \end{aligned} \tag{8.21}$$

Thus, from (8.20) and (8.21) we obtain the following theorem.

Theorem 8.1. *Let* $\begin{bmatrix} \mathbf{a} \\ \mathbf{l} \end{bmatrix}$ *be a column vector in* $\mathbb{C}^6$ *whose first three components consist of a vector* $\mathbf{a} \in \mathbb{C}^3$ *that satisfies* (8.16) *and whose last three components consist of the vector* $\mathbf{l} \in \mathbb{C}^3$ *given by* (8.18). *Then*

[d] Here we restrict our attention to the case where the traction produced by $\overset{\circ}{\boldsymbol{T}}$ on $\mathbf{n}\cdot\mathbf{x} = 0$ vanishes, i.e., $\sum_{j=1}^{3} \overset{\circ}{T}_{ij} n_j = 0$ $(i = 1, 2, 3)$ on $\mathbf{n}\cdot\mathbf{x} = 0$.

the following six-dimensional eigen-relation holds:

$$\mathbf{N}\begin{bmatrix}\mathbf{a}\\ \mathbf{l}\end{bmatrix} = p\begin{bmatrix}\mathbf{a}\\ \mathbf{l}\end{bmatrix}, \tag{8.22}$$

where $\mathbf{N}$ *is the* 6×6 *real matrix defined by*

$$\mathbf{N} = \begin{bmatrix} -\mathbf{T}^{-1}\mathbf{R}^T & \mathbf{T}^{-1} \\ -\mathbf{Q} + \mathbf{R}\mathbf{T}^{-1}\mathbf{R}^T & -\mathbf{R}\mathbf{T}^{-1} \end{bmatrix}. \tag{8.23}$$

We call the eigenvalue problem (8.22) *Stroh's eigenvalue problem.*

When $v = 0$, it follows from (2) of Lemma 8.1 that the solutions p_α $(1 \leq \alpha \leq 6)$ to (8.17), i.e., the eigenvalues of $\mathbf{N}$, are not real. As v increases from $v = 0$, at some point Eq. (8.17) ceases to have only complex roots. Since we are concerned with the surface-wave solution (8.13), where $\operatorname{Im} p > 0$, we shall restrict our attention to the range of v for which all the solutions to (8.17) are complex.

Let $\widetilde{\mathbf{m}} = (\widetilde{m}_1, \widetilde{m}_2, \widetilde{m}_3)$ and $\widetilde{\mathbf{n}} = (\widetilde{n}_1, \widetilde{n}_2, \widetilde{n}_3)$ be orthogonal unit vectors in $\mathbb{R}^3$, which are obtained by rotating the orthogonal unit vectors $\mathbf{m}$ and $\mathbf{n}$ around their vector product $\mathbf{m} \times \mathbf{n}$ by an angle ϕ $(-\pi \leq \phi < \pi)$ so that

$$\widetilde{\mathbf{m}} = \widetilde{\mathbf{m}}(\phi) = \mathbf{m}\cos\phi + \mathbf{n}\sin\phi, \quad \widetilde{\mathbf{n}} = \widetilde{\mathbf{n}}(\phi) = -\mathbf{m}\sin\phi + \mathbf{n}\cos\phi. \tag{8.24}$$

Let $\mathbf{Q}(\phi), \mathbf{R}(\phi)$ and $\mathbf{T}(\phi)$ be the 3×3 real matrices given by

$$\begin{aligned}
\mathbf{Q}(\phi) &= \left(\sum_{j,l=1}^{3} B_{ijkl}\widetilde{m}_j\widetilde{m}_l\right)_{i\downarrow k\rightarrow 1,2,3} - \rho v^2 \cos^2\phi\,\mathbf{I},\\
\mathbf{R}(\phi) &= \left(\sum_{j,l=1}^{3} B_{ijkl}\widetilde{m}_j\widetilde{n}_l\right)_{i\downarrow k\rightarrow 1,2,3} + \rho v^2 \cos\phi\sin\phi\mathbf{I},\\
\mathbf{T}(\phi) &= \left(\sum_{j,l=1}^{3} B_{ijkl}\widetilde{n}_j\widetilde{n}_l\right)_{i\downarrow k\rightarrow 1,2,3} - \rho v^2 \sin^2\phi\mathbf{I},
\end{aligned} \tag{8.25}$$

where $\mathbf{I}$ is the 3×3 identity matrix. Then $\mathbf{Q}(0), \mathbf{R}(0)$, and $\mathbf{T}(0)$ are equal to $\mathbf{Q}, \mathbf{R}$, and $\mathbf{T}$ in (8.15), respectively.

As we shall see in Proposition 8.2, Definition 8.3 and Theorem 8.4, we have introduced the matrices $\mathbf{Q}(\phi), \mathbf{R}(\phi)$ and $\mathbf{T}(\phi)$ for the ultimate purpose of finding properties of the vectors $\mathbf{a}$ and $\mathbf{l}$ in (8.22) and of the surface impedance matrix $\mathbf{Z}(v)$, which is defined later through these vectors.

Definition 8.1. The limiting velocity $v_L = v_L(\mathbf{m}, \mathbf{n})$ is the lowest velocity for which the matrices $\mathbf{Q}(\phi)$ and $\mathbf{T}(\phi)$ become singular for some angle ϕ:

$$v_L = \inf\{v > 0 \,|\, \exists\phi; \det \mathbf{Q}(\phi) = 0\} = \inf\{v > 0 \,|\, \exists\phi; \det \mathbf{T}(\phi) = 0\}. \quad (8.26)$$

We will give a characterization of $v_L = v_L(\mathbf{m}, \mathbf{n})$ in terms of body waves whose direction of propagation is on the $\mathbf{m}$–$\mathbf{n}$ plane.

A solution to (8.12) of the form

$$\boldsymbol{u} = (u_1, u_2, u_3) = \mathbf{a}\, e^{-\sqrt{-1}\,k\left(\widetilde{\mathbf{m}}\cdot\mathbf{x} - c(\phi)t\right)} \quad (8.27)$$

represents a body wave with direction of propagation $\widetilde{\mathbf{m}}$, wave number k, velocity $c(\phi)$, and polarization $\mathbf{a} = \mathbf{a}(\phi)$. Substituting this into (8.12), we easily observe that $\rho c(\phi)^2$ and $\mathbf{a}(\phi)$ are an eigenvalue and an eigenvector of the positive definite matrix called the acoustic tensor

$$\left(\sum_{j,l=1}^{3} B_{ijkl}\widetilde{m}_j\widetilde{m}_l\right)_{i\downarrow k\rightarrow 1,2,3}, \quad (8.28)$$

respectively.

Let

$$\lambda_i(\phi)\ (i = 1, 2, 3), \quad 0 < \lambda_1(\phi) \leq \lambda_2(\phi) \leq \lambda_3(\phi)$$

be the eigenvalues of the acoustic tensor (8.28). Corresponding to these eigenvalues, there exist three body waves, which have direction of propagation $\widetilde{\mathbf{m}}$, phase velocity

$$c_i(\phi) = \sqrt{\frac{\lambda_i(\phi)}{\rho}}\ (i = 1, 2, 3), \quad 0 < c_1(\phi) \leq c_2(\phi) \leq c_3(\phi), \quad (8.29)$$

and polarizations $\mathbf{a}_i = \mathbf{a}_i(\phi) \in \mathbb{R}^3$ $(i = 1, 2, 3)$, which are mutually orthogonal.

Proposition 8.1.

$$v_L = v_L(\mathbf{m}, \mathbf{n}) = \min_{-\frac{\pi}{2}<\phi<\frac{\pi}{2}} \frac{c_1(\phi)}{\cos\phi}. \quad (8.30)$$

Proof. We see from the first equation in (8.25) that the eigenvalues of $\mathbf{Q}(\phi)$ are

$$\lambda_i(\phi) - \rho v^2 \cos^2 \phi, \quad i = 1, 2, 3.$$

Hence the assertion follows from (8.29) and the first equality in (8.26). $\square$

A useful construction of $v_L(\mathbf{m}, \mathbf{n})$ is in terms of the slowness section in the $\mathbf{m}$–$\mathbf{n}$ plane. The slowness section in the $\mathbf{m}$–$\mathbf{n}$ plane consists of the three closed curves generated by the radius vectors

$$\frac{1}{c_i(\phi)}\widetilde{\mathbf{m}}, \quad i = 1, 2, 3 \quad (-\pi < \phi \leq \pi).$$

The curve corresponding to the slowest velocity $c_1(\phi)$ defines the silhouette of the slowness section and is called the outer profile.

It follows from (8.30) that

$$v_L^{-1} = \max_{-\frac{\pi}{2}<\phi<\frac{\pi}{2}} \frac{1}{c_1(\phi)} \cos\phi. \tag{8.31}$$

Since ϕ is the angle of rotation about $\mathbf{m} \times \mathbf{n}$ between the $\mathbf{m}$-axis and the vector $\widetilde{\mathbf{m}}$,

$$\frac{1}{c_1(\phi)} \cos\phi$$

is the projection on the $\mathbf{m}$-axis of the point on the outer profile. Then by (8.31), the limiting slowness $v_L(\mathbf{m}, \mathbf{n})^{-1}$ is the absolute maximum of the set of such projections. Thus, we obtain the following corollary.

Corollary 8.1. *In the* $\mathbf{m}$–$\mathbf{n}$ *plane, let* L *be a line parallel to the* $\mathbf{n}$-*axis approaching the slowness section from the right and making the first tangential contact with the outer profile at some point* T. *The limiting slowness* $v_L(\mathbf{m}, \mathbf{n})^{-1}$ *is the projection of* T *on the* $\mathbf{m}$-*axis. Let* $\hat{\phi}$ *be the angle between* $\overrightarrow{OT}$ *and the* $\mathbf{m}$-*axis* $(-\frac{\pi}{2} < \hat{\phi} < \frac{\pi}{2})$. *Then*

$$v_L^{-1} = \max_{-\frac{\pi}{2}<\phi<\frac{\pi}{2}} \frac{1}{c_1(\phi)} \cos\phi = \frac{1}{c_1(\hat{\phi})} \cos\hat{\phi} \tag{8.32}$$

(*see Fig.* 8.1). *The corresponding body wave represented by* (8.27) *propagates in the direction of the radius vector* $\overrightarrow{OT}$ *with the velocity* $c_1(\hat{\phi})$ *and its polarization is an eigenvector of the acoustic tensor* (8.28) *at* $\phi = \hat{\phi}$ *pertaining to the smallest eigenvalue* $\rho c_1(\hat{\phi})^2$.

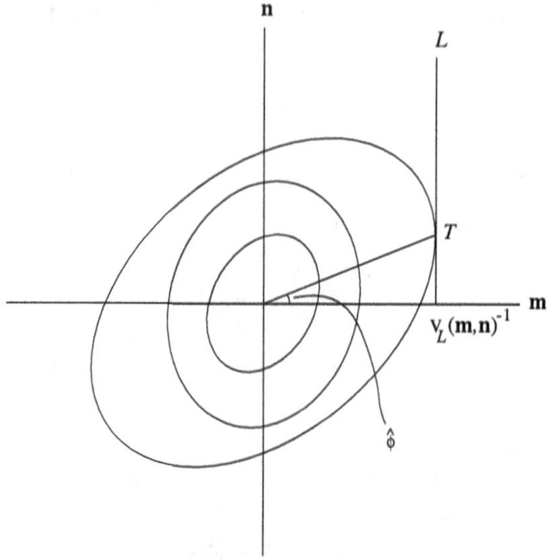

Fig. 8.1. Slowness section and limiting velocity.

Example 8.1. The effective elastic coefficients of a material in an unstressed and isotropic state are given by

$$B_{ijkl} = \lambda\delta_{ij}\delta_{kl} + \mu(\delta_{ik}\delta_{jl} + \delta_{il}\delta_{kj}), \tag{8.33}$$

where λ and μ are the Lamé constants. Then it follows that[e]

$$\begin{aligned} \mathbf{Q}(\phi) &= (\lambda+\mu)\widetilde{\mathbf{m}} \otimes \widetilde{\mathbf{m}} + (\mu - \rho v^2 \cos^2\phi)\,\mathbf{I}, \\ \det \mathbf{Q}(\phi) &= (\mu - \rho v^2 \cos^2\phi)^2(\lambda + 2\mu - \rho v^2\cos^2\phi). \end{aligned}$$

Taking into account that $\mu < \lambda + 2\mu$, which is guaranteed by the strong convexity condition for (8.33) (cf. the paragraph which follows (8.10)), we immediately see that v_L for the material in question is given by $v_L^{\mathrm{Iso}} = \sqrt{\frac{\mu}{\rho}}$ for any $\mathbf{m}$ and $\mathbf{n}$. The acoustic tensor (8.28) of this material is given by

$$(\lambda+2\mu)\widetilde{\mathbf{m}} \otimes \widetilde{\mathbf{m}} + \mu(\widetilde{\mathbf{n}} \otimes \widetilde{\mathbf{n}} + \boldsymbol{\ell} \otimes \boldsymbol{\ell}),$$

where $\boldsymbol{\ell} = \mathbf{m}\times\mathbf{n} = \widetilde{\mathbf{m}}\times\widetilde{\mathbf{n}}$; its eigenvalues are $\lambda+2\mu$ (simple) and μ (double). Hence the slowness sections of this material are three circles centered at

[e]Henceforth we will use the notations $\otimes$ and $\times$ to denote the tensor product and the vector product, respectively.

the origin, two of which have the same radius $\sqrt{\frac{\rho}{\mu}}$ and the other has radius $\sqrt{\frac{\rho}{\lambda+2\mu}}$.

The interval $0 < v < v_L$ is called the subsonic range.[f]

We present below several fundamental properties of the matrices $\mathbf{Q}(\phi), \mathbf{T}(\phi)$, $\mathbf{N}$ and the sextic equation (8.17) in the subsonic range. The proof of the following lemma is given, for example, in [12, Lemma 3.2].

Lemma 8.2.

(1) *The symmetric matrices* $\mathbf{Q}(\phi)$ *and* $\mathbf{T}(\phi)$ *are positive definite for all* ϕ *if and only if* $0 \leq v < v_L$.

(2) *When* $0 \leq v < v_L$, *the solutions* p_α $(1 \leq \alpha \leq 6)$ *to the sextic equation* (8.17), *i.e., the eigenvalues of* $\mathbf{N}$, *are not real and they occur in complex conjugate pairs.*

Henceforth we will restrict our attention to the subsonic range $0 \leq v < v_L$ and take

$$\operatorname{Im} p_\alpha > 0, \quad \alpha = 1, 2, 3. \tag{8.34}$$

Now let $0 \leq v < v_L$, and let $\begin{bmatrix} \mathbf{a}_\alpha \\ \mathbf{l}_\alpha \end{bmatrix}$ $(\alpha = 1, 2, 3)$ be linearly independent eigenvectors of the eigenvalue problem (8.22) pertaining to the eigenvalues p_α $(\alpha = 1, 2, 3, \ \operatorname{Im} p_\alpha > 0)$, respectively. Motivated by (8.13) and (8.19), we observe that the general form[g] of the surface-wave solution is given by

$$\boldsymbol{u} = \sum_{\alpha=1}^{3} c_\alpha \mathbf{a}_\alpha \, e^{-\sqrt{-1}\, k(\mathbf{m}\cdot\mathbf{x} + p_\alpha \mathbf{n}\cdot\mathbf{x} - vt)} \tag{8.35}$$

and the corresponding traction on the surface $\mathbf{n} \cdot \mathbf{x} = 0$ produced by the solution above is given by

$$\boldsymbol{t} = -\sqrt{-1} k \sum_{\alpha=1}^{3} c_\alpha \mathbf{l}_\alpha e^{-\sqrt{-1} k(\mathbf{m}\cdot\mathbf{x} - vt)}, \tag{8.36}$$

[f] The reader should not consider "subsonic" and "ultrasonic" to be related adjectives. "Ultrasonic" refers to stress waves with frequencies above 2×10^4 Hz, the upper limit of human hearing. Rayleigh waves, which have a phase velocity in the subsonic range, can have ultrasonic frequencies.

[g] We have used the term "general" because $\mathbf{a}_\alpha$ $(\alpha = 1, 2, 3)$ are linearly independent in $\mathbb{C}^3$, which we shall see soon.

where c_α $(1 \le \alpha \le 3)$ are arbitrary complex constants. The arrangement (8.34) guarantees that the surface-wave solution (8.35) in the subsonic range decays exponentially as $\mathbf{n} \cdot \mathbf{x} \longrightarrow -\infty$.

Remark 8.1. When the six-dimensional eigenvalue problem (8.22) does not have six linearly independent eigenvectors, generalized eigenvector(s) must be introduced. Let $\begin{bmatrix} \mathbf{a}_\alpha \\ \mathbf{l}_\alpha \end{bmatrix} \in \mathbb{C}^6$ $(\alpha = 1, 2, 3)$ be linearly independent eigenvector(s) and generalized eigenvector(s) of $\mathbf{N}$ associated with the eigenvalues p_α $(\alpha = 1, 2, 3)$. Then the form of the surface-wave solution (8.35) must be modified according to the degeneracy of the eigenvalue problem (8.22) (see, e.g., [12, Sec. 3.2]). However, the corresponding displacements on the surface $\mathbf{n} \cdot \mathbf{x} = 0$ have the same form

$$\boldsymbol{u} = \sum_{\alpha=1}^{3} c_\alpha \mathbf{a}_\alpha e^{-\sqrt{-1}k(\mathbf{m}\cdot\mathbf{x} - vt)}, \tag{8.37}$$

and the corresponding tractions on $\mathbf{n} \cdot \mathbf{x} = 0$ all have the same form

$$\boldsymbol{t} = -\sqrt{-1}\, k \sum_{\alpha=1}^{3} c_\alpha \mathbf{l}_\alpha e^{-\sqrt{-1}k(\mathbf{m}\cdot\mathbf{x} - vt)}. \tag{8.38}$$

Let us turn to the rotated orthogonal unit vectors (8.24) in the $\mathbf{m}$–$\mathbf{n}$ plane and to the matrices (8.25) defined through these rotated vectors. For $0 \le v < v_L$, let $\mathbf{N}(\phi)$ be the 6×6 real matrix defined by

$$\mathbf{N}(\phi) = \begin{bmatrix} -\mathbf{T}(\phi)^{-1}\mathbf{R}(\phi)^T & \mathbf{T}(\phi)^{-1} \\ -\mathbf{Q}(\phi) + \mathbf{R}(\phi)\mathbf{T}(\phi)^{-1}\mathbf{R}(\phi)^T & -\mathbf{R}(\phi)\mathbf{T}(\phi)^{-1} \end{bmatrix}. \tag{8.39}$$

Note that $\mathbf{N}(0)$ is equal to $\mathbf{N}$ defined by (8.23).

The following property of Stroh's eigenvector problem is fundamental to the Stroh formalism. It serves as a basis for the derivation of the Barnett–Lothe integral formalism and is also elegant in itself.

Theorem 8.2. *For* $0 \le v < v_L$, *let* $\begin{bmatrix} \mathbf{a} \\ \mathbf{l} \end{bmatrix}$ *be an eigenvector of* $\mathbf{N}(0)$ *associated with the eigenvalue* $p = p_0$. *Then*

$$\mathbf{N}(\phi) \begin{bmatrix} \mathbf{a} \\ \mathbf{l} \end{bmatrix} = p(\phi) \begin{bmatrix} \mathbf{a} \\ \mathbf{l} \end{bmatrix} \tag{8.40}$$

for all ϕ, *and the eigenvalue* $p(\phi)$ *of* $\mathbf{N}(\phi)$ *satisfies the Riccati equation*

$$\frac{d}{d\phi}p = -1 - p^2 \tag{8.41}$$

with $p(0) = p_0$.

Various proofs of this theorem can be found in several papers listed at the beginning of this section. Here, by avoiding the introduction of "dynamical elastic coefficients", we present a slight improvement of the proof given in [12].

Proof. We denote the differentiation $\frac{d}{d\phi}$ by $'$. From (8.24) we get

$$\widetilde{\mathbf{m}}' = -\mathbf{m}\sin\phi + \mathbf{n}\cos\phi = \widetilde{\mathbf{n}}, \quad \widetilde{\mathbf{n}}' = -\mathbf{m}\cos\phi - \mathbf{n}\sin\phi = -\widetilde{\mathbf{m}}.$$

Then from (8.25) it follows that

$$\begin{aligned}
\mathbf{Q}(\phi)' &= \left(\sum_{j,l=1}^{3} B_{ijkl}\widetilde{n}_j\widetilde{m}_l + \sum_{j,l=1}^{3} B_{ijkl}\widetilde{m}_j\widetilde{n}_l\right)_{i\downarrow k\to 1,2,3} \\
&\quad + 2\rho v^2\cos\phi\sin\phi\mathbf{I} = \mathbf{R}(\phi) + \mathbf{R}(\phi)^T, \\
\mathbf{R}(\phi)' &= \left(\sum_{j,l=1}^{3} B_{ijkl}\widetilde{n}_j\widetilde{n}_l - \sum_{j,l=1}^{3} B_{ijkl}\widetilde{m}_j\widetilde{m}_l\right)_{i\downarrow k\to 1,2,3} \\
&\quad + \rho v^2(\cos^2\phi - \sin^2\phi)\mathbf{I} = \mathbf{T}(\phi) - \mathbf{Q}(\phi), \\
\mathbf{T}(\phi)' &= -\left(\sum_{j,l=1}^{3} B_{ijkl}\widetilde{m}_j\widetilde{n}_l + \sum_{j,l=1}^{3} B_{ijkl}\widetilde{n}_j\widetilde{m}_l\right)_{i\downarrow k\to 1,2,3} \\
&\quad - 2\rho v^2\cos\phi\sin\phi\mathbf{I} = -\mathbf{R}(\phi) - \mathbf{R}(\phi)^T.
\end{aligned} \tag{8.42}$$

Put

$$\mathbf{H}(\phi) = [\mathbf{Q} + p(\mathbf{R} + \mathbf{R}^T) + p^2\mathbf{T}](\phi).$$

From (8.42) we have

$$\begin{aligned}
\mathbf{H}'(\phi) &= [\mathbf{R} + \mathbf{R}^T + p'(\mathbf{R} + \mathbf{R}^T) + 2p(\mathbf{T} - \mathbf{Q}) + 2pp'\mathbf{T} - p^2(\mathbf{R} + \mathbf{R}^T)](\phi) \\
&= [-2p\mathbf{Q} + (1 + p' - p^2)(\mathbf{R} + \mathbf{R}^T) + 2p(1 + p')\mathbf{T}](\phi).
\end{aligned}$$

Suppose that $p(\phi)$ satisfies the Riccati equation (8.41). Then the equation above becomes

$$\mathbf{H}'(\phi) = -2p[\mathbf{Q} + p(\mathbf{R} + \mathbf{R}^T) + p^2\mathbf{T}](\phi) = -2p(\phi)\mathbf{H}(\phi). \tag{8.43}$$

Now put

$$\mathbf{h}(\phi) = [\mathbf{Q} + p(\mathbf{R} + \mathbf{R}^T) + p^2\mathbf{T}](\phi)\mathbf{a} = \mathbf{H}(\phi)\mathbf{a},$$

where $\mathbf{a}$ satisfies (8.16). Then we get

$$\mathbf{h}(0) = \mathbf{0}. \tag{8.44}$$

It follows from (8.43) that

$$\mathbf{h}'(\phi) = -2p(\phi)\mathbf{H}(\phi)\mathbf{a} = -2p(\phi)\mathbf{h}(\phi). \tag{8.45}$$

The solution to (8.41) is $p(\phi) = \tan(\phi_0 - \phi)$ with $\tan(\phi_0) = p_0$ and $\mathrm{Im}\, p_0 \neq 0$ or $p(\phi) \equiv \pm\sqrt{-1}$, both of which are bounded smooth functions of ϕ. Thus from (8.44) and (8.45) we obtain

$$\mathbf{h}(\phi) = [\mathbf{Q} + p(\mathbf{R} + \mathbf{R}^T) + p^2\mathbf{T}](\phi)\mathbf{a} = \mathbf{0} \tag{8.46}$$

for all ϕ.

Now put

$$\boldsymbol{l}(\phi) = [\mathbf{R}^T + p\mathbf{T}](\phi)\mathbf{a}.$$

Then, by the assumption of the theorem, we have

$$\boldsymbol{l}(0) = \mathbf{l}.$$

It follows from (8.41) and (8.42) that

$$\begin{aligned}\boldsymbol{l}'(\phi) &= [\mathbf{T} - \mathbf{Q} + p'\mathbf{T} - p(\mathbf{R} + \mathbf{R}^T)](\phi)\mathbf{a} \\ &= -[\mathbf{Q} + p(\mathbf{R} + \mathbf{R}^T) + p^2\mathbf{T}](\phi)\mathbf{a} = -\mathbf{h}(\phi),\end{aligned}$$

which is equal to zero by (8.46). Hence we obtain

$$\boldsymbol{l}(\phi) = [\mathbf{R}^T + p\mathbf{T}](\phi)\mathbf{a} = \boldsymbol{l}(0) = \mathbf{l} \tag{8.47}$$

for all ϕ. In the same manner that (8.16) and (8.18) lead to the eigenrelation (8.22), from (8.46) and (8.47) we obtain (8.40).[h] □

[h]In the proof, we have assumed that $p(\phi)$ satisfies the Riccati equation (8.41). However, it follows that $p(\phi)$, being the eigenvalue of $\mathbf{N}(\phi)$ in (8.40), *must be* the solution to (8.41), because an eigenvalue that corresponds to the same eigenvector is unique.

When $\mathbf{N}(0)$ has generalized eigenvectors, they do not have the simple invariance as in Theorem 8.2, but there is a rule that describes their dependence on ϕ, for which we refer to the literature given at the beginning of this section.

We now proceed to the Barnett–Lothe integral formalism.

Definition 8.2. For $0 \leq v < v_L$, we define the 6×6 real matrix $\mathbf{S} = \mathbf{S}(v)$ to be the angular average of the 6×6 matrix $\mathbf{N}(\phi)$ over $[-\pi, \pi]$:

$$\mathbf{S} = \begin{bmatrix} \mathbf{S}_1 & \mathbf{S}_2 \\ \mathbf{S}_3 & \mathbf{S}_1^T \end{bmatrix} = \frac{1}{2\pi} \int_{-\pi}^{\pi} \mathbf{N}(\phi) d\phi, \tag{8.48}$$

where $\mathbf{S}_1 = \mathbf{S}_1(v), \mathbf{S}_2 = \mathbf{S}_2(v)$ and $\mathbf{S}_3 = \mathbf{S}_3(v)$ are 3×3 real matrices defined by

$$\begin{aligned} \mathbf{S}_1 &= \frac{1}{2\pi} \int_{-\pi}^{\pi} -\mathbf{T}(\phi)^{-1}\mathbf{R}(\phi)^T d\phi, \quad \mathbf{S}_2 = \frac{1}{2\pi} \int_{-\pi}^{\pi} \mathbf{T}(\phi)^{-1} d\phi, \\ \mathbf{S}_3 &= \frac{1}{2\pi} \int_{-\pi}^{\pi} -\mathbf{Q}(\phi) + \mathbf{R}(\phi)\mathbf{T}(\phi)^{-1}\mathbf{R}(\phi)^T \, d\phi, \end{aligned} \tag{8.49}$$

and $\mathbf{Q}(\phi)$, $\mathbf{R}(\phi)$ and $\mathbf{T}(\phi)$ are given by (8.25).

By (1) of Lemma 8.2, the matrices $\mathbf{S}_2$ and $\mathbf{S}_3$ are symmetric and $\mathbf{S}_2$ is positive definite for $0 \leq v < v_L$.

Now we take the angular average of Stroh's eigenvalue problem.

Theorem 8.3. *For $0 \leq v < v_L$, let $\begin{bmatrix} \mathbf{a}_\alpha \\ \mathbf{l}_\alpha \end{bmatrix}$ be an eigenvector or generalized eigenvector of $\mathbf{N}(0)$ corresponding to the eigenvalues p_α $(\alpha = 1, 2, 3)$ with $\operatorname{Im} p_\alpha > 0$. Then for $0 \leq v < v_L$,*

$$\mathbf{S} \begin{bmatrix} \mathbf{a}_\alpha \\ \mathbf{l}_\alpha \end{bmatrix} = \sqrt{-1} \begin{bmatrix} \mathbf{a}_\alpha \\ \mathbf{l}_\alpha \end{bmatrix}. \tag{8.50}$$

Proof. When $\begin{bmatrix} \mathbf{a}_\alpha \\ \mathbf{l}_\alpha \end{bmatrix}$ is an eigenvector of $\mathbf{N}(0)$ corresponding to the eigenvalues p_α $(\alpha = 1, 2, 3)$ with $\operatorname{Im} p_\alpha > 0$, we take the angular average of both sides of (8.40). For $p(\phi) = \tan(\phi_0 - \phi)$ with $\tan(\phi_0) = p_0$, $\operatorname{Im} p_0 > 0$ and for $p(\phi) \equiv \pm\sqrt{-1}$ it follows that $\frac{1}{2\pi} \int_{-\pi}^{\pi} p(\phi) d\phi = \sqrt{-1}$, which, combined with (8.48), leads to (8.50).

When $\begin{bmatrix} \mathbf{a}_\alpha \\ \mathbf{l}_\alpha \end{bmatrix}$ is a generalized eigenvector of $\mathbf{N}(0)$, the proof of (8.50) is a little complicated: refer to the literature given at the beginning of this section. $\square$

Proposition 8.2. *For* $0 \le v < v_L$, *let* $\begin{bmatrix} \mathbf{a}_\alpha \\ \mathbf{l}_\alpha \end{bmatrix}$ $(\alpha = 1, 2, 3)$ *be linearly independent eigenvector(s) or generalized eigenvector(s) of* $\mathbf{N}(0)$ *corresponding to the eigenvalues* p_α $(\alpha = 1, 2, 3)$ *with* $\operatorname{Im} p_\alpha > 0$. *Then their displacement parts* $\mathbf{a}_\alpha$ $(\alpha = 1, 2, 3)$ *are linearly independent.*

Proof. The first three rows of the system (8.50) are written, by using the notation in (8.49), as

$$\mathbf{S}_1 \mathbf{a}_\alpha + \mathbf{S}_2 \mathbf{l}_\alpha = \sqrt{-1}\mathbf{a}_\alpha, \quad \alpha = 1, 2, 3.$$

Then we get

$$\mathbf{S}_2 \mathbf{l}_\alpha = (\sqrt{-1}\mathbf{I} - \mathbf{S}_1)\mathbf{a}_\alpha, \quad \alpha = 1, 2, 3.$$

Since $\mathbf{S}_2$ is invertible, multiplying both sides by $\mathbf{S}_2^{-1}$, we obtain

$$\mathbf{l}_\alpha = (\sqrt{-1}\mathbf{S}_2^{-1} - \mathbf{S}_2^{-1}\mathbf{S}_1)\mathbf{a}_\alpha, \quad \alpha = 1, 2, 3. \tag{8.51}$$

Suppose that $\mathbf{a}_\alpha (\alpha = 1, 2, 3)$ are linearly dependent. Then there exists a set of complex numbers $(c_1, c_2, c_3) \neq (0, 0, 0)$ such that

$$\sum_{\alpha=1}^{3} c_\alpha \mathbf{a}_\alpha = \mathbf{0}.$$

Then from (8.51) it follows that

$$\sum_{\alpha=1}^{3} c_\alpha \mathbf{l}_\alpha = \mathbf{0},$$

and therefore

$$\sum_{\alpha=1}^{3} c_\alpha \begin{bmatrix} \mathbf{a}_\alpha \\ \mathbf{l}_\alpha \end{bmatrix} = \mathbf{0}.$$

This contradicts the assumption that $\begin{bmatrix} \mathbf{a}_\alpha \\ \mathbf{l}_\alpha \end{bmatrix}$ $(\alpha = 1, 2, 3)$ are linearly independent. □

Now we can define the surface impedance matrix, which was first introduced by Ingebrigtsen and Tonning [22] and later given by Lothe and Barnett [8] in the framework of the Stroh formalism.

Definition 8.3. For $0 \le v < v_L$, let $\begin{bmatrix} \mathbf{a}_\alpha \\ \mathbf{l}_\alpha \end{bmatrix}$ $(\alpha = 1, 2, 3)$ be linearly independent eigenvector(s) or generalized eigenvector(s) of $\mathbf{N}(0)$ corresponding to the eigenvalues p_α ($\alpha = 1,2,3$) with $\operatorname{Im} p_\alpha > 0$. The surface impedance matrix $\mathbf{Z}(v)$ is the 3×3 matrix given by

$$\mathbf{Z}(v) = -\sqrt{-1}[\mathbf{l}_1, \mathbf{l}_2, \mathbf{l}_3][\mathbf{a}_1, \mathbf{a}_2, \mathbf{a}_3]^{-1}, \tag{8.52}$$

where $[\mathbf{l}_1, \mathbf{l}_2, \mathbf{l}_3]$ and $[\mathbf{a}_1, \mathbf{a}_2, \mathbf{a}_3]$ denote 3×3 matrices, which consist of the column vectors $\mathbf{l}_\alpha$ and $\mathbf{a}_\alpha$, respectively.

Therefore, $\mathbf{Z}(v)$ expresses a linear relationship between (i) the displacements given at the surface $\mathbf{n} \cdot \mathbf{x} = 0$ of the form (8.37), which pertain to the surface-wave solution of the form (8.35)[i] propagating in the direction of $\mathbf{m}$ with the phase velocity v, and (ii) the tractions needed to sustain them at that surface of the form (8.38).

Remark 8.2. There is an arbitrariness in the choice of the linearly independent eigenvectors and generalized eigenvectors. However, Theorem 8.4, below, implies that the arbitrariness is canceled out in the product of the two matrices in (8.52), and hence $\mathbf{Z}(v)$ is well defined.

Theorem 8.4. *For* $0 \le v < v_L$,

$$\mathbf{Z}(v) = \mathbf{S}_2^{-1} + \sqrt{-1}\,\mathbf{S}_2^{-1}\mathbf{S}_1, \tag{8.53}$$

where the real matrices $\mathbf{S}_1$ *and* $\mathbf{S}_2$ *are given by* (8.49).

Proof. It is obvious from (8.51) and (8.52). □

Let $\begin{bmatrix} \mathbf{a}_\alpha \\ \mathbf{l}_\alpha \end{bmatrix}$ ($\alpha = 1,2,3$) be linearly independent eigenvectors or generalized eigenvectors of $\mathbf{N}(0)$ corresponding to the eigenvalues p_α ($\alpha = 1,2,3$) with $\operatorname{Im} p_\alpha > 0$. Then from Theorem 8.3 it follows that

$$\mathbf{S}^2 \begin{bmatrix} \mathbf{a}_\alpha \\ \mathbf{l}_\alpha \end{bmatrix} = - \begin{bmatrix} \mathbf{a}_\alpha \\ \mathbf{l}_\alpha \end{bmatrix} \quad (\alpha = 1,2,3). \tag{8.54}$$

We see from (8.22) that the complex conjugates of $\begin{bmatrix} \mathbf{a}_\alpha \\ \mathbf{l}_\alpha \end{bmatrix}$ ($\alpha = 1,2,3$) are linearly independent eigenvectors or generalized eigenvectors of $\mathbf{N}(0)$ corresponding to the complex conjugates of the eigenvalues p_α ($\alpha = 1,2,3$).

[i]When $\mathbf{N}(0)$ has generalized eigenvectors, the form of (8.35) will have to be modified slightly. cf. Remark 8.1.

Hence we take the complex conjugate of both sides of (8.54) to see that (8.54) holds for six linearly independent eigenvectors or generalized eigenvectors of $\mathbf{N}(0)$. Therefore, we get

$$\mathbf{S}^2 = -\mathbf{I}_6 \quad (0 \leq v < v_L), \tag{8.55}$$

where $\mathbf{I}_6$ denotes the 6×6 identity matrix. Then the blockwise expression of (8.55) obtained from (8.48) gives

$$\mathbf{S}_1^2 + \mathbf{S}_2\mathbf{S}_3 = -\mathbf{I}, \quad \mathbf{S}_1\mathbf{S}_2 + \mathbf{S}_2\mathbf{S}_1^T = \mathbf{0} \quad (0 \leq v < v_L). \tag{8.56}$$

Since $\mathbf{S}_2$ is symmetric and invertible for $0 \leq v < v_L$, it follows from the second equality in (8.56) that

$$\mathbf{S}_2^{-1}\mathbf{S}_1 = -(\mathbf{S}_2^{-1}\mathbf{S}_1)^T \quad (0 \leq v < v_L), \tag{8.57}$$

which implies that $\mathbf{S}_2^{-1}\mathbf{S}_1$ is antisymmetric. Hence from (8.53) we obtain the following corollary.

Corollary 8.2. *The surface impedance matrix* $\mathbf{Z}(v)$ *is Hermitian for* $0 \leq v < v_L$.

8.4. Rayleigh Waves in Anisotropic Materials

Rayleigh waves are elastic surface waves, which propagate along the traction-free surface $\mathbf{n} \cdot \mathbf{x} = 0$ with a phase velocity in the subsonic range $0 < v < v_L$, and whose amplitude decays exponentially with depth below that surface.

Let $\mathbf{m}$ and $\mathbf{n}$ be orthogonal unit vectors in $\mathbb{R}^3$. Following the setting of Section 8.3, we consider Rayleigh waves that propagate along the surface $\mathbf{n} \cdot \mathbf{x} = 0$ in the direction of $\mathbf{m}$ with the phase velocity v_R satisfying $0 < v_R < v_L$, and whose amplitude decays exponentially as $\mathbf{n} \cdot \mathbf{x} \longrightarrow -\infty$, and which produce no tractions on $\mathbf{n} \cdot \mathbf{x} = 0$. Here $v_L = v_L(\mathbf{m}, \mathbf{n})$ is the limiting velocity in Definition 8.1.

We take the $\mathbb{C}^3$-vectors $\mathbf{a}_\alpha$ and $\mathbf{l}_\alpha$ $(\alpha = 1, 2, 3)$ so that $\begin{bmatrix} \mathbf{a}_\alpha \\ \mathbf{l}_\alpha \end{bmatrix} (\alpha = 1, 2, 3)$ are linearly independent eigenvector(s) or generalized eigenvector(s) of $\mathbf{N}(0)$ at $v = v_R$ associated with the eigenvalues p_α $(\alpha = 1, 2, 3, \operatorname{Im} p_\alpha > 0)$. Then the existence of Rayleigh waves implies that the corresponding traction on $\mathbf{n} \cdot \mathbf{x} = 0$ given by (8.38) vanishes for $v = v_R$. In other words, there exists a set of complex numbers $(c_1, c_2, c_3) \neq (0, 0, 0)$ such that

$$\sum_{\alpha=1}^{3} c_\alpha \mathbf{l}_\alpha = \mathbf{0} \quad \text{at } v = v_R, \tag{8.58}$$

which is equivalent to

$$\det[\mathbf{l}_1, \mathbf{l}_2, \mathbf{l}_3] = 0 \quad \text{at } v = v_R. \tag{8.59}$$

We will give a characterization of the Rayleigh-wave velocity v_R in terms of the surface impedance matrix $\mathbf{Z}(v)$ defined by (8.52), or equivalently by (8.53), which we shall use in investigating the perturbation of v_R in the next section.

Theorem 8.5. *A necessary and sufficient condition for the existence of Rayleigh waves in the half-space* $\mathbf{n} \cdot \mathbf{x} \leq 0$, *which propagate along the surface* $\mathbf{n} \cdot \mathbf{x} = 0$ *in the direction of* $\mathbf{m}$ *with the phase velocity* v_R *in the subsonic range* $0 < v < v_L$, *is*

$$\det \mathbf{Z}(v) = 0 \quad \textit{at } v = v_R \quad (0 < v_R < v_L). \tag{8.60}$$

Proof. Recall that Rayleigh waves produce no tractions on $\mathbf{n} \cdot \mathbf{x} = 0$. The theorem is obvious from (8.59), Proposition 8.2, and (8.52). □

The surface impedance matrix $\mathbf{Z}(v)$ $(0 \leq v < v_L)$ has the following fundamental properties:

(1) $\mathbf{Z}(v)$ is Hermitian for $0 \leq v < v_L$ (Corollary 8.2).
(2) $\mathbf{Z}(0)$ is positive definite[j] (cf. [8], Sec. 7.D of [9], Sec. 6.6 of [11], and [23, 24]).
(3) The Hermitian matrix $\frac{d}{dv}\mathbf{Z}(v)$ is negative definite for $0 < v < v_L$ (cf. [8, 25, 26]).

Proofs of assertions (2) and (3) are long: refer to the literature mentioned above.

Using assertion (3), we obtain (cf. Theorem 7 of [10], [12]):

Lemma 8.3. *For* $0 \leq v < v_L$, *the eigenvalues of* $\mathbf{Z}(v)$ *are monotonic decreasing functions of* v.

[j]This assertion holds when $(B_{ijkl})_{i,j,k,l=1,2,3}$ satisfies the strong convexity condition. However, since we have assumed that a deviation of the medium from its unstressed state caused by the initial stress is small (see the comment just after (8.10)), the assertion remains valid.

Therefore, the eigenvalues of $\mathbf{Z}(v)$ decrease monotonically with v in the interval $0 \leq v < v_L$ from their positive values at $v = 0$, and one of them becomes zero when v equals the Rayleigh-wave velocity v_R (Theorem 8.5).

Using the integral expression (8.53) for $\mathbf{Z}(v)$, we get (cf. the comments after Theorem 7 of [10], [12]) the following lemma.

Lemma 8.4. *At most one eigenvalue of* $\mathbf{Z}(v)$ *can be negative at* $v = v_L$.

The last two lemmas, combined with Theorem 8.5, imply the following corollary.

Corollary 8.3. *For given orthogonal unit vectors* **m** *and* **n**, *the phase velocity of a Rayleigh wave is uniquely determined if the Rayleigh wave exists.*[k]

Example 8.2. The effective elastic coefficients of a material in an unstressed and isotropic state are given by (8.33). Algebraic manipulations of (8.52) give

$$\mathbf{Z}(v) = -\sqrt{-1}\left[\mu p_1 \boldsymbol{\ell} \otimes \boldsymbol{\ell} + \frac{V p_3}{1 + p_1 p_3}\mathbf{m} \otimes \mathbf{m} + \frac{V p_1}{1 + p_1 p_3}\mathbf{n} \otimes \mathbf{n} \right.$$
$$\left. + \left(2\mu - \frac{V}{1 + p_1 p_3}\right)(\mathbf{m} \otimes \mathbf{n} - \mathbf{n} \otimes \mathbf{m})\right], \tag{8.61}$$

where

$$V = \rho v^2, \quad p_1 = p_2 = \sqrt{-1}\sqrt{\frac{\mu - V}{\mu}}, \quad p_3 = \sqrt{-1}\sqrt{\frac{\lambda + 2\mu - V}{\lambda + 2\mu}},$$

$\boldsymbol{\ell} = \mathbf{m} \times \mathbf{n}$ is the vector product, and $\otimes$ denotes the tensor product. Here we have used the explicit expressions of $\mathbf{a}_\alpha$ and $\mathbf{l}_\alpha$ ($\alpha = 1, 2, 3$) given in (3.36) and (3.158) of [12].

[k]The polarization vector $\sum_{\alpha=1}^{3} c_\alpha \mathbf{a}_\alpha$ in (8.37) of the Rayleigh wave at the surface $\mathbf{n} \cdot \mathbf{x} = 0$ is also uniquely determined up to a constant multiplicative factor.

When $v = v_L^{\mathrm{Iso}} = \sqrt{\frac{\mu}{\rho}}$,

$$p_1 = 0, \quad p_3 = \sqrt{-1}\sqrt{\frac{\lambda+\mu}{\lambda+2\mu}}.$$

Hence

$$\mathbf{Z}(v_L^{\mathrm{Iso}}) = \mu\sqrt{\frac{\lambda+\mu}{\lambda+2\mu}}\mathbf{m}\otimes\mathbf{m} - \sqrt{-1}\mu(\mathbf{m}\otimes\mathbf{n} - \mathbf{n}\otimes\mathbf{m}),$$

whose eigenvalues are 0 and the two real roots of the quadratic equation of p:

$$p^2 - \mu\sqrt{\frac{\lambda+\mu}{\lambda+2\mu}}\,p - \mu^2 = 0.$$

Obviously, this has two roots of opposite sign, which implies that one of the eigenvalues of $\mathbf{Z}(v)$ is negative at $v = v_L^{\mathrm{Iso}}$. Therefore, a Rayleigh wave exists for any $\mathbf{m}$ and $\mathbf{n}$. As an example, we show in Fig. 8.2 the variation of the eigenvalues of $\mathbf{Z}(v)$ $(0 \le v \le v_L)$ in (8.61), which pertains to a specific unstressed and isotropic material.

The phase velocity $v_R^{\mathrm{Iso}}(< v_L^{\mathrm{Iso}})$ of the Rayleigh wave is obtained through (8.60) and (8.61) from the equation

$$\frac{Vp_1}{1+p_1p_3}\frac{Vp_3}{1+p_1p_3} + \left(2\mu - \frac{V}{1+p_1p_3}\right)^2 = 0,$$

which is equivalent to the cubic equation

$$V^3 - 8\mu V^2 + \frac{8\mu^2(3\lambda+4\mu)}{\lambda+2\mu}V - \frac{16\mu^3(\lambda+\mu)}{\lambda+2\mu} = 0.$$

In the next section, using Theorem 8.5 we derive an equation for the phase velocity of Rayleigh waves (i.e., a secular equation) that propagate on the free surface of a material in an unstressed and orthotropic or transversely isotropic state and study the perturbation of the phase velocity caused by the presence of the initial stress and/or caused by the deviation of the incremental elasticity tensor from its comparative orthotropic or transversely isotropic state.

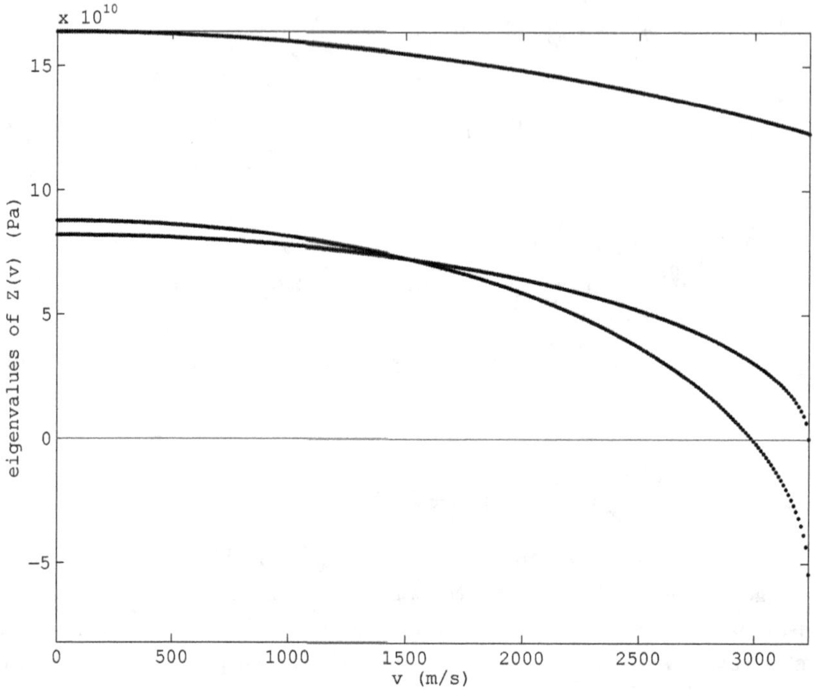

Fig. 8.2. Variation of the three eigenvalues of $\mathbf{Z}(v)$ in the interval $0 \leq v < v_L$ (black curves) which pertains to mild steel in an unstressed and isotropic state whose effective elastic coefficients are given by (8.33) with $\lambda = 107.4$, $\mu = 81.9$ (GPa) and whose density is given by $\rho = 7837$ kg/m^3. In this figure, the subsonic range is divided into 300 subintervals of equal length and the three eigenvalues are computed at each partition point. The figure was produced by a Matlab program written under the first author's joint work with Samuli Siltanen at University of Helsinki in the spring of 2014.

8.5. Perturbation of the Phase Velocity of Rayleigh Waves in Prestressed Anisotropic Media When the Base Material is Orthotropic

Let us turn to the constitutive equation (8.2). As a base material we take an orthotropic medium[1] with elasticity tensor $\mathbb{C}^{\text{Orth}}$. Suppose the base material can be such that it occupies the half-space $x_3 \leq 0$ and its planes of reflectional symmetry coincide with the three coordinate planes. Then

[1]We say that an elastic material is orthotropic if there exist three mutually orthogonal planes such that (8.6) holds for any orthogonal tensor $\boldsymbol{Q} = (Q_{ij})_{i,j=1,2,3}$ pertaining to a reflection with respect to one of these planes.

the elasticity tensor $\mathbb{C}^{\text{Orth}}$ is expressed with the Voigt notation as

$$\mathbb{C}^{\text{Orth}} = (C_{rs}^{\text{Orth}}) = \begin{pmatrix} C_{11} & C_{12} & C_{13} & 0 & 0 & 0 \\ & C_{22} & C_{23} & 0 & 0 & 0 \\ & & C_{33} & 0 & 0 & 0 \\ & & & C_{44} & 0 & 0 \\ & & & & C_{55} & 0 \\ Sym. & & & & & C_{66} \end{pmatrix}. \tag{8.62}$$

Suppose that the incremental elasticity tensor $\mathbb{L}$ is composed of an orthotropic part $\mathbb{C}^{\text{Orth}}$ and a perturbative part $\mathbb{A}$, the latter of which expresses a deviation of the material from its orthotropic unperturbed state. For the perturbative part $\mathbb{A}$, we assume the major and minor symmetries

$$a_{ijkl} = a_{klij} = a_{jikl}, \quad i, j, k, l = 1, 2, 3$$

but do not assume any material symmetry. In the Voigt notation, $\mathbb{A}$ can be written as

$$\mathbb{A} = (a_{rs}) = \begin{pmatrix} a_{11} & a_{12} & a_{13} & a_{14} & a_{15} & a_{16} \\ & a_{22} & a_{23} & a_{24} & a_{25} & a_{26} \\ & & a_{33} & a_{34} & a_{35} & a_{36} \\ & & & a_{44} & a_{45} & a_{46} \\ & & & & a_{55} & a_{56} \\ Sym. & & & & & a_{66} \end{pmatrix} \tag{8.63}$$

and the 21 components in the upper triangular part of matrix (8.63) are generally all independent. In this setting, $\mathbb{L}$ can be written as a fourth-order tensor on symmetric tensors $\boldsymbol{E}$ in the form

$$\mathbb{L}[\boldsymbol{E}] = \mathbb{C}^{\text{Orth}}[\boldsymbol{E}] + \mathbb{A}[\boldsymbol{E}]; \tag{8.64}$$

here the comparative orthotropic part $\mathbb{C}^{\text{Orth}}$ and the perturbative part $\mathbb{A}$ are also treated as fourth-order tensors on $\boldsymbol{E}$. In this section, we shall investigate how $\mathbb{A}$ affects the phase velocity of Rayleigh waves.

We also consider the change of the phase velocity of Rayleigh waves due to the presence of the initial stress. We assume that the surface $x_3 = 0$ of the half-space is free of traction. Since the initial stress is here taken to be homogeneous, zero traction at the boundary $x_3 = 0$ implies that the components $\overset{\circ}{T}_{i3}$ $(i = 1, 2, 3)$ of $\overset{\circ}{\boldsymbol{T}}$ must vanish. Then the second-order tensor $\overset{\circ}{\boldsymbol{T}}$ is represented by a 3×3 matrix as

$$\overset{\circ}{\boldsymbol{T}} = \begin{pmatrix} \overset{\circ}{T}_{11} & \overset{\circ}{T}_{12} & 0 \\ \overset{\circ}{T}_{12} & \overset{\circ}{T}_{22} & 0 \\ 0 & 0 & 0 \end{pmatrix}. \tag{8.65}$$

Thus, in our constitutive equation (8.2) with (8.64), $\mathbb{A}$ and $\overset{\circ}{\boldsymbol{T}}$ express the deviation of the medium in question from its comparative orthotropic and unstressed state and then we consider what influence $\mathbb{A}$ and $\overset{\circ}{\boldsymbol{T}}$ exert upon the phase velocity of Rayleigh waves that propagate along the surface of the material half-space.

Remark 8.3. As we mentioned in Section 8.2, $\mathbb{L}$ generally depends on $\overset{\circ}{\boldsymbol{T}}$. Hence $\mathbb{A}$ also may depend on $\overset{\circ}{\boldsymbol{T}}$. Throughout this chapter, except for the model studied in Sec. 8.6 we will keep the dependence of $\mathbb{L}$ and $\mathbb{A}$ on $\overset{\circ}{\boldsymbol{T}}$ implicit. Thus, in Theorem 8.6, Remark 8.4 and below, when we refer to the effect of $\overset{\circ}{\boldsymbol{T}}$ on the perturbation formulas, we mean only the contribution of the initial stress through the $\boldsymbol{H}\overset{\circ}{\boldsymbol{T}}$ term in (8.2).

For an orthotropic base material whose elasticity tensor is given by (8.62), when there exist Rayleigh waves propagating along the surface of the half-space $x_3 \leq 0$ in the direction of the 2-axis, the phase velocity v_R^{Orth} satisfies the secular equation

$$R^{\mathrm{Orth}}(v) = 0, \tag{8.66}$$

where

$$R^{\mathrm{Orth}}(v) = C_{33}C_{44}(C_{22} - V)V^2 - (C_{44} - V)\left(C_{33}(C_{22} - V) - C_{23}^2\right)^2 \tag{8.67}$$

and $V = \rho v^2$. Equations (8.66) and (8.67) follow from (8.59) (cf. [27]).

As mentioned in the introduction of this chapter, our purpose is to derive a perturbation formula that shows how $\mathbb{A}$ and $\overset{\circ}{\boldsymbol{T}}$ affect the phase velocity of Rayleigh waves from its comparative orthotropic and unstressed value v_R^{Orth}. Now we state a theorem that gives the perturbation formula.

Theorem 8.6. *In a prestressed medium whose incremental elasticity tensor* $\mathbb{L}$ *and initial stress* $\overset{\circ}{\boldsymbol{T}}$ *are given by* (8.64) *and* (8.65) *respectively,* *the phase velocity of Rayleigh waves which propagate along the surface of the half-space $x_3 \le 0$ in the direction of the 2-axis can be written, up to terms linear in the perturbative part $\mathbb{A}$ of $\mathbb{L}$ and the initial stress $\overset{\circ}{\boldsymbol{T}}$, as*

$$v_R = v_R^{\mathrm{Orth}} - \frac{1}{2\rho v_R^{\mathrm{Orth}}}\Big[\gamma_{22}(v_R^{\mathrm{Orth}})a_{22} + \gamma_{23}(v_R^{\mathrm{Orth}})a_{23} + \gamma_{33}(v_R^{\mathrm{Orth}})a_{33} + \gamma_{44}(v_R^{\mathrm{Orth}})a_{44} - \overset{\circ}{T}_{22}\Big], \tag{8.68}$$

where

$$\gamma_{ij}(v) = \frac{N_{ij}(v)}{D(v)} \quad (\{ij\} = \{22\}, \{23\}, \{33\}, \{44\}), \tag{8.69}$$

$$\begin{aligned}
N_{22}(v) &= C_{33}[-2C_{44}(C_{22}C_{33} - C_{23}^2) + 2(C_{22}C_{33} - C_{23}^2 + C_{33}C_{44})V \\
&\quad + (C_{44} - 2C_{33})V^2], \\
N_{23}(v) &= 4C_{23}(C_{44} - V)(C_{22}C_{33} - C_{23}^2 - C_{33}V), \\
N_{33}(v) &= (C_{22} - V)[-2C_{44}(C_{22}C_{33} - C_{23}^2) + 2(C_{22}C_{33} - C_{23}^2 + C_{33}C_{44})V \\
&\quad + (C_{44} - 2C_{33})V^2] = \frac{C_{22} - V}{C_{33}} N_{22}(v), \\
N_{44}(v) &= \frac{-V}{C_{44}}(C_{22}C_{33} - C_{23}^2 - C_{33}V)^2, \\
D(v) &= (C_{22}C_{33} - C_{23}^2)(C_{22}C_{33} - C_{23}^2 + 2C_{33}C_{44}) \\
&\quad + 2C_{33}[C_{22}C_{44} - 2(C_{22}C_{33} - C_{23}^2) - C_{33}C_{44}]V \\
&\quad + 3C_{33}(C_{33} - C_{44})V^2, \\
V &= \rho v^2.
\end{aligned}$$

Remark 8.4. Only four components a_{22}, a_{23}, a_{33} and a_{44} of the perturbative part $\mathbb{A}$ of $\mathbb{L}$ and one component $\overset{\circ}{T}_{22}$ of the initial stress $\overset{\circ}{\boldsymbol{T}}$ in the $\boldsymbol{H}\overset{\circ}{\boldsymbol{T}}$ term of (8.2) affect the first-order perturbation of the phase velocity v_R. This is also true for the case where the base material is unstressed and isotropic [13]. When the base material is generally anisotropic, Song and Fu [28] obtained a formula on the first-order perturbation of the phase velocity of Rayleigh waves; that formula involves the eigenvalues and the eigenvectors of Stroh's eigenvalue problem for the base material. Also, they applied their formula to the case where the base material is monoclinic. There they asserted that for Rayleigh waves polarized in a symmetry plane of the monoclinic material, which we take to be the 2–3 plane in this instance, the first-order perturbation of v_R will not involve any components of $\mathbb{A}$ in which suffix 1 appears at least once, i.e., no components of $\mathbb{A}$ other than $a_{22}, a_{23}, a_{24}, a_{33}, a_{34}$ and a_{44} will affect the first-order perturbation of v_R. For more discussion on work [28] refer to [17].

When the base material is transversely isotropic[m] and the axis of symmetry coincides with the 3-axis, its elasticity tensor $\mathbb{C}^{\mathrm{Trans}}$ is expressed in the Voigt notation as

$$\mathbb{C}^{\mathrm{Trans}} = (C_{rs}^{\mathrm{Trans}}) = \begin{pmatrix} C_{11} & C_{12} & C_{13} & 0 & 0 & 0 \\ & C_{11} & C_{13} & 0 & 0 & 0 \\ & & C_{33} & 0 & 0 & 0 \\ & & & C_{44} & 0 & 0 \\ & & & & C_{44} & 0 \\ Sym. & & & & & (C_{11}-C_{12})/2 \end{pmatrix}. \tag{8.70}$$

Corollary 8.4. *In a prestressed medium whose incremental elasticity tensor* $\mathbb{L}$ *and initial stress* $\overset{\circ}{\boldsymbol{T}}$ *are given by* $\mathbb{L} = \mathbb{C}^{\mathrm{Trans}} + \mathbb{A}$ *and* (8.65), *respectively, the first-order perturbation formula for the phase velocity of Rayleigh waves that propagate along the surface of the half-space* $x_3 \leq 0$ *in the direction of the 2-axis is given by* (8.68) *and* (8.69) *with* C_{22} *and* C_{23} *there replaced by* C_{11} *and* C_{13} *of* (8.70), *respectively, and with* v_R^{Orth} *replaced by the phase velocity* v_R^{Trans} *of Rayleigh waves propagating on the surface of the half-space* $x_3 \leq 0$ *of the comparative transversely isotropic and unstressed medium, which solves* (8.66) *under the aforementioned replacements of the components of the elasticity tensor.*

Remark 8.5. Note that in Theorem 8.6 the 2-axis is an axis of two-fold rotational symmetry of the orthotropic base material. Hence formula (8.68) and its obvious modification are applicable only to two propagation directions in the $x_3 = 0$ plane. On the other hand, when the base material is transversely isotropic with the 3-axis being the ∞-axis, any propagation direction in the $x_3 = 0$ plane can be used to define the 2-axis. Thus Corollary 8.4 is applicable to any propagation direction of Rayleigh waves.

[m] We say that an elastic material is transversely isotropic if there exists a unit vector $\boldsymbol{n}$ such that (8.6) holds for any orthogonal transformation $\boldsymbol{Q} = (Q_{ij})_{i,j=1,2,3}$ which satisfies $\boldsymbol{Qn} = \boldsymbol{n}$ or $\boldsymbol{Qn} = -\boldsymbol{n}$.

For a base material with $\overset{\circ}{\boldsymbol{T}} = \mathbf{0}$, $\mathbb{A} = \mathbb{O}$ and $\mathbb{L} = \mathbb{C}^{\text{Orth}}$, the limiting velocity v_L^{Orth} of the surface waves that propagate along the surface of the half-space $x_3 \leq 0$ in the direction of the 2-axis is given by

$$\rho(v_L^{\text{Orth}})^2 = \min(C_{66},\ V^*), \tag{8.71}$$

where

$$V^* = \sup\{0 < V \leq \min(C_{22}, C_{44}) \mid \sqrt{C_{33}(C_{22} - V)} + \sqrt{C_{44}(C_{44} - V)} \geq |C_{23} + C_{44}|\},$$

and the corresponding surface impedance matrix is given by

$$\mathbf{Z}^{\text{Orth}}(v) = (Z_{ij})_{i\downarrow j\to 1,2,3} = \overline{\mathbf{Z}^{\text{Orth}}(v)}^T, \tag{8.72}$$

$$Z_{11} = \sqrt{C_{55}(C_{66} - V)}, \quad Z_{12} = Z_{13} = 0, \quad Z_{22} = \frac{\sqrt{C_{44}(C_{22} - V)}H}{G},$$

$$Z_{23} = \frac{-\sqrt{-1}\sqrt{C_{44}}J}{G}, \quad Z_{33} = \frac{\sqrt{C_{33}(C_{44} - V)}H}{G},$$

where

$$H = H(V) = \sqrt{\left(\sqrt{C_{33}(C_{22} - V)} + \sqrt{C_{44}(C_{44} - V)}\right)^2 - (C_{23} + C_{44})^2},$$

$$G = G(V) = \sqrt{C_{33}(C_{22} - V)} + \sqrt{C_{44}(C_{44} - V)},$$

$$J = J(V) = \sqrt{C_{33}C_{44}(C_{22} - V)} - C_{23}\sqrt{C_{44} - V}$$

and $V = \rho v^2$ (cf. [11, Sec. 12-10] and [12, Sec. 3.8]).

From Theorem 8.5 it can be proved (cf. [11, 12]) that Rayleigh waves that propagate along the surface of the half-space $x_3 \leq 0$ in the direction of the 2-axis exist if $V^* \leq C_{66}$ and that the secular equation for the phase velocity of the Rayleigh waves in the base material is given by $Z_{22}Z_{33} + Z_{23}^2 = 0$, which is found to be equivalent to (8.66).

Sketch of Proof of Theorem 8.6. Below we shall briefly sketch a derivation of the perturbation formula (8.68) in Theorem 8.6. We will investigate the effects of $\mathbb{A}$ and $\overset{\circ}{\boldsymbol{T}}$ on the surface impedance matrix $\mathbf{Z}(v)$ pertaining to the surface waves that propagate in the direction of the 2-axis along the surface of the half-space $x_3 \leq 0$ of an elastic medium whose incremental elasticity tensor and the initial stress have the form (8.64) and (8.65), respectively.

Since we are concerned with the terms in v_R up to those linear in the perturbative part $\mathbb{A}$ of $\mathbb{L}$ and the initial stress $\overset{\circ}{\boldsymbol{T}}$, suppose that we can write the surface impedance matrix $\mathbf{Z}(v)$, up to terms linear in $\mathbb{A}$ and $\overset{\circ}{\boldsymbol{T}}$, as

$$\mathbf{Z}(v) \approx \mathbf{Z}^{\mathrm{Orth}}(v) + \mathbf{Z}^{\mathrm{Ptb}}(v).$$

We will use the notation $\approx$ to indicate that we are retaining terms up to those linear in the perturbative part $\mathbb{A}$ of $\mathbb{L}$ and the initial stress $\overset{\circ}{\boldsymbol{T}}$ and are neglecting the higher-order terms. Note that each component of $\mathbf{Z}^{\mathrm{Ptb}}(v)$ is a linear function of $\mathbb{A}$ and $\overset{\circ}{\boldsymbol{T}}$. From (8.72) we can write

$$\mathbf{Z}(v) \approx \begin{pmatrix} Z_{11} + \zeta_{11} & \zeta_{12} & \zeta_{13} \\ \overline{\zeta_{12}} & Z_{22} + \zeta_{22} & Z_{23} + \zeta_{23} \\ \overline{\zeta_{13}} & -Z_{23} + \overline{\zeta_{23}} & Z_{33} + \zeta_{33} \end{pmatrix},$$

where

$$\mathbf{Z}^{\mathrm{Ptb}}(v) = \overline{\mathbf{Z}^{\mathrm{Ptb}}(v)}^{T} = (\zeta_{ij})_{i\downarrow j\to 1,2,3}.$$

Hence it follows that

$$\begin{aligned} \det\ \mathbf{Z}(v) &\approx (Z_{11} + \zeta_{11})(Z_{22} + \zeta_{22})(Z_{33} + \zeta_{33}) \\ &\quad - (Z_{11} + \zeta_{11})(Z_{23} + \zeta_{23})(-Z_{23} + \overline{\zeta_{23}}) \\ &\approx (Z_{11} + \zeta_{11})(Z_{22}Z_{33} + Z_{23}^2 + Z_{33}\zeta_{22} + Z_{22}\zeta_{33} + Z_{23}(\zeta_{23} - \overline{\zeta_{23}})). \end{aligned}$$

We see from $Z_{11} = \sqrt{C_{55}(C_{66} - V)}$ and (8.71) that $Z_{11} > 0$ in the subsonic range $0 < v < v_L^{\mathrm{Orth}}$ of surface waves that propagate along the surface of the half-space $x_3 \leq 0$ in the direction of the 2-axis of the base material. Since $\overset{\circ}{\boldsymbol{T}}$ and $\mathbb{A}$ are sufficiently small, $Z_{11} + \zeta_{11} > 0$ in the subsonic range of the surface waves in question. Thus we obtain from Theorem 8.5 an approximate secular equation for v_R:

$$\Delta(v) = 0, \tag{8.73}$$

where

$$\Delta(v) = Z_{22}Z_{33} + Z_{23}^2 + Z_{33}\zeta_{22} + Z_{22}\zeta_{33} + Z_{23}(\zeta_{23} - \overline{\zeta_{23}}).$$

Lemma 8.5. *The effects of the initial stress* $\overset{\circ}{\boldsymbol{T}}$ *and the perturbative part* $\mathbb{A}$ *of the incremental elasticity tensor* $\mathbb{L}$ *on the approximate secular equation* (8.73), *more precisely, on the components* ζ_{22} *and* ζ_{33} *of* $\mathbf{Z}^{\mathrm{Ptb}}(v)$

and $\zeta_{23} - \overline{\zeta_{23}}$, *to first order of* $\overset{\circ}{\boldsymbol{T}}$ *and* $\mathbb{A}$, *come only from* $a_{22}, a_{23}, a_{33}, a_{44}$ *and* $\overset{\circ}{T}_{22}$.

To prove this lemma we use the integral representation (8.53) of $\mathbf{Z}(v)$. The matrix $\mathbf{S}_1$ can be written as

$$\mathbf{S}_1 \approx \mathbf{S}_1^{\mathrm{Orth}} + \mathbf{S}_1^{\mathrm{Ptb}}. \tag{8.74}$$

Here

$$\mathbf{S}_1^{\mathrm{Orth}} = \mathbf{S}_1^{\mathrm{Orth}}(v) = \frac{1}{2\pi}\int_{-\pi}^{\pi}(-\mathbf{T}_v^{\mathrm{Orth}}(\phi)^{-1}\mathbf{R}_v^{\mathrm{Orth}}(\phi)^T)d\phi$$

is of zeroth order in $\overset{\circ}{\boldsymbol{T}}$ and $\mathbb{A}$, where

$$\mathbf{R}_v^{\mathrm{Orth}}(\phi) = \left(\sum_{j,l=1}^{3} C_{ijkl}^{\mathrm{Orth}}\,\widetilde{m}_j\widetilde{n}_l\right)_{i\downarrow k\to 1,2,3} + \rho\, v^2\cos\phi\sin\phi\;\mathbf{I},$$

$$\mathbf{T}_v^{\mathrm{Orth}}(\phi) = \left(\sum_{j,l=1}^{3} C_{ijkl}^{\mathrm{Orth}}\widetilde{n}_j\widetilde{n}_l\right)_{i\downarrow k\to 1,2,3} - \rho v^2\sin^2\phi\mathbf{I},$$

and

$$\begin{aligned}\mathbf{S}_1^{\mathrm{Ptb}} = \mathbf{S}_1^{\mathrm{Ptb}}(v) = \frac{1}{2\pi}\int_{-\pi}^{\pi}&(-\mathbf{T}_v^{\mathrm{Orth}}(\phi)^{-1}\mathbf{R}^{\mathrm{Ptb}}(\phi)^T \\ &+\mathbf{T}_v^{\mathrm{Orth}}(\phi)^{-1}\mathbf{T}^{\mathrm{Ptb}}(\phi)\mathbf{T}_v^{\mathrm{Orth}}(\phi)^{-1}\mathbf{R}_v^{\mathrm{Orth}}(\phi)^T)d\phi\end{aligned} \tag{8.75}$$

is of first order in $\overset{\circ}{\boldsymbol{T}}$ and $\mathbb{A}$, where

$$\mathbf{R}^{\mathrm{Ptb}}(\phi) = \sum_{j,l=1}^{3}\overset{\circ}{T}_{jl}\widetilde{m}_j\widetilde{n}_l\mathbf{I} + \left(\sum_{j,l=1}^{3} a_{ijkl}\widetilde{m}_j\widetilde{n}_l\right)_{i\downarrow k\to 1,2,3},$$

$$\mathbf{T}^{\mathrm{Ptb}}(\phi) = \sum_{j,l=1}^{3}\overset{\circ}{T}_{jl}\widetilde{n}_j\widetilde{n}_l\;\mathbf{I} + \left(\sum_{j,l=1}^{3} a_{ijkl}\widetilde{n}_j\widetilde{n}_l\right)_{i\downarrow k\to 1,2,3}.$$

We also have

$$\mathbf{S}_2 \approx \mathbf{S}_2^{\mathrm{Orth}} + \mathbf{S}_2^{\mathrm{Ptb}},$$

where

$$\mathbf{S}_2^{\text{Orth}} = \mathbf{S}_2^{\text{Orth}}(v) = \frac{1}{2\pi}\int_{-\pi}^{\pi} \mathbf{T}_v^{\text{Orth}}(\phi)^{-1} d\phi$$

is of zeroth order in $\overset{\circ}{\boldsymbol{T}}$ and $\mathbb{A}$ and

$$\mathbf{S}_2^{\text{Ptb}} = \mathbf{S}_2^{\text{Ptb}}(v) = \frac{-1}{2\pi}\int_{-\pi}^{\pi} \mathbf{T}_v^{\text{Orth}}(\phi)^{-1}\mathbf{T}^{\text{Ptb}}(\phi)\mathbf{T}_v^{\text{Orth}}(\phi)^{-1} d\phi$$

is of first order in $\overset{\circ}{\boldsymbol{T}}$ and $\mathbb{A}$. Hence we can deduce from

$$\mathbf{Z}(v) = \mathbf{S}_2^{-1} + \sqrt{-1}\mathbf{S}_2^{-1}\mathbf{S}_1 \approx \mathbf{Z}^{\text{Orth}}(v) + \mathbf{Z}^{\text{Ptb}}(v)$$

that

$$\begin{aligned}\mathbf{Z}^{\text{Ptb}}(v) &= -(\mathbf{S}_2^{\text{Orth}})^{-1}\mathbf{S}_2^{\text{Ptb}}(\mathbf{S}_2^{\text{Orth}})^{-1} \\ &\quad + \sqrt{-1}[-(\mathbf{S}_2^{\text{Orth}})^{-1}\mathbf{S}_2^{\text{Ptb}}(\mathbf{S}_2^{\text{Orth}})^{-1}\mathbf{S}_1^{\text{Orth}} + (\mathbf{S}_2^{\text{Orth}})^{-1}\mathbf{S}_1^{\text{Ptb}}] \\ &= -(\mathbf{S}_2^{\text{Orth}})^{-1}\mathbf{S}_2^{\text{Ptb}}\,\mathbf{Z}^{\text{Orth}}(v) + \sqrt{-1}(\mathbf{S}_2^{\text{Orth}})^{-1}\mathbf{S}_1^{\text{Ptb}}. \end{aligned} \tag{8.76}$$

Looking carefully at the components ζ_{22} and ζ_{33} of (8.76) and $\zeta_{23} - \overline{\zeta_{23}}$, we can confirm the lemma. For details refer to [17].

From the lemma we immediately obtain the following proposition.

Proposition 8.3. *The effects of the initial stress* $\overset{\circ}{\boldsymbol{T}}$ *(cf. Remark* 8.3) *and the perturbative part* $\mathbb{A}$ *of the incremental elasticity tensor* $\mathbb{L}$ *on the phase velocity* v_R*, to first order of* $\overset{\circ}{\boldsymbol{T}}$ *and* $\mathbb{A}$*, come only from* $a_{22}, a_{23}, a_{33}, a_{44}$ *and* $\overset{\circ}{T}_{22}$.

This proposition allows us to reduce the case where the perturbative part $\mathbb{A}$ is generally anisotropic and the initial stress $\overset{\circ}{\boldsymbol{T}}$ is generally given by (8.65) to the orthotropic case with uniaxial stress in the propagation direction, which provides a highly simplified derivation of the perturbation formula (8.68). The argument is similar to that which is given at the end of Section 8.7 for deriving perturbation formula (8.101) in Theorem 8.7 from Proposition 8.5. For details see [17].

8.6. An Inverse Problem on Recovery of Initial Stress

Consider the body of a composite material, which occupies the half-space $x_3 \leq 0$ and, after homogenization, is transversely isotropic (with the 3-axis defined by the unit vector $\boldsymbol{e}_3$ as the ∞-axis) except for the presence of a prestress $\overset{\circ}{\boldsymbol{T}}$ given by (8.65). In this section, we will investigate what information about the prestress $\overset{\circ}{\boldsymbol{T}}$ could be inferred from boundary measurements of the phase velocities of Rayleigh waves that propagate along the free surface.

In the following we will choose the 1- and 2-axes of the Cartesian coordinate system in question arbitrarily. We assume that the constitutive equation of the composite material is of the form (cf. [29] for a different but similar setting)

$$\boldsymbol{S} = \overset{\circ}{\boldsymbol{T}} + \boldsymbol{H}\overset{\circ}{\boldsymbol{T}} + \mathbb{L}(\overset{\circ}{\boldsymbol{T}})[\boldsymbol{E}] = \overset{\circ}{\boldsymbol{T}} + \boldsymbol{H}\overset{\circ}{\boldsymbol{T}} + \mathbb{C}[\boldsymbol{E}] + \mathbb{D}[\overset{\circ}{\boldsymbol{T}}, \boldsymbol{E}], \tag{8.77}$$

where the elasticity tensor $\mathbb{C}$ and the acoustoelastic tensor $\mathbb{D}$ satisfy

$$\boldsymbol{Q}\mathbb{C}[\boldsymbol{E}]\boldsymbol{Q}^T = \mathbb{C}[\boldsymbol{Q}\boldsymbol{E}\boldsymbol{Q}^T], \quad \boldsymbol{Q}\mathbb{D}[\overset{\circ}{\boldsymbol{T}}, \boldsymbol{E}]\boldsymbol{Q}^T = \mathbb{D}[\boldsymbol{Q}\overset{\circ}{\boldsymbol{T}}\boldsymbol{Q}^T, \boldsymbol{Q}\boldsymbol{E}\boldsymbol{Q}^T], \tag{8.78}$$

respectively, for all orthogonal transformations $\boldsymbol{Q}$ that obey $\boldsymbol{Q}\boldsymbol{e}_3 = \boldsymbol{e}_3$ or $\boldsymbol{Q}\boldsymbol{e}_3 = -\boldsymbol{e}_3$. Restriction $(8.78)_1$ dictates that $\mathbb{C}$ is given by $\mathbb{C}^{\mathrm{Trans}}$ or by (8.62) with the additional conditions

$$C_{11} = C_{22}, \quad C_{13} = C_{23}, \quad C_{44} = C_{55}, \quad C_{66} = \frac{1}{2}(C_{11} - C_{12}). \tag{8.79}$$

In the following, for simplicity we will adopt the special assumption that the bilinear function $\mathbb{D}[\overset{\circ}{\boldsymbol{T}}, \boldsymbol{E}]$ is isotropic, i.e., it satisfies $(8.78)_2$ for each orthogonal transformation $\boldsymbol{Q}$. With the understanding that $\boldsymbol{E}$ and $\overset{\circ}{\boldsymbol{T}}$ are written as E_{kl} and $\overset{\circ}{T}_{mn}$, respectively, we then have [21]

$$\begin{aligned} D_{ijklmn} &= \beta_1\delta_{ij}\delta_{kl}\delta_{mn} + \frac{1}{2}\beta_2(\delta_{ik}\delta_{jl} + \delta_{il}\delta_{jk})\delta_{mn} \\ &\quad + \frac{1}{2}\beta_3((\delta_{im}\delta_{jn} + \delta_{in}\delta_{jm})\delta_{kl} + (\delta_{km}\delta_{ln} + \delta_{kn}\delta_{lm})\delta_{ij}) \\ &\quad + \frac{1}{4}\beta_4(\delta_{ik}\delta_{lm}\delta_{jn} + \delta_{il}\delta_{km}\delta_{jn} + \delta_{ik}\delta_{ln}\delta_{jm} \\ &\quad + \delta_{il}\delta_{kn}\delta_{jm} + \delta_{im}\delta_{jl}\delta_{kn} + \delta_{im}\delta_{jk}\delta_{ln} + \delta_{in}\delta_{jl}\delta_{km} \\ &\quad + \delta_{in}\delta_{jk}\delta_{lm}), \end{aligned} \tag{8.80}$$

where β_1, β_2, β_3 and β_4 are material constants. Assumption (8.80), while simplistic in the sense that it could not be expected to be adequate for a transversely isotropic composite that is strongly anisotropic, serves the purpose of illustrating all the main issues pertaining to the inverse problem in question. We refer the reader to Remark 8.6 for further comments and to [17] for a full account of the case where $\mathbb{D}$ is transversely isotropic.

Here the acoustoelastic tensor $\mathbb{D}$ plays the role of the perturbative part $\mathbb{A}$ in (8.64). It follows from (8.80) that the parameters a_{22}, a_{23}, a_{33} and a_{44} in (8.68) are given by the following formulas:

$$a_{22} = \sum_{m,n=1}^{3} D_{2222mn}\mathring{T}_{mn} = (\beta_1 + \beta_2)(\mathring{T}_{11} + \mathring{T}_{22}) + 2\beta_3\mathring{T}_{22} + 2\beta_4\mathring{T}_{22}, \tag{8.81}$$

$$a_{23} = \sum_{m,n=1}^{3} D_{2233mn}\mathring{T}_{mn} = \beta_1(\mathring{T}_{11} + \mathring{T}_{22}) + \beta_3\mathring{T}_{22}, \tag{8.82}$$

$$a_{33} = \sum_{m,n=1}^{3} D_{3333mn}\mathring{T}_{mn} = (\beta_1 + \beta_2)(\mathring{T}_{11} + \mathring{T}_{22}), \tag{8.83}$$

$$a_{44} = \sum_{m,n=1}^{3} D_{2323mn}\mathring{T}_{mn} = \frac{1}{2}\beta_2(\mathring{T}_{11} + \mathring{T}_{22}) + \frac{1}{2}\beta_4\mathring{T}_{22}. \tag{8.84}$$

Note that $\mathring{T}_{12}$ does not appear in these equations.

Let σ_1 and σ_2 be the principal stresses in the 1–2 plane, and let the 1′- and 2′-axes define the corresponding principal directions of the initial stress. Let ψ be the angle between the 1′-axis and the 2-axis. We choose the labels 1′ and 2′ such that $0 < \psi \leq \pi/2$. The non-trivial components of the initial stress are then given by

$$\begin{aligned} &\mathring{T}_{11} = \frac{1}{2}(\sigma_1 + \sigma_2) - \frac{1}{2}(\sigma_1 - \sigma_2)\cos 2\psi, \quad \mathring{T}_{12} = \frac{1}{2}(\sigma_1 - \sigma_2)\sin 2\psi, \\ &\mathring{T}_{22} = \frac{1}{2}(\sigma_1 + \sigma_2) + \frac{1}{2}(\sigma_1 - \sigma_2)\cos 2\psi. \end{aligned} \tag{8.85}$$

Substituting the expressions for $\mathring{T}_{11}$ and $\mathring{T}_{22}$ into Eqs. (8.81)–(8.84), we have

$$a_{22} = (\beta_1 + \beta_2 + \beta_3 + \beta_4)(\sigma_1 + \sigma_2) + (\beta_3 + \beta_4)(\sigma_1 - \sigma_2)\cos 2\psi, \tag{8.86}$$

$$a_{23} = \left(\beta_1 + \frac{1}{2}\beta_3\right)(\sigma_1 + \sigma_2) + \frac{1}{2}\beta_3(\sigma_1 - \sigma_2)\cos 2\psi, \tag{8.87}$$

$$a_{33} = (\beta_1 + \beta_2)(\sigma_1 + \sigma_2), \tag{8.88}$$

$$a_{44} = \left(\frac{1}{2}\beta_2 + \frac{1}{4}\beta_4\right)(\sigma_1 + \sigma_2) + \frac{1}{4}\beta_4(\sigma_1 - \sigma_2)\cos 2\psi. \tag{8.89}$$

Substituting the preceding expressions and that for $\overset{\circ}{T}_{22}$ into (8.68), we obtain for Rayleigh waves propagating in the 2-direction the formula

$$v_R = v_R^{\text{Trans}} + A(\sigma_1 + \sigma_2) + B(\sigma_1 - \sigma_2)\cos 2\psi, \tag{8.90}$$

where v_R^{Trans} is the solution of the secular equation (8.66) for the unstressed medium, and

$$A = -\frac{1}{2\rho v_R^{\text{Trans}}}\left(\gamma_{22}(\beta_1 + \beta_2 + \beta_3 + \beta_4) + \gamma_{23}\left(\beta_1 + \frac{1}{2}\beta_3\right)\right.$$
$$\left. + \gamma_{33}(\beta_1 + \beta_2) + \gamma_{44}\left(\frac{1}{2}\beta_2 + \frac{1}{4}\beta_4\right) - \frac{1}{2}\right), \tag{8.91}$$

$$B = -\frac{1}{2\rho v_R^{\text{Trans}}}\left(\gamma_{22}(\beta_3 + \beta_4) + \frac{1}{2}\beta_3\gamma_{23} + \frac{1}{4}\beta_4\gamma_{44} - \frac{1}{2}\right). \tag{8.92}$$

Now consider Rayleigh waves with propagation direction $\boldsymbol{m} = (\cos\theta, \sin\theta, 0)$ or θ for simplicity, and let $v_R(\theta)$ be the phase velocity. Let φ be the angle of rotation about the 3-axis that will bring the 1- and 2-axes to the 1′- and 2′-directions, respectively. Consider a new Cartesian coordinate system $OX''Y''$ where $\boldsymbol{m}$ is in the direction of the 2″-axis. Since $\psi = \theta - \varphi$ (see Fig. 8.3) is the angle between the 1′-axis and the 2″-axis, Eq. (8.90) dictates that $v_R(\theta)$ is given by the formula

$$v_R(\theta) = v_R^{\text{Trans}} + A(\sigma_1 + \sigma_2) + B(\sigma_1 - \sigma_2)\cos 2(\theta - \varphi). \tag{8.93}$$

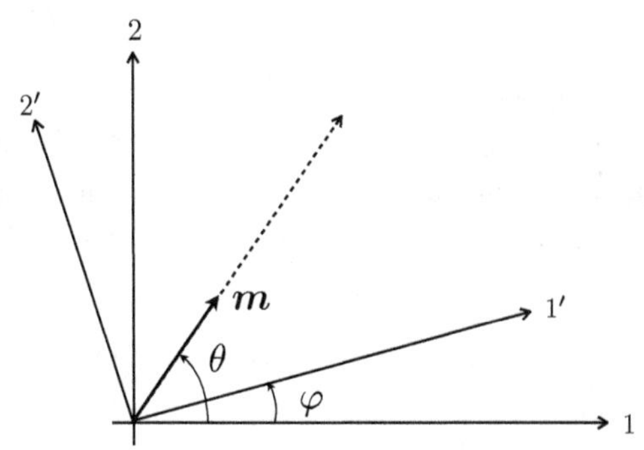

Fig. 8.3. The angle between the 1′-axis and the propagation direction is $\theta - \varphi$.

The recovery of φ (i.e., the principal directions of $\overset{\circ}{\boldsymbol{T}}$), of $\sigma_1 - \sigma_2$, and of $\sigma_1 + \sigma_2$ from boundary measurements of Rayleigh-wave phase velocities have different degrees of difficulty. We will discuss these problems in turn below:

(1) **Recovery of the principal directions of $\overset{\circ}{\boldsymbol{T}}$:** By measuring $v_R(\theta)$ for various θ, we can determine φ without any knowledge of v_R^{Trans}, A, B, σ_1, σ_2.

(2) **Recovery of $\boldsymbol{\sigma_1 - \sigma_2}$:** Measurement of $v_R(\theta)$ alone will not suffice for determination of $\sigma_1 - \sigma_2$. We will also have to ascertain the value of the acoustoelastic constant B. If it is feasible to conduct experiments in which $v_R(\theta)$ is measured for various θ while the sample is subjected to additional (known) applied stresses, then B can be determined. Another possibility is to estimate C_{22}, C_{23}, C_{33}, C_{44}, β_3, and β_4 through micromechanical modeling or some homogenization scheme and then use (8.92) to calculate B.

(3) **Recovery of $\boldsymbol{\sigma_1 + \sigma_2}$:** Suppose we can perform experiments such as those described in item (2) above. After the values of φ, B, and $\sigma_1 - \sigma_2$ are ascertained, we can determine the acoustoelastic constant A. But we still cannot recover $\sigma_1 + \sigma_2$ from measurement of $v_R(\theta)$ unless v_R^{Trans} has already been found. If the initial stress $\overset{\circ}{\boldsymbol{T}}$ is a result of the manufacturing process of the composite material, then it may not be possible to prepare unstressed samples with the same $\mathbb{C}$ as that of the composite material. In that case, getting theoretical estimates of C_{22}, C_{23}, C_{33}, and C_{44} through micromechanical modeling or homogenization would be the only way that might lead us out of this difficulty. Since we have been treating the effects of the initial stress $\overset{\circ}{\boldsymbol{T}}$ as a first-order perturbation, we have assumed *a priori* that the size of the term $A(\sigma_1 + \sigma_2)$ is at least an order of magnitude smaller than v_R^{Trans}. For the approach outlined here to work, the measured value of $v_R(\theta)$ and the theoretical estimate of v_R^{Trans} should be sufficiently accurate and the size of the acoustoelastic constant A sufficiently large so that the errors involved do not mask the contribution of $\sigma_1 + \sigma_2$ when this sum is at significant levels.

Remark 8.6. As mentioned above, for transversely isotropic composites that are strongly anisotropic, the isotropic assumption (8.80) for $\mathbb{D}$ will be inadequate. It should be replaced by its transversely isotropic counterpart, which has 16 material constants (instead of four for the isotropic case).

The procedure that leads to formula (8.90) for v_R and the formula itself, however, remain valid, although the parameters A and B are now given by somewhat more complicated formulas that involve, besides elastic coefficients from $\mathbb{C}$, seven material constants from $\mathbb{D}$ (see [17] for details). The discussion above on the recovery of the principal directions of $\overset{\circ}{\boldsymbol{T}}$, of $\sigma_1 - \sigma_2$, and of $\sigma_1 + \sigma_2$ from boundary measurements of Rayleigh-wave phase velocities requires only minor modifications that result from the fact that the formulas for A and B now involve more material constants. In fact, the sheer number of such constants might make their experimental determination arduous or even unfeasible. A potentially more practical approach is to design experiments based on formula (8.90) to measure the acoustoelastic constants A and B directly. Alternatively, one may consider developing micromechanical models or homogenization schemes by which the material parameters of the homogenized material that appear in $\mathbb{C}$ and $\mathbb{D}$ can be estimated from the structure of the composite and the mechanical properties of its constituents.

8.7. Perturbation of the Polarization Ratio of Rayleigh Waves in Prestressed Anisotropic Media When the Base Material is Orthotropic

We consider Rayleigh waves which propagate along the traction-free surface of the half-space $\mathbf{n} \cdot \mathbf{x} \leq 0$ in the direction of $\mathbf{m}$ with the phase velocity v_R. The displacement field $\boldsymbol{u}$ of such Rayleigh waves at the surface $\mathbf{n} \cdot \mathbf{x} = 0$ is written through (8.37) as

$$\boldsymbol{u} = \mathbf{a}_{\text{pol}}\, e^{-\sqrt{-1}\,k(\mathbf{m}\cdot\mathbf{x} - v_R\, t)}, \quad \mathbf{a}_{\text{pol}} = \sum_{\alpha=1}^{3} c_\alpha \mathbf{a}_\alpha, \tag{8.94}$$

where $c_\alpha \in \mathbb{C}$ ($\alpha = 1, 2, 3$) are the coefficients which satisfy (8.58) and the vectors $\mathbf{a}_\alpha \in \mathbb{C}^3$ ($\alpha = 1, 2, 3$) become the first three components of linearly independent eigenvector(s) or generalized eigenvector(s) of $\mathbf{N}(0)$ at $v = v_R$ associated with the eigenvalues of positive imaginary parts (cf. the paragraph that includes formula (8.58)). The vector $\mathbf{a}_{\text{pol}} \in \mathbb{C}^3$ is called the polarization vector of the Rayleigh waves at $\mathbf{n} \cdot \mathbf{x} = 0$.

The Stroh formalism allows us to express the polarization vector $\mathbf{a}_{\text{pol}}$ in terms of the Rayleigh-wave velocity v_R and the real matrices $\mathbf{S}_1(v)$ and $\mathbf{S}_3(v)$ in (8.49).

Proposition 8.4 ([9]). *Let $\mathbf{S}_1(v)$ and $\mathbf{S}_3(v)$ be the 3×3 real matrices in (8.49), and let $\mathbf{e}_1$ and $\mathbf{e}_2$ be orthogonal unit vectors in $\mathbb{R}^3$ such that*

$$\mathbf{S}_3(v_R)\mathbf{e}_1 \neq \mathbf{0}, \quad \big(\mathbf{S}_3(v_R)\mathbf{e}_1\big) \times \mathbf{e}_2 \neq \mathbf{0}, \tag{8.95}$$

where the symbol $\times$ denotes the vector product. Then the Rayleigh waves have at the surface $\mathbf{n} \cdot \boldsymbol{x} = 0$ the polarization vector

$$\mathbf{a}_{\text{pol}} = \big(\mathbf{S}_3(v_R)\mathbf{e}_1\big) \times \mathbf{e}_2 - \sqrt{-1}\, \mathbf{S}_1(v_R)[\big(\mathbf{S}_3(v_R)\mathbf{e}_1\big) \times \mathbf{e}_2]. \tag{8.96}$$

The real part $\mathbf{a}^+$ and the imaginary part $\mathbf{a}^-$ of $\mathbf{a}_{\text{pol}}$ define the plane to which the paths of surface particles are confined, so that their displacements are expressed by taking the real part of (8.94), i.e.,

$$\mathbf{a}^+ \cos k(\mathbf{m} \cdot \boldsymbol{x} - v_R t) + \mathbf{a}^- \sin k(\mathbf{m} \cdot \boldsymbol{x} - v_R t). \tag{8.97}$$

It is proved (see [9, Sec. 7.B]) that $\mathbf{a}^+$ and $\mathbf{a}^-$ are linearly independent.

Let us turn to the setting in Section 8.5; the medium has the incremental elasticity tensor $\mathbb{L}$ (8.64), carries the initial stress $\mathring{\boldsymbol{T}}$ (8.65), and the constitutive equation is given by (8.2). We consider Rayleigh waves which propagate along the traction-free surface of the half-space $x_3 \leq 0$ in the direction of the 2-axis with the phase velocity v_R. Using the component-wise expression

$$\mathbf{a}^+ = \begin{bmatrix} a_1^+ \\ a_2^+ \\ a_3^+ \end{bmatrix} \quad \text{and} \quad \mathbf{a}^- = \begin{bmatrix} a_1^- \\ a_2^- \\ a_3^- \end{bmatrix},$$

we rewrite (8.97) as

$$\begin{bmatrix} a_1^+ \\ a_2^+ \\ a_3^+ \end{bmatrix} \cos k(x_2 - v_R t) + \begin{bmatrix} a_1^- \\ a_2^- \\ a_3^- \end{bmatrix} \sin k(x_2 - v_R t). \tag{8.98}$$

Since the 2-axis agrees with the propagation direction, the longitudinal component of (8.98) is

$$\begin{aligned} &a_2^+ \cos k(x_2 - v_R t) + a_2^- \sin k(x_2 - v_R t) \\ &= \sqrt{(a_2^+)^2 + (a_2^-)^2} \sin\big(k(x_2 - v_R t) + \alpha\big), \end{aligned}$$

and since Rayleigh waves propagate along the surface $x_3 = 0$, the normal component of (8.98) is

$$a_3^+ \cos k(x_2 - v_R t) + a_3^- \sin k(x_2 - v_R t)$$
$$= \sqrt{(a_3^+)^2 + (a_3^-)^2} \sin (k(x_2 - v_R t) + \beta),$$

where α and β are the angles determined by

$$\tan\alpha = \frac{a_2^+}{a_2^-}, \quad \tan\beta = \frac{a_3^+}{a_3^-}, \quad \left(-\frac{\pi}{2} \le \alpha, \beta \le \frac{\pi}{2}\right).$$

The polarization ratio r_R of the Rayleigh waves on the surface $x_3 = 0$ is defined by the ratio of the maximum longitudinal component to the maximum normal component of the displacements on $x_3 = 0$:

$$r_R = \frac{\sqrt{(a_2^+)^2 + (a_2^-)^2}}{\sqrt{(a_3^+)^2 + (a_3^-)^2}}. \tag{8.99}$$

In the comparative orthotropic and unstressed medium defined by $\mathbb{L}[\boldsymbol{E}] = \mathbb{C}^{\text{Orth}}[\boldsymbol{E}]$, $\mathbb{A}[\boldsymbol{E}] = \boldsymbol{0}$ and $\overset{\circ}{\boldsymbol{T}} = \boldsymbol{0}$, the polarization ratio is given by

$$r_R^{\text{Orth}} = \sqrt{\frac{C_{33}V_R^{\text{Orth}}}{C_{22}C_{33} - C_{23}^2 - C_{33}V_R^{\text{Orth}}}}, \tag{8.100}$$

where $V_R^{\text{Orth}} = \rho\left(v_R^{\text{Orth}}\right)^2$ and v_R^{Orth} satisfies (8.66) (cf. [27])[n].

Our purpose in this section is to derive a perturbation formula that shows how $\mathbb{A}$ in (8.64) and $\overset{\circ}{\boldsymbol{T}}$ of the term $\boldsymbol{H}\overset{\circ}{\boldsymbol{T}}$ in (8.2) affect the polarization ratio of Rayleigh waves from its value r_R^{Orth} for the comparative orthotropic and unstressed base material (cf. Fig. 8.4).

[n]The fractional expression under the square-root sign on the right-hand side of (8.100) is positive. In fact, the strong convexity condition for $\mathbb{C}^{\text{Orth}}$ implies that $C_{22}, C_{33}, C_{44} > 0$, $C_{22}C_{33} - C_{23}^2 > 0$, which leads to $R^{\text{Orth}}(0) = -C_{44}\left(C_{22}C_{33} - C_{23}^2\right)^2 < 0$ (see (8.67)). Suppose that there exists a velocity $\tilde{v}$ $(0 < \tilde{v} \le v_R^{\text{Orth}})$ such that $C_{22}C_{33} - C_{23}^2 - C_{33}\widetilde{V} = 0$, where $\widetilde{V} = \rho(\tilde{v})^2$. Since (8.71) implies that $\widetilde{V} \le V_R^{\text{Orth}} < \rho\left(v_L^{\text{Orth}}\right)^2 \le C_{22}$, it then follows that $R^{\text{Orth}}(\tilde{v}) > 0$. Hence there exists $\hat{v}$ $(0 < \hat{v} < \tilde{v})$ such that $R^{\text{Orth}}(\hat{v}) = 0$, which contradicts Corollary 8.3. Therefore, $C_{22}C_{33} - C_{23}^2 - C_{33}V$ decreases monotonically with $V \ge 0$ from its positive value at $V = 0$, but remains positive as long as V moves between 0 and V_R^{Orth}.

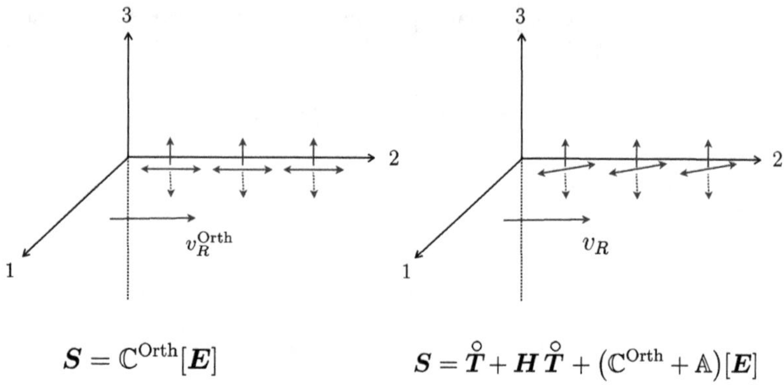

Fig. 8.4. Change of the polarization vector due to the effects of $\mathbb{A}$ and the initial stress.

Theorem 8.7. *In a prestressed medium whose incremental elasticity tensor* $\mathbb{L}$ *and initial stress* $\overset{\circ}{\boldsymbol{T}}$ *are given by* (8.64) *and* (8.65), *respectively, the polarization ratio of Rayleigh waves which propagate along the surface of the half-space* $x_3 \leq 0$ *in the direction of the* 2*-axis can be written, to within terms linear in the perturbative part* $\mathbb{A}$ *of* $\mathbb{L}$ *and the initial stress* $\overset{\circ}{\boldsymbol{T}}$, *as*

$$r_R = r_R^{\text{Orth}} + \eta_{22}(v_R^{\text{Orth}})\, a_{22} + \eta_{23}(v_R^{\text{Orth}})\, a_{23} + \eta_{33}(v_R^{\text{Orth}})\, a_{33} + \eta_{44}(v_R^{\text{Orth}})\, a_{44}, \tag{8.101}$$

where

$$\begin{aligned}
\eta_{22}(v) &= \frac{-C_{33}C_{44}\sqrt{C_{44}-V}}{2E(v)}[(C_{22}C_{33}-C_{23}^2)^2 - 2(C_{22}C_{33}-C_{23}^2)C_{33}V \\
&\quad + (C_{33}^2 - C_{23}^2)V^2], \\
\eta_{23}(v) &= \frac{C_{23}\sqrt{C_{44}-V}}{E(v)}[(C_{22}C_{33}-C_{23}^2)^2C_{44} - 2(C_{22}C_{33}-C_{23}^2)C_{33}C_{44}V \\
&\quad + (C_{33}-C_{22})C_{33}C_{44}V^2], \\
\eta_{33}(v) &= \frac{\sqrt{C_{44}-V}}{2C_{33}E(v)}[2C_{22}(C_{22}C_{33}-C_{23}^2)^2C_{33}C_{44} - (C_{22}C_{33}-C_{23}^2) \\
&\quad \times \big((C_{22}C_{33}-C_{23}^2)(2C_{22}C_{33}+C_{23}^2) + 4C_{22}C_{33}^2C_{44}\big)V \\
&\quad + C_{33}\big(2(C_{22}C_{33}-C_{23}^2)(2C_{22}C_{33}+C_{23}^2) \\
&\quad - C_{22}(C_{22}C_{33}+C_{23}^2-2C_{33}^2)C_{44}\big)V^2 \\
&\quad - C_{33}\big(C_{22}C_{33}(2C_{33}-C_{44}) + C_{23}^2(C_{33}-2C_{44})\big)V^3],
\end{aligned}$$

$$\eta_{44}(v) = \frac{C_{22}C_{33} - C_{23}^2}{2C_{44}D(v)}\sqrt{C_{33}(C_{22}C_{33} - C_{23}^2 - C_{33}V)V},$$

$$E(v) = \sqrt{C_{44}(C_{22} - V)(C_{22}C_{33} - C_{23}^2 - C_{33}V)VD(v)V},$$

and $D(v)$ *is given in* (8.69) *and* $V = \rho v^2$.

Remark 8.7. Only four components a_{22}, a_{23}, a_{33} and a_{44} of the perturbative part $\mathbb{A}$ of $\mathbb{L}$ can affect the first-order perturbation of the polarization ratio r_R. To first order, the $\boldsymbol{H}\overset{\circ}{\boldsymbol{T}}$ term in (8.2) has no perturbative effect on r_R.

Proof. First we make use of (8.96) to obtain an expression of $\mathbf{a}_{\mathrm{pol}}$ which is correct up to terms linear in the perturbative part $\mathbb{A}$ of $\mathbb{L}$ and the initial stress $\overset{\circ}{\boldsymbol{T}}$. In a parallel way in which we have derived (8.74), we get

$$\mathbf{S}_3 \approx \mathbf{S}_3^{\mathrm{Orth}} + \mathbf{S}_3^{\mathrm{Ptb}}. \tag{8.102}$$

Here

$$\mathbf{S}_3^{\mathrm{Orth}} = \mathbf{S}_3^{\mathrm{Orth}}(v) = \frac{1}{2\pi}\int_{-\pi}^{\pi}(-\mathbf{Q}_v^{\mathrm{Orth}}(\phi) + \mathbf{R}_v^{\mathrm{Orth}}(\phi)\mathbf{T}_v^{\mathrm{Orth}}(\phi)^{-1}\mathbf{R}_v^{\mathrm{Orth}}(\phi)^T)\,d\phi$$

is of zeroth order in $\overset{\circ}{\boldsymbol{T}}$ and $\mathbb{A}$, where

$$\mathbf{Q}_v^{\mathrm{Orth}}(\phi) = \left(\sum_{j,l=1}^{3} C_{ijkl}^{\mathrm{Orth}}\widetilde{m}_j\widetilde{m}_l\right)_{i\downarrow k\to 1,2,3} - \rho v^2\cos^2\phi\,\mathbf{I},$$

and

$$\begin{aligned}\mathbf{S}_3^{\mathrm{Ptb}} = \mathbf{S}_3^{\mathrm{Ptb}}(v) = \frac{1}{2\pi}\int_{-\pi}^{\pi}&(-\mathbf{Q}^{\mathrm{Ptb}}(\phi) + \mathbf{R}^{\mathrm{Ptb}}(\phi)\mathbf{T}_v^{\mathrm{Orth}}(\phi)^{-1}\mathbf{R}_v^{\mathrm{Orth}}(\phi)^T \\ &+ \mathbf{R}_v^{\mathrm{Orth}}(\phi)\mathbf{T}_v^{\mathrm{Orth}}(\phi)^{-1}\mathbf{R}^{\mathrm{Ptb}}(\phi)^T \\ &- \mathbf{R}_v^{\mathrm{Orth}}(\phi)\mathbf{T}_v^{\mathrm{Orth}}(\phi)^{-1}\mathbf{T}^{\mathrm{Ptb}}(\phi)\mathbf{T}_v^{\mathrm{Orth}}(\phi)^{-1}\mathbf{R}_v^{\mathrm{Orth}}(\phi)^T)\,d\phi\end{aligned}$$

is of first order in $\overset{\circ}{\boldsymbol{T}}$ and $\mathbb{A}$, where

$$\mathbf{Q}^{\mathrm{Ptb}}(\phi) = \sum_{j,l=1}^{3}\overset{\circ}{T}_{jl}\widetilde{m}_j\widetilde{m}_l\,\mathbf{I} + \left(\sum_{j,l=1}^{3} a_{ijkl}\widetilde{m}_j\widetilde{m}_l\right)_{i\downarrow k\to 1,2,3},$$

and the matrices $\mathbf{R}_v^{\mathrm{Orth}}(\phi)$, $\mathbf{T}_v^{\mathrm{Orth}}(\phi)$, $\mathbf{R}^{\mathrm{Ptb}}(\phi)$, and $\mathbf{T}^{\mathrm{Ptb}}(\phi)$ have been defined in the proof of Lemma 8.5.

The Rayleigh waves in question propagate along the surface of the half-space $x_3 \leq 0$ in the direction of the 2-axis. Hence we take

$$\mathbf{m} = (0, 1, 0) \quad \text{and} \quad \mathbf{n} = (0, 0, 1).$$

Then Theorem 8.4 and formula (8.72) imply that

$$\left(\mathbf{S}_2^{\text{Orth}}\right)^{-1} = \begin{pmatrix} Z_{11} & 0 & 0 \\ 0 & Z_{22} & 0 \\ 0 & 0 & Z_{33} \end{pmatrix}, \quad \mathbf{S}_1^{\text{Orth}} = \begin{pmatrix} 0 & 0 & 0 \\ 0 & 0 & T_{23} \\ 0 & T_{32} & 0 \end{pmatrix}, \tag{8.103}$$

where

$$T_{23} = T_{23}(v) = \frac{-1}{\sqrt{C_{22} - V}} \frac{J}{H},$$

$$T_{32} = T_{32}(v) = \sqrt{\frac{C_{44}}{C_{33}(C_{44} - V)}} \frac{J}{H}, \quad V = \rho v^2,$$

which, together with the first identity of (8.56), gives

$$\mathbf{S}_3^{\text{Orth}} = -\left(\mathbf{S}_2^{\text{Orth}}\right)^{-1}\left(\mathbf{I} + \left(\mathbf{S}_1^{\text{Orth}}\right)^2\right) = \begin{pmatrix} S_{11} & 0 & 0 \\ 0 & S_{22} & 0 \\ 0 & 0 & S_{33} \end{pmatrix}, \tag{8.104}$$

where

$$S_{11} = S_{11}(v) = -\sqrt{C_{55}(C_{66} - V)},$$

$$S_{22} = S_{22}(v) = \frac{\sqrt{C_{44}}}{\sqrt{C_{33}(C_{44} - V)}\, GH} \times \left(\sqrt{C_{44}}\, J^2 - \sqrt{C_{33}(C_{22} - V)(C_{44} - V)} H^2\right),$$

$$S_{33} = S_{33}(v) = \frac{\sqrt{C_{33}(C_{44} - V)}}{\sqrt{C_{44}(C_{22} - V)}} S_{22}, \quad V = \rho v^2.$$

Let $t_{ij} = t_{ij}(v)$ and $s_{ij} = s_{ij}(v)$ be the (i, j) components of the matrices $\mathbf{S}_1^{\text{Ptb}}(v)$ and $\mathbf{S}_3^{\text{Ptb}}(v)$, respectively. They are linear functions of $\mathbb{A}$ and $\mathring{\boldsymbol{T}}$. We see from (8.74) and the second formula in (8.103) that

$$\mathbf{S}_1(v) \approx \mathbf{S}_1^{\text{Orth}}(v) + \mathbf{S}_1^{\text{Ptb}}(v) = \begin{pmatrix} t_{11} & t_{12} & t_{13} \\ t_{21} & t_{22} & T_{23} + t_{23} \\ t_{31} & T_{32} + t_{32} & t_{33} \end{pmatrix}, \tag{8.105}$$

and from (8.102) and (8.104) that

$$\mathbf{S}_3(v) \approx \mathbf{S}_3^{\mathrm{Orth}}(v) + \mathbf{S}_3^{\mathrm{Ptb}}(v) = \begin{pmatrix} S_{11}+s_{11} & s_{12} & s_{13} \\ s_{12} & S_{22}+s_{22} & s_{23} \\ s_{13} & s_{23} & S_{33}+s_{33} \end{pmatrix}. \tag{8.106}$$

Let us put $\mathbf{e}_1 = (1,0,0)$ and $\mathbf{e}_2 = (0,1,0)$ in the formula (8.96). From (8.106) it follows that

$$\mathbf{S}_3(v)\mathbf{e}_1 \approx (\mathbf{S}_3^{\mathrm{Orth}} + \mathbf{S}_3^{\mathrm{Ptb}})(v)\mathbf{e}_1 = (S_{11}+s_{11}, s_{12}, s_{13})^T$$

and

$$(\mathbf{S}_3(v)\mathbf{e}_1) \times \mathbf{e}_2 \approx \big((\mathbf{S}_3^{\mathrm{Orth}} + \mathbf{S}_3^{\mathrm{Ptb}})(v)\mathbf{e}_1\big) \times \mathbf{e}_2 = (-s_{13}, 0, S_{11}+s_{11})^T.$$

Since $\mathbb{A}$ and $\overset{\circ}{\boldsymbol{T}}$ are sufficiently small, v_R is close to v_R^{Orth} and $s_{ij}(v_R)$ ($1 \le i \le j \le 3$) are sufficiently small. From (8.71) and $v_R^{\mathrm{Orth}} < v_L^{\mathrm{Orth}}$ it follows that $S_{11}(v_R^{\mathrm{Orth}}) < 0$ and $S_{11}(v_R) < 0$. Hence the preceding two vectors do not vanish at $v = v_R$, which guarantees the assumptions (8.95). Thus, by (8.105) and (8.106), the formula (8.96) can be written, to within terms linear in $\mathbb{A}$ and $\overset{\circ}{\boldsymbol{T}}$, as

$$\begin{aligned}
\mathbf{a}_{\mathrm{pol}} &= (\mathbf{I} - \sqrt{-1}\,\mathbf{S}_1(v_R))\,[(\mathbf{S}_3(v_R)\mathbf{e}_1) \times \mathbf{e}_2] \\
&\approx \Big(\mathbf{I} - \sqrt{-1}(\mathbf{S}_1^{\mathrm{Orth}}(v_R) + \mathbf{S}_1^{\mathrm{Ptb}}(v_R))\Big)\,[((\mathbf{S}_3^{\mathrm{Orth}}(v_R) + \mathbf{S}_3^{\mathrm{Ptb}}(v_R))\mathbf{e}_1) \times \mathbf{e}_2] \\
&= \begin{pmatrix} 1-\sqrt{-1}\,t_{11} & -\sqrt{-1}\,t_{12} & -\sqrt{-1}\,t_{13} \\ -\sqrt{-1}\,t_{21} & 1-\sqrt{-1}\,t_{22} & -\sqrt{-1}\,(T_{23}+t_{23}) \\ -\sqrt{-1}\,t_{31} & -\sqrt{-1}\,(T_{32}+t_{32}) & 1-\sqrt{-1}\,t_{33} \end{pmatrix}_{v=v_R} \\
&\quad \times \begin{pmatrix} -s_{13} \\ 0 \\ S_{11}+s_{11} \end{pmatrix}_{v=v_R} \\
&\approx \begin{pmatrix} -s_{13} \\ 0 \\ S_{11}+s_{11} \end{pmatrix}_{v=v_R} + \sqrt{-1}\begin{pmatrix} -S_{11}\,t_{13} \\ -T_{23}\,(S_{11}+s_{11}) - S_{11}\,t_{23} \\ -S_{11}\,t_{33} \end{pmatrix}_{v=v_R}.
\end{aligned}$$

Hence a real form of the displacements on the surface $x_3 = 0$ of the Rayleigh waves is given from (8.97), to first order in $\mathbb{A}$ and $\mathring{\boldsymbol{T}}$, as

$$\begin{pmatrix} -s_{13}(v_R) \\ 0 \\ S_{11}(v_R) + s_{11}(v_R) \end{pmatrix} \cos k(x_2 - v_R t) + \begin{pmatrix} -S_{11}(v_R)\, t_{13}(v_R) \\ -T_{23}(v_R)\,(S_{11} + s_{11})\,(v_R) - S_{11}(v_R)\, t_{23}(v_R) \\ -S_{11}(v_R)\, t_{33}(v_R) \end{pmatrix} \sin k(x_2 - v_R t). \tag{8.107}$$

From (8.98), (8.99) and (8.107) it follows that

$$r_R \approx \frac{|T_{23}(v_R)\,(S_{11} + s_{11})\,(v_R) + S_{11}(v_R)\, t_{23}(v_R)|}{|S_{11}(v_R) + s_{11}(v_R)|}. \tag{8.108}$$

We recall that $S_{11}(v_R) < 0$ when $\mathbb{A}$ and $\mathring{\boldsymbol{T}}$ are sufficiently small. On the other hand, from (8.71) and $v_R^{\text{Orth}} < v_L^{\text{Orth}}$ it follows that $H(V_R^{\text{Orth}}) > 0$, where $V_R^{\text{Orth}} = \rho\,(v_R^{\text{Orth}})^2$. Moreover, $\sqrt{C_{33}(C_{22} - V_R^{\text{Orth}})} > C_{23}$ (see footnote "n") and $\sqrt{C_{44}} > \sqrt{C_{44} - V_R^{\text{Orth}}}$ imply that $J(V_R^{\text{Orth}}) > 0$. Hence $|T_{23}(v_R^{\text{Orth}})|$ is positive.[o] Thus we obtain from (8.108)

$$r_R \approx \left| T_{23}(v_R) + \frac{S_{11}(v_R)\, t_{23}(v_R)}{S_{11}(v_R) + s_{11}(v_R)} \right| \approx |T_{23}(v_R) + t_{23}(v_R)|. \tag{8.109}$$

Recalling that $t_{23}(v)$ is a linear function of $\mathbb{A}$ and $\mathring{\boldsymbol{T}}$, we get

$$t_{23}(v_R) \approx t_{23}(v_R^{\text{Orth}}). \tag{8.110}$$

Moreover, it follows from the Taylor expansion of $T_{23}(v)$ around $v = v_R^{\text{Orth}}$ that

$$T_{23}(v_R) \approx T_{23}(v_R^{\text{Orth}}) + T_{23}'(v_R^{\text{Orth}})\,(v_R - v_R^{\text{Orth}}). \tag{8.111}$$

Therefore, we obtain

$$r_R \approx \left| T_{23}(v_R^{\text{Orth}}) + T_{23}'(v_R^{\text{Orth}})\,(v_R - v_R^{\text{Orth}}) + t_{23}(v_R^{\text{Orth}}) \right|. \tag{8.112}$$

---

[o]Using (8.66), we easily check that $|T_{23}(v_R^{\text{Orth}})| = r_R^{\text{Orth}}$.

Proposition 8.5. *The effects of the initial stress* $\overset{\circ}{\boldsymbol{T}}$ *and the perturbative part* $\mathbb{A}$ *of the incremental elasticity tensor* $\mathbb{L}$ *on the polarization ratio* r_R, *to first order of* $\overset{\circ}{\boldsymbol{T}}$ *and* $\mathbb{A}$, *come from* $a_{22}, a_{23}, a_{33}, a_{44}$ *and* $\overset{\circ}{T}_{22}$.

Proof. Theorem 8.6 implies that the effects of $\overset{\circ}{\boldsymbol{T}}$ and $\mathbb{A}$ on $v_R - v_R^{\text{Orth}}$, to first order of $\overset{\circ}{\boldsymbol{T}}$ and $\mathbb{A}$, come from $a_{22}, a_{23}, a_{33}, a_{44}$ and $\overset{\circ}{T}_{22}$. Also, by the argument that follows (55) of [17], the effects of $\overset{\circ}{\boldsymbol{T}}$ and $\mathbb{A}$ on $t_{23}(v)$, to first order of $\overset{\circ}{\boldsymbol{T}}$ and $\mathbb{A}$, come from $a_{22}, a_{23}, a_{33}, a_{44}$ and $\overset{\circ}{T}_{22}$. □

This proposition allows us to reduce the case where the perturbative part $\mathbb{A}$ is generally anisotropic and the initial stress $\overset{\circ}{\boldsymbol{T}}$ is generally given by (8.65) to the orthotropic case with uniaxial stress in the propagation direction, which provides a highly efficient derivation of the perturbation formula (8.101).

Indeed, consider another prestressed elastic half-space $x_3 \leq 0$ (II) with incremental elasticity tensor

$$\mathbb{L}^{\text{Orth}}[\boldsymbol{E}] = \mathbb{C}^{\text{Orth}}[\boldsymbol{E}] + \mathbb{A}^{\text{Orth}}[\boldsymbol{E}], \tag{8.113}$$

where the principal part $\mathbb{C}^{\text{Orth}}$ is exactly the same as that of the given prestressed medium (I), and the perturbative part $\mathbb{A}^{\text{Orth}}$ is orthotropic with the coordinate planes $x_i = 0$ $(i = 1, 2, 3)$ as the planes of reflection symmetry and has its non-trivial components exactly the same as the corresponding ones in the perturbative part $\mathbb{A}$ of the given medium (I). Clearly $\mathbb{L}^{\text{Orth}}$ is orthotropic and has the coordinate planes $x_i = 0$ $(i = 1, 2, 3)$ as the planes of reflection symmetry. Furthermore, we assume that the initial stress in the half-space (II) has the form $\overset{\circ}{\boldsymbol{T}} = \text{diag}(0, \overset{\circ}{T}_{22}, 0)$ and consider Rayleigh waves that propagate along its free surface. The following proposition is a simple extension of [27] (see also [19, Proposition 6.4]).

Proposition 8.6. *The polarization ratio of Rayleigh waves which propagate in the direction of the* 2*-axis along the surface of the prestressed orthorhombic half-space* (II) *is given by*

$$r_R^{\mathbb{L}\text{-Orth}} = \sqrt{\frac{L_{33}(V_R^{\mathbb{L}\text{-Orth}} - \overset{\circ}{T}_{22})}{L_{22}L_{33} - L_{23}^2 - L_{33}(V_R^{\mathbb{L}\text{-Orth}} - \overset{\circ}{T}_{22})}}, \tag{8.114}$$

where $V_R^{\mathbb{L}\text{-Orth}} = \rho(v_R^{\mathbb{L}\text{-Orth}})^2$. *Here* $v_R^{\mathbb{L}\text{-Orth}}$, *the phase velocity of the Rayleigh waves described above, satisfies the secular equation*

$$R^{\mathbb{L}\text{-Orth}}(v) = 0, \tag{8.115}$$

where

$$R^{\mathbb{L}\text{-Orth}}(v) = L_{33}L_{44}(L_{22} - (V - \mathring{T}_{22}))(V - \mathring{T}_{22})^2 \\ - (L_{44} - (V - \mathring{T}_{22}))(L_{33}(L_{22} - (V - \mathring{T}_{22})) - L_{23}^2)^2 \tag{8.116}$$

and $V = \rho v^2$.

The components of the perturbative part $\mathbb{A}^{\text{Orth}}$ and of the initial stress $\mathring{\boldsymbol{T}}$ included in $r_R^{\mathbb{L}\text{-Orth}}$ are, through (8.113), $a_{22}, a_{23}, a_{33}, a_{44}$ and $\mathring{T}_{22}$. By Proposition 8.5, these are exactly the same components that affect the first-order perturbation of the polarization ratio r_R of the given prestressed medium (I). Hence it should also be possible to obtain from (8.114) the perturbation formula for the polarization ratio of the Rayleigh waves in question, i.e., Rayleigh waves which propagate in the direction of the 2-axis on the surface of the given prestressed elastic half-space $x_3 \leq 0$ (I) whose incremental elasticity tensor and the initial stress have the forms (8.64) and (8.65), respectively.

In fact, observing that

$$r_R^{\mathbb{L}\text{-Orth}}\Big|_{\mathbb{A}^{\text{Orth}}=\mathbf{0},\mathring{\boldsymbol{T}}=\mathbf{0}} = r_R^{\text{Orth}},$$

from the Taylor expansion of r_R around $(\mathbb{A}^{\text{Orth}}, \mathring{\boldsymbol{T}}) = (\mathbf{0}, \mathbf{0})$ we get

$$r_R \approx r_R^{\text{Orth}} + \sum_{r,s} \frac{\partial r_R^{\mathbb{L}\text{-Orth}}}{\partial a_{rs}}\Big|_{\mathbb{A}^{\text{Orth}}=\mathbf{0},\mathring{\boldsymbol{T}}=\mathbf{0}} a_{rs} + \frac{\partial r_R^{\mathbb{L}\text{-Orth}}}{\partial \mathring{T}_{22}}\Big|_{\mathbb{A}^{\text{Orth}}=\mathbf{0},\mathring{\boldsymbol{T}}=\mathbf{0}} \mathring{T}_{22}, \tag{8.117}$$

where the summation $\sum_{r,s}$ on the right-hand side is taken for the indices $(r, s) = (2, 2), (2, 3), (3, 3)$ and $(4, 4)$[p].

[p] $r_R^{\mathbb{L}\text{-Orth}}$ is independent of $\mathring{T}_{22}$, because (8.115) and (8.116) imply that $V_R^{\mathbb{L}\text{-Orth}} - \mathring{T}_{22}$ is independent of $\mathring{T}_{22}$, and by (8.114), so is $r_R^{\mathbb{L}\text{-Orth}}$. Hence the last term on the right-hand side of (8.117) vanishes.

We see from (8.64) that

$$\left.\frac{\partial r_R^{\mathbb{L}\text{-Orth}}}{\partial a_{rs}}\right|_{\mathbb{A}^{\text{Orth}}=\mathbf{0},\mathring{\boldsymbol{T}}=\mathbf{0}} = \left.\frac{\partial r_R^{\mathbb{L}\text{-Orth}}}{\partial L_{rs}}\right|_{\mathbb{A}^{\text{Orth}}=\mathbf{0},\mathring{\boldsymbol{T}}=\mathbf{0}} + \left.\frac{\partial r_R^{\mathbb{L}\text{-Orth}}}{\partial V_R^{\mathbb{L}\text{-Orth}}}\frac{\partial V_R^{\mathbb{L}\text{-Orth}}}{\partial a_{rs}}\right|_{\mathbb{A}^{\text{Orth}}=\mathbf{0},\mathring{\boldsymbol{T}}=\mathbf{0}}. \tag{8.118}$$

On the other hand, it follows that

$$\left.\frac{\partial V_R^{\mathbb{L}\text{-Orth}}}{\partial a_{rs}}\right|_{\mathbb{A}^{\text{Orth}}=\mathbf{0},\mathring{\boldsymbol{T}}=\mathbf{0}} = -\gamma_{rs}, \quad (r,s) = (2,2),(2,3),(3,3) \quad \text{and} \quad (4,4),$$

where γ_{rs} are the coefficients of the perturbation formula for the phase velocity given by (8.69). Since $r_R^{\mathbb{L}\text{-Orth}}$ is given by (8.114) explicitly, we can calculate directly its derivatives in (8.118). Therefore, from (8.117) and (8.118) we obtain Theorem 8.7. □

Remark 8.8. A comparison of the formulas (8.68) in Theorem 8.6 and (8.101) in Theorem 8.7 reveals that in the context of Section 8.6, in parallel to (8.93) for angular dependence of Rayleigh-wave velocity, we have the formula

$$r_R(\theta) = r_R^{\text{Trans}} + C(\sigma_1+\sigma_2) + D(\sigma_1-\sigma_2)\cos 2(\theta-\varphi) \tag{8.119}$$

for angular dependence of polarization ratio, where

$$C = \eta_{22}(\beta_1+\beta_2+\beta_3+\beta_4) + \eta_{23}\left(\beta_1+\frac{1}{2}\beta_3\right) + \eta_{33}(\beta_1+\beta_2) + \eta_{44}\left(\frac{1}{2}\beta_2+\frac{1}{4}\beta_4\right), \tag{8.120}$$

$$D = \eta_{22}(\beta_3+\beta_4) + \frac{1}{2}\beta_3\eta_{23} + \frac{1}{4}\beta_4\eta_{44}. \tag{8.121}$$

Formula (8.119) can be used instead of (8.93) in non-destructive evaluation of stress. In fact, doing so may be advantageous for materials where r_R is more sensitive than v_R to changes in initial stress (cf. the discussion at the end of the introduction). On the other hand, (8.119) and (8.93) are formulas with the same structure. The discussion and caveats given in Section 8.6 on using (8.93) for recovery of initial stress (see the paragraph that precedes Remark 8.6) applies almost verbatim to using (8.119) for that purpose.

Acknowledgments

We are grateful to Prof Vladislav Mantič, who kindly invited us to submit the first version of this chapter to the first edition of the book and who encouraged us to revise the chapter for the second edition. He also gave us many helpful suggestions about the literature on composite materials. We also thank the reviewers who read our manuscript carefully and gave us many comments. The work of Tanuma was partly supported by Grant-in-Aid for Scientific Research (C) (Nos. 22540111 & 26400157 & 19K03559), Society for the Promotion of Science, Japan. The research efforts of Man were supported in part by a grant (No. DMS-0807543) from the US National Science Foundation.

References

[1] Rose J.L., 1999. *Ultrasonic Waves in Solid Media*, Cambridge University Press, Cambridge.

[2] Datta S.K., 2000. Wave propagation in composite plates and shells, in *Comprehensive Composite Materials*, Vol. 1, A. Kelly, C. Zweben, (eds.), Elsevier, Amsterdam, pp. 511–558.

[3] Ostoja-Starzewski M., 2008. *Microstructural Randomness and Scaling in Mechanics of Materials*, Chapman & Hall/CRC, Taylor & Francis Group.

[4] Datta S.K., Shah A.H., 2009. *Elastic Waves in Composite Media and Structures: With Applications to Ultrasonic Nondestructive Evaluation*, CRC Press, Taylor & Francis Group.

[5] Datta S.K., Ledbetter H.M., 1983. Elastic constants of fiber-reinforced boron-aluminum: Observation and theory, *Int. J. Solids Struct.*, 19, 885–894.

[6] Stroh A.N., 1958. Dislocations and cracks in anisotropic elasticity, *Philos. Mag.*, 3, 625–646.

[7] Stroh A.N., 1962. Steady state problems in anisotropic elasticity, *J. Math. Phys.*, 41, 77–103.

[8] Lothe J., Barnett D.M., 1976. On the existence of surface-wave solutions for anisotropic elastic half-spaces with free surface, *J. Appl. Phys.*, 47, 428–433.

[9] Chadwick P., Smith G.D., 1977. Foundations of the theory of surface waves in anisotropic elastic materials, *Adv. Appl. Mech.*, 17, 303–376.

[10] Barnett D.M., Lothe J., 1985. Free surface (Rayleigh) waves in anisotropic elastic half-spaces: the surface impedance method, *Proc. R. Soc. Lond. A*, 402, 135–152.

[11] Ting T.C.T., 1996. *Anisotropic Elasticity*, Oxford University Press, New York.

[12] Tanuma K., 2007. Stroh formalism and Rayleigh waves, *J. Elasticity*, 89, 5–154.

[13] Tanuma K., Man C.-S., 2006. Perturbation formula for phase velocity of Rayleigh waves in prestressed anisotropic media, *J. Elasticity*, 85, 21–37.

[14] Wu T.-T., Chai J.-F., 1994. Propagation of surface waves in anisotropic solids: Theoretical calculation and experiment, *Ultrasonics*, 32, 21–29.

[15] Nayfeh A.H., 1995. *Wave Propagation in Layered Anisotropic Media with Applications to Composites*, North-Holland, Amsterdam.

[16] Wu T.-T., Liu Y.-H., 1999. On the measurement of anisotropic elastic constants of fiber-reinforced composite plate using ultrasonic bulk wave and laser generated Lamb wave, *Ultrasonics*, 37, 405–412.

[17] Tanuma K., Man C.-S., Du W., 2013. Perturbation of phase velocity of Rayleigh waves in pre-stressed anisotropic media with orthorhombic principal part, *Math. Mech. Solids*, 18, 301–322.

[18] Junge M., Qu J., Jacobs L.J., 2006. Relationship between Rayleigh wave polarization and state of stress, *Ultrasonics*, 44, 233–237.

[19] Tanuma K., Man C.-S., 2008. Perturbation formulas for polarization ratio and phase shift of Rayleigh waves in prestressed anisotropic media, *J. Elasticity*, 92, 1–33.

[20] Man C.-S., Lu W.Y., 1987. Towards an acoustoelastic theory for measurement of residual stress, *J. Elasticity*, 17, 159–182.

[21] Man C.-S., 1998. Hartig's law and linear elasticity with initial stress, *Inverse Problems*, 14, 313–319.

[22] Ingebrigtsen K.A., Tonning A., 1969. Elastic surface waves in crystals, *Phys. Rev.*, 184, 942–951.

[23] Chadwick P., Ting T.C.T., 1987. On the structure and invariance of the Barnett–Lothe tensors, *Quart. Appl. Math.*, 45, 419–427.

[24] Ting T.C.T., 1988. Some identities and the structure of $\mathbf{N}_i$ in the Stroh formalism of anisotropic elasticity, *Quart. Appl. Math.*, 46, 109–120.

[25] Mielke A., Fu Y.B., 2004. Uniqueness of the surface-wave speed: a proof that is independent of the Stroh formalism, *Math. Mech. Solids*, 9, 5–15.

[26] Ito H., Construction and existence criterion of Rayleigh and Stoneley waves by means of factorization of matrix polynomials, preprint.

[27] Royer D., Dieulesaint E., 1984. Rayleigh wave velocity and displacement in orthorhombic, tetragonal, hexagonal, and cubic crystals, *J. Acoust. Soc. Am.*, 76, 1438–1444.

[28] Song Y.Q., Fu Y.B., 2007. A note on perturbation formulae for the surface-wave speed due to peturbations in material properties, *J. Elasticity*, 88, 187–192.

[29] Paroni R., Man C.-S., 1999. Constitutive equations of elastic polycrystalline materials, *Arch. Ration. Mech. Anal.*, 150, 153–177.

https://doi.org/10.1142/9781800611887_0009

Chapter 9

Advanced Mathematical Models and Efficient Numerical Simulation in Composite Processes

Emmanuelle Abisset-Chavanne[*,§], **Anais Barasinski**[†,¶], **and Francisco Chinesta**[‡,||]

[*]I2M Lab @ Arts et Métiers Institute of Technology, Esplanade des Arts et Métiers, 33400 Talence, France

[†]Universite de Pau et des Pays de l'Adour, E2S UPPA, CNRS, IPREM, 64000 Pau, France

[‡]ESI Chair & PIMM Lab @ Arts et Métiers Institute of Technology, 151 Boulevard de l'Hôpital, 75013 Paris, France

[§]emmanuelle.abisset-chavanne@ensam.eu

[¶]anais.barasinski@univ-pau.fr

[||]francisco.chinesta@ensam.eu

Abstract

Over the last decades, an increasing number of functional and structural parts, made so far with metals, has been progressively reengineered by replacing metallic materials by polymers, reinforced polymers and composites. The motivation for this substitution may be the weight reduction, the simpler, cheaper or faster forming process, or the ability to exploit additional functionalities. The fillers usually employed cover a broad range involving many scales: (i) the nanometer scale (e.g., carbon nanotubes, graphene, fullerene, nanodiamonds); (ii) the micrometer to the millimeter scale (particles and short fibers); (iii) the centimeter scale of fibers used in SMC and BMC composite processes; and finally (iv) the macroscopic scale where fibrous reinforcements are made of continuous fibers arranged in bundles. When load-bearing capacities are especially looked for, continuous fiber-reinforcement polymers are selected. In that case, the impregnation of the reinforcement with a low viscosity polymer involves the flow of a Newtonian or non-Newtonian fluid in the complex multi-scale microstructure related to the fiber and tow arrangement. Reinforced polymers are selected instead of high performance polymers of equivalent properties since the latter are generally more expensive. When looking for functional properties, the use of nano-charges opens a wide spectrum of possibilities but also raises new challenges, such as dispersion of charges into the polymer matrix and occurrence of aggregation and disaggregation mechanisms. Suspensions of practical interest involve many scales and many concentration regimes, the latter ranging from dilute to highly concentrated. In the present chapter, we survey modern developments related to the multi-scale modeling and simulation of reinforced polymers and composites.

Contents

9.1. Introduction

Composites manufacturing processes involve many different physics and length and timescales. When considering reinforced polymers the main issue is related to the flow induced anisotropy, whose prediction and control is needed in order to optimize both, the processes and the manufactured parts. Standard reinforced polymers flow models were developed under a number of strong hypotheses. Most of them are based on the fact of considering dilute suspensions with the reinforcement (consisting of nano or micro particles) immersed into a Newtonian fluid that flows in complex geometries. Moreover, those particles are assumed rigid enough to neglect eventual deformations and their kinematics assumed unperturbed by any eventual confinement effect occurring when the characteristic particles length is of the same order than the gap in which the suspension is flowing. The first part of the present chapter focuses on the state of the art and the most recent developments of flow involving reinforced polymers.

When considering long-fibers composites, an important issue concerns the nature of the macroscopic models defined in plate or shell domains characterized by having a dimension (the thickness) several orders of magnitude lower than the other representative in-plane dimensions. This fact, even if it is not a major conceptual issue, is a real handicap for simulation purposes. This situation is not new; plate and shell theories were successfully developed many years ago and they were intensively used in structural mechanics. These theories make use of some kinematic and mechanic hypotheses to reduce the 3D nature of mechanical models to 2D reduced models defined in the shell or plate middle surface. In the case of elastic behaviors, the derivation of such reduced models is quite simple and it constitutes the foundations of classical plate and shell theories. Today, most commercial codes for structural mechanics applications propose different type of plate and shell finite elements, even in the case of multilayered composites plates or shells.

However, in composites manufacturing processes the physics encountered in such multilayered plate or shell domains is much more rich, because usually it involves chemical reactions, crystallization and strongly coupled thermomechanical behaviors. The complexity of the involved physics makes impossible the introduction of pertinent hypotheses for reducing the dimensionality of the model from 3D to 2D. In that case a fully 3D modeling is compulsory, and because the richness of the thickness description (many coupled physics and many plies differently oriented) the approximation of the fields involved in the models needs thousands of nodes distributed along the thickness direction. Thus, fully 3D descriptions involve thousands of millions of degrees of freedom that should be solved many times because of the history-dependent thermomechanical behavior. Moreover, when we are considering optimization or inverse identification, many direct problems have to be solved in order to reach the minimum of a certain cost function. In the case of inverse analysis, such cost function is the difference between the predicted and measured fields.

Today, the solution of such fully 3D models involved in composites manufacturing processes remains intractable despite the impressive progresses reached in mechanical modeling, numerical analysis, discretization techniques and computer science during the last decade. New numerical techniques are needed for approaching such complex scenarios, able to proceed to the solution of fully 3D multiphysics models in geometrically complex parts (e.g., the whole aircraft). The well-experienced mesh-based discretizations techniques fail because the excessive number of degrees of freedom involved in the fully 3D discretizations where very fine meshes are required in the thickness direction (despite its reduced dimension) and also in the in-plane directions to avoid too distorted meshes and also because some processes (e.g., tape placement) implies thermomechanical loads moving on the plate or shell requiring fine enough meshes of the plate or shell surfaces. We are in a real impasse. The only getaway is to explore new discrezation strategies able to circumvent or at least alleviate the drawbacks related to mesh-based discretizations of fully 3D models defined in plate or shell domains, as well as the complex plates and shells assemblages usually encountered in composite structures.

The second part of the present chapter focuses on the modeling and efficient simulation of numerous models encountered in composites engineering, requiring rich enough representations (most of time 3D) and whose models are defined in this kind of degenerated domains (plates and shells).

Finally, when considering laminates, special attention must be paid in the appropriate and fine enough representation of physics at the interfaces where thermal resistances, molecular diffusion, and squeeze flow during consolidation, occur. The third and last part of the chapter focuses on the physics encountered during consolidation of composites laminates with special emphasis on the efficient treatment of physics encountered at the interfaces level.

9.2. Reinforced Polymers

Fiber suspensions can be described at different scales: (i) the microscopic scale, related to individual fibers, (ii) the mesoscopic scale, which considers a population of fibers within a local representative volume, and (iii) the macroscopic scale related to the forming process and the final part itself. For an overview, see the recent monograph [1].

The kinematics of an ellipsoidal particle oriented in direction $\mathbf{p}$ (with $\|\mathbf{p}\| = 1$) and immersed in a Newtonian fluid flow characterized by the velocity gradient $\nabla \mathbf{v}$ are given by the Jeffery equation [2]

$$\dot{\mathbf{p}} = \boldsymbol{\Omega} \cdot \mathbf{p} + \mathcal{F}\left(\mathbf{D} \cdot \mathbf{p} - \nabla \mathbf{v} : (\mathbf{p} \otimes \mathbf{p})\mathbf{p}\right), \tag{9.1}$$

where $\mathbf{D} = 1/2\left(\nabla \mathbf{v} + (\nabla \mathbf{v})^T\right)$, $\boldsymbol{\Omega} = 1/2\left(\nabla \mathbf{v} - (\nabla \mathbf{v})^T\right)$, and the shape factor $\mathcal{F} = \frac{r^2-1}{r^2+1}$ depends on the ellipsoid aspect ratio r (major to minor axes ratio). Note that due to the symmetry of $(\mathbf{p} \otimes \mathbf{p})$, we have $\nabla \mathbf{v} : (\mathbf{p} \otimes \mathbf{p}) = \mathbf{D} : (\mathbf{p} \otimes \mathbf{p})$.

Mesoscopic kinetic theory models result from the coarsening of microscopic descriptions. In kinetic theory models, the individuality of the particles is lost in favor of a statistical description that substitutes the microscopic entities with a series of conformation coordinates [3, 4]. For example, when considering a suspension of rigid rods, the mesoscopic description consists in giving the fraction of rods that at position $\mathbf{x}$ and time t are oriented along direction $\mathbf{p}$. This information is contained in the probability distribution function – pdf – $\psi(\mathbf{x}, t, \mathbf{p})$, whose evolution is governed by the Fokker–Planck equation

$$\frac{\partial \psi}{\partial t} + \nabla_x \cdot (\mathbf{v}\psi) + \nabla_p \cdot (\dot{\mathbf{p}}\psi) = 0. \tag{9.2}$$

The Fokker–Planck equation being highly-dimensional, it cannot be solved by means of standard mesh-based discretization techniques. This issue is known as the curse of dimensionality. Solution procedures based on the use of particles at the mesoscopic scale have been extensively developed by many authors [5–12]. On the other hand, there are few works on the solution of the Fokker–Planck equation with standard discretization techniques [13, 14]. We have proposed in [15, 16] a new solution technique called Proper Generalized Decomposition (PGD) that allows for the direct solution of the Fokker–Planck equation in conformation spaces of high dimension [17, 18].

Finally, at the macroscopic scale, the pdf is substituted with some of its moments [19, 20], e.g., the second and fourth-order moments, $\mathbf{a}$ and $\mathbf{A}$, that read respectively

$$\mathbf{a} = \int_{\mathcal{S}} \mathbf{p} \otimes \mathbf{p} \psi \, d\mathbf{p}, \tag{9.3}$$

and

$$\mathbf{A} = \int_{\mathcal{S}} \mathbf{p} \otimes \mathbf{p} \otimes \mathbf{p} \otimes \mathbf{p} \psi \, d\mathbf{p}, \tag{9.4}$$

where $\mathcal{S}$ is the surface of the unit sphere where the orientation vector $\mathbf{p}$ is defined.

Here, the level of detail and the involved physics are sacrificed in favor of computational efficiency. The equations governing the time evolution of $\mathbf{a}$ are obtained by taking the time derivative of Eq. (9.3), expressing $\dot{\psi}$ from Eq. (9.2) and then integrating by parts. This procedure yields

$$\dot{\mathbf{a}} = \int_{\mathcal{S}} (\dot{\mathbf{p}} \otimes \mathbf{p} + \mathbf{p} \otimes \dot{\mathbf{p}}) \, \psi \, d\mathbf{p}, \tag{9.5}$$

which, in view of Eq. (9.1), leads to

$$\dot{\mathbf{a}} = \nabla \mathbf{v} \cdot \mathbf{a} + \mathbf{a} \cdot (\nabla \mathbf{v})^T - 2 \, \mathbf{A} : \nabla \mathbf{v}. \tag{9.6}$$

In order to close the problem, $\mathbf{A}$ must be expressed as a function of the lower-order moment $\mathbf{a}$, using one of the available approximate closure relations [21–24]. The impact of closure relations on the computed solutions can be significant [25, 26], and it is *a priori* unpredictable in most of cases.

Theoretical suspension models not only depend on the chosen scale of description, but also on the concentration regime considered and on the nature of the suspending fluid.

9.2.1. Fiber suspensions in Newtonian fluids

For Newtonian suspending fluids, the different concentration regimes (dilute, semi-dilute, semi-concentrated and concentrated) have been extensively analyzed from the modeling, simulation and experimental viewpoints.

In the dilute regime, fiber–fiber interactions are neglected altogether. For the semi-dilute regime, these interactions are usually taken into account in the form of a phenomenological randomizing mechanism, i.e., one adds a diffusion term in the Fokker–Planck equation to obtain

$$\frac{\partial \psi}{\partial t} + \nabla_x \cdot (\mathbf{v} \psi) + \nabla_p \cdot (\dot{\mathbf{p}} \psi) = \nabla_p (D_r \nabla_p \psi), \tag{9.7}$$

where D_r is a diffusion coefficient. At the macroscopic scale, this leads to

$$\dot{\mathbf{a}} = \nabla\mathbf{v}\cdot\mathbf{a} + \mathbf{a}\cdot(\nabla\mathbf{v})^T - 2\mathbf{A} : \nabla\mathbf{v} - 6D_r\left(\mathbf{a} - \frac{\mathbf{I}}{3}\right). \tag{9.8}$$

Randomizing effects result from Brownian effects due to thermal agitation within the suspending fluid and from rod–rod interactions in the semi-dilute regime. Brownian effects were taken into account in [9, 27, 28], within the microscopic framework. When randomizing effects are due to rod–rod interactions, direct numerical simulations [29] suggest that the diffusion coefficient D_r scales linearly with the strain rate $\dot{\gamma}$, i.e., the second invariant of the rate of strain tensor. Thus, $D_r \approx C_I\dot{\gamma}$. Introducing this linear scaling of the diffusion coefficient into Eq. (9.8) gives the so-called Folgar–Tucker model [30]:

$$\dot{\mathbf{a}} = \nabla\mathbf{v}\cdot\mathbf{a} + \mathbf{a}\cdot(\nabla\mathbf{v})^T - 2\cdot\mathbf{A} : \mathbf{D} - 6C_I\dot{\gamma}\left(\mathbf{a} - \frac{\mathbf{I}}{3}\right). \tag{9.9}$$

There is a wide literature on dilute and semi-dilute suspensions, addressing modeling [31–35], flows [20, 36–38] and rheology [39, 40]. These models describe quite well the experimental observations.

When the concentration further increases, intense fiber–fiber interactions occur which must be taken into account appropriately, as for example, in the model proposed in [41, 42]. Recent experiments suggest that short fibers in concentrated suspensions align more slowly as a function of strain than what models based on Jeffery's equation predict [43]. In order to address this issue, Wang *et al.* [43] proposed the use of a strain reduction factor, but this approach violates the principle of objectivity. Later, the same authors proposed an objective model by decoupling the time evolution of both the eigenvalues and the eigenvectors of the second-order orientation tensor [44]. In [45], an anisotropic rotary diffusion is proposed that accounts for fiber–fiber interactions; the model parameters were selected by matching the experimental steady-state orientation in simple shear flow and by requiring stable steady states and physically realizable solutions.

The most complex scenario is that of the concentrated flow regime involving entangled suspensions or dense clusters immersed into the suspending fluid, exhibiting specific kinematics and complex aggregation/disaggregation mechanisms [46]. The first natural question is how to describe such systems. At the macroscopic scale, one could try to fit some power-law constitutive equation, however, this description does not hold for the microstructure. At the microscopic scale, direct numerical simulations describing complex fiber–fiber interactions can be carried out in small-enough representative volumes [29, 47–49].

A first attempt at describing dilute suspensions composed of rigid and deformable clusters from a micromechanical point of view was proposed in [50]. Later, kinematic predictions for rigid and deformable clusters were compared with direct numerical simulations in [51]. An enriched description of the kinematics of rigid clusters within a multi-scale framework was addressed in [52].

In [52], we considered the kinematics of rigid clusters composed of rods. It is assumed that hydrodynamic forces act on the N beads of the rods involved in the cluster. The location of each bead $\mathcal{B}_i$ with respect to the cluster center of gravity $\mathcal{G}$ is given by $L_i\mathbf{p}_i$, where $\mathbf{p}_i$ is the unit vector pointing from $\mathcal{G}$ to $\mathcal{B}_i$, as shown in Fig. 9.1.

A balance of momenta yields the cluster rotary velocity. By defining the cluster conformation tensor $\mathbf{c}$ as follows,

$$\mathbf{c} = \frac{\sum_{i=1}^{N} L_i^2(\mathbf{p}_i \otimes \mathbf{p}_i)}{\sum_{i=1}^{N} L_i^2}, \tag{9.10}$$

the cluster rotary velocity $\boldsymbol{\omega}$ is given by [52]:

$$\boldsymbol{\omega} = (\mathbf{I} - \mathbf{c})^{-1}(\boldsymbol{\epsilon} : (\nabla\mathbf{v} \cdot \mathbf{c})), \tag{9.11}$$

where $\boldsymbol{\epsilon}$ the Levi-Civita permutation tensor.

Entangled systems involving moderately long fibers are generally described by using some *ad hoc* adaptations of the Folgar–Tucker model [30]. This model, however, has its origins in the Jeffery equation whose validity is by construction restricted to the dilute regime. For this reason, the relevance of this approach must be confirmed from both the theoretical

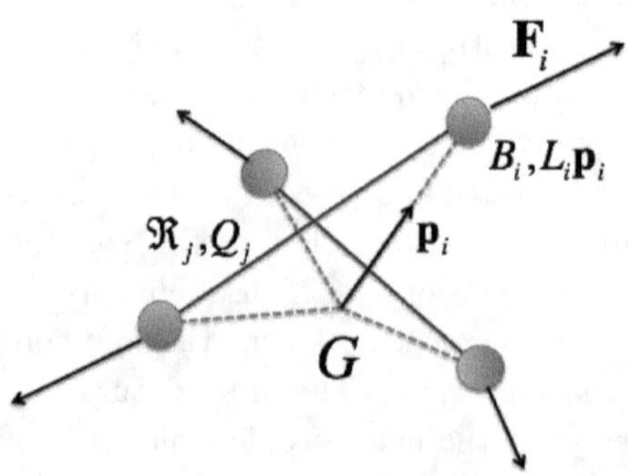

Fig. 9.1. Rigid cluster composed of rods.

and experimental viewpoints. Moreover, when the fiber length and the number of interactions per fiber increase, fiber bending mechanisms are activated. Attempts at modeling rod bending have been published. Some consider the rod composed of rigid segments connected by springs activated by bending [8, 9, 53, 54], while others consider richer dumbbell models [55]. Finally, we addressed in [56] the issue of confinement and highlighted the related limitations of moment-based descriptions.

9.2.2. Fiber suspensions in non-Newtonian fluids

As mentioned previously, the vast majority of available models are based on the Jeffery equation [2] that describes the motion of an ellipsoidal particle immersed in a Newtonian fluid. Nowadays, there is no general Jeffery counterpart available for non-Newtonian suspending fluids. Different theoretical studies exist that consider particles immersed in non-Newtonian fluids. Many of them concern the motion of particles (e.g., sedimentation) [57–64], others experimental studies on the motion and orientation of fibers in viscoelastic fluids [65, 66], or theoretical work for second-order fluids [67, 68]. However, to our knowledge, a multi-scale modeling framework is lacking that could be exploited in the simulation of forming processes involving reinforced polymers. Leal [68] and Brunn [67] published important results for second-order viscoelastic fluids, in the limit of low Weissenberg numbers. They derived the equations governing the motion of rods (Leal [68]) and transversely isotropic particles (Brunn [67]) that could be viewed as the counterparts of the Jeffery equation for the fluids and flow regimes considered in their derivation. In view of their relevance, we briefly detail the main ingredients of these two pioneering works.

The study of the kinematics of transversely isotropic particles suspended in a viscoelastic medium has first been addressed in the work of Leal [68]. The rheology of the suspending fluid is described by the Rivlin–Ericksen second-order fluid model. In this case, the dimensionless Cauchy stress reads

$$\boldsymbol{\sigma} = -P\mathbf{I} + \mathbf{A}_1 + \epsilon_0 \left[(\mathbf{A}_1)^2 + \epsilon_1 \mathbf{A}_2\right], \tag{9.12}$$

where P is the isotropic pressure, while the first and second-order Rivlin–Ericksen tensors $\mathbf{A}_1$ and $\mathbf{A}_2$ are given, respectively, by

$$\mathbf{A}_1 = 2\mathbf{D}, \tag{9.13}$$

and

$$\mathbf{A}_2 = \frac{\partial \mathbf{D}}{\partial t} + \mathbf{v} \cdot \nabla \mathbf{D} + (\nabla \mathbf{v})^T \cdot \mathbf{D} + \mathbf{D} \cdot \nabla \mathbf{v}. \tag{9.14}$$

In Eq. (9.12), the material constants ϵ_0 and ϵ_1 are related to the suspending fluid normal stress coefficients. The flow is considered rheologically slow, in the sense that the flow characteristic timescale is large compared to the fluid intrinsic relaxation time. Thus, the dimensionless number ϵ_0 remains close to zero. Under this assumption, the perturbation of the velocity field due to presence of the particle differs only slightly from the one corresponding to a Newtonian suspending fluid. This perturbation is estimated by performing an asymptotic expansion of the velocity field $\mathbf{v}$ around the particle in terms of powers of ϵ_0:

$$\mathbf{v} = \mathbf{v}^{(0)} + \sum_{i=1}^{\infty} \epsilon_0^i \mathbf{v}^{(i)}, \tag{9.15}$$

and retaining only terms up to first order in ϵ_0. The zeroth-order term corresponds to the flow field obtained in the Newtonian case. It is determined from a line Stokeslet distribution along the particle axis of symmetry.

For slender particles, the intensity of the Stokeslet distribution can be determined analytically as a power series of the particle aspect ratio, by enforcing the appropriate no-slip condition at the particle surface. In [69], it was proven that the dominant terms of the series are independent of the particle geometry.

In the work of Leal [68], only the first-order non-Newtonian contributions to the flow field generated by these geometry-independent terms of the Stokeslet distribution are considered. By means of the Lorentz reciprocal theorem, the resultant force and torque acting on the particle were calculated without explicitly solving for the velocity and stress field perturbations. Finally, by enforcing that the particle remains force and torque-free, the particle rotary velocity $\boldsymbol{\omega}$ is determined. However, no explicit expression for $\boldsymbol{\omega}$ is given, but only some particularizations to specific flows, as for example, the particle rotation in a simple shear flow.

In [67], Brunn later extended Leal's work to general, transversally isotropic particles, by introducing the concept of material tensors depending on the particle geometry and fluid properties. In that work, the fluid rheology is also described by a second-order fluid model, i.e., that proposed by Giesekus in [70]. In dimensionless form, this model reads:

$$\boldsymbol{\sigma} = -P\mathbf{I} + 2\mathbf{D} + 2\left[\kappa_0^{(1)}\left(\frac{\partial \mathbf{D}}{\partial t} + \mathbf{v}\cdot\nabla\mathbf{D} - \boldsymbol{\Omega}\cdot\mathbf{D} + \mathbf{D}\cdot\boldsymbol{\Omega}\right) + \kappa_0^{(2)}\mathbf{D}\cdot\mathbf{D}\right], \tag{9.16}$$

with the material coefficients $\kappa_0^{(1)}$ and $\kappa_0^{(2)}$ given in [67].

In Brunn's work, a general expression for the particle rotary velocity is derived. It consists of a Newtonian contribution enriched with a first-order non-Newtonian correction. The particle orientation, described by the unit vector $\mathbf{p}$ defining the particle axis, thus evolves according to

$$\begin{aligned}\dot{\mathbf{p}} &= \boldsymbol{\Omega} \cdot \mathbf{p} + \mathcal{F}\left(\mathbf{D} \cdot \mathbf{p} - (\mathbf{D} : (\mathbf{p} \otimes \mathbf{p}))\,\mathbf{p}\right) \\ &\quad - (\mathbf{I} - \mathbf{p} \otimes \mathbf{p}) \cdot \mathbf{D} \cdot (H_1 \mathbf{D} \cdot \mathbf{p} - H_2\,(\mathbf{D} : (\mathbf{p} \otimes \mathbf{p}))\,\mathbf{p}), \qquad (9.17)\end{aligned}$$

where the scalar $\mathcal{F}$ depends, in the case of an ellipsoid or a tri-dumbbell, on the particle aspect ratio r (i.e., the length ratio of major to minor particle axes), $\mathcal{F} = \frac{r^2-1}{r^2+1}$, and the scalar parameters H_1 and H_2 depend on the material tensors associated to the particle and fluid system [67].

The particular case of rigid tri-dumbbells is considered in [67], establishing the link between tri-dumbbells and ellipsoids. When applied to a single dumbbell composed of two non-interacting but rigidly-interconnected beads, emulating rod-like particles, Brunn's model reduces to that proposed by Leal [68].

It is important to note that the first term in Eq. (9.17), $\boldsymbol{\Omega} \cdot \mathbf{p} + \mathcal{F}\left(\mathbf{D} \cdot \mathbf{p} - (\mathbf{D} : (\mathbf{p} \otimes \mathbf{p}))\,\mathbf{p}\right)$, coincides with the classical Jeffery equation for ellipsoids of aspect ratio r in a Newtonian fluid [2].

Although Brunn's model provides a detailed description of particle orientation in a viscoelastic fluid a low Weissenberg number, the first-order correction term introduces higher powers of $\mathbf{p}$, with important consequences on the related macroscopic descriptions, as discussed in [71], where it was proved that a nice compromise consists of assuming standard models (valid in the case of Newtonian suspending fluids) and considering an effective velocity gradient that consists in the one related to a Newtonian fluid flow and an extra-term accounting for the non-Newtonian perturbation. In that paper, both microscopic and macroscopic descriptions were derived and it was proved that in simple shear flows of second-order fluids ellipsoid tends to orient along the vorticity direction instead of aligning along the shear direction as is the case when considering Newtonian fluids.

9.3. Multi-Physics in Laminates

Plates and shells are very common in nature and thus they inspired engineers that used both them from the very beginning of structural mechanics. Nowadays, plates and shells parts are massively present in most engineering applications.

This type of structural elements involves homogeneous and heterogeneous materials, isotropic and anisotropic, linear and nonlinear. The appropriate design of such parts consists not only in the structural analysis of the parts for accommodating the design loads, but also in the analysis of the associated manufacturing processes because many properties in the final parts are induced by the forming process itself (e.g., flow-induced microstructures). Thus, fine analyses concern both, the structural parts and their associated forming processes.

In general, the whole design requires the solution of some mathematical models governing the evolution of the quantities of interest. These models consist of a set of partial differential equations combining general balance equations (mass, energy and momentum) and some specific constitutive equations depending on the considered physics, the last involving different material parameters. These complex equations (in general nonlinear and strongly coupled) must be solved in the domain of interest.

When addressing plate or shell geometries the domains in which the mathematical models must be solved become degenerated because one of its characteristic dimensions (the thickness in the present case) is much lower than the other characteristic dimensions. We will understand the consequences of such degeneracy later. When analytical solutions are neither available nor possible because of the geometrical or behavior complexities, the solution must be calculated by invoking any of the available numerical techniques (finite elements, finite differences, finite volumes, methods of particles, ...).

In the numerical framework the solution is only obtained in a discrete number of points, usually called nodes, distributed in the domain. From the solution at those points, it can be interpolated at any other point in the domain. In general regular nodal distributions are preferred because they offer better accuracies. In the case of degenerated plate or shell domains one could expect that if the solution evolves significantly in the thickness direction, a large enough number of nodes must be distributed along the thickness direction to ensure the accurate representation of the field evolution in that direction. In that case, a regular nodal distribution in the whole domain will imply the use of an extremely large number of nodes with the consequent impact on the numerical solution efficiency.

When simple behaviors and domains were considered, plate and shell theories were developed in the structural mechanics framework allowing, through the introduction of some hypotheses, reducing the 3D complexity to a 2D one related to the problem now formulated by considering the in-plane coordinates.

In the case of fluid flows this dimensionality reduction is known as lubrication theory and it allows efficient solutions of fluids flows taking place in plate or shell geometries for many type of fluids, linear (Newtonian) and nonlinear. The interest of this type of flows, taking place in plate and shell geometries, is not only due to the fact that it is involved in the manufacturing processes of plate and shell parts, but also due to the fact that many tests for characterizing material behaviors involve it.

However, as soon as richer physics are concerned by the models and the considered geometries differ of those ensuring the validity of the different reduction hypotheses, simplified simulations are compromised and they fail in their predictions.

In these circumstances, the reduction from the 3D model to a 2D simplified one is not obvious, and 3D simulations appear many times as the only valid route for addressing such models, that despite the fact of being defined in degenerated geometries (plate or shell) they seem requiring a fully 3D solution. However in order to integrate such a calculation (fully 3D and implying an impressive number of degrees of freedom) in usual design procedures, a new efficient (fast and accurate) solution procedure is needed.

A new discretization technique based on the use of separated representations was proposed some years ago for addressing multidimensional models suffering the so-called curse of dimensionality, where standard mesh-based techniques fail [15]. The curse of dimensionality was circumvented thanks to those separated representations that transformed the solution of a multidimensional problem into a sequence of lower dimensional problems. The interested reader can refer to the recent reviews [17, 72, 73] and the references therein.

A direct consequence was separating the physical space. Thus in plate domains an in-plane–out-of-plane decomposition was proposed for solving 3D flows occurring in RTM (Resin Transfer Moulding) processes [17], then for solving elasticity problems in plates [74] and shells [75] and coupled multiphysics problems [50]. In those cases, the 3D solution was obtained from the solution of a sequence of 2D problems (the ones involving the in-plane coordinates) and 1D problems (the ones involving the coordinate related to the plate thickness).

It is important emphasizing the fact that these approaches are radically different from standard ones. We propose a 3D solver able to compute the different unknown 3D fields without the necessity of introducing any hypothesis. The most outstanding advantage is that 3D solutions can be obtained with a computational cost characteristic of standard 2D solutions. The separated representations constructor (solver) is based on the Proper Generalized Decomposition (PGD).

9.3.1. PGD at a glance

Most of the existing model reduction techniques proceed by projecting the problem solution onto a reduced basis (this constitutes the wide class of *projection-based model order reduction* methods). Therefore, the construction of the reduced basis usually constitutes the first step in the solution procedure, giving rise to a second important distinction when classifying Model Order Reduction (MOR) techniques: *a posteriori* versus *a priori* model order reduction. One must be careful on the suitability of a particular reduced basis when employed for representing the solution of a particular problem, particularly if it was obtained through *snapshots* of slightly different problems. This difficulty (at least partially) disappears if the reduced basis is constructed at the same time that the problem is solved (in other words: a priori with no need for snapshots of different problems). Thus, each problem has its associated basis in which its solution is expressed. One could consider few vectors in the basis, leading to a reduced representation, or all the terms needed for approximating the solution up to a certain accuracy level. The Proper Generalized Decomposition (PGD), which is described in general terms in what follows proceeds in this manner.

When calculating the transient solution of a generic problem, say $u(x,t)$, we usually consider a given basis of space functions $N_i(x)$, $i = 1, \ldots, N_n$, the so-called shape functions within the finite element framework, being N_n the number of nodes. They approximate the problem solution as

$$u(x,t) \approx \sum_{i=1}^{N_n} a_i(t) N_i(x). \qquad (9.18)$$

This implies a space-time separated representation where the time-dependent coefficients $a_i(t)$ are unknown at each time instant (when proceeding incrementally) and the space functions $N_i(x)$ are given "a priori", e.g., piece-wise polynomials. POD (Proper Orthogonal Decomposition) and Reduced Basis methodologies consider a set of global, reduced basis $\phi_i(x)$

for approximating the solution instead of the generic, but local, finite element functions $N_i(x)$. The former are expected to be more adequate to approximate the problem at hand. Thus, it results

$$u(x,t) \approx \sum_{i=1}^{R} b_i(t)\phi_i(x), \tag{9.19}$$

where it is expected that $R \ll N_n$. Again, Eq. (9.19) represents a space-time separated representation where the time-dependent coefficient must be calculated at each time instant during the incremental solution procedure.

Inspired from these results, one could consider the general space-time separated representation

$$u(x,t) \approx \sum_{i=1}^{N} X_i(x) \cdot T_i(t), \tag{9.20}$$

where now neither the time-dependent functions $T_i(t)$ nor the space functions $X_i(x)$ are *a priori* known. Both will be computed on the fly when solving the problem.

As soon as one postulates that the solution of a transient problem can be expressed in the separated form (9.20), whose approximation functions $X_i(x)$ and $T_i(t)$ will be determined during the problem solution, one could make a step forward and assume that the solution of a multidimensional problem $u(x_1, \ldots, x_d)$ could be found in the separated form

$$u(x_1, x_2, \ldots, x_d) \approx \sum_{i=1}^{N} X_i^1(x_1) \cdot X_i^2(x_1) \cdot \ldots \cdot X_i^d(x_d), \tag{9.21}$$

and even more, expressing the 3D solution $u(x,y,z)$ as a finite sum decomposition involving low-dimensional functions

$$u(x,y,z) \approx \sum_{i=1}^{N} X_i(x) \cdot Y_i(y) \cdot Z_i(z), \tag{9.22}$$

or

$$u(x,y,z) \approx \sum_{i=1}^{N} X_i(x,y) \cdot Z_i(z). \tag{9.23}$$

Equivalently, the solution of a parametric problem $u(\mathbf{x}, t, p_1, \ldots, p_\wp)$ could be approximated as

$$u(\mathbf{x}, t, p_1, \ldots, p_\wp) \approx \sum_{i=1}^{N} X_i(\mathbf{x}) \cdot T_i(t) \cdot \prod_{k=1}^{\wp} P_i^k(p_k). \tag{9.24}$$

9.3.2. Heat transfer in laminates

In what follows, we are illustrating the construction of the Proper Generalized Decomposition of a generic model defined in a plate domain $\Xi = \Omega \times \mathcal{I}$ with $\Omega \subset \mathbb{R}^2$ and $\mathcal{I} = [0, H] \subset \mathbb{R}$. For the sake of simplicity, we consider the model related to the steady-state heat conduction equation:

$$\nabla \cdot (\mathbf{K} \cdot \nabla u) = 0, \tag{9.25}$$

in a plate geometry that contains P plies in the plate thickness. Each ply is characterized by its conductivity tensor $\mathbf{K}_i(x, y)$ which is assumed constant through the ply thickness. Moreover, without any loss of generality, we assume the same thickness h for the different plies constituting the laminate. Thus, we can define a characteristic function representing the position of each ply $i = 1, \ldots, P$:

$$\chi_i(z) = \begin{cases} 1 & z_i \leq z \leq z_{i+1}, \\ 0 & \text{otherwise}, \end{cases} \tag{9.26}$$

where $z_i = (i-1)h$ defines the location of the ith-ply in the laminate thickness. Now, the laminate conductivity can be given in the following separated form:

$$\mathbf{K}(x, y, z) = \sum_{i=1}^{i=P} K_i(\mathbf{x}) \cdot \chi_i(z), \tag{9.27}$$

where $\mathbf{x}$ denotes the in-plane coordinates, i.e., $\mathbf{x} = (x, y) \in \Omega$.

The weak form of Eq. (9.25), with appropriate boundary conditions, writes:

$$\int_\Xi \nabla u^* \cdot (\mathbf{K} \cdot \nabla u) \, d\Xi = 0, \tag{9.28}$$

with the test function u^* defined in an appropriate functional space. The solution $u(x, y, z)$ is searched under the separated form:

$$u(\mathbf{x}, z) \approx \sum_{j=1}^{j=N} X_j(\mathbf{x}) \cdot Z_j(z). \tag{9.29}$$

In what follows, we are illustrating the construction of such a decomposition. For this purpose, we assume that at enrichment step $n < N$ the solution $u^n(\mathbf{x}, z)$ is already known:

$$u^n(\mathbf{x}, z) = \sum_{j=1}^{j=n} X_j(\mathbf{x}) \cdot Z_j(z), \tag{9.30}$$

and that at the present step $n + 1$ we look for the solution enrichment $R(\mathbf{x}) \cdot S(z)$:

$$u^{n+1}(\mathbf{x}, z) = u^n(\mathbf{x}, z) + R(\mathbf{x}) \cdot S(z). \tag{9.31}$$

The algorithm starts by finding the first couple of unknown functions, that is $n = 0$.

The test function involved in the weak form is searched under the form:

$$u^*(\mathbf{x}, z) = R^*(\mathbf{x}) \cdot S(z) + R(\mathbf{x}) \cdot S^*(z). \tag{9.32}$$

By introducing Eqs. (9.31) and (9.32) into Eq. (9.28) it results:

$$\int_\Xi \left(\begin{pmatrix} \tilde{\nabla} R^* \cdot S \\ R^* \cdot \dfrac{dS}{dz} \end{pmatrix} + \begin{pmatrix} \tilde{\nabla} R \cdot S^* \\ R \cdot \dfrac{dS^*}{dz} \end{pmatrix} \right) \cdot \left(\mathbf{K} \cdot \begin{pmatrix} \tilde{\nabla} R \cdot S \\ R \cdot \dfrac{dS}{dz} \end{pmatrix} \right) d\,\Xi$$
$$= - \int_\Xi \left(\begin{pmatrix} \tilde{\nabla} R^* \cdot S \\ R^* \cdot \dfrac{dS}{dz} \end{pmatrix} + \begin{pmatrix} \tilde{\nabla} R \cdot S^* \\ R \cdot \dfrac{dS^*}{dz} \end{pmatrix} \right) \cdot \mathbf{Q}^n \, d\,\Xi, \tag{9.33}$$

where $\tilde{\nabla}$ denotes the plane component of the gradient operator, i.e., $\tilde{\nabla}^T = \left(\frac{\partial}{\partial x}, \frac{\partial}{\partial y}\right)$ and $\mathbf{Q}^n$ denotes the flux at iteration n:

$$\mathbf{Q}^n = \mathbf{K} \cdot \sum_{j=1}^{j=n} \begin{pmatrix} \tilde{\nabla} X_j(\mathbf{x}) \cdot Z_j(z) \\ X_j(\mathbf{x}) \cdot \dfrac{dZ_j(z)}{dz} \end{pmatrix}. \tag{9.34}$$

Now, as the enrichment process is nonlinear we propose to search the couple of functions $R(\mathbf{x})$ and $S(z)$ by applying the alternating direction fixed point algorithm. Thus, assuming $R(\mathbf{x})$ known, we compute $S(z)$, and then we update $R(\mathbf{x})$. The process continues until reaching convergence. The converged solutions allow defining the next term in the finite sum decomposition: $R(\mathbf{x}) \to X_{n+1}(\mathbf{x})$ and $S(z) \to Z_{n+1}(z)$. We are illustrating each one of the just referred steps. The global enrichment procedure stops when the norm of the just compared functions have becomes small enough or when more advanced stopping criteria apply [18].

9.3.2.1. *Computing $R(\mathbf{x})$ from $S(z)$*

When $S(z)$ is known, the test function reduces to:

$$u^*(\mathbf{x}, z) = R^*(\mathbf{x}) \cdot S(z), \tag{9.35}$$

and the weak form (9.33) reduces to:

$$\int_\Xi \begin{pmatrix} \tilde{\nabla} R^* \cdot S \\ R^* \cdot \dfrac{dS}{dz} \end{pmatrix} \cdot \left(\mathbf{K} \cdot \begin{pmatrix} \tilde{\nabla} R \cdot S \\ R \cdot \dfrac{dS}{dz} \end{pmatrix} \right) d\Xi = - \int_\Xi \begin{pmatrix} \tilde{\nabla} R^* \cdot S \\ R^* \cdot \dfrac{dS}{dz} \end{pmatrix} \cdot \mathbf{Q}^n \, d\Xi. \tag{9.36}$$

Now, as all the functions involving the coordinate z are known, they can be integrated over $\mathcal{I} = [0, H]$. Thus, if we consider:

$$\mathbf{K} = \begin{pmatrix} \mathbb{K} & \mathbf{k} \\ \mathbf{k}^T & \kappa \end{pmatrix}, \tag{9.37}$$

with

$$\mathbb{K} = \begin{pmatrix} \mathbf{K}_{xx} & \mathbf{K}_{xy} \\ \mathbf{K}_{xy} & \mathbf{K}_{yy} \end{pmatrix}, \tag{9.38}$$

$$\mathbf{k} = \begin{pmatrix} \mathbf{K}_{xz} \\ \mathbf{K}_{yz} \end{pmatrix} \tag{9.39}$$

and $\kappa = \mathbf{K}_{zz}$, then we can define:

$$\mathbf{K}^x = \begin{pmatrix} \int_\mathcal{I} \mathbb{K} \cdot S^2 \, dz & \int_\mathcal{I} \mathbf{k} \cdot \dfrac{dS}{dz} \cdot S \, dz \\ \int_\mathcal{I} \mathbf{k}^T \cdot \dfrac{dS}{dz} \cdot S \, dz & \int_\mathcal{I} \kappa \cdot \left(\dfrac{dS}{dz} \right)^2 dz \end{pmatrix}, \tag{9.40}$$

and

$$(\mathbf{Q}^x)^n = \sum_{j=1}^{j=n} \begin{pmatrix} \int_{\mathcal{I}} \mathbb{K} \cdot S \cdot Z_j \, dz & \int_{\mathcal{I}} \mathbf{k} \cdot \frac{dZ_j}{dz} \cdot S \, dz \\ \int_{\mathcal{I}} \mathbf{k}^T \cdot \frac{dS}{dz} \cdot Z_j \, dz & \int_{\mathcal{I}} \kappa \cdot \frac{dS}{dz} \cdot \frac{dZ_j}{dz} \, dz \end{pmatrix} \cdot \begin{pmatrix} \tilde{\nabla} X_j(\mathbf{x}) \\ X_j(\mathbf{x}) \end{pmatrix}, \tag{9.41}$$

that allows writing Eq. (9.36) into the form:

$$\int_\Omega \begin{pmatrix} \tilde{\nabla} R^* \\ R^* \end{pmatrix} \cdot \left(\mathbf{K}^x \cdot \begin{pmatrix} \tilde{\nabla} R \\ R \end{pmatrix} \right) d\Omega = - \int_\Omega \begin{pmatrix} \tilde{\nabla} R^* \\ R^* \end{pmatrix} \cdot (\mathbf{Q}^x)^n \, d\Omega \tag{9.42}$$

that defines an elliptic 2D problem defined in Ω.

9.3.2.2. *Computing $S(z)$ from $R(\mathbf{x})$*

When $R(\mathbf{x})$ is known the test function writes:

$$u^*(\mathbf{x}, z) = R(\mathbf{x}) \cdot S^*(z) \tag{9.43}$$

and the weak form (9.33) reduces to:

$$\int_\Xi \begin{pmatrix} \tilde{\nabla} R \cdot S^* \\ R \cdot \frac{dS^*}{dz} \end{pmatrix} \cdot \left(\mathbf{K} \cdot \begin{pmatrix} \tilde{\nabla} R \cdot S \\ R \cdot \frac{dS}{dz} \end{pmatrix} \right) d\,\Xi = - \int_\Xi \begin{pmatrix} \tilde{\nabla} R \cdot S^* \\ R \cdot \frac{dS^*}{dz} \end{pmatrix} \cdot \mathbf{Q}^n \, d\,\Xi. \tag{9.44}$$

Now, as all the functions involving the in-plane coordinates $\mathbf{x} = (x, y)$ are known, they can be integrated over Ω. Thus, using the previous notation we can define:

$$\mathbf{K}^z = \begin{pmatrix} \int_\Omega (\tilde{\nabla} R) \cdot (\mathbb{K} \cdot \tilde{\nabla} R) \, d\Omega & \int_\Omega (\tilde{\nabla} R) \cdot \mathbf{k} \cdot R \, d\Omega \\ \int_\Omega (\tilde{\nabla} R) \cdot \mathbf{k} \cdot R \, d\Omega & \int_\Omega \kappa \cdot R^2 \, d\Omega \end{pmatrix}, \tag{9.45}$$

and

$$(\mathbf{Q}^z)^n = \sum_{j=1}^{j=n} \begin{pmatrix} \int_\Omega (\tilde{\nabla} R) \cdot (\mathbb{K} \cdot \tilde{\nabla} X_j) \, d\Omega & \int_\Omega (\tilde{\nabla} R) \cdot \mathbf{k} \cdot X_j \, d\Omega \\ \int_\Omega (\tilde{\nabla} X_j) \cdot \mathbf{k} \cdot R \, d\Omega & \int_\Omega \kappa \cdot X_j \cdot R \, d\Omega \end{pmatrix} \cdot \begin{pmatrix} Z_j(z) \\ \frac{dZ_j}{dz}(z) \end{pmatrix}, \tag{9.46}$$

that allows writing Eq. (9.44) into the form:

$$\int_{\mathcal{I}} \begin{pmatrix} S^* \\ \frac{dS^*}{dz} \end{pmatrix} \cdot \left(\mathbf{K}^z \cdot \begin{pmatrix} S \\ \frac{dS}{dz} \end{pmatrix} \right) dz = - \int_{\mathcal{I}} \begin{pmatrix} S^* \\ \frac{dS^*}{dz} \end{pmatrix} \cdot (\mathbf{Q}^z)^n \, dz \tag{9.47}$$

that defines a one-dimensional boundary value problem (BVP).

9.3.3. 3D RTM

We now illustrate in some detail the application of the PGD to the modeling of resin transfer moulding processes. We consider the flow within a porous medium in a plate domain $\Xi = \Omega \times \mathcal{I}$ with $\Omega \subset \mathbb{R}^2$ and $\mathcal{I} = [0, H] \subset \mathbb{R}$. The governing equation is obtained by combining Darcy's law, which relates the fluid velocity to the pressure gradient,

$$\mathbf{v} = -\mathbf{K} \cdot \nabla p, \tag{9.48}$$

and the incompressibility constraint,

$$\nabla \cdot \mathbf{v} = 0. \tag{9.49}$$

Introduction of Eq. (9.48) into Eq. (9.49) yields a single equation for the pressure field:

$$\nabla \cdot (\mathbf{K} \cdot \nabla p) = 0. \tag{9.50}$$

The mould contains a laminate preform composed of P different anisotropic plies of thickness h, each one characterized by a permeability tensor $\mathbf{K}_i(x, y)$ that is assumed constant through the ply thickness. We define a characteristic function

$$\chi_i(z) = \begin{cases} 1 & z_i \leq z \leq z_{i+1}, \\ 0 & \text{otherwise}, \end{cases} \tag{9.51}$$

where $z_i = (i-1)h$ is the location of the ith-ply in the plate thickness. The laminate's permeability is thus given in separated form as follows:

$$\mathbf{K}(x, y, z) = \sum_{i=1}^{P} \mathbf{K}_i(\mathbf{x}) \cdot \chi_i(z), \tag{9.52}$$

where $\mathbf{x}$ denotes the in-plane coordinates, i.e., $\mathbf{x} = (x, y) \in \Omega$.

The weak form of Eq. (9.50) reads:

$$\int_{\Xi} \nabla p^* \cdot (\mathbf{K} \cdot \nabla p) \, d\Xi = 0, \tag{9.53}$$

for all test functions p^* selected in an appropriate functional space. Dirichlet boundary conditions are imposed for the pressure at the inlet and outlet of the flow domain $p(\mathbf{x} \in \Gamma_D) = p_g(\mathbf{x})$, while zero flux (i.e., no flow) $\nabla p \cdot \mathbf{n} = 0$ is imposed elsewhere ($\mathbf{n}$ being the unit outwards vector defined on the domain boundary) as a weak boundary condition. We seek an approximate solution $p(x, y, z)$ in the PGD form:

$$p(\mathbf{x}, z) \approx \sum_{j=1}^{N} X_j(\mathbf{x}) \cdot Z_j(z). \tag{9.54}$$

The PGD algorithm then proceeds as follows. Assume that the first n functional products have been computed, i.e.,

$$p^n(\mathbf{x}, z) = \sum_{j=1}^{n} X_j(\mathbf{x}) \cdot Z_j(z), \tag{9.55}$$

is a known quantity. We must now perform an enrichment step to obtain

$$p^{n+1}(\mathbf{x}, z) = p^n(\mathbf{x}, z) + R(\mathbf{x}) \cdot S(z). \tag{9.56}$$

The test function involved in the weak form is given by

$$p^*(\mathbf{x}, z) = R^*(\mathbf{x}) \cdot S(z) + R(\mathbf{x}) \cdot S^*(z). \tag{9.57}$$

Introducing Eqs. (9.56) and (9.57) into Eq. (9.53), we obtain

$$\begin{aligned}
&\int_{\Xi} \left(\begin{pmatrix} \tilde{\nabla} R^* \cdot S \\ R^* \cdot \dfrac{dS}{dz} \end{pmatrix} + \begin{pmatrix} \tilde{\nabla} R \cdot S^* \\ R \cdot \dfrac{dS^*}{dz} \end{pmatrix} \right) \cdot \left(\mathbf{K} \cdot \begin{pmatrix} \tilde{\nabla} R \cdot S \\ R \cdot \dfrac{dS}{dz} \end{pmatrix} \right) d\,\Xi \\
&= - \int_{\Xi} \left(\begin{pmatrix} \tilde{\nabla} R^* \cdot S \\ R^* \cdot \dfrac{dS}{dz} \end{pmatrix} + \begin{pmatrix} \tilde{\nabla} R \cdot S^* \\ R \cdot \dfrac{dS^*}{dz} \end{pmatrix} \right) \cdot \mathbf{Q}^n \, d\,\Xi,
\end{aligned} \tag{9.58}$$

where $\tilde{\nabla}$ denotes the plane component of the gradient operator, i.e., $\tilde{\nabla}^T = \left(\frac{\partial}{\partial x}, \frac{\partial}{\partial y} \right)$ and $\mathbf{Q}^n$ is a flux term known at step n:

$$\mathbf{Q}^n = \mathbf{K} \cdot \sum_{j=1}^{n} \begin{pmatrix} \tilde{\nabla} X_j(\mathbf{x}) \cdot Z_j(z) \\ X_j(\mathbf{x}) \cdot \dfrac{dZ_j(z)}{dz} \end{pmatrix}. \tag{9.59}$$

As discussed previously, each enrichment step of the PGD algorithm is a nonlinear problem which must be performed by means of a suitable iterative process. Here, we compute the unknown functions $R(\mathbf{x})$ and $S(z)$ by applying an alternating direction fixed point algorithm. Thus, assuming $R(\mathbf{x})$ known, we compute $S(z)$, and then we update $R(\mathbf{x})$. The process continues until reaching convergence. The converged solutions provide the next functional product of the PGD: $R(\mathbf{x}) \rightarrow X_{n+1}(\mathbf{x})$ and $S(z) \rightarrow Z_{n+1}(z)$. Both calculations are exactly the same that were considered in the thermal model just described.

9.3.4. The elastic problem defined in plate domains

We proposed in [74] and original in-plane–out-of-plane decomposition of the 3D elastic solution in a plate geometry. The elastic problem was defined in a plate domain $\Xi = \Omega \times \mathcal{I}$ with $(x, y) \in \Omega$, $\Omega \subset \mathbb{R}^2$ and $z \in \mathcal{I}$, $\mathcal{I} = [0, H] \subset \mathbb{R}$, being H the plate thickness. The separated representation of the displacement field $\mathbf{u} = (u_1, u_2, u_3)$ reads:

$$\mathbf{u}(x,y,z) = \begin{pmatrix} u_1(x,y,z) \\ u_2(x,y,z) \\ u_3(x,y,z) \end{pmatrix} \approx \sum_{i=1}^{N} \begin{pmatrix} P_1^i(x,y) \cdot T_1^i(z) \\ P_2^i(x,y) \cdot T_2^i(z) \\ P_3^i(x,y) \cdot T_3^i(z) \end{pmatrix}, \tag{9.60}$$

where P_k^i, $k = 1, 2, 3$, are functions of the in-plane coordinates (x, y) whereas T_k^i, $k = 1, 2, 3$, are functions involving the thickness coordinate z. In [74] we compared the first modes of such separated representations with the kinematic hypotheses usually considered in plate theories. Similar behavior was noticed in the case of elastic solutions in shell domains with respect to classical shell theories.

Expression (9.60) can be written in a more compact form by using the Hadamard (component-to-component) product:

$$\mathbf{u}(x,y,z) \approx \sum_{i=1}^{N} \mathbf{P}^i(x,y) \circ \mathbf{T}^i(z), \tag{9.61}$$

where vectors $\mathbf{P}^i$ and $\mathbf{T}^i$ contain functions P_k^i and T_k^i, respectively.

Because neither the number of terms in the separated representation of the displacement field nor the dependence on x_3 of functions T_k^i are assumed *a priori*, the approximation is flexible enough for representing the fully 3D solution, being obviously more general than theories assuming particular *a priori* evolutions in the thickness direction z.

Let us consider a linear elasticity problem on a plate domain $\Xi = \Omega \times \mathcal{I}$. The weak form using the so-called Voigt notation reads:

$$\int_{\Xi} \epsilon(\mathbf{u}^*)^T \cdot \mathbf{K} \cdot \epsilon(\mathbf{u})\, d\mathbf{x} = \int_{\Xi} \mathbf{u}^* \cdot \mathbf{f}_d\, d\mathbf{x} + \int_{\Gamma_N} \mathbf{u}^* \cdot \mathbf{F}_d\, d\mathbf{x}, \quad \forall \mathbf{u}^*, \tag{9.62}$$

where $\mathbf{K}$ is the generalized 6×6 Hooke tensor, $\mathbf{f}_d$ represents the volumetric body forces while $\mathbf{F}_d$ represents the traction applied on the boundary Γ_N.

The separation of variables introduced in Eq. (9.60) yields the following expression for the derivatives of the displacement components u_i, $i = 1, 2, 3$:

$$\frac{\partial u_i}{\partial x} \approx \sum_{k=1}^{k=N} \frac{\partial P_i^k}{\partial x} \cdot T_i^k, \tag{9.63}$$

$$\frac{\partial u_i}{\partial y} \approx \sum_{k=1}^{k=N} \frac{\partial P_i^k}{\partial y} \cdot T_i^k, \tag{9.64}$$

and

$$\frac{\partial u_i}{\partial z} \approx \sum_{k=1}^{k=N} P_i^k \cdot \frac{\partial T_i^k}{\partial z}, \tag{9.65}$$

from which we can obtain the separated vector form of the strain tensor $\boldsymbol{\epsilon}$:

$$\boldsymbol{\epsilon}(\mathbf{u}(x,y,z)) \approx \sum_{k=1}^{N} \begin{pmatrix} \frac{\partial P_1^k}{\partial x} \cdot T_1^k \\ \frac{\partial P_2^k}{\partial y} \cdot T_2^k \\ P_3^k \cdot \frac{\partial T_3^k}{\partial z} \\ \frac{\partial P_1^k}{\partial y} \cdot T_1^k + \frac{\partial P_2^k}{\partial x} \cdot T_2^k \\ \frac{\partial P_3^k}{\partial x} \cdot T_3^k + P_1^k \cdot \frac{\partial T_1^k}{\partial z} \\ \frac{\partial P_3^k}{\partial y} \cdot T_3^k + P_2^k \cdot \frac{\partial T_2^k}{\partial z} \end{pmatrix}. \tag{9.66}$$

Depending on the number of non-zero elements in the $\mathbf{K}$ matrix, the development of $\boldsymbol{\epsilon}(\mathbf{u}^*)^T \cdot \mathbf{K} \cdot \boldsymbol{\epsilon}(\mathbf{u})$ involves different number of terms, 21 in the case of an isotropic material and 41 in the case of general anisotropic behaviors.

The separated representation constructor proceeds by computing a term of the sum at each iteration. Assuming that the first $n-1$ modes (terms of the finite sum) of the solution were already computed, $\mathbf{u}^{n-1}(x,y,z)$ with $n \geq 1$, the solution enrichment reads:

$$\mathbf{u}^n(x,y,z) = \mathbf{u}^{n-1}(x,y,z) + \mathbf{P}^n(x,y) \circ \mathbf{T}^n(z), \tag{9.67}$$

where both vectors $\mathbf{P}^n$ and $\mathbf{T}^n$ containing functions P_i^n and T_i^n ($i = 1, 2, 3$) depending on (x,y) and z, respectively, are unknown at the present iteration. The test function $\mathbf{u}^*$ reads $\mathbf{u}^* = \mathbf{P}^* \circ \mathbf{T}^n + \mathbf{P}^n \circ \mathbf{T}^*$.

The introduction of Eq. (9.67) into (9.62) results in a nonlinear problem. We proceed by considering the simplest linearization strategy, an alternated directions fixed point algorithm, that proceeds by calculating $\mathbf{P}^{n,k}$ from $\mathbf{T}^{n,k-1}$ and then by updating $\mathbf{T}^{n,k}$ from the just calculated $\mathbf{P}^{n,k}$ where k refers to the step of the nonlinear solver. The iteration procedure continues until convergence, that is, until reaching the fixed point $\|\mathbf{P}^{n,k} \circ \mathbf{T}^{n,k} - \mathbf{P}^{n,k-1} \circ \mathbf{T}^{n,k-1}\| < \epsilon$, that results in the searched functions $\mathbf{P}^{n,k} \rightarrow \mathbf{P}^n$ and $\mathbf{T}^{n,k} \rightarrow \mathbf{T}^n$. Then, the enrichment step continues by looking for the next mode $\mathbf{P}^{n+1} \circ \mathbf{T}^{n+1}$. The enrichment stops when the model residual becomes small enough.

When $\mathbf{T}^n$ is assumed known, we consider the test function $\mathbf{u}^\star$ given by $\mathbf{P}^\star \circ \mathbf{T}^n$. By introducing the trial and test functions into the weak form and then integrating in $\mathcal{I}$ because all the functions depending on the thickness coordinate are known, we obtain a 2D weak formulation defined in Ω whose discretization (by using a standard discretization strategy, e.g., finite elements) allows computing $\mathbf{P}^n$.

Analogously, when $\mathbf{P}^n$ is assumed known, the test function $\mathbf{u}^\star$ is given by $\mathbf{P}^n \circ \mathbf{T}^\star$. By introducing the trial and test functions into the weak form and then integrating in Ω because all the functions depending on the in-plane coordinates (x, y) are at present known, we obtain a 1D weak formulation defined in $\mathcal{I}$ whose discretization (using any technique for solving standard ODE equations) allows computing $\mathbf{T}^n$.

As discussed in [74] this separated representation allows computing 3D solutions while keeping a computational complexity characteristic of 2D solution procedures. If we consider a hexahedral domain discretized using a regular structured grid with N_1, N_2 and N_3 nodes in the x_1, x_2 and x_3 directions respectively, usual mesh-based discretization strategies imply a challenging issue because the number of nodes involved in the model scales with $N_1 \cdot N_2 \cdot N_3$, however, by using the separated representation and assuming that the solution involves N modes, one must solve about N 2D problems related to the functions involving the in-plane coordinates (x, y) and the same number of 1D problems related to the functions involving the thickness coordinate z. The computing time related to the solution of the one-dimensional problems can be neglected with respect to the one required for solving the two-dimensional ones. Thus, the resulting complexity scales as $N \cdot N_1 \cdot N_2$. By comparing both complexities we can notice that as soon as $N_3 \gg N$ the use of separated representations leads to impressive computing time savings, making possible the solution of models never until now solved, and even using light computing platforms.

9.3.5. 3D elastic problem in a shell domain

In this section, we generalize the rationale just described for the solution of elastic problems defined in shell domains.

9.3.5.1. *Shell representation*

The shell domain Ω^S, assumed with constant thickness, can be described from a reference surface $\mathbf{X}$, that in what follows will be identified to the shell middle surface but that in the general case could be any other one, parametrized by the coordinates ξ, η, that is $\mathbf{X}(\xi, \eta)$, where:

$$\mathbf{X}(\xi,\eta) = \begin{pmatrix} X_1(\xi,\eta) \\ X_2(\xi,\eta) \\ X_3(\xi,\eta) \end{pmatrix}. \tag{9.68}$$

Being $\mathbf{n}$ the unit vector normal to the middle surface, the shell domain Ω^S can be parametrized from:

$$\mathbf{x}(\xi,\eta,\zeta) = \mathbf{X}(\xi,\eta) + \zeta \cdot \mathbf{n}. \tag{9.69}$$

The geometrical transformation $(\xi, \eta, \zeta) \rightarrow (x_1, x_2, x_3)$ involves [75]

$$\tilde{\mathbf{F}} = \left[\frac{\partial \mathbf{x}}{\partial \xi} \frac{\partial \mathbf{x}}{\partial \eta} \mathbf{n} \right]. \tag{9.70}$$

The inverse transformation $(x_1, x_2, x_3) \rightarrow (\xi, \eta, \zeta)$, described by $\tilde{\mathbf{F}}^{-1}$ can be also easily obtained [75].

9.3.5.2. *Weak form*

The weak form of the elastic problem defined in the shell domain Ω^S using again the Voigt notation writes:

$$\int_{\Omega^S} \epsilon(\mathbf{u}^*)^T \cdot \mathbf{K} \cdot \epsilon(\mathbf{u})\, d\mathbf{x} = \int_{\Omega^S} \mathbf{u}^* \cdot \mathbf{f}_d\, d\mathbf{x} + \int_{\Gamma_N^S} \mathbf{u}^* \cdot \mathbf{F}_d\, d\mathbf{x}. \tag{9.71}$$

Now we are considering the coordinates transformation introduced in the previous section mapping $\mathbf{x} \in \Omega^S$ into $(\xi, \eta, \zeta) \in \Xi = \Omega \times \mathcal{I}$, with $(\xi, \eta) \in \Omega \subset \mathbb{R}^2$ and $\zeta \in \mathcal{I} \subset \mathbb{R}$.

The geometric transformation requires to transform the differential operator as well as the different volume and surface elements. Knowing that under the small displacements and strains assumption, the strain tensor consists of the symmetric part of the gradient of displacement tensor, i.e.,

$$\epsilon(\mathbf{u}) = \frac{1}{2}\left(\nabla\mathbf{u} + (\nabla\mathbf{u})^T\right), \tag{9.72}$$

it can be transformed taking into account the transformation of the gradient differential operator

$$\nabla(\cdot) = \nabla_\xi(\cdot)\cdot\tilde{\mathbf{F}}^{-1}, \tag{9.73}$$

where $\nabla_\xi(\cdot)$ denotes the gradient in the parametric space.

The volume element involved in the integral in Ω^S writes [75]

$$d\mathbf{x} = dx_1 \cdot dx_2 \cdot dx_3 = \sqrt{a}\cdot(1 - 2\cdot H\cdot\zeta + K\cdot\zeta^2)\cdot d\xi\cdot d\eta\cdot d\zeta, \tag{9.74}$$

where a is the determinant of the metric tensor related to the middle surface mapping and H and K the curvatures [75].

9.3.5.3. *In-plane–out-of-plane separated representation*

With the weak form defined in $\Xi = \Omega \times \mathcal{I}$ the situation is quite similar to the one encountered in the analysis of elastic problems in plate geometries, that was addressed in [74] and that we just summarized in the previous section.

In what follows, we use the following matrix notation: If $\mathbf{a}$ and $\mathbf{b}$ are vectors of the same dimension, vector $\mathbf{c}$, defined from $\mathbf{c} = \mathbf{a}\circ\mathbf{b}$, has as components $c_i = a_i \cdot b_i$. If $\mathbf{a}$ and $\mathbf{b}$ are second-order tensor with the same size, tensor $\mathbf{c}$, defined from $\mathbf{c} = \mathbf{a}\circ\mathbf{b}$, has components $c_{ij} = a_{ij}\cdot b_{ij}$ (no sum with respect to the repeated indexes). In this case it results $\mathbf{a}:\mathbf{b} = c$, with the scalar c given by $c = a_{ij}\cdot b_{ij}$ considering sum with respect to the repeated indexes (Einstein's summation convention).

We could perform an in-plane–out-of-plane separated representation of the displacement field, similar to (9.60) but now involving the coordinates (ξ, η, ζ)

$$\mathbf{u}(\xi,\eta,\zeta) = \begin{pmatrix} u_1(\xi,\eta,\zeta) \\ u_2(\xi,\eta,\zeta) \\ u_3(\xi,\eta,\zeta) \end{pmatrix} \approx \sum_{i=1}^{N} \begin{pmatrix} P_1^i(\xi,\eta)\cdot T_1^i(\zeta) \\ P_2^i(\xi,\eta)\cdot T_2^i(\zeta) \\ P_3^i(\xi,\eta)\cdot T_3^i(\zeta) \end{pmatrix}, \tag{9.75}$$

or in a more compact form

$$\mathbf{u}(\xi,\eta,\zeta) \approx \sum_{i=1}^{N} \mathbf{P}^i(\xi,\eta)\circ\mathbf{T}^i(\zeta). \tag{9.76}$$

As explained in the previous section, the construction of such a separated representation is performed sequentially, thus assuming known the solution at iteration $n-1$, the solution at iteration n is sought as

$$\mathbf{u}^n(\xi,\eta,\zeta) = \mathbf{u}^{n-1}(\xi,\eta,\zeta) + \mathbf{P}^n(\xi,\eta) \circ \mathbf{T}^n(\zeta). \tag{9.77}$$

By introducing (9.77) in the weak form and using the alternated directions fixed point algorithm we can calculate $\mathbf{P}^n(\xi,\eta)$ by assuming $\mathbf{T}^n(\zeta)$ known and then updated $\mathbf{T}^n(\zeta)$ from the just calculated $\mathbf{P}^n(\xi,\eta)$. The iteration continues until reaching the convergence (the fixed point) that determines both functions $\mathbf{P}^n(\xi,\eta)$ and $\mathbf{T}^n(\zeta)$.

However, the decomposition in a problem defined in Ω for calculating function $\mathbf{P}^n(\xi,\eta)$, obtained by integrating the weak form in $\mathcal{I}$, and in another problem defined in $\mathcal{I}$ for calculating function $\mathbf{T}^n(\zeta)$, obtained by integrating the weak form in Ω, requires the separated representation of all the operators, variables, coefficients and functions involved in the weak form.

For the displacement (the trial $\mathbf{u}$ and the test $\mathbf{u}^*$ displacements), we just indicated the separated in-plane–out-of-plane representation. This representation allows defining a separated representation of the associated strain tensors $\boldsymbol{\epsilon}(\mathbf{u})$ and $\boldsymbol{\epsilon}(\mathbf{u}^*)$ as illustrated in Eq. (9.66) but for this purpose we must define a separated representation of the transformation gradient involved in Eq. (9.73) $\tilde{\mathbf{F}}^{-1}$. This issue was deeply considered in [75].

9.3.6. Squeeze flow in composite laminates

The in-plane–out-of-plane separated representation allows the solution of full 3D flow models defined in plate geometries with a computational complexity characteristic of 2D simulations. In the present case, the 3D velocity field reads

$$\mathbf{v}(\mathbf{x},z) = \begin{pmatrix} u(\mathbf{x},z) \\ v(\mathbf{x},z) \\ w(\mathbf{x},z) \end{pmatrix} \approx \begin{pmatrix} \sum\limits_{i=1}^{N} P_i^1(\mathbf{x}) \cdot T_i^1(z) \\ \sum\limits_{i=1}^{N} P_i^2(\mathbf{x}) \cdot T_i^2(z) \\ \sum\limits_{i=1}^{N} P_i^3(\mathbf{x}) \cdot T_i^3(z) \end{pmatrix}, \tag{9.78}$$

which leads to a separated representation of the strain rate, when introduced into the flow problem weak form allows the calculation of functions $P_i(x, y)$ by solving the corresponding 2D equations and functions $T_i(z)$ by solving the associated 1D equations, as described later.

Equation (9.78) can be rewritten in the compact form

$$\mathbf{v}(\mathbf{x}, z) \approx \sum_{i=1}^{N} \mathbf{P}_i(\mathbf{x}) \circ \mathbf{T}_i(z). \tag{9.79}$$

Using notation in (9.78), the velocity gradient $\nabla\mathbf{v}(\mathbf{x}, z)$ can be written as:

$$\nabla\mathbf{v} = \begin{pmatrix} \frac{\partial u}{\partial x} & \frac{\partial u}{\partial y} & \frac{\partial u}{\partial z} \\ \frac{\partial v}{\partial x} & \frac{\partial v}{\partial y} & \frac{\partial v}{\partial z} \\ \frac{\partial w}{\partial x} & \frac{\partial w}{\partial y} & \frac{\partial w}{\partial z} \end{pmatrix} \approx \sum_{i=1}^{N} \begin{pmatrix} \frac{\partial P_i^1}{\partial x} & \frac{\partial P_i^1}{\partial y} & P_i^1 \\ \frac{\partial P_i^2}{\partial x} & \frac{\partial P_i^2}{\partial y} & P_i^2 \\ \frac{\partial P_i^3}{\partial x} & \frac{\partial P_i^3}{\partial y} & P_i^3 \end{pmatrix}$$

$$\circ \begin{pmatrix} T_i^1 & T_i^1 & \frac{\partial T_i^1}{\partial z} \\ T_i^2 & T_i^2 & \frac{\partial T_i^2}{\partial z} \\ T_i^3 & T_i^3 & \frac{\partial T_i^3}{\partial z} \end{pmatrix} = \sum_{i=1}^{N} \mathbb{P}_i(\mathbf{x}) \circ \mathbb{T}_i(z). \tag{9.80}$$

The solution of different 3D flow problems within the in-plane–out-of-plane separated representation is revisited in the next sections.

9.3.6.1. *Stokes model*

The Stokes flow model is defined in $\Xi = \Omega \times \mathcal{I}$, $\Omega \subset \mathbb{R}^2$ and $\mathcal{I} \subset \mathbb{R}$, and for an incompressible fluid, in absence of inertia and mass terms reduces to:

$$\begin{cases} \nabla \cdot \boldsymbol{\sigma} = \mathbf{0}, \\ \boldsymbol{\sigma} = -p\mathbf{I} + 2\eta\mathbf{D}, \\ \nabla \cdot \mathbf{v} = 0, \end{cases} \tag{9.81}$$

where $\boldsymbol{\sigma}$ is the Cauchy's stress tensor, $\mathbf{I}$ the unit tensor, η the fluid viscosity, p the pressure (Lagrange multiplier associated with the incompressibility constraint) and the rate of strain tensor $\mathbf{D}$ defined as

$$\mathbf{D} = \frac{\nabla\mathbf{v} + (\nabla\mathbf{v})^T}{2}. \tag{9.82}$$

A penalty formulation is used to circumvent the issue related to stable mixed formulations (LBB give full form of LBB conditions) within the separated representation, which to this day is an open issue. The mass balance is modified by introducing a penalty coefficient λ whose value is usually very small. The penalty formulation can be written as

$$\nabla \cdot \mathbf{v} + \lambda p = 0, \tag{9.83}$$

or more explicitly

$$p = -\frac{\nabla \cdot \mathbf{v}}{\lambda} = -\frac{\mathrm{Tr}(\mathbf{D})}{\lambda}, \tag{9.84}$$

where Tr( ) refers the trace operator. The trace of a tensor, and in particular the trace of the rate of strain, can be written as $\mathrm{Tr}(\mathbf{D}) = \mathbf{D} : \mathbf{I}$.

The weak form of the penalized Stokes problem, for a test velocity $\mathbf{v}^*$ vanishing on the boundary in which the velocity is prescribed, and assuming null tractions in the remaining part of the domain boundary, can be written as

$$\int_{\Omega \times \mathcal{I}} \left\{ \frac{1}{\lambda} \mathrm{Tr}(\mathbf{D}^*) \mathrm{Tr}(\mathbf{D}) + 2\eta \mathbf{D}^* : \mathbf{D} \right\} d\mathbf{x}\, dz = 0, \tag{9.85}$$

where as proved in [76] can be expressed from

$$2\eta \mathbf{D}^* : \mathbf{D} \approx \frac{\eta}{2} \sum_{j=1}^{N} \sum_{k=1}^{4} \left(\mathbb{A}_{jk}^*(\mathbf{x}) : \mathbb{B}_{jk}(z) + \mathbb{A}_{jk}(\mathbf{x}) : \mathbb{B}_{jk}^*(z) \right), \tag{9.86}$$

and

$$\frac{1}{\lambda} \mathrm{Tr}(\mathbf{D}^*) \cdot \mathrm{Tr}(\mathbf{D}) \approx \frac{1}{\lambda} \sum_{j=1}^{N} \left(\mathbb{F}_j^*(\mathbf{x}) : \mathbb{G}_j(z) + \mathbb{F}_j(\mathbf{x}) : \mathbb{G}_j^*(z) \right), \tag{9.87}$$

where $\mathbf{D}^*$ is the strain rate related to the test field $\mathbf{v}^*$.

The construction of the solution separated representation is performed incrementally, a term of the sum at each iteration. Thus, supposing that at iteration $n-1$, $n \geq 1$, the first $n-1$ terms of the velocity separated representation were already computed

$$\mathbf{v}^{n-1}(\mathbf{x}, z) = \sum_{i=1}^{n-1} \mathbf{P}_i(\mathbf{x}) \circ \mathbf{T}_i(z), \tag{9.88}$$

the terms involved in the weak form (9.85) are

$$\mathbf{D}^* : \mathbf{D}^{n-1} = \frac{1}{4}\sum_{j=1}^{n-1}\sum_{k=1}^{4}\left(\mathbb{A}^*_{jk}(\mathbf{x}) : \mathbb{B}_{jk}(z) + \mathbb{A}_{jk}(\mathbf{x}) : \mathbb{B}^*_{jk}(z)\right), \tag{9.89}$$

and

$$\mathrm{Tr}(\mathbf{D}^*) \cdot \mathrm{Tr}(\mathbf{D}^{n-1}) = \sum_{j=1}^{n-1}\left(\mathbb{F}^*_j(\mathbf{x}) : \mathbb{G}_j(z) + \mathbb{F}_j(\mathbf{x}) : \mathbb{G}^*_j(z)\right), \tag{9.90}$$

respectively.

When looking for the improved velocity field $\mathbf{v}^n(\mathbf{x}, z)$ at iteration n

$$\mathbf{v}^n(\mathbf{x}, z) = \sum_{i=1}^{n} \mathbf{P}_i(\mathbf{x}) \circ \mathbf{T}_i(z) = \mathbf{v}^{n-1}(\mathbf{x}, z) + \mathbf{P}_n(\mathbf{x}) \circ \mathbf{T}_n(z), \tag{9.91}$$

we consider the test function $\mathbf{v}^*(\mathbf{x}, z)$

$$\mathbf{v}^* = \mathbf{P}^* \circ \mathbf{T}_n + \mathbf{P}_n \circ \mathbf{T}^*. \tag{9.92}$$

The fixed point algorithm for solving the resulting nonlinear problem proceeds by calculating functions $\mathbf{T}_n$ by assuming known functions $\mathbf{P}_n$, and then updating the first. The iteration continues until reaching the fixed point.

Consider a laminate composed of P layers in which each layer involves a linear and isotropic viscous fluid of viscosity η_i, thus the extended Stokes flow problem in its weak form involves the dependence of the viscosity along the thickness direction.

If H is the total laminate thickness, and assuming for the sake of simplicity and without loss of generality that all the plies have the same thickness h, it results $h = \frac{H}{P}$. Now, from the characteristic function of each ply $\chi_i(z)$, $i = 1, \ldots, P$:

$$\chi_i(z) = \begin{cases} 1 & \text{if } (i-1)h \le z < ih, \\ 0 & \text{elsewehere}, \end{cases} \tag{9.93}$$

the viscosity reads

$$\eta(\mathbf{x}, z) = \sum_{i=1}^{P} \eta_i \cdot \chi_i(z), \tag{9.94}$$

where it is assumed, again without loss of generality, that the viscosity does not evolve in the plane, i.e., $\eta_i(\mathbf{x}) = \eta_i$.

This decomposition is fully compatible with the velocity separated representation (9.78) and with the in-plane–out-of-plane decomposition considered for solving the associated weak form.

9.3.6.2. *Power-law fluid*

The Stokes model extended to power-law fluids reads:

$$\begin{cases} \nabla p = \nabla \cdot \mathbf{T}, \\ \nabla \cdot \mathbf{v} = 0, \end{cases} \tag{9.95}$$

where the extra-stress tensor for power-law fluids writes:

$$\mathbf{T} = 2KD_{\mathrm{eq}}^{n-1}\mathbf{D}, \tag{9.96}$$

with the equivalent strain rate D_{eq} given by

$$D_{\mathrm{eq}} = \sqrt{2(\mathbf{D} : \mathbf{D})}, \tag{9.97}$$

where ":" denotes the tensor product twice contracted.

The solution is again carried out by using a penalty formulation. Moreover, at each iteration of the nonlinear solver we must evaluate the equivalent strain rate, and write it in a separated form for enhancing the efficiency of the separated representation solver. The simplest way for performing such decomposition

$$D_{\mathrm{eq}}^{n-1} \approx \sum_{i=1}^{i=M} F_i(x, y) \cdot G_i(z), \tag{9.98}$$

consists of using a singular value decomposition. This decomposition is optimal but it requires a 3D reconstruction and data storage that can be expensive from the computational point of view.

As previously discussed this rationale can be easily extended for considering multi-layered domains with different behaviors, from an adequate in-plane–out-of-plane expression of the viscosity along the domain thickness.

9.3.6.3. *Brinkman's model*

In composite manufacturing processes resin located between fibers in the reinforcement layers also flows. A usual approach for evaluating the resin flow in such circumstances consists of solving the associated Darcy's model. It is well known that Darcy–Stokes coupling at the interlayers generates numerical instabilities because the localized boundary layers whose accurate description requires very rich representations (very fine meshes along the laminate thickness).

In [77] we proposed to use the Brinkman model that allows representing in a unified manner both the Darcy and the Stokes behaviors. In order to avoid numerical inaccuracies a very fine representation along the thickness direction was considered and for circumventing the exponential increase in the number of degrees of freedom that such a fine representation would imply when extended to the whole laminate domain, we considered again the in-plane–out-of-plane separated representation previously introduced.

The Brinkman's model is defined by

$$\nabla p = \mu \cdot \mathbf{K}^{-1} \cdot \mathbf{v} + \eta \cdot \Delta \mathbf{v}, \tag{9.99}$$

where μ is the dynamic viscosity, $\mathbf{K}$ the layer permeability tensor and η the dynamic effective viscosity.

In the zones where Stokes model applies (resin layers) we assign a very large isotropic permeability $\mathbf{K} = \mathbf{I}$ (units in the metric system and $\mathbf{I}$ being the unit tensor) whereas in the ones occupied by the reinforcement, the permeability is assumed anisotropic, being several orders of magnitude lower, typically 10^{-8}. Thus the Darcy's component in Eq. (9.99) does not perturb the Stokes flow in the resin layers, and it becomes dominant in the reinforcement layers. Additionally by choosing this outstanding difference in permeability, representative of the one observed in Liquid Resin Infusion processes when highly porous distribution media are used, we also want to give the evidence that this type of problem can be addressed by the proposed approach.

9.3.6.4. *Squeeze flow of multiaxial laminates*

The case of a prepreg ply reinforced by continuous fibers oriented along direction $\mathbf{p}^T = (p_x, p_y, 0)$, $\|\mathbf{p}\| = 1$ (form now on $\mathbf{p}$ denotes the reinforcement direction whereas p represents the scalar pressure field), is analyzed here. It is assumed that the thermoplastic resin exhibits Newtonian behavior. Thus the velocity $\mathbf{v}(\mathbf{x}, z)$ of the equivalent anisotropic fluid must satisfy the incompressibility and inextensibility constraints

$$\nabla \cdot \mathbf{v} = 0, \tag{9.100}$$

and

$$\mathbf{p}^T \cdot \nabla \mathbf{v} \cdot \mathbf{p} = 0, \tag{9.101}$$

respectively. Expression (9.101) can be rewritten using tensor notation as $\nabla \mathbf{v} : \mathbf{a} = 0$, where the second-order orientation tensor $\mathbf{a}$ is defined from $\mathbf{a} = \mathbf{p} \cdot \mathbf{p}^T = \mathbf{p} \otimes \mathbf{p}$.

The orientation tensor $\mathbf{a}$ has only planar components (the out-of-plane fiber orientation can be neglected in the case of laminates), it is symmetric and of unit trace, i.e.,

$$\mathbf{a} = \begin{pmatrix} a_{xx} & a_{xy} & 0 \\ a_{yx} & a_{yy} & 0 \\ 0 & 0 & 0 \end{pmatrix} = \begin{pmatrix} \mathcal{A} & \mathbf{0} \\ \mathbf{0}^T & 0 \end{pmatrix}, \tag{9.102}$$

where $\mathcal{A}$ represents the plane component of the orientation tensor $\mathbf{a}$, $a_{xy} = a_{yx}$ (i.e., $\mathcal{A} = \mathcal{A}^T$) and $a_{yy} = 1 - a_{xx}$.

The simplest expression of the Ericksen's constitutive equation [78] can be written in the compact form as follows:

$$\boldsymbol{\sigma} = -p\mathbf{I} + T\mathbf{a} + 2\eta_T \mathbf{D} + 2(\eta_L - \eta_T)(\mathbf{D} \cdot \mathbf{a} + \mathbf{a} \cdot \mathbf{D}), \tag{9.103}$$

and introduced in the linear momentum balance

$$\nabla \cdot \boldsymbol{\sigma} = \mathbf{0}, \tag{9.104}$$

allows the flow kinematics calculation.

In Eq. (9.103), p and T represent, respectively, the Lagrange multipliers related to the incompressibility and inextensibility, and η_L and η_T the longitudinal and transverse shear viscosities, respectively.

By introducing the incompressibility and inextensibility constraints from a penalty formulation it results:

$$\nabla \cdot \mathbf{v} + \lambda p = 0, \tag{9.105}$$

and

$$\nabla \mathbf{v} : \mathbf{a} + \xi T = 0, \tag{9.106}$$

with λ and ξ small enough, leading to:

$$p = -\frac{\nabla \cdot \mathbf{v}}{\lambda} = -\frac{1}{\lambda}\mathrm{Tr}(\mathbf{D}), \tag{9.107}$$

and

$$T = -\frac{\nabla \mathbf{v} : \mathbf{a}}{\xi} = -\frac{1}{\xi}\mathbf{D} : \mathbf{a}, \tag{9.108}$$

where the fact that $\mathbf{a}$ is symmetric, implying $\nabla \mathbf{v} : \mathbf{a} = \mathbf{D} : \mathbf{a}$, is used.

The weak form for a test velocity $\mathbf{v}^*(\mathbf{x}, z)$ vanishing at the boundary in which velocity is prescribed and assuming null tractions in the remaining part of the domain boundary can be expressed as

$$\int_{\Omega \times \mathcal{I}} \mathbf{D}^* : \boldsymbol{\sigma} \, d\mathbf{x} \, dz = 0. \tag{9.109}$$

By introducing the Ericksen constitutive equation (9.103) as well as both penalty expressions (9.105) and (9.106), it can be written as

$$\int_{\Omega \times \mathcal{I}} \mathbf{D}^* : \boldsymbol{\sigma} \, d\mathbf{x} \, dz = \int_{\Omega \times \mathcal{I}} \mathbf{D}^* : \left(\frac{\mathrm{Tr}(\mathbf{D})}{\lambda} \mathbf{I} - \frac{\mathbf{D} : \mathbf{a}}{\xi} \mathbf{a} + \eta_T \mathbf{D} \right. \\ \left. + \tilde{\eta} \left(\mathbf{D} \cdot \mathbf{a} + \mathbf{a} \cdot \mathbf{D} \right) \right) d\mathbf{x} \, dz = 0, \tag{9.110}$$

with $\tilde{\eta} = \eta_L - \eta_T$. This integral form can be rewritten as

$$\int_{\Omega \times \mathcal{I}} \left\{ \frac{\mathrm{Tr}(\mathbf{D}^*) \cdot \mathrm{Tr}(\mathbf{D})}{\lambda} - \frac{(\mathbf{D}^* : \mathbf{a}) \cdot (\mathbf{D} : \mathbf{a})}{\xi} \right. \\ \left. + \eta_T \mathbf{D}^* : \mathbf{D} + \tilde{\eta} \mathbf{D}^* : (\mathbf{D} \cdot \mathbf{a} + \mathbf{a} \cdot \mathbf{D}) \right\} d\mathbf{x} \, dz = 0. \tag{9.111}$$

At this stage, the in-plane–out-of-plane separated representation constructor of $\mathbf{v}(\mathbf{x}, z)$ proceeds as described in the previous sections.

The following remarks can be addressed:

- If $\mathbf{a} = \mathbf{0}$ this formulation reduced to one related to the Stokes flow problem.
- Laminates can be addressed by associating with each ply the planar fiber orientation $\mathbf{p}_i(\mathbf{x})$, with its out-of-plane component vanishing, from which the associated orientation tensor $\mathbf{a}_i(\mathbf{x})$ results in $\mathbf{a}_i(\mathbf{x}) = \mathbf{p}_i(\mathbf{x}) \otimes \mathbf{p}_i(\mathbf{x})$. Using again the characteristic function of the i-ply, $\chi_i(z)$, $i = 1, \ldots, P$, the orientation tensor in the laminate, $\mathbf{a}(\mathbf{x}, z)$, can be expressed as

$$\mathbf{a}(\mathbf{x}, z) = \sum_{i=1}^{P} \mathbf{a}_i(\mathbf{x}) \chi_i(z). \tag{9.112}$$

- If the fiber orientation is constant in each plane, then the laminate orientation tensor can be expressed as

$$\mathbf{a}(z) = \sum_{i=1}^{P} \mathbf{a}_i \chi_i(z). \tag{9.113}$$

9.3.7. Electromagnetic models in laminates

Composites parts tend to represent an increasing volume of production in transport industry (aeronautic and automotive). This is due to their combination of high mechanical properties and low mass. Although one aspect is still a disadvantage, this concerns their long cycle time.

Conventional processing methods for producing polymer composite parts usually involve the application of heat to the material by convection or conductive heating through elements, which depend on surface heat transfer. Microwave (MW) technology relies on volumetric heating, that means thermal energy is transferred through electromagnetic fields to materials that can absorb it at specific frequencies. Volumetric heating enables better process temperature control and less overall energy use, which can results in shorter processing cycles. Furthermore, comparable mechanical properties are shown between parts made with the MW technology and parts made with a traditional curing system. These virtues of the MW technology have attracted interest in developing the method and adopting it for the production of thermoset as well as thermoplastic composite materials.

The main drawback of this technology today is that the complex physics involved in the conversion of electromagnetic energy to thermal energy (heating) is not entirely understood and controlled. These models will simulate the way electromagnetic energy is propagated within the material volume and the various interfaces, then converted to thermal energy. The main challenge concerns the high-resolution description of the electromagnetic and thermal fields in a composite laminate, that involve plies whose characteristic in-plane dimension is orders of magnitude higher than the ones related to the thickness (typical aspect ratio are of tens of thousands). In that situation, the use of in-plane–out-of-plane separated representations within the Proper Generalized Decomposition (PGD) framework seems an appealing and valuable route for solving 3D models, very rich in both, the in-plane and the out-of-plane directions, while ensuring a computational complexity of standard 2D models.

The physical model consists of the Maxwell equations, that after some manipulations and assuming an harmonic electric field, reduces to

$$\nabla^2 \mathbf{E} = \gamma^2 \mathbf{E}, \tag{9.114}$$

with $\gamma^2 = i\omega\mu(\sigma + i\omega\epsilon)$, where μ, σ and ϵ are respectively the permeability, conductivity and permittivity, all them depending on the considered material and ω the wave pulsation.

Within the Proper Generalized Decomposition (PGD) framework, allows writing the electric field in the 3D separated form

$$\mathbf{E}(\mathbf{x}, z) \approx \sum_{j=1}^{N} \mathbf{X}_j(\mathbf{x}) \circ \mathbf{Z}_j(z). \tag{9.115}$$

If H is the total laminate thickness, and assuming for the sake of simplicity and without loss of generality that the P plies have the same thickness h, it results $h = \frac{H}{P}$. Now, from the characteristic function of each ply $\chi_i(z)$, $i = 1, \ldots, P$:

$$\chi_i(z) = \begin{cases} 1 & \text{if } (i-1)h \leq z < ih, \\ 0 & \text{elsewehere}, \end{cases} \tag{9.116}$$

the expression of γ^2 reads

$$\gamma^2(\mathbf{x}, z) = \sum_{i=1}^{P} \gamma_i^2(\mathbf{x}) \cdot \chi_i(z), \tag{9.117}$$

where it is assumed that in each ply γ^2 does not evolve along its thickness. When it is not the case a SVD must be applied in order to separate the electromagnetic properties involved in γ^2.

9.4. Coupled Physics at Interfaces

The use of thermoplastic composites, more and more considered for a variety of industrial applications, requires specific processes because the high viscosity of thermoplastic resins limits the use of standard liquid moulding manufacturing processes.

Pre-impregnated thermoplastic composites (pre-preg) are being widely used for manufacturing complex parts. In these processes *in situ* consolidation is envisaged and for that, manufacturing processes must ensure an adequate degree of intimate contact as well as the molecular diffusion across the contact interface. Both mechanisms are strongly influenced by the thermal history. From one side, the polymer viscosity that depends on the temperature, determine the squeeze flow occurring at the interfaces when pressure is applied on the laminate. This flow determines the progressive reduction of the asperities initially present on the pre-preg surfaces and the increase of the degree of intimate contact. As soon as surfaces enter in contact molecular diffusion starts, but again the molecular mobility depends strongly on the existing temperature through the so-called temperature-dependent reptation time.

Since the intimate contact at the interface level is imperfect due to the surface roughness, the heat transfer is perturbed and for quantifying its effect thermal contact resistances (TCR) were considered in our former works for performing accurate macroscopic thermal analyses in ATP (Automated Tape Placement) processes [79, 80]. The TCR is a direct macroscopic consequence of the imperfect intimate contact originated by the surface roughness, that involves as many scales as the ones present in the representation of the rough surface.

From a conceptual viewpoint there are no major difficulties: surfaces could be measured without neglecting the minimum detail and then, a high-resolution thermal model could be solved on the actual surface geometry, using the nowadays advanced experimental and computational facilities. However, even if such fine enough thermal models able to capture the microscopic events (surface asperities) do not require the use of TCR, the resulting numerical complexity for attaining the finest scales involved in the roughness representation becomes unaffordable even for the nowadays computational capabilities. Thus, the use of the TCR concept seems justified from a practical point of view. In this context, the study of its dependence on the geometrical parameters defining the composite interfaces seems of major interest for understanding the physics involved in pre-preg *in situ* consolidation, and more specifically the evolution of it during the consolidation, where due to the compression the interface evolves due to the squeeze flow at the level of the surface asperities.

9.4.1. Surface representation

The characterization of random surfaces and the effect of roughness on physics defined at surface level is a recurrent issue widely addressed in many works. Two usual surface descriptors are as follows: (i) the Structure Function $S(\tau)$ and (ii) the Power Spectral Density $P(\omega)$, both defined for a random process $z(t)$ as

$$S(\tau) = \langle (z(t) - z(t+\tau))^2 \rangle, \tag{9.118}$$

where $\langle \bullet \rangle$ refers to the average, and

$$P(\omega) = |\mathcal{F}(\omega)|^2, \tag{9.119}$$

with $\mathcal{F}(\omega)$ the Fourier's transform of $z(t)$

$$\mathcal{F}(\omega) = \int_{-\infty}^{\infty} z(t) e^{-i\omega t} dt. \tag{9.120}$$

From these two functions one can extract, as described and discussed in [81], two useful surface descriptors, the directional topothesy $C(\theta)$, quantifying the roughness amplitude, and the directional fractal dimension $D(\theta)$, quantifying the self-affinity through the space scales. As also proved in [81], when considering one-directional pre-impregnated tapes, the resulting surfaces exhibit high anisotropy, justifying the consideration of a 2D analysis along the transverse direction to the fibers.

Having been proved the fractal nature of the surfaces, they can be assimilated to a Cantor set. Even if such representation does not exactly describe the actual surface, it has two main advantages: (i) it makes possible to reduce the computational cost due to the fact that the geometry is very much simplified and composed of rectangular elements representing the asperities, making it possible, as described in the next section, the use of efficient separated representations within the proper generalized decomposition framework; and (ii) it presents the possibility of modifying easily the different fractal parameters in order to perform sensibility and parametric analyses.

The Cantor interface is created by removing rectangles, whose sides lengths evolve at each generation n_g, as illustrated in Fig. 9.2, according to:

$$Y_{n_g+1} = \left(\frac{1}{f_y}\right)^{n_g+1} Y_0, \tag{9.121}$$

and

$$Z_{n_g+1} = \left(\frac{1}{f_z}\right)^{n_g+1} Z_0, \tag{9.122}$$

where f_y and f_z are the scaling factors in each direction, Y_0 is the total length of the Cantor interface and Z_0 is the height of the deepest interface asperity. When increasing the number of generations, the asperities characteristic size decreases. The maximum number of generations is determined by the fiber size, which determines the finest scale of analysis. At each generation s rectangles result from each parent.

Thus, the interface is fully determined by using four parameters, the scaling factors f_y and f_z, the number of asperities s created at each generation and the considered number of generations n_g, the last, as just

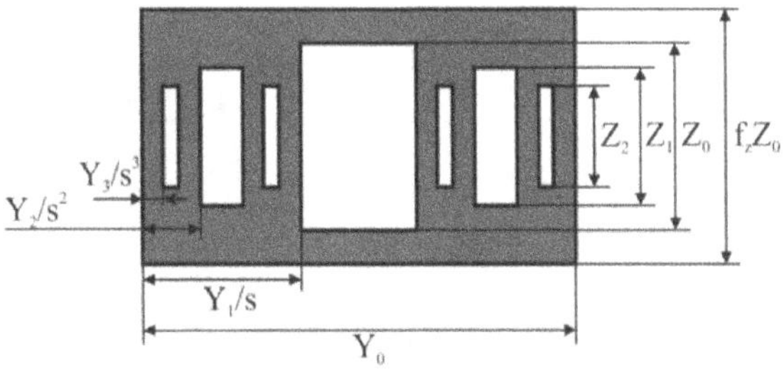

Fig. 9.2. Cantor interface with $n_g = 3$ generations and $s = 2$.

argued, given by the finest scale present in the interface, the fiber diameter in the case study here addressed. These four parameters can be easily extracted from the data provided by a profilometer.

If the domain occupied by the composite tape cross section is denoted by $\Omega_{tp} = \{(y, z),\ y \in [0, Y_0],\ z \in [0, Z(y)]\}$, the intersection of the line $z = u$, $u \in [0, \max_y Z(y)]$ with Ω_{tp}, $\Omega_{tp} \cap \{z = u\}$, measures $Y_u = |\Omega_{tp} \cap \{z = u\}|$. Obviously if $Y_u = 0$, the intersection $\Omega_{tp} \cap \{z = u\}$ has a null measure. On the contrary, when $Y_u = Y_0$ the line $z = u$ is fully contained in the composite. The evolution with u of Y_u is related to the fractal descriptors. In [81] authors illustrate the identification of the fractal parameters from the surface geometrical information of a tape obtained from a profilometer.

9.4.2. High-resolution numerical solution

One of the main advantages of representing the surface by a Cantor fractal is that it becomes described from a population of rectangular elements of different sizes as depicted in Fig. 9.2. In those circumstances one could perform high-resolution thermal simulations by transforming the 2D heat conduction problem (defined in the tapes cross section) into a sequence of 1D heat transfer problems along directions y and z when assuming a separated representation of both the temperature field and the material thermal properties within the Proper Generalized Decomposition (PGD) framework.

In that framework the temperature field $T(y, z)$ is searched in a separated form

$$T(y, z) \approx \sum_{i=1}^{M} F_i(y) \cdot G_i(z), \tag{9.123}$$

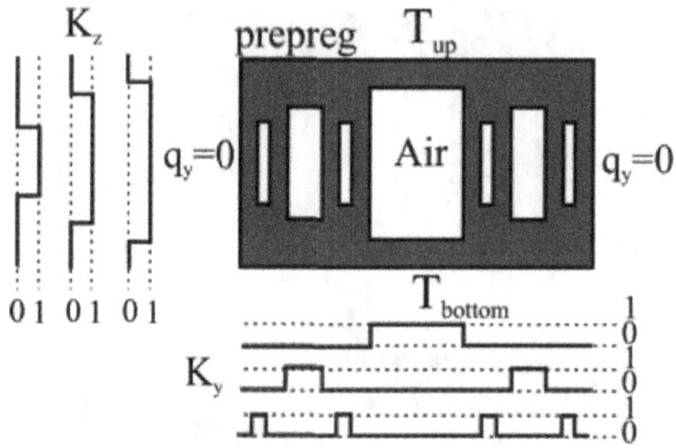

Fig. 9.3. Separated representation of the thermal conductivity.

and the thermal properties, the in-plane thermal conductivity in the present case, assumed with respect to Fig. 9.3, as

$$k(y,z) = \left(1 - \sum_{j=1}^{n_g} K_y^j(y) \cdot K_z^j(z)\right) k_c + \left(\sum_{j=1}^{n_g} K_y^j(y) \cdot K_z^j(z)\right) k_a, \tag{9.124}$$

where k_a and k_c are, respectively, the thermal conductivities of air and composite (in the transverse directions to the fibers arrangement), n_g is the number of fractal generations and K_y^j and K_z^j are the characteristic functions related to the rectangles occupied by the air at generation $j \leq n_g$ (according to Fig. 9.3).

The heat conduction problem is defined in the rectangular domain Ω illustrated in Fig. 9.3, with prescribed temperatures at the top and bottom boundaries, T_u and T_b, respectively ($T_u > T_b$), and null heat fluxes enforced on the lateral boundaries. The domain Ω is expressed from $\Omega = [0, L] \times [0, H]$, that in reference to the fractal characterization illustrated in Fig. 9.2, $L = Y_0$ and $H = f_z Z_0$.

The heat problem weak form writes: *Find $T(y,z) \in H^1(\Omega)$ verifying the Dirichlet boundary conditions $T(y,0) = T_b$ and $T(y,H) = T_u$ such that*

$$\int_\Omega \nabla T^* k(y,z) \nabla T \, d\mathbf{x} = 0 \quad \forall T^*(y,z) \in H_0^1(\Omega), \tag{9.125}$$

where $H^1(\Omega)$ and $H_0^1(\Omega)$ are the usual Sobolev functional spaces.

Now, injecting the conductivity separated form (9.124) into the weak form (9.125) and assuming that at iteration m, $m < M$, we have already computed the first m terms of the separated representation (9.123), i.e.,

$$T^m(y,z) \approx \sum_{i=1}^{m} F_i(y) \cdot G_i(z), \tag{9.126}$$

the present iteration looks for $T^{m+1}(y,z)$

$$T^{m+1}(y,z) = T^m(y,z) + F_{m+1}(y) \cdot G_{m+1}(z), \tag{9.127}$$

whose associated test function $T^*(y,z)$ reads

$$T^*(y,z) = F^*(y) \cdot G_{m+1}(z) + F_{m+1}(y) \cdot G^*(z). \tag{9.128}$$

Thus, with $\Omega = \Omega_y \times \Omega_z$ ($\Omega_y = [0,L]$ and $\Omega_z = [0,H]$) the weak form results:

$$\int_{\Omega_y \times \Omega_z} \nabla(F^*(y)G_{m+1}(z) + F_{m+1}(y)G^*(z))k(y,z)\nabla(F_{m+1}(y)G_{m+1}(z))\,d\mathbf{x}$$
$$= \int_{\Omega_y \times \Omega_z} \nabla(F^*(y)G_{m+1}(z) + F_{m+1}(y)G^*(z))k(y,z)$$
$$\nabla T^m(y,z)\,d\mathbf{x}. \tag{9.129}$$

As the problem defined by Eq. (9.129) becomes nonlinear, $F_{m+1}(y)$ and $G_{m+1}(z)$ being unknown at present iteration, the use of a linearization strategy becomes compulsory. We consider the simplest one, the one considering alternated search directions, that computes $F_{m+1}(y)$ by assuming $G_{m+1}(z)$ known (the one coming from the previous iteration of the nonlinear solution procedure) and then updating $G_{m+1}(z)$ from the just calculated $F_{m+1}(y)$. The iteration process continues until reaching the fixed point that results in the searched functions $F_{m+1}(y)$ and $G_{m+1}(z)$. Then the enrichment iteration continues for calculating $F_{m+2}(y)$ and $G_{m+2}(z)$ from $T^{m+1}(y,z)$ an so on. The enrichment iteration stops when the norm of the undated temperature becomes small enough $\|T^M(y,z) - T^{M-1}(y,z)\| < \epsilon$, even if more sophisticated error indicators exist [18].

Thus, the solution of problem (9.129) results in a sequence of one-dimensional problems, one for calculating $F_{m+1}(y)$ from $G_{m+1}(z)$

assumed known

$$\int_{\Omega_y \times \Omega_z} \nabla(F^*(y)G_{m+1}(z))k(y,z)\nabla(F_{m+1}(y)G_{m+1}(z))\,d\mathbf{x}$$
$$= \int_{\Omega_y \times \Omega_z} \nabla(F^*(y)G_{m+1}(z))k(y,z)\nabla T^m(y,z)\,d\mathbf{x}, \qquad (9.130)$$

and the other for updating $G_{m+1}(z)$ from the just calculated $F_{m+1}(y)$

$$\int_{\Omega_y \times \Omega_z} \nabla(F_{m+1}(y)G^*(z))k(y,z)\nabla(F_{m+1}(y)G_{m+1}(z))\,d\mathbf{x}$$
$$= \int_{\Omega_y \times \Omega_z} \nabla(F_{m+1}(y)G^*(z))\; k(y,z)\; \nabla T^m(y,z)\; d\mathbf{x}, \qquad (9.131)$$

with $k(y,z)$ in Eqs. (9.130) and (9.131) given by Eq. (9.124).

Because in Eq. (9.130) all the functions depending on the coordinate z are known, integrals in Ω_z can be performed in order to obtain the one-dimensional weak form, that after discretization leads to $F_{m+1}(y)$. The same rationale applied to Eq. (9.131) allows obtaining the corresponding problem for calculating $G_{m+1}(z)$. For more details on the computational implementation the interested reader can refer to [18] and the references therein.

As soon as the solution $T(y,z)$ is obtained we can compute the net heat flux Q by integrating the heat flux along any section $z = C \in [0,H]$ (being zero the heat fluxes on the lateral domain boundaries):

$$Q = \int_{\Omega_y} k(y, z = C)\; \left.\frac{\partial T(y,z)}{\partial z}\right|_{y,z=C} dy, \qquad (9.132)$$

from which we can obtain the thermal contact resistance (TCR)

$$TCR = \frac{T_u - T_b}{Q} = \frac{\Delta T}{Q}. \qquad (9.133)$$

9.4.3. Surface evolution during the *in-situ* consolidation

We consider the flow of a rheo-thinning fluid under the lubrication hypotheses to derive the squeeze flow equations, to be applied on each rectangular element involved in the Cantor fractal representation of the composite surface.

We consider the linear momentum and mass balance equations

$$\nabla \cdot \boldsymbol{\sigma} = \mathbf{0}, \tag{9.134}$$

and

$$\nabla \cdot \mathbf{v} = 0, \tag{9.135}$$

respectively, with $\boldsymbol{\sigma}$ and $\mathbf{v}$ the Cauchy's stress and the fluid velocity respectively. For a power-law incompressible fluid the Cauchy's stress reads:

$$\boldsymbol{\sigma} = -p\mathbf{I} + \boldsymbol{\tau} = -p\mathbf{I} + 2\kappa D_{\text{eq}}^{n-1}\mathbf{D}, \tag{9.136}$$

where p is the pressure, that can be interpreted as the Lagrange multiplayer associated with the incompressibility constraint, κ the material consistency, n the power index ($n = 1$ results in a Newtonian fluid for which κ becomes its viscosity), D_{eq} the equivalent strain rate usually given by second invariant of the strain rate tensor $\mathbf{D}$ (symmetric component of the velocity gradient $\nabla\mathbf{v}$)

$$\mathbf{D} = \frac{1}{2}\left(\nabla\mathbf{v} + (\nabla\mathbf{v})^T\right), \tag{9.137}$$

and

$$D_{\text{eq}} = \sqrt{2\mathbf{D} : \mathbf{D}}, \tag{9.138}$$

where $(\bullet : \bullet)$ refers to the tensor product twice contracted.

In what follows the flow equations are considered in $2D$ (y, z) with the components of the velocity field given by $\mathbf{v}^T = (v, w)$. Figure 9.4 illustrates the problem geometry.

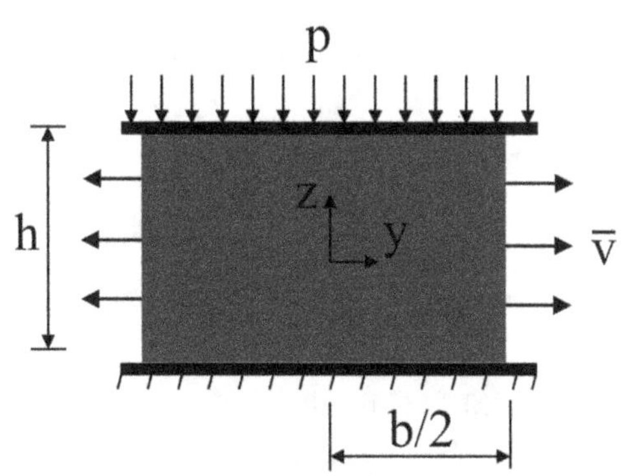

Fig. 9.4. Squeeze flow problem geometry.

Now, by assuming that the flow takes place in a very thin gap, the following kinematic hypotheses (lubrication hypotheses) apply:

$$\begin{cases} \dfrac{\partial v}{\partial z} \gg \dfrac{\partial v}{\partial y}, \\ w \approx 0, \end{cases} \tag{9.139}$$

that introduced into the linear momentum balance results

$$\frac{\partial p}{\partial y} = \kappa \frac{\partial}{\partial z}\left(D_{\text{eq}}^{n-1}\frac{\partial v}{\partial z}\right), \tag{9.140}$$

and

$$\frac{\partial p}{\partial z} \approx 0, \tag{9.141}$$

where in the present case

$$D_{\text{eq}} = \left|\frac{\partial v}{\partial z}\right|. \tag{9.142}$$

From Eq. (9.141) we conclude that the pressure does not depend on the z-coordinate, i.e., $p = p(y)$, that allows the integration of Eq. (9.140) with respect to the z-coordinate

$$D_{\text{eq}}^{n-1}\frac{\partial v}{\partial z} = \frac{1}{\kappa}\frac{\partial p}{\partial y} z + \mathcal{C}, \tag{9.143}$$

where the integration constant $\mathcal{C}$ vanishes because the velocity gradient vanishes at $z = 0$ (flow symmetry condition). Thus, it results

$$D_{\text{eq}}^{n-1}\frac{\partial v}{\partial z} = \frac{1}{\kappa}\frac{\partial p}{\partial y} z, \tag{9.144}$$

whose square power, taking into account relation (9.142), results

$$D_{\text{eq}}^{2n} = \frac{z^2}{\kappa^2}\left(\frac{\partial p}{\partial y}\right)^2, \tag{9.145}$$

or

$$D_{\text{eq}} = \left(\frac{|z|}{\kappa}\left|\frac{\partial p}{\partial y}\right|\right)^{\frac{1}{n}}, \tag{9.146}$$

that introduced into Eq. (9.144) results

$$\left(\frac{|z|}{\kappa}\left|\frac{\partial p}{\partial y}\right|\right)^{\frac{n-1}{n}}\frac{\partial v}{\partial z} = \frac{1}{\kappa}\frac{\partial p}{\partial y} z. \tag{9.147}$$

By defining $\alpha = \frac{1}{\kappa}\left|\frac{\partial p}{\partial y}\right|$, the integration of Eq. (9.147) in the z-coordinate assuming $v(y, z = \pm h/2) = 0$ finally results

$$v(y,z) = \frac{1}{\kappa}\frac{\partial p}{\partial y}\alpha^{\frac{1-n}{n}}\frac{n}{n+1}\left(|z|^{\frac{1+n}{n}} - \left(\frac{h}{2}\right)^{\frac{1+n}{n}}\right), \tag{9.148}$$

from which the mean velocity $\overline{v}(y)$ reads

$$\overline{v} = \frac{2\int_0^{h/2} v(y,z)dz}{h} = -\frac{n}{1+2n}\frac{\alpha^{\frac{1-n}{n}}}{\kappa}2^{-\frac{1+n}{n}}h^{\frac{n+1}{n}}\frac{\partial p}{\partial y}. \tag{9.149}$$

The mass balance writes

$$h\frac{\partial \overline{v}(y)}{\partial y} + \frac{dh}{dt} = 0, \tag{9.150}$$

that taking into account expression (9.149) and the definition of α, the last considered for $\frac{\partial p}{\partial y} \geq 0$, results

$$h\frac{\partial}{\partial y}\left(\frac{\partial p}{\partial y}\left(\frac{1}{\kappa}\frac{\partial p}{\partial y}\right)^{\frac{1-n}{n}}\right)\mathcal{D}_1 + \frac{dh}{dt} = 0, \tag{9.151}$$

with $\mathcal{D}_1 = -\frac{n}{1+2n}\frac{1}{\kappa}2^{-\frac{1+n}{n}}h^{\frac{n+1}{n}}$.

Developing the derivative with respect to the y-coordinate in the previous equation it results

$$\mathcal{D}_2\left(\frac{\partial p}{\partial y}\right)^{\frac{1-n}{n}}\frac{\partial^2 p}{\partial y^2} + \frac{dh}{dt} = 0, \tag{9.152}$$

with $\mathcal{D}_2 = \frac{h\mathcal{D}_1}{n}\left(\frac{1}{\kappa}\right)^{\frac{1-n}{n}}$. The previous equation can be rewritten for $\frac{\partial p}{\partial y} \neq 0$ as

$$\frac{\partial^2 p}{\partial y^2} + \frac{dh}{dt}\frac{1}{\mathcal{D}_2}\left(\frac{\partial p}{\partial y}\right)^{\frac{n-1}{n}} = 0, \tag{9.153}$$

that making use of the change of variable $q = \frac{\partial p}{\partial y}$ results

$$\frac{\partial q}{\partial y} + \frac{dh}{dt}\frac{1}{\mathcal{D}_2}q^{\frac{n-1}{n}} = 0, \tag{9.154}$$

whose integral results

$$q^{\frac{1}{n}} = -\frac{1}{n\mathcal{D}_2}\frac{dh}{dt}y + \mathcal{C}_1, \tag{9.155}$$

where the integration constant $\mathcal{C}_1$ vanishes if $q = \left.\frac{\partial p}{\partial y}\right|_{y=0} = 0$ (symmetry condition). Thus, the pressure derivative $\frac{\partial p}{\partial y} = q$ results

$$\frac{\partial p}{\partial y} = \left(-\frac{1}{n\mathcal{D}_2}\frac{dh}{dt}\right)^n y^n, \tag{9.156}$$

that integrating again results

$$p = \left(-\frac{1}{n\mathcal{D}_2}\frac{dh}{dt}\right)^n \frac{y^{n+1}}{n+1} + \mathcal{C}_2, \tag{9.157}$$

where $\mathcal{C}_2$ is obtained by assuming that at $y = \pm b/2$ the pressure is p_0. Thus, it results finally

$$p - p_0 = \frac{1}{n+1}\left(-\frac{1}{n\mathcal{D}_2}\frac{dh}{dt}\right)^n \left(y^{n+1} - \left(\frac{b}{2}\right)^{n+1}\right). \tag{9.158}$$

The total force f originating the thickness reduction $\frac{dh}{dt}$ can be obtained from

$$f = \int_{-\frac{b}{2}}^{\frac{b}{2}} (p - p_0)dy, \tag{9.159}$$

and from it $\frac{dh}{dt}$, that taking into account the mass conservation $bh = b_0h_0$ (where the index $(\bullet)_0$ refers to values at the initial time) results

$$\frac{dh}{dt} = -\frac{h^{2+\frac{1}{n}}\left(\frac{1}{\kappa}\right)^{\frac{1}{n}} n \left(\frac{(2+n)h^2 f\left(\frac{b_0h_0}{h}\right)^{-n}}{b_0^2h_0^2}\right)^{\frac{1}{n}}}{1+2n}. \tag{9.160}$$

The thickness evolution rate given by Eq. (9.160) depends on the fluid rheology through n and κ, the asperity initial geometry b_0 and h_0, the present height at time t, h, and the applied force f.

If F denotes the total force applied on the tape, the resulting force acting on a rectangle of generation n_g results

$$f = \frac{F}{s^{n_g}}. \tag{9.161}$$

Now, if h_{n_g} denotes the height of a rectangle of generation n_g at time t, Eq. (9.160) reads

$$\frac{dh_{n_g}}{dt} = -\frac{h_{n_g}^{2+\frac{1}{n}}\left(\frac{1}{\kappa}\right)^{\frac{1}{n}} n \left(\frac{(2+n)h_{n_g}^2 \frac{F}{s^{n_g}}\left(\frac{b_0 h_0}{h_{n_g}}\right)^{-n}}{b_0^2 h_0^2}\right)^{\frac{1}{n}}}{1+2n} = \mathcal{F}(h_{h_g}). \quad (9.162)$$

Thus, the rectangle height time evolution results from integrating the previous equation:

$$\int_{h_{n_g,0}}^{h_{n_g,f}} \frac{1}{\mathcal{F}(h_{n_g})} dh_{n_g} = (\Delta t)_{n_g}, \quad (9.163)$$

where $(\Delta t)_{n_g}$ represents the time needed for removing rectangles of generation n_g and $h_{n_g,0}$ and $h_{n_g,f}$ respectively the initial and final height of rectangles of that generation.

In order to obtain the initial and final heights the mass conservation is again invoked, that as illustrated in Fig. 9.5, reads:

$$h_{n_g,f}\left(\frac{1}{f_y}\right)^{n_g-1} Y_0 = h_{n_g,0}\left(\frac{1}{f_y}\right)^{n_g} Y_0, \quad (9.164)$$

that results in

$$h_{n_g,f} = \frac{h_{n_g,0}}{f_y}. \quad (9.165)$$

Thus, being Z_{n_g} the thickness of the n_g generation rectangles, it results

$$h_{n_g,0} = Z_{n_g} - \Delta h_{n_g} \quad (9.166)$$

with $\Delta h_{n_g} = Z_{n_g+1} - h_{n_g+1,f}$.

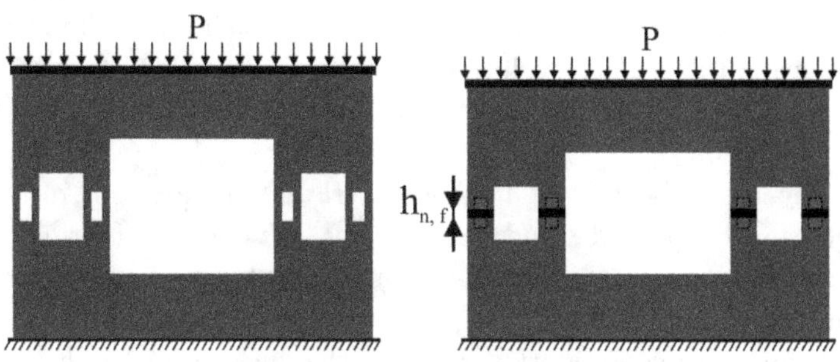

Fig. 9.5. Initial and final configurations.

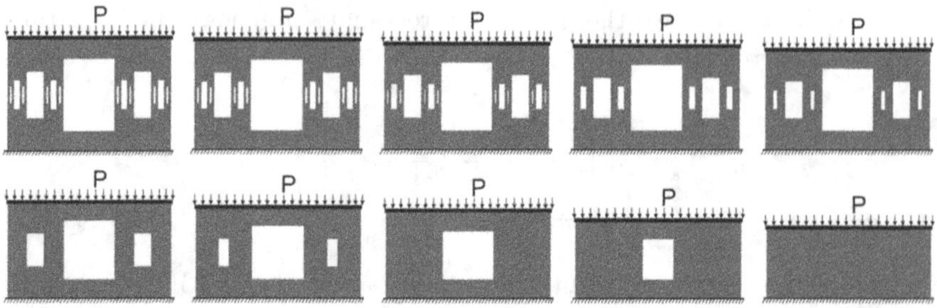

Fig. 9.6. Evolution of the surface Cantor representation during the consolidation. During tape placement air entrapped in the voids associated with the fractal surface representation can scape in the third direction, the one perpendicular to the depicted cross section.

Figure 9.6 depicts a consolidation sequence from the evolution of the surface fractal representation.

9.4.4. Consolidation simulation strategy

As described in the previous section the evolution of the geometry initially described from a Cantor fractal can be predicted taking into account the squeeze flow of the different rectangular elements involved in the fractal description of the surface.

The rectangles of the last generation are removed the first when compressing, then it is the turn for the ones of the previous generation and so on. The evolution does not correspond with a sequence of the same fractal by removing one generation at each compression step, because as proved in the previous section (see e.g., Eq. (9.165)) as the rectangle volume does not coincide with the contiguous void rectangles, the rectangles height of a particular generation is affected by the squeeze flow of the next generations. For this reason Eq. (9.163) must be carefully integrated for evaluating the surface evolution that remains all along the simulation composed by rectangles, but different to the ones associated to the fractal initial description.

In any case, the fact of having a description based on rectangular elements makes possible and very easily solving the thermal model while using a separated representation of both the temperature field and the thermal conductivity as previously described. Thus, at different times during the consolidation process the integration of Eq. (9.163) allows updating the surface description, and then the solution of the heat conduction equation in the updated surface representation allows calculating the equivalent TCR. At each thermal calculation, the thermal conductivity must be expressed in a separated form from the characteristic functions describing the deformed rectangles of all the generations existing at that instant.

9.5. Conclusions

In this chapter, we addressed three main issues encountered in composites forming processes. The first concerned the modeling of reinforced polymers with special emphasis on the suspending fluid rheology (Newtonian versus non-Newtonian) as well as to the concentration regimes (dilute, semi-dilute/semi-concentrated and concentrated). Then, when moving to continuous fiber composites the issue related to the efficient simulation of multi-physics problems in laminates was addressed. The third part of the chapter focused in the physics encountered at the interfaces level, and particularly to the ones taking place during thermoplastics laminates consolidation.

References

[1] Binetruy C., Chinesta F., Keunings R., 2015. *Flows in Polymers, Reinforced Polymers and Composites: A Multiscale Approach*, Springerbriefs, Springer.
[2] Jeffery G.B., 1922. The motion of ellipsoidal particles immersed in a viscous fluid, *Proc. R. Soc. London A* 102, 161–179.
[3] Bird R.B., Curtiss C.F., Armstrong R.C., Hassager O., 1987. *Dynamics of Polymeric Liquids, Volume 2: Kinetic Theory*, Wiley.
[4] Doi M., Edwards S.F., 1987. *The Theory of Polymer Dynamics*, Clarendon Press, Oxford.
[5] Ammar A., Chinesta F., 2005. A particle strategy for solving the Fokker–Planck equation governing the fiber orientation distribution in steady recirculating flows involving short fiber suspensions, in M. Griebel, M.A. Schweitzer (eds.), *Meshfree Methods for Partial Differential Equations II*, Springer, pp. 1–16.
[6] Chaubal C.V., Srinivasan A., Egecioglu O., Leal L.G., 1997. Smoothed particle hydrodynamics techniques for the solution of kinetic theory problems, *J. Non-Newton. Fluid. Mech.*, 70, 125–154.
[7] Chinesta F., Chaidron G., Poitou A., 2003. On the solution of the Fokker–Planck equation in steady recirculating flows involving short fiber suspensions, *J. Non-Newton. Fluid. Mech.*, 113, 97–125.
[8] Cruz C., Illoul L., Chinesta F., Regnier G., 2010. Effects of a bent structure on the linear viscoelastic response of Carbon Nanotube diluted suspensions, *Rheol. Acta*, 49, 1141–1155.

[9] Cruz C., Chinesta F., Regnier G., 2012. Review on the Brownian dynamics simulation of bead-rod-spring models encountered in computational rheology, *Arch. Comput. Method. Eng.*, 19, 227–259.

[10] Öttinger H.C., Laso M., 1992. Smart polymers in finite-element calculations, in P. Moldenaers and R. Keunings (eds.), *Theoretical and Applied Rheology*, Vol. 1, Proc. XIth Int. Congr. on Rheology, Elsevier, Amsterdam, pp. 286–288.

[11] Wapperom P., Keunings R., Legat V., 2000. The Backward-Tracking Lagrangian Particle Method for transient viscoelastic flows, *J. Non-Newton. Fluid. Mech.*, 91, 273–295.

[12] Wapperom P., Keunings R., 2001. Numerical simulation of branched polymer melts in transient complex flows using pom-pom models, *J. Non-Newton. Fluid. Mech.*, 97, 267–281.

[13] Chauviere C., Lozinski A., 2004. Simulation of dilute polymer solutions using a Fokker–Planck equation, *Comput. Fluids*, 33, 687–696.

[14] Lozinski A., Chauviere C., 2003. A fast solver for Fokker–Planck equation applied to viscoelastic flows calculations: 2D FENE model, *J. Comput. Phys.*, 189, 607–625.

[15] Ammar A., Mokdad B., Chinesta F., Keunings R., 2006. A new family of solvers for some classes of multidimensional partial differential equations encountered in kinetic theory modeling of complex fluids, *J. Non-Newton. Fluid. Mech.*, 139, 153–176.

[16] Ammar A., Mokdad B., Chinesta F., Keunings R., 2007. A new family of solvers for some classes of multidimensional partial differential equations encountered in kinetic theory modeling of complex fluids. Part II: transient simulation using space-time separated representations, *J. Non-Newton. Fluid. Mech.*, 144, 98–121.

[17] Chinesta F., Ammar A., Leygue A., Keunings R., 2011. An overview of the Proper Generalized Decomposition with applications in computational rheology, *J. Non-Newton. Fluid. Mech.*, 166, 578–592.

[18] Chinesta F., Keunings R., Leygue A., 2014. *The Proper Generalized Decomposition for Advanced Numerical Simulations. A Primer*, Springerbriefs, Springer.

[19] Advani S., Tucker Ch., 1987. The use of tensors to describe and predict fiber orientation in short fiber composites, *J. Rheol.*, 31, 751–784.

[20] Advani S. (ed.), 1994. *Flow and Rheology in Polymer Composites Manufacturing*, Elsevier.

[21] Advani S., Tucker Ch., 1990. Closure approximations for three-dimensional structure tensors, *J. Rheol.*, 34, 367–386.

[22] Dupret F., Verleye V., 1999. Modelling the flow of fiber suspensions in narrow gaps, in *Advances in the Flow and Rheology of Non-Newtonian Fluids*, Rheology Series, Elsevier, pp. 1347–1398.

[23] Kroger M., Ammar A., Chinesta F., 2008. Consistent closure schemes for statistical models of anisotropic fluids, *J. Non-Newton. Fluid. Mech.*, 149, 40–55.

[24] Pruliere E., Ammar A., El Kissi N., Chinesta F., 2009. Recirculating flows involving short fiber suspensions: Numerical difficulties and efficient advanced micro-macro solvers, *Arch. Comput. Method. Eng.*, 16, 1–30.

[25] Chiba K., Ammar A., Chinesta F., 2005. On the fiber orientation in steady recirculating flows involving short fibers suspensions, *Rheol. Acta*, 44, 406–417.

[26] Keunings R., 1997. On the Peterlin approximation for finitely extensible dumbbells, *J. Non-Newton. Fluid. Mech.*, 68, 85–100.

[27] Grassia P., Hinch J., Nitsche L.C., 1995. Computer simulations of brownian motion of complex systems, *J. Fluid Mech.*, 282, 373–403.

[28] Grassia P., Hinch J., 1996. Computer simulations of polymer chain relaxation via brownian motion, *J. Fluid Mech.*, 308, 255–288.

[29] Mezher R., Abisset-Chavanne E., Ferec J., Ausias G., Chinesta F., 2015. Direct simulation of concentrated fiber suspensions subjected to bending effects, *Model. Simul. Mater. Sci. Eng.*, 23, 055007.

[30] Folgar F., Tucker Ch., 1984. Orientation behavior of fibers in concentrated suspensions, *J. Reinf. Plast. Comp.*, 3, 98–119.

[31] Batchelor G.K., 1970. The stress system in a suspension of force-free particles, *J. Fluid Mech.*, 41, 545–570.

[32] Hand G.L., 1962. A theory of anisotropic fluids, *J. Fluid Mech.*, 13, 33–62.

[33] Hinch J., Leal G., 1972. The effect of Brownian motion on the rheological properties of a suspension of non-spherical particles. *J. Fluid Mech.*, 52, 683–712.

[34] Hinch J., Leal G., 1975. Constitutive equations in suspension mechanics. Part I. *J. Fluid Mech.*, 71, 481–495.

[35] Hinch J., Leal G., 1976. Constitutive equations in suspension mechanics. Part II, *J. Fluid Mech.*, 76, 187–208.

[36] Azaiez J., Chiba K., Chinesta F., Poitou A., 2002. State-of-the-art on numerical simulation of fiber-reinforced thermoplastic forming processes, *Arch. Comput. Method. Eng.*, 9, 141–198.

[37] Cueto E., Monge R., Chinesta F., Poitou A., Alfaro I., Mackley M., 2010. Rheological modeling and forming process simulation of CNT nanocomposites, *Int. J. Mater. Form.*, 3, 1327–1338.

[38] Tucker Ch., 1991. Flow regimes for fiber suspensions in narrow gaps, *J. Non-Newton. Fluid. Mech.*, 39, 239–268.

[39] Ma A., Chinesta F., Mackley M., 2009. The rheology and modelling of chemically treated Carbon Nanotube suspensions, *J. Rheol.*, 53, 547–573.

[40] Petrie C., 1999. The rheology of fiber suspensions, *J. Non-Newton. Fluid. Mech.*, 87, 369–402.

[41] Ferec J., Ausias G., Heuzey M.C., Carreau P., 2009. Modeling fiber interactions in semiconcentrated fiber suspensions, *J. Rheol.*, 53, 49–72.

[42] Ferec J., Abisset-Chavanne E., Ausias G., Chinesta F., 2014. The use of interaction tensors to describe and predict rod interactions in rod suspensions, *Rheol. Acta*, 53, 445–456.

[43] Wang J., Silva C.A., Viana J.C., van Hattum F.W.J., Cunha A.M., Tucker Ch., 2008. Prediction of fiber orientation in a rotating compressing and expanding mold, *Polym. Eng. Sci.*, 1405–1413.

[44] Wang J., O'Gara J., Tucker Ch., 2008. An objective model for slow orientation kinetics in concentrated fiber suspensions: Theory and rheological evidence, *J. Rheol.*, 52, 1179–1200.

[45] Phelps J., Tucker Ch., 2009. An anisotropic rotary diffusion model for fiber orientation in short and long fiber thermoplastics, *J. Non-Newton. Fluid. Mech.*, 156, 165–176.

[46] Ma A., Chinesta F., Ammar A., Mackley M., 2008. Rheological modelling of Carbon Nanotube aggregate suspensions, *J. Rheol.*, 52, 1311–1330.

[47] Ausias G., Fan X.J., Tanner R., 2006. Direct simulation for concentrated fiber suspensions in transient and steady state shear flows, *J. Non-Newton. Fluid. Mech.*, 135, 46–57.

[48] Le Corre S., Caillerie D., Orgéas L., Favier D., 2004. Behavior of a net of fibers linked by viscous interactions: theory and mechanical properties, *J. Mech. Phys. Solids*, 52, 395–421.

[49] Le Corre S., Dumont P., Orgéas L., Favier D., 2005. Rheology of highly concentrated planar fiber suspensions, *J. Rheol.*, 49, 1029.

[50] Chinesta F., 2013. From single-scale to two-scales kinetic theory descriptions of rods suspensions, *Arch. Comput. Method. Eng.*, 20, 1–29.

[51] Abisset-Chavanne E., Mezher R., Le Corre S., Ammar A., Chinesta F., 2013. Kinetic theory microstructure modeling in concentrated suspensions, *Entropy*, 15, 2805–2832.

[52] Abisset-Chavanne E., Chinesta F., Ferec J., Ausias G., Keunings R., 2015. On the multiscale description of dilute suspensions of non-Brownian rigid clusters composed of rods, *J. Non-Newton. Fluid. Mech.*, 222, 34–44.

[53] Ortman K., Baird D., Wapperom P., Whittington A., 2012. Using startup of steady shear flow in a sliding plate rheometer to determine material parameters for the purpose of predicting long fiber orientation, *J. Rheol.*, 56, 955–981.

[54] Strautins U., Latz A., 2007. Flow-driven orientation dynamics of semiflexible fiber systems, *Rheol. Acta*, 46, 1057–1064.

[55] Abisset-Chavanne E., Ferec J., Ausias G., Cueto E., Chinesta F., Keunings R., 2015. A second-gradient theory of dilute suspensions of flexible rods in a Newtonian fluid, *Arch. Comput. Method. Eng.*, 22, 511–527.

[56] Perez M., Scheuer A., Abisset-Chavanne E., Chinesta F., Keunings R., 2016. A multi-scale description of orientation in confined suspensions involving rods, *J. Non-Newton. Fluid. Mech.*, 233, 61–74.

[57] Choi Y.J., Hulsen M.A., Meijer H.E.H., 2010. An extended finite element method for the simulation of particulate viscoelastic flows, *J. Non-Newton. Fluid. Mech.*, 165, 607–624.

[58] D'Avino G., Maffettone P.L., 2015. Particle dynamics in viscoelastic liquids, *J. Non-Newton. Fluid. Mech.*, 215, 80–104.

[59] Harlen O., 2002. The negative wake behind a sphere sedimenting through a viscoelastic fluid, *J. Non-Newton. Fluid. Mech.*, 108, 411–430.
[60] Singh P., Joseph D.D., Hesla T.I., Glowinski R., Pan T.-W., 2000. A distributed Lagrange multiplier/fictitious domain method for viscoelastic particulate flows, *J. Non-Newton. Fluid. Mech.*, 91, 165–188.
[61] Tanner R., Housiadas K.D., Qi F., 2014. Mechanism of drag increase on spheres in viscoelastic cross-shear flows, *J. Non-Newton. Fluid. Mech.*, 203, 51–53.
[62] Villone M.M., D'Avino G., Hulsen M.A., Greco F., Maffettone P.L., 2013. Particle motion in square channel flow of a viscoelastic liquid: Migration vs. secondary flows, *J. Non-Newton. Fluid. Mech.*, 195, 1–8.
[63] Yu Z., Phan-Thien N., Fan Y., Tanner R., 2002. Viscoelastic mobility problem of a system of particles, *J. Non-Newton. Fluid. Mech.*, 104, 87–124.
[64] Yu Z., Wachs A., Peysson Y., 2006. Numerical simulation of particle sedimentation in shear-thinning fluids with a fictitious domain method, *J. Non-Newton. Fluid. Mech.*, 136, 126–139.
[65] Gunes D.Z., Scirocco R., Mewis J., Vermant J., 2008. Flow-induced orientation of non-spherical particles: Effect of aspect ratio and medium rheology, *J. Non-Newtonian Fluid Mech.*, 155, 39–50.
[66] Iso Y., Cohen C., Koch D.L., 1996. Orientation in simple shear flow of semi-dilute fiber suspensions 2. Highly elastic fluids, *J. Non-Newton. Fluid. Mech.*, 62, 135–153.
[67] Brunn P., 1977. The slow motion of a rigid particle in a second-order fluid, *J. Fluid Mech.*, 82, 529–547.
[68] Leal L.G., 1975. The slow motion of slender rod-like particles in a second-order fluid, *J. Fluid Mech.*, 69, 305–337.
[69] Batchelor G.K., 1970. Slender-body theory for particles of arbitrary cross-section in Stokes flow, *J. Fluid Mech.*, 44, 419–440.
[70] Giesekus H., 1963. Die simultane translations-und rotations bewegung einer kugel in einer elastovisken flussigkeit, *Rheol. Acta*, 3, 59–71.
[71] Borzacchiello D., Abisset-Chavanne E., Chinesta F., Keunings R., 2016. Orientation kinematics of short fibers in a second-order viscoelastic fluid, *Rheol. Acta*, 55, 397–409.
[72] Chinesta F., Ammar A., Cueto E., 2010. Recent advances and new challenges in the use of the Proper Generalized Decomposition for solving multidimensional models, *Arch. Comput. Method. Eng.*, 17, 327–350.
[73] Chinesta F., Ladeveze P., Cueto E., 2011. A short review in model order reduction based on Proper Generalized Decomposition, *Arch. Comput. Method. Eng.*, 18, 395–404.
[74] Bognet B., Leygue A., Chinesta F., Poitou A., Bordeu F., 2012. Advanced simulation of models defined in plate geometries: 3D solutions with 2D computational complexity, *Comput. Method. Appl. Mech. Eng.*, 201, 1–12.
[75] Bognet B., Leygue A., Chinesta F., 2014. Separated representations of 3D elastic solutions in shell geometries, *Adv. Model. Simul. in Eng. Sci.*, 1, 4.

[76] Ghnatios Ch., Abisset-Chavanne E., Binetruy Ch., Chinesta F., Advani S., 2016. 3D modeling of squeeze flow of multiaxial laminates, *J. Non-Newton. Fluid. Mech.*, 234, 188–200.

[77] Ghnatios Ch., Chinesta F., Binetruy Ch., 2015. The squeeze flow of composite laminates, *Int. J. Mater. Form.*, 8, 73–83.

[78] Ericksen J.L., 1959. Anisotropic fluids, *Arch. Ration. Mech. Anal.*, 231–237.

[79] Barasinski A., Leygue A., Soccard E., 2014. Identification of non-uniform thermal contact resistance in automated tape placement process, *Int. J. Mater. Form.*, 7, 479–486.

[80] Chinesta F., Leygue A., Bognet B., Ghnatios Ch., Poulhaon F., Bordeu F., Barasinski A., Poitou A., Chatel S., Maison-Le-Poec S., 2014. First steps towards an advanced simulation of composites manufacturing by automated tape placement, *Int. J. Mater. Form.*, 7, 81–92.

[81] Leon A., Barasinski A., Nadal E., Chinesta F., 2015. High-resolution thermal analysis at thermoplastic pre-impregnated composite interfaces, *Compos. Interface.*, 22, 767–777.

https://doi.org/10.1142/9781800611887_0010

Chapter 10

Modeling Fracture and Complex Crack Networks in Laminated Composites

Carlos G. Dávila[*,‡], **Cheryl A. Rose**[*,§], **Endel V. Iarve**[†,¶], **and Frank A. Leone**[*,||]

[*]*NASA Langley Research Center, Hampton, VA 23681, USA*

[†]*University of Texas at Arlington, Arlington, TX 76019, USA*

[‡]*carlos.g.davila@nasa.gov*

[§]*cheryl.a.rose@nasa.gov*

[¶]*endel.iarve@uta.edu*

[||]*frank.leone@nasa.gov*

Abstract

Recent advances in computational methods and numerous published demonstrations of successful representations of the propagation of composite damage mechanisms indicate that the day is imminent when reliable tools for the virtual testing of composite structures will replace some mechanical testing in the design and certification process. Given these rapid developments and the apparent diversity of the proposed approaches, it is necessary to formulate the conditions under which a given model can be expected to work and when it will cease to be adequate. In this chapter, we examine the fundamental concepts that are required for predicting damage in composites with the intent of providing a basis to help select the idealizations that are necessary, physically reasonable, and computationally tractable. Issues of the objectivity of fracture propagation with continuum damage mechanics models are discussed and the application of the extended finite element method to avoid these difficulties is explored.

Contents

10.1. Introduction: Damage Idealization and Scale

Fracture in a composite structure is the result of the evolution of discrete damage events, such as fiber/matrix debonding, matrix cracking, delamination between plies, and fiber failure. These damage modes evolve in various combinations that depend on the stacking sequence and ply thicknesses and cause redistributions of stresses in the failing composite. Some combinations of damage may reduce local stress concentrations, while others may precipitate a structural collapse. Therefore, a methodology capable of predicting structural strength must take into account damage initiation and propagation.

However, the details of the mechanisms that lead to failure are not fully understood due to the complexity of the idealization of the individual constituent responses and their interactions. The presence of

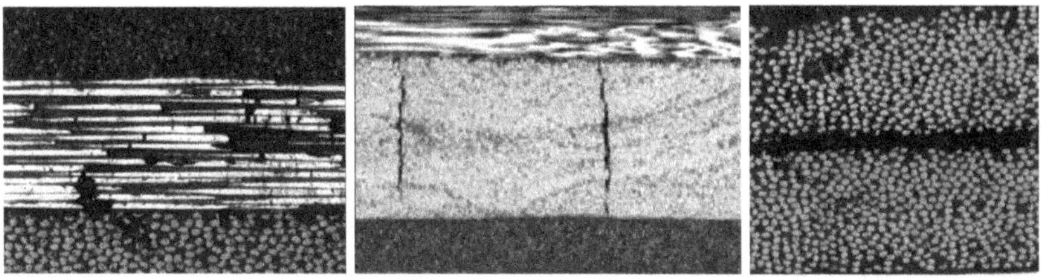

(a) Fiber fracture (b) Matrix transverse crack (c) Delamination

Fig. 10.1. Damage mechanisms in laminated composites.

two constituents, the fiber and the matrix, and the extreme anisotropy in both stiffness and strength properties result in damage mechanisms at different levels. The damage mechanisms can be divided into intralaminar and interlaminar damage. As shown in Fig. 10.1, intralaminar damage mechanisms correspond to fiber fracture and matrix cracking, whereas interlaminar damage mechanisms correspond to the interfacial separation of the plies (delamination).

The formulation of the governing physical principles of damage evolution depends on the scale of the idealization of the damage process, which may span from molecular dynamics to structural mechanics, and includes the intermediate scales of micro- and mesomechanics. The damage models shown in Fig. 10.2 illustrate four typical scales of damage idealization. The micromechanical scale model shown in Fig. 10.2(a) represents what is normally the smallest scale of composite damage idealization, in which detailed matrix energy-dissipating mechanisms such as matrix plasticity and damage and fiber/matrix interface cracking are represented [1]. The representation of damage at this level is typically based on a reduction of the material stiffness. Hence, a fracture is represented as a band of localized volumetric stiffness reduction, referred to as a "weak discontinuity," as opposed to a "strong discontinuity" in which voids are represented by displacement discontinuities in the model. Due to computational constraints, micromechanical models are typically two-dimensional (2D) and represent domains much smaller than a ply thickness. Consequently, they are useful for representing a composite hardening response before cracks localize at either the ply level or on a larger scale.

The discrete damage mechanics (DDM) model shown in Fig. 10.2(b) represents a new class of analysis methods in which the plies are represented at the mesoscale level, i.e., the material is assumed to be homogeneous with orthotropic properties, and where enriched finite elements [2] are used to insert cracks and delaminations in locations that are independent of

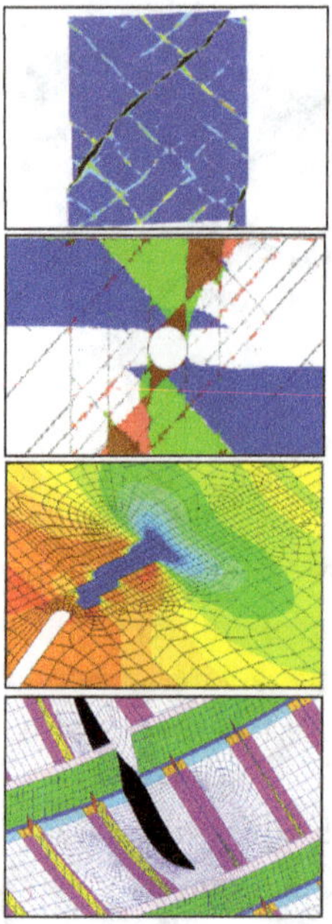

Scale of idealization	Damage type	Typical approaches
(a) **Microscale (Weak discontinuity)**	Fiber/matrix interface, matrix plasticity and damage	• Representation volume element (RVE) models (unit cell)
(b) **Mesoscale (Strong discontinuity)**	Discrete damage	• x-FEM + Cohesive laws
(c) **Mesoscale (Weak discontinuity)**	Intralaminar damage	• Continuum damage mechanics (CDM)
(d) **Structural scale (Strong discontinuity)**	Through-the-thickness crack or delamination	• Fracture mechanics and modifications • Strain softening, cohesive laws

Fig. 10.2. Levels of damage idealization from microscale to structural scale.

mesh orientation. DDM methods are classified as either element enrichment methods (e.g., [3]) or nodal enrichment methods, such as the extended finite element method (x-FEM) [4, 5]. DDM methods can be used to represent complex networks of transverse matrix cracks and delaminations using a single cohesive law [6]. A form of x-FEM, the regularized extended finite element method (Rx-FEM), is discussed in Section 10.7.

A similar level of kinematic fidelity can also be obtained without the use of element enrichment when the crack directions are known *a priori*. Since the direction of transverse matrix cracks in a unidirectional tape ply laminate is parallel to the fiber direction, finite element meshes can be constructed that are aligned with the fiber direction in the ply and in which the crack paths are pre-defined. Bouvet *et al.*, for instance, model individual plies with narrow strips of elements tied together with cohesive elements [7]. This modeling technique can represent networks of matrix cracks and delaminations without requiring the numerical complexity of discrete damage methods.

The most common idealization for composite damage is the continuum damage mechanics (CDM) model at the mesoscale level, in which plies are assumed to be composed of a homogeneous material with orthotropic properties and intraply damage modes, such as fiber fracture, fiber kinking, and matrix cracking, are represented as a reduction in the corresponding stiffnesses [8, 9]. In CDM models, localized damage is therefore represented as a weak discontinuity, as opposed to the strong discontinuities used in DDM models. An example of a mesoscale model of a notched wing skin subjected to compression loads is shown in Fig. 10.2(c).

Mesoscale approaches such as DDM and CDM have several computational advantages compared to micromechanical-level models. Mesoscale models use material properties that can be determined from lamina-level characterization tests, and imperfections and variabilities that must be considered at lower scales can be disregarded. To account for micromechanical effects such as fiber/matrix debonding, some mesoscale methods perform concurrent analyses on idealized unit-cell models that represent typical distribution patterns of fibers in the matrix. Other mechanisms, such as transverse matrix cracking or fiber kinking, which depend on mesoscale data such as ply thickness, are usually taken into account with failure criteria and *in situ* strengths. These failure criteria are used in conjunction with material degradation schedules to soften the material properties associated with a particular mode of failure. The delamination failure mode is represented at the mesoscale level as a strong discontinuity between plies, which is generally modeled using either cohesive elements or nonlinear springs [10].

The mesoscale CDM approach is one of the most widely used approaches to calculate the damage tolerance of structural components. However, the relatively diffused representation of damage in the CDM approach may lack sufficient resolution to capture some important damage interactions at the micromechanical scale. In particular, mesoscale CDM models have difficulty predicting the correct direction of propagation of matrix cracks parallel to the fibers. Furthermore, the damage models for intralaminar damage may not interact correctly with the delamination models, e.g., [11, 12].

Finally, the model of a through-crack in a fuselage panel illustrated in Fig. 10.2(d) is an example of a structural-level damage model. The crack is represented as a strong discontinuity, and prediction of the propagation of the crack could be based on a strain-softening law or a criterion based on the critical energy release rate [13, 14]. However, any structural-level crack propagation criterion is strongly dependent on the material system and laminate configuration and consequently must be determined for each new material system and laminate stacking sequence. In addition, structural-level semi-empirical fracture models cannot address the characteristics of the crack-tip damage zone nor the complex interactions between micro- and macro-failures associated with the crack-extension process. Instead, the crack-tip damage zone is simulated as some "effective" notch-tip damage zone that is assumed to grow in a self-similar manner. In many cases, self-similar crack growth is not observed, and the lack of resolution in the damage mechanisms often renders structural-level damage models inaccurate after a short propagation of damage.

It is clear from the preceding overview of typical damage modeling strategies that the conceptual idealization of damage, i.e., the identification, characterization, and formulation of the governing physical mechanisms that constitute damage evolution, are different at each scale of idealization. Damage idealizations at lower structural scales have higher resolution and kinematic freedom and can capture multiple damage mechanisms with a separate damage law for each mechanism. These damage laws are likely to be simpler and require lower fracture toughnesses than the damage laws used in coarser higher-level models to represent the same global response. For instance, mesoscale DDM models may only need one simple cohesive law to represent a variety of matrix damage patterns, while CDM models may need multiple empirical stiffness degradation laws and interacting activation functions to represent the softening of the material.

The choice of a modeling approach is further complicated by the fact that there is a vast number of published demonstrations of different successful representations of the propagation of damage. A variety of damage models have been championed and, given the complexity of these methods, their differences, strengths, and shortcomings are unclear. The objective of the present chapter is to examine the common features and basic principles of well-established approaches for predicting damage propagation in composites to establish which damage mechanisms are important and how to select the idealizations that are necessary, physically reasonable, and computationally tractable.

In the following sections, we first examine the capabilities of cohesive laws to represent crack initiation and propagation. Then, the concepts of the continuum representation of composite material response are discussed and the intrinsic limitations of continuum damage models for laminated composites are outlined. Finally, a modeling technique based on x-FEM, which overcomes many of the limitations of continuum damage models is presented and the capabilities of the x-FEM methodology are illustrated with some examples.

10.2. Crack Initiation and Propagation

10.2.1. Linear elastic fracture of composites

The ability to predict crack propagation in composites emerged four decades ago with the development of computational methods based on the theory of linear elastic fracture mechanics (LEFM). In particular, the virtual crack closure technique [15] is a computationally simple procedure with which to calculate the energy release rate (ERR) for each mode of fracture along a crack front. According to the Griffith criterion [16], when the sum of these ERR values exceeds the corresponding critical fracture toughness, the crack should propagate. However, LEFM is limited to applications in which the fracture process zone is confined to the immediate neighborhood of the crack tip itself, so it cannot be applied to a number of important cracking problems involving some of the tougher, more ductile structural materials and adhesives sometimes described as quasi-brittle [17], that fracture after extensive nonlinear deformation.

For many fracture processes in composite materials and structures, the fracture process zone can be relatively large compared to other structural dimensions. The fracture process zone is characterized by plastic deformations and progressive material softening due to nonlinear material deformations, such as microcracking, void formation, and fiber/matrix pullout. The size of the fracture process zone is dependent on the type of material softening and it must be considered in many situations of crack growth in composite structures. For example, the development of a process zone gives rise to stable growth and crack growth resistance. The apparent fracture toughness increases with crack growth — an effect called the *R-curve* — until the process zone is fully developed. In addition, the effects of material strength, the R-curve, and structural size on the residual strength of structures made of these materials may be misunderstood, disregarded, or perceived as solely statistical if the fracture process zone is not taken into account in fracture calculations.

Reduced-singularity criteria such as the Mar–Lin criterion [13, 18] or Sun's two-parameter fracture criterion [19] were devised for extending the classical linear theory to situations where the stress distribution ahead of the crack tip does not follow a square-root singularity. However, these criteria can easily be supplanted by nonlinear fracture mechanics (NLFM). NLFM provides a framework for characterizing crack growth resistance and for analyzing initial amounts of stable crack growth.

The cohesive crack model is an NLFM methodology that was developed to simulate the nonlinear fracture response near a crack tip. Cohesive crack models have the ability to describe the process of void nucleation from inclusions. Therefore, they can be applied to an initially un-cracked structure and can describe the entire fracture process, from no crack to complete structural failure. In the following sections, the formulation of cohesive laws is presented. We examine the differences in the failure/ fracture of a quasi-brittle material subjected to either a uniform stress field or a "singular" stress field caused by a crack tip. Next, we examine the ability of cohesive laws to account for the size effect on structural strength and different aspects of composite failure related to material characteristics. Finally, the effect of material softening on the size of the fracture process zone and the R-curve effect is demonstrated.

10.2.2. Cohesive laws

Cohesive crack models are based on kinematic descriptions that use strong discontinuities in the displacement field. Cohesive interfacial laws are phenomenological mechanical relations between the traction σ and the interfacial separation δ such that with increasing interfacial separation the tractions across the interface reach a maximum and then decrease and vanish when complete decohesion occurs. It can be shown by performing a J-integral calculation along a contour surrounding a notch tip that the resulting work of interfacial separation is related to Griffith's fracture criterion [16]:

$$G_c = \int_0^{\delta_f} \sigma(\delta) d\delta, \tag{10.1}$$

where G_c is the critical energy release rate (ERR) and δ_f is the critical interfacial separation.

The complete representation of a fracture in a continuum requires an ability to model both the initiation and growth of the fracture. The principles of the representation of a fracture in a continuum can be illustrated by considering a quasi-brittle bar of length L and cross-section A, as shown in Fig. 10.3(a). To model the fracture of the bar, a cohesive interface is introduced. The crack is assumed to open according to the

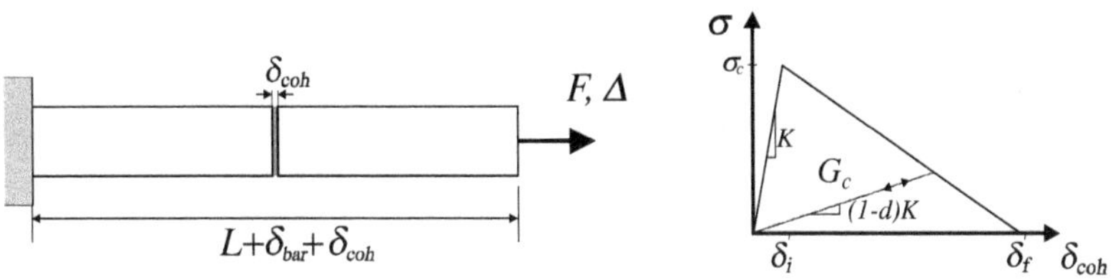

(a) (b)

Fig. 10.3. (a) Elastic bar with a cohesive crack. (b) Bilinear cohesive law.

softening law shown in Fig. 10.3(b). Initially, the crack is assumed to be elastic and the crack closing forces are related to the interfacial displacement jump δ_{coh} by a high penalty stiffness K. If the displacement jump exceeds a critical value δ_i the crack closing forces are assumed to soften linearly such that the area under the traction-displacement curve is equal to the fracture toughness G_c. Complete separation is achieved when the displacement jump exceeds δ_f. Therefore, the bilinear cohesive law can be expressed in two parts:

$$\sigma = \begin{cases} K\delta, & \delta < \delta_i, \\ (1-d)K\delta, & \delta_i \leq \delta < \delta_f, \end{cases} \tag{10.2}$$

where the damage variable d is a function of the displacement jump and accounts for the reduction in the load-carrying ability of the material as a result of damage.

The response of the bar can be obtained from the compatibility and equilibrium equations:

$$\Delta = \delta_{\text{bar}} + \delta_{\text{coh}} \quad \text{and} \quad \sigma = \sigma_{\text{bar}} = \sigma_{\text{coh}}. \tag{10.3}$$

The deformation of the bulk material is $\delta_{\text{bar}} = \sigma L/E$ where E is Young's modulus. If the bar is undamaged ($\sigma \leq \sigma_c$), the opening of the cohesive crack is $\delta_{\text{coh}} = \sigma/K$. Therefore, the force-displacement response of the bar before damage is

$$F = A\sigma = EA\frac{\Delta}{L + \frac{E}{K}}. \tag{10.4}$$

It can be observed that as long as $K \gg E/L$ the compliance introduced by the cohesive law can be neglected. It is also clear from Eq. (10.4) that the maximum load that the bar can withstand, $F_c = A\sigma_c$, and the response of the bar before failure depend on strength alone and not on a fracture criterion.

For a bilinear cohesive law, the relationship between the damage variable d and the displacement jump δ_{coh} has the form:

$$d = \frac{\delta_f(\delta_{\text{coh}} - \delta_i)}{\delta_{\text{coh}}(\delta_f - \delta_i)}, \tag{10.5}$$

where $\delta_i = \frac{\sigma_c}{K}$ and $\delta_f = \frac{2G_c}{\sigma_c}$. Therefore, the response of the bar after damage initiation is obtained by substituting d, Δ, δ_i, and δ_f into Eq. (10.4), which results in:

$$F = A\sigma = EA\frac{\Delta - \frac{2G_c}{\sigma_c}}{L - \frac{2EG_c}{\sigma_c^2} + \frac{E}{K}}. \tag{10.6}$$

The response of the bar after damage initiation is stable under displacement control only if $\frac{\partial F}{\partial \Delta} \leq 0$ which, assuming that $K \gg E/L$, gives:

$$L \leq \frac{2EG_c}{\sigma_c^2}. \tag{10.7}$$

The right-hand side of Eq. (10.7) is a characteristic dimension associated with material properties that is typically much smaller than the structural dimensions. Consequently, the failure of a uniformly stressed problem, such as the present bar, is typically unstable, i.e., no quasistatic equilibrium solution exists, and the failure load is independent of the fracture toughness G_c. As a result of the unstable fracture, the shape of the softening response is of no consequence. On the other hand, when cohesive models are used to predict the propagation of crack fronts, the fracture toughness G_c is the dominant material property and the effect of the strength σ_c may be minor.

In the following section, the fundamental concepts in the analysis of crack propagation using cohesive laws and LEFM are compared and some principles related to the effect of size on structural strength are outlined.

10.2.2.1. *Length of the fracture process zone*

LEFM assumes that the material is elastic and that the mechanisms that consume the fracture energy act at the crack tip. NLFM was initiated by Irwin [20] with a model for ductile solids based on an elastic/perfectly plastic material response to describe the effect of plastic material behavior in the vicinity of the crack tip on fracture propagation. By assuming that

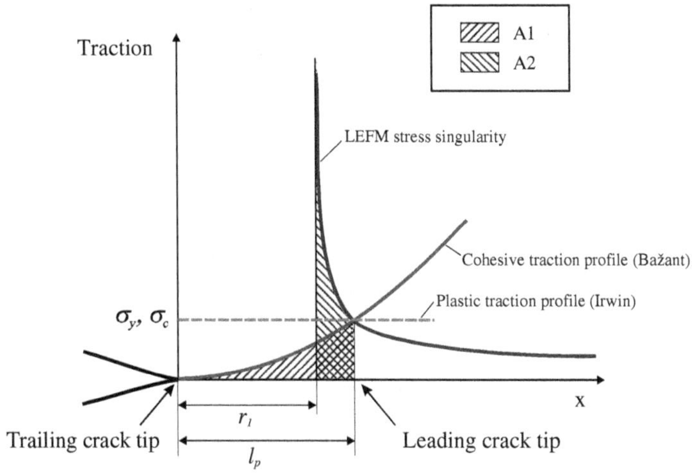

Fig. 10.4. Stress profile ahead of the crack tip [21].

plasticity affects the stress field only in the vicinity of the crack tip, Irwin estimated the size of the plastic region by equating the yield strength to the stress of the elastic field ahead of the leading crack tip (see Fig. 10.4). A generalization of Irwin's model for material softening due to damage was proposed by Bažant and Planas [21]. In the Irwin–Bažant model, the traction profile in the inelastic zone ahead of the trailing crack tip follows a general expression:

$$\sigma = \sigma_C \left(\frac{x}{l_p} \right)^{\beta}, \tag{10.8}$$

where l_p is the size of the cohesive zone (or the fracture process zone), x is the distance from the trailing crack tip, and β is a parameter that describes the stress field in the process zone.

In the elastic zone, the traction profile follows the expression given by the LEFM solution. Accounting for an offset r_1 to be determined, the stress singularity at the crack tip given by the linear elastic solution is

$$\sigma = \sqrt{\frac{EG_C}{2\pi(x - r_1)}}. \tag{10.9}$$

The length l_p and the offset r_1 are obtained by assuming that the traction given by Eqs. (10.8) and (10.9) are equal at $x = l_p$, and that the areas A1 and A2 shown in Fig. 10.4 are also equal. The crack propagates when the energy release rate G is equal to the critical value G_c. Therefore, the size of the cohesive zone when the crack is propagating in a self-similar, steady-state fashion can be solved using the previous equations, resulting in:

$$l_p^\infty = \frac{\beta + 1}{\pi} \frac{EG_c}{\sigma_c^2}. \tag{10.10}$$

The notation l_p^∞ indicates that Eq. (10.10) is valid only when the structural dimensions are infinite compared to the crack length, in which case the elastic stress field is represented accurately by Eq. (10.9). Finite element analysis results indicate that the steady-state length of the process zone for a bilinear law, such as the one represented in Fig. 10.3(b), is approximated well using $\beta = 1$ in Eq. (10.10) (see [22]).

In many situations of crack growth in composite structures, the process zone length l_p may be relatively large compared to other structural dimensions. The fracture process zone length is generally considered to represent an intrinsic characteristic of a material response and the fact that it can range from about 10 nm for a silicon wafer to several meters for concrete underlines the diversity in the response of different materials [23]. However, when the structural dimensions are small, the boundary conditions can influence the stress distribution ahead of the crack tip and empirical corrections must be applied to Eq. (10.10). This can be the case, for instance, in delamination, where the process zone length is shorter than the l_p^∞ estimated by Eq. (10.10), especially for thin adherends [22]. Conversely in a notched specimen whose height is less than the crack length, the non-singular stresses become significant and induce a process zone that is longer than l_p^∞ [19]. The development of analytical expressions for predicting the length of the process zone accurately for general configurations is the subject of ongoing research.

For many fracture processes in composite materials and structures, the length of the process zone must be considered because its formation is responsible for an increase in fracture toughness with crack growth, a response denoted as the resistance curve or the R-curve. In the presence of an R-curve, the toughness measured during crack propagation typically increases monotonically until reaching a steady-state value. In the case of delamination, the increase in toughness with crack growth is attributed mostly to fiber bridging across the delamination plane. Since it is generally assumed that fiber bridging only occurs in unidirectional test specimens and not in general laminates, the toughness of the material is taken as the initial toughness and the toughness for steady-state propagation is ignored [24]. However, recent experimental work on the delamination between plies with different orientations indicates that R-curve effects are always present in delamination such that the steady-state values of the critical ERR is typically between four and five times greater than the initiation values [25–27]. In a through-the-thickness fracture of composite laminates (Fig. 10.2(d)), the R-curve effect is caused by a combination of damage mechanisms leading the process zone, and fiber bridging in the trailing region of the process zone. This R-curve response makes it difficult to predict the effect of structural size on strength using LEFM, even in the case of self-similar propagation, as described in the following section. Test results for notched laminated panels of different sizes indicate that their strength cannot normally be predicted using a constant fracture toughness [13].

Finally, knowledge of the length of the process zone l_p is also useful for determining the finite element mesh requirements for a given material since it is necessary to use more than three elements in the process zone [28].

10.2.2.2. *Size effects*

The goal of any damage theory is to predict the effect of size, i.e., the change in strength when the spatial dimensions are scaled up from the coupon to the structure. No viable physical theory can be postulated without a clear understanding of the effects of scaling. The LEFM theory, in which all the fracture processes are assumed to occur at the crack tip, exhibits the strongest possible size effect. LEFM scaling predicts that the nominal strength is inversely proportional to the square root of structure size, as shown in the following example. Consider the notched specimen shown in Fig. 10.5(a) with an initial crack length a_0 and subjected to a tensile stress σ_u. Let D, b, and t be the width, the length, and the thickness of the specimen, respectively. The elastic solution can be approximated as a stress field consisting of two regions: one which is loaded elastically, V_{el}, and another which is unloaded, V_{ul}. The elastic energy in the specimen for a crack length a is then

$$U_e = \frac{\sigma_u^2}{2E} V_{\text{el}} = \frac{\sigma_u^2}{2E}(Db - ka^2)t, \tag{10.11}$$

where E is the elastic modulus and the slope k of the boundary between V_{el} and V_{ul} is a constant that is independent of the specimen dimension D and crack length a. Griffith's fracture criterion states that crack propagation

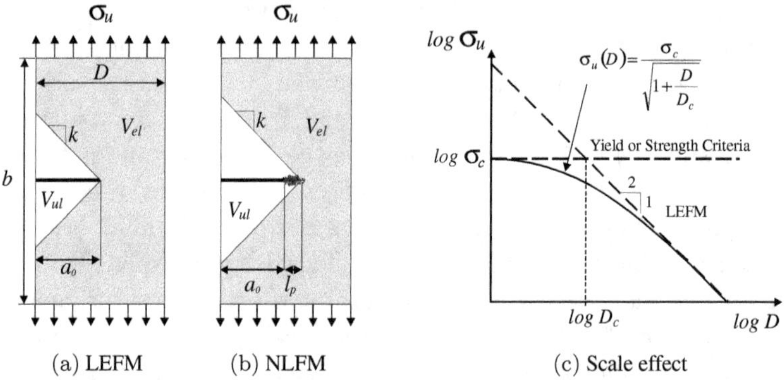

Fig. 10.5. Brittle and quasi-brittle crack propagation in a uniformly-loaded specimen.

occurs when the change in elastic energy per new surface area created is equal to the fracture toughness G_c. Therefore,

$$-\frac{1}{t}\left[\frac{\partial U_e}{\partial a}\right]_{\sigma_u} = \frac{\sigma_u^2}{E}ka = G_c. \tag{10.12}$$

The failure strength σ_u is obtained by solving Eq. (10.12), which gives $\sigma_u = \sqrt{EG_c/ka}$. Since the failure strength decreases with the crack length a, the maximum strength for the specimen corresponds to the initial crack length a_0:

$$\sigma_u^{\mathrm{MAX}} = \sqrt{\frac{EG_c}{ka_0}}. \tag{10.13}$$

Consider a series of experiments conducted on specimens with the same material and the same geometric proportions but with different dimensions. On a logarithmic plot, the strength in Eq. (10.13) scales as a line with slope $-1/2$, as shown in Fig. 10.5(c).

In contrast with the LEFM assumptions, a crack in a quasi-brittle material can propagate in a stable manner until a damage zone is formed behind the crack tip. The effect of the damage zone is equivalent to an increment l_p to the crack length, as shown in Fig. 10.5(b). Therefore, the failure strength can be expressed as

$$\sigma_u = \sqrt{\frac{EG_c}{k(a_0 + l_p)}}. \tag{10.14}$$

Since a_0 is proportional to D, the strength represented by Eq. (10.14) results in the scaling law

$$\sigma_u(D) = \frac{\sigma_c}{\sqrt{1 + \frac{D}{D_c}}}, \tag{10.15}$$

which can be plotted on a logarithmic scale as shown in Fig. 10.5(c). Equation (10.15) is known as Bažant's scaling law [17]. It can be observed that the failure of small specimens is governed by strength considerations, while that of large specimens is governed by LEFM. Furthermore, the degree to which size effects on structural strength can be predicted by LEFM depends on the laminate notch sensitivity, which is a function of both the laminate material and the notch length. The notch sensitivity can be expressed by the dimensionless ratio η of the notch length a over l_p:

$$\eta = \frac{a}{l_p} \begin{cases} \eta < 5 & \text{ductile damage, plasticity,} \\ 5 \leq \eta \leq 100 & \text{quasi-brittle fracture mechanics,} \\ \eta > 100 & \text{brittle.} \end{cases} \tag{10.16}$$

The load-carrying capability of notch-ductile components is dictated by strength; that of notch-brittle parts is dictated by fracture toughness; and for the range in between these extremes, both strength and toughness play a role. The process zone length for polymeric composites is of the order of a few millimeters, so notched panels with notch lengths greater than approximately 1–10 cm are notch-brittle [8].

Specimens that do not exhibit localized fracture planes, such as in the case of brooming, splitting, etc., may exhibit much enhanced notch-ductility. For instance, notch-ductility is enhanced by multiple cracking and matrix splitting along the fiber direction. It has been shown that the fracture toughness of a composite laminate increases with fiber strength and decreases with fiber/matrix shear strength [29].

Since cohesive laws are defined in terms of strength and fracture toughness, they can represent equally well the propagation of cracks as the softening of a material, as in the example of the tension-loaded quasi-brittle bar, which gives them the ability to produce the entire range of structural scaling represented by Eq. (10.15) and shown in Fig. 10.5(c).

10.2.2.3. *Softening law and the R-curve effect*

The relationship between the functional form of the material softening law, the length of the process zone, and the shape of the R-curve has not received much attention from the computational mechanics community. As described in the previous section, the physics of stable crack growth should be viewed as the gradual development of a fracture process zone behind the crack tip that produces a stabilizing influence on crack growth, characterized by a rising R-curve. For a bilinear softening curve such as the one shown in Fig. 10.3(b), the R-curve can be approximated by the expression [30]:

$$G_R(\Delta a) = \begin{cases} G_c \dfrac{\Delta a}{l_p}\left(2 - \dfrac{\Delta a}{l_p}\right) & \text{for } \Delta a < l_p, \\ G_c & \text{for } \Delta a \geq l_p, \end{cases} \tag{10.17}$$

where l_p can be estimated from Eq. (10.10) with $\beta = 1$.

In composite materials, more than one physical phenomenon is often involved in the fracture process. Some phenomena act at small opening displacements, which are confined to correspondingly small distances from the crack tip, and others act at higher displacements, which extend further into the crack wake. In these situations, more complicated softening laws than the bilinear softening law may be necessary to capture the correct crack growth response. For example, when fiber bridging or friction effects are present, a softening law may be required that has a peak at low crack displacements to represent the tip process zone, and a long tail at high crack displacements to represent the bridging in the wake of the crack.

In this case, multilinear softening laws can be obtained by combining two or more bilinear cohesive laws, as illustrated in Fig. 10.6. The two underlying linear responses may be seen as representing different phenomena, such as a quasi-brittle delamination fracture characterized by a small critical opening displacement δ_{c1}, and fiber bridging characterized by a lower peak stress and a larger critical opening displacement δ_{c2}. A multilinear cohesive law can provide a more accurate approximation of the process zone length and a more accurate approximation of an experimentally determined R-curve.

Consider a trilinear cohesive law such as that shown in Fig. 10.6. To describe such a trilinear law, it is convenient to consider the superposition of two bilinear cohesive laws that peak at the same opening displacement. Two bilinear softening responses are used for convenience and do not necessarily correspond to two distinct failure modes, which could peak at different opening displacements. In fact, the bridging strength does not typically contribute to the peak strength, which is associated with the intrinsic fracture process prior to the bridging process. Consequently,

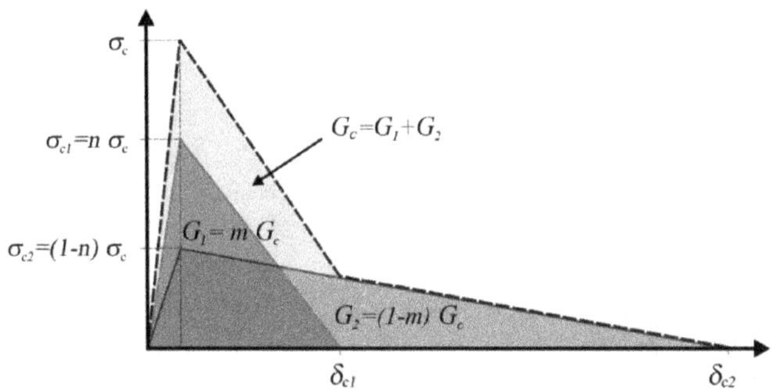

Fig. 10.6. Trilinear cohesive law obtained by the superposition of two bilinear laws.

a trilinear cohesive law can be described by the proportions: $\sigma_{c1} = n\sigma_c$, $\sigma_{c2} = (1-n)\sigma_c$, $G_1 = mG_c$, and $G_2 = (1-m)G_c$ with $0 \leq n,\ m \leq 1$, so that

$$G_c = G_1 + G_2 \quad \text{and} \quad \sigma_c = \sigma_{c1} + \sigma_{c2}. \tag{10.18}$$

A procedure for determining the strength ratio n and the toughness ratio m that approximate an experimentally determined R-curve is presented below. On the basis of Eq. (10.17), an expression for an R-curve that results from the sum of two bilinear cohesive laws [30] is defined as:

$$G_R(\Delta a) = \underbrace{n\,G_c\,\frac{\Delta a}{l_{p1}}\left(2 - \frac{n}{m}\frac{\Delta a}{l_{p1}}\right)}_{=G_1 \text{ if } \Delta a \geq \frac{m}{n} l_{p1}} + \underbrace{(1-n)G_c\,\frac{\Delta a}{l_{p1}}\left(2 - \frac{(1-n)}{(1-m)}\frac{\Delta a}{l_{p1}}\right)}_{=G_2 \text{ if } \Delta a \geq \frac{1-m}{1-n} l_{p1}}, \tag{10.19}$$

where $l_{p1} = \frac{2}{\pi}\frac{E\,G_c}{\sigma_c^2}$ is the length of the process zone for a single bilinear cohesive law. If the two superposed bilinear cohesive laws are ordered such that $m/n \leq (1-m)/(1-n)$, then the process zone length for the resulting trilinear law is

$$l_{p2} = \frac{1-m}{1-n} l_{p1}. \quad \text{Otherwise,} \quad l_{p2} = \frac{m}{n} l_{p1}. \tag{10.20}$$

Consequently, $l_{p2} \geq l_{p1}$, i.e., the process zone length for a trilinear cohesive law is longer than that of the corresponding bilinear cohesive law. For example, consider the problem examined in [30] of a crack with an initial length $a_0 = 25\,\text{mm}$ in a material with modulus $E = 70\,\text{GPa}$, fracture toughness $G_c = 180\,\text{kJ/m}^2$, and strength $\sigma_c = 2000\,\text{MPa}$. The length of

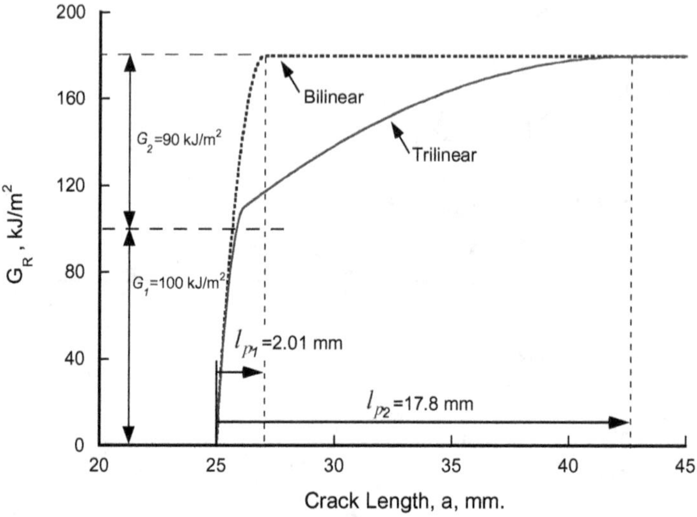

Fig. 10.7. R-curves for bilinear and trilinear cohesive laws.

the process zone is equal to $l_{p1} = 2.01$ mm, and the associated R-curve obtained from Eq. (10.17) is shown in Fig. 10.7. The R-curve for a trilinear cohesive law defined by $m = 0.556$ and $n = 0.944$ obtained from Eq. (10.20) is equal to $l_{p2} = 17.8$ mm, as shown in Fig. 10.7.

10.2.2.4. *Mixed-mode cohesive laws*

In composites, the direction of propagation of matrix cracks and delaminations is typically constrained by the presence of fibers. Therefore, crack propagation is likely to occur under mixed-mode loading, and a general formulation for cohesive laws must address mixed-mode fracture. A mixed-mode cohesive law can be illustrated in a single three-dimensional (3D) map by representing Mode I on the 2–3 plane, and Mode II in the 1–3 plane, as shown in Fig. 10.8. The triangle $O - Y - \delta_I^F$ is the bilinear material response in pure Mode I and $O - S - \delta_{II}^F$ is the bilinear material response in pure Mode II. It can be observed that the tensile strength Y is lower than the shear strength S, and the ultimate displacement in shear can be larger than in tension. In this 3D map, any point on the 0–1–2 plane represents a mixed-mode relative displacement. Under mixed-mode loading, damage initiates at δ^0 and complete fracture is reached at δ^F. Consequently, the tractions for Mode I and Mode II under mixed-mode loading follow the

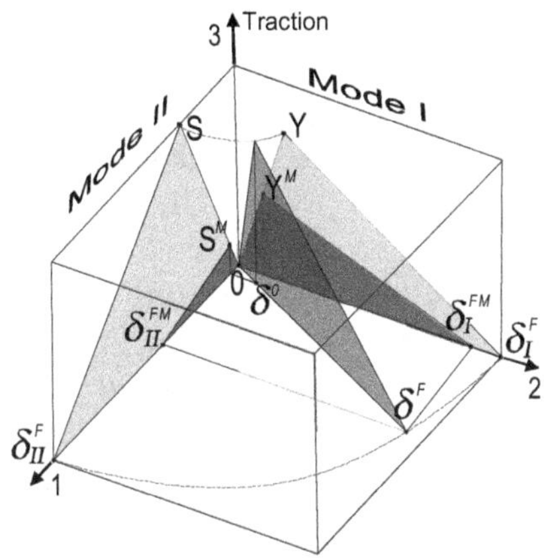

Fig. 10.8. Mixed-mode cohesive law.

reduced curves $O - Y^M - \delta_I^{FM}$ and $O - S^M - \delta_{II}^{FM}$, respectively. The areas under these two curves represent the fracture energies under mixed-mode loading. In the model proposed by Turon *et al.* [31], the initial strength and the critical value of the ERR in the mixed-mode cohesive interface damage law are functions of the mode-mixity parameter B:

$$B = \frac{G_{II}}{G_{II} + G_I} = \frac{K_{II}\delta_{II}^2}{K_{II}\delta_{II}^2 + K_I\delta_I^2}, \tag{10.21}$$

where G_I and G_{II} are the Mode I and II critical ERR values, K_I and K_{II} are the Mode I and II penalty stiffnesses, and δ_I and δ_{II} are the displacement jumps in the directions normal to and along the crack surface. The critical strength and fracture toughness values for mixed-mode loading are defined as

$$\begin{aligned} \sigma_c^2 &= Y^2 + (S^2 - Y^2)B^\eta, \\ G_c &= G_{Ic} + (G_{IIc} - G_{Ic})B^\eta, \end{aligned} \tag{10.22}$$

where Y and S are the transverse tensile strength and shear strength, respectively, of a composite ply, and η is a numerical parameter for fitting experimental mixed-mode data. Similar mixed-mode models have been adopted by a number of authors [32–36] and have been extended to trilinear cohesive laws by Hansen *et al.* [37].

Despite the maturity of cohesive laws, some issues regarding the prediction of crack propagation under mixed-mode conditions remain unresolved. According to any of the mode-mixity measures defined in the references cited above, the mode ratio is rarely if ever constant during fracture. Even in specimens such as the mixed-mode bending (MMB) specimen, where from the LEFM point of view the mode ratio is constant during propagation, it can be observed that the opening displacements at damage initiation are dominated by Mode II, and that immediately before complete separation the displacement jumps are mostly in Mode I. Sørensen *et al.* [38] and Turon *et al.* [39] have observed that the ratio of interlaminar strengths affects the prediction of delamination propagation, even when the crack length is long and the propagation should be according to LEFM. By enforcing the condition of a non-negative damage rate under variable mode mixity, Turon obtained a relationship between interlaminar strengths, fracture toughnesses, and penalty stiffnesses that has the form:

$$K_{II} = K_I \frac{G_{Ic}}{G_{IIc}} \left(\frac{S}{Y}\right)^2 . \tag{10.23}$$

When the penalty stiffnesses are selected according to Eq. (10.23) and the LEFM assumptions are valid, then the load–displacement curve for propagation predicted from a mixed-mode cohesive formulation is equivalent to that predicted using LEFM. However, additional research is needed for a full understanding of mixed-mode crack propagation.

10.3. Continuum Representation of Material Response

10.3.1. Distributed damage vs. localization of fracture

All materials exhibit irreversible nonlinearities, which can be due to plasticity, damage, and fracture. When a material's constitutive tangent relationship between the strain increments and the stresses is positive definite, as shown in Fig. 10.9, the response is said to be of a hardening type. Hardening is a macroscopic, distributed, and irreversible material response that locally redistributes stress concentrations and eliminates any stress singularities. Examples of hardening damage include plasticity and distributed damage mechanisms at the microstructural scale. Different material nonlinearities can be identified by comparing the unloading paths to the loading paths. Unloading path A is a typical linear elastic material response. Path B has the same slope as path A, but with an additional strain offset, indicating that the material has undergone plastic deformation. Path C differs from path A in terms of both slope and offset, indicating the presence of additional nonlinearities such as cracking. The constitutive tangential stiffness for loading path D is non-positive definite and the

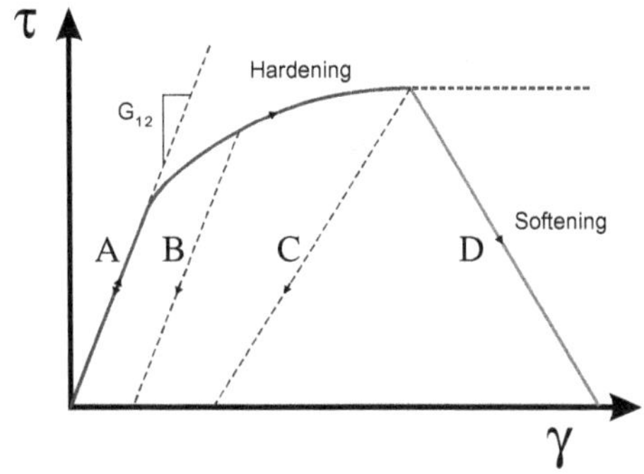

Fig. 10.9. Typical shear stress–strain response exhibiting multiple nonlinearities.

response is said to be of a softening type, which corresponds to the development and coalescence of voids and microcracks. As a consequence of softening, damage localizes along a fracture surface while material adjacent to the fracture surface unloads elastically.

A number of models have been proposed to represent material nonlinearities. The hardening of composites can be modeled with local constitutive models, i.e., models in which the stresses at a given point depend uniquely on the history of the strains up to that point. These models rely on the implicit assumption that the material can be treated as a continuum at any arbitrarily small scale. These models are described by differential equations and lack the notion of characteristic length.

Alternatively, the constitutive response of a material can be approximated using a spatially periodic representative volume element (RVE) to represent the micromechanical response of individual constituents and their interactions [40, 41]. The degree of complexity of the system is often reduced by considering a 2D approximation of the 3D continuum, assuming plane-strain conditions for a 2D model of the material. It is assumed that each RVE deforms in a repetitive way, identical to its neighbors. Periodic boundary conditions are imposed on each RVE in order to ensure compatibility of the deformation field along the boundaries. RVEs must be large enough to reflect the stochastic fluctuations of material properties on the pertinent scale while the computational requirements call for an RVE to be as small as possible. However, even when the RVE is large, extending the analysis to a highly nonlinear material response that leads to localization of damage renders the periodicity in the boundary conditions, and consequently the RVE approach, unsatisfactory [42].

Until Bažant and others developed the concept of crack bands [43], the idea of using strain softening to represent cracking in a continuum was controversial [44]. It was often argued that materials with a non-positive definite tangential moduli tensor do not exist. The point in the deformation history where the tangential stiffness of the constitutive model loses its positive definite properties indicates the formation of discontinuities. From the mathematical point of view, the loss of ellipticity of the governing differential equation induces numerical difficulties related to the ill-posedness of the boundary-value problem [45, 46]. Since an infinitesimal change in the data can cause a finite change in the solution, ill-posedness is manifested by the pathological sensitivity of numerical results to finite element discretization.

The problem can be easily illustrated by considering a simple example of a quasi-brittle bar loaded in monotonic tension, as shown in Fig. 10.10.

The constitutive damage model is a function of the strain given by the following expression:

$$\sigma(\varepsilon) = \begin{cases} E\varepsilon, & \varepsilon \leq \varepsilon_i, \\ (1-d)E\varepsilon, & \varepsilon_i < \varepsilon < \varepsilon_f, \quad d = \dfrac{\varepsilon_f(\varepsilon - \varepsilon_i)}{\varepsilon(\varepsilon_f - \varepsilon_i)}, \\ 0, & \varepsilon > \varepsilon_f, \end{cases} \tag{10.24}$$

where d is a scalar damage variable and E is the elastic modulus of the material. While the strains do not exceed ε_i, the force–displacement relationship of the bar is

$$F = A\sigma = \frac{EA}{L}\delta. \tag{10.25}$$

The material properties and the geometry of a real bar cannot be exactly uniform. Assuming that the strength of a small region of the bar Ω_B

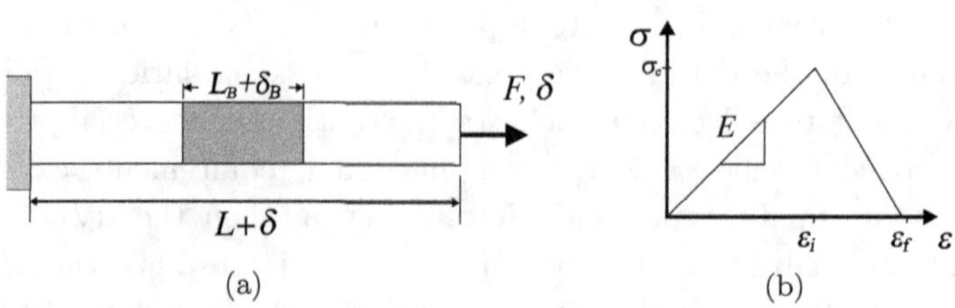

Fig. 10.10. (a) Bar under tensile load. (b) Constitutive response with linear softening.

is lower than the strength of the remaining portion of the bar Ω_A, damage localizes in the region Ω_B. Consequently, the material in Ω_B softens and in order to satisfy the equilibrium conditions the material in Ω_A unloads elastically. Equilibrium dictates that the stress along the bar is

$$\sigma = \sigma_A = \sigma_B, \tag{10.26}$$

which yields

$$\frac{\sigma}{E} = \varepsilon_A = \varepsilon_i \frac{\varepsilon_f - \varepsilon_B}{\varepsilon_f - \varepsilon_i}. \tag{10.27}$$

Compatibility of the displacements gives

$$(L - L_B)\varepsilon_A + L_B \varepsilon_B = \delta. \tag{10.28}$$

Using Eqs. (10.27) and (10.28) to solve for ε_A and ε_B gives the force–displacement relationship during the damage process:

$$F = A\sigma = EA \frac{\delta - L_B \varepsilon_f}{L - L_B \frac{\varepsilon_f}{\varepsilon_i}}. \tag{10.29}$$

As Eq. (10.29) indicates, the force–displacement relationship for damage depends on the length of damage localization L_B, which can take any value between zero and L. Consequently, the problem has infinitely many solutions, as the post-peak solutions illustrated in Fig. 10.11 indicate, and it is not clear which of these solutions is correct. In addition, some post-peak responses, such as that for $L_B/L = 0.01$, exhibit an artificial contraction once the critical strain is exceeded in Ω_B, which is not a physical behavior.

In finite element analyses, the length L_B is related to the element length L_e. Consider a model of the bar composed of N_e linear beam elements of equal length. The length of the localized zone is $L_B = L_e = L/N_e$. The post-peak solution given by Eq. (10.29) is therefore strongly dependent on the number of elements N_e. As the number of elements tends to infinity, the post-peak response approaches the unloading response of the initial (linear elastic) solution. Furthermore, the energy dissipated in the localized zone is calculated as $\int_V \int_0^{\varepsilon_f} \sigma d\varepsilon dV = 1/2\sigma_c \varepsilon_f A L_e$, which depends on the element length L_e. At the limit, when the element size tends to zero, the computational model predicts failure without any energy being dissipated, a physically unacceptable result. If the crack-band model does not permit reducing the element size to zero, convergence cannot be defined and the

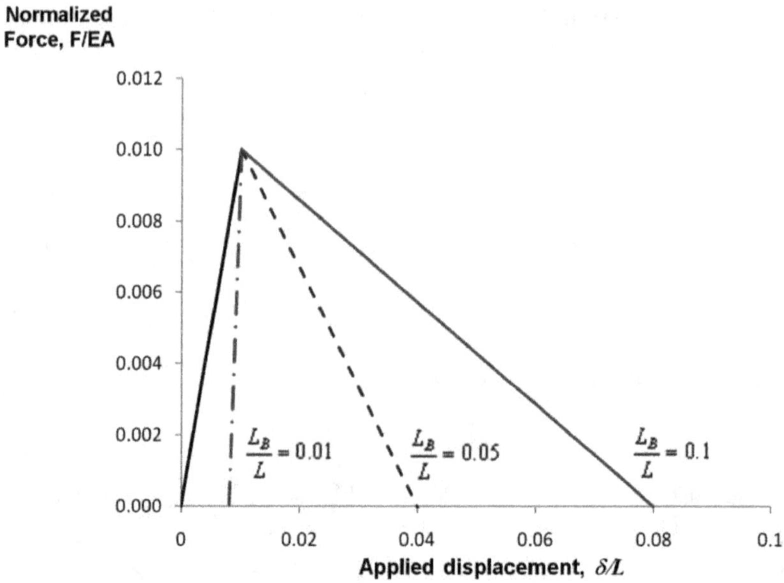

Fig. 10.11. Force–displacement response of a bar loaded in tension for different numerical discretizations.

boundary-value problem becomes ill-posed. Somehow, the boundary-value problem has to be regularized.

In the context of continuum mechanics, so-called non-local techniques are available to resolve mesh sensitivity issues and to retain the objectivity of the numerical response. Most of these techniques introduce spatial interaction terms that have a smearing effect on the deformation fields, and thus preclude localization in a plane [47, 48]. A similar smearing can be obtained in a computationally simpler technique using rate-dependent (viscous) properties [49]. The material properties necessary for crack smearing, or strain softening, are chosen such that the width of a localizing diffuse crack band in a continuum is equal to the characteristic length, which is associated with the material response.

Alternatively, the objectivity of the numerical solution can be simply achieved by adjusting the post-peak material response using a characteristic element length L_e. This technique, proposed by Bažant and Oh [43], consists of ensuring that the computed dissipated energy due to the fracture process is constant and equal to the product between the fracture energy G_c and the crack surface A. Solving the resulting equation for the ultimate strain gives:

$$\int_V \int_0^{\varepsilon_f} \sigma d\varepsilon dV = G_c A \Rightarrow AL_e \frac{E\varepsilon_i \varepsilon_f}{2} = G_c A \Rightarrow \varepsilon_f = \frac{2G_c}{L_e E \varepsilon_i}. \quad (10.30)$$

When Eq. (10.30) is substituted into (10.29), the response of the bar becomes

$$F = EA \frac{\delta - \frac{2G_c}{\sigma_c}}{L - \frac{2EG_c}{\sigma_c^2}}. \quad (10.31)$$

The response of the bar represented by Eq. (10.31) is independent of the crack band or discretization length L_e. In addition, the response given by Eq. (10.31) is identical to the response obtained from (10.6) using a cohesive crack model, provided that the penalty stiffness of the cohesive law is a very large number: $K \gg E/L$.

It can be observed that the adjusted constitutive model takes into account a size effect, since the response of the bar depends on the length, L: a longer bar has a more brittle post-peak response. It can be shown that under displacement control, the post-failure response described by Eq. (10.31) is stable only if $L \leq 2EG_c/\sigma_c^2$.

Another important aspect in the simulation of fracture using crack-band models is that there exists a maximum size of the finite elements that can be used in the simulation of a crack band. In order to avoid snap-back in the constitutive model illustrated in Fig. 10.12, the ultimate strain cannot be less than $\varepsilon_i = \sigma_c/E$. Under this circumstance, the maximum characteristic size of the finite element is

$$AL_e^{\mathrm{Max}} \frac{\sigma_c^2}{2E} = G_c A \Rightarrow L_e^{\mathrm{Max}} = \frac{2EG_c}{\sigma_c^2}. \quad (10.32)$$

Note that Eq. (10.32) has the same form as the equation governing the stability of the bar loaded in tension. The same principle applies at both

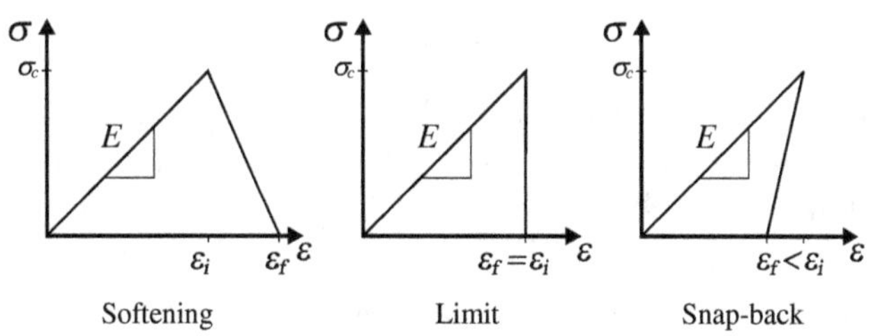

Fig. 10.12. Snap-back at constitutive level.

scales: unstable failure occurs when the elastic strain energy in the bar is greater than the fracture energy required to fail it, once initiated. While this is physically valid in the case of the bar, if the element size is greater than the critical size in Eq. (10.32), the element will always fail unstably and a model composed of such elements will be unable to predict any stable fracture process.

The implications of Eq. (10.32) are important: the use of an element size larger than L_e^{Max}, which for some composite damage modes is a fraction of a millimeter, results in the overestimation of the critical ERR. When it is computationally impractical to use a sufficiently fine mesh, a workaround consists of artificially reducing the strength in order to preserve the correct value of G_c [28, 50]. Reducing the strength in a damage model to increase the maximum acceptable element size should be done only for problems where the solution is expected to be fracture dependent and not strength dependent.

10.3.2. Idealization of damage modes in composite materials

The complex damage mechanisms occurring in advanced composite materials result in additional difficulties in the numerical simulation of failure. While some intralaminar damage mechanisms such as transverse matrix cracking (Fig. 10.1(b)) and delamination (Fig. 10.1(c)) occur in easily identifiable fracture planes that are parallel to the fiber direction, others such as fiber failure are more difficult to idealize. These complexities make the representation of intralaminar damage using a kinematic description based on strong discontinuities a formidable task. A more computationally tractable approach consists of using failure criteria based on the homogenized stress or strain state to idealize the mechanisms of failure.

10.3.3. Failure criteria and strength

Many failure criteria have been proposed to predict the onset of matrix cracking and fiber fracture, some of which are described and compared in the worldwide failure exercise (WWFE) [51, 52]. However, few criteria can represent several relevant aspects of the failure process of laminated composites, e.g., the increase of apparent shear strength when applying moderate values of transverse compression, or the detrimental effect of the in-plane shear stresses in failure by fiber kinking. The LaRC03 failure criteria [53] and subsequent evolutions [54] address some of the limitations of other failure criteria as identified from the WWFE. For example, the LaRC criteria account for the effect of ply thickness, fiber misalignment in compression, and the effect of shear nonlinearity on fiber kinking and *in situ* strength.

10.3.4. Crack tunneling and *in situ* strength

Transverse matrix cracking is often considered a benign mode of failure because it normally causes such a small reduction in the overall stiffness of a structure that it is difficult to detect during a test. However, transverse matrix cracks can have a strong effect on the development of damage. Work by Green *et al.* [55] and others indicates that scaling effects in the failure of the matrix produce different modes of failure. Thicker plies were found to crack and cause delaminations at relatively low loads, while thinner plies resulted in more brittle failure mechanisms with cleaner through-thickness fracture surfaces.

To predict matrix cracking in a laminate subjected to in-plane shear and transverse tensile stresses, a failure criterion must account for the *in situ* strengths. The *in situ* effect, originally detected in Parvizi *et al.*'s [56] tensile tests of cross-ply glass-fiber-reinforced plastics, is characterized by higher transverse tensile and shear strengths of a ply when it is constrained by plies with different fiber orientations in a laminate, compared with the strength of the same ply in a unidirectional laminate. The *in situ* strength also depends on the number of plies clustered together and on the fiber orientations of the constraining plies. The results of Wang's [57] tests of $[0/90_n/0]$ carbon/epoxy laminates indicate that thinner plies exhibit a higher transverse tensile strength.

Both experimental and analytical methods [58] have been proposed to determine the *in situ* strengths. The *in situ* strengths can be calculated using fracture mechanics solutions for the propagation of cracks in a constrained ply. For typical ply thicknesses, it can be assumed that defects exist in a ply and that these material defects span the thickness of the ply, as shown in Fig. 10.13. Cracking of the ply can be assumed to occur when the ply is loaded above the load required to propagate the slit crack in the fiber direction (tunneling). It can be shown that the stress required for

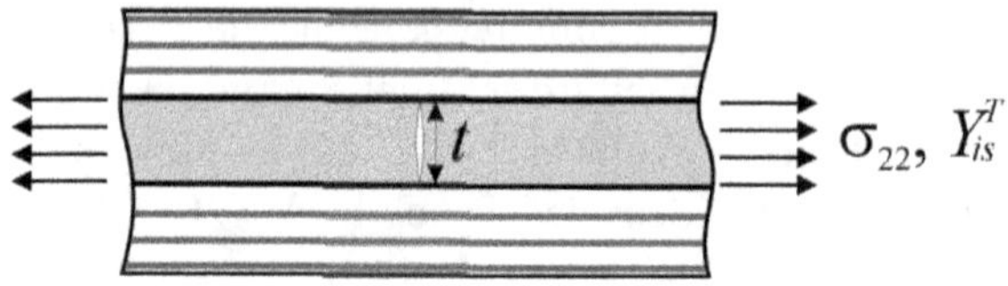

Fig. 10.13. *In situ* strength determined from the propagation of a slit crack in a ply.

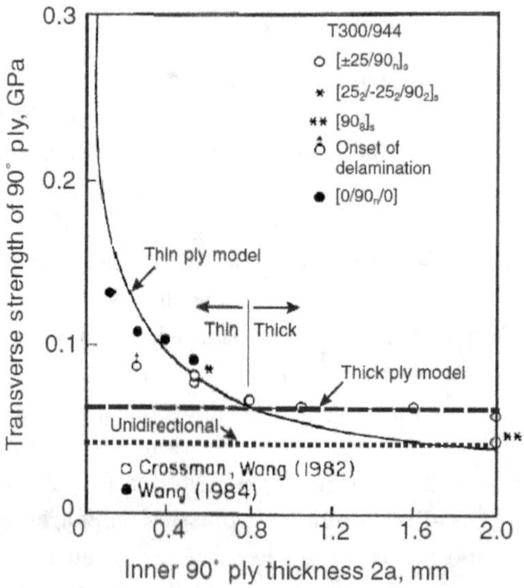

Fig. 10.14. Transverse tensile strength as a function of ply thickness [57].

tunneling (or the *in situ* strength) can be approximated as [58]

$$Y_{is}^T = \sqrt{\frac{8G_{Ic}}{\pi t \Lambda_{22}^0}}, \tag{10.33}$$

where G_{Ic} is the Mode 1 fracture toughness of the matrix, t is the ply thickness, and Λ_{22}^0 is an elastic modulus of the material. Predicted *in situ* strengths as a function of the ply thickness and some experimental values of *in situ* strength are shown in Fig. 10.14 for a T300/944 graphite/epoxy. A similar approach is followed to calculate the *in situ* shear strength.

In multidirectional composite laminates subjected to uniform stress states, cracks accumulate during the loading process. As the loading on an individual ply is increased, new cracks suddenly appear, initially at rather random locations, and then with a progressively more uniform crack spacing. Eventually, new sets of cracks appear deterministically equally spaced between the original cracks. Finally, the crack density reaches a saturation value at which a different event occurs, such as delamination or fiber failure in an adjacent ply.

As the density of cracks in each ply of the laminate increases, networks of cracks are formed, which can link up through the thickness of the laminate by inducing delaminations. A simplified damage progression sequence of coupled transverse matrix cracking and interlaminar delamination is shown in Figs. 10.15(b)–(d) for the case of a laminated plate subjected to a tensile load. Initially, the laminate is undamaged, as shown in Fig. 10.15(a). As a result of the load application, transverse matrix cracks form in different plies of the laminate, as shown in Fig. 10.15(b). In the absence of a stress concentration, the locations of the initial matrix cracks are random, and

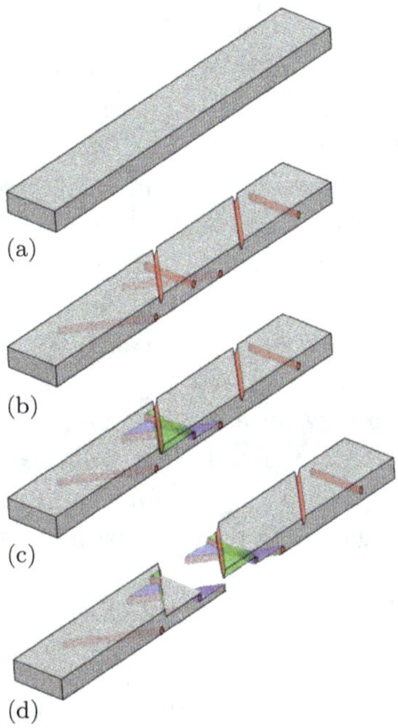

Fig. 10.15. Idealized damage progression sequence in a laminated composite plate subjected to tensile loading: (a) initial stage without damage, (b) matrix cracking stage, (c) delamination stage, linking up matrix cracks in various plies, and (d) specimen fracture.

cannot be known *a priori*. At some value of the applied load, delaminations initiate from the matrix cracks, Fig. 10.15(c). These delaminations can connect matrix cracks in adjacent plies, which can cause the disintegration of the laminate, Fig. 10.15(d).

10.3.5. Continuum damage models for composite materials

Continuum damage mechanics models for composite materials were pioneered by Ladevèze and LeDantec [59], Matzenmiller *et al.* [60], and others based on previous work by Kachanov [61], Lemaître *et al.* [62], and others. In these composite damage models, a distinction is made between the different failure modes, especially between fiber and matrix failure. These models include a progressive softening of the material response, with internal damage variables describing the softening response. However, a softening response causes a localization of the strains along a surface known as the failure surface and these strain-softening models typically do not include a characteristic length. Therefore, they exhibit pathological dependencies on element size. Upon reducing the mesh size to zero, such analyses predict that failure would occur with zero energy dissipation.

To resolve this lack of objectivity with respect to element size, a characteristic length must be inserted into the constitutive model. The evolution of intralaminar damage in laminated composites can be represented by softening laws that define the evolution of damage in terms of the fracture energy dissipated in each damage mode. Most damage models, such as the progressive damage model for composites provided in Abaqus® [34] and typical cohesive elements [31, 63], represent the evolution of damage with linear softening laws that are described by a maximum traction and a critical energy release rate.

Using the LaRC03 failure criteria [53] as damage activation functions F_M, it is possible to formulate a continuum damage model to predict the propagation of M damage mechanisms occurring at the intralaminar level. Each damage activation function predicts one type of damage mechanism using the following equations:

$$F_M = \phi_M(\varepsilon^t) - r_M^t \leq 0, \tag{10.34}$$

where r_M^t are internal variables (equal to 1 at time $t = 0$), and the functions $\phi_M(\varepsilon^t)$ of the strains ε^t correspond to the LaRC03 failure criteria. When a damage activation function is satisfied, $F_M \geq 0$, the associated damage variable d_M takes on a positive, non-zero value less than or equal to 1, and the ply compliance tensor is affected by the presence of damage. Using the model proposed by Maimí [9], the compliance matrix of a damaged ply is defined as:

$$H = \begin{bmatrix} \dfrac{1}{(1-d_1)E_1} & -\dfrac{\upsilon_{12}}{E_2} & 0 \\ -\dfrac{\upsilon_{12}}{E_2} & \dfrac{1}{(1-d_2)E_2} & 0 \\ 0 & 0 & \dfrac{1}{(1-d_6)G_{12}} \end{bmatrix}, \tag{10.35}$$

where d_1 is the damage variable associated with fiber fracture, d_2 is the damage state variable associated with matrix cracking, and d_6 is a damage variable associated with both damage mechanisms.

In addition to the damage activation functions and damaged compliance tensor, it is necessary to define the evolution laws for the damage variables d_M. The damage evolution laws need to ensure that the computed energy dissipated is independent of the refinement of the mesh.

A complete definition of a continuum damage mechanics model for the simulation of intralaminar damage can be found in [9, 50]. The algorithm for the integration of the constitutive damage model was implemented in an Abaqus® UMAT subroutine. The CDM model simulates localized intralaminar damage using strain softening constitutive models. In order to avoid mesh-dependent solutions, energy dissipation is regularized for each damage mechanism using a modification of the crack-band model. To avoid physically unacceptable snap-backs of the material response, the maximum allowable element size is determined using closed-form equations [50]. If the element size exceeds this maximum for any damage mode, the corresponding strength is automatically reduced to preserve the correct ERR.

10.4. CDM: Limitations

To predict the ultimate strength of composite structures, it is necessary to have an accurate numerical representation of all damage modes and their interactions. Some of the most complex damage models available rely on CDM to represent the intralaminar damage modes (e.g., transverse matrix cracking and fiber failure) and use cohesive zone models to capture delamination between ply interfaces. Some of the combined CDM/cohesive models, such as the impact models of Lopes *et al.* [64], rely on extremely fine meshes with one or more elements through the thickness of a ply, while others use stacks of shell elements to represent sublaminates within larger structures [65, 66]. However, despite advances in progressive damage modeling, recent studies (e.g., van der Meer and Sluys [67]) indicate that CDM models coupled with cohesive zone models may not always represent laminate failure sequences properly. These deficiencies are particularly evident when the observed fracture mode exhibits matrix splitting and pullouts [55] or when the fracture is characterized by a strong coupling between transverse matrix cracking and delamination [12, 67].

The deficiencies of the predictive capabilities consist of several issues, including the incorrect prediction of the damage zone size normal to the fracture direction when using crack-band models and the inability of local CDM models to reliably predict matrix crack paths. These limitations are mostly due to the fact that CDM models are usually implemented as "local" rather than "non-local" models [42], i.e., the evolution of damage in a local CDM model is evaluated at individual integration points without consideration of the state of damage at neighboring locations. The following discussion pertains mostly to such local implementations, since non-local damage models are less widely used due to the difficulty in implementing them within the finite element method.

The premise of the crack-band approach for regularizing CDM models is that damage localizes into a band with a width equivalent to the element dimension. If the element size is either significantly larger or smaller than the damage process zone, the crack-band approach may not correctly predict the width of the damage zone nor the local stress field. The size of the elements within the CDM crack band affects the severity of the stress concentration produced by the CDM crack. Consequently, the stress redistribution resulting from damage development may be inaccurately predicted and can potentially result in inaccurate representation of damage mode interactions and failure sequences for problems where the results are sensitive to primarily the strength inputs or to both the strength and fracture toughness inputs.

As a result of homogenization and damage localization, CDM models have difficulties predicting crack paths. Since homogenization eliminates the physical distinction between the fibers and the matrix, a strain softening CDM model does not distinguish between cracks that propagate along the fiber direction from those that cross fibers [67]. In CDM models implemented with damage localization, the damage state at any integration point in the model depends only on the stress field at that point rather than the damage state of neighboring points. Therefore, the direction of damage evolution is driven only by the instantaneous local stress distribution. In other words, the local direction of cracking may be predicted correctly by the failure criteria, but if the morphology and the kinematics of the cracked material are not properly accounted for in the damage model, the sequence of failures that eventually defines the path of a crack at a macroscopic level may be predicted incorrectly.

The potential inability of CDM models to determine the correct direction of propagation is particularly evident when the stress field is dominated by shear. Consider two different plies in a laminate with a notch that is subjected to shear, as shown in Fig. 10.16. In both situations, the stress level required to initiate matrix microcracks is correctly predicted by the failure criterion. Furthermore, both situations would result in an identical sequence of failures, since the stress field is identical. However, it is clearly not the same to propagate a crack in a sequence of linked microcracks (Fig. 10.16(a)) as it is to propagate a crack band across fibers (Fig. 10.16(b)). Matrix cracking in a shear band running parallel to the fibers is a relatively brittle failure mechanism, whereas matrix cracking normal to the fibers produces a damage band that requires much more work to propagate.

The sensitivity of CDM predictions to the finite element mesh orientation also contributes to the difficulty in predicting the crack path. Although the objectivity of the solution with respect to element size is addressed with the crack-band approach described in the previous section, the predicted damage may be dependent on mesh orientation and element

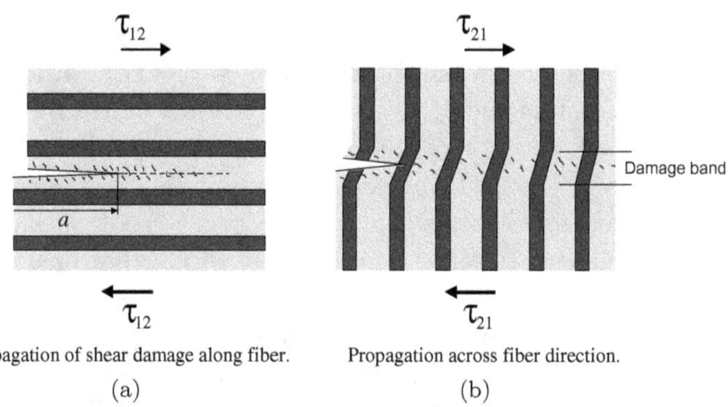

Propagation of shear damage along fiber. (a)

Propagation across fiber direction. (b)

Fig. 10.16. Idealized propagation of shear damage [67].

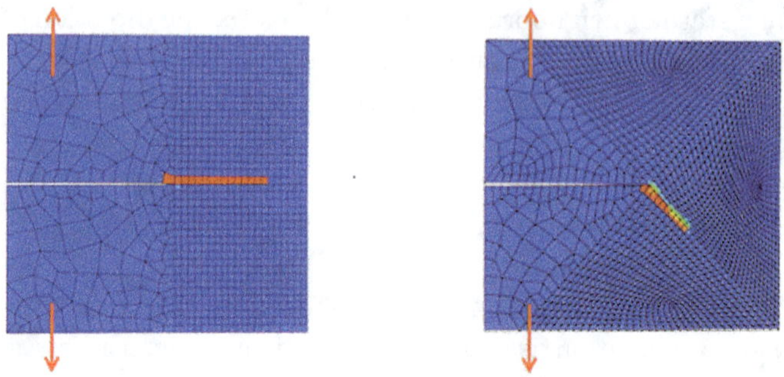

Mesh aligned with crack direction.
(a)

Mesh inclined to crack direction.
(b)

Fig. 10.17. Effect of mesh orientation on crack path in a unidirectional CT specimen.

shape. When strain-softening constitutive models are used in a finite element simulation, damage tends to propagate along preferred directions, consisting of either element edges or element diagonals. A demonstration of the sensitivity of simulation results to mesh orientation is provided in Fig. 10.17 for a unidirectional compact tension (CT) specimen with fibers oriented at 90° to the load direction. Results are presented for simulations obtained with a mesh oriented parallel to the fiber direction (Fig. 10.17(a)). and with an inclined mesh in front of the crack tip (Fig. 10.17(b)). The crack should propagate along the fiber direction. However, the results show directional bias, and the simulated crack band propagates in the direction of the element alignment.

The tendency for damage to localize along mesh lines can be partially attributed to shear locking [68]. In the CDM methodology, a crack or displacement discontinuity is represented by a degradation of the corresponding terms in the constitutive stiffness. As the crack opens, the stiffness degradation is such that stress should not be transferred across the crack faces. However, such unloading may not occur due to in-plane shear locking. Shear locking here refers to inappropriate shear stress transfer across a widely open smeared crack, which occurs when an element cannot shear without inducing tensile strains. In a ply with orthotropic properties, a matrix crack is typically represented by setting the transverse shear modulus G_{12} and the transverse Young's modulus E_{22} to zero. However, quadrilateral elements have been shown to exhibit coupling between γ_{12} and ε_{11} unless the element edges are aligned with the softening band or are oriented at 45° to the band [69]. Furthermore, the tendency of the element to lock is dependent on the order of integration of the element: fully integrated elements are more susceptible to pathological in-plane shear locking than reduced-integration elements.

The shear stress transfer across an open smeared cracked caused by shear locking can result in inaccurate prediction of stress redistribution after damage development. Iarve *et al.* [70] demonstrated shear locking for a simple case of a unidirectional $[0_8]$ graphite/epoxy open-hole tension specimen. Experiments show that this specimen exhibits splitting cracks parallel to the load and tangential to the hole. Splitting near holes and notches reduces the stress concentration, so predicting their effect is essential in obtaining the ultimate strength with any accuracy. Iarve's progressive damage analyses were conducted using a radial-type mesh pattern with different levels of mesh refinement, where a radial mesh typically consists of a pattern of elements radiating from a circular hole and ending at a rectangular boundary. Iarve observed that it is not possible to predict the stress relaxation with a radial mesh due to longitudinal splitting unless the fiber-direction modulus E_{11} is also set to zero.

When damage localizes and a fracture path is known *a priori*, a relatively simple approach to circumvent some of the limitations of CDM noted above consists of aligning the mesh with the direction of fracture [71]. Mesh alignment can be used to force a matrix crack to localize along the fiber direction. The benefit of mesh alignment is demonstrated in Fig. 10.18 for an open-hole tension specimen. Predictions were obtained for an $[0_8]$ IM7-8552 laminate using the continuum damage model provided within the

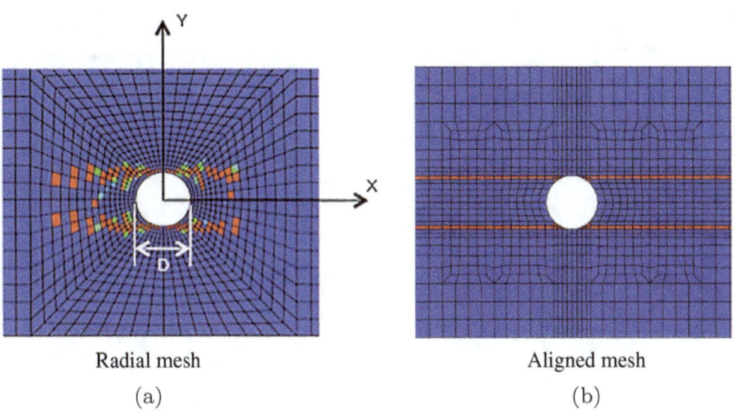

Fig. 10.18. Splitting damage predicted using a radial mesh and an aligned mesh.

Abaqus® finite element code [34]. Each ply was modeled using quadrilateral reduced integration shell elements, S4R, to minimize the tendency for locking. Analyses were obtained using a radial mesh and an aligned mesh, as shown in Fig. 10.18. The loading direction and fiber direction are parallel to the X-axis.

The damage zones predicted in the region of the hole using the radial and aligned meshes are shown in Fig. 10.18. The results indicate that both models predict longitudinal splitting tangential to the hole. The load vs. end displacement response obtained with the two models is shown in Fig. 10.19(a). The radial mesh model predicts failure of the specimen at 65% of the load predicted by the model with the aligned mesh. The failure load is severely underestimated by the radial mesh due to the inability of rotated elements to represent shearing along the axial split.

The tensile stresses along the width (Y direction) of the specimen are shown in Fig. 10.19(b). It can be seen in Fig. 10.19(b) that the results obtained with the aligned mesh have a stress concentration at the hole that is substantially less severe than the stress concentration in the radial mesh model. Shear locking in the damaged elements of the radial mesh model transfer shear load across the splitting cracks even though the matrix-dependent moduli G_{12} and E_{22} have been set to zero. This load transfer across the splitting cracks causes premature failure of the fiber ligaments on either side of the hole.

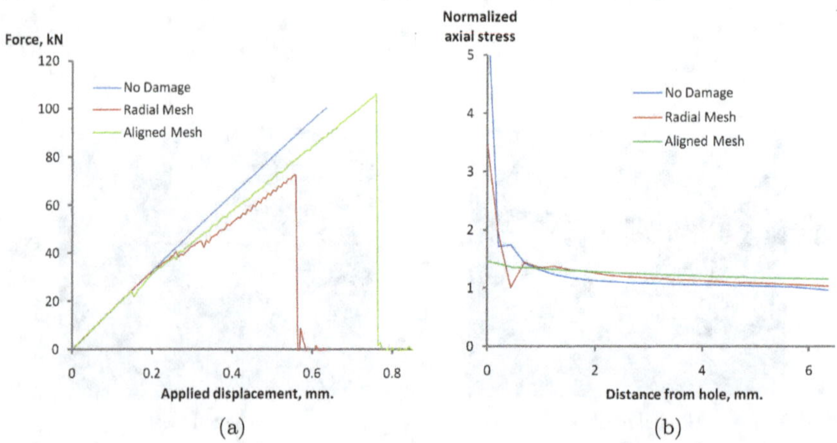

Fig. 10.19. Effect of mesh type on predicted ultimate failure and fiber-stress relaxation due to splitting in a $[0_8]$ laminate. (a) Load vs. end displacement and (b) normalized axial stress distribution.

Using an aligned mesh in the open hole model, however, does not entirely remove the stress concentration from the edge of the hole. A stress concentration of approximately 1.6 persists, as shown in Fig. 10.19(b). Large shear deformations in strain softening CDM models can also cause load transfer across cracks, even while using fiber-aligned meshes. This additional source of shear load transfer is related to inaccurately tracking the orientations of the cracked material and the deformed continuum. When a cracked material is subjected to simple shear deformation, neither the material nor the crack face should rotate despite the rotation of the current frame of the continuum. To avoid nonphysical load transfer across the crack, it is necessary to accurately define the orientation of the crack throughout the deformation history of the material.

10.5. Bridging the Gap between DDM and CDM

In order to address some of the limitations of CDM discussed in the previous section, developers have started to incorporate features of higher-fidelity DDM methods into CDM methods, where possible. Cohesive zone modeling (Section 10.2.2) and CDM methods (Section 10.3.5) are closely related in that the former uses traction-displacement laws to represent the opening of discrete cracks, while the latter uses stress-strain laws to do the same for a smeared, cracked continuum. Rather than developing separate, parallel methods for intralaminar and interlaminar matrix cracks, Camanho *et al.* [72] proposed a new smeared crack CDM model in which intralaminar matrix cracks within the bulk material are represented using embedded cohesive laws. These intralaminar cohesive cracks can be inserted in the bulk material with crack surface normals with any orientation in the 2–3 plane. An inserted matrix crack with a crack normal oriented at α degrees away from the 2-direction is shown in Fig. 10.20. The exact angle α of the crack normal depends on the stress state that is acting on the crack at the time of initiation. Additive strain decomposition is used to separate the total strain into bulk material strain ε^e and cracking strain ε^c components:

$$\varepsilon = \varepsilon^e + \varepsilon^c = \varepsilon^e + R_{\mathrm{cr}}\varepsilon^c_{\mathrm{cr}}R^T_{\mathrm{cr}}, \tag{10.36}$$

where R_{cr} defines the crack coordinate system in terms of the crack normal angle α. The cracking strain ε^c is transformed into the coordinate system of the crack to yield $\varepsilon^c_{\mathrm{cr}}$. Conventional cohesize zone modeling approaches are applied to determine the state of stress and damage on the crack in

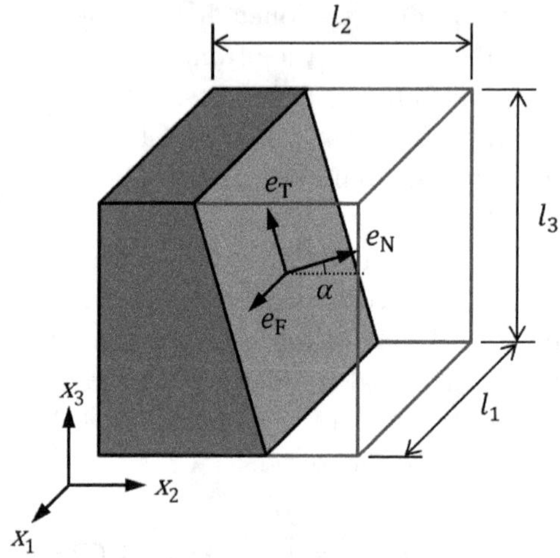

Fig. 10.20. Orientation of an inserted cohesive matrix crack in the methods of Camanho *et al.* [72] and Leone [74].

terms of $\varepsilon^c_{\mathrm{cr}}$. Hooke's law is applied to determine the state of stress in the elastic material in terms of ε^e. Equilibrium is enforced between the elastic material and cracking regions in order to solve for the two states of deformation, and the overall state of stress. A similar approach, the continuum-decohesive finite element (CDFE) of Prabhakar and Waas [73], utilizes an embedded cohesive law to represent the formation and growth of intraply cracks in fiber-reinforced materials. The CDFE method is a finite element formulation that involves applying the principle of virtual work to a cracked continuum, deriving an enriched set of displacements for real and internal dummy nodes, and generating an equivalent element stiffness for the cracked element through static condensation. These approaches unify the theories used for inter- and intralaminar matrix damage, simplifying the overall composite damage modeling approach for fiber-reinforced composites.

In progressive damage analyses, damaged material points may have to stretch to several times their characteristic length L_e as part of complicated, multi-mode failure processes. Under these large deformations, it is important that damaged material points accurately represent the kinematics of a crack. Misrepresenting the kinematics of a crack under large deformation can lead to non-physical load transfer across the crack. Leone [74] expanded on the smeared crack CDM model of Camanho for problems involving geometric nonlinearity and large shear deformations.

Rather than additively decomposing the strain, the deformation gradient decomposition (DGD) method is applied to track the orientation of cracks throughout their deformation history. In this method, the deformation gradient tensor of the continuum F is decomposed into cohesive displacement jump vector δ and bulk material deformation gradient tensor F_B components:

$$F^{(2)} = F_B^{(2)} + R_{\text{cr}}^T \delta \frac{1}{L_e}, \tag{10.37}$$

where the superscript $^{(2)}$ represents the second column of a tensor, corresponding to the matrix-direction. A schematic representation of the decomposition of the deformation gradient tensor in the 1–2 plane is shown in Fig. 10.21. In the DGD approach, the cohesive crack surface normal e_N in the current, deformed configuration is tracked using F_B through:

$$e_N = F_B^{-T} \begin{bmatrix} 0 \\ \cos(\alpha) \\ \sin(\alpha) \end{bmatrix}. \tag{10.38}$$

As a result, the crack coordinate system R_{cr} is a function of the current bulk material deformation and the fracture angle α. As in the approach of Camanho *et al.* [72], conventional cohesize zone modeling approaches are applied to determine the state of stress and damage on the crack. The Green–Lagrange strain and second Piola–Kirchhoff stress are calculated for the bulk material using F_B. Equilibrium is enforced between the bulk

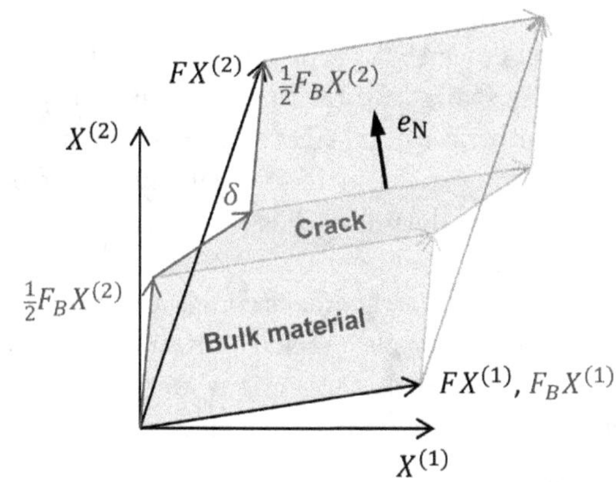

Fig. 10.21. Decomposition of the deformation gradient tensor of the continuum into a cohesive displacement jump and a bulk material deformation gradient tensor.

material and the cohesive crack in order to solve for the two states of deformation, and the overall state of stress.

While the methods of Camanho *et al.* [72] and Leone [74] use cohesive laws to soften damaged material, both methods have lower fidelity than discretely inserting cohesive elements into the original mesh. In both of these CDM methods, a single integration point must represent the deformation and stress state of a crack and the material on either side of the crack. While equilibrium can be assumed across a crack face, assuming that the material on either side of a crack has equal stresses along the crack face is a simplification which may introduce load transfer across the crack and smearing of the overall stress state.

10.6. Regularized x-FEM (Rx-FEM) Framework

The extended finite element method (x-FEM) is a technique that can be used to predict the location and evolution of matrix cracks in composites while avoiding the aforementioned limitations associated with CDM models. x-FEM is a mesh enrichment technique based on a pioneering concept by Moës *et al.* [4], which facilitates the introduction of displacement discontinuities such as cracks at locations and along directions that are independent of the underlying finite element mesh. Although most of the research on x-FEM is devoted to arbitrary crack propagation in isotropic materials, recent applications to composite materials include delamination modeling and textile composite architecture representation [5]. Huynh *et al.* [75] provide a review of contemporary developments in x-FEM as well as novel applications to interfacial crack analysis in 2D and 3D problems.

Modeling a matrix crack that propagates parallel to the fiber direction in a ply is conceptually straightforward using x-FEM. However, it is more difficult to model networks of matrix cracks in a laminate where the fracture planes of matrix cracks in individual plies intersect at common interfaces and can cause delaminations that link the cracks through the thickness. Within the traditional x-FEM approach, the difficulty in modeling networks of linked matrix cracks could be addressed by developing a special enrichment for multiple crack situations or by connecting two enriched/cracked elements. Such connections were recently accomplished in a quasi-2D formulation, for example, by van der Meer and Sluys [76] and Ling *et al.* [77].

Another direction in which x-FEM is being developed is the regularized extended finite element method (Rx-FEM) proposed by Iarve [78–80].

In Rx-FEM, the step function approaches to describe the crack surface, which are used in x-FEM, are replaced by a continuous function. Displacement shape functions are used to approximate the step function, and the Gauss integration can be retained for element stiffness matrix computation regardless of the orientation of the crack. The cohesive connection between two plies in which cracks have been introduced can be established by computing integrals of the products of the shape functions at the ply interface. Therefore, with the Rx-FEM technique a kinematically powerful model of crack networks can be constructed in which transverse matrix cracks are inserted parallel to the fiber direction at locations determined using a failure criterion.

A simulation begins without any initial matrix cracks. As the loading is increased, matrix cracks are inserted according to the LaRC03 failure criterion [53]. The criterion is evaluated at each integration point and if the criterion is exceeded a matrix crack oriented in the fiber direction is added. The crack is inserted using the displacement enrichment necessary to model the displacement jump. The magnitude of the jump is initially zero and is controlled by an interface cohesive law (Turon *et al.* [31]). The same cohesive law is used at the ply interfaces to represent potential delamination surfaces. A Newton–Raphson procedure is applied to find the equilibrium solution at each load step of the implicit incremental solution.

The following section describes the formulation of Rx-FEM. The goal here is to highlight the concepts. More detailed discussion of the formulation can be found in [78, 80]. Then, a few examples that illustrate the application of Rx-FEM to predict matrix cracking-induced delamination failure in unnotched laminated composites are considered. The ability of the model to predict the effects of ply thickness and ply orientation on matrix cracking and delamination are discussed as well.

10.6.1. Matrix crack modeling using Rx-FEM

A discontinuous displacement field over a crack surface Γ_α can be represented using two continuous displacement fields $\boldsymbol{u}_1$ and $\boldsymbol{u}_2$ and the Heaviside step function H as follows [4]:

$$\boldsymbol{u} = H(f_\alpha)\boldsymbol{u}_1 + (1 - H(f_\alpha))\boldsymbol{u}_2, \tag{10.39}$$

where f_α is a signed distance function of the crack surface Γ_α. This function is defined for an arbitrary point as the distance from the point to the crack surface. The signed distance function is positive if the point is located in the direction of the normal to the crack surface and negative if it is located in the direction opposite to the normal. The strain field and subsequently the stress field are computed similarly on each side of the crack from the continuous displacement fields $\boldsymbol{u}_1$ and $\boldsymbol{u}_2$ as

$$\varepsilon = H(f_\alpha)\varepsilon_1 + (1 - H(f_\alpha))\varepsilon_2, \tag{10.40}$$

$$\sigma = H(f_\alpha)\sigma_1 + (1 - H(f_\alpha))\sigma_2. \tag{10.41}$$

The calculation of the strain energy of a volume, i.e., an element containing a crack Γ_α, is more difficult because the approximations given in Eqs. (10.39)–(10.41) are discontinuous across the surface Γ_α. Therefore, separate computations of the stiffness matrix are required on each side of the crack using complex element subdivisions and the associated integration points. Nevertheless, the strain energy of the volume of interest V can be cast in the following form:

$$W = \int_V (H(f_a)W_1 + (1 - H(f_a))W_2)dV, \tag{10.42}$$

where W_1 and W_2 are computed from the strain and stress on each side of the crack and the aforementioned integration detail is hidden by the presence of the step function. The cohesive energy over the crack surface S_α can be written using the cohesive law provided in Eq. (10.2) as

$$M = \int_{S_\alpha} \left(\int_0^\delta \sigma(\delta)d\delta \right) ds, \tag{10.43}$$

where the displacement jump $\delta = \|\boldsymbol{u}_1 - \boldsymbol{u}_2\|$ and the normal vector to the crack surface can be readily computed. Performing the surface integration is not straightforward because the crack surface inside the volume V has to be discretized in order to perform the surface integration. On the other hand, if there were a practical way of dealing with the step function and its derivatives, the surface integral in Eq. (10.43) would be calculated as a volume integral using

$$M = \int_V |\nabla H| \left(\int_0^\delta \sigma(\delta)d\delta \right) dV. \tag{10.44}$$

Equation (10.44) can be readily understood since

$$\nabla H(f_\alpha) = \mathcal{D}(f_\alpha)\nabla f_\alpha, \tag{10.45}$$

where $\mathcal{D}(x)$ is the Dirac delta function, and the gradient of the signed distance function is by definition the unit normal vector to the crack surface so that $|\nabla H(f_\alpha)| = \mathcal{D}(f_\alpha)$. However, the transition from the volume integral to a surface integral in Eq. (10.44) is intuitive, especially considering the 1D case where the "volume" is a line and the "crack surface" is a point. Clearly, evaluation of Eq. (10.44) within the standard x-FEM framework has no practical application and needs to be evaluated by surface discretization.

In the regularized x-FEM formulation, the step function $H(f_\alpha)$ is replaced with a continuous function $\tilde{H}(\mathbf{x})$. In the formulation proposed by Iarve [78], the function $\tilde{H}(\mathbf{x})$ is expressed as a superposition of the displacement approximation functions. This function is equal to 0 or 1 everywhere except in the vicinity of the crack surface. If the displacement approximation functions are $X_i(\mathbf{x})$, then

$$\tilde{H}(\mathbf{x}) = \sum h_i X_i(\mathbf{x}) \tag{10.46}$$

and the coefficients h_i are obtained as

$$h_i = \frac{\int_V X_i^{\langle f_\alpha \rangle} f_\alpha^{dV}}{\int_V X_i dV} \quad \text{or} \quad h_i = 0.5 + 0.5\frac{\int_V X_i f_\alpha dV}{\int_V X_i |f_\alpha| dV}. \tag{10.47}$$

As indicated in Eq. (10.47), the coefficient h_i is equal to 0 or 1 if the signed distance function does not change sign in the support domain of the shape function X_i. On the other hand, when the crack crosses the support domain of the shape function, $0 < h_i < 1$, the evaluation of Eqs. (10.47) requires determination of the signed distance function from the crack surface, which is a trivial task in the case of a straight crack that is normal to the ply interfaces and has a prescribed orientation.

The strain energy, Eq. (10.42), and the cohesive energy, Eq. (10.44), can be evaluated in the Rx-FEM formulation, where $H(f_\alpha)$ is replaced by $\tilde{H}(\mathbf{x})$, using standard Gauss quadratures for a crack of arbitrary direction and location, which is fully defined by the h_i coefficients. The price for such simplicity and robustness is the non-zero width of the crack surface, which is associated with the width of the gradient zone $|\nabla \tilde{H}| > 0$. The width of this zone is a function of the mesh size. In other words, standard x-FEM encloses the crack inside the element and Rx-FEM smears the crack surface over possibly more than one element. Energetically, however, the cohesive energy of Rx-FEM will approach that of the regular crack with mesh refinement. Reducing the mesh size, i.e., reducing the size of the support

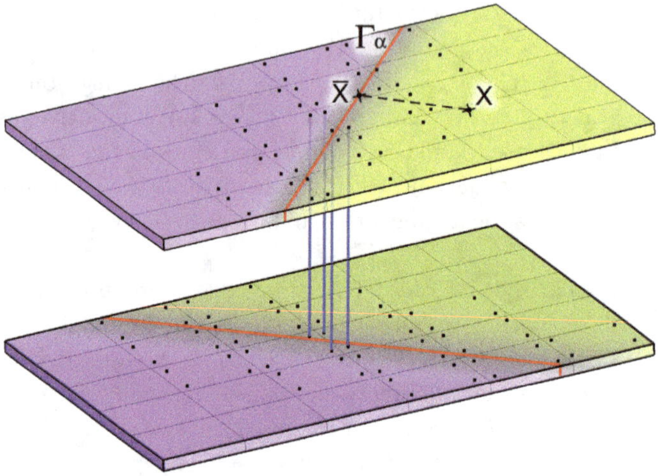

Fig. 10.22. Two plies with different cracking directions. The cracking direction in each ply and the shaded level plot of the regularized step function $\tilde{H}$ are shown.

domain of the shape functions, reduces the size of the band where X_i is not constant. Therefore, the difference between the step function representation of the crack surface $H(f_\alpha)$ and the continuous representation $\tilde{H}(\mathbf{x})$ is reduced.

The ability to maintain the Gauss quadratures in "cracked" elements also allows for simple calculation of the cross products of shape functions on the surfaces between plies with arbitrary fiber and matrix cracking directions. This is illustrated in Fig. 10.22, where two plies with different crack directions are shown, and the regularized step functions $\tilde{H}$ associated with the crack in each ply are shown by shading. Enriched functions in each ply are obtained by multiplying the shape functions by the respective step functions. Since the step functions are smooth functions and do not alter the support domains for the shape functions, the cross integrals of the enriched functions in the two plies can be easily computed by the Gauss quadrature.

10.7. Rx-FEM Simulations

Numerical results are presented in the following sections to verify and illustrate the proposed methodology. First, we consider in detail the initiation of delamination from a given matrix crack in a transverse crack tension (TCT) specimen. Next, we simulate failure in multilayered composites and evaluate the ply thickness and lamination effects on matrix cracking and delamination initiation and propagation.

10.7.1. Transverse crack tension test

The TCT specimen was designed to measure Mode II interlaminar fracture toughness [81]. It consists of three unidirectional ($\theta = 0$) plies with thicknesses t, $2t$, and t and where the middle ply is cut at the specimen mid-length prior to curing. The specimen is subjected to axial tensile loading, and when a critical load is exceeded, delaminations between the middle ply and the top and bottom plies develop suddenly. These delaminations continue to propagate in a stable manner with increasing load until the delaminations reach the grips, at which point the load is carried by the two undamaged outer plies.

A TCT specimen was analyzed using a conventional finite element (FE) model, shown in Fig. 10.23(a), and an Rx-FEM model, shown in Fig. 10.23(b), to illustrate the ability of the Rx-FEM method to insert transverse matrix cracks at arbitrary locations and orientations, and to demonstrate the ability to represent the correct interactions between transverse matrix cracks and the corresponding delaminations. Although Fig. 10.23 shows the entire laminate thickness, only half of the laminate thickness was modeled and symmetry conditions were applied to the mid-surface of the laminate. The shaded red regions in the figures show the initial middle ply crack. In the case of the conventional FE model, the crack in

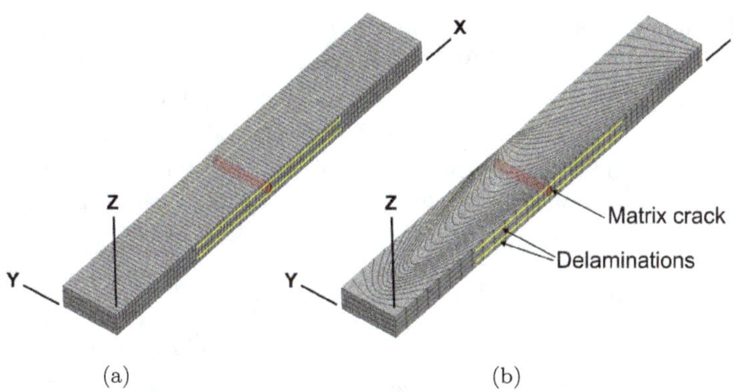

Fig. 10.23. (a) Conventional finite element (FE) model and (b) Rx-FEM model of the TCT specimen.

Table 10.1. Material properties used in the analyses.

		T300/914C [18]	T300/976 [20]
E_{11}	(GPa)	139.9	138
E_{22}, E_{33}	(GPa)	10.1	10.3
G_{23}	(GPa)	3.7	3.1
$G_{12,13}$	(GPa)	4.6	5.5
ν_{23}		0.436	0.66
ν_{12}, ν_{13}		0.3	0.3
α_{11}	$(1/^{0}C)$	—	0.4×10^{-6}
α_{22}, α_{33}	$(1/^{0}C)$	—	2.54×10^{-5}
$T - T_0$	(^{0}C)	—	–125
Y_t	(MPa)	80	37.9
Y_c	(MPa)	300	200
S	(MPa)	100	100
G_{IC}	(J/m^2)	120	157
G_{IIC}	(J/m^2)	500	315

the middle ply is aligned with a mesh line and is simply modeled by using unconnected double nodes.

For the Rx-FEM model, a curved non-uniform mesh was used to demonstrate the mesh independence of the approach. In the Rx-FEM model, the middle ply crack was inserted at the start of the analysis and is not aligned with the mesh cell boundaries. The total number of elements in the longitudinal direction is 120 for both models. However, the local density of the Rx-FEM mesh near the delamination crack tip is not uniform due to the irregularity of the mesh. In both models, the delaminations between the plies were modeled using the cohesive technique described in Section 10.2.2 and one element was used through the thickness of each ply (due to symmetry half of the middle ply was represented by 1 element). The material properties for T300/914C from [81] are summarized in Table 10.1. The thermal prestress is not considered since all plies have the same orientation and there is no mismatch of thermal expansion properties between plies. The load vs. applied displacement curves predicted by the two models are shown in Fig. 10.24. The two responses are nearly identical.

10.7.2. Effect of ply thickness

One of the key factors affecting the matrix cracking and delamination failure modes in laminated composites is the ply thickness. A systematic experimental study of delamination failure as a function of ply thickness was conducted by Crossman and Wang [82]. A T300/934 $[\pm 25/90_n]_s$

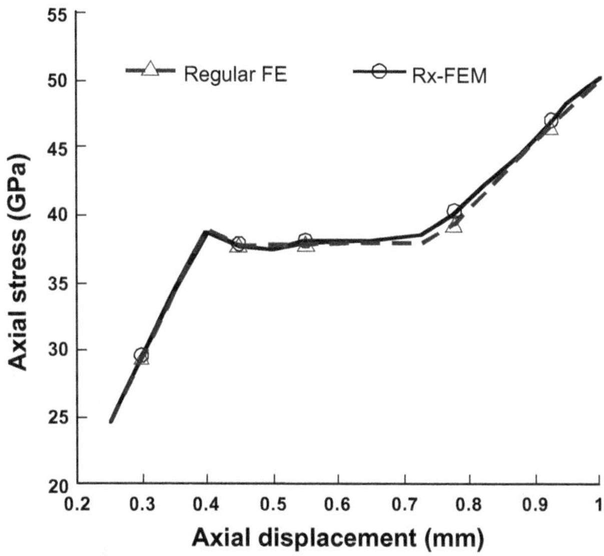

Fig. 10.24. Load vs. displacement curves of the TCT specimen predicted using the conventional FE model and the Rx-FEM model.

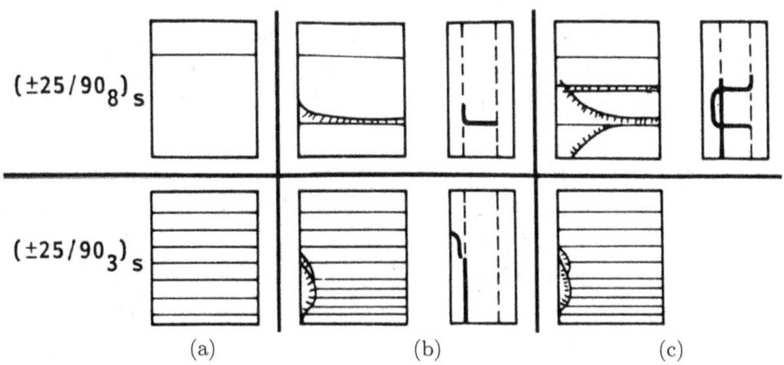

Fig. 10.25. Fracture sequence in $(25/-25/90_n)_s$ laminates: (a) just prior to delamination; (b) subsequent to delamination; (c) just prior to final failure. Reprinted with permission from [82].

laminate family with $n = 1, \ldots, 8$ was subjected to uniaxial tensile loading, perpendicular to the 90° ply, and failure loads and patterns were carefully documented. The delamination patterns (hatched lines) and crack densities (spacing between horizontal lines) can be observed in Fig. 10.25 for three load levels and for two thicknesses of the 90° ply, namely $n = 3$ and 8.

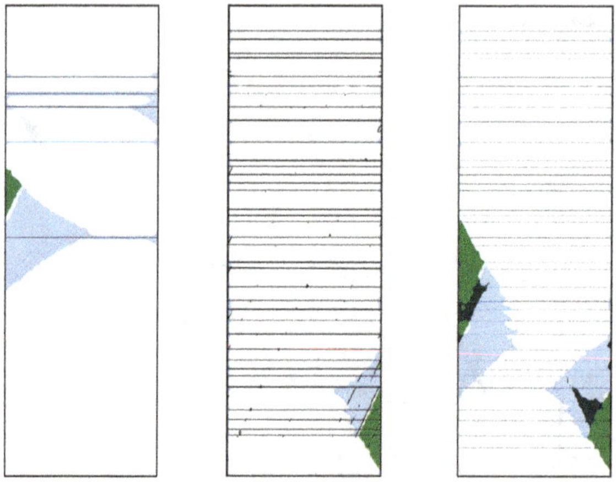

Fig. 10.26. Predicted cracking and delamination patterns in $[\pm 25/90_8]_s$ and $[\pm 25/90_3]_s$ laminates. Blue areas correspond to predicted delaminations at the $90/-25$ interface and green areas correspond to predicted delaminations at the $25/-25$ interface.

The differences observed in the shapes of the delaminations in the two laminates are evident: in the $n = 8$ case, the delamination is funneling off the individual matrix cracks, whereas for the thinner plies, the delamination spreads over multiple transverse matrix cracks, and is referred to as "oyster shaped" in [82]. A significant difference between the two cases is also seen in the 90° ply transverse crack densities.

The results of simulating the tensile loading of these two laminates are presented in Fig. 10.26. The material properties for T300/936 used in the model are shown in Table 10.1. To illustrate the damage development process, damage variable contours for both transverse matrix cracks and delaminations are plotted on the undeformed geometry. The areas of delamination correspond to interfaces where the value of the damage variable d exceeds 0.995. The Rx-FEM transverse matrix cracks in each ply correspond to surfaces where the discontinuity function $\tilde{H}$ is equal to 0.5.

Predicted matrix cracking and delamination patterns for the laminates with $n = 8$ and $n = 3$ are shown in Fig. 10.26. Blue areas correspond to predicted delaminations at the $90/-25$ interface, and green areas correspond to predicted delaminations at the $25/-25$ interface. The state of cracking and delamination immediately before the complete failure of the

$[\pm 25/90_8]_s$ laminate is shown in Fig. 10.26(a). The delaminations in the $[\pm 25/90_3]_s$ laminate evolve extremely rapidly before failure. A sequence of two states of delamination at nearly identical loads is shown in Figs. 10.26(b) and (c). It can be observed that the predicted density of matrix cracking for $n = 8$ is significantly lower than predicted for the thinner $n = 3$ case. In addition, the shape of the delamination in Fig. 10.26(a) is very similar to the experimental funnel-type delamination shown in Fig. 10.25. In both the experimental observations and the predictions, thin delamination areas accompany all matrix cracks.

The delaminations predicted for $n = 3$ (Figs. 10.26(b) and (c)) cover multiple transverse cracks and have shapes consistent with the experimental results shown in Fig. 10.25. The extent of the delamination in Fig. 10.26(b) is very similar to that in Fig. 10.25. It is likely that the larger predicted delamination in Fig. 10.26(c) corresponds to an unstable equilibrium state, which is unlikely to be caught in the experiment. Since the predicted extent of the delaminations and the crack density is in good agreement with the experimental observations for both ply thicknesses, it is concluded that the effect of ply thickness on matrix cracking and delamination evolution is well represented by the Rx-FEM model.

10.7.3. Internal delamination vs. edge delamination

In the case of the $[\pm 25/90_8]_s$ laminates considered above, the delamination evolution process initiates from the intersection of matrix cracks and the free edges, leading to the eventual disintegration of the laminate. It is of interest to evaluate the present methodology for characterizing the process of matrix crack-induced damage accumulation in laminates with different ply orientations, where the delamination and matrix cracking evolution and interaction patterns may vary. A number of angle-ply laminate configurations were experimentally and analytically investigated by Johnson and Chang [83]. The T300/976 graphite fiber material system (see Table 10.1 for ply level properties) was used. Tensile failure of a $[\pm 45/90]_s$ laminate and a $[\pm 60_2]_s$ laminate, both considered in [83], with a ply thickness of 0.127 mm are considered below. These laminates do not contain any 0° plies and completely lose their load-carrying capacity as a result of matrix cracking and delamination.

Predicted matrix crack and delamination damage evolution patterns for a $[\pm 45/90]_s$ laminate and for a $[\pm 60_2]_s$ laminate are shown in Fig. 10.27. Damage patterns are shown at three load levels, including the load level

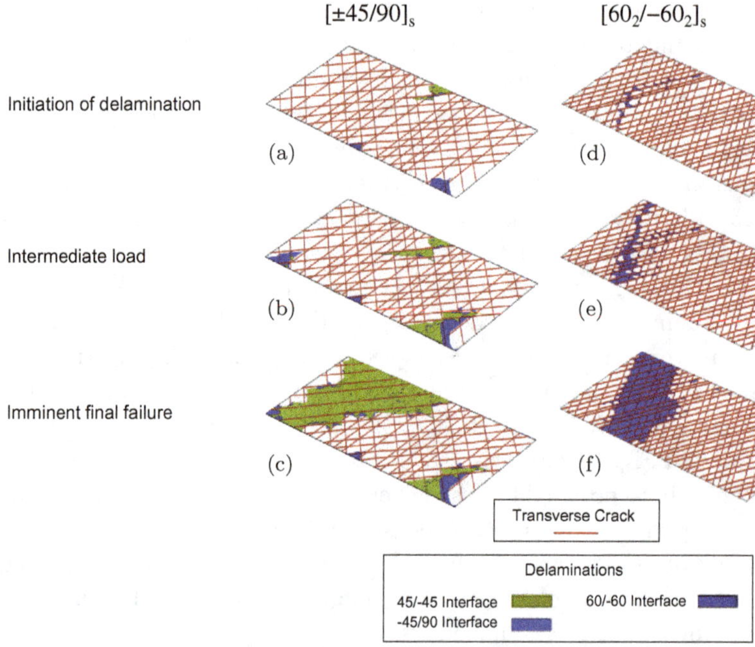

Fig. 10.27. Predicted damage at three increasing load levels in a $[45/-45/90]_s$ laminate (a,b,c) and in a $[60_2/-60_2]_s$ laminate (d,e,f).

immediately preceding the simulated final failure. For both laminates considered, a few cracks develop in the very early stages of loading (not shown). All of the matrix cracks then quickly grow though the width of the specimen.

The general damage evolution process of the $[\pm 45/90]_s$ specimen is similar to the edge delamination initiated process seen before in the $[\pm 25/90_8]_s$ laminates. Triangular-shaped delaminations initiate in multiple locations on the $+45/-45$ and $-45/90$ interfaces at the matrix crack and free edge intersections, as shown in Figs. 10.27(a), (b), and (c). As the loading increases, the delaminations grow inwardly and expand in size until they connect the two edges and the interfaces via matrix cracks, at which point the specimen fails.

The failure process in the $[60_2/-60_2]_s$ specimen is starkly different compared to the previous laminate. Delamination initiation and propagation are not anchored around the outer edges of the specimen as in the $[\pm 45/90]_s$ specimen. Delamination in this case initiates in the interior of the specimen at the matrix crack intersections, as shown in Fig. 10.27(d). As the load is increased, the delamination grows in the interior of the specimen. Figure 10.27(e) shows a delamination band of almost uniform length through the entire width of the specimen, which then extends and allows the matrix cracks to separate the plies (Fig. 10.27(f)). This difference in failure mechanisms between the two laminates has been observed experimentally [83].

The ability to address various failure mechanisms arising in non-traditional composite laminates without modifying the analysis framework and/or mesh is a critical advantage of x-FEM technology. Such capability is becoming increasingly important with aerospace companies focused on increasing the structural efficiency of composites and breaking away from traditional laminate design.

Significant developments and applications of Rx-FEM have taken place since the first edition of this book. Basic results, including finite element implementation and its application to failure prediction in laminated composites with open holes, were described in [84–86], which emphasize the issue of the interaction between matrix cracking and delamination. Fiber-dominated composite failure mechanisms were considered in [87, 88]. Progressive fiber failure is modeled by using a CDM approach with a trilinear stress–strain response. The fiber-direction softening response is characterized by two main physical quantities, the strength σ_c and the fracture toughness G_c as well as the knee-over characteristics defined by m and n. Determination of these characteristics for failure in the direction perpendicular to fibers is an open topic of experimental research, especially in the case of compression failure [89]. The application of DDM to compression loading problems was recently presented in [87]. While the Rx-FEM modeling of cracking and delamination is independent of loading direction, models of progressive fiber failure under compressive loading are significantly less mature than in tension. Therefore, errors in the open hole compression strength prediction are significantly higher than in tension, even though the predicted damage distribution at 90% of the failure load correlated with X-ray CT measurements [87]. Significant research effort is required to establish robust and reliable DDM compression failure prediction methodology and demonstrate similar level of accuracy as achieved in tensile failure prediction in [87, 88].

10.8. Conclusions

Vast numbers of new computational models capable of predicting the damage processes in composites are continuously advocated. Given the increasing complexity of these numerical methods, the task of distilling technological breakthroughs by sorting through vastly different demonstrations of successful representations of the propagation of damage mechanisms is daunting. In the present chapter, an overview of the fundamentals of the idealization of damage in composites was presented with an emphasis on identifying the issues associated with the scale of damage idealization and size effects. It was shown that the ability of damage models to predict the initiation and propagation of damage is related to the shape of the softening law and its corresponding characteristic length, which depends on the material properties and on the scale (resolution) of the idealization selected. The capabilities of advanced continuum damage mechanics (CDM) models were reviewed and their pathological deficiencies were discussed. In particular, the conditions were demonstrated under which cohesive laws and continuum damage models can achieve objectivity with respect to the mesh size, and how crack propagation using CDM models is constrained by mesh orientation and damage localization. Finally, the use of an extended finite element technique to model damage propagation by inserting cohesive cracks in arbitrary directions was presented as an emerging technology that avoids some of the limitations of CDM models.

Acknowledgments

The authors are grateful to Prof Pedro P. Camanho for helpful discussions and some unpublished material that was used in this chapter. The authors thank Dr. David Mollenhauer, AFRL, for his technical contributions. Funding under NASA NRA NNX08AB05A and the AFRL contract FA8650-05-D-5052 with the University of Dayton Research Institute are gratefully acknowledged by Endel Iarve.

References

[1] González C., Llorca J., 2007. Mechanical behavior of unidirectional fiber-reinforced polymers under transverse compression: Microscopic mechanisms and modeling, *Compos. Sci. Technol.*, 67, 2795–2806.

[2] Oliver J., Huespe A.E., Sánchez P.J., 2006. A comparative study on finite elements for capturing strong discontinuities: E-FEM vs X-FEM, *Comp. Meth. Appl. Mech. Eng.*, 195, 4732–4752.

[3] Rudraraju S. Salvi A., Garikipati K., Waas A.M., 2012. Predictions of crack propagation using a variational multiscale approach and its application to fracture in laminated fiber reinforced composites, *Compos. Struct.*, 94, 3336–3346.

[4] Moës N., Dolbow J., Belytschko T., 1999. A finite element method for crack growth without remeshing, *Int. J. Num. Meth. Eng.*, 46, 131–150.

[5] Belytschko T., Parimi C., Moës N., Sukumar N., Usui S., 2003. Structured extended finite element methods for solids defined by implicit surfaces, *Int. J. Num. Meth. Eng.*, 56, 609–635.

[6] Mollenhauer D.H., Iarve E.V., Putthanarat S., Hallett S.R., Li X., 2010. Application of discrete damage modeling to laminated composite overheight compact tension specimens, *Proc. 41st SDM Conf.*, Orlando, FL.

[7] Bouvet C., Rivallant S., Barrau J.J., 2012. Low velocity impact modeling in composite laminates capturing permanent indentation, *Compos. Sci. Technol.*, 72, 1977–1988.

[8] Camanho P.P., Maimí P., Dávila C.G., 2007. Prediction of size effects in notched laminates using continuum damage mechanics, *Compos. Sci. Technol.*, 67, 2715–2727.

[9] Maimí P., Camanho P.P., Mayugo J.A., Dávila C.G., 2007. A continuum damage model for composite laminates: Part I — constitutive model, *Mech. Mater.*, 39, 897–908.

[10] Mabson G., Lyle R.D., Dopker B., Hoyt D.M., Baylor J.S., Graesser D.L., 2007. Fracture interface elements for static and fatigue analysis, *Proc. 16th Int. Conf. on Composite Materials*, Kyoto, Japan.

[11] Dávila C.G., Camanho P.P., de Moura M.F.S.F., 2001. Progressive damage analyses of skin/stringer debonding, *Proc. ASC 16th Tech. Conf.*, Blacksburg, VA.

[12] Allix O., Guedra-Degeorges D., Guinard S., Vinet A., 2000. Analyse de la tenue aux impacts à faible vitesse et faible énergie des stratifiés composites par la mécanique de l'endommagement, *Méc. Ind.*, 1, 27–35.

[13] MIL-HDBK-17-3, 2005. Materials usage, design and analysis, rev e., *DoD Single Stock Point (DODSSP)*, Philadelphia, PA.

[14] Bergan A.C., Dávila C.G., Leone F.A., Awerbuch J., Tan T.-M., 2016. A mode I cohesive law characterization procedure for through-the-thickness crack propagation in composite laminates, *Comp. B*, 94, 338–349.

[15] Rybicki E.F., Kanninen M.F., 1977. A finite element calculation of stress intensity factors by a modified crack closure integral, *Eng. Fract. Mech.*, 9, 931–939.

[16] Griffith A.A., 1921. The phenomena of rupture and flow in solids, *Phil. Trans. Roy. Soc. of London A*, 221, 163–197.

[17] Bažant Z.P., Chen E.-P., 1996. Scaling of structural failure, Sandia Report SAND96–2948, Sandia National Laboratories, Albuquerque, NM.

[18] Mar J.W., Lin K.Y., 1977. Fracture of boron/aluminum composites with discontinuities, *J. Compos. Mater.*, 11, 405–421.

[19] Sun C.T., Qian H., 2009. Brittle fracture beyond the stress intensity factor, *J. Mech. Mat. Str.*, 4, 743–753.

[20] Irwin G.R., 1970. Fracture strength of relatively brittle structures and materials, *J. Franklin Inst.*, 290, 513–521.

[21] Bažant Z.P., Planas J., 1998. *Fracture and Size Effect in Concrete and Other Quasibrittle Materials*, CRC, Boca Raton, FL.

[22] Turon A., Costa J., Camanho P.P., Maimí P., 2008. Analytical and numerical investigation of the length of the cohesive zone in delaminated composite materials in P.P. Camanho *et al.* (eds), *Mechanical Response of Composites*, Springer, Berlin.

[23] Bažant Z.P., 2002. Concrete fracture models: Testing and practice, *Eng. Fract. Mech.*, 69, 165–205.

[24] ASTM D5528-01, 2002. Standard test method for mode I interlaminar fracture toughness of unidirectional fiber-reinforced polymer matrix composites, *Annual Book of ASTM Standards*, American Society for Testing and Materials, West Conshohocken, PA.

[25] Pereira A.B., de Morais A.B., 2004. Mode I interlaminar fracture of carbon/ epoxy multidirectional laminates, *Compos. Sci. Technol.*, 64, 2261–2270.

[26] Schön J., Nyman T., Blom A., Ansell H., 2000. A numerical and experimental investigation of delamination behaviour in the DCB specimen, *Compos. Sci. Technol.*, 60, 173–184.

[27] Wagner W., Balzani C., 2008. Simulation of delamination in stringer stiffened fiber-reinforced composite shells, *Comput. Struct.*, 86, 930–939.

[28] Turon A., Dávila C.G., Camanho P.P., Costa J., 2007. An engineering solution for mesh size effects in the simulation of delamination using cohesive zone models, *Eng. Fract. Mech.*, 74, 1665–1682.

[29] Bao G., Suo Z., 1992. Remarks on crack-bridging concepts, *Appl. Mech. Rev.*, 45, 355–366.

[30] Dávila C.G., Rose C.A., Camanho P., 2009. A procedure for superposing linear cohesive laws to represent multiple damage mechanisms in the fracture of composites, *Int. J. Fracture*, 158, 211–223.

[31] Turon A., Camanho P.P., Costa J., Dávila C.G., 2006. A damage model for the simulation of delamination in advanced composites under variable-mode loading, *Mech. Mater.*, 38, 1072–1089.

[32] Dávila C.G., Camanho P.P., 2001. Decohesion elements using two and three-parameter mixed-mode criteria, *Proc. AHS Structures Specialists Meeting*, Williamsburg, VA.

[33] Turon A., Camanho P.P., Costa J., Dávila C.G., 2004. An interface damage model for the simulation of delamination under variable-mode ratio in composite materials, NASA/TM-2004–213277, Hampton, VA, 32.

[34] 2008. *Abaqus 6.8 User's Manual*, Dassault Systèmes, Providence, RI, USA.

[35] Jiang W.-G., Hallett S.R., Green B.G., Wisnom M.R., 2007. A concise interface constitutive law for analysis of delamination and splitting in composite materials and its application to scaled notched tensile specimens, *Int. J. Num. Meth Eng.*, 69, 1982–1995.

[36] Pinho S.T., Iannucci L., Robinson P., 2006. Formulation and implementation of decohesion elements in an explicit finite element code, *Comp. A*, 37, 778–789.

[37] Hansen A.L., Lund E., Sørensen B.F., 2007. Simulation of delamination including fiber bridging using cohesive zone models, *Proc. ECCOMAS Thematic Conference on Mechanical Response of Composites* (P.P. Camanho *et al.*, eds), Porto, Portugal.

[38] Sørensen B.F., Goutianos S., Jacobsen T.K., 2009. Strength scaling of adhesive joints in polymer-matrix composites, *Int. J. Solids Struct.*, 46, 741–761.

[39] Turon A., Camanho P.P., Costa J., Renart J., 2010. Accurate simulation of delamination growth under mixed-mode loading using cohesive elements: Definition of interlaminar strengths and elastic stiffness, *Compos. Struct.*, 92, 1857–1864.

[40] Chamis C.C., Murthy P.L.N., Minnetyan L., 1996. Progressive fracture of polymer matrix composite structures, *Theor. App. Fract. Mech.*, 25, 1–15.

[41] Paley M., Aboudi J., 1992. Micromechanical analysis of composites by the generalized cells model, *Mech. Mater.*, 14, 127–139.

[42] Jirásek M., 1998. Nonlocal models for damage and fracture: Comparison of approaches, *Int. J. Solids Struct.*, 35, 4133–4145.

[43] Bažant Z.P., Oh B., 1983. Crack band theory for fracture of concrete, *Mater. Struct.*, 16, 155–177.

[44] Bažant Z.P., 2002. Reminiscences on four decades of struggle and progress in softening damage and size effects, *Concrete J.*, 40, 16–28.

[45] Jirásek M., 2007. Mathematical analysis of strain localization, *Rev. Euro. Génie Civil.*, 11, 977–991.

[46] William K.J., 2002. Constitutive models for engineering materials, in R.A. Meyers (ed), *Encyclopedia of Physical Science and Technology, 3rd edn.*, Academic Press, pp. 603–633.

[47] Peerlings R.H.J., de Borst R., Brekelmans W.A.M., Geers M.G.D., 2002. Localisation issues in local and nonlocal continuum approaches to fracture, *Euro. J. Mech. A*, 21, 175–189.

[48] Simo J.C., Ju J.W., 1987. Strain- and stress-based continuum damage models — I. Formulation, *Int. J. Solids Struct.*, 23, 821–840.

[49] Bažant Z.P., Jirásek M., 2002. Nonlocal integral formulations of plasticity and damage: Survey of progress, *J. Eng. Mech.*, 128, 1119–1149.

[50] Maimí P., Camanho P.P., Mayugo J.A., Dávila C.G., 2007. A continuum damage model for composite laminates: Part II — Computational implementation, *Mech. Mater.*, 39, 909–919.

[51] Hinton M.J., Kaddour A.S., Soden P.D., 2002. A comparison of the predictive capabilities of current failure theories for composite laminates, judged against experimental evidence, *Compos. Sci. Technol.*, 62, 1725–1797.

[52] Soden P.D., Hinton M.J., Kaddour A.S., 1998. A comparison of the predictive capabilities of current failure theories for composite laminates, *Compos. Sci. Technol.*, 58, 1225–1254.

[53] Dávila C.G., Camanho P.P., Rose C.A., 2005. Failure criteria for FRP laminates, *J. Compos. Mater.*, 39, 323–345.

[54] Pinho S.T., Dávila C.G., Camanho P.P., Iannucci L., Robinson P., 2005. Failure models and criteria for FRP under in-plane or three-dimensional stress states including shear nonlinearity, NASA/TM-2005-213530, Hampton, VA.

[55] Green B.G., Wisnom M.R., Hallett S.R., 2007. An experimental investigation into the tensile strength scaling of notched composites, *Comp. A*, 38, 867–878.

[56] Parvizi A., Garrett K., Bailey J., 1978. Constrained cracking in glass fibre-reinforced epoxy cross-ply laminates, *J. Mat. Sci.*, 13, 195–201.

[57] Wang A.S.D., 1984. Fracture mechanics of sublaminate cracks in composite materials, *Comp. Tech. Rev.*, 6, 45–62.

[58] Camanho P.P., Dávila C.G., Pinho S.T., Iannucci L., Robinson P., 2006. Prediction of *in situ* strengths and matrix cracking in composites under transverse tension and in-plane shear, *Comp. A*, 37, 165–176.

[59] Ladevèze P., LeDantec E., 1992. Damage modelling of the elementary ply for laminated composites, *Compos. Sci. Technol.*, 43, 257–267.

[60] Matzenmiller A., Lubliner J., Taylor R.L., 1995. A constitutive model for anisotropic damage in fiber-composites, *Mech. Mater.*, 20, 125–152.

[61] Kachanov L.M., 1986. *Introduction to Continuum Damage Mechanics*, Martinus Nijhoff, Dordrecht, NL.

[62] Lemaitre J., Desmorat R., Sauzay M., 2000. Anisotropic damage law of evolution, *Euro. J. Mech. A*, 19, 187–208.

[63] Camanho P.P., Dávila C.G., de Moura M.F.S.F., 2003. Numerical simulation of mixed-mode progressive delamination in composite materials, *J. Compos. Mater.*, 37, 1415–1438.

[64] Lopes C.S., Camanho P.P., Gürdal Z., Maimí P., González E.V., 2009. Low-velocity impact damage on dispersed stacking sequence laminates. Part II: Numerical simulations, *Compos. Sci. Technol.*, 69, 937–947.

[65] Overgaard L.C.T., Lund E., Camanho P.P., 2010. A methodology for the structural analysis of composite wind turbine blades under geometric and material induced instabilities, *Comput. & Struct.*, 88, 1092–1109.

[66] Bisagni C., Vescovini R., Dávila C. G., 2011. Single-stringer compression specimen for the assessment of damage tolerance of postbuckled structures, *J. Aircraft*, 48, 495–502.

[67] van der Meer F.P., Sluys L.J., 2009. Continuum models for the analysis of progressive failure in composite laminates, *J. Compos. Mater.*, 43, 2131–2156.

[68] Jirásek, M. and Zimmermann, T., 1998. Analysis of rotating crack model. *J. Eng. Mech.*, 124(8), 842–851.

[69] Jirásek M., 2001. *Modeling of Localized Damage and Fracture in Quasibrittle Materials*, Springer, Berlin.

[70] Iarve E.V., Mollenhauer D., Kim R., 2005. Theoretical and experimental investigation of stress redistribution in open hole composite laminates due to damage accumulation, *Comp. A*, 36, 163–171.

[71] Laš V., Zemčík R., 2008. Progressive damage of unidirectional composite panels, *J. Compos. Mater.*, 42, 25–44.

[72] Camanho P.P., Bessa M.A., Catalanotti G., Vogler M., Rolfes R., 2013. Modeling the inelastic deformation and fracture of polymer composites — Part II: Smeared crack model, *Mech. Mater.*, 59, 36–49.

[73] Prabhaker P., Waas A.M., 2013. A novel continuum-decohesive finite element for modeling in-plane fracture in fiber reinforced composites, *Compos. Sci. Technol.*, 83, 1–10.

[74] Leone F.A., 2015. Deformation gradient decomposition for representing matrix cracks in fiber-reinforced materials, *Comp. A*, 76, 334–341.

[75] Huynh D.B.P., Belytschko T., 2009. The extended finite element method for fracture in composite materials, *Int. J. Num. Meth. Eng.*, 77, 214–239.

[76] van der Meer F., Sluys L., 2009. A phantom node formulation with mixed mode cohesive law for splitting in laminates, *Int. J. Fracture*, 158, 107–124.

[77] Ling D., Yang Q., Cox B., 2009. An augmented finite element method for modeling arbitrary discontinuities in composite materials, *Int. J. Fracture*, 156, 53–73.

[78] Iarve E.V., 2003. Mesh independent modelling of cracks by using higher order shape functions, *Int. J. Num. Meth. Eng.*, 56, 869–882.

[79] Patzák B., Jirásek M., 2003. Process zone resolution by extended finite elements, *Eng. Fract. Mech.*, 70, 957–977.

[80] Benvenuti E., 2008. A regularized XFEM framework for embedded cohesive interfaces, *Comp. Meth. Appl. Mech. Eng.*, 197, 4367–4378.

[81] König M., Krüger R., Kussmaul K., von Alberti M., Gädke M., 1997. Characterizing static and fatigue interlaminar fracture behaviour of a first generation graphite/epoxy composite, in S.J. Hooper (ed), *Composite Materials: Testing and Design*, ASTM, West Conshohocken, PA, pp. 60–81.

[82] Crossman S.W., Wang A.S.D., 1982. The dependence of transverse cracking and delamination on ply thickness in graphite/epoxy laminates, in *Damage in Composite Materials*, ASTM International, pp. 118–139.

[83] Johnson P., Chang F.-K., 2001. Characterization of matrix crack-induced laminate failure. Part I: Experiments, *J. Compos. Mater.*, 35, 2009–2035.

[84] Iarve E.V., Gurvich M.R., Mollenhauer D.H., Rose C.A., Dávila C.G., 2011. Mesh-independent matrix cracking and delamination modeling in laminated composites, *Int. J. Num. Meth. Eng.*, 88, 749–777.

[85] Swindeman M.J., Iarve E.V., Brockman R.A., Mollenhauer D.H., 2013. Strength prediction in open hole composite laminates by using discrete damage modeling, *AIAA J.*, 51, 936–945.

[86] Iarve E.V., Mollenhauer D.H., 2015. Mesh independent matrix cracking and delamination modelling in advanced composite materials, in *Numerical modeling of failure in advanced composites*, Woodhead Publishing, pp. 227–264.

[87] Hoos K., Iarve E.V., Braginsky M., Zhou E., Mollenhauer D.H., 2017. Static strength prediction in laminated composites by using discrete damage modeling, *J. Compos. Mater*, 51, 1473–1492.

[88] Mollenhauer D.H., Ward L., Iarve E.V., Putthanarat S., Hoos K., Hallett S., Li X., 2012. Simulation of discrete damage in composite overheight compact tension specimens, *Comp. A*, 43, 1667–1679.

[89] Laffan M.J., Pinho S.T., Robinson P.R., Ianucci L., McMillan A.J., 2012. Measurement of the fracture toughness associated with the longitudinal fibre compressive failure mode of laminated composites, *Comp. A*, 43, 1930–1038.

https://doi.org/10.1142/9781800611887_0011

Chapter 11

Delamination and Adhesive Contacts, Their Mathematical Modeling and Numerical Treatment

Tomáš Roubíček[*,†,††], **Martin Kružík**[‡,§,‡‡], **Jan Zeman**[§,‡,§§], **Christos G. Panagiotopoulos**[¶,¶¶], **Roman Vodička**[‖,‖‖], and **Vladislav Mantič**[**,***]

[*]*Mathematical Institute, Charles University, Sokolovská 83, CZ-186 75 Praha 8, Czech Republic*
[†]*Institute of Thermomechanics, Czech Academy of Sciences, Dolejškova 5, CZ-182 00 Praha 8, Czech Republic*
[‡]*Institute of Information Theory and Automation, Czech Academy of Sciences, Pod vodárenskou věží 4, CZ-182 08 Praha 8, Czech Republic*
[§]*Faculty of Civil Engineering, Czech Technical University, Thákurova 7, CZ-166 29 Praha 6, Czech Republic*
[¶]*Department of Music Technology and Acoustics, Hellenic Mediterranean University, GR-71410 Heraklion, Crete, Greece*
[‖]*Faculty of Civil Engineering, Technical University of Košice, Vysokoškolská 4, SK-042 00 Košice, Slovakia*
[**]*Escuela Técnica Superior de Ingeniería, Universidad de Sevilla, Camino de los Descubrimientos s/n, 41092 Seville, Spain*
[††]*tomas.roubicek@mff.cuni.cz*
[‡‡]*kruzik@utia.cas.cz*
[§§]*jan.zeman@cvut.cz*
[¶¶]*pchr@hmu.gr*
[‖‖]*roman.vodicka@tuke.sk*
[***]*mantic@us.es*

Abstract

This chapter reviews mathematical approaches to inelastic processes on the surfaces of elastic bodies. We mostly consider a quasistatic and rate-independent evolution at small strains. Various concepts of solutions are introduced and applied (including their comparison), e.g., to elastic-brittle delamination, cohesive contact problems, and to delamination in various fracture modes, or combined with friction. Besides the theoretical treatment, numerical experiments are also presented. Several implicit time discretization schemes are exploited. Finally, generalizations to dynamic and thermodynamic processes are outlined, together with an extension to the homogenization of composite materials with debonding phases.

Contents

11.1. Introduction

This chapter is devoted to formulation, mathematical, and numerical treatments of inelastic processes on surfaces. Through this chapter, with the exception of Sections 11.6.6 and 11.6.7, these inelastic processes are considered to be *quasistatic* (no inertia is taken into account) and *rate-independent* (no internal timescale is considered). We will address only small-strain models. Our general framework will solely be based on the hypothesis that the evolution is governed by a time-dependent Gibbs-type *stored energy* functional $\mathscr{E}$ (involving external loading) and a *dissipation potential* $\mathscr{R}$, which, being degree-1 positively homogeneous, reflects the rate-independence of the process (i.e., invariance under any monotone rescaling of time). Both functionals are defined on a suitable state space considered here in the form $\mathscr{U} \times \mathscr{Z}$. With only a small loss of generality, we usually assume, with some exceptions in Sections 11.6 and 11.7, that $\mathscr{R}$ involves just the z-component of a state $q = (u, z) \in \mathscr{U} \times \mathscr{Z}$. This distinguishes z as a "slow" variable, while u is a "fast" variable because its velocity is not controlled by any dissipation.

As a prototypical application of the energetic framework in modeling composite materials and structures, we now briefly introduce a simplified *elastic delamination model* (sometimes also called *debonding*) treated in more detail later in Section 11.5. Here, we restrict our attention to two elastic domains Ω_1 and Ω_2, sharing an interface $\Gamma_{\rm C}$. The structure is subjected to a time-dependent hard-device loading with a Dirichlet boundary condition imposed by displacements $w_{\rm D}(t)$ acting at a part of the external boundary $\Gamma_{\rm D}$ (with $u_{\rm D}(t)$ denoting an extension of $w_{\rm D}(t)$ to bodies Ω_1 and Ω_2).

In this context, $u + u_{\rm D}$ is used to denote the *displacement field*. Moreover, we will work with an *internal variable* z (possibly vectorial) describing inelastic delamination processes on the boundary $\Gamma_{\rm C}$. Various possibilities are presented in this chapter. The simplest scenario works with a scalar-valued *delamination* (also called *damage*) *variable* considered as a function of $x \in \Gamma_{\rm C}$, with $z = 1$ and $z = 0$ corresponding to undamaged and fully damaged interfacial points x, respectively. The time-independent set $\mathscr{U}$ consists of the kinematically admissible displacements, satisfying the homogeneous Dirichlet boundary conditions on $\Gamma_{\rm D}$ and frictionless contact conditions on $\Gamma_{\rm C}$:

$$u = 0 \text{ on } \Gamma_{\rm D} \quad \text{and} \quad [\![u]\!]_{\rm n} \geq 0 \text{ on } \Gamma_{\rm C}. \tag{11.1a}$$

Analogously, the set of admissible internal variables $\mathscr{Z}$ is defined as

$$0 \leq z \leq 1 \quad \text{on } \Gamma_{\rm C}. \tag{11.1b}$$

The notation $[\![u]\!]_{\rm n}$ in (11.1a) stands for the normal component of the displacement jump $[\![u]\!]$.

Now, given admissible u and z, the stored energy functional reads

$$\mathscr{E}(t,u,z) := \sum_{i=1}^{2} \int_{\Omega_i} \mathbb{C}^{(i)} e(u){:}e\left(\frac{1}{2}u + u_{\mathrm{D}}(t)\right) \mathrm{d}x + \int_{\Gamma_{\mathrm{C}}} \frac{z}{2}\mathbb{K}[\![u]\!]\cdot[\![u]\!]\,\mathrm{d}S, \tag{11.2}$$

where $e(u)$ denotes the small-strain tensor associated with displacement u, $\mathbb{C}^{(i)}$ stores the stiffness tensor of the ith domain, and $\mathbb{K}$ is the tensor of elastic interfacial stiffnesses. The dissipation rate is defined as

$$\mathscr{R}(\dot{z}) := \begin{cases} \displaystyle\int_{\Gamma_{\mathrm{C}}} a|\dot{z}|\,\mathrm{d}S & \text{if } \dot{z} \le 0 \text{ on } \Gamma_{\mathrm{C}}, \\ \infty & \text{otherwise,} \end{cases} \tag{11.3}$$

with a denoting the interface fracture energy, representing the energy dissipated by complete delamination of a unit area of interface. The value ∞ in (11.3) is used to ensure unidirectionality of the delamination phenomena, i.e., no healing of the interface is admissible during the loading process. Sometimes only a finitely valued $\mathscr{R}$ is considered, which means that *healing* (also called *rebonding*) is allowed, cf., e.g., [1–3]. Such models are, however, mathematically less difficult in some aspects than (11.3) and their interpretation is rather limited (as the configuration, possibly shifted after complete delamination, has a tendency to remember its initial state after healing if not combined with an interfacial plasticity like in [4]), and will not be particularly addressed in this chapter.

After the energy functionals (11.2) and (11.3) have been specified, the tools and techniques presented in the remainder of this chapter will allow us to study the delamination evolution rigorously, including theoretically supported numerical simulations. Note that the generality of the energetic framework makes it easy to incorporate more realistic interfacial constitutive laws naturally and to couple the delamination phenomena with other inelastic processes such as plasticity, damage, or phase transformations. In addition, the energy functional (11.2) can easily be adapted to the periodic homogenization theory, which makes the framework directly applicable to the analysis of, e.g., fiber-matrix debonding in fibrous composites, see also Section 11.8 for a concrete example.

In comparison to the first edition of this chapter [5], we introduce a wider variety of solution concepts and their mutual comparison, and suppress the role of energy conservation in merely rate-independent models, present more numerical experiments (exploiting various time-discretization schemes beside bare backward Euler one), reflecting also the research pursued by our wider team during the past five years. Moreover, some recent references of this intensively developing area have been added, although we do not claim that the list is complete in any sense.

11.2. Concepts in Quasistatic Rate-Independent Evolution

We will consider a Banach space[a] $\mathscr{X} \supset \mathscr{Z}$ on which $\mathscr{R}$ is defined as having a domain with a nonempty interior and being coercive, and degree-1 homogeneous in the sense $\mathscr{R}(\lambda z) = \lambda\mathscr{R}(z)$ for any $\lambda \geq 0$. The set $\mathscr{X}$ must be a Banach space (to define the degree-1 homogeneity), but here also $\mathscr{U}$ and $\mathscr{Z}$ will always be Banach spaces.

Formally, for $\mathscr{E} : [0,T]\times\mathscr{U}\times\mathscr{Z} \to \mathbb{R}\cup\{\infty\}$ and $\mathscr{R} : \mathscr{Z} \to [0,\infty]$, the rate-independent evolution we have in mind is governed by the following system of *doubly nonlinear* degenerate *parabolic/elliptic variational inclusions*:

$$\partial_u\mathscr{E}(t,u,z) \ni 0 \qquad \text{and} \qquad \partial\mathscr{R}(\dot{z}) + \partial_z\mathscr{E}(t,u,z) \ni 0, \tag{11.4}$$

where $\dot{z} := \frac{\mathrm{d}z}{\mathrm{d}t}$ and the symbol "∂" refers to a (partial) subdifferential,[b] and $\mathscr{R}(\cdot)$, $\mathscr{E}(t,\cdot,z)$, and $\mathscr{E}(t,u,\cdot)$ are convex functionals in all specific models considered in this chapter.

In mechanics, internal variables and the inclusion (11.4), sometimes also called Biot's equation [6], are used for so-called generalized standard materials [7], cf. also [8] specifically for adhesive contact. There are sound variational principles supporting the abstract model (11.4), although their applicability should not be overestimated.

The first inclusion in (11.4) expresses the *minimum-energy principle*, asserting that at any time t the displacement u minimizes $u \mapsto \mathscr{E}(t,u,z(t))$.

Assuming for simplicity $\mathscr{E}$ smooth and denoting the partial differentials by $\mathscr{E}'_t$, $\mathscr{E}'_u$, and $\mathscr{E}'_z$, one can postulate a so-called Lagrangian in the form

$$\begin{aligned}\mathscr{L}(t,u,z,\dot{z}) &:= \frac{\mathrm{d}}{\mathrm{d}t}\mathscr{E} + \mathscr{R} \\ &= \mathscr{E}'_t(t,u,z) + \langle\mathscr{E}'_u(t,u,z),\dot{u}\rangle + \langle\mathscr{E}'_z(t,u,z),\dot{z}\rangle + \mathscr{R}(\dot{z}) \\ &= \mathscr{E}'_t(t,u,z) + \langle\mathscr{E}'_z(t,u,z),\dot{z}\rangle + \mathscr{R}(\dot{z}),\end{aligned} \tag{11.5}$$

[a] A normed linear space which is complete is called a Banach space. Recall that a normed space X is complete if every Cauchy sequence in X converges to a limit in X.

[b] Recall that the subdifferential $\partial f(x)$ of a convex function $f\colon X \to \mathbb{R}\cup\{\infty\}$ at a point x is defined as the convex closed subset $\partial f(x) := \{x^* \in X^*;\ \forall v \in X\colon f(x)+\langle x^*, v-x\rangle \leq f(v)\}$ of the dual space X^*; conventionally, $\langle\cdot,\cdot\rangle : X^* \times X \to \mathbb{R}$ denotes the duality pairing between the Banach space X and its dual $X^* := \{x^*\colon X \to \mathbb{R}$, linear and continuous$\}$.

where we also used the first inclusion in (11.4). Then the second inclusion in (11.4) can be derived as the first-order optimality condition for

$$\dot{z} \mapsto \mathscr{L}(t, u, z, \dot{z}) \quad \text{is minimal for any time } t. \tag{11.6}$$

One refers to (11.6) as a *minimum dissipation-potential principle*, cf. [9–12].

The degree-1 homogeneity of $\mathscr{R}$ still allows for further interpretation of the flow rule, i.e., the second inclusion in (11.4). Defining the convex "elastic domain" $K := \partial\mathscr{R}(0)$, the second inclusion in (11.4) only means $\langle \omega - \mathfrak{z}, w - \dot{z} \rangle \geq 0$ for any w and any $\omega \in \partial\mathscr{R}(w)$, where we introduced the so-called thermodynamic *driving force* $\mathfrak{z} \in -\partial_z\mathscr{E}(t, u, z)$. As the set $\partial_z\mathscr{E}$ is not a singleton in general, we will call $\mathfrak{z}$ rather an *actual driving force*, in order to distinguish it from the all formally *available driving forces* from $-\partial_z\mathscr{E}(t, u, z)$. The adjective "available" is sometimes used in fracture mechanics, referring to the energy release rate. In particular, for $w = 0$ one obtains

$$\langle \mathfrak{z}, \dot{z} \rangle = \max_{\omega \in K} \langle \omega, \dot{z} \rangle. \tag{11.7}$$

To derive (11.7), we used that $\mathfrak{z} \in \partial\mathscr{R}(\dot{z}) \subset \partial\mathscr{R}(0) = K$ thanks to the degree-1 homogeneity of $\mathscr{R}(\cdot)$, so that always $\langle \mathfrak{z}, \dot{z} \rangle \leq \max_{\omega \in K} \langle \omega, \dot{z} \rangle$. The identity (11.7) means that the dissipation due to the actual driving force $\mathfrak{z}$ is maximal provided that the order-parameter rate $\dot{z}$ is kept fixed, while a possible driving force ω varies freely over the *set of all admissible driving forces* K. This resembles the so-called Hill's *maximum-dissipation principle* [13], cf. also [14–17].

Let us also observe that the set $K = \partial\mathscr{R}(0)$ determines $\mathscr{R}$ because $\mathscr{R} = \delta_K^*$ with δ_K denoting the so-called indicator function[c] of the set K and δ_K^* is its Legendre–Fenchel conjugate.[d] In terms of K, by standard convex-analysis calculus [18], Eq. (11.7) can also be written as

$$\dot{z} \in [\partial\delta_K^*]^{-1}(\mathfrak{z}) = \partial[[\delta_K^*]^*](\mathfrak{z}) = \partial\delta_K(\mathfrak{z}) = N_K(\mathfrak{z}), \tag{11.8}$$

where $N_K(\mathfrak{z})$ denotes the normal cone[e] to K at $\mathfrak{z}$. This is the well-known principle from plasticity theory, called *normality condition*, stating that the rate of the internal parameter z (representing, as a special case in plasticity

[c]This means that $\delta_K(v) = 0$ if $v \in K$ and $\delta_K(v) = \infty$ if $v \notin K$.

[d]The Legendre–Fenchel conjugate $f^* : X^* \to \mathbb{R} \cup \{\infty\}$ of a function $f : X \to \mathbb{R} \cup \{\infty\}$ is defined as $f^*(x^*) := \sup_{x \in X} \langle x^*, x \rangle - f(x)$.

[e]Recall that the normal cone $N_K(x)$ to a convex set $K \subset X$ at $x \in X$ is defined as $N_K(x) := \{x^* \in X^*;\ \forall v \in K :\ \langle x^*, v - x \rangle \leq 0\}$. It is a generalization of the notion "outer normal vector" to ∂K.

theory, the plastic-deformation rate) belongs to the cone of outward normals to the elasticity domain, see also [14, 19] or [20, Sec. 3.2], [21, Sec. 2.4.4] or [22, Sec. 2.6].

11.3. Mathematical Concepts to Solve the System (11.4)

Initialized by David Hilbert as the 20th of his famous problems [23], it was recognized that the classical formulation of (initial-)boundary-value problems in terms of derivatives (or here subdifferentials) is not natural and rather some integral variants are to be devised. An initiative pursued during the whole 20th century led to various more suitable concepts. The most conventional are so-called weak solutions. Here, as far as static or incremental problems are concerned, we will rely on respective potentials and, instead of classical differential equations or inequalities, we will use the variational structure involving the underlying functionals $\mathscr{E}$ and $\mathscr{R}$, and possible some others, see [24] for a brief introduction to variational methods. Here, in addition, the rate independence is a certain simplification which, however, makes the class of weak solutions (also called local solutions) very wide and needs various refinements to be carefully interpreted, cf. in particular [25, 26]. We will present the relevant concepts and solutions first on a rather abstract level, not counting with the concrete form of $\mathscr{E}$ and $\mathscr{R}$.

11.3.1. General weak solutions — Local solutions

To design the concept of a weak solution to (11.4), we use $\mathfrak{z}$ as in Section 11.2, and use the standard definition of the convex subdifferential as already mentioned above to write the three subdifferentials in (11.4) as the system of three inequalities

$$\forall(t,\tilde{u},\tilde{z}): \quad \mathscr{R}(\tilde{z}) - \langle \mathfrak{z}(t), \tilde{z} - \dot{z}(t)\rangle \geq \mathscr{R}(\dot{z}(t)), \tag{11.9a}$$

$$\mathscr{E}(t,\tilde{u},z(t)) \geq \mathscr{E}(t,u(t),z(t)), \tag{11.9b}$$

$$\mathscr{E}(t,u(t),\tilde{z}) + \langle \mathfrak{z}(t), \tilde{z} - z(t)\rangle \geq \mathscr{E}(t,u(t),z(t)), \tag{11.9c}$$

where the first and third inequalities mean that $\mathfrak{z} \in \partial\mathscr{R}(\dot{z})$ and $-\mathfrak{z} \in \partial_z\mathscr{E}(t,u(t),z(t))$, respectively, while (11.9b) states that $u(t)$ minimizes the energy $\mathscr{E}(t,\cdot,z(t))$. As $\mathscr{R}$ is homogeneous only of degree 1, $\dot{z}$ can be expected to be bounded only in the Bochner–Lebesgue space $L^1(0,T;\mathscr{X})$,[f] or rather in the corresponding space of measures. Thus, the expression $\langle \mathfrak{z}, \dot{z}\rangle$ in (11.9a) might not be well defined or not allow for a mathematical treatment regarding various limit procedures. Thus, it is desirable to convert it to a more suitable form. For this, we use the formal chain rule

$$\mathscr{E}(T,u(T),z(T)) - \mathscr{E}(0,u_0,z_0) = \int_0^T \frac{\mathrm{d}}{\mathrm{d}t}\mathscr{E}(t,u,z)\,\mathrm{d}t$$
$$= \int_0^T \big(\mathscr{E}_t'(t,u,z) - \langle \mathfrak{z}, \dot{z}\rangle\big)\,\mathrm{d}t; \qquad (11.10)$$

cf. (11.5). We integrate (11.9a) over $[0,T]$ and substitute for $\int_0^T \mathscr{R}(\dot{z})\,\mathrm{d}t$ the total variation

$$\mathrm{Diss}_{\mathscr{R}}(z,[0,T]) := \sup \sum_{j=1}^{N} \mathscr{R}(z(t_j) - z(t_{j-1})), \qquad (11.11)$$

where the supremum is taken over all partitions $0 \le t_0 < t_1 < \cdots < t_{N-1} \le t_N \le T$ of $[0,T]$; indeed, if z is absolutely continuous, then $\mathrm{Diss}_{\mathscr{R}}(z,[0,T]) = \int_0^T \mathscr{R}(\dot{z})\,\mathrm{d}t$. This turns (11.9a) into

$$\mathscr{E}(T,u(T),z(T)) + \mathrm{Diss}_{\mathscr{R}}(z,[0,T])$$
$$- \int_0^T \mathscr{E}_t'(t,u,z)\,\mathrm{d}t - \mathscr{E}(0,u_0,z_0) \le \int_0^T \mathscr{R}(\tilde{z}) - \langle \mathfrak{z}, \tilde{z}\rangle\,\mathrm{d}t. \qquad (11.12)$$

Note that (11.12) implies the energy inequality, more specifically for $\tilde{z} = 0$ we get

$$\underbrace{\mathscr{E}(T,u(T),z(T))}_{\substack{\text{stored energy}\\ \text{at time } t=T}} + \underbrace{\mathrm{Diss}_{\mathscr{R}}(z,[0,T])}_{\substack{\text{energy dissipated}\\ \text{during } [0,T]}} \le \underbrace{\int_0^T \mathscr{E}_t'(t,u,z)\,\mathrm{d}t}_{\substack{\text{work done by}\\ \text{mechanical load}}} + \underbrace{\mathscr{E}(0,u_0,z_0)}_{\substack{\text{stored energy}\\ \text{at time } t=0}}. \qquad (11.13)$$

We will use the notation $\mathrm{B}([0,T];\mathscr{U})$ for the Banach space of bounded measurable functions $[0,T] \to \mathscr{U}$ defined everywhere, and $\mathrm{BV}([0,T];\mathscr{X})$

[f] The notation $L^p(\cdot)$ stands for the Banach space of measurable functions whose p-power is integrable on the indicated domain, here $[0,T]$. Such spaces are called Lebesgue spaces. If the functions take values in a general Banach space $\mathscr{X}$ then one applies a natural generalization of measurability due to Bochner.

for functions $[0,T] \to \mathscr{X}$ with bounded variation;[g] recall that $\mathscr{X} \supset \mathscr{Z}$ is a Banach space on which $\mathscr{R}$ is coercive, i.e., $\mathscr{R}(z) \geq \varepsilon\|z\|$ for some $\varepsilon > 0$.

Definition 11.1 (Weak solutions). The process $(u, z, \mathfrak{z}) : [0,T] \to \mathscr{U} \times \mathscr{Z} \times \mathscr{Z}^*$ is called a weak solution of the initial-value problem given by $(\mathscr{U} \times \mathscr{Z}, \mathscr{E}, \mathscr{R})$ and the initial condition (u_0, z_0) if $u \in \mathrm{B}([0,T]; \mathscr{U})$, $z \in \mathrm{B}([0,T]; \mathscr{Z}) \cap \mathrm{BV}([0,T]; \mathscr{X})$, $\mathfrak{z} \in \mathrm{B}([0,T]; \mathscr{Z}^*)$ and

(i) the inequality (11.12) with (11.11) holds for a.a. t_1 and t_2, where $0 \leq t_1 < t_2 \leq T$, and for any $\tilde{z} \in \mathscr{Z}$, i.e.,

$$\mathscr{E}(t_2, u(t_2), z(t_2)) + \mathrm{Diss}_{\mathscr{R}}(z, [t_1, t_2]) - \mathscr{E}(t_1, u(t_1), z(t_1)) \leq \int_{t_1}^{t_2} \mathscr{E}_t'(t, u, z) + \mathscr{R}(\tilde{z}) - \langle \mathfrak{z}, \tilde{z} \rangle \, \mathrm{d}t, \tag{11.14}$$

(ii) (11.9b) holds for any $\tilde{u} \in \mathcal{U}$ and a.a. $t \in [0,T]$,
(iii) (11.9c) holds for any $\tilde{z} \in \mathcal{Z}$ and a.a. $t \in [0,T]$,
(iv) the initial conditions $u(0) = u_0$ and $z(0) = z_0$ hold.

The advantage of the above definition is that it is completely derivative free, i.e., no time derivative $\dot{z}$ and no (sub)differentials of $\mathscr{E}$ or $\mathscr{R}$ occur explicitly in Definition 11.1. In fact, if $\mathscr{E}(t, \cdot, z)$ or $\mathscr{E}(t, u, \cdot)$ are not convex and thus (11.4) loses meaning, Definition 11.1 still yields a certain generalized solution. If a weak solution $(u, z, \mathfrak{z})$ is such that $\dot{z}$ is absolutely continuous (i.e., $\dot{z}$ is not a measure but an L^1-function) and if $\mathscr{E}(t, \cdot, z)$ and $\mathscr{E}(t, u, \cdot)$ are convex, then (u, z) solves the original problem (11.4) for a.e. time $t \in [0,T]$. This justifies the above definition.

An inconvenience is also the involvement of the driving-force field $\mathfrak{z}$ valued in $\mathscr{Z}^*$, which obviously does not bear any generalization for spaces $\mathscr{Z}$ lacking a linear structure, like the Griffith-delamination problem (11.48)–(11.49), below. Thus one may be tempted to make further generalizations. As $\mathscr{R}$ is degree-1 homogeneous and convex, $\partial\mathscr{R}(\dot{z}) \subset \partial\mathscr{R}(0)$ and thus $-\mathscr{E}_z'(t, u(t), z(t)) \in \partial\mathscr{R}(\dot{z})$ implies $\mathscr{R}(\tilde{z}) + \langle \mathscr{E}_z'(t, u(t), z(t)), \tilde{z} \rangle \geq \mathscr{R}(0) = 0$. The convexity of $\mathscr{E}(t, u, \cdot)$ and $\langle \mathscr{E}_u'(t, u, z), \tilde{u} \rangle$ then further imply $\mathscr{R}(\tilde{z}) + \mathscr{E}(t, u(t), \tilde{z} + z(t)) - \mathscr{E}(t, u(t), z(t)) \geq 0$. After making a substitution $\tilde{z} := \tilde{z} - z(t)$ we arrive to the so-called *semi-stability*

$$\forall\, \tilde{z} \in \mathscr{Z} : \qquad \mathscr{E}(t, u(t), z(t)) \leq \mathscr{E}(t, u(t), \tilde{z}) + \mathscr{R}(\tilde{z} - z(t)). \tag{11.15}$$

---

[g] Recall that the variation of $z : [0,T] \to \mathscr{X}$ is defined as $\sup \sum_{j=1}^{N} \|z(t_j) - z(t_{j-1})\|$, where $\|\cdot\|$ is the norm on $\mathscr{X}$ and the supremum is taken over all partitions of $[0,T]$.

Definition 11.2 (Local solutions). The process $(u, z, \mathfrak{z}) : [0, T] \to \mathscr{U} \times \mathscr{Z} \times \mathscr{Z}^*$ is called a local solution[h] of the initial-value problem given by $(\mathscr{U} \times \mathscr{Z}, \mathscr{E}, \mathscr{R})$ and the initial condition (u_0, z_0) if the energy inequality (i.e., (11.14) for $\tilde{z} = 0$) holds for a.a. $[t_1, t_2] \subset [0, T]$, and if for a.a. $t \in [0, T]$, (11.9b) holds for any $\tilde{u} \in \mathcal{U}$ together with (11.15) and again the initial conditions $u(0) = u_0$ and $z(0) = z_0$ hold.

This definition was introduced for a special crack problem in [27], and as a general concept under the name "dissipative trajectories" in [28, Definition 6.1], and further investigated in the general theory of rate-independent processes in [25]. In fact, Definition 11.1 yields slightly more general solutions than Definition 11.2, but in the some cases when $\partial\mathscr{R}$ or $\partial_z\mathscr{E}$ are (locally) bounded (here, e.g., healing in delamination or some cohesive contacts) these definitions are equivalent to each other, cf. [26, Proposition 3.3.5] or [29, Proposition 2.3].

If $\mathscr{E}(t, \cdot, \cdot)$ is convex, then the class of local (or weak) solution is fairly small (and even uniqueness of such solution holds under additional smoothness qualification, cf. [30] or also [26, Sec. 3.4.4]). Yet, convex stored energies cannot yield the pursued phenomena typically observed in adhesive contact mechanics (as activated, sudden rupture) and various careful refinements of these wide concepts are highly desired.

11.3.2. Weak solutions conserving energy: Energetic solutions

As said above, the drawback of the above two definitions is their rather low selectivity.[i] An attractive temptation (although not always physically relevant) is the requirement of a conservation of the mechanical energy in the sense an equality in (11.13). Relying again on the degree-1 homogeneity and convexity of $\mathscr{R}$ and now even on joint convexity of $\mathscr{E}(t, \cdot, \cdot)$, we can see that $-\mathscr{E}_z'(t, u, z) \in \partial\mathscr{R}(\dot{z})$ implies $\mathscr{R}(\tilde{z}) + \mathscr{E}(t, \tilde{u} + u, \tilde{z} + z) - \mathscr{E}(t, u, z) \geq 0$, which is obviously just (11.16) below. Thus we arrive at the following definition.

[h] More precisely, local solutions requires the energy inequality to hold for all $[t_1, t_2] \subset [0, T]$, while solutions defined here are rather called a.e.-local solutions.

[i] For example, weak solutions to the brittle delamination problem in the formulation (11.54) do not necessarily have the so-called Griffith property and do not recover the original problem (11.48)–(11.49), through the formula (11.55), in contrast to the energetic solutions (see Definition 11.3), which enjoy this property.

Definition 11.3 (Energetic solutions, [30–32]). The process (u,z): $[0,T]\to\mathscr{U}\times\mathscr{Z}$ is called an energetic solution to the initial-value problem given by $(\mathscr{U}\times\mathscr{Z},\mathscr{E},\mathscr{R})$ and the initial condition (u_0,z_0) if $u\in\mathrm{B}([0,T];\mathscr{U})$, $z\in\mathrm{B}([0,T];\mathscr{Z})\cap\mathrm{BV}([0,T];\mathscr{X})$, and

(i) the energy inequality (11.13) holds,
(ii) the following *stability* inequality holds for any $t\in[0,T]$:

$$\forall(\tilde{u},\tilde{z})\in\mathscr{U}\times\mathscr{Z}:\quad \mathscr{E}(t,u(t),z(t))\le\mathscr{E}(t,\tilde{u},\tilde{z})+\mathscr{R}(\tilde{z}-z(t)),\tag{11.16}$$

(iii) the initial conditions $u(0)=u_0$ and $z(0)=z_0$ hold.

In fact, any energetic solution satisfies (11.13) as an equality and thus, more conventionally, the definition of energetic solutions instead employs the energy equality. If an energetic solution exists it is also a weak or local solution. Indeed, if (11.13) holds then (11.12) follows because $\mathscr{R}\ge 0$ and $\langle\mathfrak{z},\tilde{z}\rangle\ge 0$ due to the fact that $0\in\partial\mathscr{R}(0)$. Taking $\tilde{z}:=z$ in (11.16) gives (11.9b). Finally, setting $\tilde{u}:=u$ in (11.16) and exploiting the convexity of $\mathscr{E}(t,u,\cdot)$ proves (11.9c).

An important step is to apply Definition 11.3 in cases when $\mathscr{E}(t,\cdot,\cdot)$ non-convex, which is just a typical situation in the modeling of quasistatic delamination processes.

An efficient theoretical tool to prove the existence of energetic solutions is the *implicit time discretization* of (11.4) by the so-called *backward Euler formula*. Being constructive, it simultaneously suggests a conceptual numerical algorithm; cf. Remark 11.3 below. Considering, for simplicity, an equidistant partition of $[0,T]$ with a time step $\tau>0$, it formally leads to the recursive problem

$$\partial_u\mathscr{E}(k\tau,u_\tau^k,z_\tau^k)\ni 0\quad\text{and}\quad\partial\mathscr{R}\left(\frac{z_\tau^k-z_\tau^{k-1}}{\tau}\right)+\partial_z\mathscr{E}(k\tau,u_\tau^k,z_\tau^k)\ni 0\tag{11.17}$$

for $k=1,2,\ldots,T/\tau$, starting from $u_\tau^0=u_0$ and $z_\tau^0=z_0$. The values (u_τ^k,z_τ^k) are to approximate the values of (some of) exact solutions $(u(t),z(t))$ at time $t=k\tau$ with $k=1,2,\ldots,T/\tau$, and to define an approximate solution (u_τ,z_τ), cf. (11.22) below. Note that, in fact, $\partial\mathscr{R}$ is homogeneous of degree 0 so that the factor $1/\tau$ in (11.17) can be omitted. The potential structure of the problem allows for a conceptually constructive way to obtain a solution to (11.17), namely by solving the *incremental global-minimization problem*, also referred to as *incremental variational problem*,

$$\left.\begin{array}{ll}\text{minimize} & (u,z)\mapsto \mathscr{E}(k\tau,u,z)+\mathscr{R}(z-z_\tau^{k-1})\\ \text{subject to} & (u,z)\in \mathscr{U}\times\mathscr{Z}.\end{array}\right\} \tag{11.18}$$

This energy minimization provides a unified approach to both the continuum thermodynamics of inelastic solids, e.g., [10, 12], as well as to computational inelasticity, [9, 11, 33]. In addition, rigorous mathematical theory has been established to study the time-continuous behavior of the problem (11.18) (corresponding to the limit $\tau \to 0$), see [25, 26, 30, 34, 35] and Section 11.3 below.

Note that (11.17), the time-discretized analog of (11.4), is a first-order optimality condition for any solution to (11.18). This is also called the *direct method* for solving (11.17), i.e., no approximate problem is in principle needed to ensure the existence of a solution to (11.17), see, e.g., [36] or [24] for a brief survey. Of course, if $\mathscr{E}(t,\cdot,\cdot)$ is not convex, as is usual in delamination problems, (11.17) might admit multiple solutions. The problem (11.18), referring to global (not only local) minimizers, encompasses the energetic-solution concept, which in turn is easily amenable to mathematical and numerical treatment. The following assertion uses quite minimal hypotheses:

Proposition 11.1 (Existence of time-discrete solutions). *If* $\mathscr{E}(t,\cdot,\cdot)$ *is lower semicontinuous and coercive*[j] *on* $\mathscr{U}\times\mathscr{Z}$ *and also* $\mathscr{R}\geq 0$ *is lower semicontinuous, then the incremental problem* (11.18) *possesses a solution.*

Let us consider a solution $(u_\tau^k, z_\tau^k)\in\mathscr{U}\times\mathscr{Z}$ of the incremental problem (11.18) at the level k. Comparing the energy value of (11.18) for a solution in the time step k with energy at arbitrary $(\tilde u,\tilde z)$, we obtain the discrete stability:

$$\begin{aligned}\mathscr{E}(k\tau,u_\tau^k,z_\tau^k) &\leq \mathscr{E}(k\tau,\tilde u,\tilde z)+\mathscr{R}(\tilde z-z_\tau^{k-1})-\mathscr{R}(z_\tau^k-z_\tau^{k-1})\\ &\leq \mathscr{E}(k\tau,\tilde u,\tilde z)+\mathscr{R}(\tilde z-z_\tau^k),\end{aligned} \tag{11.19}$$

where we also used the degree-1 homogeneity and the convexity of $\mathscr{R}$, which yields the triangle inequality $\mathscr{R}(\tilde z-z_\tau^{k-1})\leq\mathscr{R}(z_\tau^k-z_\tau^{k-1})+\mathscr{R}(\tilde z-z_\tau^k)$.

[j]This essentially means that the sub-level sets of $\mathscr{E}(t,\cdot,\cdot)$, i.e., $\{(u,z)\in\mathscr{U}\times\mathscr{Z};\ \mathscr{E}(t,u,z)\leq c\}$, are, for any $c\in\mathbb{R}$ that makes them non-empty, compact in some topology of $\mathscr{U}\times\mathscr{Z}$ which makes $\mathscr{E}(t,\cdot,\cdot)$ and $\mathscr{R}(\cdot)$ lower semicontinuous.

Comparing the energy value of a solution at the level k with that for a solution $(u_\tau^{k-1}, z_\tau^{k-1})$ of the incremental problem (11.18) at the level $k-1$ gives $\mathscr{E}(k\tau, u_\tau^k, z_\tau^k) + \mathscr{R}(z_\tau^k - z_\tau^{k-1}) \le \mathscr{E}(k\tau, u_\tau^{k-1}, z_\tau^{k-1}) + \mathscr{R}(z_\tau^{k-1} - z_\tau^{k-1}) = \mathscr{E}(k\tau, u_\tau^{k-1}, z_\tau^{k-1})$, which yields an upper estimate of the energy balance in the kth step:

$$
\begin{aligned}
&\mathscr{E}(k\tau, u_\tau^k, z_\tau^k) + \mathscr{R}(z_\tau^k - z_\tau^{k-1}) - \mathscr{E}((k-1)\tau, u_\tau^{k-1}, z_\tau^{k-1}) \\
&\quad \le \mathscr{E}(k\tau, u_\tau^{k-1}, z_\tau^{k-1}) - \mathscr{E}((k-1)\tau, u_\tau^{k-1}, z_\tau^{k-1}) \\
&\quad = \int_{(k-1)\tau}^{k\tau} \mathscr{E}_t'(t, u_\tau^{k-1}, z_\tau^{k-1})\,\mathrm{d}t. \qquad (11.20)
\end{aligned}
$$

Eventually, writing the stability (11.19) at the level $k-1$ and testing it by $(\tilde{u}, \tilde{z}) = (u_\tau^k, z_\tau^k)$ gives a lower estimate of the energy balance in the kth step:

$$
\begin{aligned}
&\mathscr{E}(k\tau, u_\tau^k, z_\tau^k) + \mathscr{R}(z_\tau^k - z_\tau^{k-1}) - \mathscr{E}((k-1)\tau, u_\tau^{k-1}, z_\tau^{k-1}) \\
&= \mathscr{E}((k-1)\tau, u_\tau^k, z_\tau^k) + \int_{(k-1)\tau}^{k\tau} \mathscr{E}_t'(t, u_\tau^k, z_\tau^k)\mathrm{d}t + \mathscr{R}(z_\tau^k - z_\tau^{k-1}) \\
&\quad -\mathscr{E}((k-1)\tau, u_\tau^{k-1}, z_\tau^{k-1}) \ge \int_{(k-1)\tau}^{k\tau} \mathscr{E}_t'(t, u_\tau^k, z_\tau^k)\mathrm{d}t. \qquad (11.21)
\end{aligned}
$$

It is convenient to introduce the notation for the piecewise constant interpolants $\overline{u}_\tau$ and $\underline{u}_\tau$, defined by

$$
\overline{u}_\tau(t) := u_\tau^k \qquad \text{for } t \in ((k-1)\tau, k\tau], \qquad (11.22a)
$$

$$
\underline{u}_\tau(t) := u_\tau^{k-1} \quad \text{for } t \in [(k-1)\tau, k\tau). \qquad (11.22b)
$$

The notation $\overline{z}_\tau$ and $\underline{z}_\tau$ has an analogous meaning. Moreover, we define

$$
\overline{\mathscr{E}}_\tau(t, u, z) := \mathscr{E}(k\tau, u, z) \quad \text{for } t \in ((k-1)\tau, k\tau]. \qquad (11.22c)
$$

In terms of these interpolants, one can write (11.19), (11.20), and (11.21) summed over k in a compact form (11.23)–(11.24):

Proposition 11.2 (Stability and two-sided energy estimate). *Let $\mathscr{R}$ be degree-1 positively homogeneous and let $\mathscr{E}_t'(\cdot, u, z) \in L^1(0,T)$ for any (u, z). Then the discrete stability*

$$
\forall (\tilde{u}, \tilde{z}) \in \mathscr{U} \times \mathscr{Z}: \quad \overline{\mathscr{E}}_\tau(t, \overline{u}_\tau(t), \overline{z}_\tau(t)) \le \overline{\mathscr{E}}_\tau(t, \tilde{u}, \tilde{z}) + \mathscr{R}(\tilde{z} - \overline{z}_\tau(t)) \quad (11.23)
$$

holds for any $t \in [0,T]$, *and, for any* $s = k\tau \in [0,T]$, $k \in \mathbb{N}$, *the following two-sided energy inequality holds*:

$$\begin{aligned}
&\int_0^s \mathscr{E}_t'(t,\overline{u}_\tau(t),\overline{z}_\tau(t))\,\mathrm{d}t \\
&\qquad \le \overline{\mathscr{E}}_\tau(s,\overline{u}_\tau(s),\overline{z}_\tau(s)) + \mathrm{Diss}_{\mathscr{R}}(\overline{z}_\tau,[0,s]) - \overline{\mathscr{E}}_\tau(0,u_0,z_0) \\
&\qquad \le \int_0^s \mathscr{E}_t'(t,\underline{u}_\tau(t),\underline{z}_\tau(t))\,\mathrm{d}t. \qquad (11.24)
\end{aligned}$$

The two-sided energy estimate may facilitate the numerical solution of the global-optimization problem (11.18). It should be emphasized that in all the applications considered in Sections 11.4–11.6, the incremental problem (11.18) involves a non-convex functional, whose minimization is therefore very delicate, and iterative procedures need good starting points. Various back-tracking strategies based only on the two-sided energy estimate (11.24) have been designed and tested in [37–39] for similar kinds of problems, see also Section 11.5.2 for additional details.

Let us assume, with some restriction of generality but still covering all problems presented here, that:

$$\exists\epsilon>0 \qquad \forall t,u,z: \quad \mathscr{E}(t,u,z) \ge \epsilon(\|u\|_{\mathscr{U}}^2 + \|z\|_{\mathscr{Z}}^2) - 1/\epsilon, \qquad (11.25a)$$

$$\exists\gamma\in L^1(0,T)\ \forall t,u,z: \quad |\mathscr{E}_t'(t,u,z)| \le \gamma(t)(1+\|u\|_{\mathscr{U}}), \qquad (11.25b)$$

$$\exists\epsilon>0 \qquad \forall z: \quad \mathscr{R}(z) \ge \epsilon\|z\|_{\mathscr{X}}. \qquad (11.25c)$$

Proposition 11.3 (Convergence of discrete solutions). *Let* (11.25) *hold,* u_0 *be stable, and* $\mathscr{E}(t,\cdot,z)$ *be strictly convex. For* $\tau \to 0$, *there is a subsequence of the sequence of approximate solutions* $\{(\overline{u}_\tau,\overline{z}_\tau)\}_{\tau>0}$, *which converges to some* (u,z) *in the sense*

$$\overline{u}_\tau(t) \to u(t) \ \text{ in } \mathscr{U} \qquad \text{for any } t\in[0,T], \qquad (11.26a)$$

$$\overline{z}_\tau(t) \to z(t) \ \text{ in } \mathscr{Z} \qquad \text{for any } t\in[0,T], \qquad (11.26b)$$

$$\mathrm{Diss}_{\mathscr{R}}(\overline{z}_\tau,[0,t]) \to \mathrm{Diss}_{\mathscr{R}}(z,[0,t]) \qquad \text{for any } t\in[0,T], \qquad (11.26c)$$

$$\mathscr{E}_t'(\cdot,\overline{u}_\tau(\cdot),\overline{z}_\tau(\cdot)) \to \mathscr{E}_t'(\cdot,u(\cdot),z(\cdot)) \quad \text{in } L^1(0,T). \qquad (11.26d)$$

Moreover, every (u,z) *obtained by such a limit process is an energetic solution to the problem* $(\mathscr{E},\mathscr{R},z_0)$.

The proof of Proposition 11.3 conventionally relies on the following steps:

(1) *A priori* estimates, derived from the upper energy estimate in (11.24) by using the coercivity/growth assumption (11.25) and Gronwall's inequality: namely one gets

$$\|\overline{u}_\tau\|_{\mathrm{B}([0,T];\mathscr{U})} \le C, \tag{11.27a}$$

$$\|\overline{z}_\tau\|_{\mathrm{B}([0,T];\mathscr{Z})\cap\mathrm{BV}([0,T];\mathscr{X})} \le C. \tag{11.27b}$$

(2) The selection of convergent subsequences using Banach's and Helly's principles; the latter is used for the z-component, which has a bounded variation. Using in addition the strict convexity of $\mathscr{E}(t,\cdot,z)$, one can show that the u-component converges at each time t.

(3) Passage to the limit of the discrete stability (11.23) by finding a so-called mutual-recovery sequence [40], i.e.,
$\forall$ stable sequence[k] $(t_k,u_k,z_k)\to(t,u,z)$ $\forall(\tilde{u},\tilde{z})$ $\exists(\tilde{u}_k,\tilde{z}_k)$:

$$\limsup_{k\to\infty}(\mathscr{E}(t_k,\tilde{u}_k,\tilde{z}_k)+\mathscr{R}(\tilde{z}_k-z_k)-\mathscr{E}(t_k,u_k,z_k)) \le \mathscr{E}(t,\tilde{u},\tilde{z})+\mathscr{R}(\tilde{z}-z)-\mathscr{E}(t,u,z). \tag{11.28}$$

(4) Passage to the limit by weak lower-semicontinuity in the upper energy estimate, i.e., in the second inequality in (11.24).

Merging Propositions 11.1 and 11.3, one obtains the following corollary.

Corollary 11.1 (Existence of energetic solutions). *Under the assumptions of Propositions* 11.1 *and* 11.3, *energetic solutions in the sense of Definition* 11.3 *do exist.*

Remark 11.1 (Special unidirectional processes). In many delamination models $\mathscr{R}$ has the special form

$$\mathscr{R}(\dot{z}) = \delta_K(\dot{z}) + \langle a,\dot{z}\rangle \tag{11.29}$$

with some cone $K\subset\mathscr{Z}$ and some $a\in\mathscr{Z}^*$ non-negative in the sense that $\langle a,z\rangle\ge 0$ for any $z\in K$. Then one can evaluate $\mathrm{Diss}_{\mathscr{R}}(z,[0,T])$ in (11.12)

[k] A sequence $\{(t_k,u_k,z_k)\}_{k\in\mathbb{N}}$ is called stable if $\sup_{k\in\mathbb{N}}\mathscr{E}(t_k,u_k,z_k)<\infty$ and if $\mathscr{E}(t_k,u_k,z_k)\le\mathscr{E}(t_k,\tilde{u},\tilde{z})+\mathscr{R}(\tilde{z}-z_k)$ for all $(\tilde{u},\tilde{z})\in\mathscr{U}\times\mathscr{Z}$.

very explicitly. Indeed, any solution satisfying (11.12) must have $\dot{z} \in K$, hence $z(t_j) - z(t_{j-1}) \in K$ for any $t_j \geq t_{j-1}$, so that (11.11) gives

$$\begin{aligned} \mathrm{Diss}_{\mathscr{R}}(z,[0,T]) &= \sup \sum_{j=1}^{N} \langle a, z(t_j) - z(t_{j-1})\rangle = \sup \langle a, z(t_N) - z(t_0)\rangle \\ &= \langle a, z(T) - z(0)\rangle = \mathscr{R}(z(T) - z(0)). \end{aligned} \tag{11.30}$$

In this particular case, one can equivalently consider the dissipation potential $\mathscr{R}_0 : \mathscr{Z} \to \{0, \infty\}$ as $\mathscr{R}_0(\dot{z}) = \delta_K(\dot{z})$, if one augments the stored energy by the term $\langle a, z\rangle$. In view of (11.4), this is obvious when writing

$$\begin{aligned} \partial\mathscr{R}(\dot{z}) + \partial_z\mathscr{E}(t,u,z) &= \partial[\mathscr{R}_0(\dot{z}) + \langle a, \dot{z}\rangle] + \partial_z\mathscr{E}(t,u,z) \\ &= \partial\mathscr{R}_0(\dot{z}) + a + \partial_z\mathscr{E}(t,u,z) \\ &= \partial\mathscr{R}_0(\dot{z}) + \partial_z\mathscr{E}_0(t,u,z) \end{aligned}$$

for $\mathscr{E}_0(t,u,z) := \mathscr{E}(t,u,z) + \langle a, z\rangle$. The philosophy behind this formula is that the contribution to the stored energy via a unidirectional process can never be gained back, and it is thus stored forever, which means that it is dissipated. This alternative setting has been considered, e.g., in [41–44]. It should be emphasized that this purely mechanical alternative is no longer equivalent in the full thermodynamical context when the dissipated energy contributes to heat production, in contrast to the stored energy, cf. Section 11.6.7.

Remark 11.2 (More general dissipation). Sometimes it is useful to consider $\mathscr{R} = \mathscr{R}(u, z, \dot{z})$, and then the inclusion (11.4) modifies to

$$\partial_u\mathscr{E}(t,u,z) \ni 0, \quad \partial_{\dot{z}}\mathscr{R}(u,z,\dot{z}) + \partial_z\mathscr{E}(t,u,z) \ni 0. \tag{11.31}$$

A priori estimates based on the test by $\dot{z}$ are the same. Now in general, $\mathrm{Diss}_{\mathscr{R}}(z;[0,T])$ in (11.13) depends also on the u-component and in terms of a placeholder $q = (u, z)$ is defined by

$$\mathrm{Diss}_{\mathscr{D}}(q;[0,T]) := \sup \sum_{i=1}^{N} \mathscr{D}\big(q(t_{i-1},\cdot), q(t_i,\cdot)\big), \tag{11.32}$$

where the supremum is taken over all partitions of the type $0 \leq t_0 < t_1 < \cdots < t_N \leq T$, $N \in \mathbb{N}$; here $\mathscr{D}$ denotes a so-called *dissipation distance* defined in [45], reflecting the minimum dissipation-potential principle (11.6), by

$$\mathscr{D}(q_1, q_2) := \inf\Big\{ \int_0^1 \mathscr{R}(u, z, \dot{z})\,\mathrm{d}t; \quad q = (u, z) \in C^1([0,1]; \mathscr{X}),\ \ q(0) = q_1,\ \ q(1) = q_2 \Big\}. \tag{11.33}$$

As before, one assumes the positive one-homogeneity of $\mathscr{R}(u, z, \cdot)$. This implies the triangle inequality

$$\forall\, q_1, q_2, q_3 \in \mathscr{U} \times \mathscr{Z}: \quad \mathscr{D}(q_1, q_3) \leq \mathscr{D}(q_1, q_2) + \mathscr{D}(q_2, q_3). \tag{11.34}$$

In terms of the dissipation distance, the incremental problem (11.18) takes the form

$$\left.\begin{array}{ll} \text{minimize} & (u, z) \mapsto \mathscr{E}(k\tau, u, z) + \mathscr{D}((u_\tau^{k-1}, z_\tau^{k-1}), (u, z)) \\ \text{subject to} & (u, z) \in \mathscr{U} \times \mathscr{Z}. \end{array}\right\} \tag{11.35}$$

An important step in the conceptual generalization is to consider the dissipation distance $\mathscr{D} \geq 0$ satisfying (11.34) without any reference to $\mathscr{R}$, and even without requiring any linear structure on $\mathscr{U} \times \mathscr{Z}$. The mutual-recovery-sequence condition (11.28) then modifies to

$$\limsup_{k\to\infty}(\mathscr{E}(t_k, \tilde{q}_k) + \mathscr{D}(q_k, \tilde{q}_k) - \mathscr{E}(t_k, q_k)) \leq \mathscr{E}(t, \tilde{q}) + \mathscr{D}(q, \tilde{q}) - \mathscr{E}(t, q). \tag{11.36}$$

Note that, if $\mathscr{R} = \mathscr{R}(\dot{z})$ and $\mathscr{Z}$ is a Banach space, formula (11.33) yields $\mathscr{D}(q_1, q_2) = \mathscr{R}(z_2 - z_1)$ with $q_i = (u_i, z_i)$, and one obtains the former case.

11.3.3. Weak solutions of stress-driven types

If $\mathscr{E}(t, \cdot, \cdot)$ is not convex or if $\mathscr{R}$ depends on (u, z), the energy-conserving solutions may be time-discontinuous even if the external loading is time-continuous. As already said, energetic solutions evolve as soon as it is energetically not disadvantageous. It should be noted that this may, however, not be exactly always in full agreement with the response of real systems where some other rate-dependent phenomena may come into play on some occasions and a rather non-physical tendency for developing too early jumps may often be observed. Typically, one may imagine the delamination of a very large elastic body which have a capacity to store large energy already under a very low stress and then, counting with Definition 11.3, the delamination process may be triggered under such a low stress, cf. also [26, Sec. 4.3.2.4]. Similarly, in mixed-mode delamination, energetic solutions may have tendency to slide to less-dissipative Mode I even in case when it is physically not expected, cf. [46].

For this reason, there are also some other concepts of solutions that are sometimes applicable and which successfully compete with energetic solutions, cf. also [25] for a comparison with other concepts in general and [47–51] in the context of crack propagation. In particular, a well-motivated class of concepts is based on adding a small viscosity into one or both inclusions in (11.4). Passing such viscosity to zero leads to so-called vanishing-viscosity solutions to the rate-independent problem (11.4). In the context of delamination, the vanishing-viscosity in the second inclusion in (11.4) was exploited in [52, 53], leading to a definition which involves a defect measure balancing the energetics.

Rather opposite concepts of solutions try to seek such solutions that cannot exhibit discontinuous responses if the driving force is not big enough. The concepts with clear physical interpretation exploit (presumably small) viscosity and study its asymptotics when it converges to zero. There are various options which may lead to various local solutions. Essentially, we can consider (different) viscosities in u, or in z, or in both. The last, most general option augments the evolution system (11.4) as

$$\epsilon_1 V_1 \dot{u} + \partial_u \mathscr{E}(t,u,z) \ni 0 \quad \text{and} \quad \epsilon_2 V_2 \dot{z} + \partial\mathscr{R}(\dot{z}) + \partial_z \mathscr{E}(t,u,z) \ni 0, \tag{11.37}$$

where $V_1 : \mathscr{U} \to \mathscr{U}^*$ and $V_2 : \mathscr{Z} \to \mathscr{Z}^*$ are linear positive definite operators having quadratic potentials $\mathscr{V}_1 : \mathscr{U} \to \mathbb{R}$ and $\mathscr{V}_2 : \mathscr{Z} \to \mathbb{R}$, respectively. We will use the shorthand notation $\epsilon = (\epsilon_1, \epsilon_2)$. Under mild assumptions on data, cf., e.g., [26, 52, 54, 55], conventional weak solutions (u_ϵ, z_ϵ) to the parabolic system (11.37) exist and, beside the a-priori estimates (11.27) which now read as

$$\|u_\epsilon\|_{\mathrm{B}([0,T];\mathscr{U})} \le C, \tag{11.38a}$$

$$\|z_\epsilon\|_{\mathrm{B}([0,T];\mathscr{Z})\cap\mathrm{BV}([0,T];\mathscr{X})} \le C, \tag{11.38b}$$

we now have also the estimates

$$\|\dot{u}_\epsilon\|_{L^2([0,T];\mathscr{U})} \le C/\sqrt{\epsilon_1}, \tag{11.38c}$$

$$\|\dot{z}_\epsilon\|_{L^2([0,T];\mathscr{Z})\cap\mathrm{BV}([0,T];\mathscr{X})} \le C/\sqrt{\epsilon_2}. \tag{11.38d}$$

Moreover, instead of the inequality (11.13), we have the energy balance

$$\mathscr{E}(t, u_\epsilon(t), z_\epsilon(t)) + \mathrm{Diss}_{\mathscr{R}}(z_\epsilon, [0,t]) \\ + \int_0^t 2\epsilon_1 \mathscr{V}_1(\dot{u}_\epsilon) + 2\epsilon_2 \mathscr{V}_2(\dot{z}_\epsilon)\,\mathrm{d}t = \int_0^t \mathscr{E}_t'(t, u_\epsilon, z_\epsilon)\,\mathrm{d}t + \mathscr{E}(0, u_0, z_0). \tag{11.39}$$

Definition 11.4 ($\mathscr{V}$-approximable semi-energetic solutions). A local solution (u,z) according Definition 11.2 is called $\mathscr{V}$-approximable semi-energetic solutions with respect to the viscosity $\mathscr{V} = \mathscr{V}(\dot{u},\dot{z}) = \mathscr{V}_1(\dot{u}) + \mathscr{V}_2(\dot{z})$ if, for some subsequence of $\epsilon = (\epsilon_1, \epsilon_2) \to (0,0)$ and for some non-negative (so-called defect) measure μ on $[0,T]$,

$$u_\epsilon(t) \to u(t) \quad \text{in } \mathscr{U} \text{ for a.a. } t \in [0,T], \quad \text{and} \tag{11.40a}$$

$$z_\epsilon(t) \to z(t) \quad \text{weakly* in } \mathscr{Z} \text{ for all } t \in [0,T], \quad \text{and} \tag{11.40b}$$

$$2\epsilon_1 \mathscr{V}_1(\dot{u}_\epsilon) + 2\epsilon_2 \mathscr{V}_2(\dot{z}_\epsilon) \to \mu \quad \text{weakly* in measures on } [0,T], \tag{11.40c}$$

and the following energy equality holds for a.a. $t \in [0,T]$:

$$\mathscr{E}(t, u(t), z(t)) + \mathrm{Diss}_{\mathscr{R}}(z, [0,t]) + \int_0^t \mu(\mathrm{d}t) \\ = \int_0^t \mathscr{E}_t'(t, u, z)\,\mathrm{d}t + \mathscr{E}(0, u_0, z_0). \tag{11.40d}$$

For a general theory and more solution concepts we refer to [54, 55] or also [26, Sec. 3.8]. Existence of $\mathscr{V}$-approximable solutions are shown by limiting (11.39) towards (11.40d). Note that, by omitting the μ-term from (11.40d) we obtain the energy inequality on a.a. time intervals $[t_1, t_2]$. This inequality is strict if μ is not zero.

Here, as we (intentionally) do not insist on energy equality, we may advantageously use the *time discretization of a fractional-step type* to decouple the system. Thus, instead of a recursive backward-Euler formula (11.17), we consider

$$\epsilon_1 V_1 \frac{u_\tau^k - u_\tau^{k-1}}{\tau} + \partial_u \mathscr{E}(k\tau, u_\tau^k, z_\tau^{k-1}) \ni 0, \tag{11.41a}$$

$$\epsilon_2 V_2 \frac{z_\tau^k - z_\tau^{k-1}}{\tau} + \partial\mathscr{R}\left(\frac{z_\tau^k - z_\tau^{k-1}}{\tau}\right) + \partial_z \mathscr{E}(k\tau, u_\tau^k, z_\tau^k) \ni 0. \tag{11.41b}$$

In contrast to (11.17), this scheme is decoupled in the sense that, at a current time level, one can solve first (11.41a) to obtain u_τ^k and then (11.41b) for z_τ^k. In view of (11.2), it is realistic to assume $\mathscr{E}(t,\cdot,\cdot)$ separately strictly convex, i.e., both $\mathscr{E}(t,u,\cdot)$ and $\mathscr{E}(t,\cdot,z)$ are strictly convex and even quadratic. Then, taking into account the constraints (11.1) and the 1-homogeneity of $\mathscr{R}$ in (11.3), both problems (11.41) lead to (possibly after a suitable Mosco-type transformation) a strictly convex linear-quadratic problems which are, after space discretization, efficiently numerically solvable. In particular, the global nonconvex minimization problem related to the fully implicit Euler formula (11.17) is thus eliminated. For fixed $\epsilon_1 \geq 0$ and $\epsilon_2 \geq 0$, the recursive system of inclusions (11.41) possesses a unique solution $\{(u_\tau^k, z_\tau^k)\}_{k=1,\ldots,T/\tau}$.

An important attribute of the scheme (11.41) is that it still complies with the upper energy estimate, i.e., the latter inequality in (11.24), provided that $\mathscr{E}(t,\cdot,\cdot)$ is separately convex. Indeed, testing (11.41a) by $u_\tau^k - u_\tau^{k-1}$ and using the convexity of $\mathscr{E}(k\tau,\cdot,z_\tau^{k-1})$, one obtains the inequality

$$2\epsilon_1\tau\mathscr{V}_1\Big(\frac{u_\tau^k - u_\tau^{k-1}}{\tau}\Big) + \mathscr{E}(k\tau, u_\tau^k, z_\tau^{k-1}) - \mathscr{E}(k\tau, u_\tau^{k-1}, z_\tau^{k-1}) \leq 0, \tag{11.42a}$$

while testing (11.41b) by $z_\tau^k - z_\tau^{k-1}$ and using the convexity of $\mathscr{E}(k\tau, u_\tau^k,\cdot)$, one obtains the inequality

$$\begin{aligned}2\epsilon_2\tau\mathscr{V}_2\Big(\frac{z_\tau^k - z_\tau^{k-1}}{\tau}\Big) + \tau\mathscr{R}\Big(\frac{z_\tau^k - z_\tau^{k-1}}{\tau}\Big)&\\ + \mathscr{E}(k\tau, u_\tau^k, z_\tau^k) - \mathscr{E}(k\tau, u_\tau^k, z_\tau^{k-1}) \leq 0.&\end{aligned} \tag{11.42b}$$

Summing it up, one exploits cancellation of the terms $\pm\mathscr{E}(k\tau, u_\tau^k, z_\tau^{k-1})$. Using again the calculus (11.20), one eventually arrives at

$$\begin{aligned}&\mathscr{E}(k\tau, u_\tau^k, z_\tau^k) + 2\epsilon_1\tau\mathscr{V}_1\Big(\frac{u_\tau^k - u_\tau^{k-1}}{\tau}\Big)\\ &\quad + 2\epsilon_2\tau\mathscr{V}_2\Big(\frac{z_\tau^k - z_\tau^{k-1}}{\tau}\Big) + \tau\mathscr{R}\Big(\frac{z_\tau^k - z_\tau^{k-1}}{\tau}\Big)\\ &\qquad \leq \mathscr{E}((k-1)\tau, u_\tau^{k-1}, z_\tau^{k-1}) + \int_{(k-1)\tau}^{k\tau} \mathscr{E}_t'(t, u_\tau^{k-1}, z_\tau^{k-1})\,\mathrm{d}t,\end{aligned} \tag{11.43}$$

which is the second inequality in (11.24) augmented also by the viscous dissipation ϵ-terms. After summing it up for $k = 1, \ldots, T/\tau$, one obtains the discrete analog of (11.39) as an inequality. This ensures numerical stability of this fractional-step method and, in qualified cases, also convergence for $\tau \to 0$. Actually, such splitting works in more than two steps when $\mathscr{E}(t, \cdot)$ is separately convex at each of (more than two) variables, and $\mathscr{R}$ is a sum of functionals of each of these variables, cf., e.g., [56, Remark 8.25].

The lower energy estimate, i.e., the first inequality in (11.24), does not hold but, on the other hand, it is not so much needed as before, because the global minimization of a non-convex functional in (11.18) as well as the backtracking algorithm in Table 11.2 below was avoided. For fixed $\epsilon_1 > 0$ and $\epsilon_2 > 0$, the scheme (11.41) converges (in terms of subsequences) for $\tau \to 0$ a conventional weak solution to the system of inclusions (11.37).

Moreover, denoting this (not unique[1]) solution by (u_ϵ, z_ϵ), the convergence (in terms of subsequences) for $\epsilon = (\epsilon_1, \epsilon_2) \to (0, 0)$ to the specific local solutions according Definitions 11.2 and 11.4 holds, cf. [26, Sec. 3.8.3]. Denoting $(u_{\epsilon,\tau}, z_{\epsilon,\tau})$ the solution interpolated in time from what is obtained by the scheme (11.41), the joint convergence for $(\epsilon, \tau) \to (0, 0)$ was proved under the condition $\epsilon/\tau \to \infty$ in [54, 55].

One may have an immediate idea for usage of the scheme (11.41) for $\epsilon_1 = 0$ and $\epsilon_2 = 0$ because the existence of approximate solutions and the a-priori estimates (11.27) hold. Although it is in conflict with the mentioned condition $\epsilon/\tau \to 0$, even such approximate solutions converge in qualified cases for $\tau \to 0$ (in the sense of subsequences) towards local solutions according to Definition 11.2, cf. [29] or also [26, Sec. 4.3.4.3] in the context of the delamination problems. The energy is (intentionally!) not conserved by such a scheme. Noteworthy, numerical experiments in particular cases in [53] show a surprisingly good coincidence with vanishing-viscosity-type of solutions which (when counting the approximated defect measure) asymptotically conserve energy. An attempt for reflection of this (currently not explained) phenomenon has been done by defining a particular sort of the (otherwise not much specific) local solutions, cf. [29]. To this goal, in [26, Definition 3.3.8], an integral variant of (11.7), namely

$$\int_0^T \mathfrak{z}(t) \mathrm{d}z(t) = \int_0^T \max_{\omega \in K} \langle \omega, \cdot \rangle \, \dot{z}(\mathrm{d}t) = \mathrm{Diss}_{\mathscr{R}}(z; [0, T]) \tag{11.44}$$

[1]Interestingly, if the delamination variable gradient would be considered, then (u_ϵ, z_ϵ) is even unique, cf. [26, Proposition 4.3.51].

for some available driving force $\mathfrak{z}(t) \in -\partial_z\mathscr{E}(t,u(t),z(t))$ has been devised, where the first integral is the Moore–Pollard modification of the lower Riemann–Stieltjes integral[m] and the second one is a variation of the measure $\dot{z}$ which is here just $\mathrm{Diss}_{\mathscr{R}}(z;[0,T])$, as used in (11.44).

Intuitively, it is obvious that, at least on the discrete level, the approximate solutions are driven by force rather than energies, and there is a good chance that they satisfy at least with a reasonable accuracy the maximum-dissipation principle (11.7), although this principle even in its integral variant (11.44) is not an ultimate dogma and even a simple counterexample can be constructed with two debonding springs of different fracture toughness, cf. [26, Example 4.3.40]. The mentioned approximation of the maximum-dissipation principle may be devised as $\int_0^T \bar{\mathfrak{z}}_\tau \mathrm{d}\bar{z}_\tau(t) \sim \mathrm{Diss}_{\mathscr{R}}(\bar{z}_\tau;[0,T])$ for some driving force $\bar{\mathfrak{z}}_\tau \in -\bar{\mathscr{E}}(t,\bar{u}_\tau(t),\bar{z}_\tau(t))$, where the approximate equality "$\sim$" is to be verified a posteriori and, if satisfied with a reasonable accuracy, one can have a certain indicator that the "inviscid" fractional-step scheme, i.e., (11.41) with $\epsilon_1 = 0$ and $\epsilon_2 = 0$, gives some stress-driven-like solution.[n]

Remark 11.3 (Numerics). One can further approximate $\mathscr{U}$ and $\mathscr{Z}$ in (11.35) by some finite-dimensional Banach subspaces $\mathscr{U}_h$ and $\mathscr{Z}_h$. Thus, in concrete situations, we obtain computationally implementable numerical strategies, as also demonstrated in Section 11.5.2. Besides, one has a convergence analysis for $h \to 0$ at one's disposal in specific cases; essentially, the proofs reduce to finding a suitable mutual-recovery sequence for conditions similar to (11.28) or (11.36), but involving also a sequence of functionals $\mathscr{E}_h$, which coincide with $\mathscr{E}$ on $[0,T] \times \mathscr{U}_h \times \mathscr{Z}_h$ while being $= \infty$ elsewhere, cf. [58]. For an example see (11.60) below.

[m] More specifically, $\int_0^T \mathfrak{z}(t)\mathrm{d}z(t) = \limsup \sum_{j=1}^N \inf_{t\in[t_{j-1},t_j]}\langle\mathfrak{z}(t), z(t_j)-z(t_{j-1})\rangle$ where "limsup" is taken over all partitions $0 \le t_0 < t_1 < \cdots < t_N \le T$ with $N \in \mathbb{N}$ ordered (directed) by inclusion. If z is absolutely continuous and $\mathfrak{z}$ is bounded and in duality with $\dot{z}$ (which cannot be expected in general for delamination problems we have in mind, however), then this definition coincides with the expected Lebesgue integral $\int_0^T \langle\mathfrak{z}(t),\dot{z}(t)\rangle\,\mathrm{d}t$. Actually, [29, 46, 57] use mistakenly the Riemann–Stieltjes definition which uses "sup" in place of "limsup" and which does not work unless z is scalar-valued non-decreasing.

[n] Unfortunately, although this scheme is simple and very efficient in many (or even most of) applications, one cannot say more if this approximate integral maximum-dissipation principle is not satisfied and one cannot make a limit passage towards (11.44).

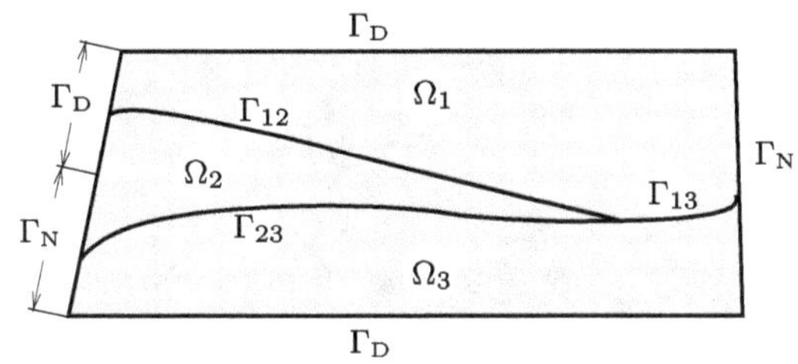

Fig. 11.1. Illustration of the geometry and the notation.

11.4. Quasistatic Brittle Delamination, The Griffith Concept

We now present models of quasistatic delamination that can be covered by the general abstract ansatz (11.4) or (11.31). We start, in this section, with models of brittle delamination.

Let $\Omega \subset \mathbb{R}^d$ $(d = 2, 3)$ be a bounded Lipschitz domain,[o] and let us consider its decomposition into a finite number of mutually disjoint Lipschitz subdomains Ω_i, $i = 1, \ldots, N$. Further denote $\Gamma_{ij} = \partial\Omega_i \cap \partial\Omega_j$ the (possibly empty) boundary between Ω_i and Ω_j. Thus, Γ_{ij} represents a prescribed $(d-1)$-dimensional surface, which may undergo delamination. We assume that the boundary $\partial\Omega$ is the union of two disjoint subsets $\Gamma_{\rm D}$ and $\Gamma_{\rm N}$, with

$$\mathscr{L}^{d-1}(\partial\Omega_i \cap \Gamma_{\rm D}) > 0, \quad i = 1, \ldots, N, \tag{11.45}$$

where $\mathscr{L}^{d-1}$ denotes the $(d-1)$-dimensional Lebesgue measure. On the Dirichlet part of the boundary $\Gamma_{\rm D}$, we impose a time-dependent boundary displacement $w_{\rm D}(t)$, while the remaining part $\Gamma_{\rm N}$ is assumed to be free (Fig. 11.1). Therefore, any admissible displacement $u : \bigcup_{i=1}^N \Omega_i \to \mathbb{R}^d$ has to be equal to a prescribed "hard-device" loading $w_{\rm D}(t)$ on $\Gamma_{\rm D}$.

We consider here the case of linear elasticity determined, on each subdomain Ω_i, by the elastic-moduli tensor $\mathbb{C}^{(i)}$. Moreover, we take into account the local non-interpenetration of matter by requiring, for the displacement u, that $[\![u]\!]_{ij} \cdot \nu_{ij} \geq 0$ on Γ_{ij}, where ν_{ij} denotes the unit normal to Γ_{ij} oriented from Ω_j to Ω_i. Here, $[\![u]\!]_{ij}$ denotes the jump $u|_{\Omega_i} - u|_{\Omega_j}$,

[o]Recall that a domain is called Lipschitz if its boundary can be covered by a finite number of graphs of Lipschitz functions. Roughly speaking, it excludes corners of 0° or 360°, otherwise most domains considered in engineering are Lipschitz.

with $u|_{\Omega_i}$ being the trace on $\partial\Omega_i$ of the restriction of u to Ω_i. For

$$A \subset \Gamma_{\rm C} := \bigcup_{i<j} \Gamma_{ij}, \tag{11.46}$$

we consider the stored-energy functional $\mathscr{E}$ in the form

$$\mathscr{E}(t,u,A) := \begin{cases} \displaystyle\sum_{i=1}^{N} \frac{1}{2} \int_{\Omega_i} \mathbb{C}^{(i)} e(u){:}e(u)\,{\rm d}x & \text{if } u = w_{\rm D}(t) \text{ on } \Gamma_{\rm D}, \\ & [\![u]\!]_{\rm n} \geq 0 \text{ on } \Gamma_{\rm C}, \\ & [\![u(x)]\!] = 0 \text{ for } x \in A, \\ \infty & \text{elsewhere,} \end{cases} \tag{11.47}$$

where $e(u) = \frac{1}{2}(\nabla u)^\top + \frac{1}{2}\nabla u$ is the small-strain tensor, and where $[\![u]\!]$ stands for the particular jump $[\![u]\!]_{ij}$ on the whole union $\Gamma_{\rm C}$ of Γ_{ij}s, and where we write for brevity $[\![u]\!]_{\rm n} \geq 0$ on $\Gamma_{\rm C}$ instead of $[\![u]\!]_{ij} \cdot \nu_{ij} \geq 0$ on Γ_{ij}, $i<j$. This means, in particular, that the part A of the contact boundary $\Gamma_{\rm C}$ is perfectly glued while the rest $\Gamma_{\rm C} \backslash A$ is completely delaminated and a frictionless unilateral, so-called *Signorini contact* takes place there. This unilateral condition is important for preventing an unphysical delamination by a mere compression of the surface.

During the time-dependent loading $w_{\rm D}$, the glued part $A = A(t)$ possibly evolves. In the simplest model, this process is considered unidirectional, i.e., healing is not allowed so that $t \mapsto A(t)$ is non-increasing, and for activation of delamination one needs (and thus dissipates) a specific energy $a : \Gamma_{\rm C} \to \mathbb{R}^+$ (in joules per unit area). The dissipated energy (understood also as the so-called dissipation distance) is then

$$\mathscr{D}(A_1, A_2) := \begin{cases} \displaystyle\int_{A_2 \backslash A_1} a(x){\rm d}S & \text{if } A_1 \subset A_2 \subset \Gamma_{\rm C}, \\ \infty & \text{otherwise.} \end{cases} \tag{11.48}$$

Specifically, the dissipated energy does not depend on particular fracture modes; cf. Sec. 11.6.2 below for a refinement of this model. The philosophy of such a quasistatic evolution is related to the *Griffith fracture criterion* [59], which states that a crack grows as soon as the energy release is more than the fracture toughness, here determined by a in (11.48). This "geometrical" framework was used in the small-strain setting in [27, 48, 51] and also large strains in [50] for polyconvex materials,[p] and in [60] for quasiconvex materials.[q]

The definition of the energetic solutions (11.13) involves the time derivative of the stored energy, which is hardly defined for (11.47) unless $w_{\rm D}$ is constant in time. Therefore, using the additive shift $u - u_{\rm D}(t)$ (where $u_{\rm D}$ is a suitable extension of the formerly defined $w_{\rm D}$) we resort to time-dependent Dirichlet boundary conditions. Thus, up to an irrelevant time-dependent constant, (11.47) transforms to

$$\mathscr{E}(t,u,A) := \begin{cases} \displaystyle\sum_{i=1}^{N} \int_{\Omega_i} \mathbb{C}^{(i)} e(u) : e\Big(\frac{u}{2} + u_{\rm D}(t)\Big)\,{\rm d}x & \text{if } u = 0 \text{ on } \Gamma_{\rm D}, \\ & [\![u]\!]_{\rm n} \geq 0 \text{ on } \Gamma_{\rm C}, \\ & [\![u(x)]\!] = 0 \text{ for } x \in A, \\ \infty & \text{elsewhere}, \end{cases} \tag{11.49}$$

if one assumes that the Dirichlet loading allows for an extension such that

$$u_{\rm D}|_{\Gamma_{\rm C}} = w_{\rm D}. \tag{11.50}$$

Note that, with a slight abuse of notation, the symbol u in (11.49) and thereafter denotes the shifted displacement field satisfying $u = 0$ on $\Gamma_{\rm D}$. After such a shift of the Dirichlet boundary conditions, $\mathscr{E}_t'$ does exist and one has the following simple formula for it:

$$\mathscr{E}_t'(t,u,A) = \sum_{i=1}^{N} \int_{\Omega_i} \mathbb{C}^{(i)} e(u) : e\big(\dot{u}_{\rm D}(t)\big)\,{\rm d}x. \tag{11.51}$$

Defining[r]

$$\mathscr{U} := \{u \in W^{1,2}(\Omega \backslash \Gamma_{\rm C}; \mathbb{R}^d);\ u|_{\Gamma_{\rm D}} = 0\}, \tag{11.52a}$$

$$\mathscr{Z} := \{A \subset \Gamma_{\rm C};\ A \text{ measurable}\} \tag{11.52b}$$

and adopting a generalization from Remark 11.2, one can claim:

---

[p]Materials whose stored energy density is a convex function of the deformation gradient, its cofactor, and determinant.

[q]Materials whose stored energy density f is quasiconvex, i.e., $f(F)|\Omega| \leq \int_\Omega f(F + \nabla u(x))\,{\rm d}x$ for all smooth mappings $u : \Omega \to \mathbb{R}^3$ vanishing at $\partial\Omega$.

[r]Notation $W^{k,p}(\Omega)$ stands for the Banach space of functions on Ω whose kth derivatives belong to L^p-space.

Proposition 11.4. *Let the prescribed boundary displacement $w_{\rm D}$ belong to $W^{1,1}(I;W^{1/2,2}(\Gamma_{\rm D};\mathbb{R}^d))$ with $W^{1/2,2}$ denoting a Sobolev–Slobodetskiĭ space*[s] *and allow for an extension $u_{\rm D}\in W^{1,1}(I;W^{1,2}(\Omega;\mathbb{R}^d))$ satisfying* (11.50) *and let $A_0 \subset \Gamma_{\rm C}$ be measurable. Then the problem $(\mathscr{U}\times\mathscr{Z},\mathscr{E},\mathscr{D},A_0)$ defined by* (11.48), (11.49), *and* (11.52), *possesses at least one energetic solution in the sense of Definition* 11.3.

Having such an energetic solution (u,A), it is natural to consider the shifted $(u+u_{\rm D},A)$ as a solution to the original problem with $\mathscr{E}$ from (11.47) even if $\mathscr{E}_t'$ is not well defined. Also, let us note that there is no explicit linear structure for the As that would allow us to write first-order optimality conditions like (11.4) but, nevertheless, the concept of energetic solutions still works.

Further, it is convenient to reformulate this problem in a way that $\mathscr{Z}$ is a subset of a Banach space. We introduce a so-called delamination parameter $z:\Gamma_{\rm C}\to[0,1]$, meaning a fraction of fixed adhesive: $z(x)=0$ means complete delamination and $z(x)=1$ means 100% perfect bonding, while $z(x)=\frac{1}{2}$ means that 50% of the adhesive is debonded at $x\in\Gamma_{\rm C}$. Here it is appropriate to consider the model

$$[\![u(x)]\!]=0 \quad \text{for a.e. } x\in\Gamma_{\rm C} \text{ such that } \ z(x)>0, \tag{11.53}$$

expressing that delamination can occur only if the adhesive is completely debonded, i.e., only if $z(x)=0$. Thus, instead of (11.49), we now consider

$$\mathscr{E}(t,u,z):=\begin{cases}\displaystyle\sum_{i=1}^{N}\int_{\Omega_i}\mathbb{C}^{(i)}e(u):e\Big(\frac{u}{2}+u_{\rm D}(t)\Big)\,{\rm d}x & \text{if } u=0 \text{ on } \Gamma_{\rm D},\\ & [\![u]\!]_{\rm n}\geq 0 \text{ on } \Gamma_{\rm C},\\ & z[\![u]\!]=0 \text{ on } \Gamma_{\rm C},\\ & 0\leq z\leq 1 \text{ on } \Gamma_{\rm C},\\ \infty & \text{elsewhere,}\end{cases} \tag{11.54a}$$

the dissipation potential

$$\mathscr{R}(\dot z):=\begin{cases}\displaystyle\int_{\Gamma_{\rm C}}a|\dot z|\,{\rm d}S & \text{if } \dot z\leq 0 \text{ on } \Gamma_{\rm C},\\ \infty & \text{otherwise,}\end{cases} \tag{11.54b}$$

[s]The space $W^{1/2,2}(\Gamma_{\rm D})$, involving fractional derivatives, is just the space of traces on $\Gamma_{\rm D}$ of all functions from $W^{1,2}(\Omega)$.

and, instead of (11.52b),

$$\mathscr{Z} := L^\infty(\Gamma_{\rm C}). \tag{11.54c}$$

Now we can use Definition 11.3.

Proposition 11.5. *Let $w_{\rm D} \in W^{1,1}(I; W^{1/2,2}(\Gamma_{\rm D}))$ and $0 \leq z_0 \leq 1$ be measurable. The problem $(\mathscr{U} \times \mathscr{Z}, \mathscr{E}, \mathscr{R}, z_0)$ defined by* (11.52a) *and* (11.54) *possesses energetic solutions in the sense of Definition* 11.3.

The relation between the previous "geometrical" concept used in Proposition 11.4 and this "functional" concept is that, if z takes only values 0 or 1, i.e., always $z = \chi_A$ for some $A \subset \Gamma_{\rm C}$,[t] then

$$\mathscr{D}(A_1, A_2) = \mathscr{R}(z_2 - z_1) \quad \text{with } z_1 = \chi_{A_1}, \quad z_2 = \chi_{A_2}. \tag{11.55}$$

It has been proved in [61] that any energetic solution (u, z) to the brittle delamination problem, whose existence was stated in Proposition 11.5, is of the Griffith type in the sense that z takes only values 0 or 1. Thus, in particular, Proposition 11.4 is proved if Proposition 11.5 is proved. The latter essentially relies on the explicit construction of the *mutual-recovery sequence* for condition (11.28) from [62], namely

$$\tilde{u}_k := u_k \quad \text{and} \quad \tilde{z}_k := \begin{cases} z_k\tilde{z}/z & \text{where } z > 0 \\ 0 & \text{where } z = 0. \end{cases} \tag{11.56}$$

Note that $0 \leq \tilde{z}_k \leq z_k$ always and, if one considers $u_k \to u$ weakly in $W^{1,2}(\Omega; \mathbb{R}^d)$ and $z_k \to z$ weakly* in $L^\infty(\Gamma_{\rm C}; \mathbb{R})$,[u] then also $\tilde{u}_k \to \tilde{u}$ weakly and $\tilde{z}_k \to \tilde{z}$ weakly*.

As the adhesive does not exhibit any elastic response in model (11.47), we refer to it as a *brittle delamination.*

The classical formulation corresponding to (11.4) with $\mathscr{E}$ from (11.54a) and $\mathscr{R}$ from (11.54b) consists in the equilibrium of forces on each subdomain Ω_i and several *complementarity problems.* Using a simplified notation

---

[t]Here χ_A denotes the characteristic function of a set A, i.e., $\chi_A(x) = 1$ if $x \in A$ while $\chi_A(x) = 0$ otherwise.

[u]The adjective "weak*" refers to testing by functions from a so-called pre-dual space. Here, as $L^\infty(\Gamma_{\rm C}; \mathbb{R}) = L^1(\Gamma_{\rm C}; \mathbb{R})^*$, weak* $L^\infty(\Gamma_{\rm C}; \mathbb{R})$-convergence means that $\lim_{k\to\infty} \int_{\Gamma_{\rm C}} z_k \cdot \varphi \,{\rm d}S = \int_{\Gamma_{\rm C}} z \cdot \varphi \,{\rm d}S$ for every $\varphi \in L^1(\Gamma_{\rm C}; \mathbb{R})$.

$\mathbb{C} = \mathbb{C}(x) = \mathbb{C}^{(i)}$ if $x \in \Omega_i$, at a current time t, we can write it as:

$$\operatorname{div}\sigma = 0, \quad \sigma = \mathbb{C}^{(i)}e(u) \qquad \text{in } \Omega_i,\ i = 1,\dots,N, \tag{11.57a}$$

$$u = w_{\rm D}(t,\cdot) \qquad \text{on } \Gamma_{\rm D}, \tag{11.57b}$$

$$\sigma\nu = 0 \qquad \text{on } \Gamma_{\rm N}, \tag{11.57c}$$

$$\left.\begin{array}{l} [\![\sigma]\!]\nu = 0 \\ [\![u]\!]_{\rm n} \geq 0, \quad \sigma_{\rm n}(u)[\![u]\!]_{\rm n} = 0 \\ \sigma_{\rm n}(u) \leq 0 \ \text{ wherever } z(t,\cdot) = 0 \\ z[\![u]\!] = 0 \\ \dot{z} \leq 0, \quad \xi \leq a, \quad \dot{z}(\xi - a) = 0 \\ \xi \in N_{[0,1]}(z) + \partial_z I([\![u]\!], z) \end{array}\right\} \quad \text{on } \Gamma_{\rm C}, \tag{11.57d}$$

where $N_{[0,1]}(\cdot) : \mathbb{R} \rightrightarrows \mathbb{R}$ is the normal-cone mapping, I denotes the indicator function of the constraint $z[\![u]\!] = 0$, ξ is the driving "force" for the delamination, and $\sigma\nu := \mathbb{C}^{(i)}e(u)|_\Gamma\nu$ is the traction stress on $\Gamma = \Gamma_{\rm C}$ or $\Gamma = \Gamma_{\rm N}$ adjacent to Ω_i. Moreover, its normal and tangential components on $\Gamma_{\rm C}$ are denoted $\sigma_{\rm n}(u) = (\sigma\nu)\cdot\nu$ and $\sigma_{\rm t}(u) = \sigma\nu - ((\sigma\nu)\cdot\nu)\nu$, respectively, so that we have the decomposition $\sigma\nu = \sigma_{\rm n}\nu + \sigma_{\rm t}$. Notice that, since by our choice ν turns out to be the inner unit normal on Ω_1, $\sigma_{\rm n}$ in (11.57d) is non-positive.

11.5. Elastic-Brittle Delamination

In contrast with Section 11.4, we will consider that the adhesive has an elastic response, which is called *elastic-brittle delamination.*

11.5.1. The model and its asymptotics to brittle delamination

Assuming a linear response of the adhesive, the possible modification of (11.54a) is

$$\mathscr{E}_{\mathbb{K}}(t,u,z) := \begin{cases} \displaystyle\sum_{i=1}^{N} \int_{\Omega_i} \mathbb{C}^{(i)}e(u){:}e\left(\frac{u}{2} + u_{\rm D}(t)\right) \mathrm{d}x & \\ \displaystyle\qquad + \int_{\Gamma_{\rm C}} \frac{1}{2} z\mathbb{K}[\![u]\!]\cdot[\![u]\!]\,\mathrm{d}S & \text{if } u = 0 \text{ on } \Gamma_{\rm D}, \\ & [\![u]\!]_{\rm n} \geq 0 \text{ on } \Gamma_{\rm C}, \\ & 0 \leq z \text{ on } \Gamma_{\rm C}, \\ \infty & \text{elsewhere,} \end{cases} \tag{11.58}$$

with $\mathbb{K}$ being a positive-definite matrix representing the elastic response of the adhesive. Note that the constraint $z \le 1$ cannot be active if the initial condition z_0 satisfies it, therefore it is not explicitly involved in $\mathscr{E}_{\mathbb{K}}$.

The classical formulation corresponding to (11.4) with $\mathscr{E} = \mathscr{E}_{\mathbb{K}}$ from (11.58) and $\mathscr{R}$ from (11.54b) consists of an equilibrium of forces on each subdomain Ω_i and three complementarity problems on $\Gamma_{\rm C}$, corresponding to the three subdifferentials in functionals $\mathscr{E}_{\mathbb{K}}$ and $\mathscr{R}$. Before shifting the Dirichlet conditions, it is (ρ is the Lagrange multiplier to the constraint $z \ge 0$):

$$\operatorname{div}\sigma = 0, \quad \sigma = \mathbb{C}^{(i)}e(u) \qquad \text{in } \Omega_i,\ i = 1,\dots,N, \tag{11.59a}$$

$$u = w_{\rm D}(t,\cdot) \qquad \text{on } \Gamma_{\rm D}, \tag{11.59b}$$

$$\sigma\nu = 0 \qquad \text{on } \Gamma_{\rm N}, \tag{11.59c}$$

$$\left.\begin{array}{l}
[\![\sigma]\!]\nu = 0 \\
\sigma\nu + z\mathbb{K}[\![u]\!] = 0 \\
[\![u]\!]_{\rm n} \ge 0, \quad \sigma_{\rm n}(u) \le 0, \quad \sigma_{\rm n}(u)[\![u]\!]_{\rm n} = 0 \\
\dot z \le 0, \quad \frac{1}{2}\mathbb{K}[\![u]\!]\cdot[\![u]\!] + \rho \le a \\
\dot z\left(\frac{1}{2}\mathbb{K}[\![u]\!]\cdot[\![u]\!] + \rho - a\right) = 0 \\
z \ge 0, \quad \rho \le 0, \quad \rho z = 0
\end{array}\right\} \text{ on } \Gamma_{\rm C}. \tag{11.59d}$$

Mathematically speaking, elastic-brittle delamination is a regularization of brittle delamination. In fact, the condition $z[\![u]\!] = 0$ in (11.54a) can obviously be modified to $\sqrt{z}[\![u]\!] = 0$ with entirely the same effect, and then, after a penalization using a quadratic penalty with an L^2-type (possibly anisotropic) norm, $(\int_{\Gamma_{\rm C}}\mathbb{K}u\cdot u\,{\rm d}S)^{1/2}$ yields exactly (11.58). Thus, one can expect convergence[v] for $\mathbb{K} \to \infty$ to brittle delamination. This has been proved in [62].

For a computer implementation, one also needs a spatial discretization. The simplest choice is P1-finite elements for u and P0-finite elements for z, assuming that all Ω_i are polyhedral and triangulated consistently on the joint boundary $\Gamma_{\rm C}$. The *mutual-recovery sequence*, cf. Remark 11.3 above,

[v]The shorthand notation $\mathbb{K} \to \infty$ means that the minimal eigenvalue of $\mathbb{K}$ goes to ∞.

can be taken as:

$$\tilde{u}_h := \Pi_{\mathscr{U},h}\tilde{u}, \qquad \tilde{z}_h := z_h \Pi_{\mathscr{Z},h}(\tilde{z}/z), \tag{11.60}$$

where $\tilde{z}(x)/z(x)$ is defined 0 if $z(x) = 0$, $h = 1/k$ denotes the mesh size, and $\Pi_{\mathscr{U},h}$ and $\Pi_{\mathscr{Z},h}$ are standard projectors on the finite-element subspace, the latter making element-wise constant averages.[w] Merging it with a time discretization, the general result follows [40]. If we consider P1-elements for z with $\Pi_{Z,h}$ the corresponding projector, we can take $\tilde{z}_h = \Pi_{Z,h}(\tilde{z} - \|z_h - z\|_{L^\infty(\Gamma_{\rm C})})^+$, where $^+$ denotes the positive part. Merging the convergence to the brittle limit $\mathbb{K} \to \infty$ with numerical approximation $(\tau, h) \to (0, 0)$ seems possible only in a rather implicit way, cf. also [61, 63] for this type of result.[x]

Proposition 11.6. *Assume* (11.52a) *and* (11.54c). *Let* $(u_\mathbb{K}, z_\mathbb{K})$ *denote the energetic solution to the problem* $(\mathscr{E}_\mathbb{K}, \mathscr{R})$ *from* (11.58) *and* (11.54b). *Let* $(u_{\mathbb{K},\tau}, z_{\mathbb{K},\tau})$ *stand for the approximate solutions obtained by the implicit semidiscretization in time (with a time step* τ*), and* $(u_{\mathbb{K},\tau,h}, z_{\mathbb{K},\tau,h})$ *for its numerical approximation constructed by this implicit time discretization and the above finite element method (FEM) discretization in space (with* h *a mesh parameter). Also let the above qualification of* $w_{\rm D}$ *and* z_0 *be satisfied. Then:*

(i) *If* $\mathbb{K} \to \infty$, *then* $(u_\mathbb{K}, z_\mathbb{K})$ *converges (in terms of subsequences) to energetic solutions to the brittle problem* $(\mathscr{E}, \mathscr{R})$ *from* (11.54a) *and* (11.54b) *in the sense* (11.26). *The same holds also for* $(u_{\mathbb{K},\tau}, z_{\mathbb{K},\tau})$ *for* $\mathbb{K} \to \infty$ *and* $\tau \to 0$.

[w] Note that the product of element-wise constant functions z_h and $\Pi_{\mathscr{Z},h}(\tilde{z}/z)$ is again element-wise constant, hence $z_h \in \mathscr{Z}_h$. As $0 \le \Pi_{\mathscr{Z},h}(\tilde{z}/z) \le 1$, we have also $0 \le \tilde{z}_h \le z_h$, hence $\mathscr{R}(\tilde{z}_h - z_h) < \infty$. As $\Pi_{\mathscr{Z},h}(\tilde{z}/z) \to \tilde{z}/z$ in any $L^p(\Gamma_{\rm C})$, $p < +\infty$, and $z_h \to z$ weakly, from (11.60) we have $\tilde{z}_h \to z(\tilde{z}/z) = \tilde{z}$ weakly* in fact in $L^\infty(\Gamma_{\rm C})$ due to the *a priori* bound of values in $[0, 1]$.

[x] More explicitly H occurring in Proposition 11.6(iii) might be supported by a local Lipschitz continuity of $(z, u) \mapsto z[\![u]\!]^2 \colon L^2(\Gamma_{\rm C}) \times W^{1/2,2}(\Omega) \to L^1(\Gamma_{\rm C})$ and by a rate of approximation by finite element (FE) discretization in these norms, cf. [58, Proposition 3.3]. If $d = 3$, this continuity is due to $\|z_1[\![u_1]\!]^2 - z_2[\![u_2]\!]^2\|_{L^1(\Gamma_{\rm C})} \le \|z_1 - z_2\|_{L^2(\Gamma_{\rm C})}\|[\![u_1]\!]\|^2_{L^4(\Gamma_{\rm C})} + 2\|z_2\|_{L^\infty(\Gamma_{\rm C})} (\|[\![u_1]\!]\|_{L^4(\Gamma_{\rm C})} + \|[\![u_2]\!]\|_{L^4(\Gamma_{\rm C})}) \|[\![u_1 - u_2]\!]\|_{L^{4/3}(\Gamma_{\rm C})}$, and then due to the continuity of the trace operator $W^{1/2,2}(\Omega) \to L^2(\Gamma_{\rm C})$. To get the rate of convergence, it seems inevitable to use a gradient theory for z. Then $H(\mathbb{K}) \sim o(|\mathbb{K}|^{-1/2})$ is expected.

(ii) *For* $\mathbb{K}$ *fixed and for* $(\tau, h) \to (0,0)$, $(u_{\mathbb{K},\tau,h}, z_{\mathbb{K},\tau,h})$ *converges (in terms of subsequences) to energetic solutions of* $(\mathscr{E}_{\mathbb{K}}, \mathscr{R})$.

(iii) *There* $H : \mathbb{R}^{d\times d} \to \mathbb{R}^+$ *converging to* 0 *sufficiently fast for* $\mathbb{K} \to \infty$ *such that the "stability criterion"* $h \leq H(\mathbb{K})$ *ensures the convergence of* $(u_{\mathbb{K},\tau,h}, z_{\mathbb{K},\tau,h})$ *(in terms of subsequences) to energetic solutions of the brittle problem* $(\mathscr{E}, \mathscr{R})$.

The different solution concepts are particularly important for the brittle limit when $\mathbb{K} \to \infty$. Lumped-parameter (i.e., zero-dimensional) examples show that the energetic solutions qualitatively differ from stress-driven type (i.e., vanishing-viscosity) solutions in the sense that the scalling of the activation threshold must be different to obtain a resonable brittle limit, cf. [64] or also [26, Sec. 4.3.2.4]. In contrast to the scaling in Proposition 11.6 which uses constant fracture toughness a in (11.54b) for energetic-solution concept and which we also used in Fig. 11.5, the mentioned examples suggest that the stress-driven-solution concepts should use the scaling

$$a \sim \frac{1}{|\mathbb{K}|} \quad \text{for the brittle limit for } \mathbb{K} \to \infty. \tag{11.61}$$

In fact, this scaling has already been investigated numerically in engineering literature for static problems close to the onset of rupture, cf. [65, Formula (16)] or [66, Formula (7)]. The rigorous proof of convergence for $\mathbb{K} \to \infty$ is however very technical and, in fact, works only in a $W^{1,1}$-regularized variant for z valued in $[0,1]$, cf. [64].

11.5.2. Numerical implementation

The theoretical developments presented up to this point provide a convenient framework for an implementable numerical scheme by discretizing the time-incremental formulation (11.18) in the space variables by standard finite element methods, recall Remark 11.3. Hence, each domain Ω_i is triangulated using elements with a mesh size h, cf. Remark 11.3. Recall that we use conforming discretizations, i.e., that two interfacial nodes belonging to the adjacent domains Ω_i and Ω_j are geometrically identical, and that the same mesh is used to approximate variables u and z. Note that, in the following, we denote by boldface letters nodal discretized variables and omit the subscripts $\mathbb{K}$ and h. Now, the finite element discretization with a suitable numbering of nodes yields a discrete incremental problem in the form

$$\left.\begin{array}{ll}\text{minimize} & (\boldsymbol{u},\boldsymbol{z}) \mapsto \mathscr{E}_h(k\tau,\boldsymbol{u},\boldsymbol{z}) + \mathscr{R}_h(\boldsymbol{z}-\boldsymbol{z}_\tau^{k-1}) \\ \text{subject to} & \boldsymbol{B}_{\rm E}\boldsymbol{u} = \boldsymbol{0}, \quad \boldsymbol{B}_{\rm I}\boldsymbol{u} \geq \boldsymbol{0}, \quad \boldsymbol{z}_\tau^{k-1} \geq \boldsymbol{z} \geq \boldsymbol{0}.\end{array}\right\} \tag{11.62}$$

Here, $\boldsymbol{u} \in \mathbb{R}^{n_u}$ stores the nodal displacements for individual subdomains and $\boldsymbol{z} \in \mathbb{R}^{n_z}$ designates the delamination parameters associated with interfacial element edges. The discretized stored energy functional has the form related to (11.58),

$$\mathscr{E}_h(t,\boldsymbol{u},\boldsymbol{z}) = \boldsymbol{u}^\top \boldsymbol{K}\left(\frac{1}{2}\boldsymbol{u} + \boldsymbol{w}_{\rm D}(t)\right) + \frac{1}{2}[\![\boldsymbol{u}]\!]^\top \boldsymbol{k}(\boldsymbol{z})[\![\boldsymbol{u}]\!], \tag{11.63}$$

where $\boldsymbol{K} = \operatorname{diag}(\boldsymbol{K}_1, \boldsymbol{K}_2, \ldots, \boldsymbol{K}_N)$ is a symmetric positive semi-definite block-diagonal stiffness matrix of order n_u (derived from $\mathbb{C}^{(i)}$), $[\![\boldsymbol{u}]\!] \in \mathbb{R}^{n_k}$ stores the displacement jumps at interfacial nodes, and $\boldsymbol{k}$ is a symmetric positive-definite interfacial stiffness matrix of order n_k, which is derived from $\mathbb{K}$ and depends linearly on $\boldsymbol{z}$. The discrete dissipation potential is expressed as

$$\mathscr{R}_h(\boldsymbol{z}) = -\boldsymbol{a}^\top \boldsymbol{z}, \tag{11.64}$$

where the entries of $\boldsymbol{a} \in \mathbb{R}^m$ store the amount of energy dissipated by the complete delamination of an interfacial element; see [67] for additional details. The constraints in problem (11.62) consist of the homogeneous Dirichlet boundary conditions prescribed at nodes specified by a full-rank $m_E \times n_u$ Boolean matrix $\boldsymbol{B}_{\rm E}$, nodal non-penetration conditions specified by a full-rank matrix $\boldsymbol{B}_{\rm I} \in \mathbb{R}^{m_I \times n_u}$ storing the corresponding components of the normal vector, and the box constraints on the internal variable.

The discrete incremental problem (11.62) represents a large-scale non-convex program (due to the $\boldsymbol{k}(\boldsymbol{z})$ term), which is very difficult to solve using a monolithic approach. Nevertheless, it can be observed that the problem is separately convex with respect to the variables $\boldsymbol{u}$ and $\boldsymbol{z}$. This directly suggests using the *alternating minimization* algorithm, proposed by Bourdin *et al.* [68] for variational models of fracture. In the current context, the algorithm is summarized in Table 11.1.

The individual sub-problems of the alternating minimization algorithm can efficiently be resolved using specialized solvers. In particular, step (11.65) now becomes a quadratic programming problem, for which optimal duality-based solvers have been recently developed [69, 70]. Owing

Table 11.1. Conceptual implementation of the alternating minimization algorithm.

(1) Set $j = 0$ and $\boldsymbol{z}^{(0)} = \boldsymbol{z}_\tau^{k-1}$

(2) Repeat

(a) Set $j = j + 1$
(b) Solve for $\boldsymbol{u}^{(j)}$:

$$\left.\begin{array}{ll}\text{minimize} & \boldsymbol{u} \mapsto \mathscr{E}_h(k\tau, \boldsymbol{u}, \boldsymbol{z}^{(j-1)}) \\ \text{subject to} & \boldsymbol{B}_{\mathrm{E}}\boldsymbol{u} = \boldsymbol{w}_{\mathrm{D}}(k\tau) \quad \boldsymbol{B}_{\mathrm{I}}\boldsymbol{u} \geq \boldsymbol{0}\end{array}\right\} \tag{11.65}$$

(c) Solve for $\boldsymbol{z}^{(j)}$:

$$\left.\begin{array}{ll}\text{minimize} & \boldsymbol{z} \mapsto \mathscr{E}_h(k\tau, \boldsymbol{u}^{(j)}, \boldsymbol{z}) + \mathscr{R}_h(\boldsymbol{z} - \boldsymbol{z}_\tau^{k-1}) \\ \text{subject to} & \boldsymbol{z}_\tau^{k-1} \geq \boldsymbol{z} \geq \boldsymbol{0}\end{array}\right\} \tag{11.66}$$

(d) Until $\|\boldsymbol{z}^{(j)} - \boldsymbol{z}^{(j-1)}\| < \eta$

(3) Set $\boldsymbol{u}_\tau^k = \boldsymbol{u}^{(j)}$ and $\boldsymbol{z}_\tau^k = \boldsymbol{z}^{(j)}$

to the piecewise constant approximation of the delamination parameters, problem (11.66) can be solved locally element-by-element in a closed form.

Note that this method allows for a non-constant $\mathbb{C}^{(i)}$ and works equally well for non-homogeneous materials. Let us, however, mention that if all $\mathbb{C}^{(i)}$ are independent of x, one can alternatively (and more efficiently) apply boundary element methods (BEM) thus combining recent advances in BEM-based solvers for the Signorini problem [71] with developments in computational materials science [72–75].

Even though the alternating minimization algorithm performs well for a wide range of computational examples, it only converges to a local minimizer of the objective function (11.62), cf. [76, 77], whereas the energetic solution concept relies on global energy minimization. To overcome this discrepancy, Mielke *et al.* [39] proposed a heuristic *back-tracking algorithm* based on the two-sided energy inequality (11.24). The resulting algorithm proceeds as shown in Table 11.2.

It should be noted that there is generally no guarantee that the algorithm will locate the global optimum of the objective function (11.62); nevertheless computational experiments suggest that it delivers solutions with lower energies than the basic alternating minimization scheme [39, 76]. An alternative approach is offered by stochastic optimization techniques [37, 38], but this comes at the expense of a substantial increase of computational cost.

Table 11.2. Conceptual implementation of an energy-based back-tracking-in-time procedure.

(1) Set $k = 1$, $\boldsymbol{z}^0 = \boldsymbol{z}^{(0)} = \mathbf{1}$

(2) Repeat

(a) Determine $\boldsymbol{z}_\tau^k$ using the alternating minimization algorithm for time t_k and initial value $\boldsymbol{z}^{(0)}$

(b) Set $\boldsymbol{z}^{(0)} = \boldsymbol{z}_\tau^k$

(c) If

$$\int_{t_\tau^{k-1}}^{t_\tau^k} \partial_t \mathscr{E}_h(t, \boldsymbol{u}_\tau^k, \boldsymbol{z}_\tau^k)\mathrm{d}t \leq \mathscr{E}_h(t_k, \boldsymbol{u}_\tau^k, \boldsymbol{z}_\tau^k) + \mathscr{R}_h(\boldsymbol{z}_\tau^{k-1} - \boldsymbol{z}_\tau^k) - \mathscr{E}_h(t_{k-1}, \boldsymbol{u}_\tau^{k-1}, \boldsymbol{z}_\tau^{k-1}) \leq \int_{t_\tau^{k-1}}^{t_\tau^k} \partial_t \mathscr{E}_h(t, \boldsymbol{u}_\tau^{k-1}, \boldsymbol{z}_\tau^{k-1})\mathrm{d}t,$$

set $k = k + 1$

(d) Else set $k = k - 1$

(e) Until $k > T/\tau$

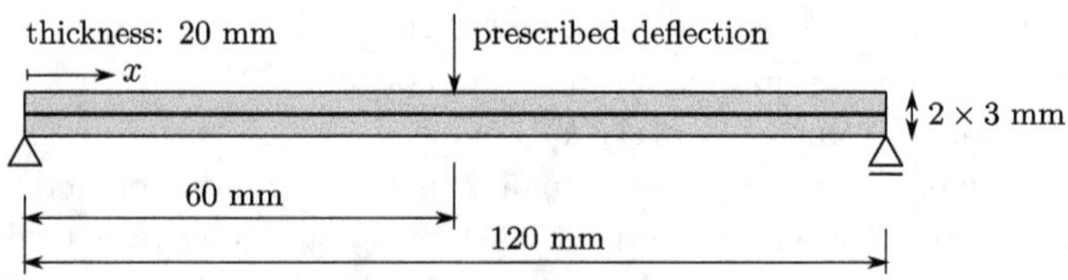

Fig. 11.2. Setup of the flexure test.

11.5.3. Illustrative examples

The basic features of the proposed solution concepts and their numerical treatment will be demonstrated by means of several two-dimensional examples.

First, the energetic-solution concept will be applied to the response of a two-layer beam in bending, imposed by a vertical displacement at the mid-span. The geometrical details of the experiment, adapted from [78], are shown in Fig. 11.2; the thickness refers to a plane-stress model used in the calculations. The elastic properties of the bulk material are characterized[y] by Young's modulus of 75 GPa and Poisson's ratio of 0.3 (corresponding to

[y]This means we use an isotropic material with $\mathbb{C}$ determined by $\mathbb{C}e{:}e = \lambda|\mathrm{trace}\, e|^2 + 2\mu|e(u)|^2$ with the so-called Lamé constants $\lambda = \nu E/((1+\nu)(1-2\nu))$ and $\mu = E/(2+2\nu)$, when E denotes Young's modulus and ν Poisson's ratio.

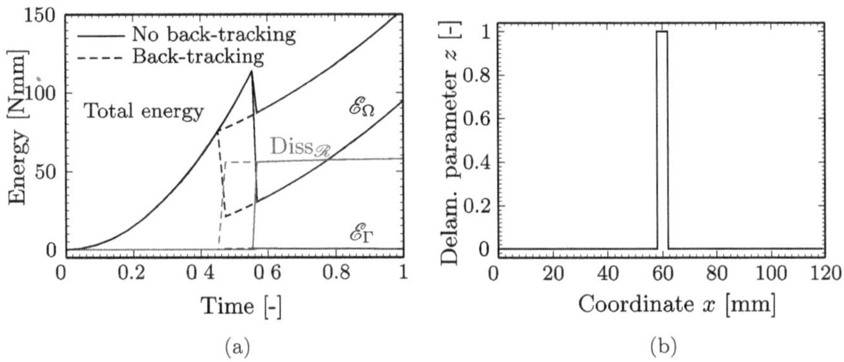

Fig. 11.3. (a) Illustration of the back-tracking procedure, (b) distribution of delamination parameter at $T = 1$; $h = 1\,\mathrm{mm}$, $\tau = 0.025$ and $\mathbb{K} = 10^5\mathbb{I}\,\mathrm{GPa/m}$; $\mathscr{E}_\Omega$ = energy stored in bulk, $\mathscr{E}_\Gamma$ = the interfacial contribution.

aluminum), the interfacial fracture toughness is set to $a = 25\,\mathrm{Jm}^{-2}$ and the maximum vertical displacement amounts to 1 mm at $T = 1$. The problem is discretized by identical isosceles right triangles with side length h and uniform time step τ. The energetics of the delamination process is shown in Fig. 11.3(a), highlighting the difference between local energy minimization and the time back-tracking scheme. In particular, the local scheme predicts initially elastic behavior, followed by almost complete delamination of the two layers at $t \approx 0.56$, accompanied by interfacial energy dissipation. However, exactly at this step the two-sided inequality is violated, as detected by the back-tracking algorithm. Inductively using this solution as the initial guess of the alternating iterative scheme, the algorithm returns to the original elastic path, thereby predicting a response leading to a lower value of the total energy for $t \in [0.46, 0.56]$. During the whole time interval, the contribution of the stored interfacial energy remains relatively small, owing to the large value of the interfacial stiffness. Notice that a small part of the interface remains intact even at $T = 1$, see Fig. 11.3(b), due to the presence of compressive traction at the mid-span. This explains why the dissipated energy in Fig. 11.3(a) saturates at a slightly smaller value than 60 Nmm, which corresponds to complete delamination. Figure 11.4(a) demonstrates the convergence of the approximate solutions for $h \to 0$. The results confirm that the overall energetic picture is almost independent of the spatial discretization, and that no spurious numerical oscillations are observed. The same conclusion holds for the force-displacement diagrams, shown in Fig. 11.4(b). Finally, the convergence of the debonding process

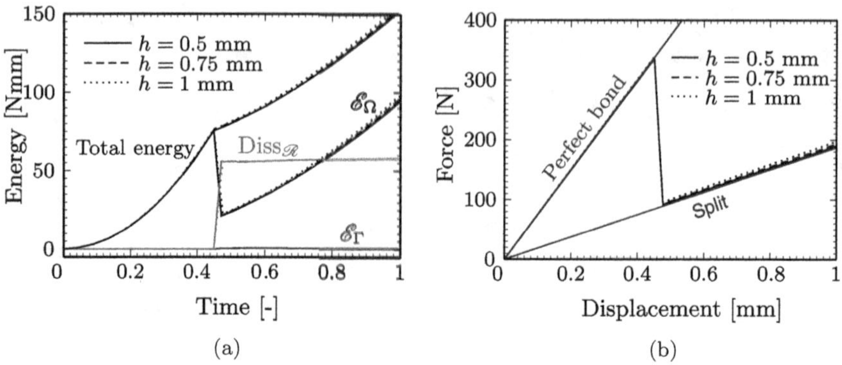

Fig. 11.4. Convergence for $h \to 0$: (a) energetics and (b) force-displacement diagram; $\tau = 0.025$ and $\mathbb{K} = 10^5\mathbb{I}$ GPa/m; $\mathscr{E}_\Omega$ = energy stored in bulk, $\mathscr{E}_\Gamma$ = the interfacial contribution.

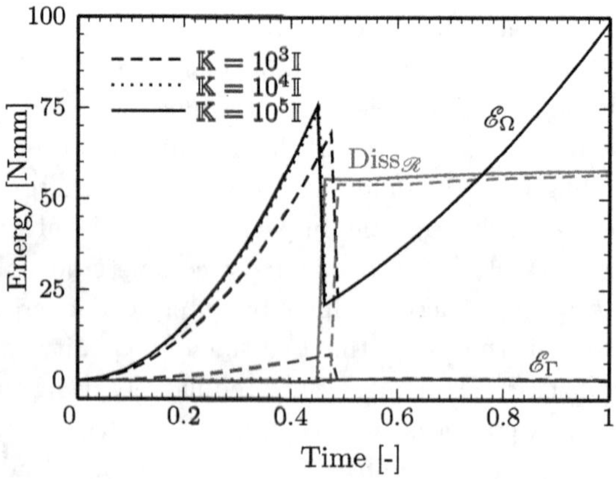

Fig. 11.5. Convergence for $\mathbb{K} \to \infty$ towards the brittle model devised in [62] based on the energetic-solution concept, justified in Proposition 11.6; $\tau = 0.025$ and $h = 0.5$ mm; $\mathscr{E}_\Omega$ = energy stored in bulk, $\mathscr{E}_\Gamma$ = the interfacial contribution.

as $\mathbb{K} \to \infty$ is illustrated in Fig. 11.5. Notice that, as with $h \to 0$, the energetics appears to be only mildly dependent on the interfacial stiffness and that, already for $\mathbb{K} \approx 10^4\mathbb{I}$ GPa/m, the FE-based solution accurately approximates the Griffith-type behavior discussed in Section 11.4.

The second example will illustrate the vanishing-viscosity approach and, in particular, the (approximate) defect measure arising from the dissipation through the (vanishing) viscosity. Let us remind that our adhesive model yields a discontinuous response of the mechanical stress

$\sigma := \partial_{[\![u]\!]}(\frac{z}{2}\mathbb{K}[\![u]\!]\cdot[\![u]\!]) = z\mathbb{K}[\![u]\!]$ within a displacement-controlled experiment. Namely, starting from an unstressed configuration, the stress linearly increases with a prescribed $[\![u]\!]$ until the driving force $\mathfrak{z} := \partial_z(\frac{z}{2}\mathbb{K}[\![u]\!]\cdot[\![u]\!]) = \frac{1}{2}\mathbb{K}[\![u]\!]\cdot[\![u]\!]$ reaches the activation threshold a used in (11.54b), then z jumps to zero and the mechanical stress jumps to zero, too; Fig. 11.6 depicts the isotropic case $\mathbb{K} = \kappa\mathbb{I}$ with some $\kappa > 0$.

The philosophy of the stress-driven-type solutions as, e.g., the vanishing-viscosity ones (in contrast to energetic solutions) is to follow the diagrams in Fig. 11.6. Typically, in our concrete delamination problem, the abstract "viscosity" operators occurring in (11.37) are considered as $V_1 = \mathscr{V}_1'$ with the potential $\mathscr{V}_1(\dot{u}) = \sum_{i=1}^{2}\int_{\Omega_i}\frac{1}{2}\mathbb{D}^{(i)}e(\dot{u}){:}e(\dot{u})\,dx$, which gives rise to the visco-elastic material in the Kelvin–Voigt rheology, and $V_2 = \mathscr{V}_2'$ with $\mathscr{V}_2(\dot{z}) = \int_{\Gamma_C}\frac{1}{2}\dot{z}^2\,\mathrm{d}S$. Vanishing viscosity both in the bulk and in the adhesive was considered in [79]. Here, for numerical experiments, we confine ourselves to the bulk (vanishing) viscosity, while the adhesive is considered merely elastic as in the first example, i.e., $\epsilon_1 > 0$ while $\epsilon_2 = 0$ in (11.37). A two-dimensional isotropic viscoelastic specimen is glued on Γ_C to the rigid obstacle, as shown in Fig. 11.7. In contrast to μ from the abstract Definition 11.4, we can now obtain a more interesting defect measure on $\bar{\Omega} \times [0, T]$ as a weak* limit of the time-and-space distribution

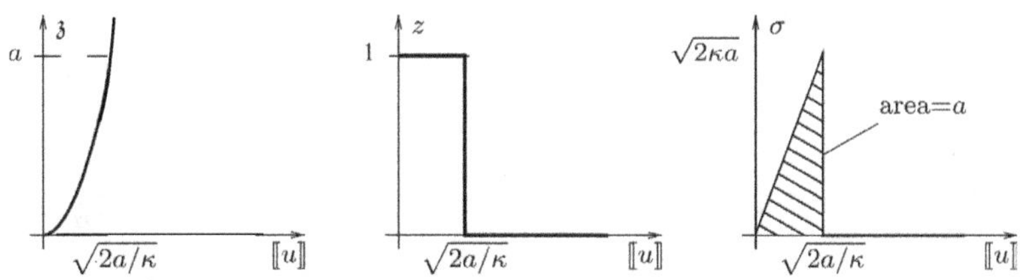

Fig. 11.6. Schematic illustration of the response of the driving force $\mathfrak{z}$, the delamination z, and the mechanical stress σ in model (11.58) with $\mathbb{K} = \kappa\mathbb{I}$ and (11.54b) during the displacement-controlled experiment; in fact, after rupture occurs, $\mathfrak{z}$ displayed schematically on the left diagram is only a selection of an available driving force from the set-valued mapping $-\partial_z\mathscr{E}(t; u, 0)$.

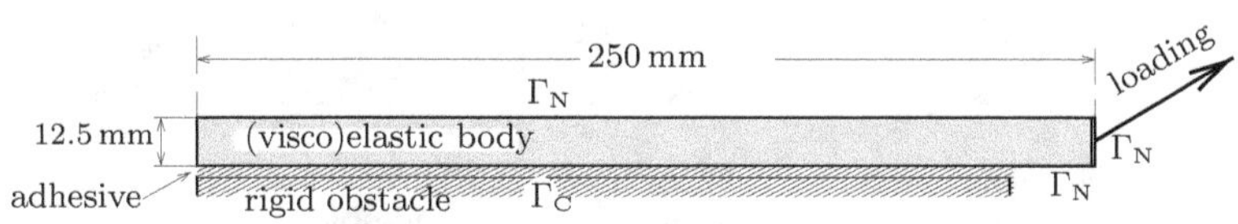

Fig. 11.7. Geometry of a two-dimensional rectangular-shaped specimens subjected to the linearly increasing loading on the right-hand side of Γ_N.

of the viscous-dissipation rate $\mu_\epsilon := \epsilon\mathbb{C}e(\dot{u}_\epsilon){:}e(\dot{u}_\epsilon)$. Evolution in time (as an integral $\int_0^t \mu_\epsilon(x,t)\mathrm{d}t$) of the spatial distribution over the specimen Ω is depicted in Fig. 11.8 until the delamination is completed. We use the fractional-step scheme (11.41) and the time step τ controlled in such a way that the energy equality (for a chosen small $\epsilon > 0$) is satisfied with a good accuracy while ϵ was made successively smaller until a reasonably good convergence of the distribution in Fig. 11.8 and also the other quantities was observed. For more results concerning this and other experiments, we refer to [53].

Computationally, the model for big $\mathbb{K}$ becomes very "stiff" and calculations based on the global minimization arising from the concept of energetic solutions become troublesome. The stress-driven concepts and

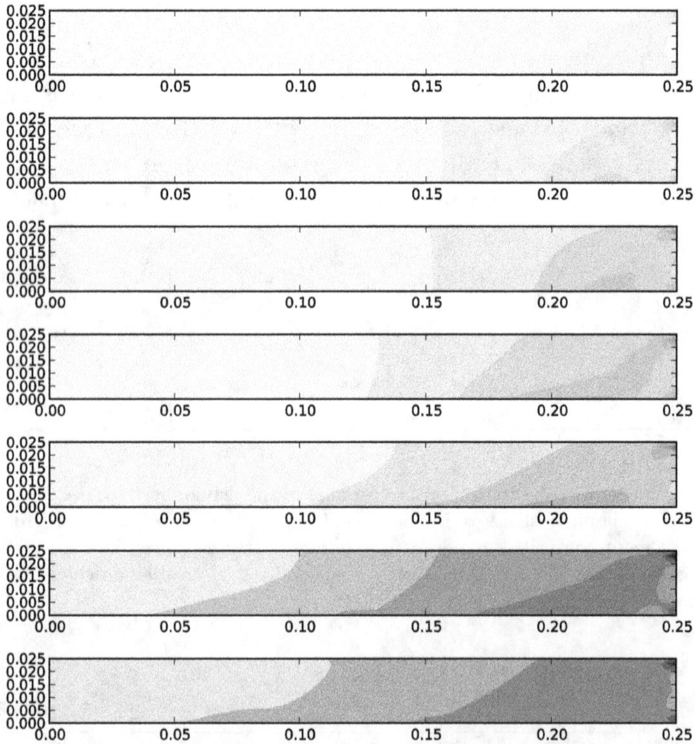

Fig. 11.8. The (approximate) spatial density of the defect measure, i.e., the distribution of the energy dissipated by (even very small) viscosity over the time interval $[0,t]$, i.e., $\int_0^t \epsilon\mathbb{C}e(\dot{u}_\epsilon){:}e(\dot{u}_\epsilon)\,\mathrm{d}t$, depicted in a gray scale in seven snaphots selected at (equi-distantly distributed) time instants t.

fractional-step discretization then become attractive both from physical/modeling as well as computational viewpoints. Some two-dimensional experiments for the geometry as in Fig. 11.23 are presented in [64], illustrating also a nice convergence under the scaling (11.61).

11.6. Various Refinements and Enhancements

The models introduced so far represent a very basic scenario, which was intentionally simplified to make the explanation of the underlying concepts easier. Engineers, however, deal with various advanced ideas not so far discussed. The goal of this section is to demonstrate that they too can be involved in this theory.

11.6.1. Cohesive contacts

The adhesive model presented in Section 11.5 yields a discontinuous response of the mechanical stress, cf. Fig. 11.6. The engineering literature often considers instead the continuous response of the mechanical stress, however. It is referred to as a *cohesive*-type *contact* and urges some modification of the above model. One simple option is to modify $\mathscr{E}_{\mathbb{K}}$ from (11.58) as follows:

$$\mathscr{E}_{\mathbb{K}_1,\mathbb{K}_2}(t,u,z) := \begin{cases} \displaystyle\sum_{i=1}^{N}\int_{\Omega_i}\mathbb{C}^{(i)}e(u){:}e\Big(\frac{u}{2}+u_{\rm D}(t)\Big)\,\mathrm{d}x \\ \displaystyle\qquad +\int_{\Gamma_{\rm C}}\frac{z\mathbb{K}_1[\![u]\!]+z^2\mathbb{K}_2[\![u]\!]}{2}\cdot[\![u]\!] \\ \displaystyle\qquad\qquad +\frac{\kappa_0}{r}|\nabla_{\rm S} z|^r\,\mathrm{d}S & \text{if } u=0 \text{ on } \Gamma_{\rm D}, \\ & [\![u]\!]_{\rm n}\ge 0 \text{ on } \Gamma_{\rm C}, \\ & 0\le z\le 1 \text{ on } \Gamma_{\rm C}, \\ \infty & \text{elsewhere,} \end{cases} \tag{11.67}$$

where we used a $(d-1)$-dimensional "surface" gradient[z] $\nabla_{\rm S}$ and assume $r > d-1$ and for mathematical reasons $\kappa_0 > 0$. This last term has a similar "non-local" effect as in the frequently used gradient theory in damage; also the analysis and especially the constructions of a mutual-recovery sequence in the sense of [40] are the same as in damage models. In particular,

[z]The notation $\nabla_{\rm S} z$ of the surface gradient stands for $\nabla z - \nu(\nu\cdot\nabla z)$.

for $1 < r \leq d-1$ one has to use a sophisticated construction from [80, 81], otherwise a simpler construction from [82] works, too. In the context of delamination, gradient theory has been used, e.g., in [41, Chapter 14] or [83, 84].

To demonstrate the response of this model, consider the isotropic adhesive response $\mathbb{K}_1 = \kappa_1 \mathbb{I}$ and $\mathbb{K}_2 = \kappa_2 \mathbb{I}$. Then, the mechanical stress σ: $= \partial_{[\![u]\!]}(\frac{1}{2}(\kappa_1 z + \kappa_2 z^2)|[\![u]\!]|^2) = (\kappa_1 z + \kappa_2 z^2)[\![u]\!]$ within a pulling experiment again linearly increases with $[\![u]\!]$ until the driving force $\mathfrak{z} := \partial_z(\frac{1}{2}(\kappa_1 z + \kappa_2 z^2)|[\![u]\!]|^2) = (\frac{1}{2}\kappa_1 + \kappa_2 z)|[\![u]\!]|^2$ reaches a, which happens for $|[\![u]\!]| = \sqrt{2a/(\kappa_1 + 2\kappa_2)}$, and then z starts evolving while holding $\mathfrak{z} = a$, i.e., $z = \frac{a}{\kappa_2}|[\![u]\!]|^{-2} - \frac{\kappa_1}{2\kappa_2}$, until it arrives at 0, which happens for $|[\![u]\!]| = \sqrt{2a/\kappa_1}$; thus the mechanical stress decays as $\sigma = (\kappa_1 z + \kappa_2 z^2)[\![u]\!] = (\frac{a^2}{\kappa_2}|[\![u]\!]|^{-4} - \frac{\kappa_1^2}{4\kappa_2})[\![u]\!]$ to zero; see Fig. 11.9. The continuous response of σ (Fig. 11.9, right) is addressed as a *cohesive-zone model*, cf., e.g., [85, 86]. As (11.58), a feature of (11.67) is that it is separately quadratic both in the u- and the z-variable, so one can advantageously use alternating minimization algorithms to solve the incremental minimization problems of the type (11.18) as in [39, 68, 77].

More generally, one can consider a continuous, increasing function $\phi: [0,1] \to \mathbb{R}^+$ with $\phi(0) = 0$, and replace the z-term under the surface integral in (11.58) by $\phi(z)|[\![u]\!]|^2$. Repeating the previous arguments, the response in a tensile experiment starting from $z = 1$ exhibits a quadratic dependence on the driving force $\mathfrak{z} = \phi'(z)|[\![u]\!]|^2$ until it reaches the activation threshold a, which happens for $|[\![u]\!]| = \sqrt{a/\phi'(1)}$. Then z starts evolving while holding $\mathfrak{z} = a$, which yields $z = [\phi']^{-1}(|[\![u]\!]|^2/a)$ and the actual stress $\sigma = 2\phi([\phi']^{-1}(|[\![u]\!]|^2/a))[\![u]\!]$, until it arrives at 0.

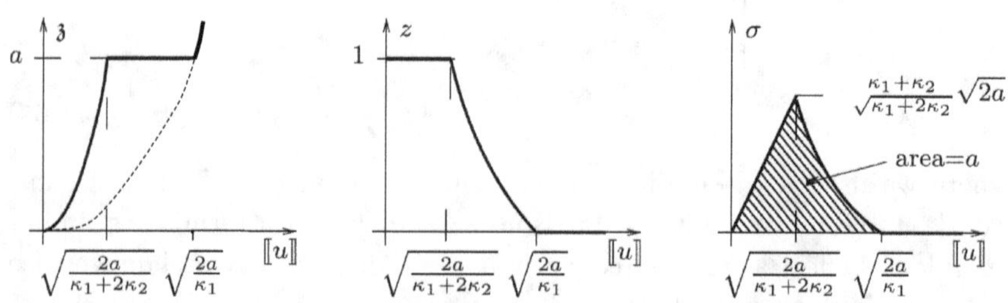

Fig. 11.9. Schematic illustration of the response of the refined model (11.67) with $\mathbb{K}_1 = \kappa_1 \mathbb{I}$ and $\mathbb{K}_2 = \kappa_2 \mathbb{I}$ and (11.54b) under the pulling experiment; again, like in Fig. 11.6-left, after the rupture is completed, the left diagram shows only schematically a selection of an available driving force rather than the actual driving force $\mathfrak{z}$.

Up to the gradient term, the equivalent effect can be obtained by a substitution of $\phi(z)$ by a new delamination variable, say ζ. This leads to the stored-energy term $\zeta|[\![u]\!]|^2$ and the dissipation distance $a|\phi^{-1}(\widetilde{\zeta})-\phi^{-1}(\zeta)|$ if $\widetilde{\zeta}\le\zeta$, which corresponds to the dissipation metric $a|\dot{\zeta}|/\phi'(\phi^{-1}(\zeta))$. Thus, in terms of this new variable, we obtain the situation

$$\text{effective activation} = \frac{a}{\phi'(\phi^{-1}(\zeta))}, \qquad \text{stress} = 2\zeta\,[\![u]\!], \tag{11.68}$$

while the driving force is $|[\![u]\!]|^2$. In fact, having an "optically" *non-associative*[aa] *model* like (11.68), one can conversely explicitly construct the dissipation metric, which is instead an exceptional situation due to the one-dimensionality and uni-directionality of the considered delamination process.

In practical applications, frequently used *cohesive zone models* (CZM) are *bilinear* or *exponential* (also called Ortiz-Pandolfi), cf. [87–89].[bb] Both of them can be obtained in the present formulation by using various functions ϕ, see [90], the following choices have simple analytical expressions:

$$\text{Bilinear:} \qquad \phi(z) = \frac{1}{2}\kappa\frac{\beta z}{1+\beta-z}, \qquad \text{with } \beta>0, \tag{11.69a}$$

$$\text{Exponential:} \qquad \phi(z) = \frac{1}{2}\kappa\mathrm{e}^{-\omega}, \qquad \text{with } z=\mathrm{e}^{-\omega}\left(1+\omega+\frac{1}{2}\omega^2\right). \tag{11.69b}$$

In the bilinear case, κ is the undamaged interface stiffness, so that in the weakening part of the traction-displacement diagram the slope is $-\beta\kappa$. Thus, for the model parameter $\beta\to+\infty$, the bilinear CZM reverts to the elastic-brittle model. Similarly, in the exponential case, κ is the original interface stiffness, i.e., the slope of the traction-displacement diagram at the origin. The relations for driving force, damage or stress can be derived as above for the formulation (11.67), their graphical representations,

[aa] Here "non-associative" means that there is no unique activation threshold associated with the dissipation mechanism. Sometimes the adjective "non-associative" instead means that the dissipative forces do not have any potential.

[bb] The adjectives of the CZMs typically refer to the resulting traction-separation law (in particular k-linear means piece-wise linear dependence of the traction on the displacement jump with k pieces). In the context of the damage model represented by the function ϕ, these adjectives may be confusing, e.g., the bilinear CZM is defined by a rational function ϕ, see (11.69a).

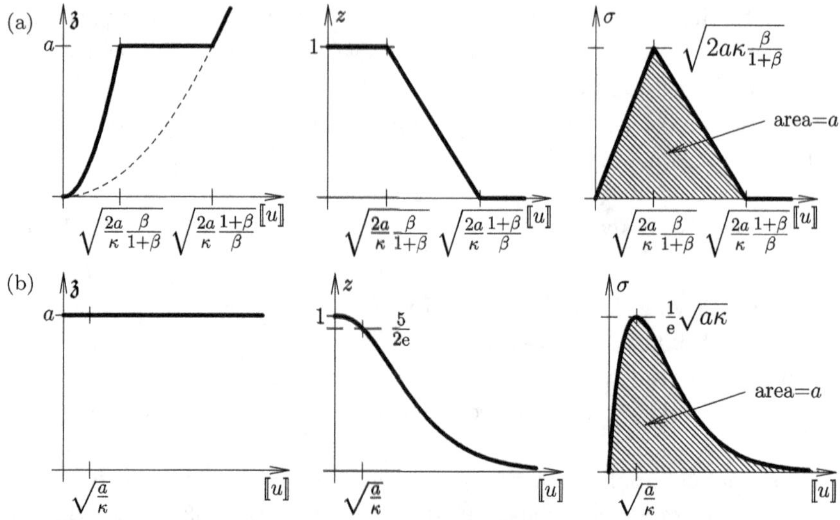

Fig. 11.10. Schematic illustration of the (a) bilinear and (b) exponential CZMs defined by functions $\phi(z)$ in (11.69). The response of the driving force $\mathfrak{z}$, the delamination z, and the mechanical stress σ during the displacement-controlled experiment. As before, after rupture occurs, $\mathfrak{z}$ displayed schematically in the left diagram is only a selection of an available driving force similarly as in Fig. 11.9.

corresponding to Fig. 11.9, are shown in Fig. 11.10. Thus, it is clear that in the exponential CZM the damage starts from the very beginning of loading ($\mathfrak{z} = a$) which may be unrealistic, moreover, it may cause troubles in numerical solutions, so that its modified version should be used as defined in [90].

Some applications of the presented CZMs for the analysis of fiber-reinforced composites at micro-scale are shown in Section 11.7.

11.6.2. Delamination in Modes I, II and mixed modes

Dissipation in the so-called Mode I (delamination by opening) is smaller than in the so-called Mode II (delamination by shearing); sometimes the difference may be tens or even hundreds of percent and under general loading it depends on the so-called *fracture-mode-mixity* angle, cf. [91–94]. Microscopically, the additional dissipation in Mode II may be explained by a plastic process both in the adhesive itself and in a narrow bulk vicinity of the delamination surface before the actual delamination starts, cf. [92, 95], or by some rough structure of the interface, cf. [96]. These plastic processes are less relevant in Mode I if the plastic strain is valued in $\mathbb{R}^{d\times d}_{\rm dev}$ = the "incompressible" (=trace free) symmetric strain, as usually considered. Modeling the narrow plastic strip around $\Gamma_{\rm C}$ is computationally difficult, and thus the various simplified phenomenological models are worth considering.

An immediate reflection of the standard engineering approach as, e.g., in [66, 97–99] is to modify (11.54b) by an activation threshold $a = a(\psi)$ depending on the so-called *fracture-mode-mixity angle* ψ. For $\mathbb{K} = {\rm diag}(\kappa_{\rm n}, \kappa_{\rm t}, \kappa_{\rm t})$, this angle can be defined as

$$\psi = \psi(\llbracket u \rrbracket) := \arctan\sqrt{\frac{\kappa_{\rm t}|\llbracket u \rrbracket_{\rm t}|^2}{\kappa_{\rm n}|\llbracket u \rrbracket_{\rm n}|^2}}, \tag{11.70}$$

where $\llbracket u \rrbracket_{\rm t}$ and $\llbracket u \rrbracket_{\rm n}$ stand for the tangential and the normal displacement jump, arising in the decomposition $\llbracket u \rrbracket = \llbracket u \rrbracket_{\rm n}\nu + \llbracket u \rrbracket_{\rm t}$ with $\llbracket u \rrbracket_{\rm n} = \llbracket u \rrbracket \cdot \nu$ with ν a unit normal to $\Gamma_{\rm C}$. Typical phenomenology is that $\kappa_{\rm t} < \kappa_{\rm n}$ (usually reaching no more than 80% of $\kappa_{\rm n}$). A typical, phenomenological form of $a(\cdot)$ used in engineering [97] is, e.g.,

$$a(\psi) := a_{\rm I}\big(1 + \tan^2((1-\lambda)\psi)\big), \tag{11.71}$$

where $a_{\rm I} = a(0)$ is the activation threshold for fracture Mode I and λ is the so-called *fracture-mode sensitivity* parameter. For example, for moderately strong fracture-mode sensitivity, which means the ratio $a_{\rm II}/a_{\rm I}$ is about 5–10 (with $a_{\rm II} = a(90°)$ being the activation threshold for the pure fracture Mode II), one has λ about 0.2–0.3; cf. [98]. Therefore, this model uses the dissipation rate from Remark 11.2 in a general form, namely

$$\mathscr{R}(u, \dot z) := \begin{cases} \displaystyle\int_{\Gamma_{\rm C}} a(\psi(\llbracket u \rrbracket))|\dot z|\,{\rm d}S & \text{if } \dot z \le 0 \text{ on } \Gamma_{\rm C}, \\ \infty & \text{otherwise.} \end{cases} \tag{11.72}$$

An immediate idea is to use a semi-implicit time discretization, leading to a modification of the incremental minimization problem (11.18) as follows:

$$\left.\begin{array}{ll} \text{minimize} & (u,z) \mapsto \mathscr{E}_{\mathbb{K}}(k\tau, u, z) + \mathscr{R}(u_\tau^{k-1}, z - z_\tau^{k-1}) \\ \text{subject to} & (u,z) \in \mathscr{U} \times \mathscr{Z}. \end{array}\right\} \tag{11.73}$$

The convergence of this method is indeed guaranteed in some cases [100], in particular when the stored energy is uniformly convex. This, however, is not the case for $\mathscr{E}_{\mathbb{K}}$. The discrete stability inequality (11.19) is modified to

$$\mathscr{E}(k\tau, u_\tau^k, z_\tau^k) \leq \mathscr{E}(k\tau, \tilde{u}, \tilde{z}) + \mathscr{R}(u_\tau^{k-1}, \tilde{z} - z_\tau^k) \tag{11.74}$$

and the following energy inequalities hold:

$$\begin{aligned}\mathscr{E}(k\tau, u_\tau^k, z_\tau^k) &+ \mathscr{R}(u_\tau^{k-1}, z_\tau^k - z_\tau^{k-1}) - \mathscr{E}((k-1)\tau, u_\tau^{k-1}, z_\tau^{k-1}) \\ &\leq \int_{(k-1)\tau}^{k\tau} \mathscr{E}_t'(t, u_\tau^{k-1}, z_\tau^{k-1})\,\mathrm{d}t\end{aligned} \tag{11.75}$$

and

$$\begin{aligned}\mathscr{E}(k\tau, u_\tau^k, z_\tau^k) &+ \mathscr{R}(u_\tau^{k-2}, z_\tau^k - z_\tau^{k-1}) - \mathscr{E}((k-1)\tau, u_\tau^{k-1}, z_\tau^{k-1}) \\ &\geq \int_{(k-1)\tau}^{k\tau} \mathscr{E}_t'(t, u_\tau^k, z_\tau^k)\,\mathrm{d}t.\end{aligned} \tag{11.76}$$

The first inequality follows from the minimality of (u_τ^k, z_τ^k) when compared with $(u_\tau^{k-1}, z_\tau^{k-1})$ while the second is implied by the discrete stability (11.74) of $(u_\tau^{k-1}, z_\tau^{k-1})$ with $(\tilde{u}, \tilde{z}) = (u_\tau^k, z_\tau^k)$. We then get only $W^{1,1}$ bounds on piecewise affine interpolants of $\{z_\tau^k\}$. Hence, concentrations of $\dot{z}$ can appear in the limit z, which is thus a function of the bounded variation only. To pass to the limit in the dissipation term one would need to enhance the sophisticated techniques developed in [101, 102] and then only get the energy inequality. There is, however, an obvious peculiarity in direct application of the previous concepts from Remark 11.2 because the dissipation distance $\mathscr{D} = \mathscr{D}_{\mathscr{R}}$ defined implicitly by (11.33) here can be evaluated explicitly as $\mathscr{D}(q_1, q_2) = \int_{\Gamma_{\rm C}} \min_{0 \leq \tilde{\psi} \leq \pi/2} a(\tilde{\psi})|z_1 - z_2|\mathrm{d}S$ if $z_2 \leq z_1$ a.e. on $\Gamma_{\rm C}$, otherwise it is infinite. The existence of an energetic solution of the model determined by $(\mathscr{E}_{\mathbb{K}}, \mathscr{D}_{\mathscr{R}})$ defined by (11.58) and (11.72) can be conventionally shown, but such solutions do not distinguish particular modes at all. This is a quite well-known effect in non-associative models, indicating that sometimes other concepts for solutions are more relevant, cf. [25, 54]. Altogether, the analysis of (11.73) (or e.g., a fully implicit modification of it) is not entirely clear. Moreover, a question remains as to whether one can indeed model the desired mode-mixity-sensitive effect in all situations in such a way. It is likely that the higher gradients in u are needed to control $[\![u]\!]$ in $C(\bar{\Gamma};\mathbb{R}^d)$ to give a good meaning to $\int_0^T \mathscr{R}(u,\dot{z})\,\mathrm{d}t$ with $\mathscr{R}(u,\dot{z})$ from (11.72), i.e., to $\int_{\bar{\Gamma}_{\mathrm{C}}} a(\psi([\![u]\!]))|\dot{z}|(\mathrm{d}S)$.[cc]

To overcome these drawbacks, but still considering an additional dissipation in Mode II, one can use the concept of viscosity from Section 11.3.3 for which the existence of solutions based on the fractional-step discretization of the type (11.41) is relatively simple,[dd] although a limit passage as in Definition 11.4 meets analytical difficulties. Again, the decoupled scheme (11.41) avoids global minimization of non-convex functionals. The efficiency of this model and its discretization is demonstrated on a two-dimensional experiment from Fig. 11.11. The response as far deformations and dissipation is concerned is depicted in Fig. 11.12. For detailed data concerning the material and more results concerning this experiments we refer to [103]. For convergence of the inviscid fractional-step time discretization we refer to [57] or also to [26, Proposition 4.3.49]. Numerical tests with appropriately formulated approximate maximum-dissipation principle are presented in [46, 57].

Alternatively, one can consider an additional inelastic process on Γ_{C}. For this, we may introduce a dissipative variable representing the "plastic" *tangential slip* s_{p} on Γ_{C}, and devise a plastic-like model with kinematic-type hardening for it, namely

$$\mathscr{Z} := L^\infty(\Gamma_{\mathrm{C}}) \times L^2(\Gamma_{\mathrm{C}};\mathbb{R}^{d-1}), \quad \mathscr{X} := L^1(\Gamma_{\mathrm{C}}) \times L^1(\Gamma_{\mathrm{C}};\mathbb{R}^{d-1}), \quad (11.77\text{a})$$

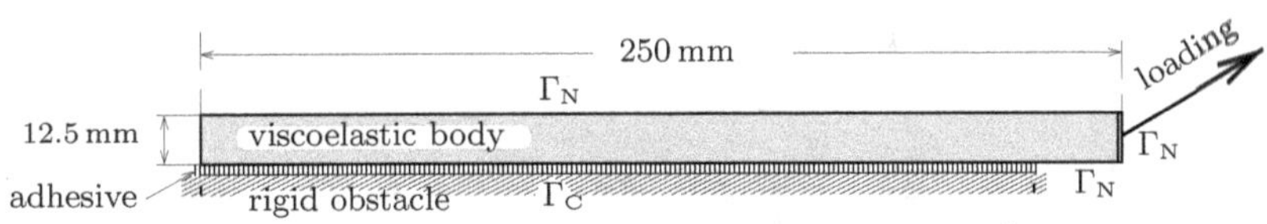

Fig. 11.11. Geometry of a two-dimensional rectangular-shaped specimen subjected to the linearly increasing loading on the right-hand side of Γ_{N} for the mixed-mode delamination experiment.

[cc]In fact, this scenario indeed works in the viscous Kelvin–Voigt rheology from Section 11.6.6, as demonstrated in [103] and in the anisothermal situation in [104].

[dd]The rate-independent delamination requires the stored energy $\mathscr{E}(t,\cdot,z)$ to have a polynomial growth bigger than space dimension (i.e., 2 or 3) which, on the other hand, does not allow for a rigorous proof of the energy conservation like (11.39), cf. [103] for details.

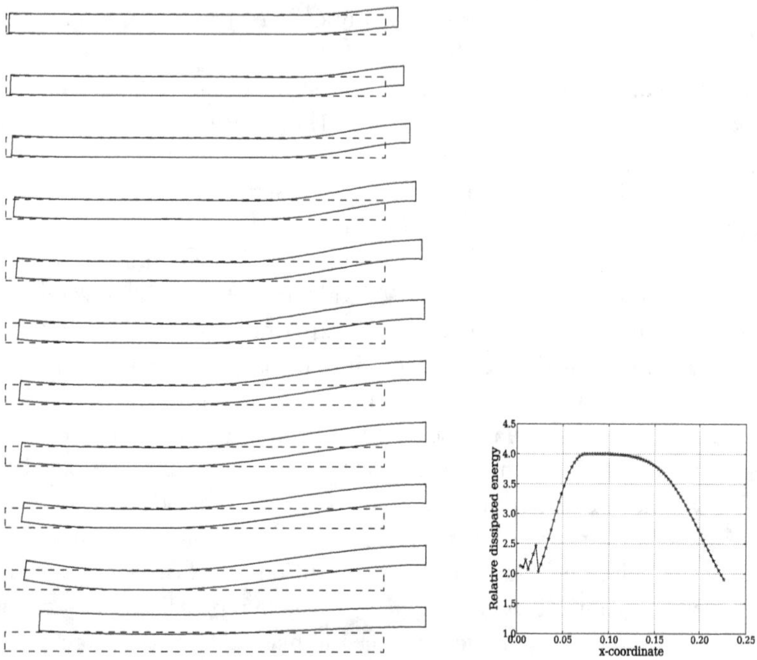

Fig. 11.12. Deformed configurations (magnified 100×) of the specimen from Fig. 11.11 depicted in selected (not equi-distributed) 11 snapshots and the final distribution along Γ_C of the (relative) dissipated energy, documenting mode variation (1 = pure Mode I, 4 = pure Mode II). Interestingly, at the end of the process, the delamination also starts rather in Mode I from the left-hand side which is opposite to the loaded right-hand side.

$$
\mathscr{E}(t,u,z,s_{\mathrm{p}}) := \begin{cases} \displaystyle\sum_{i=1}^{N}\int_{\Omega_i}\mathbb{C}^{(i)}e(u){:}e\Big(\frac{u}{2}+u_{\mathrm{D}}(t)\Big)\,\mathrm{d}x \\ \displaystyle\qquad +\int_{\Gamma_{\mathrm{C}}} z\Big(\frac{\kappa_{\mathrm{n}}}{2}|[\![u]\!]_{\mathrm{n}}|^2+\frac{\kappa_{\mathrm{t}}}{2}|[\![u]\!]_{\mathrm{t}}-s_{\mathrm{p}}|^2\Big) \\ \displaystyle\qquad +\frac{\kappa_{\mathrm{H}}}{2}|s_{\mathrm{p}}|^2+\frac{\kappa_0}{r}|\nabla_{\mathrm{S}} z|^r\,\mathrm{d}S & \text{if } u=0 \text{ on } \Gamma_{\mathrm{D}}, \\ & [\![u]\!]_{\mathrm{n}}\geq 0 \text{ on } \Gamma_{\mathrm{C}}, \\ & 0\leq z\leq 1 \text{ on } \Gamma_{\mathrm{C}}, \\ \infty & \text{otherwise,} \end{cases} \tag{11.77b}
$$

$$
\mathscr{R}(\dot z,\dot s_{\mathrm{p}}) \;\; := \;\; \begin{cases} \displaystyle\int_{\Gamma_{\mathrm{C}}} a_1|\dot z|+a_2|\dot s_{\mathrm{p}}|\mathrm{d}S & \text{if } \dot z\leq 0 \text{ a.e. on } \Gamma_{\mathrm{C}}, \\ \infty & \text{otherwise,} \end{cases} \tag{11.77c}
$$

with $[\![u]\!] = [\![u]\!]_\mathrm{n}\nu + [\![u]\!]_\mathrm{t}$, $[\![u]\!]_\mathrm{n} = [\![u]\!]\cdot\nu$ and ν a unit normal to Γ_C, while $\mathscr{U}$ is again from (11.52a). Rigorously, $s_\mathrm{p} \in L^2(\Gamma_\mathrm{C};\mathbb{R}^{d-1})$ is considered as a $(d-1)$-dimensional vector field embedded in $\mathbb{R}^d$ space to give meaning to the expression $[\![u]\!]_\mathrm{t} - s_\mathrm{p}$. As in (11.67), we again use a gradient theory for z with $r > d-1$ and $\kappa_0 > 0$ to facilitate the construction of the mutual-recovery sequence.[ee] When $d = 3$, the physical dimensions are: $[a_\mathrm{I}] = \mathrm{J/m^2}, [a_2] = \mathrm{J/m^3}$, and $[\kappa_\mathrm{t}] = [\kappa_\mathrm{n}] = [\kappa_\mathrm{H}] = \mathrm{J/m^4}$. The activation criterion to trigger delamination is now

$$\frac{1}{2}\left(\kappa_\mathrm{n}\Big|[\![u]\!]_\mathrm{n}\Big|^2 + \kappa_\mathrm{t}\Big|[\![u]\!]_\mathrm{t} - s_\mathrm{p}\Big|^2\right) \le a_\mathrm{I}. \tag{11.78}$$

Starting from the initial conditions $s_{\mathrm{p},0} = 0$ and $z_0 = 1$, the response for pure Mode I is essentially the same as in Fig. 11.6 (right), because no evolution of s_p is triggered by for opening. To analyze the response pure Mode II, realize that the tangential stress σ_t is a derivative of $\mathscr{E}$ with respect to $[\![u]\!]_\mathrm{t}$, and thus $\sigma_t(u, s_\mathrm{p}) = \kappa_\mathrm{t}([\![u]\!]_\mathrm{t} - s_\mathrm{p})$ if $z = 1$. In analogy with the plasticity, the slope of evolution of s_p under hardening is $\kappa_\mathrm{t}/(\kappa_\mathrm{t} + \kappa_\mathrm{H})$. From (11.78), one can see that delamination is triggered if $\frac{1}{2}\kappa_\mathrm{t}|[\![u]\!]_\mathrm{t} - s_\mathrm{p}|^2 = \frac{1}{2}\sigma_\mathrm{t}^2/\kappa_\mathrm{t}$ reaches the threshold a_I, i.e., if the tangential stress σ_t achieves the threshold $\sqrt{2a_\mathrm{I}\kappa_\mathrm{t}}$, as depicted in Fig. 11.13 (right). The delamination in Mode II is thus triggered under the tangential displacement

$$s_\mathrm{II} = \frac{\sqrt{2\kappa_\mathrm{t}^3 a_\mathrm{I}} - a_2\kappa_\mathrm{t} + \sqrt{2\kappa_\mathrm{t}\kappa_\mathrm{H}^2 a_\mathrm{I}}}{\kappa_\mathrm{t}\kappa_\mathrm{H}} \tag{11.79a}$$

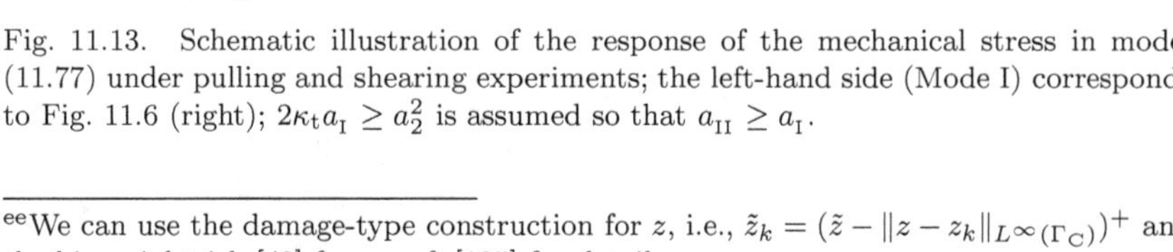

Fig. 11.13. Schematic illustration of the response of the mechanical stress in model (11.77) under pulling and shearing experiments; the left-hand side (Mode I) corresponds to Fig. 11.6 (right); $2\kappa_\mathrm{t}a_\mathrm{I} \ge a_2^2$ is assumed so that $a_\mathrm{II} \ge a_\mathrm{I}$.

[ee] We can use the damage-type construction for z, i.e., $\tilde{z}_k = (\tilde{z} - \|z - z_k\|_{L^\infty(\Gamma_\mathrm{C})})^+$ and the binomial trick [40] for s_p; cf. [105] for details.

and, after some algebra, one can see that the overall dissipated energy is

$$a_{II} = a_I + \tfrac{a_2}{\kappa_H}(\sqrt{2\kappa_t a_I} - a_2), \tag{11.79b}$$

provided $2\kappa_t a_I \geq a_2^2$. The *fracture-mode sensitivity* a_{II}/a_I is then indeed more than 1, namely $1 + a_2(\sqrt{2\kappa_t a_I} - a_2)/(\kappa_H a_I)$. The surface plastic slip stops evolving after delamination and, as used for (11.79b), only if after delamination the driving stress $\kappa_H s_{II}$ has a magnitude less than a_2. In view of (11.79a), it needs $\kappa_t a_I < 2a_2^2$. Thus, to produce the desired effects, our model should work with parameters satisfying

$$\frac{1}{2}\kappa_t a_I < a_2^2 \leq 2\kappa_t a_I. \tag{11.80}$$

The validity of this model has been tested numerically in [105, 106].

An interesting open problem is the limit passage of this model under a suitable scaling to a brittle model as in Proposition 11.6.

Let us note that in both of the mode-mixity-sensitive models considered in this section the stored energy involves z linearly. However, it is not difficult to combine *cohesive-zone*-type *models* from Section 11.6.1 with these mode-mixity-sensitive models. Thus, e.g., (11.72) can be generalized to

$$\mathscr{R}(u, z, \dot{z}) := \begin{cases} \displaystyle\int_{\Gamma_C} a(\psi([\![u]\!]), z)|\dot{z}|\,\mathrm{d}S & \text{if } \dot{z} \leq 0 \text{ on } \Gamma_C, \\ \infty & \text{otherwise.} \end{cases} \tag{11.81}$$

To facilitate the mathematical analysis, one again needs the stored energy to be augmented by the delamination gradient.

The influence of the mixed-mode behavior will be illustrated with an example of the mixed-mode flexure test [78, Sec. 3.2] shown in Fig. 11.14. The material properties of the bulk material are the same as in Section 11.5.3, the elastic-brittle interface is now characterized by the stiffnesses $\kappa_n = 810\,\text{GPa/m}$ and $\kappa_t = 760\,\text{GPa/m}$. The Mode I and Mode II fracture energies are set to $a_I = 200\,\text{Jm}^{-2}$ and $a_{II} = 900\,\text{Jm}^{-2}$, in order

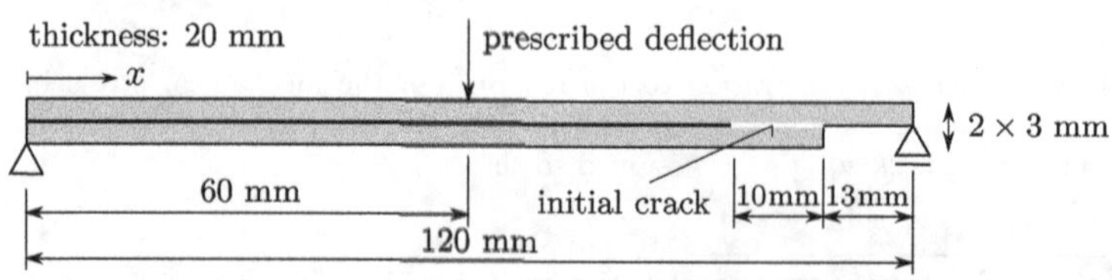

Fig. 11.14. Setup of the mixed-mode flexure test.

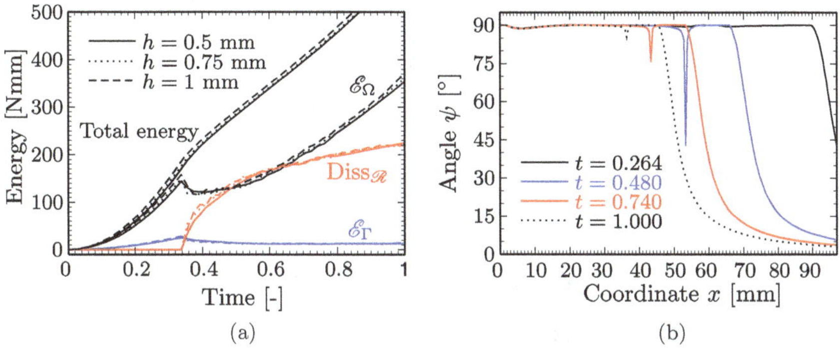

Fig. 11.15. (a) Energetics of mixed mode flexure test ($\mathscr{E}_\Omega$ = energy stored in bulk, $\mathscr{E}_\Gamma$ = interfacial contribution) and convergence for $h \to 0$, and (b) evolution of mode-mixity angles calculated for $h = 0.5\,\mathrm{mm}$ and $\tau = 0.025$ by using the energetic-solution concept.

to achieve a more ductile structural response. The prescribed mid-span displacement equals 2.5 mm at $T = 1$. The non-associative model (11.72) with (11.58) is used.

Figure 11.15(a) summarizes energetics of the delamination evolution. After the initial elastic regime, delamination initiates in a combined normal and shear mode, see Fig. 11.15(b). This is accompanied by a high increase of the dissipated energy. With the increasing load, however, the mode mixity gradually changes towards the opening mode. The production of dissipated energy decreases and the interfacial stored energy almost vanishes; see Fig. 11.16 for an illustration. The response remains almost independent of the mesh size h. Moreover, the back-tracking algorithm remained inactive for the whole loading range, which confirms the energy stability of the delamination evolution. Note that the peaks in the mode-mixity angles in Fig. 10(b) are related to the changes of the sign of the tangential slip $[\![u]\!]_\mathrm{t}$, recall Eq. (11.70).

Both the models (11.54a)–(11.72) and (11.77) are conceptually quite different, although they aim to cope with the same phenomenon of the mode-dependent delamination. For a comparison and mutual fitting of these two models when used the stress-driven solution concept we refer to [107].

Remark 11.4 (Mode III). Delamination by twisting (i.e., Mode III) exhibits specific behavior and is often also considered, though we not consider this sort of model here. In fact, it would be possible to model such regimes by making the activation threshold dependent on the angle between

Fig. 11.16. Ten snapshots of delamination evolution during the flexure test of the specimen from Fig. 11.14; displacements are magnified five times.

$\nabla_{\rm S} z$ and the tangential stress. Obviously, there needs to be compactness in terms of $\nabla_{\rm S} z$, which would have to occur "nonlinearly" in the model, so that an even higher gradient of z is involved in $\mathscr{E}$.

11.6.3. Multi-threshold delamination

Some of the engineering literature incorporates parallel breakable springs with different elastic and inelastic properties, cf. [66, 98]. On the continuum-mechanical level, this idea can be reflected by a generalization of the previous model by considering J different adhesives acting simultaneously on $\Gamma_{\rm C}$:

$$\mathscr{E}(t,u,z_1,\dots,z_J) := \begin{cases} \displaystyle\sum_{i=1}^{N}\int_{\Omega_i}\mathbb{C}^{(i)}e(u){:}e\Big(\frac{u}{2}+u_{\rm D}(t)\Big)\,\mathrm{d}x \\ \displaystyle +\int_{\Gamma_{\rm C}}\frac{\sum_{j=1}^{J}z_j\mathbb{K}_j[\![u]\!]}{2}\cdot[\![u]\!]\,\mathrm{d}S \\ \qquad\qquad\qquad \text{if } u=0 \text{ on } \Gamma_{\rm D}, \\ \qquad\qquad\qquad [\![u]\!]_{\rm n}\ge 0 \text{ on } \Gamma_{\rm C}, \\ \qquad\qquad\qquad 0\le z_j\le 1 \text{ on } \Gamma_{\rm C}, \\ \infty \qquad\qquad\quad \text{otherwise,} \end{cases} \tag{11.82a}$$

$$\mathscr{R}(\dot{z}_1,\dots,\dot{z}_I) \quad := \quad \begin{cases} \sum_{j=1}^{J} \int_{\Gamma_C} a_j |\dot{z}_j| \, dS & \text{if } \max_{j=1,\dots,J} \dot{z}_j \leq 0 \text{ on } \Gamma_C, \\ \infty & \text{otherwise.} \end{cases} \tag{11.82b}$$

Again, the advantage of the energetic formulation is that there is no problem in combining multi-threshold models with cohesive-zone models from Section 11.6.1 and/or the mode-mixity-sensitive models from Section 11.6.2. Also, instead of parameterizing the various adhesives by a discrete parameter $j = 1,\dots,J$, one could use a "continuous" parameter.

An example for implementation of multi-threshold damage uses a double cantilever beam adopted from [108] whose dimensions are shown in Fig. 11.17(a). The material parameters of the beam are $E = 176.6\,\text{GPa}$ and $\nu = 0.34$.

The initial crack of length ℓ_{ini} is situated in the specimen midplane. To predict crack propagation along the interface Γ_C, a CZM with a trilinear stress-separation law (a generalization of the bilinear one), similar to that in [108], is used to model the interface behavior. The considered stress-separation law is shown in Fig. 11.17(b). This also includes consideration of the "cohesive" term over Γ_C, i.e., replacement of the pertinent integral in (11.82a) by that from (11.67), or specifically by its equivalent with ϕ from (11.69a): $\int_{\Gamma_C} (\phi_1(z_1)[\![u]\!] + \phi_2(z_2)[\![u]\!]) \cdot [\![u]\!] \, dS$. In fact, both ϕ_1 and ϕ_2 are chosen as a scalar (damage dependent, according to the bilinear CZM (11.69a)) multiple of the initial stiffness $\mathbb{K}$, see [109].

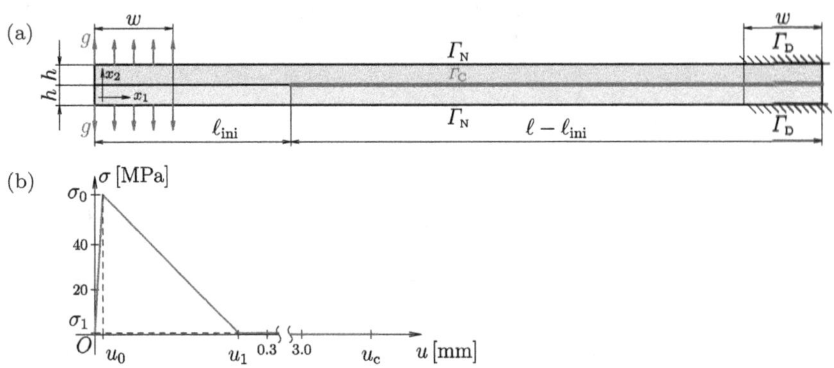

Fig. 11.17. (a) Set-up for a double cantilever beam test. Dimensions are: $\ell = 190\,\text{mm}$, $\ell_{\text{ini}} = 55\,\text{mm}$, $w = 20\,\text{mm}$, $h = 5\,\text{mm}$. (b) The stress response of the interface given by a two-threshold delamination model corresponding to a trilinear CZM. The parameters are: $u_0 = 0.014\,\text{mm}$, $u_1 = 0.25\,\text{mm}$, $u_c = 4\,\text{mm}$, $\sigma_0 = 62\,\text{MPa}$, $\sigma_1 = 0.67\,\text{MPa}$.

The adopted split of fracture energy according to (11.82b) considers $a_1 = 7.745\,\mathrm{kJm}^{-2}$ and $a_2 = 1.34\,\mathrm{kJm}^{-2}$. Such a split corresponds to crack initiation near the crack tip and subsequent crack propagation with a large process zone, as justified, e.g., for fiber composites by fiber bridging in [108].

The calculated deformations, stresses and two-threshold damages are shown in Fig. 11.18 for various time instants of the applied displacement load, where the stress-driven type solution of Section 11.3.3 was used. After initiating both levels of damage, the stress distribution close to the crack tip (point where z_1 and z_2 are still equal to one) remains almost the same. But there is a difference at least at the part where z_2 is non-vanishing. A detailed plot within the last snap-shot focuses on the stress distribution corresponding to z_2 evolution which is small but non-zero according to the stress-displacement graph in Fig. 11.17. The graphs also show that the damage parameter z_2 is initiated after z_1 has reached zero.

11.6.4. Combinations with other inelastic processes in bulk

The definite advantage of the energetic formulation (i.e., in terms of functionals $\mathscr{E}$ and $\mathscr{R}$) is that one can easily combine the above delamination models with other inelastic processes, like damage or plasticity in the bulk.

Let us illustrate this by a simple example, augmenting the model from Section 11.5 by linearized, single-threshold *plasticity* with kinematic *hardening*. The additional variable is then the plastic strain π valued in $\mathbb{R}^{d\times d}_{\mathrm{dev}} := \{\pi \in \mathbb{R}^{d\times d}, \pi^\top = \pi, \mathrm{tr}\,\pi = 0\}$, and the plastic response is determined by the convex "elasticity" domain $S^{(i)} \subset \mathbb{R}^{d\times d}_{\mathrm{dev}}$ and by a hardening tensor $\mathbb{H}^{(i)}$ on each Ω_i. After shifting the Dirichlet conditions, the functionals are as follows:

$$\mathscr{E}_{\mathbb{K}}(t,u,z,\pi) := \begin{cases} \displaystyle\sum_{i=1}^{N} \int_{\Omega_i} \mathbb{C}^{(i)}(e(u){-}\pi) : \Big(e\Big(\frac{u}{2} + u_{\mathrm{D}}(t)\Big) - \frac{\pi}{2}\Big) + \mathbb{H}^{(i)}\pi{:}\pi \,\mathrm{d}x & \\ \displaystyle\qquad + \int_{\Gamma_{\mathrm{C}}} \frac{1}{2} z\mathbb{K}[\![u]\!]\cdot[\![u]\!]\,\mathrm{d}S & \text{if } u = 0 \text{ on } \Gamma_{\mathrm{D}}, \\ & [\![u]\!]_{\mathrm{n}} \geq 0 \text{ on } \Gamma_{\mathrm{C}}, \\ & 0 \leq z \text{ on } \Gamma_{\mathrm{C}}, \\ \infty & \text{otherwise}, \end{cases} \tag{11.83a}$$

$$\mathscr{R}(\dot z,\dot\pi) := \begin{cases} \displaystyle\int_{\Gamma_{\mathrm{C}}} a|\dot z|\,\mathrm{d}S + \sum_{i=1}^{N} \int_{\Omega_i} \delta^*_{S^{(i)}}(\dot\pi)\,\mathrm{d}x & \text{if } \dot z \leq 0 \text{ on } \Gamma_{\mathrm{C}}, \\ \infty & \text{otherwise}, \end{cases} \tag{11.83b}$$

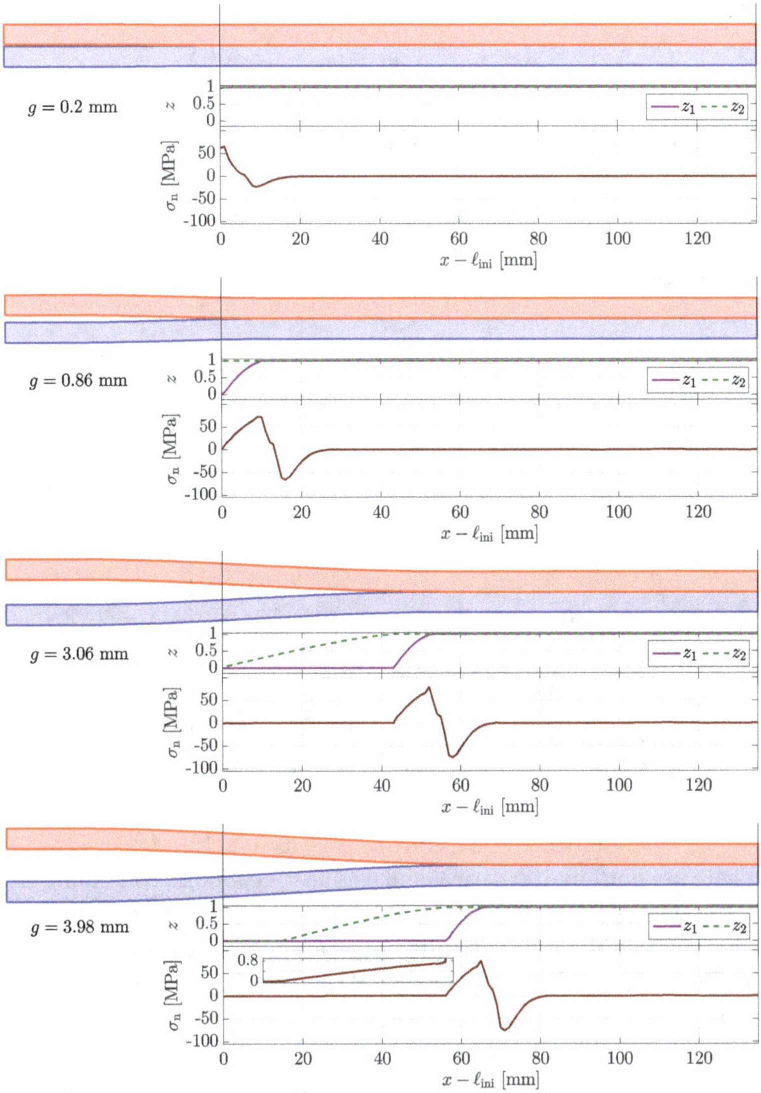

Fig. 11.18. Deformations of the double-cantilever beam, the interfacial damage evolution, and the normal stress distribution in the cracked interface at selected time instants corresponding to the prescribed displacement g, $\ell_{\rm ini}$ is the initial crack length, cf. Fig. 11.17. The last snapshot includes also a detail of the stress distribution in the part where z_1 is already zero but z_2 is positive, according to the prescribed trilinear CZM in Fig. 11.17, the maximal stress at this region being 0.67 MPa.

and, instead of (11.52b),

$$\mathscr{Z} := L^\infty(\Gamma_{\rm C}) \times L^2(\Omega; \mathbb{R}^{d\times d}_{\rm dev}). \tag{11.83c}$$

The classical formulation corresponding to (11.4) with $\mathscr{E}$ and $\mathscr{R}$ from (11.54b) is an equilibrium of forces on each subdomain Ω_i and four complementarity problems on $\Gamma_{\rm C}$, corresponding to the four subdifferentials occurring in the involved functionals $\mathscr{E}_{\mathbb{K}}$ and $\mathscr{R}$. Before shifting the Dirichlet conditions, this formulation is

$$\left.\begin{array}{l}\operatorname{div}\sigma = 0, \quad \sigma = \mathbb{C}^{(i)}(e(u) - \pi)\\ \dot\pi \in N_{S^{(i)}}(\mathbb{H}^{(i)}\pi - \sigma)\end{array}\right\} \quad \text{in } \Omega_i, \tag{11.84a}$$

$$u = w_{\rm D}(t,\cdot) \qquad \text{on } \Gamma_{\rm D}, \tag{11.84b}$$

$$\sigma\nu = 0 \qquad \text{on } \Gamma_{\rm N}, \tag{11.84c}$$

$$\left.\begin{array}{l}[\![\sigma]\!]\nu = 0\\ \sigma\nu + z\mathbb{K}[\![u]\!] = 0\\ [\![u]\!]_{\rm n} \ge 0, \quad \sigma_{\rm n}(u) \le 0, \quad \sigma_{\rm n}(u)[\![u]\!]_{\rm n} = 0\\ \dot z \le 0, \quad \frac{1}{2}\mathbb{K}[\![u]\!]\cdot[\![u]\!] + \rho \le a\\ \dot z\Big(\frac{1}{2}\mathbb{K}[\![u]\!]\cdot[\![u]\!] + \rho - a\Big) = 0\\ z \ge 0, \quad \rho \le 0, \quad \rho z = 0\end{array}\right\} \quad \text{on } \Gamma_{\rm C}, \tag{11.84d}$$

where $N_{S^{(i)}}$ denotes the normal cone to $S^{(i)}$, as in (11.8), and $i = 1, \ldots, N$.

There are both mathematical and engineering studies combining elasto-plasticity with cracks, cf., e.g., [110, 111]; this is a highly non-trivial problem because the crack path is not *a priori* prescribed, in contrast to delamination and perfect (the so-called Prandtl–Reuss) plasticity, which has been considered in [111]. A combination with damage models in the bulk is also easily possible. Recent studies [61, 80] reveal that the delamination model from Section 11.4 can be obtained as the limit when, instead of a surface $\Gamma_{\rm C}$ undergoing delamination, one considers a narrow strip of material undergoing damage and the thickness of that strip goes to zero. A challenging conjecture is whether the refined models from Section 11.6 can be justified in this way, e.g., whether considering a narrow strip of a material undergoing damage and plasticity with kinematic hardening might recover the mode-mixity-dependent model (11.77) under a suitable scaling. This could support former engineering studies as, e.g., in [95, 112].

In the static or quasi-static case, the bulk damage can approximate cracks with not a-priori prescribed paths by using the *Ambrosio–Tortorelli functional* [113]

$$\mathscr{E}(u,z)=\int_\Omega \frac{\epsilon^2+z^2}{2}\mathbb{C}e(u){:}e(u)+\frac{a_0}{4\epsilon}(1-z)^2+\epsilon a_0|\nabla z|^2\,\mathrm{d}x \tag{11.85}$$

provided $\epsilon\to 0$, imitating the philosophy that fracture is in fact a bulk damage which is eventually complete but localized on very small volumes along evolving surfaces where the fracture propagates, cf. [68, 76, 114, 115].

11.6.5. Another inelastic process on the surface: Friction

Another inelastic process that can be considered directly on the contact boundary is dry friction. It is relatively easy in bilateral contact (i.e., always in Mode II) but, if combined with the Signorini unilateral contact, it is recognized as extremely difficult in general[ff] and therefore various regularizations (often with a reasonable mechanical interpretation) are adopted.

Likely the most popular model for unilateral frictional contact in engineering employs the so-called *normal-compliance* concept, cf., e.g., [1, 2, 8, 26, 116–119]. The model including Coulomb friction then provides the force equilibrium on the contact interface $\Gamma_{\rm C}$ as follows:

$$[\![\sigma]\!]\vec{n}=0\quad\text{ with }\ \sigma=\mathbb{C}e(u), \tag{11.86a}$$

$$\sigma_{\rm n}-2\phi_{\rm n}(z)[\![u]\!]_{\rm n}-\sigma_{\rm c}=0,\ \text{ with }\ \sigma_{\rm c}:=-\kappa_{\rm g}[\![u]\!]_{\rm n}^-, \tag{11.86b}$$

$$|\sigma_{\rm f}|<-\mu(z)\sigma_{\rm c}\ \ \Rightarrow\ \ [\![\dot u]\!]_{\rm t}=0, \tag{11.86c}$$

$$|\sigma_{\rm f}|=-\mu(z)\sigma_{\rm c}\ \ \Rightarrow\ \ \exists\lambda\ge 0:\ \ \sigma_{\rm f}=\lambda[\![\dot u]\!]_{\rm t}, \tag{11.86d}$$

$$\sigma_{\rm f}:=\sigma_{\rm t}-2\phi_{\rm t}(z)[\![u]\!]_{\rm t}, \tag{11.86e}$$

with

$$\sigma_{\rm t}:=\sigma\vec{n}-\sigma_{\rm n}\vec{n}\quad\text{with}\quad\sigma_{\rm n}:=\vec{n}^\top\sigma\vec{n},$$

and the flow rule for delamination on $\Gamma_{\rm C}$:

$$a(\psi\,([\![u]\!])\,)+\phi_{\rm t}'(z)[\![u]\!]_{\rm t}^2+\phi_{\rm n}'(z)[\![u]\!]_{\rm n}^2+N_{[0,1]}(z)\ni{\rm div}_{\rm s}(\kappa_2\nabla_{\rm s}z) \tag{11.86f}$$

[ff] Often, results are available only for small friction coefficient (with the goal to prevent interesting jump regimes and without specifying what "small" means) or for a given friction (independent of the normal force), etc. Particular difficulties occur in combination of the Signorini unilateral contact with inertia as considered in Section 11.6.6).

as described in [120]. The weak formulation requires at least an additional dissipation term for Coulomb friction on Γ_C, say in dissipation functional from (11.54b) or (11.72). Thus, we may devise the following model:

$$\mathscr{E}(t,u,z) := \begin{cases} \displaystyle\sum_{i=1}^{N} \int_{\Omega_i} \mathbb{C}^{(i)} e(u){:}e\Big(\frac{u}{2} + u_D(t)\Big)\,\mathrm{d}x & \\ \displaystyle\quad + \int_{\Gamma_C} \phi_n(z)|\,[\![u]\!]_n|^2 + \phi_t(z)|\,[\![u]\!]_t|^2 & \\ \displaystyle\quad + \frac{\kappa_g}{2}|\,[\![u]\!]_n^-|^2 + \frac{\kappa_0}{r}|\nabla_S z|^r\,\mathrm{d}S & \text{if } u=0 \text{ on } \Gamma_D, \\ & 0 \le z \le 1 \text{ on } \Gamma_C, \\ \infty & \text{otherwise,} \end{cases} \tag{11.87a}$$

$$\mathscr{R}(u,z;\dot z) := \begin{cases} \displaystyle\int_{\Gamma_C} \mu(z)\kappa_g|\,[\![u]\!]_n^-|\,\|\,[\![\dot u]\!]_t\| & \\ \displaystyle\qquad + a\left(\psi\left([\![u]\!]_n\right)\right)|\dot z|\,\mathrm{d}S & \text{if } \dot z \le 0 \text{ a.e. on } \Gamma_C, \\ \infty & \text{otherwise.} \end{cases} \tag{11.87b}$$

Notice that the normal compliance penalization form (introducing the parameter κ_g) of the normal contact is used and that the friction coefficient μ is considered to be damage dependent, cf. [120]. Actually, the rigorous analysis needs still to consider a suitable viscoelastic material, in particular in the Kelvin–Voigt rheology with $\sigma = \mathbb{D}^{(i)} e(\dot u) + \mathbb{C}^{(i)} e(u)$ as in (11.88) below and then $\mathscr{R}$ in (11.87b) augments by the term $\sum_{i=1}^{N} \int_{\Omega_i} \frac{1}{2}\mathbb{D}^{(i)} e(\dot u){:}e(\dot u)\,\mathrm{d}x$.

Another regularization exploits the model (11.77): taking κ_t very large, $[\![u]\!]_t$ is then approximately equal to s_p and the activation threshold a_2 is in the position of the friction, cf. [120].

11.6.6. Dynamical adhesive contact in visco-elastic materials

So far, we have considered only quasistatic models, which have relatively broad applicability. In some situations, additional effects must be taken into account, however. In particular, even under very slow loading, spontaneous rupture of weak surfaces Γ_C may lead to the emission of elastic waves in the bulk, which may backward interact with the rate-independent delamination hosted on Γ_C. Thus inertial effects must be considered. It is natural to take into account also attenuation in the bulk. Considering the Kelvin–Voigt rheology, the simplest model from Section 11.5 thus modifies to:

$$\varrho^{(i)}\ddot{u} - \operatorname{div}\sigma = 0, \quad \sigma = \mathbb{C}^{(i)}e(u) + \mathbb{D}^{(i)}e(\dot{u}) \qquad \text{in } \Omega_i, \tag{11.88a}$$

$$u = w_{\rm D}(t,\cdot) \qquad \text{on } \Gamma_{\rm D}, \tag{11.88b}$$

$$\sigma\nu = 0 \qquad \text{on } \Gamma_{\rm N}, \tag{11.88c}$$

$$\left.\begin{aligned} &[\![\sigma]\!]\nu = 0\\ &\sigma\nu + z\mathbb{K}[\![u]\!] = 0\\ &[\![u]\!]_{\rm n} \geq 0, \quad \sigma_{\rm n}(u) \leq 0, \quad \sigma_{\rm n}(u)[\![u]\!]_{\rm n} = 0\\ &\dot{z} \leq 0, \quad \frac{1}{2}\mathbb{K}[\![u]\!]\cdot[\![u]\!] + \rho \leq a\\ &\dot{z}\Big(\frac{1}{2}\mathbb{K}[\![u]\!]\cdot[\![u]\!] + \rho - a\Big) = 0\\ &z \geq 0, \quad \rho \leq 0, \quad \rho z = 0 \end{aligned}\right\} \quad \text{on } \Gamma_{\rm C}, \tag{11.88d}$$

where $\varrho^{(i)} > 0$ is a mass density and $\mathbb{D}^{(i)}$ is a fourth-order symmetric positive definite tensor determining the attenuation of the material occupying the domain Ω_i, $i = 1,\ldots,N$. This model is analyzed by a semidiscretization in time leading to a recursive increment formula. We refer to [121] for more details when the backward Euler formula is used, or to [26, Chapter 5] or [122, 123] when the backward Euler formula possibly combined with the fractional-step strategy and a smoothening of the unilateral contact is used; in particular, the inertial term $\varrho^{(i)}\ddot{u}$ is approximated by the second-order difference $\varrho^{(i)}(u_\tau^k - 2u_\tau^{k-1} + u_\tau^{k-2})/\tau^2$.

For a slow loading, one expects that the solution (after a corresponding rescaling of time) will converge to a vanishing-viscosity solution of the rate-independent model,[gg] cf. [124] where a viscosity in the adhesive has been also considered.

For an analogous model but with viscous (instead of activated rate-independent) adhesion we refer to [2, Chapter 5] or, e.g., [1, 125], and, with plasticity in the bulk, to [126].

The typical application of dynamic adhesive contact is in the modeling of spontaneous rupture on lithospheric faults (i.e., weak surfaces in the language of mechanical engineers) with the emission of elastic waves having the capability to trigger, e.g., another possible rupture on an adjacent fault and inelastic damage, examples of which occur on the Earth's surface

[gg] The proof of this intuitive phenomenon is however not trivial due to the inertial term interacting with a non-convex potential $\mathscr{E}(t,\cdot)$, as emphasized already in [26, Sec. 5.1.2.2].

as *earthquakes*. Typical rupture processes run in pure Mode II because the enormous gravity pressures on the faults deep under the Earth's surface exclude Mode I. Thus, Signorini contact might even be *a priori* simplified to the condition $[\![u]\!]_n = 0$ on Γ_C. An important phenomenon that facilitates spontaneous rupture within earthquake modeling is the so-called *slip weakening*; cf. models and discussion, e.g., in [127–133]. There are several options for how to describe the weakening phenomena within the models presented above. One can, e.g., modify model (11.77) by considering the activation threshold a_I in (11.77c) dependent on the plastic slip s_p or, perhaps more physically, on $\int_0^t |\dot{s}_p| dt$, which then makes the dissipation non-associative. This option is capable of modeling one particular rupturing. For repeated rupturing (combined with healing), a dependence of both a_1, and a_2 in (11.77c) on delamination parameter z can be considered instead, cf. also [4].

Another possibility arises if $\mathscr{E}(t, u, \cdot)$ is concave. In model (11.67) in Section 11.6.1 with $\mathbb{K}_1 = \kappa_1 \mathbb{I}$ and $\mathbb{K}_2 = \kappa_2 \mathbb{I}$, this occurs if κ_2 is small, namely $0 > \kappa_2 > -\kappa_1/2$. Then Fig. 11.6 is relevant but the rupture happens under the mechanical stress $\sigma = (\kappa_1 + \kappa_2)\sqrt{2a/(\kappa_1 + 2\kappa_2)}$ when $|[\![u]\!]| = \sqrt{2a/(\kappa_1 + 2\kappa_2)}$ as in Fig. 11.9. Thus, one can model *delamination weakening*. Note that the initial stress σ leading to delamination may be made very large by sending $\kappa_2 \downarrow -\kappa_1/2$, even if the total dissipated energy may be independently moderate. One can see the weakening effect also from (11.68): if $\kappa_2 < 0$ then ϕ is strictly concave hence ϕ' is decreasing and thus the effective activation threshold (11.68) gets smaller if η decreases (ranging over a monotone branch of ϕ of course).

The weakening phenomenon may also occur in the multi-threshold models from Section 11.6.3: consider $\mathbb{K}_1 > \mathbb{K}_2 > \cdots$ (=ordering of positive definite matrices) and $a_1 > a_2 > \cdots$; initially the first adhesive layer can withstand high stress but when debonded the next adhesive layers can withstand smaller and smaller stresses.

It is well known that the mentioned backward Euler discretization of the inertial term $\varrho\ddot{u}$ exhibits unacceptably large artificial numerical attenuation

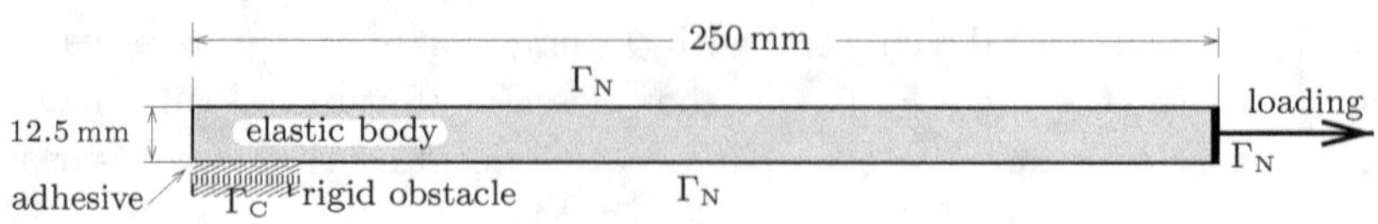

Fig. 11.19. Geometry of a two-dimensional rectangular-shaped specimen subjected to the loading $f_1 = f_1(t)$ on the right-hand side of Γ_N monotonically increasing until rupture is completed.

and is mainly useful for theoretical analysis. Computationally, efficient approaches for so-called transient or wave propagation problems (i.e. low-frequency vibrations or high-frequency waves, respectively) use various implicit or explicit schemes. An interesting implicit scheme conserving energy relies on replacing the second-order equation (11.88a) by the system

$$\dot{u} = v, \quad \varrho^{(i)}\dot{v} - \operatorname{div}\sigma = 0, \quad \sigma = \mathbb{C}^{(i)}e(u) + \mathbb{D}^{(i)}e(v) \qquad \text{in } \Omega_i \tag{11.89}$$

which is then discretized in time by the *Crank-Nicolson scheme*,[hh] originally designed for a single parabolic equation [136]. When combined with the fractional-step splitting, this results to

$$\frac{u_\tau^k - u_\tau^{k-1}}{\tau} = \frac{v_\tau^k + v_\tau^{k-1}}{2} \quad \text{and} \quad \varrho^{(i)}\frac{v_\tau^k - v_\tau^{k-1}}{\tau} = \operatorname{div}\frac{\sigma_\tau^k + \sigma_\tau^{k-1}}{2} \tag{11.90a}$$

with $\sigma_\tau^k = \mathbb{C}^{(i)}e(u_\tau^k) + \mathbb{D}^{(i)}e(v_\tau^k)$, accompanied by the boundary conditions on Γ_{C} using z_τ^{k-1}, i.e.,

$$\frac{\sigma_\tau^k + \sigma_\tau^{k-1}}{2}\nu + z_\tau^{k-1}\mathbb{K}\left[\!\!\left[\frac{u_\tau^k + u_\tau^{k-1}}{2}\right]\!\!\right] = 0 \tag{11.90b}$$

and, when (u_τ^k, v_τ^k) is obtained, to be followed by the second step to obtain z_τ^k in the same way as used for the fractional-step splitting before, cf. (11.41b). Combining the binomial formula relying on the quadratic structure of $\mathscr{E}(t,\cdot,z)$ and $\mathscr{V}_1$ as well as of the kinetic energy with the cancellation effect as we already said for (11.42)–(11.43), the upper-energy estimate always holds. In "perfectly" cohesive (like (11.67) with $\mathbb{K}_1 = 0$) bilateral contact analogous to the bulk model (11.85) where no constraints would be needed, we would obtain even the full energy conservation in this discrete scheme. In our adhesive unilateral contact when $\mathscr{E}(t,\cdot,\cdot)$ involves constraints (11.1) violating the separate-quadratic structure, one can make a penalty-like approximation (i.e., in particular the normal-compliance approximation of the unilateral contact) and then time discretization by using difference quotients as in [137], which again yield the full energy conservation in this discrete scheme, cf. [138] for details.

A numerical experiment with a two-dimensional rectangular-shaped specimen from Fig. 11.19 can show a wave emitted by the rupture, as displayed in Fig. 11.20. This wave is then reflected on the right-hand side of the specimen and propagates back to the left-hand side where it is again

[hh] In fact, this is a special case of a popular Hilber–Hughes–Taylor scheme [134] which generalizes an older scheme by Newmark [135].

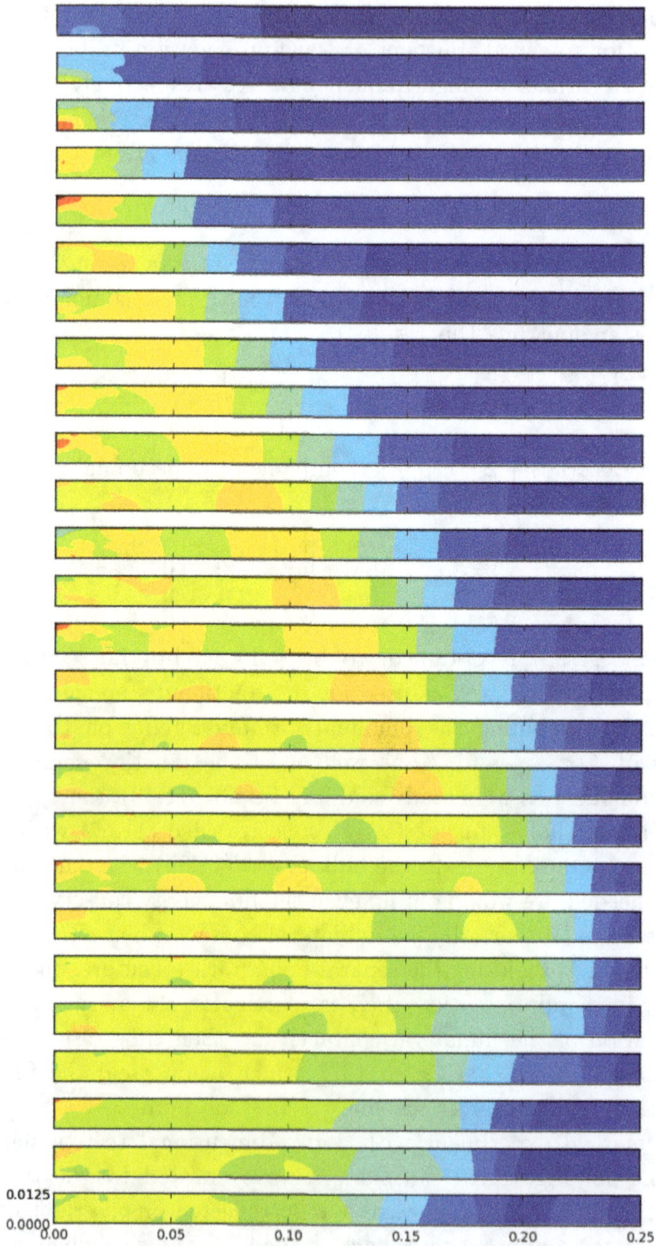

Fig. 11.20. A wave emitted by the sudden adhesive rupture of the adhesive boundary in Fig. 11.19 and propagating through the specimen and then reflected on the right-hand side back. Magnitude of the velocity is depicted at 25 snapshots.

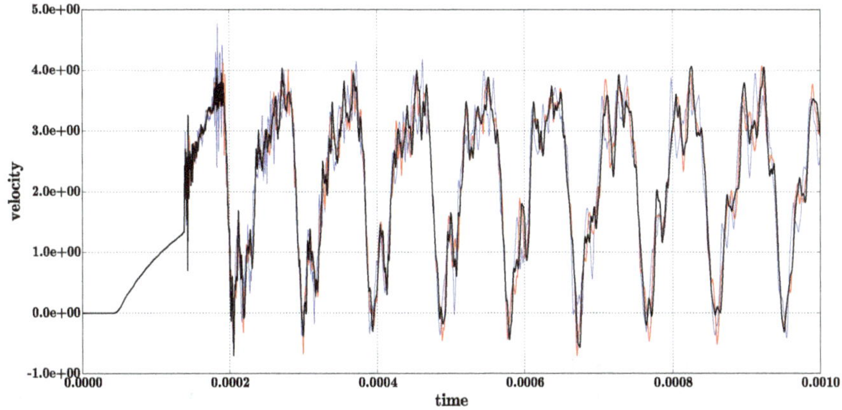

Fig. 11.21. Non-attenuating vibrations after the rupture showing both the lowest eigenfrequency of the bar from Fig. 11.19 superposed by various higher-frequency vibrations. The time-evolution of the velocity of the left-hand side of the specimen is depicted.

reflected and so on. The time varying velocity of the left-hand side of the specimen from Fig. 11.19 is displayed in Fig. 11.21. For other experiments as well as for numerical stability and convergence of the scheme (11.90) combined with the fractional-step splitting of the delamination we refer to [138].

The fractional-step Crank–Nicolson scheme (11.90) can advantageously be used for the separately-quadratic model (11.85) for fixed $\epsilon > 0$ combined both with the inertia, cf. [139] where it was further combined with some diffusive processes and heat transfer.

Another phenomenon in the bulk that can easily be modeled is *creep*. We again consider an internal strain variable π and the stored energy $\mathscr{E}_{\mathbb{K}}$ from (11.83a) but, instead of $\mathscr{R}$ from (11.83b), we take

$$
\mathscr{R}(\dot{u},\dot{z},\dot{\pi}) = \sum_{i=1}^{N} \int_{\Omega_i} \Big(\mathbb{D}^{(i)}(e(\dot{u})-\dot{\pi}) : (e(\dot{u})-\dot{\pi}) + \mathbb{D}_{\mathrm{M}}^{(i)}\dot{\pi}:\dot{\pi} \Big)\,\mathrm{d}x
+ \begin{cases} \displaystyle\int_{\Gamma_{\mathrm{C}}} a|\dot{z}|\,\mathrm{d}S & \text{if } \dot{z} \le 0 \text{ on } \Gamma_{\mathrm{C}}, \\ \infty & \text{otherwise}, \end{cases} \tag{11.91}
$$

with some positive definite tensors $\mathbb{D}_{\mathrm{M}}^{(i)}$ which determines so-called *Maxwell rheology*. In fact, the model (11.83a) and (11.91) combines Kelvin–Voigt and Maxwell rheologies, which is sometimes called a Jeffreys viscoelastic rheology. Without inertia and in homogeneous isotropic cases where, in addition $\mathbb{D}^{(i)} = \chi\mathbb{C}^{(i)}$ and $\mathbb{D}_{\mathrm{M}}^{(i)} = \chi_{\mathrm{M}}\mathbb{C}^{(i)}$ for some fixed relaxation-time constants $\chi > 0$ and $\chi_{\mathrm{M}} > 0$, after a suitable computationally "cheap" transformation, one can use conventional BEM methods for spatial discretization, cf. [140] also for some other viscoelastic rheologies.

11.6.7. Thermodynamics of adhesive contacts

Interesting features might be triggered in non-isothermal situations. Mechanical stresses in thermally expanding materials can be created by spatially varying temperature profiles. Also, merging materials with different thermal expansion coefficients (as is typical in laminated composites) creates mechanical stresses even with a spatially equilibrated temperature. Such a thermomechanical load may lead to delamination. This may naturally influence heat transfer through the delaminated surfaces. Hence, besides the usual thermomechanical coupling due to viscous dissipation and thermal expansion in the bulk, coupling by delamination also occurs.

Focusing again on the Kelvin–Voigt rheology as in Section 11.6.6, the thermodynamically consistent model, naturally involving the additional variable θ in the bulk as temperature, is

$$\left.\begin{aligned} &\varrho\ddot{u} - \operatorname{div}\sigma = 0\\ &\sigma = \mathbb{C}^{(i)}(e(u) - \theta\mathbb{E}^{(i)}) + \mathbb{D}^{(i)}e(\dot{u})\\ &c^{(i)}(\theta)\dot{\theta} - \operatorname{div}(\mathbb{L}^{(i)}\nabla\theta) = (\mathbb{D}^{(i)}e(\dot{u}) - \theta\mathbb{C}^{(i)}\mathbb{E}^{(i)}){:}e(\dot{u}) \end{aligned}\right\}\ \text{in } \Omega_i, \tag{11.92a}$$

$$u = w_{\mathrm{D}}(t,\cdot) \qquad \text{on } \Gamma_{\mathrm{D}}, \tag{11.92b}$$

$$\sigma\nu = 0 \qquad \text{on } \Gamma_{\mathrm{N}}, \tag{11.92c}$$

$$\left.\begin{aligned} &[\![\sigma]\!]\nu = 0\\ &\sigma\nu + z\mathbb{K}[\![u]\!] = 0\\ &[\![u]\!]_{\mathrm{n}} \geq 0, \quad \sigma_{\mathrm{n}}(u) \leq 0, \quad \sigma_{\mathrm{n}}(u)[\![u]\!]_{\mathrm{n}} = 0\\ &\dot{z} \leq 0, \quad \frac{1}{2}\mathbb{K}[\![u]\!]\cdot[\![u]\!] + \rho \leq a = a_0 + a_1\\ &\dot{z}\Big(\frac{1}{2}\mathbb{K}[\![u]\!]\cdot[\![u]\!] + \rho - a\Big) = 0\\ &z \geq 0, \quad \rho \leq 0, \quad \rho z = 0\\ &[\![\mathbb{L}\nabla\theta]\!]\cdot\nu = -a_1\dot{z}\\ &\langle\langle\mathbb{L}\nabla\theta\cdot\nu\rangle\rangle + \eta([\![u]\!], z)[\![\theta]\!] = 0 \end{aligned}\right\}\ \text{on } \Gamma_{\mathrm{C}}, \tag{11.92d}$$

where $\langle\langle\cdot\rangle\rangle$ denotes the average of traces from both sides of $\Gamma_{\rm C}$, and where $\mathbb{E}^{(i)}$ is the matrix of thermal expansion coefficient, which may depend on Ω_i as $\mathbb{C}^{(i)}$ and $\mathbb{D}^{(i)}$. Further $c^{(i)} = c^{(i)}(\theta)$ is the heat capacity, and $\mathbb{L}^{(i)}$ is the positive-definite heat-conductivity tensor, and $\eta = \eta([\![u]\!], z)$ is the heat-transfer coefficient through the delaminating boundary $\Gamma_{\rm C}$.

Note that only a part $a_1/a = a_1/(a_1 + a_0)$ of the mechanical energy dissipated during delamination contributes to heat production on $\Gamma_{\rm C}$, while the rest a_0/a is irreversibly stored (=dissipated) in the debonded adhesive without contributing to the heat balance.

Mode II causes considerably more heating than Mode I, as experimentally documented in [141]. For example, bearing in mind (11.71), one may consider the splitting

$$a_0(\psi) := a_{\rm I}, \quad a_1(\psi) := a_{\rm I} \tan^2((1-\lambda)\psi), \tag{11.93}$$

which indicates that Mode I delamination does not contribute to heat production at all, and only the additional dissipation for Mode II contributes to heat production on the delaminating surface.

For model (11.77), it would be natural to involve the dissipation via $a_2|\dot{s}_{\rm d}|$ as a measure-valued heat source acting on $\Gamma_{\rm C}$ while the dissipation by pure opening along the surface, i.e., $a_{\rm I}|\dot{z}|$, would contribute instead to the stored energy. Mathematical analysis of such a problem, as well as numerical experiments, seem challenging. For mode-mixity-independent dissipation, this model has been analyzed in [142]; for Griffith-type rate-dependent adhesion see also [143]. Recently, mode-mixity-sensitive dissipation (e.g., of the type (11.71)) was also scrutinized in [104] by implementing the concept of non-simple materials. The limit passage from adhesive to brittle delamination in the spirit of Proposition 11.6 but in the context of thermo-visco-elasticity like (11.92) was analyzed in [144] without inertia effects.

In many situations, the bulk is nearly isothermal while a substantial variation of temperature occurs only on the adhesive contact or its vicinity. Then it is computationally advantageous to introduce a "surface temperature" in addition to the bulk temperature, cf. [143, 145] or also [26, Sec. 5.3.3.3]. This idea occurs also in physics in the context of surfaces arising as shear bands during plastification [146]. If the bulk "reservoir" is big, the bulk temperature can possibly be even ignored, being considered constant, and then advantageously BEM can be exploited, cf. [120]. An application of this approach can be for the popular Dieterich–Ruina rate-and-state-dependent friction model of fault slip during earthquakes

[147, 148] where the delamination variable can undergo healing and is interpreted as aging, while the important weakening effect can alternatively be achieved when introducing the surface temperature [149] that can be related to the so-called flash heating during large earthquakes.

11.7. Applications to Fiber-Reinforced Composites

Although the focus of this chapter is on the modeling of macroscopic delamination problems, the presented framework can easily be adapted, e.g., to the homogenization of composites with debonding interfaces. In the analysis, we use both the energetic solution described in Section 11.3.2 and the stress-driven solution of Section 11.3.3. Moreover, for the interface, the models with cohesive contacts described in Section 11.6.1 are implemented.

First, we use the energetic-solution concept. In the context of two-scale homogenization, the stored energy associated with the unit cell problem is [150, 151]:

$$\mathscr{E}_{\mathbb{K}}(t,u,z) := \begin{cases} \displaystyle\sum_{i=1}^{N}\frac{1}{2}\int_{\Omega_i}\mathbb{C}^{(i)}(e(u)+E(t)):(e(u)+E(t))\,\mathrm{d}x \\ \displaystyle +\int_{\Gamma_{\mathrm{C}}}\frac{\phi(z)}{2}\mathbb{K}[\![u]\!]\cdot[\![u]\!]+\frac{\kappa_0}{r}|\nabla_{\mathrm{s}}z|^r\,\mathrm{d}S & \text{if } u \text{ is } \Omega\text{-periodic}, \\ & \displaystyle\int_{\Omega}u\,\mathrm{d}x=0, \\ & [\![u]\!]_{\mathrm{n}}\geq 0 \text{ on } \Gamma_{\mathrm{C}}, \\ & 0\leq z \text{ on } \Gamma_{\mathrm{C}}, \\ \infty & \text{otherwise}, \end{cases}$$

where $E(t)$ designates the macroscopic strain tensor, u now denotes a periodic microscopic displacement, and the function ϕ is used to model the cohesive contact with a piecewise linear traction-separation law as described e.g., by Fig. 11.10(a). The dissipative potential (11.54b) remains unchanged, as well as the numerical treatment of the incremental problem. The periodic boundary conditions and the macroscopic strain are incorporated by the Lagrange multipliers technique introduced, e.g., in [152]. As an example, we consider a cross-section of a unit cell of a fiber-reinforced composite like that shown in Fig. 11.22(a) but finite one, subject to a bi-axial macroscopic stretching $E_{11} = E_{22} = 1.5\%$ at $T = 1$. The geometry of the problem is defined by the fiber volume fraction equal to 50% and the diameter of the fiber is taken as 10 µm. The material data for individual phases and

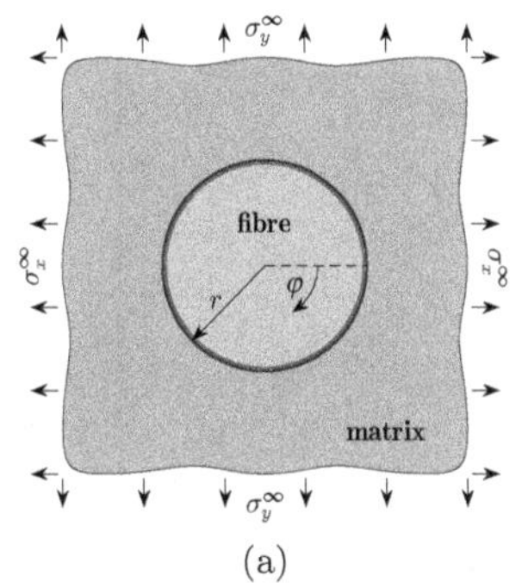

Fig. 11.22. (a) A single-fiber debonding problem under a biaxial far-field load, and (b) a fiber-bundle debonding problem (representing local geometry of fibers in a fiber-reinforced composite) under a far-field tension load.

Table 11.3. Material data.

Matrix Young's modulus	1 MPa
Matrix Poisson's ratio	0.4
Fiber Young's modulus	150 MPa
Fiber Poisson's ratio	0.3
Interfacial fracture energy, a	$0.02\ \mathrm{J/m^2}$
Interfacial elastic stiffnesses, $\kappa_{\rm n} = \kappa_{\rm t}$	0.5 GPa/m
Interfacial cohesive contact function	$\phi(z) = z/(1 - z + 10^{-4})$

interfaces appear in Table 11.3; the gradient term was neglected by setting $\kappa_0 = 0$.

The energetics of the progressive debonding is shown in Fig. 11.23, together with representative snapshots of the debonding evolution. Due to the prescribed cohesive law, we capture the gradual transition from a stiff elastic interface, i.e., when the highest values of stress exists in the fiber, to the completely debonded configuration. In this situation, the whole load is carried by the matrix phase and the stored interfacial energy drops to zero.

This simple study is complemented with a numerical simulation of debonding evolution in a complex 20-particle unit cell subject to macroscopic shear $E_{12} = 1\%$. The results in Fig. 11.24 confirm that the energetic approach, combined with robust duality-based solvers, captures the complex mechanisms of multiple contact, sliding, and gradual debonding between fibers and matrix in geometrically complicated real-world material samples.

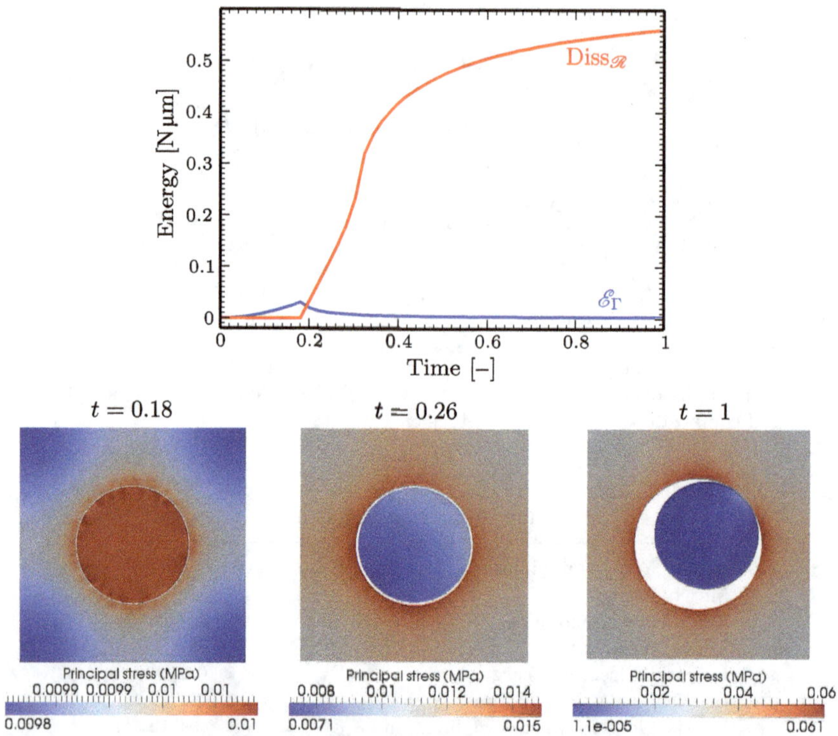

Fig. 11.23. Energetics of single-fiber debonding and three selected snapshots of displacement (magnified 20×) showing the spatial distribution of stress.

The stress-driven type of solution from Section 11.3.3 is used in the next couple of tests typical for fiber-reinforced composites. The results will be presented for the two variants shown in Fig. 11.22, see [99, 153, 154], accordingly the material properties of a common glass(1)/epoxy(2) composite are: $E_1 = 70.8\,\text{GPa}$, $E_2 = 2.79\,\text{GPa}$, $\nu_1 = 0.22$, $\nu_2 = 0.33$, fiber radius $r = 7.5\,\mu\text{m}$. The properties of the interface which is considered to obey the bilinear CZM are: the fracture energy $a = 2\,\text{Jm}^{-2}$, the maximal stress $\sigma_\text{c} = 90\,\text{MPa}$, see Fig. 11.10(a), and the parameter $\beta = 0.1$. In order to achieve a more ductile structural response, the data in the solved example also include dependency on the fracture-mode-mixity as it is shown in Section 11.6.2 Eqs. (11.70), (11.71), (11.72). Thus, in fact a is a_I and σ_c is $\sigma_{\text{n}\,\text{c}}$. Additionally, we need to define also the fracture energy in Mode II $a_\text{II} = 13.66\,\text{Jm}^{-2}$ and the initial tangential stiffness $\kappa_\text{t} = 0.25\kappa_\text{n}$.

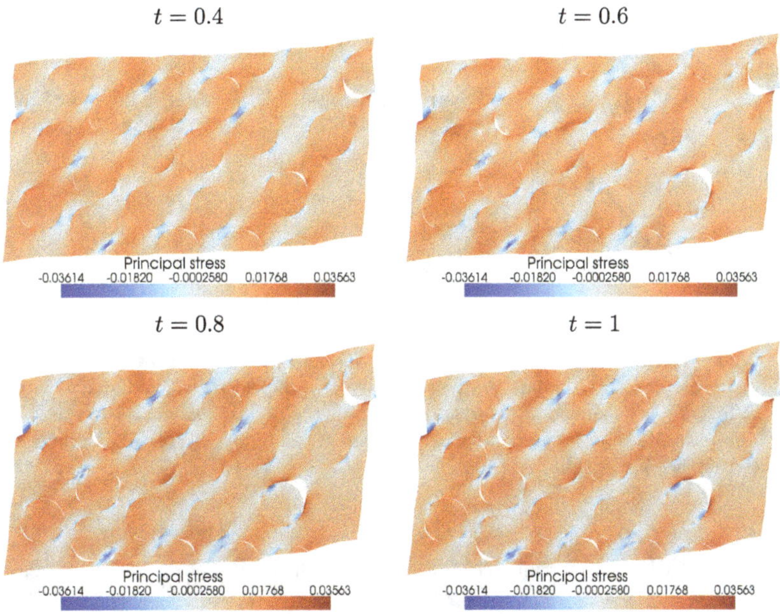

Fig. 11.24. Debonding in a fiber-reinforced composite: selected snapshots of a gradually loaded representative cell containing 20 fibers with depicted displacements (magnified 20×) showing the spatial distribution of stress.

In the case of the single fiber, the dependence of the damage initiation on the ratio of the magnitudes of two applied far-field loads σ_x^∞ and σ_y^∞ was observed. The failure curves in Fig. 11.25 are plotted for various achieved levels of damage. In this way, the first initiation of damage is referred by the value 0.9999, while the value 0 reflects that z equals to zero at least at one point. The presented deformations, corresponding to the failure curve '0' show how and where the damage appears. It is also observed that the present formulation predicts symmetric cracking of the interface, different conditions should be defined to obtain a non-symmetric one as presented in [155]. Finally, the figure shows that having the compressive loading prevailing, the Mode II becomes an important part of solution.

In the case of the multi-fiber problem of Fig. 11.22, the far-field tensile stress σ_y^∞ is applied. The material parameters are the same, except for the interface behavior is defined by the exponential CZM, with a modification according to [90], where, we keep the fracture energies and maximal cohesive stresses the same as in the previous example. The deformations

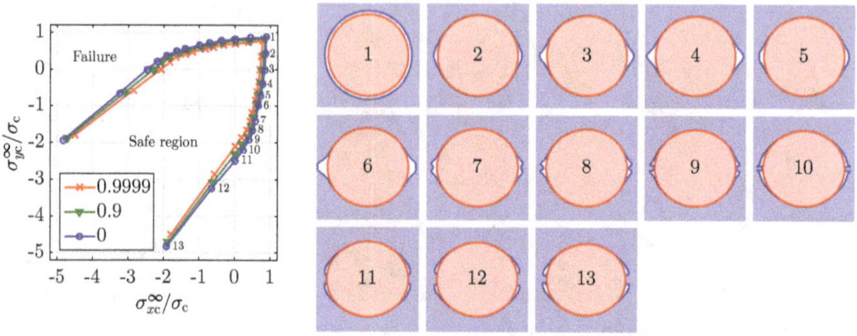

Fig. 11.25. Failure curves for the single-fiber model in the case of the biaxial far-field loading: each curve corresponds to different level of achieved interface damage (minimal z in the legend). The shown opening gaps (magnified 15×) are associated to the damage level 0 for each of the numbered calculated points of the failure curve. The bilinear CZM was used in the calculation.

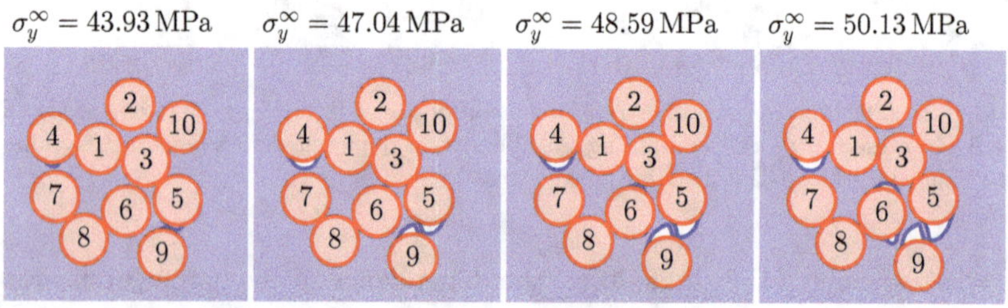

Fig. 11.26. Debonding in a fiber-bundle: selected snapshots of a gradually loaded representative cell containing 10 fibers with depicted opening gaps (magnified 5×).

in the vicinity of the fibers are shown at several instants of loading in Fig. 11.26, where the initiation and subsequent propagation of the fiber-matrix interface cracks can be observed. Additionally, the current damage state at one of the selected instants is shown in Fig. 11.27. The angle φ is clockwise measured at each fiber from its eastern point, cf. Fig. 11.22(a). The state of damage can be easily related to the obtained deformations, the fibers unaffected by the load still have $z \approx 1$ along the whole perimeter.

Another phenomenon that can be considered in a fiber-matrix debonding interface is friction [1, 8, 119] as suggested in (11.86) combined with viscoelastic materials as mentioned in Section 11.6.5. The difference between the frictional and frictionless cases for the problem of uniaxial compression of a single fiber problem solved above is shown in Fig. 11.28.

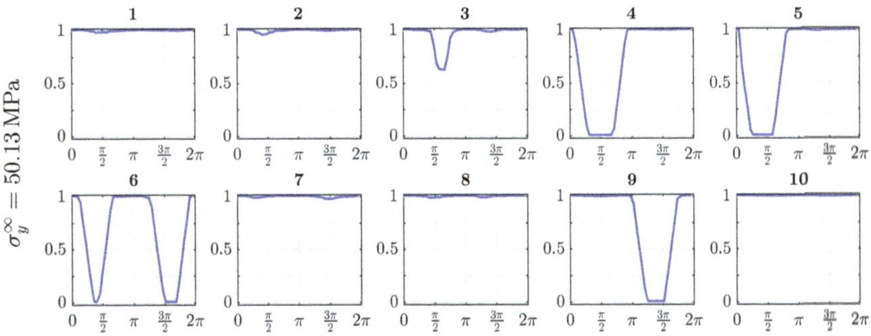

Fig. 11.27. Debonding in a fiber-bundle: the damage state at selected prescribed load corresponding to the last snap-shot in Fig. 11.26. The numbers of the fibers coincide with those in Figs. 11.22 and 11.26.

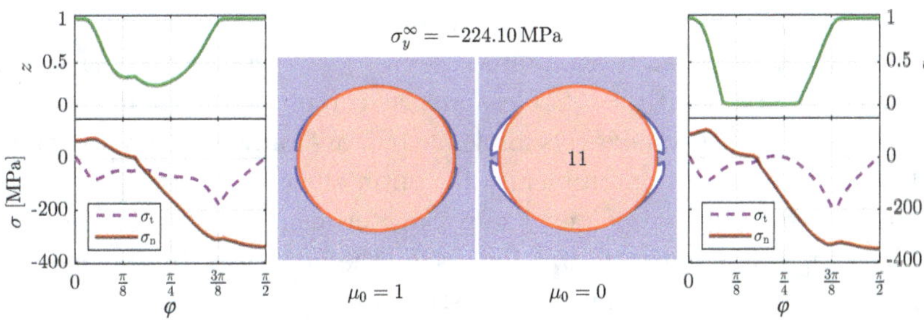

Fig. 11.28. Influence of the friction in the case of uniaxial compression case of the single-fiber problem from Figs. 11.22(a), 11.25: fiber debonding (magnification 15×), damage and stress distribution along a part of the interface (lower right quarter, φ starts at the eastern point). Value '11' refers to the particular number of the load case used in Fig. 11.25.

In the solution, we used the damage-dependent friction in the form $\mu(z) = \mu_0(1-z)^4$ and we plotted the solution at the time instant, where frictionless case has already reached full damage ($z = 0$) at a part of the interface, cf. Fig. 11.25. It should also be noted that the stress-driven type solution of Section 11.3.3 was applied.

11.8. Conclusion

In this chapter, we have surveyed some existing models and proposed a menagerie of new ones for delamination under small strains. The main purpose has been to cover them using the unified concept of quasistatic evolution of the form (11.4) or (11.31) — also to pursue their energetics —

and to outline the rigorous mathematical and numerical analysis based on the concept of the so-called energetic solutions. Additionally, another solution concept of the form (11.37) was outlined for a stress-driven type of solution which intentionally does not necessarily conserve energy but avoids some tendencies to non-physically early jumps. The particular solution strategies are important parts of the model in rate-independent cases. These approaches suggest efficient computational algorithms and allow for mathematically supported simulations. They also allow us to combine relatively easily and routinely various mutually competing inelastic processes both on the delaminating surface or in the bulk, and to devise advanced mathematically supported models and launch numerical simulations of such non-trivial processes. In addition, under certain conditions the approaches may also be combined with rate-dependent processes in the bulk, such as viscosity (for the solution concept based on (11.37) it is natural) in a sufficiently dissipating rheology (e.g., Kelvin–Voigt), inertia, and even thermal processes as, e.g., thermo-visco-plasticity.

When a model with a reliable and fast computable response (which depends continuously on data) is implemented, one can think about dealing with various optimization problems, like optimal control or optimal-shape design or some inverse problems like parameter identification. The desired attribute in this context is uniqueness of the response. In the field of adhesive contacts, it is related with stress-driven solution concepts and the related discretization schemes that avoid global minimization of non-convex problems and provide a unique discrete solution continuously dependent on the data, in contrast to the energetic-solution concept which makes any optimization very complicated and hardly implementable, cf. [156]. Non-smoothness of the control-to-state response due to both the activated character of the delamination process and unilateral contacts needs evaluation of suitable generalized gradients to facilitate usage of efficient optimization algorithms, cf. [157] for a problem of identification of fracture toughness and adhesive elasticity in the model (11.59).

Except a note in Section 11.4, we entirely omitted models with large strains. Let us, however, point out that the advantage of the concept of energetic solutions is that the quasistatic delamination models can be relatively easily formulated in large strains, cf. also [26, Sec. 4.2.3], in contrast to other weak-solution concepts or to dynamical viscous models of the type (11.88).

Finally, the authors would like to point out that further intensive theoretical and computational research in the interface damage and contact mechanics has continued since the revision and acceptance of this chapter in 2017. An amendment [158–187] of the References added in 2022 is surely not complete and only briefly documents ongoing activity in this field.

Acknowledgments

The authors warmly thank Ctirad Matyska, Alexander Mielke, Riccarda Rossi, and Marita Thomas for many fruitful discussions and comments, and Pavel Gruber for performing the numerical simulations presented in Figs. 11.15–11.16 and 11.23–11.24, as well as to two anonymous referees whose valuable remarks and suggestions helped us to improve the clarity of the original manuscript of the first edition [5]. The authors also acknowledge partial support from the grants 16-03823S and I 2375-N32 of the Czech Science Foundation, PGC2018-099197-B-I00 of the Spanish Ministry of Science, Innovation and Universities and European Regional Development Fund, and VEGA 1/0078/16 of the Ministry of Education, Science, Research and Sport of the Slovak Republic.

References

[1] Cocou M., Schryve M., Raous M., 2010. A dynamic unilateral contact problem with adhesion and friction in viscoelasticity, *Zeits. Angew. Math. Phys.*, 61, 721–743.

[2] Sofonea M., Han W., Shillor M., 2006. *Analysis and Approximation of Contact Problems and Adhesion or Damage*, Chapman & Hall, New York.

[3] Shillor M., Sofonea M., Telega J., 2004. *Models and Analysis of Quasistatic Contact*, Springer, Berlin.

[4] Roubíček T., Souček O., Vodička R. 2013. A model of rupturing lithospheric faults with reoccurring earthquakes, *SIAM J. Appl. Math.*, 73, 1460–1488.

[5] Roubíček T., Kružík, M., Zeman, J., 2014. Delamination and adhesive, contact models and their mathematical analysis and numerical treatment. Chap. 9 in V. Mantič (ed.), *Mathematical Methods and Models in Composites*, Imperial College Press, pp. 349–400.

[6] Biot M., 1965. *Mechanics of Incremental Deformations*, Wiley, New York.

[7] Halphen B., Nguyen Q., 1975. Sur les matériaux standards généralisés, *J. Mécanique*, 14, 39–63.

[8] Del Piero G., Raous M., 2010. A united model for adhesive interfaces with damage, viscosity, and friction, *European J. Mech. A/Solids*, 29, 496–507.

[9] Carstensen C., Hackl K., Mielke A., 2002. Non-convex potentials and microstructures in finite-strain plasticity, *Proc. Royal Soc. London Ser. A*, 458, 299–317.

[10] Hackl K., Fischer F.D., 2007. On the relation between the principle of maximum dissipation and inelastic evolution given by dissipation potential, *Proc. Royal Soc. A*, 464, 117–132.

[11] Ortiz M., Repetto E., 1999. Nonconvex energy minimization and dislocation structures in ductile single crystals, *J. Mech. Phys. Solids*, 47, 397–462.
[12] Petryk H., 2003. Incremental energy minimization in dissipative solids, *C. R. Mechanique*, 331, 469–474.
[13] Hill R., 1948. A variational principle of maximum plastic work in classical plasticity, *Q. J. Mech. Appl. Math.*, 1, 18–28.
[14] Lubliner J., 1984. A maximum dissipation principle in generalized plasticity, *Acta Mech.*, 52, 225–237.
[15] Rajagopal K., Srinivasa A., 1998. Mechanics of inelastic behavior of materials. Part I and II, *Int. J. Plasticity*, 14, 945–968, 969–998.
[16] Svoboda J., Turek I., 1991. On diffusion-controlled evolution of closed solid-state thermodynamic systems at constant temperature and pressure, *Philos. Mag. B*, 64, 749–759.
[17] Ziegler H., Wehrli C., 1987. On a principle of maximal rate of entropy production, *J. Non-Equilib. Thermodyn.*, 12, 229–243.
[18] Ekeland I., Temam R., 1999. *Convex Analysis and Variational Problems*, SIAM, Philadelphia.
[19] Simo J., 1988. A framework for finite strain elasto-plasticity based on maximum plastic dissipation and the multiplicative decomposition, *Comput. Methods Appl. Mech. Engrg.*, 66, 199–219.
[20] Han W., Reddy B., 1999. *Plasticity (Mathematical Theory and Numerical Analysis)*, Springer, New York.
[21] Maugin G., 1992. *The Thermomechanics of Plasticity and Fracture*, Cambridge University Press, Cambridge.
[22] Simo J., Hughes T., 1998. *Computational Inelasticity*, Springer, New York.
[23] Hilbert, D., 1901. Mathematische Probleme. *Archiv d. Math. u. Physik* 1, 44–63, 213–237. Engl. transl.: *Bull. Amer. Math. Soc.* 8 (1902), 437–479.
[24] Roubíček T., 2014. Calculus of variations, Chap. 17 in M. Grinfeld (ed.), *Mathematical Tools for Physicists*, Wiley, Weinheim, pp. 551–588. http://www.wiley-vch.de/books/sample/3527411887_c17.pdf
[25] Mielke A., 2010. Differential, energetic and metric formulations for rate-independent processes, in L. Ambrosio and G. Savaré (eds.), *Nonlinear PDEs and Applications*, Springer, Heidelberg, pp. 87–170.
[26] Mielke A., Roubíček T., 2015. *Rate-Independent Systems — Theory and Application*, Springer, New York.
[27] Toader R., Zanini C., 2009. An artificial viscosity approach to quasistatic crack growth, *Boll. Unione Matem. Ital.*, 2, 1–36.
[28] Stefanelli U., 2009. A variational characterization of rate-independent evolution. *Math. Nach.*, 282, 1492–1512.
[29] Roubíček T., 2015. Maximally-dissipative local solutions to rate-independent systems and application to damage and delamination problems, *Nonlin. Anal., Theory Meth. Appl.* 113, 33–50.
[30] Mielke A., Theil, F., 2004. On rate-independent hysteresis models, *Nonl. Diff. Eqns. Appl.* (*NoDEA*), 11, 151–189.
[31] Mielke A., Theil F., 1999. A mathematical model for rate-independent phase transformations with hysteresis, in H.-D. Alber, R.M. Balean, R. Farwig

(eds.), *Models of Continuum Mechanics in Analysis and Engineering*, Shaker-Verlag, Aachen, pp. 117–129.

[32] Mielke A., Theil F., Levitas V.I., 2002. A variational formulation of rate-independent phase transformations using an extremum principle, *Arch. Ration. Mech. Anal.*, 162, 137–177.

[33] Ortiz M., Stainier L., 1999. The variational formulation of visco-plastic constitutive updates, *Comput. Methods Appl. Mech. Engrg.*, 171, 419–444.

[34] Mielke A., 2005. Evolution in rate-independent systems, in C. Dafermos and E. Feireisl (eds.), *Handbook of Differential Equations, Evolutionary Equations* Elsevier B.V., Amsterdam, Vol. 2, pp. 461–559.

[35] Francfort G., Mielke A., 2006. Existence results for a class of rate-independent material models with nonconvex elastic energies, *J. Reine Angew. Math.*, 595, 55–91.

[36] Dacorogna B., 2008. *Direct Methods in the Calculus of Variations*, 2nd edn., Springer, New York.

[37] Benešová B., 2009. Modeling of shape-memory alloys on the mesoscopic level, in P. Sittner *et al.* (eds.), *Proc. ESOMAT 2009* EDP Sci., pp. 1–7.

[38] Benešová B., 2011. Global optimization numerical strategies for rate-independent processes, *J. Global Optim.*, 50, 197–220.

[39] Mielke A., Roubíček T., Zeman J., 2010. Complete damage in elastic and viscoelastic media and its energetics, *Computer Methods Appl. Mech. Engr.*, 199, 1242–1253.

[40] Mielke A., Roubíček T., Stefanelli U., 2008. Γ-limits and relaxations for rate-independent evolutionary problems, *Calc. Var., P.D.E.*, 31, 387–416.

[41] Frémond M., 2002. *Non-Smooth Thermomechanics*, Springer, Berlin.

[42] Point N., 1988. Unilateral contact with adherence, *Math. Methods Appl. Sci.*, 10, 367–381.

[43] Point N., Sacco E., 1996. A delamination model for laminated composites, *Math. Methods Appl. Sci.*, 33, 483–509.

[44] Point N., Sacco E., 1998. Mathematical properties of a delamination model, *Math. Comput. Modelling*, 28, 359–371.

[45] Mielke A., 2002. Finite elasto-plasticity, Lie groups and geodesics on SL(d), in P. Newton, A. Weinstein, P.J. Holmes (eds.), *Geometry, Dynamics, and Mechanics*, Springer-Verlag, New York, pp. 61–90.

[46] Vodička R., Mantič V., Roubíček T., 2014. Energetic versus maximally-dissipative local solutions of a quasi-static rate-independent mixed-mode delamination model, *Meccanica* 49, 2933–2963.

[47] Cagnetti F., 2009. A vanishing viscosity approach to fracture growth in a cohesive zone model with prescribed crack path, *Math. Models Meth. Appl. Sci.*, 18, 1027–1071.

[48] Knees D., Mielke A., Zanini C., 2008. On the inviscid limit of a model for crack propagation, *Math. Mod. Meth. Appl. Sci.*, 18, 1529–1569.

[49] Knees D., Schroder A., 2013. Computational aspects of quasi-static crack propagation, *Disc. Cont. Dynam. Syst. S*, 6, 63–99.

[50] Knees D., Zanini C., Mielke A., 2010. Crack growth in polyconvex materials, *Physica D*, 239, 1470–1484.

[51] Negri M., Ortner C., 2008. Quasi-static crack propagation by Griffith's criterion, *Math. Mod. Meth. Appl. Sci.*, 18, 1895–1925.
[52] Roubíček T., 2013. Adhesive contact of visco-elastic bodies and defect measures arising by vanishing viscosity, *SIAM J. Math. Anal.*, 45, 101–126.
[53] Roubíček T., Panagiotopoulos C.G., Mantič V., 2013. Quasistatic adhesive contact of visco-elastic bodies and its numerical treatment for very small viscosity, *Zeit. Angew. Math. Mech.*, 93, 823–840.
[54] Mielke A., Rossi R., Savaré G., 2012. BV solutions and viscosity approximations of rate-independent systems, *ESAIM Control Optim. Calc. Var.*, 18, 36–80.
[55] Mielke A., Rossi R., Savaré G., 2016. Balanced-viscosity (BV) solutions to infinite-dimensional rate-independent systems, *J. Europ. Math. Soc.*, 18, 2107–2165.
[56] Roubíček T., 2013. *Nonlinear Partial Differential Equations with Applications*. 2nd edn., Birkhäuser, Basel.
[57] Roubíček T., Panagiotopoulos C.G., Mantič V., 2015. Local-solution approach to quasistatic rate-independent mixed-mode delamination, *Math. Mod. Meth. Appl. Sci.*, 25, 1337–1364.
[58] Mielke A., Roubíček T., 2009. Numerical approaches to rate-independent processes and applications in inelasticity, *Math. Model. Numer. Anal.*, 43, 399–428.
[59] Griffith A., 1921. The phenomena of rupture and flow in solids, *Philos. Trans. Royal Soc. London Ser. A. Math. Phys. Eng. Sci.*, 221, 163–198.
[60] Dal Maso G., Zanini C., 2007. Quasi-static crack growth for a cohesive zone model with prescribed crack path, *Proc. R. Soc. Edinb., Sect. A. Math.*, 137(2), 253–279.
[61] Mielke A., Roubíček T., Thomas M., 2012. From damage to delamination in nonlinearly elastic materials at small strains, *J. Elast.*, 109, 235–273.
[62] Roubíček T., Scardia L., Zanini C., 2009. Quasistatic delamination problem, *Cont. Mech. Thermodynam.*, 21, 223–235.
[63] Bartels S., Roubíček T., 2011. Thermo-visco-elasticity with rate-independent plasticity in isotropic materials undergoing thermal expansion, *Math. Model. Numer. Anal.*, 45, 477–504.
[64] Roubíček T., Thomas M., Panagiotopoulos C.G., 2015. Stress-driven local-solution approach to quasistatic brittle delamination, *Nonlinear Anal., Real World Appl.* 22, 645–663.
[65] Lenci S., 2001. Analysis of a crack at a weak interface, *Int. J. Fracture*, 108, 275–290.
[66] Távara L., Mantič V., Graciani E., Cañas J., París F., 2010. Analysis of a crack in a thin adhesive layer between orthotropic materials: An application to composite interlaminar fracture toughness test, *CMES-Comp. Model. Eng. Sci.*, 58, 247–270.
[67] Kočvara M., Mielke A., Roubíček T., 2006. A rate-independent approach to the delamination problem, *Math. Mech. Solids*, 11, 423–447.
[68] Bourdin B., Francfort G.A., Marigo J.-J., 2000. Numerical experiments in revisited brittle fracture, *J. Mech. Phys. Solids.*, 48, 797–826.

[69] Dostál Z. 2009. *Optimal Quadratic Programming Algorithms: With Applications to Variational Inequalities,* Springer, New York.
[70] Dostál Z., Kozubek T., Vondrák V., Brzobohatý T., Markopoulos A., 2010. Scalable TFETI algorithm for the solution of multibody contact problems of elasticity, *Int. J. Numer. Meth. Eng.*, 82, 1384–1405.
[71] Bouchala J., Dostál Z., Sadowská M., 2009. Scalable total BETI based algorithm for 3D coercive contact problems of linear elasto-statics, *Computing*, 85(3), 189–217.
[72] Blázquez A., Vodička R., París F., Mantič V., 2002. Comparing the conventional displacement BIE and the BIE formulations of the first and second kind in frictionless contact problems, *Eng. Anal. Bound. Elem.*, 26, 815–826.
[73] Graciani E., Mantič V., París F., Blázquez A., 2005. Weak formulation of axi-symmetric frictionless contact problems with boundary elements. Application to interface crack, *Comput. Struct.* 83, 836–855.
[74] Blázquez A., Mantič V., París F., 2006. Application of BEM to generalized plane problems for anisotropic elastic materials in presence of contact, *Eng. Anal. Bound. Elem.*, 30, 489–502.
[75] Graciani E., Mantič V., París F., Varna J., 2009. Numerical analysis of debond propagation in the single fibre fragmentation test, *Compos. Sci. Technol.*, 69, 2514–2520.
[76] Bourdin B., 2007. Numerical implementation of the variational formulation for quasi-static brittle fracture, *Interface. Free Bound.*, 9, 411–430.
[77] Burke S., Ortner C., Süli E., 2010. An adaptive finite element approximation of a variational model of brittle fracture, *SIAM J. Numer. Anal.*, 48, 980–1012.
[78] Valoroso N., Champaney L., 2006. A damage-mechanics-based approach for modelling decohesion in adhesively bonded assemblies, *Eng. Fract. Mech.*, 73, 2774–2801.
[79] Mielke A., Rossi R., Savaré G., 2016. Balanced-Viscosity solutions for multi-rate systems, *J. Physics, Conf. Series*, 727, 012010.
[80] Thomas M., 2010. Rate-independent damage processes in nonlinearly elastic materials, PhD thesis, Inst. f. Math., Humboldt-Universität zu Berlin.
[81] Thomas M., Mielke A., 2010. Damage of nonlinearly elastic materials at small strain — Existence and regularity results, *Zeits. Angew. Math. Mech.*, 90, 88–112.
[82] Mielke A., Roubíček T., 2006. Rate-independent damage processes in nonlinear elasticity, *Math. Models Meth. Appl. Sci.*, 16, 177–209.
[83] Bonetti E., Bonfanti G., Rossi R., 2007. Well-posedness and long-time behaviour for a model of contact with adhesion, *Indiana Univ. Math. J.*, 56, 2787–2820.
[84] Bonetti E., Bonfanti G., Rossi R., 2008. Global existence for a contact problem with adhesion, *Math. Meth. Appl. Sci.*, 31, 1029–1064.
[85] Turon A., Davila C., Camanho P., Costa J., 2007. An engineering solution for mesh size effects in the simulation of delamination using cohesive zone models, *Eng. Fract. Mech.*, 74, 1665–1682.

[86] Xie D., Waas A., 2006. Discrete cohesive zone model for mixed-mode fracture using finite element analysis, *Eng. Fract. Mech.*, 73, 1783–1796.

[87] Ortiz M., Pandolfi A., 1999. Finite-deformation irreversible cohesive elements for three dimensional crack propagation analysis, *Int. J. Num. Meth. Engrg.*, 44, 1267–1283.

[88] Park K., Paulino G.H., 2011. Cohesive zone models: A critical review of traction-separation relationships across fracture surfaces, *Appl. Mech. Rev.*, 64(6), 060802.

[89] Távara L., Mantič V., Salvadori A., Gray L.J., París F., 2013. Cohesive-zone-model formulation and implementation using the symmetric Galerkin boundary element method for homogeneous solids, *Comput. Mech.*, 51, 535–551.

[90] Vodička R., 2016. A quasi-static interface damage model with cohesive cracks: SQP-SGBEM implementation. *Eng. Anal. Bound. Elem.* 62, 123–140.

[91] Banks-Sills L., Askenazi D., 2000. A note on fracture criteria for interface fracture, *Int. J. Fracture*, 103, 177–188.

[92] Liechti K., Chai Y., 1992. Asymmetric shielding in interfacial fracture under in-plane shear, *J. Appl. Mech.*, 59, 295–304.

[93] Mantič V., 2008. Discussion on the reference length and mode mixity for a bimaterial interface, *J. Eng. Mater. Technology*, 130, 045501–1–2.

[94] Swadener J., Liechti K., deLozanne A., 1999. The intrinsic toughness and adhesion mechanism of a glass/epoxy interface, *J. Mech. Phys. Solids*, 47, 223–258.

[95] Tveergard W., Hutchinson J., 1993. The influence of plasticity on mixed mode interface toughness, *J. Mech. Phys. Solids*, 41, 1119–1135.

[96] Evans A., Rühle M., Dalgleish B., Charalambides P., 1990. The fracture energy of bimaterial interfaces, *Metall. Mater. Trans. A-Phys. Metall. Mater. Sci.*, 21A, 2419–2429.

[97] Hutchinson J.W., Suo Z., 1992. Mixed mode cracking in layered materials, *Adv. in Appl. Mech.*, 29, 63–191.

[98] Távara L., Mantič V., Graciani E., París F., 2011. BEM analysis of crack onset and propagation along fibre-matrix interface under transverse tension using a linear elastic-brittle interface model, *Eng. Anal. Bound. Elem.*, 35, 207–222.

[99] Mantič V., Távara L., Blázquez A., Graciani E., París F., 2015. A linear elastic-brittle interface model: application for the onset and propagation of a fibre-matrix interface crack under biaxial transverse loads, *Int. J. Fract*, 195, 15–38.

[100] Mielke A., Rossi R., 2007. Existence and uniqueness results for a class of rate-independent hysteresis problems, *Math. Models Meth. Appl. Sci.*, 17, 81–123.

[101] Balder E., 1984. A general approach to lower semicontinuity and lower closure in optimal control theory, *SIAM J. Control. Optim.*, 22, 570–598.

[102] Rossi R., Savaré G., 2006. Gradient flows of non convex functionals in Hilbert spaces and applications, *ESAIM Control Optim. Calc. Var.*, 12, 564–614.

[103] Kružík M., Panagiotopoulos C.G., Roubíček T., 2015. Quasistatic adhesive contact delaminating in mixed mode and its numerical treatment, *Math. Mech. Solids*, 20, 582–599.

[104] Rossi R., Roubíček T., 2013. Adhesive contact delaminating at mixed mode, its thermodynamics and analysis, *Interface. Free Bound.*, 14, 1–37.

[105] Roubíček T., Mantič V., Panagiotopoulos C.G., 2013. A quasistatic mixed-mode delamination model, *Disc. Cont. Dynam. Syst. S*, 6, 591–610.

[106] Panagiotopoulos C.G., Mantič V., Roubíček T., 2013. BEM solution of delamination problems using an interface damage and plasticity model, *Comput. Mech.*, 51, 505–521.

[107] Panagiotopoulos C.G., Mantič V., Roubíček T., 2018. Two adhesive-contact models for quasistatic mixed-mode delamination problems, *Math. Comp. Simulation*, 145, 18–33.

[108] Gutkin R., Laffan M.L., Pinho S.T.,Robinson P., Curtis P.T., 2011. Modelling the R-curve effect and its specimen-dependence, *Int. J. Solids Struct.*, 48, 1767–1777.

[109] Vodička R., Mantič V., 2017. An energy based formulation of a quasi-static interface damage model with a multilinear cohesive law, *Disc. Cont. Dynam. Syst. S*, 10, 1539–1561.

[110] Bigoni D., Radi E., 1993. Mode I crack propagation in elastic-plastic pressure-sensitive materials, *Int. J. Solids Struct.*, 30, 899–919.

[111] Dal Maso G., Toader R., 2010. Quasistatic crack growth in elasto-plastic materials: The two-dimensional case, *Arch. Rat. Mech. Anal.*, 196, 867–906.

[112] Wey Y., Hutchinson J., 1997. Steady-state crack growth and work of fracture for solids characterized by strain gradient plasticity, *J. Mech. Phys. Solids*, 45, 1253–1273.

[113] Ambrosio L., Tortorelli V.M., 1992. On the approximation of free discontinuity problems, *Bollettino Unione Mat. Italiana*, 7, 105–123.

[114] Giacomini A., 2005. Ambrosio-Tortorelli approximation of quasi-static evolution of brittle fractures, *Calc. Var. Partial Diff. Eqs.*, 22, 129–172.

[115] Knees D., Negri M., 2017. Convergence of alternate minimization schemes for phase field fracture and damage, *Math. Mod. Meth. Appl. Sci.* 27, 1743–1794.

[116] Han W., Shillor M., Sofonea M., 2001. Variational and numerical analysis of a quasistatic viscoelastic problem with normal compliance, friction and damage, *J. Comput. Appl. Math.*, 137, 377–398.

[117] Klarbring A., Mikelič A., Shillor M., 1988. Frictional contact problems with normal compliance, *Intl. J. Engng. Sci.*, 26, 811–832.

[118] Raous M., Cangémi L., Cocu M., 1999. A consistent model coupling adhesion, friction, and unilateral contact. *Comput. Methods Appl. Mech. Engrg.* 177, 383–399.

[119] Vodička R., Mantič V., Roubíček T., 2017. Quasistatic normal-compliance contact problem of visco-elastic bodies with Coulomb friction implemented by QP and SGBEM, *J. Comp. Appl. Math.*, 315, 249–272.

[120] Roubíček T., 2019. A general thermodynamical model for adhesive frictional contacts between viscoelastic or poro-viscoelastic bodies at small strains, *Interface. Free Bound.*, 21, 169-–198.

[121] Roubíček T., 2009. Rate independent processes in viscous solids at small strains, *Math. Methods Appl. Sci.*, 32, 825–862. Erratum p. 2176.

[122] Rossi R., Thomas M., 2017. Coupling rate-independent and rate-dependent processes: existence results, *SIAM J. Math. Anal.*, 49(2), 1419—1494.

[123] Scala R., 2017. A weak formulation for a rate-independent delamination evolution with inertial and viscosity effects subjected to unilateral constraint, *Interface. Free Bound.*, 19, 79—107.

[124] Scala R., 2017. Limit of viscous dynamic processes in delamination as the viscosity and inertia vanish, *ESAIM: COCV*, 23, 593–625.

[125] Rossi R., Thomas M., 2017. From adhesive to brittle delamination in visco-elastodynamics, *Math. Mod. Meth. Appl. Sci.*, 27, 1489–1546.

[126] Jarušek J., Sofonea M., 2009. On the solvability of dynamic elastic-visco-plastic contact problems with adhesion, *Ann. Acad. Romanian Sci.*, 1, 191–214.

[127] Abercrombie R., Rice J., 2005. Can observations of earthquake scaling constrain slip weakening? *Geophys. J. Int.*, 162, 406–424.

[128] Bizzarri A., Cocco M., 2005. 3D dynamic simulations of spontaneous rupture propagation governed by different constitutive laws with rake rotation allowed, *Ann. Geophys.*, 48, 279–299.

[129] Burjánek J., 2006. *Dynamic Stress Field of Kinematic Earthquake Source Models*, PhD thesis, Math.-Phys. Faculty, Charles Univ., Prague.

[130] Das S., 2003. Spontaneous complex earthquake rupture propagation, *Pure Appl. Geophysics*, 160, 579–602.

[131] Kostrov B., Das S., 1988. *Principles of Earthquake Source Mechanics*, Cambridge University Press, New York.

[132] Madariaga R., Olsen K., 2002. Earthquake dynamics, in W. Lee *et al.* (eds.), *International Handbook of Earthquake and Engineering Seismology*, Academic Press, London.

[133] Uenishi K., Rice R., 2003. Universal nucleation length for slip-weakening rupture instability under non-uniform fault loading, *J. Geophys. Res.*, 108, 1–14.

[134] Hilber H.M., Hughes T.J.R., Taylor R.L., 1977. Improved numerical dissipation for time integration algorithms in structural dynamics, *Earthquake Eng. Struct. Dyn.*, 5, 283–292.

[135] Newmark N.M., 1959. A method of computation for structural dynamics, *J. Eng. Mech. Div.*, 85, 67–94.

[136] Crank J., Nicolson P., 1947. A practical method for numerical evaluation of solutions of partial differential equations of the heat conduction type, *Proc. Camb. Phil. Soc.*, 43, 50–67.

[137] Armero F., Petötz E., 1998. Formulation and analysis of conserving algorithms for dynamic contact/impact problems, *Computer Meth. Appl. Mech. Engr.*, 158, 269–300.

[138] Roubíček T., Panagiotopoulos C.G. 2017. Energy-conserving time-discretisation of abstract dynamical problems with applications in continuum mechanics of solids, *Numer. Funct. Anal. Optim.*, 38, 1143–1172.

[139] Roubíček T., 2017. An energy-conserving time-discretisation scheme for poroelastic media with phase-field fracture emitting waves and heat. *Disc. Cont. Dynam. Syst. S*, 10, 867–893.

[140] Panagiotopoulos C.G., Mantič V., Roubíček T., 2014. A simple and efficient BEM implementation of quasistatic linear visco-elasticity. *Int. J. Solids Struct.* 51, 2261–2271.

[141] Rittel D., 1999. Thermomechanical aspects of dynamic crack initiation, *Int. J. Fracture*, 99, 199–209.

[142] Rossi R., Roubíček T., 2011. Thermodynamics and analysis of rate-independent adhesive contact at small strains, *Nonlinear Anal., Theor.*, 74, 3159–3190.

[143] Bonetti E., Bonfanti G., Rossi R., 2009. Thermal effects in adhesive contact: Modelling and analysis, *Nonlinearity*, 22, 2697–2731.

[144] Rossi R., Thomas M., 2015. From an adhesive to a brittle delamination model in thermo-visco-elasticity, *ESAIM Control Optim. Calc. Var.*, 21, 1–59.

[145] Bonetti E., Bonfanti G., Rossi R., 2011. Long-time behaviour of a thermomechanical model for adhesive conctact, *Disc. Cont. Dynam. Systems — Ser. S*, 4, 273–309.

[146] Bouchbinder E., Langer J.S., 2009. Nonequilibrium thermodynamics of driven amorphous materials. I. Internal degrees of freedom and volume deformation, II. Effective-temperature theory, III. Shear-transformation-zone plasticity, *Phys. Rev. E.*, 80, 031131, 031132, 031133.

[147] Dieterich J.H., 1979. Modelling of rock friction. Part 1: Experimental results and constitutive equations, *J. Geophys. Res.*, 84(B5), 2161–2168.

[148] Ruina A.L., 1983. Slip instability and state variable friction laws. *J. Geophys. Res.*, 88, 10,359-10,370.

[149] Roubíček T., 2014. A note about the rate-and-state-dependent friction model in a thermodynamical framework of the Biot-type equation. *Geophysical J. Int.*, 199, 286–295.

[150] Miehe C., 2002. Strain-driven homogenization of inelastic microstructures and composites based on an incremental variational formulation, *Int. J. Numer. Meth. Engrg.*, 55, 1285–1322.

[151] Mielke A., Timofte A., 2007. Two-scale homogenization for evolutionary variational inequalities via the energetic formulation, *SIAM J. Math. Anal.*, 39, 642–668.

[152] Michel J., Moulinec H., Suquet P., 1999. Effective properties of composite materials with periodic microstructure: A computational approach, *Comput. Methods Appl. Mech. Engrg.*, 172, 109–143.

[153] Mantič V., García I.G., 2012. Crack onset and growth at the fibre-matrix interface under a remote biaxial transverse load. Application of a coupled stress and energy criterion, *Int. J. Solids Struct.*, 49, 2273–2290.

[154] Távara L., Mantič V., Graciani E., París F., 2016. Modelling interfacial debonds in unidirectional fibre-reinforced composites under biaxial transverse loads, *Compos. Struct.*, 136, 305–312.

[155] Panagiotopoulos C.G., Mantič V., 2013. Symmetric and non-symmetric debonds at fibre-matrix interface under transverse loads. An application of energetic approaches using collocation BEM. *Anales de Mecánica de la Fractura* 30, Vol. I, 125–130.

[156] Kočvara M., Outrata J.V., 2004. On the modeling and control of delamination processes. In *Control & Boundary Analysis* (J. Cagnol, J.-P. Zolezio, eds), Marcel Dekker, New York, pp. 171–190.

[157] Adam L., Outrata J.V., Roubíček T., 2017. Identification of some nonsmooth evolution systems with illustration on adhesive contacts at small strains, *Optimization*, 66, 2025–2049.

[158] Alessi R., Freddi F., 2017. Phase-field modelling of failure in hybrid laminates, *Compos. Struct.*, 181, 9–25.

[159] Bonetti E., Bonfanti G., Rossi R., 2017. Global existence for a nonlocal model for adhesive contact, *Applicable Analysis*, 97, 1315–1339.

[160] Freddi F., Iurlano F., 2017. Numerical insight of a variational smeared approach to cohesive fracture, *J. Mech. Phys. Solids*, 98, 156–171.

[161] Negri M., Scala R., 2017. A quasi-static evolution generated by local energy minimizers for an elastic material with a cohesive interface, *Nonlinear Anal. Real World Appl.*, 38, 271–305.

[162] Thomas M., Zanini C., 2017. Cohesive zone-type delamination in visco-elasticity, *Disc. Cont. Dynam. Systems-Ser. S*, 10, 1487–1517.

[163] Alessi R., Marigo J.-J., Maurini C., Vidoli S., 2018. Coupling damage and plasticity for a phase-field regularisation of brittle, cohesive and ductile fracture: One-dimensional examples, *Int. J. Mech. Sci.*, 149, 559–576.

[164] Lazzaroni G., Nardini L., 2018. On the quasistatic limit of dynamic evolutions for a peeling test in dimension one, *J. Nonlinear Sci.*, 28, 269–304.

[165] Rossi R., Thomas M., 2018. From nonlinear to linear elasticity in a coupled rate-dependent/independent system for brittle delamination, in E. Rocca, U. Stefanelli, L. Truskinovsky, A. Visintin (eds), *Trends in Applications of Mathematics to Mechanics*, Springer, Cham, pp. 127–157.

[166] Tanné E., Li T., Bourdin B., Marigo J.-J., Maurini C., 2018. Crack nucleation in variational phase-field models of brittle fracture, *J. Mech. Phys. Solids*, 110, 80–99.

[167] Thomas, M., 2018. A comparison of delamination models: modeling, properties, and applications, in P. van Meurs, M. Kimura, H. Notsu (eds.), *Mathematical Analysis of Continuum Mechanics and Industrial Applications II*, Springer, Singapore, pp. 27–38.

[168] Vodička R., Kormaníková E., Kšiňan F., 2018. Interfacial debonds of layered anisotropic materials using a quasi-static interface damage model with Coulomb friction, *Int. J. Fract.*, 211, 163–182.

[169] Alessi R., Freddi F., 2019. Failure and complex crack patterns in hybrid laminates: A phase-field approach, *Compos. B Eng.*, 179, 107256.

[170] Bacho A., Emmrich E., Mielke A., 2019. An existence result and evolutionary Γ-convergence for perturbed gradient systems, *J. Evol. Equ.*, 19, 479–522.

[171] Jirásek M., Allix O., 2019. Analysis of self-similar rate-dependent interfacial crack propagation in mode II, *Int. J. Fract.*, 220, 45–83.

[172] Rossi R., 2019. Visco-energetic solutions to some rate-independent systems in damage, delamination, and plasticity. *Math. Models Meth. Appl. Sci.*, 29, 1079–1138.

[173] Caffarelli L., Cagnetti F., Figalli A., 2020. Optimal regularity and structure of the free boundary for minimizers in cohesive zone models, *Arch. Ration. Mech. Anal.*, 237, 299–345.

[174] García-Guzmán L., Reinoso J., Távara L., París F., 2020. A consistent finite displacement and rotation formulation of the Linear Elastic Brittle Interface Model for triggering interlaminar damage in fiber-reinforced composites, *Theor. Appl. Fract. Mech.*, 108, 102644.

[175] García-Guzmán L., Reinoso J., Valverde A., Martínez-Pañeda E., Távara L., 2020. Numerical study of interface cracking in composite structures using a novel geometrically nonlinear Linear Elastic Brittle Interface Model: mixed-mode fracture conditions and application to structured interfaces, *Composite Structures*, 248, 112495.

[176] Kumar A., Bourdin B., Francfort G.A., Lopez-Pamies O., 2020. Revisiting nucleation in the phase-field approach to brittle fracture, *J. Mech. Phys. Solids*, 142, 104027.

[177] Riva F., 2020. On the approximation of quasistatic evolutions for the debonding of a thin film via vanishing inertia and viscosity. *J Nonlinear Sci.*, 30, 903–951.

[178] Asur Vijaya Kumar P.K., Dean A., Reinoso J., Paggi M., 2021. A multi phase-field-cohesive zone model for laminated composites: Application to delamination migration, *Compos. Struct.*, 276, 114471.

[179] Bonacini M., Conti S., Iurlano F., 2021. Cohesive fracture in 1D: quasi-static evolution and derivation from static phase-field models, *Arch. Ration. Mech. Anal.*, 239, 1501–1576.

[180] Bonetti E., Bonfanti G., Rossi R., 2021. A new nonlocal temperature-dependent model for adhesive contact, in E. Bonetti, C. Cavaterra, R. Natalini, M. Solci (eds.), *Mathematical Modeling in Cultural Heritage*, Springer, Cham, pp. 37–58.

[181] Bonetti E., Cavaterra C., Freddi F., Riva F., 2021. On a phase–field model of damage for hybrid laminates with cohesive interface, *Math. Meth. Appl. Sci.*, https://doi.org/10.1002/mma.7999.

[182] Dal Maso G., Rossi R., Savaré G., Toader R., 2021. Visco-energetic solutions for a model of crack growth in brittle materials, arXiv:2105.00046.

[183] Heida M., Thomas M., 2021. GENERIC for dissipative solids with bulk-interface interaction. Preprint No. 2872, WIAS, Berlin.

[184] Negri M., Scala R., 2021. Existence, energy identity, and higher time regularity of solutions to a dynamic viscoelastic cohesive interface model, *SIAM J. Math. Anal.*, 53, 5682–5730.

[185] Panagiotopoulos C.G., Tsogka C., 2021. A mixed finite element-based numerical method for elastodynamics considering adhesive interface damage for dynamic fracture, *J. Theor. Comput. Acoustics*, 29, 2150010.

[186] Riva F., Nardini L., 2021. Existence and uniqueness of dynamic evolutions for a one-dimensional debonding model with damping, *J. Evol. Equ.*, 21, 63–106.

[187] Vodička R., 2021. A new quasi-static delamination model with rate-dependence in interface damage and its operation under cyclic loading, *Int. J. Solids Struct.*, 224, 111035.

https://doi.org/10.1142/9781800611887_0012

Chapter 12

Interaction of Cracks with Interfaces

Dominique Leguillon[*,‡] **and Eric Martin**[†,§]
[*]*Institut Jean le Rond d'Alembert, CNRS UMR 7190, Sorbonne Université, 4 place Jussieu, 75005 Paris, France*
[†]*Laboratoire des Composites Thermo-Structuraux CNRS UMR 5801, Université Bordeaux 1, 3 rue La Boétie,33600 Pessac, France*
[‡]*dominique.leguillon@upmc.fr*
[§]*eric.martin@enseirb-matmeca.fr*

Abstract

The coupled criterion predicts the crack nucleation at stress concentration point. It is a twofold criterion that uses conditions for energy and tensile stress and involves both toughness and tensile strength of the material. In general, the crack jumps a finite length and then either stops or goes on growing. It has proven its effectiveness in many situations encountered in homogeneous materials like V- and U-notches and predictions fairly agree with the experimental measures. It can also be used to study specific mechanisms of degradation of composites such as delamination or fiber debonding. It has recently been used successfully to predict the initiation of delamination from a stress-free edge, the applications discussed in this chapter mainly concern the deviation of transverse cracks by an interface and the crack kinking out of an interface. A brief digression is also made in the case of interfaces showing an angle.

Contents

12.1. Introduction

Delamination is the main cause of failure of multi-materials and especially of composite laminates, the components separate leading to the total ruin or at least to a weaken structure. This topic is still subject of numerous works and a detailed and recent list of references can be found in [1] for a better overview of the problem. There are at least two important origins to delamination under static loading [2]; the first one is the classic initiation process which occurs at the intersection of a free edge with the interface between two layers. It is a zone of stress concentration described in elasticity by a singular stress field [1, 3]. The other one is less obvious because it is an internal process and therefore not directly detectable: the deflection of transverse cracks present in the most disoriented layers relatively to the tensile direction [4–6]. The classical situation in a laminate is the presence of plies 90° oriented with respect to the loading direction. Transverse microcracks are then generated which coalesce to form a transverse crack which is deflected when it reaches the interface with a 0° ply. This deflection mechanism originates a delamination crack [7]. Crack deflection at the fiber/matrix interface is also a prerequisite for the activation of toughening mechanisms like multiple matrix cracking in ceramic matrix composites (CMC) [8, 9]. Once delamination takes place, the crack can grow along the interface and at the end separate the components or it can kink into one of the adjacent materials.

We propose, in this chapter, to present the coupled criterion allowing the prediction of crack nucleation at stress concentration points [10, 11]. It combines stress and energy conditions which allow to get rid of the definition of a characteristic fracture length selected more or less arbitrarily. This criterion has been applied successfully to several situations in composite and laminated materials: delamination originating from a stress-free edge within a generalized plane strain elasticity framework [1, 12], fiber debonding [13–15]. It is illustrated herein by two mechanisms, the transverse crack deflection by an interface [16, 17] and the crack kinking out of an interface [18–21]. The analysis will be developed in plane strain and should be theoretically extended to 3D although a number of technical difficulties emerge [22, 23]. A brief digression is also made in the case of interfaces showing an angle.

12.2. The Coupled Criterion

To establish this criterion, the better generic model is the 3-point bending test on a V-notched specimen made of a homogeneous material (Fig. 12.1).

In composite materials, it applies at the intersection of the interface between two layers and a free edge for instance [1], or at the end of a transverse crack impinging on an interface as illustrated in the next sections.

The coupled criterion is based on a twofold condition to predict the nucleation of cracks in areas of stress concentrations in brittle materials: a condition on the maximum tensile or shear stress that a structure can sustain and an energy balance between the stored energy and the energy required to induce fracture [10]. The first condition refers to the tensile strength σ_c (or shear strength τ_c) while the other relies on the toughness G_c of the material (or interface). These two conditions must be satisfied simultaneously.

The form taken by these two conditions come from the theory of singularities and the asymptotic expansions of the displacement field $\underline{U}$ and the stress field $\underline{\underline{\sigma}}$ in the vicinity of the origin so-called Williams' expansion [24] (in polar coordinates with origin at the singular point formed by the notch root)

$$\begin{cases} \underline{U}(r,\theta) = \underline{C} + kr^{\lambda}\underline{u}(\theta) + \cdots, \\ \underline{\underline{\sigma}}(r,\theta) = k\, r^{\lambda-1}\underline{\underline{s}}(\theta) + \cdots, \end{cases} \tag{12.1}$$

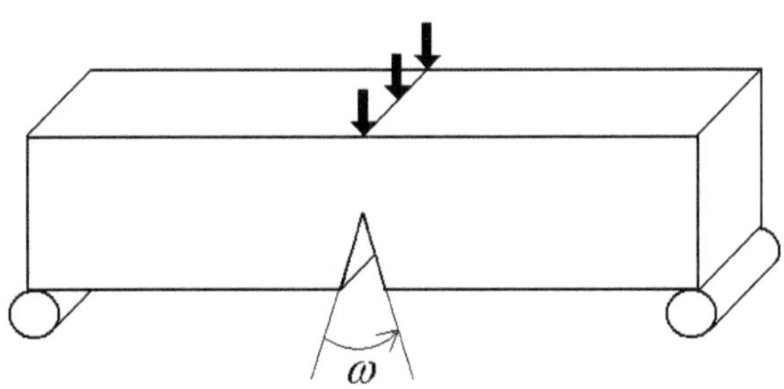

Fig. 12.1. The 3-point bending test on a V-notched homogeneous specimen.

λ is the leading singularity exponent ($1/2 \le \lambda \le 1$ for a V-notch), k (MPa.m$^{1-\lambda}$) is the generalized stress intensity factor (GSIF), $\underline{u}(\theta)$ and $\underline{\underline{s}}(\theta)$ are two angular functions and $\underline{C}$ is a constant (the rigid translation of the origin). Coefficient k depends on the whole geometry of the structure and on the remote applied load. Dots hold for less significant terms in the expansion. The exponent and the angular functions depend only on the local geometry and elastic properties; they are solutions to an eigenvalue problem and are either known analytically in some simple cases or can anyway be determined numerically using a simple algorithm. Clearly, the stress components tend to infinity as $r \to 0$, this is why it is called a singular point.

Note that here and in the following, we only address the tensile stress but extension to the shear component is straightforward.

The GSIF k can be computed using a path independent integral Ψ [25, 26], valid for any elastic fields satisfying the equilibrium to 0 (i.e., vanishing boundary conditions and balance equation within the domain surrounded by the integration path)

$$k = \frac{\Psi(\underline{U}(r,\theta), r^{-\lambda}\underline{u}^{-}(\theta))}{\Psi(r^{\lambda}\underline{u}(\theta), r^{-\lambda}\underline{u}^{-}(\theta))} \quad \text{with}$$

$$\Psi(\underline{U}, \underline{V}) = \frac{1}{2}\int_{\Gamma} (\underline{\underline{\sigma}}(\underline{U}).\underline{n}.\underline{V} - \underline{\underline{\sigma}}(\underline{V}).\underline{n}.\underline{U})\mathrm{d}s, \tag{12.2}$$

where Γ is a contour encompassing the notch root and starting and finishing on the stress-free edges of the notch, $\underline{n}$ its normal pointing toward the origin. Relation (12.2) is based on two properties

(1) If λ is an eigenvalue then there is also $-\lambda$. The so-called dual mode [25] or "super singular" function [27] $r^{-\lambda}\underline{u}^{-}(\theta)$ is a mathematical solution to the previous eigenvalue problem which has presently no special physical meaning (in particular, the elastic energy associated with this function is unbounded in the vicinity of the origin).
(2) For any pair of eigensolutions $r^{\alpha}\underline{u}_{\alpha}(\theta)$ and $r^{\beta}\underline{u}_{\beta}(\theta)$, $\beta \neq -\alpha \Rightarrow \Psi(r^{\alpha}\underline{u}_{\alpha}(\theta), r^{\beta}\underline{u}_{\alpha}(\theta)) = 0$. It is a kind of bi-orthogonality property (be careful that Ψ is not a scalar product) which allows the extraction of coefficient k. This result is a consequence of the path independence of Ψ.

Here, the only role of the dual mode is to be a mathematical extraction function, however these modes will play a greater role in the matched asymptotic expansions procedures both for the inner and outer ones (see (12.14) and (12.15) later in Section 12.3).

The stress condition (e.g., the maximum tensile stress criterion) involves the tensile component σ of the stress tensor $\underline{\underline{\sigma}}$ acting on the presupposed crack path defined by the direction θ_0 prior to its onset, it provides an upper bound to the admissible crack extension lengths a $(\lambda - 1 < 0)$

$$\sigma = kr^{\lambda-1}s(\theta_0) + \cdots \geq \sigma_c \quad \text{for} \quad 0 \leq r \leq a \Rightarrow ka^{\lambda-1}s(\theta_0) \geq \sigma_c. \qquad (12.3)$$

The coefficient $s(\theta_0)$ is a dimensionless constant derived from $\underline{\underline{s}}$ which can be normalized to $s(\theta_0) = 1$ if the failure direction (i.e., θ_0) is known [25]. Relation $(12.3)_2$ is enough to imply $\sigma \geq \sigma_c$ along the whole presupposed crack path since σ is a decreasing function of the distance to the singular point.

As will be shown in Section 12.3, expansions (12.1) can be used to define an expansion of the potential energy variation when a small crack extension appears in direction θ_0. Its leading term provides a lower bound of the crack extension length $(2\lambda - 1 > 0)$

$$G^{\text{inc}}(\theta_0) = -\frac{W(a) - W(0)}{a} = A(\theta_0)k^2a^{2\lambda-1} + \cdots \geq G_c, \qquad (12.4)$$

where $W(x)$ is the potential energy of the structure embedding a crack extension with length x. $G^{\text{inc}}(\theta_0)$ is so-called incremental energy release rate, because it depends on the increment a, emphasis is put on the fact that we do not consider the limit as $a \to 0$ as it is done in the Griffith criterion [28]. The incremental and differential criterion are identical if $\lambda = 1/2$. It is the rate of potential energy change prior to and following the onset of a new crack with length a. The scaling coefficient $A(\theta_0)$ (MPa^{-1}) is another constant depending on the local properties and on the direction θ_0 of the short crack but not on the remote applied load which occurs in (12.4) through the only coefficient k. A complete definition of A is given in the next section.

The compatibility between these two inequalities provides a characteristic length a_c at initiation (Fig. 12.2)

$$a_c = \frac{G_c}{A(\theta_0)}\left(\frac{s(\theta_0)}{\sigma_c}\right)^2. \qquad (12.5)$$

Initiation is in general (i.e., if $\lambda > 1/2$) an unstable mechanism. The crack jumps the length a_c and then go on growing or not, but a_c is not

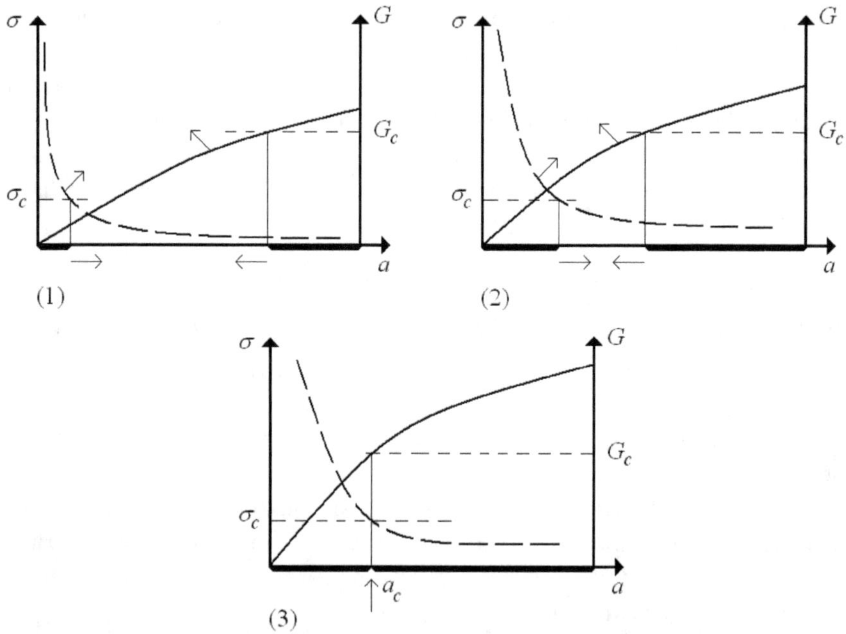

Fig. 12.2. Schematic view of the determination of a_c (12.5). (1) For a small remote load, crack extension lengths fulfilling the stress and the energy conditions are incompatible. (2) When the remote load increases, the two bounds come closer to each other. (3) Failure occurs when the two bounds merge giving a_c. Arrows indicate the motion of curves and points when the remote load increases.

defined as a crack arrest length. It means essentially that below this length the balance between the stored energy and the energy consumed during failure does not hold true, no crack smaller than a_c can be observed. This jump length is still a function of θ_0.

Then we deduce an Irwin-like [29] condition on the GSIF k which plays the classical role of the stress intensity factor (SIF) K_I

$$k \geq k_c = \left(\frac{G_c}{A(\theta_0)}\right)^{1-\lambda}\left(\frac{\sigma_c}{s(\theta_0)}\right)^{2\lambda-1}. \tag{12.6}$$

For a crack embedded in a homogeneous body, then $\lambda = 1/2$, $k = K_I$, relation (12.6) coincides with the well-known Irwin criterion. A straight edge in a homogeneous material is a limit case where there is no stress concentration, then $\lambda = 1$ and inequality (12.6) coincides with the maximum tensile stress criterion.

The direction θ_0 was assumed to be known, if not, one has to check all the possible directions and maximize the denominator in (12.6), i.e., minimize the value of k_c.

A single mode has been involved in (12.1), for a V-notch it corresponds to the symmetric case as shown in Fig. 12.1. Generalizations can be made to account for more complex loadings, it is then necessary to determine both the load causing failure and the direction of the nucleating crack [30].

The computation of the scaling coefficient $A(\theta_0)$, using matched asymptotic expansions, will be the topic of the next section. For simplicity, the dependency on θ_0 will be omitted.

Remark. Martin *et al.* [1, 31, 32] and Hebel *et al.* [33] and their coworkers operate the coupled criterion numerically without going through the semi-analytical asymptotic expansions procedure. The tensile stress σ along the presupposed crack path and the incremental energy release rate G^{inc} (12.4) are extracted from a direct finite elements (FE) computation which requires bringing a special care to the mesh refinement in the vicinity of the region where the new crack initiates. Inequalities (12.3) and (12.4) are employed without going through the calculation of λ, k, s and A and reduce to

$$\sigma = \sigma_{\text{FE}}(r) \geq \sigma_c \quad \text{for } 0 \leq r \leq a \quad \text{and}$$
$$G^{\text{inc}} = -\frac{W_{\text{FE}}(a) - W_{\text{FE}}(0)}{a} \geq G_c \tag{12.7}$$

and a is the smaller length fulfilling the two inequalities. Indeed (12.7) is the most general definition of the coupled criterion that can be used in all cases even if the functions σ_c and G^{inc} are non-monotonic. Such a situation has been encountered in [32] in the study of the fiber-matrix debonding, as well as in [34, 35] when studying different cracking mechanisms in a laminated ceramic material designed with compressive residual stresses, or in [36] about the multi-cracking phenomenon in an oxide layer, again in the presence of residual stresses.

This approach allows studying situations that cannot be taken into account in the asymptotic approach, like crack arrest after a short initiation for instance. However, it does not reveal directly (analytically) the role played by different geometric parameters of the structure like the layer thickness in case of an adhesive layer or an interphase for instance [37].

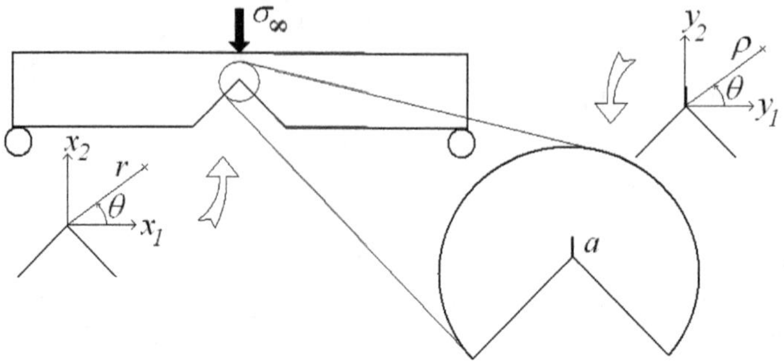

Fig. 12.3. Onset of a short crack with length a at the root of the V-notch.

12.3. Matched Asymptotic Expansions

Solving numerically an elasticity problem in a domain Ω^a embedding a short crack of length a at the root of the V-notch (Fig. 12.3) presents some difficulties because of the small size of the perturbation. Needs for drastic mesh refinements prevent to take into account too small details.

It is better trying to represent the solution in the form of an outer expansion or far field

$$\underline{U}^a(x_1, x_2) = \underline{U}^0(x_1, x_2) + \text{small correction}, \tag{12.8}$$

where $\underline{U}^0$ is solution to the same elasticity problem, but now posed on the unperturbed domain Ω^0 (Fig. 12.1) that can be considered as the limit of Ω^a as $a \to 0$ (the short crack is not visible).

It is clear that this solution $\underline{U}^0$ is a satisfying approximation of $\underline{U}^a$ away from the perturbation, i.e., outside a neighborhood of it, and thence its designation as the outer field (or far field, or remote field).

Evidently, this information is incomplete, particularly when we are interested in the fracture mechanisms. We therefore dilate the space variables by introducing the change of variables $y_i = x_i/a$. In the limit when $a \to 0$, we obtain an unbounded domain Ω^{in} (looking like the enlarge frame in Fig. 12.3) in which the length of the crack is now equal to 1.

We then search for a different representation of the solution under the form of an expansion known as inner expansion or near field

$$\underline{U}^a(x_1, x_2) = \underline{U}^a(ay_1, ay_2) = F_0(a)\underline{V}^0(y_1, y_2) + F_1(a)\underline{V}^1(y_1, y_2) + \cdots, \tag{12.9}$$

where $F_1(a)/F_0(a) \to 0$ as $a \to 0$. But conditions at infinity lack to have well-posed problems for $\underline{V}^0$ and $\underline{V}^1$. Matching rules provide these missing conditions. There must be an intermediate zone (close to the perturbation for the far field and far from it for the near field) where both inner and outer expansions are valid.

The behavior of the far field near the origin is described by the expansion in powers of r as previously encountered in Eq. (12.1)

$$\underline{U}^0(x_1, x_2) = \underline{C} + kr^\lambda \underline{u}(\theta) + \cdots . \tag{12.10}$$

Then the matching conditions can be written as follows:

$$F_0(a)\underline{V}^0(y_1, y_2) \approx \underline{C}, \quad F^1(a)\underline{V}^1(y_1, y_2) \approx ka^\lambda \rho^\lambda \underline{u}(\theta) \tag{12.11}$$

when $\rho = r/a = \sqrt{y_1^2 + y_2^2} \to \infty$ (the symbol $\approx$ means here "behaves like at infinity"), thus

$$F_0(a) = 1; \quad \underline{V}^0(y_1, y_2) = \underline{C}; \quad F_1(a) = ka^\lambda; \quad \underline{V}^1(y_1, y_2) \approx \rho^\lambda \underline{u}(\theta). \tag{12.12}$$

This matching statement is nothing else than the so-called remote load at infinity. Proceeding by superposition, it comes

$$\underline{V}^1(y_1, y_2) = \rho^\lambda \underline{u}(\theta) + \hat{\underline{V}}^1(y_1, y_2) \quad \text{with} \quad \hat{\underline{V}}^1(y_1, y_2) \approx 0. \tag{12.13}$$

More precisely, the behavior of $\hat{\underline{V}}^1(y_1, y_2)$ at infinity can be described by the dual mode $\rho^{-\lambda}\underline{u}^-(\theta)$ to $\rho^\lambda \underline{u}(\theta)$ (see Section 12.2)

$$\hat{\underline{V}}^1(y_1, y_2) = \kappa \rho^{-\lambda} \underline{u}^-(\theta) + \cdots . \tag{12.14}$$

This expansion is the analogous to (12.10) but at infinity now, κ is the GSIF and missing terms tend to 0 faster than $\rho^{-\lambda}$ at infinity. This detail is generally useless for our purpose, but it may play a role in other issues. It has been used recently to determine the length of a crack using full field measurements and digital image correlation (DIC) [38]. It allows specifying the small correction mentioned in (12.8)

$$\underline{U}^a(x_1, x_2) = \underline{U}^0(x_1, x_2) + k\kappa a^{2\lambda}(r^{-\lambda}\underline{u}^-(\theta) + \hat{\underline{U}}^1(x_1, x_2)) + \cdots . \tag{12.15}$$

Finally, eq. (12.9) rewrites

$$\underline{U}^a(x_1, x_2) = \underline{U}^a(ay_1, ay_2) = \underline{C} + ka^\lambda \underline{V}^1(y_1, y_2) + \cdots . \tag{12.16}$$

The function $\underline{V}^1(y_1, y_2)$ is computed once for all by FE in an artificially bounded (at a large distance of the perturbation) domain with either prescribed displacements or forces along the new fictitious boundary.

We have now to our disposal a description of the elastic solution prior to and following the onset of a short crack and we are able to calculate the change in potential energy $W(a) - W(0)$ which can be expressed by mean of the path independent integral Ψ already encountered in (12.2):

$$-(W(a) - W(0)) = \Psi(\underline{U}^a, \underline{U}^0). \tag{12.17}$$

Then replacing the above expansions, once for $a = 0$ and once for $a \neq 0$, into (12.17) leads to the announced expression (12.4) with $G^{\mathrm{inc}} = -(W(a) - W(0))/a$ and

$$A = \Psi(\underline{V}^1(y_1, y_2), \rho^\lambda \underline{u}(\theta)). \tag{12.18}$$

Figure 12.4 shows the dimensionless function $A^* = E^*A$ (where $E^* = E$ in plane stress and $E^* = E/(1-\nu^2)$ in plane strain, with E Young's modulus and ν Poisson's ratio of the homogeneous isotropic material) for different V-notch openings ω (Fig. 12.1) and for a crack located along the bisector (symmetric case). It can be seen as a master curve valid for any elastic isotropic material, the role of Poisson's ratio in A has been verified numerically. Note that $A^* = 2\pi$ for $\omega = 0°$ as a consequence of the normalization of the eigenmode (12.1). It is such that the tensile stress

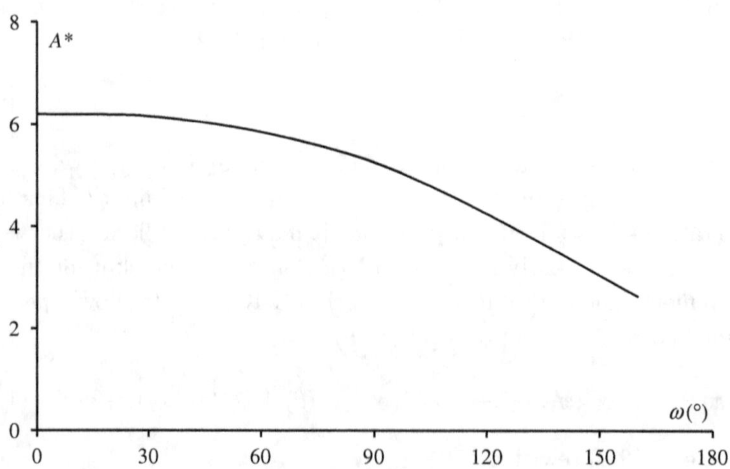

Fig. 12.4. The dimensionless coefficient A^* vs. the V-notch opening ω in the symmetric case.

$\sigma = k/r^{\lambda-1}$ along the bisector, leading to $\sigma = k/\sqrt{r}$ for $\omega = 0°$ (a crack) whereas it is usually $\sigma = k/\sqrt{2\pi r}$.

12.4. Application to the Crack Onset at a V-notch in a Homogeneous Material

Tensile tests have been carried out on PMMA V-notched specimens (E = 3250 MPa, $\nu = 0.3$, $G_c = 0.325$ MPa.mm, $\sigma_c = 75$ MPa) for different V-notch openings from 30 to 160° (Fig. 12.5) [39].

The tensile test was then numerically simulated by finite elements for an arbitrary prescribed load F_0 (note here that special care must be taken given the lack of symmetry of the specimen) and the GSIF k_0 was extracted using (12.2). A scaling with the critical value k_c (12.6) provides the corresponding force $F = F_0 \times k_c/k_0$ at failure.

A comparison between predicted and measured failure forces is illustrated in Fig. 12.6 which exhibits a fair agreement.

The above results rely on a symmetric loading. The generalization to more complex cases is straightforward. The two terms Williams' expansion (12.1) can be written as

$$\begin{cases} \underline{U}(r,\theta) = \underline{C} + k_1 r^{\lambda_1} \underline{u}^1(\theta) + k_2 r^{\lambda_2} \underline{u}^2(\theta) + \cdots \\ \underline{\underline{\sigma}}(r,\theta) = k_1\, r^{\lambda_1 - 1} \underline{\underline{s}}^1(\theta) + k_2\, r^{\lambda_2 - 1} \underline{\underline{s}}^2(\theta) + \cdots \end{cases} \quad \text{with } 1/2 \le \lambda_1 \le \lambda_2 \le 1. \tag{12.19}$$

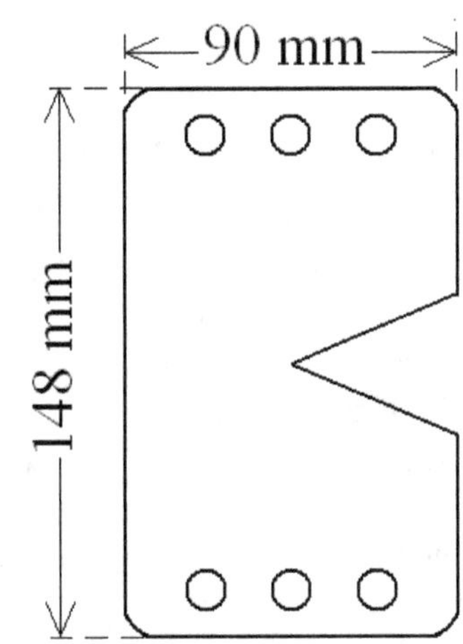

Fig. 12.5. The PMMA compact tension V-notched specimen (CTS).

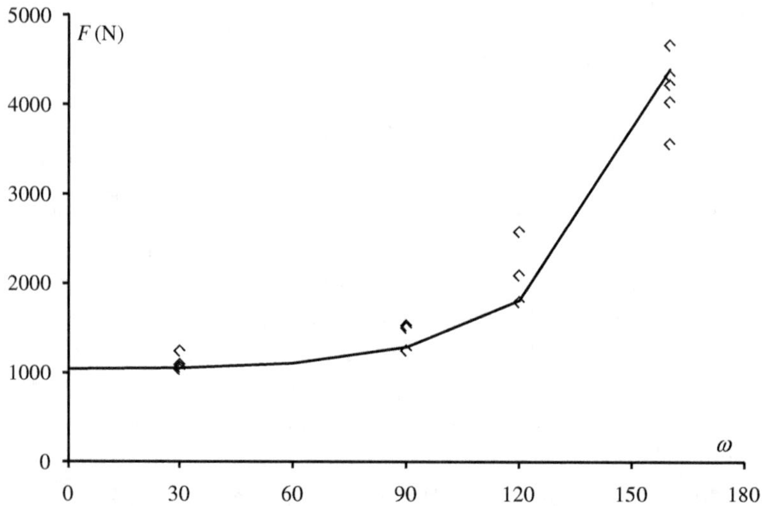

Fig. 12.6. Applied force F at failure of a V-notched specimens of PMMA as a function of the notch opening ω. Comparison between experiments (diamonds) and prediction (solid line) using the coupled criterion [39].

The leading term (previously without index and now with the index 1 in (12.19)) is associated with a symmetric mode whereas the corresponding mode of the second term is anti-symmetric (thus in the previous case $k_2 = 0$ because of symmetry). Using matched asymptotic expansions and applying the coupled criterion leads to more intricate relations taking into account a mode mix parameter [30] (see also Sections 12.7 and 12.8 on that topic).

12.5. Application to the Deflection of Transverse Cracks

We now consider a transverse crack as depicted schematically in Fig. 12.7. Despite a pre-existing crack, the singular exponent at its tip which impinges an interface is not 1/2 as usual and the situation differs from that of a crack (Fig. 12.8). For homogeneous isotropic components, if $E_2 > E_1$ (first case) then $\lambda > 1/2$ (weak singularity, it is less harmful than a crack) and vice versa if $E_2 < E_1$ (second case) then $\lambda < 1/2$ (strong singularity, more harmful than a crack) [40]. Here E_i holds for Young's modulus of ply number i (it is assumed that the two Poisson's ratios ν_1 and ν_2 equal, otherwise the rule is close to the above one but slightly altered by the contrast in ν). This obviously leads to substantially different results in terms of rupture.

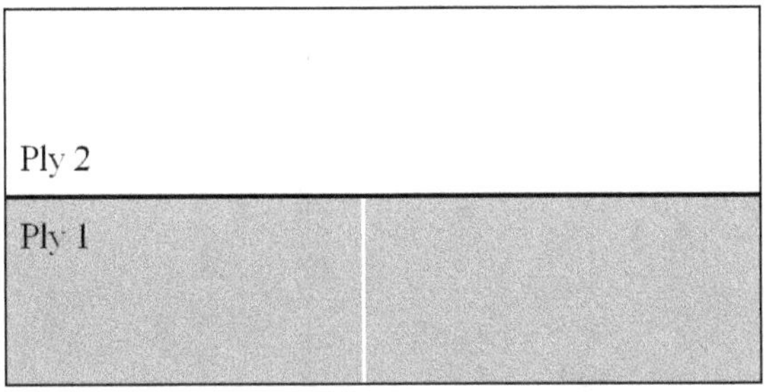

Fig. 12.7. Schematic view of a transverse crack impinging on an interface.

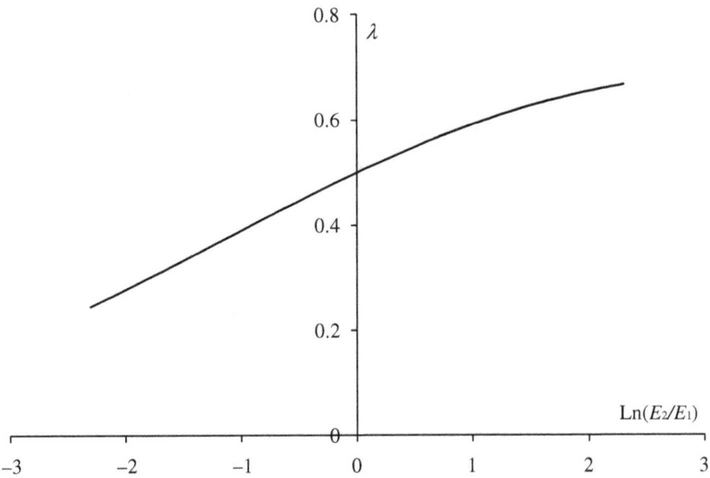

Fig. 12.8. The singular exponent vs. Young's modulus ratio of the two adjacent materials for equal Poisson's ratio $\nu_1 = \nu_2 = 0.3$ in the two materials.

We immediately notice, according to (12.4), that if $a \to 0$ then $G^{\text{inc}} \to 0$ in the first case whereas $G^{\text{inc}} \to \infty$ in the second. This property also affects the (differential) energy release rate G of a crack approaching and crossing the interface. When the crack approaches, it remains a ligament with length l between the crack tip and the interface (Fig. 12.13) and $G \to 0$ (respectively, $G \to \infty$) for a weak singularity (respectively, strong singularity) as $l \to 0$. Symmetrically, after crossing the interface the crack tip is at a distance a of it (Fig. 12.7) and G increases from 0 (weak singularity) or decreases from infinity (strong singularity) as a increases. This behavior is shown in Fig. 12.9, for different Young's moduli contrast, $E_2/E_1 = 0.1$, 0.2, 0.5, 1, 2, 5 and 10. Results are obtained using FE

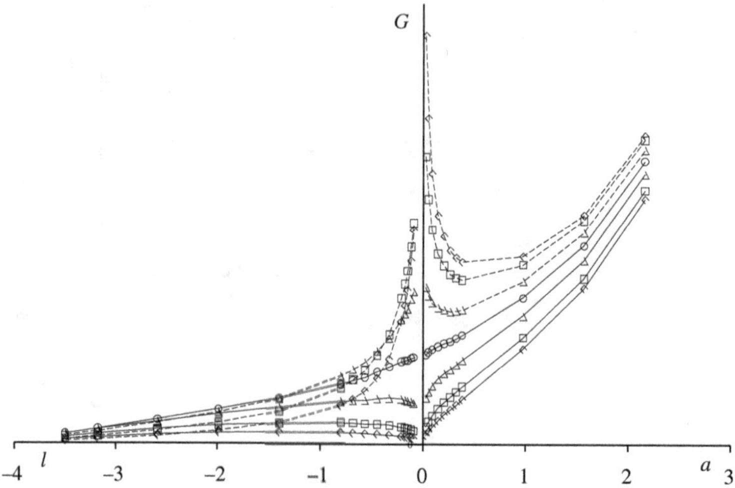

Fig. 12.9. The behavior of the energy release rate when a crack approaches an interface (left) and then crosses it (right) for different Young's moduli contrast, $E_2/E_1 = 0.1$ (dashed line and diamonds), 0.2 (dashed line and squares), 0.5 (dashed line and triangles), 1 (solid line and circles), 2 (solid line and triangles), 5 (solid line and squares) and 10 (solid line and diamonds). Units are meaningless, the emphasis is put on the general trends: increasing or decreasing functions when approaching the interface.

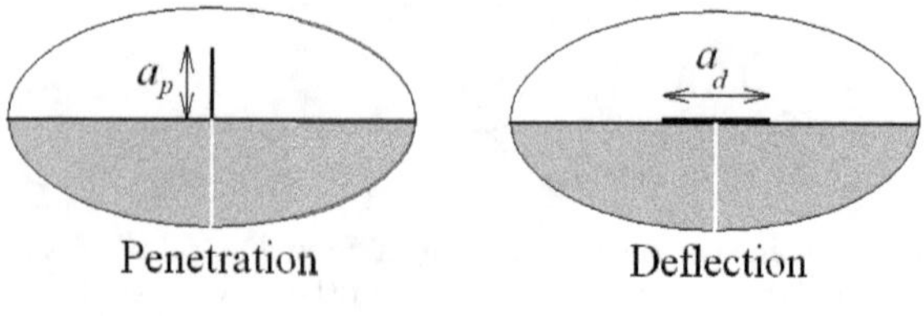

Fig. 12.10. The mechanisms of crack penetration and deflection.

and a variable crack tip location, counted negative if the crack is growing toward the interface and positive after the crossing (Fig. 12.9). Even if the energy is globally calculated at the structure level, it requires a strong mesh refinement in the area of interest to have a good geometrical description.

The question that arises now is the following: does such a transverse crack stop, penetrate material 2 or deflect along the interface to give a delamination crack (Fig. 12.10)?

Let us consider again inequality (12.4) in two cases, penetration in material 2 (index p) and deflection along the interface (index d), $G_c^{(1)}$ and

$G_c^{(2)}$ hold, respectively, for the interface and material 2 toughness

$$G_d^{\text{inc}} = A_d k^2 a_d^{2\lambda-1} + \cdots \geq G_c^{(1)} \quad \text{and} \quad G_p^{\text{inc}} = A_p k^2 a_p^{2\lambda-1} + \cdots \geq G_c^{(2)}. \tag{12.20}$$

Two cases can be considered, a double symmetric deflection (Fig. 12.10) or a single asymmetric one; the only change occurs in the coefficient A_d but does not lead to a big difference [41]. The forthcoming analysis will be carried out in the first case.

Deflection is promoted if the first inequality in (12.20) is fulfilled whereas the second one is not, then

$$\frac{G_c^{(1)}}{G_c^{(2)}} = R \leq \frac{A_d}{A_p}\left(\frac{a_d}{a_p}\right)^{2\lambda-1}. \tag{12.21}$$

The dimensionless ratio A_d/A_p is plotted in Fig. 12.11 for various material contrasts E_2/E_1 (E_i is Young's modulus of material i and $\nu_1 = \nu_2 = 0.3$).

He and Hutchinson [42] obtained a similar result (although differently) but simplified thanks to a dubious assumption [43, 44]. They consider the (differential) energy release rates G_p and G_d, respectively, at the tip of a

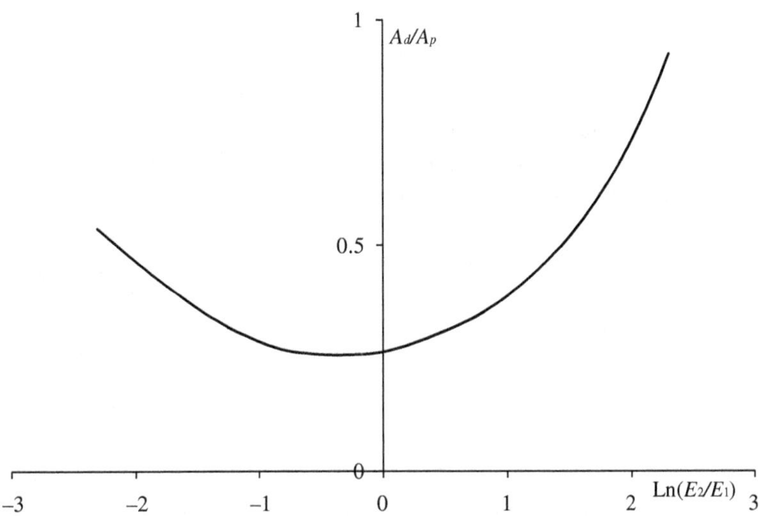

Fig. 12.11. The dimensionless ratio A_d/A_p vs. Young's moduli ratio of the two adjacent materials for a double symmetric deflection.

penetrated crack and a deflected one, the two crack extensions being equal, i.e., $a_d = a_p$, and obtain a condition on the toughness ratio of the interface and material 2 which is clearly equivalent to the ratio A_d/A_p according to (12.20) if $a_d = a_p$

$$R \leq \frac{G_d}{G_p}. \tag{12.22}$$

A discussion on this specific point can be found in [45, 46].

Clearly, it is possible to determine the two characteristic lengths a_d and a_p involved in (12.21) using the stress condition provided $\lambda > 1/2$ (otherwise both G and σ are decreasing functions of the distance to the singular point and the coupled criterion can no longer be used).

Case $\lambda > 1/2$. According to (12.5) ($\sigma_c^{(I)}$ and $\sigma_c^{(2)}$ hold for the tensile strengths of the interface and material 2 respectively) it

$$a_d = \frac{G_c^{(I)}}{A_d}\left(\frac{s_d}{\sigma_c^{(I)}}\right)^2; \quad a_p = \frac{G_c^{(2)}}{A_p}\left(\frac{s_p}{\sigma_c^{(2)}}\right)^2. \tag{12.23}$$

Thus, deflection is promoted if

$$R \leq \frac{A_d}{A_p}\left(\frac{s_d}{s_p}\frac{\sigma_c^{(2)}}{\sigma_c^{(I)}}\right)^{\frac{2\lambda-1}{1-\lambda}}. \tag{12.24}$$

The special case $\lambda = 1$ cannot be met; it would correspond to an infinitely compliant material 1 compared to material 2. Knowing that the ratio s_d/s_p remains of the same order of magnitude than 1, it is clear in (12.24) that the tensile strengths ratio plays a crucial role which can significantly alter the criterion proposed by He and Hutchinson.

As illustrated in Fig. 12.12, the deflection will be even more and more promoted as material 2 is more and more resistant (i.e., $\sigma_c^{(2)} > \sigma_c^{(1)}$). It is to be noted that Parmigiani and Thouless [45] derive the same tendency with the use of cohesive zone models (CZM).

Reference is made in both cases to the tensile stress; it is clear for penetration but less obvious for deflection. However, considering the eigenmode governing the stress field before crack propagation, one can check that the tensile component σ is equal or larger than the shear one τ, the ratio σ/τ grows from 1 to 2.8 as E_2/E_1 varies from 1 to 10. Knowing that, in addition, the shear failure is generally more difficult than in tension, it seems reasonable to consider only the tensile component.

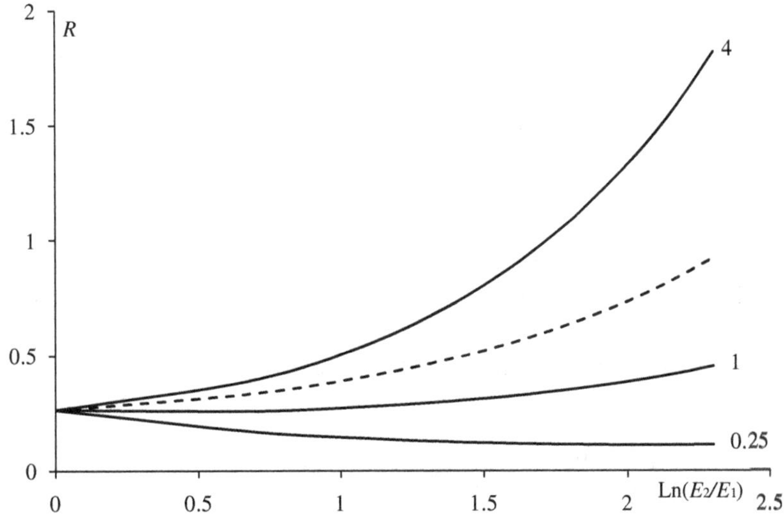

Fig. 12.12. Failure map of the criterion (12.24) for different values of the strengths ratio $\sigma_c^{(2)}/\sigma_c^{(1)} = 0.25$, 1, 4 function of the material contrast E_2/E_1. The dashed line corresponds to A_d/A_p (Fig. 12.11). Below the continuous line, conditions are favorable to a deflection along the interface, above to the penetration in material 2.

This analysis can be included in a homogenization process where the domain shown in Fig. 12.7 would be the representative volume element (RVE). Playing on the slenderness of this cell allows taking into account different densities of transverse cracks [5].

Case $\lambda < 1/2$. The coupled criterion approach treated above is no longer valid because the energy release rate is now a decreasing function of the distance to the singular point (see (12.4)) and the energy condition no longer gives any lower bound for the crack extension length [42, 47]. Under a monotonic loading the crack grows continuously, there is no crack jump. Moreover, according to (12.20), G_d and G_p tend to infinity as a_d and a_p decrease to 0 which prevent the direct use of the energy release rate at the very beginning of the crack growth process. No rigorous conclusion can be derived in this situation. He and Hutchinson [42] still propose to use (12.22) or equivalently the ratio A_d/A_p. Another approach based on the maximum dissipated energy is proposed by Leguillon *et al.* [41], however it better corresponds to the geometrical situation analyzed in the following section (Fig. 12.13).

12.6. The Cook and Gordon Mechanism

Due to the decay to 0 or the unbounded growth of the energy release rate (Fig. 12.9), it should be pointed out that the geometric situation described in Fig. 12.7 cannot be achieved by a crack growing in material 1 and approaching the interface. It can only be obtained by a mechanical action like a saw cut.

Otherwise, we have to consider a crack in material 1 with its tip at small distance l of the interface as shown in Fig. 12.13 [46, 47] and a possible debonding of the interface ahead of the crack tip as Cook and Gordon [48] studied.

Assuming a small increment $\delta l \ll l$ at the tip of this crack and according to (12.4), the (differential) energy release rate is

$$G_1 = -\lim_{\delta a_d \to 0} \frac{W(l+\delta l) - W(l)}{\delta l} = k^2 A_1 \lim_{\delta a_d \to 0} \frac{(l+\delta l)^{2\lambda} - l^{2\lambda}}{\delta l} + \cdots$$
$$= 2\lambda k^2 A_1 l^{2\lambda - 1} + \cdots . \quad (12.25)$$

A_1 is given by (12.18) where the perturbation is the small ligament with width l instead of a crack extension. Moreover, since the crack is growing in material 1

$$G_1 = G_c^{(1)} \Rightarrow k^2 l^{2\lambda-1} = \frac{G_c^{(1)}}{2\lambda A_1}. \quad (12.26)$$

This relation means that, in these conditions, knowing l or the applied load is somewhat equivalent.

Case 1: If $\lambda > 1/2$, G_1 decreases to 0 as $l \to 0$ and thus drops below $G_c^{(1)}$ (material 1 toughness). An overload must be brought for the situation to evolve. Then a competition arises: the crack still grows in material 1, the

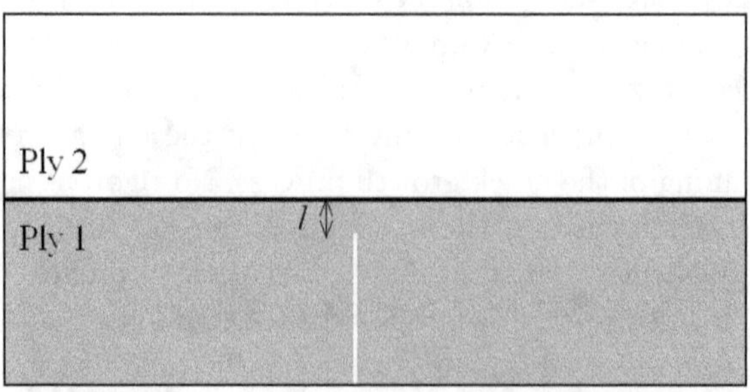

Fig. 12.13. A crack growing in material 1 and approaching the interface at a distance l.

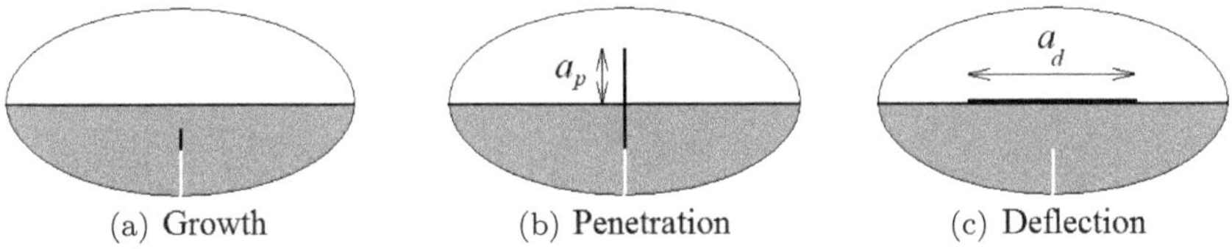

Fig. 12.14. The competition between the crack growth in material 1 (a), the crack jump in material 2 (b) and the interface debonding (c).

interface debonds ahead of the crack tip or the crack jumps and penetrates material 2 (Fig. 12.14) [46]. This latter mechanism will not be discussed here. Another mechanism called step-over, where the crack reinitiates in the second material, leaving a ligament in its wake, was discussed in [46, 49].

There are now two small parameters l and a_d (respectively, a_p) in case of debonding (respectively, penetration), which is an additional difficulty. If one is very small compared to the other, it can be neglected in a first step. If they are of the same order of magnitude the expansions can be indifferently carried out with respect to one or the other small parameter. For technical reasons it is easier to use l. By analogy with the single parameter case, the stress and energy conditions now write (we refer to [50] for the details of the proof)

$$\begin{cases} \sigma = kl^{\lambda-1}\tilde{\sigma}(\mu_d) \geq \sigma_c^{(I)}, \\ G_d^{\text{inc}} = k^2 B_d(\mu_d) l^{2\lambda-1} \geq G_c^{(I)}, \end{cases} \tag{12.27}$$

where $\mu_d = a_d/l$. The function B_d (Fig. 12.15) is an increasing function of μ_d and plays the role of A_d. It is derived from the calculation of A (2) considering three cases: the "unperturbed" one (Fig. 12.7) and successively cases illustrated in Figs. 12.13 and 12.14(b) or 12.14(c). Whereas $\tilde{\sigma}$ is a decreasing function (Fig. 12.16) and plays the role of s, it is the tensile stress associated with $\underline{V}^1$ along the presupposed crack path prior to any crack extension (i.e. with the inner term calculated on the geometry of Fig. 12.13).

The equation providing the dimensionless characteristic length μ_d derives from (12.26) and $(12.27)_2$

$$B_d(\mu_d) = 2\lambda A_1 \frac{G_c^{(I)}}{G_c^{(1)}}. \tag{12.28}$$

The dimensionless debonding length μ_d is small if the interface toughness $G_c^{(I)}$ is small.

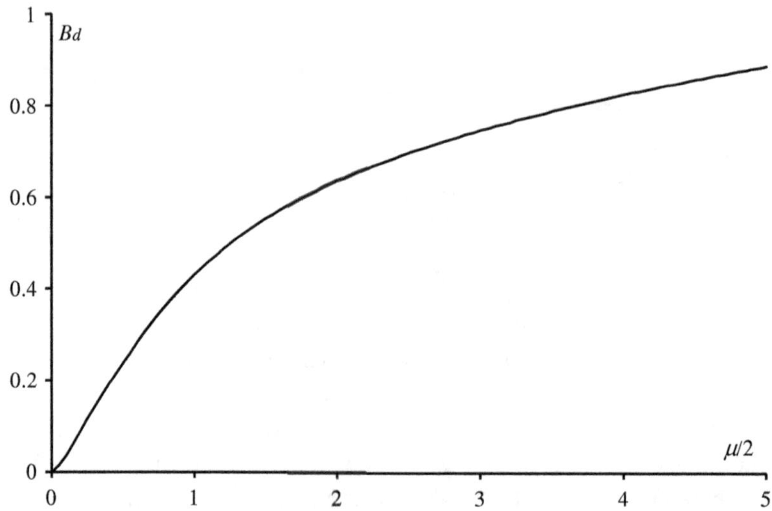

Fig. 12.15. The function $B_d(\mu)$ (MPa^{-1}) for a material contrast $E_2/E_1 = 10$.

For a material contrast $E_2/E_1 = 10$, $\lambda = 0.667$ (Fig. 12.8), $A_1 = 0.479\,\mathrm{MPa}^{-1}$, thus if $G_c^{(I)} = G_c^{(1)}$, then from (12.26) $B_d(\mu_d) = 0.639\,\mathrm{MPa}^{-1}$, $\mu_d/2 = 1.9$ (Fig. 12.15) and $\tilde{\sigma}(\mu_d) = 0.345$ (Fig. 12.16). Thus, the condition for an interface debonding ahead of the primary crack is

$$k \geq k_c = \left(\frac{G_c^{(I)}}{B_d(\mu_d)}\right)^{1-\lambda} \left(\frac{\sigma_c^{(I)}}{\tilde{\sigma}(\mu_d)}\right)^{2\lambda-1}. \tag{12.29}$$

Note that using (12.26) and $(12.27)_1$ gives l, although it is not useful to settle the criterion (12.29) which requires only μ_d

$$l = \frac{G_c^{(1)}}{2\lambda A_1}\left(\frac{\tilde{\sigma}(\mu_d)}{\sigma_c^{(I)}}\right)^2. \tag{12.30}$$

The ligament width is small if the tensile strength $\sigma_c^{(I)}$ is high. Since $a_d = \mu_d l$ and according to (12.28) and (12.30), the debond length is large for a high toughness and a small tensile strength of the interface which is often the case for polymer adhesives for instance.

The primary crack may stop or not at the distance l, it depends on how the energy release rate G_1 evolves after the onset of the debonding. Figure 12.17 shows the ratio $\gamma = G_1/G_c^{(1)}$ function of the dimensionless debonding length μ_d still for $E_2/E_1 = 10$ and $\nu = 0.3$ in both materials.

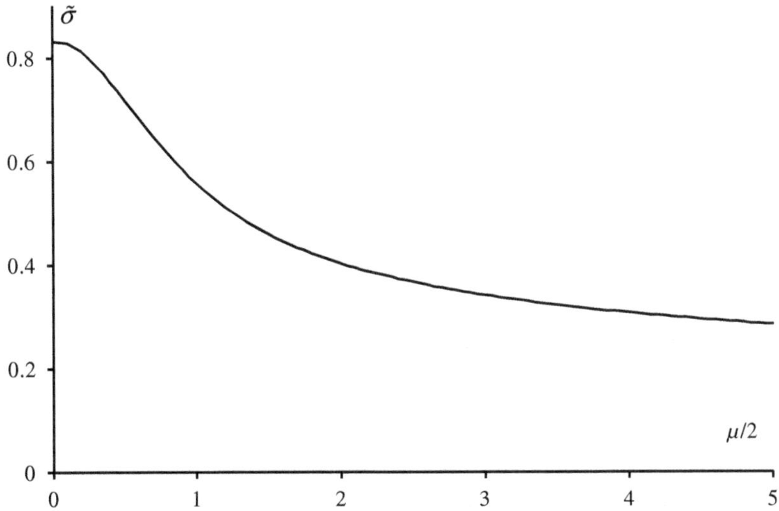

Fig. 12.16. The function $\tilde{\sigma}(\mu)$ for a material contrast $E_2/E_1 = 10$.

Obviously, on the one hand, if the characteristic debonding length μ_d is smaller than a given value (roughly $\mu_d/2 = 4$ in the present case, Fig. 12.17) the energy release rate increases as debond occurs and the primary crack restarts and breaks definitely the ligament. On the other hand, this ligament does not disappear and thus can only be observed if the debonding length is large ($\mu_d/2 = 4$ in the present case).

Case 2: If $\lambda < 1/2$, G_1 increases as $l \to 0$. For a given l, if the load (i.e., a given GSIF k) is such that (12.26) holds true then the crack accelerates toward the interface. As it impinges on the interface, there is an excess of energy in the balance

$$\Delta W_1 = kA_1 l^{2\lambda} - G_c^{(1)} l = G_c^{(1)} \left(\frac{1}{2\lambda} - 1 \right) l. \tag{12.31}$$

Then the crack deflects and the (differential) energy release rate G_d decreases as the debond length increases (it is calculated using a small increment $\delta a_d \ll a_d$ at the tip of the deflected crack and passing to the limit $\delta a_d \to 0$ like for G_1, see (12.25)). Following (12.25) and (12.26), it drops below the interface toughness at a distance a_d such that

$$G_d = 2\lambda k^2 A_d a_d^{2\lambda-1} = G_c^{(I)} \Rightarrow \left(\frac{a_d}{l} \right)^{2\lambda-1} = \frac{G_c^{(I)}}{G_c^{(1)}} \frac{A_1}{A_d}. \tag{12.32}$$

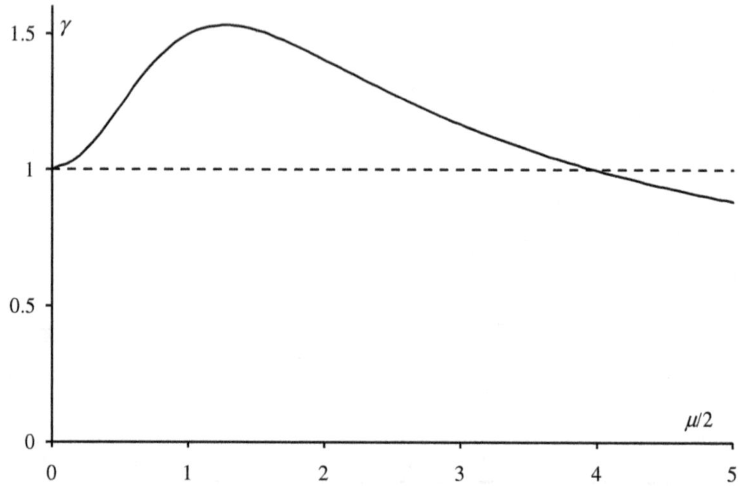

Fig. 12.17. The evolution of the energy release rate (normalized by $G_c^{(1)}$) at the tip of the primary crack after the onset of the interface debonding for a material contrast $E_2/E_1 = 10$.

At this point, the excess of energy is now

$$\Delta W = \Delta W_1 + \Delta W_d = \left(G_c^{(1)} l + G_c^{(I)} a_d\right)\left(\frac{1}{2\lambda} - 1\right). \tag{12.33}$$

Thus, the crack can go on growing on a length δa_d until it consumes this excess of energy

$$\delta a_d = \frac{\Delta W}{G_c^{(I)}} = \left(\frac{G_c^{(1)}}{G_c^{(I)}} l + a_d\right)\left(\frac{1}{2\lambda} - 1\right). \tag{12.34}$$

Of course, $a_d + \delta a_d$ is an upper bound of the delamination lengths, part of the excess of energy can be dissipated by dynamic effects, elastic waves providing noise for example.

As already mentioned, the comparison with a crack advancing in a straight line and penetrating material 2 is not considered here. The mechanism is more difficult to describe and is the subject of a work in progress.

The modeling described in Sections 12.5 and 12.6 extends to the anisotropic case provided the problem still splits into a plane and an antiplane problems. It is the case for the cross-ply laminates in CFRP, each layer being orthotropic in the appropriate basis, but not for angle-ply laminates [1].

12.7. The Interface Crack Growing Along the Interface — Delamination

As above, the analysis is restricted to 2D or generalized 2D framework. All the results described in Sections 12.2 and 12.3 can be extended to the case of the interface crack (Fig. 12.18) characterized by a complex exponent with a real part 0.5 describing the singularity at its tip. Williams' expansion in the vicinity of O (the upper bar denotes the complex conjugate) can be written as a particular case of expansion (12.19)

$$\begin{cases} \underline{U}(r,\theta) = \underline{C} + K\, r^{1/2+\mathrm{i}\varepsilon}\underline{u}(\theta) + \bar{K}\, r^{1/2-\mathrm{i}\varepsilon}\bar{\underline{u}}(\theta) + \cdots, \\ \underline{\underline{\sigma}}(r,\theta) = K\, r^{-1/2+\mathrm{i}\varepsilon}\underline{\underline{s}}(\theta) + \bar{K}\, r^{-1/2-\mathrm{i}\varepsilon}\bar{\underline{\underline{s}}}(\theta) + \cdots. \end{cases} \tag{12.35}$$

Notations are similar to that of Section 12.2, K is the complex stress intensity factor and ε is a real number function of the contrast between the adjacent materials [51, 52]. Note that, in this so-called open model, complex exponents involve oscillations and lead to some intricate formulas and even to overlapping of the two crack faces. However, in most cases of bi-materials, this overlapping zone is extremely small making the approximation admissible [52] and avoiding a much more complex formulation involving a contact problem [53].

It is convenient to use a dimensionless mode mix parameter m [52], a kind of generalization of the usual parameter $m = K_{II}/K_I$, the main difference being that now it depends on r

$$m(r) = \frac{\bar{K}}{K} r^{-2\mathrm{i}\varepsilon}. \tag{12.36}$$

Thus (12.35) can be rewritten to highlight the role of K

$$\begin{cases} \underline{U}(r,\theta) = \underline{C} + K r^{1/2+\mathrm{i}\varepsilon}\, (\underline{u}(\theta) + m(r)\bar{\underline{u}}(\theta) + \cdots), \\ \underline{\underline{\sigma}}(r,\theta) = K r^{-1/2+\mathrm{i}\varepsilon}(\underline{\underline{s}}(\theta) + m(r)\, \bar{\underline{\underline{s}}}(\theta) + \cdots). \end{cases} \tag{12.37}$$

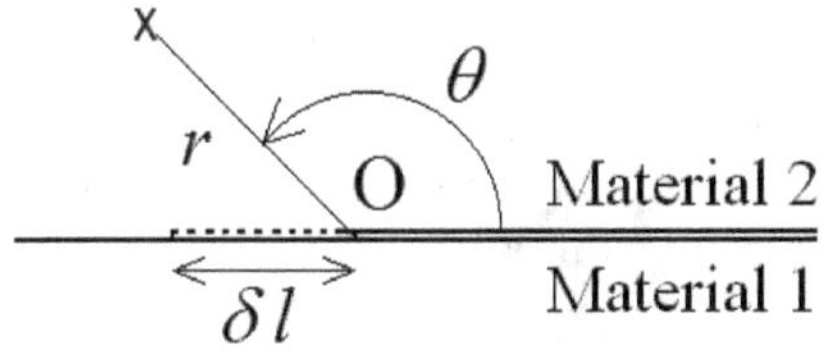

Fig. 12.18. Growth of the interface crack — Delamination.

Despite the oscillations of the displacement and stress fields, fortunately, the energy release rate does not involve oscillations

$$G = -\frac{\partial W^{\mathrm{P}}}{\partial l} = 2K\bar{K}\chi + \cdots, \tag{12.38}$$

where χ is a known real constant parameter [52], then Griffith's criterion can still be written in its usual form

$$G \geq G_c, \tag{12.39}$$

where G_{c} is the interface toughness, a "material" parameter that is difficult to determine because it may depend on the mode mix [54–56].

12.8. The Crack Kinking out of the Interface

A new difficulty arises in this case, unlike the delamination crack, the energy release rate G, i.e., the limit

$$\lim_{\delta l \to 0} \frac{W(0) - W(\delta l)}{\delta l}$$

does not exist due to oscillating terms like $\sin(\varepsilon \ln \delta l)$). To overcome this difficulty, various solutions are suggested in the literature: (i) to take into account a contact zone at the crack tip [53]; to neglect ε that is often small (≤ 0.1 for isotropic bimaterials) [18]; (iii) to measure the quantities of interest at a given fixed distance of O [52].

We propose an alternative approach: to use the T-stress (that plays a role only in the case of crack kinking) and the coupled criterion. We will develop this less traditional approach in this section.

The T-stress occurs as an additional term in the Williams expansion (12.35)

$$\begin{cases} \underline{U}(r,\theta) = \underline{C} + K\, r^{1/2+\mathrm{i}\varepsilon}\underline{u}(\theta) + \bar{K}\, r^{1/2-\mathrm{i}\varepsilon}\underline{\bar{u}}(\theta) + T\, r\underline{t}(\theta) + \cdots, \\ \underline{\underline{\sigma}}(r,\theta) = K\, r^{-1/2+\mathrm{i}\varepsilon}\underline{\underline{s}}(\theta) + \bar{K}\, r^{-1/2-\mathrm{i}\varepsilon}\underline{\underline{\bar{s}}}(\theta) + T\, \underline{\underline{\tau}}(\theta) + \cdots, \end{cases} \tag{12.40}$$

where the real function $r\underline{t}(\theta)$ is such that the associated stress field $\underline{\underline{\tau}}$ is constant and fulfills $\tau_{11} = 1$, $\tau_{12} = \tau_{22} = 0$, and where T is the corresponding real generalized stress intensity factor.

There are now two mode mix parameters, both depending on r

$$m(r) = \frac{\bar{K}}{K} r^{-2\mathrm{i}\varepsilon}; \quad M(r) = \frac{T}{K} r^{1/2-\mathrm{i}\varepsilon} \tag{12.41}$$

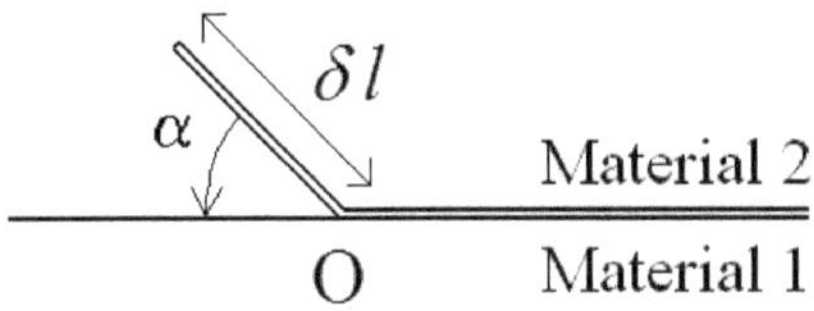

Fig. 12.19. The crack kinking out of the interface.

allowing to rewrite (12.40) as

$$\begin{cases} \underline{U}(r,\theta) = \underline{C} + Kr^{1/2+i\varepsilon}\left(\underline{u}(\theta) + m(r)\bar{\underline{u}}(\theta) + M(r)\,\underline{t}(\theta) + \cdots\right), \\ \underline{\underline{\sigma}}(r,\theta) = Kr^{-1/2+i\varepsilon}(\underline{\underline{s}}(\theta) + m(r)\,\bar{\underline{\underline{s}}}(\theta) + M(r)\,\underline{\underline{\tau}}(\theta) + \cdots) \end{cases} \tag{12.42}$$

as well as the incremental energy release rate and energy condition (α is the kink angle (Fig. 12.19))

$$G^{\text{inc}} = -\delta W/\delta l = K\bar{K}X(\alpha, m(\delta l),\ M(\delta l)) + \cdots \geq G_{\text{c}}, \tag{12.43}$$

where X is a coefficient depending on α that can be computed from a matched asymptotic expansions procedure [17].

The stress condition takes the following form:

$$\begin{aligned} &K\bar{K}Y(\alpha, m(\delta l), M(\delta l)) \geq \delta l\, \sigma_c^2 \quad \text{with} \\ &Y(\alpha, m(\delta l), M(\delta l)) = |s_{\theta\theta}(\alpha) + m(\delta l)\ \bar{s}_{\theta\theta}(\alpha) + M(\delta l)\ \tau_{\theta\theta}(\alpha) + \cdots|^2 \end{aligned} \tag{12.44}$$

leading to a complex version of the coupled criterion [57]. The compatibility between the two conditions gives an equation for the crack initiation length δl_{c} function of α

$$\delta l_{\text{c}} \frac{X(\alpha, m(\delta l_{\text{c}}), M(\delta l_{\text{c}}))}{Y(\alpha, m(\delta l_{\text{c}}), M(\delta l_{\text{c}}))} = \frac{G_{\text{c}}}{\sigma_{\text{c}}^2}. \tag{12.45}$$

The indisputable advantage of the approach is that δl_{c} is not arbitrarily chosen. Once (12.45) is solved, a condition on $|K|$ comes out for the crack initiation in the direction α

$$|K| \geq K_\alpha = \sqrt{\frac{G_c}{X(\alpha, m(\delta l_{\text{c}}), M(\delta l_{\text{c}}))}}, \quad K_{\text{f}} = \underset{\alpha}{\text{Min}}\, K_\alpha. \tag{12.46}$$

The actual kink angle α_{c} maximizes the denominator, i.e., minimizes K_α giving K_{f} (i.e., $|K|$ at failure) (generalization of the G-max criterion).

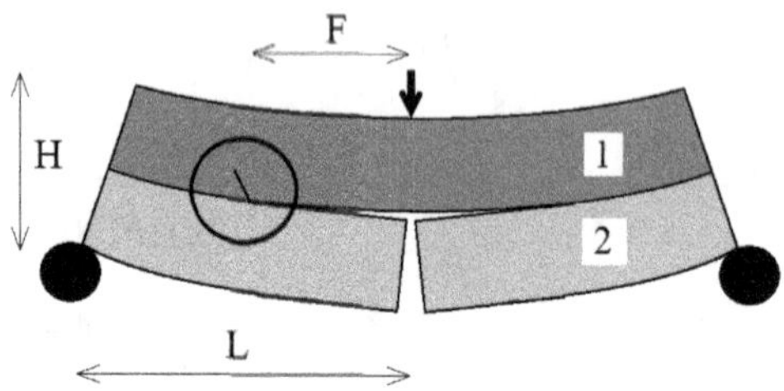

Fig. 12.20. 3-point bending test on a notched bi-layer and crack kinking out of the interface.

It must be pointed out that the T-stress together with the coupled criterion defines an initiation length that allows getting rid of the oscillation terms in the prediction of the critical load and the deflection angle of a crack kinking out of an interface.

As an application of the complex coupled criterion, the following simulation of the 3-point bending test on a bi-material (Fig. 12.20) is proposed with $F/L = 0.25$, $H/L = 0.2$.

The parameter M is Young's modulus ratio. If $M > 1$ ($M = 2, 10, 50$), the stiffer material ($E_1 = 300\,\text{GPa}$, $\sigma_{1c} = 400\,\text{MPa}$, $G_{\text{Ic}} = 0.05\,\text{MPa mm}$, $K_{\text{Ic}} = 4.06\,\text{MPa m}^{1/2}$) is in the upper position and the stiffness of the compliant material varies. Otherwise, $M < 1$ ($R = 0.5, 0.1, 0.02$), the compliant material ($E_1 = 3\,\text{GPa}$, σ_{1c} =75 MPa, $G_{\text{Ic}} = 0.35\,\text{MPa mm}$,$K_{\text{Ic}} = 1.07\,\text{MPa m}^{1/2}$) is in the upper position and the stiffness of the bottom material varies. In all cases $\nu = 0.3$.

The particular case $M = 1$ is given for comparison with

$$K = \frac{1}{2}(K_{\text{I}} + iK_{\text{II}}) \quad \text{and} \quad \underline{u}(\theta) = \underline{u}_{\text{I}}(\theta) - \mathrm{i}\underline{u}_{\text{II}}(\theta). \tag{12.47}$$

A constant force is applied in all cases, here we are only interested in the kink angle not the load at failure. Results are presented in Table 12.1. It is worth noting that the sign of ε is a convention since both + and − occur in the expansion. In the two last rows of Table 12.1, numbers in italic correspond to the same analysis but neglecting T, i.e., with $T = 0$ in the equations. Obviously, neglecting the T-stress leads in some cases to a significant discrepancy in terms of critical load and kink angle.

The influence of the T-stress on the kink angle is stronger in the stiffest material leading to a larger kink angle. Such a feature can be observed in layered ceramics [58, 59], although the schematic situation illustrated in Fig. 12.21 is slightly different from the present one (there is no long

Table 12.1. Parameters resulting of the simulation of the 3-point bending test. Numbers in italic correspond to neglecting T.

M	ε	Im(K)/Re(K)	T/Re(K) ($m^{-1/2}$)	α_c (deg.)	K_f/K_{Ic}
100	0.091	0.41	37.4	90 *30*	0.07 *0.17*
75	0.090	0.38	31.6	90 *30*	0.09 *0.21*
50	0.089	0.41	37.8	90 *40*	0.12 *0.22*
10	0.075	0.49	62.8	70 *40*	0.23 *0.28*
2	0.030	0.75	65.2	60 *50*	0.33 *0.35*
1	0.000	0.93	81.8	60 *50*	0.38 *0.39*
0.5	−0.030	−1.14	47.7	50 *50*	0.47 *048*
0.1	−0.075	−1.49	15.8	40 *40*	0.63 *0.63*
0.02	−0.089	−1.58	11.9	30 *30*	0.73 *0.73*
0.013	−0.090	−1.59	−35.4	30 *30*	0.75 *0.75*
0.01	−0.091	−1.59	−38.1	30 *30*	0.76 *0.75*

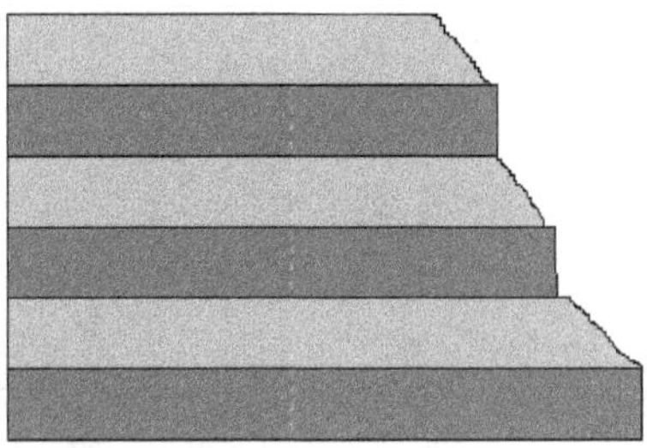

Fig. 12.21. Schematic fracture pattern in bending of a laminated ceramic made of compliant (light grey) and stiff (dark grey) layers.

interface debonding) and cannot be directly compared. The above result is still waiting for an experimental evidence.

For a complete analysis the competition between the deflection mechanism and the delamination growth should be analyzed.

12.9. The Interface Corner

This case of non-smooth interface occurs in many structures and in particular in electronic devices [60, 61]. In this latter case material 1 is for instance copper and material 2 a dielectric (a porous ceramic) (Fig. 12.22). There is a stress concentration (a singularity) at the corner whose intensity depends on the contrast between the materials [62]. The singularity exponents are real and the associated angular modes are solution to an eigenvalue problem where the usual boundary conditions in θ are

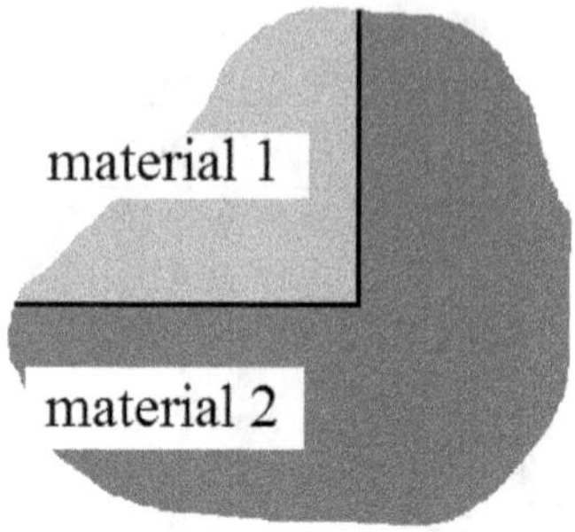

Fig. 12.22. A non-smooth interface between two materials.

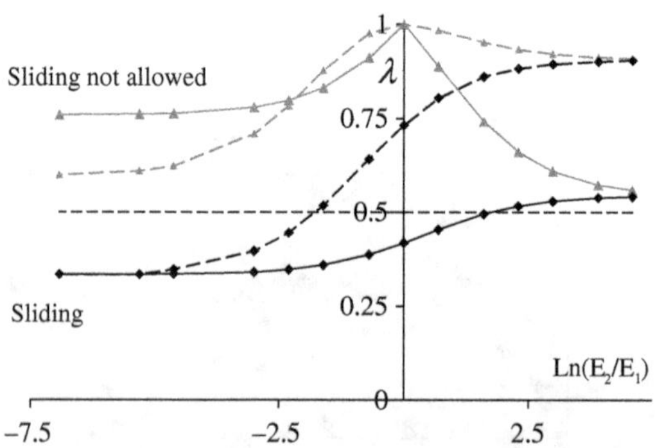

Fig. 12.23. The singularity exponents at a right-angle corner of an interface, function of the Young's modulus contrast between the materials, for a perfect bonding (in gray) and when sliding is allowed (in black). The solid line is associated with the symmetric mode and the dashed one with the antisymmetric mode.

replaced with a periodicity condition. With a perfect bonding there are two weak singularities, i.e., $0.5 \leq \lambda_1 \leq \lambda_2 \leq 1$ (Fig. 12.23).

In case of damage and debonding, depending on the applied load, either the interface opens or slides. In the first case, i.e., in case of opening, the singularity is the usual 90°. notch singularity in the matrix with $\lambda_1 = 0.545$ and $\lambda_2 = 0.906$ (independent of the matrix stiffness [24]) since there is no longer any contact between the matrix and the inclusion. When sliding is allowed along the interfaces (without friction) the singularities become strong (Fig. 12.23) if the inclusion is stiffer than the matrix. Such a property has a significant influence on the fracture mechanisms resulting of the stress concentration and the competition between different fracture mechanisms depicted in Fig. 12.24 can be studied.

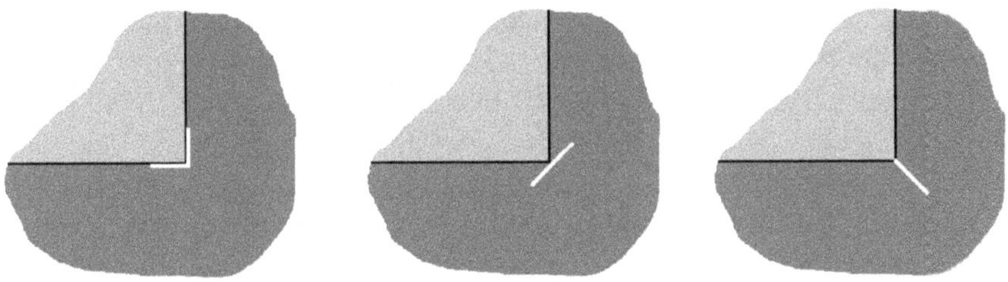

Fig. 12.24. Various scenarios of damage at the corner of an interface [63, 64].

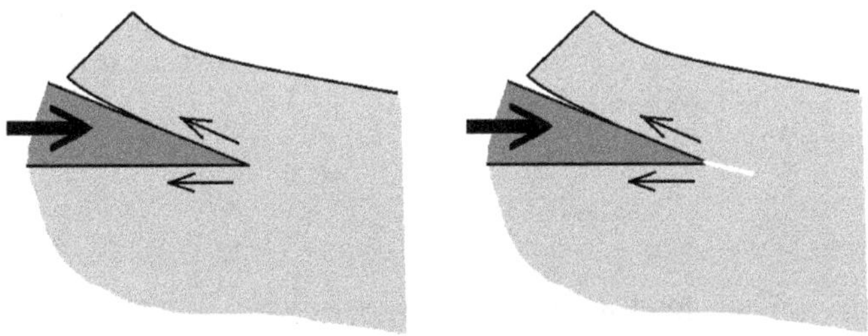

Fig. 12.25. A schematic view of a cutting tool and the formation of a crack ahead of the tool.

Among others, a possible scenario is that interface debonding occurs first (Fig. 12.24 left), then there is no longer perfect bonding and the generally strong singularities due to sliding lead to a matrix crack nucleation (Fig. 12.24 middle and right).

Another interesting example is given by a cutting tool (Fig. 12.25) where a stiff wedge is inserted into a more compliant material, with sliding along the two faces of the wedge giving rise to a strong singularity facilitating the growth of a crack.

12.10. Conclusion

Plane strain elasticity is the main assumption of this chapter but the coupled criterion can be extended without major difficulties to a generalized plane strain framework allowing in particular to analyze the initiation of delamination of angle-ply laminates [1, 12]. The analysis of the crack deflection by an interface together with the mechanism of crack kinking out of an interface developed herein, cover a wide range of problems of failure in composite laminates; especially since the anisotropy, which has not been mentioned in this chapter, do not complicate things too much as long as the assumption of plane or generalized plane elasticity still holds true.

It is obviously an extension to 3D that raises the most difficulties. There are no major conceptual changes, but everything becomes technically much more complicated. The crack extension is no longer simply described by two parameters, e.g., direction and length; the complete geometry must be taken into account. Attempts were made to predict the nucleation of small lens-shaped cracks along a straight crack front subjected to a mode III remote loading [22], and to predict delamination of an interface starting from a wedge [23]. It is worth noting also the experiments and the simulations of mode I+III crack nucleation in slanted V-notched specimens under a 4-point bending loading [65, 66]. Nevertheless, it is clear that much remains to be done in this domain.

References

[1] Martin E., Leguillon D., Carrere N., 2010. A twofold strength and toughness criterion for the onset of free-edge shear delamination in angle-ply laminates, *Int. J. Solids Struct.*, 47, 1297–1305.

[2] Sridharan S., 2008. *Delamination Behaviour of Composites.* Woodhead Publishing, Cambridge.

[3] Brewer J.C., Lagace P.A., 1998. Quadratic stress criterion for initiation of delamination, *J. Comp. Mat.*, 122, 1141–1155.

[4] Nairn J.A., Hu S., 1992. The initiation and growth of delaminations induced by matrix microcracks in laminated composites, *Int. J. Frac.*, 57, 1–24.

[5] Delisée I., 1996. *Etude de mécanismes de délaminage des composites croisés carbone/epoxy*, Thèse, Université Pierre et Marie Curie, Paris.

[6] Ladeveze P., Lubineau G., Marsal D., 2006. Towards a bridge between the micro- and mesomechanics of delamination for laminated composites, *Comp. Sci. Technol.*, 66, 698–712.

[7] Blázquez A., Mantič V., París F., McCartney L.N., 2008. Stress state characterization of delamination cracks in [0/90] symmetric laminates by BEM, *Int. J. Solids Struct.*, 45, 1632–1662.

[8] Martin E., Peters P.W.M., D. Leguillon and J.M. Quenisset (1998). Conditions for crack deflection at an interface in ceramic matrix composites, *Mat. Sci. Eng. A*, 250, 291–302.

[9] Carrère N., Martin E., Lamon J., 2000. The influence of the interphase and associated interfaces on the deflection of matrix cracks in ceramic matrix composites, *Comp. Part A*, 31, 1179–1190.

[10] Leguillon D., 2002. Strength or toughness? A criterion for crack onset at a notch, *Eur. J. of Mechanics — A/Solids*, 21, 61–72.

[11] Weissgraeber P., Leguillon D., Becker W., 2016. A review of Finite Fracture Mechanics: Crack initiation at singular and non-singular stress-raisers, *Arch. Appl. Mech.*, 86, 375–401.

[12] Pipes B.R., Pagano N.J., 1970. Interlaminar stresses in composite laminates under uniform axial extension, *J. Comp. Mat.*, 4, 538–548.

[13] Mantič V., 2009. Interface crack onset at a circular cylindrical inclusion under a remote transverse tension. Application of a coupled stress and energy criterion, *Int. J. Solids Struct.*, 46, 1287–1304.

[14] Mantič V., Garcia I.G., 2012. Crack onset and growth at the fiber-matrix interface under a remote biaxial transverse load. Application of a coupled stress and energy criterion, *Int. J. Solids Struct.*, 49, 2273–2290.

[15] Garcia I.G., Mantič V., Graciani E., 2015. Debonding at the fiber-matrix interface under remote transverse tension. One debond or two symmetric debonds, *Eur. J. Mech. A/Solids*, 53, 75–88.

[16] Leguillon D., Murer S., 2008. Crack deflection in a biaxial stress state, *Int. J. Frac.*, 150, 75–90.

[17] Leguillon D., Murer S., 2008. A criterion for crack kinking out of an interface, *Key Engineering Materials*, 385–387, 9–12.

[18] He M.Y., Hutchinson J.W., 1989. Kinking of a crack out of an interface, *ASME J. Appl. Mech.*, 56, 270–278.

[19] He M.Y., Bartlett A., Evans A.G., Hutchinson J.W., 1991. Kinking of a crack out of an interface: role of in-plane stress, *J. Am. Ceram. Soc.*, 74, 767–771.

[20] Wang T.C., C.F. Shih and Z. Suo (1992). Crack extension and kinking in laminates and bicrystals., Int. J. Solids Structures 29, 327–344.

[21] Lee W., Shin H., Yoo Y.H., 2004. Revisited criterion to prevent kinking of a crack out of a bi-material interface under the presence of in-plane residual stresses — a global approach. *Int. J. Engng. Sci.*, 42, 257–270.

[22] Leguillon D., 2003. Computation of 3D singular elastic fields for the prediction of failure at corners, *Key Eng. Mater.*, 251–252, 147–152.

[23] Leguillon D., 2014. An attempt to extend the 2D coupled criterion for crack nucleation in brittle materials to the 3D case, *Theor. Appl. Frac. Mech.*, 74, 7–17.

[24] Williams M.L., 1952. Stress singularities resulting from various boundary conditions in angular corners of plates in extension, *J. Appl. Mech.*, 19, 526–528.

[25] Leguillon D., Sanchez-Palencia E., 1987. *Computation of Singular Solutions in Elliptic Problems and Elasticity*, John Wiley & Son, New York, and Masson, Paris.

[26] Labossiere P.E.W., Dunn M.L., 1999. Stress intensities at interface corners in anisotropic bimaterials, *Engng. Frac. Mech.*, 62, 555–575.

[27] Henninger C., Roux S., Hild F., 2010. Enriched kinematic fields of cracked structures, *Int. J. Solids Struct.*, 47, 3305–3316.

[28] Griffith A.A., 1920. The phenomenon of rupture and flow in solids, *Phil. Trans. Roy. Soc. London A* 221, 163–198.

[29] Irwin G., 1958. Fracture. *Handbuch der Physics*, Springer, Berlin, VI, pp. 551–590.
[30] Yosibash, Z., Priel E., Leguillon D., 2006. A failure criterion for brittle elastic materials under mixed-mode loading, *Int. J. Frac.,* 141, 289–310.
[31] Martin E., Leguillon D., 2004. Energetic conditions for interfacial failure in the vicinity of a matrix crack in brittle matrix composites, *Int. J. Solids Struct.*, 41, 6937–6948.
[32] Martin E., Poitou B., Leguillon D., Gatt J.M., 2008. Competition between deflection and penetration at an interface in the vicinity of a main crack, *Int. J. Frac.*, 151, 247–268.
[33] Hebel J., Dieringer R., Becker W., 2010. Modelling brittle crack formation at geometrical and material discontinuities using a finite fracture mechanics approach, *Eng. Frac. Mech.*, 77, 3558–3572.
[34] Leguillon D., Sevecek O., Martin E., Bermejo R., 2015. Edge cracking due to a compressive residual stress in ceramic laminates, *C. R. Mécanique*, 343, 192–198.
[35] Leguillon D., Martin E., Sevecek O., Bermejo R., 2015. Application of the coupled stress-energy criterion to predict the fracture behaviour of layered ceramics designed with internal compressive stresses, *Eur. J. Mech. A/Solids*, 54, 94–104.
[36] Leguillon D., Lafarie-Frenot M.C., Pannier Y., Martin E., 2016. Prediction of the surface cracking pattern of an oxidized polymer induced by residual and bending stresses. *Int. J. Solids and Structures*, 91, 89–101.
[37] Leguillon D., Laurencin J., Dupeux M., 2003. Failure of an epoxy joint between two steel plates, *Eur. J. of Mech. A/Solids*, 22, 509–524.
[38] Leguillon D., 2011. Determination of the length of a short crack at a v-notch from a full filed measurement, *Int. J. Solids Struct.*, 48, 884–892.
[39] Leguillon D., Murer S., Recho N., Li J., 2009. Crack initiation at a V-notch under complex loadings — Statistical scattering, International Conference on Fracture, ICF12, Ottawa, Canada, 12–17 July 2009.
[40] Leguillon D., Sanchez-Palencia E., 1992. Fracture in heterogeneous materials, weak and strong singularities, in P. Ladevèze and O.C. Zienkiewicz (eds.), *Proc. European Conf. New Advances in Computational Structural Mechanics* Elsevier, Amsterdam, pp. 229–236.
[41] Leguillon D., Lacroix C., Martin E., 2000. Matrix crack deflection at an interface between a stiff matrix and a soft inclusion, *C. R. Acad. Sci. Paris*, Sér IIb, 328, 9–24.
[42] He M.Y., Hutchinson J.W., 1989. Crack deflection at an interface between dissimilar elastic materials, *Int. J. Solids Struct.*, 25, 1053–1067.
[43] Martinez D., Gupta V., 1994. Energy criterion for crack deflection at an interface between two orthotropic media, *J. Mech. Phys. Solids*, 42, 1247–1271.
[44] He M.Y., Evans A.G., Hutchinson J.W., 1994. Crack deflection at an interface between between dissimilar elastic materials: role of residual stresses, *Int. J. Solids Struct.*, 31, 3443–3455.

[45] Parmigiani J.P., Thouless M.D., 2006. The roles of toughness and cohesive strength on crack deflection at interfaces, *J. Mech. Phys. Solids*, 54, 266–287.
[46] Leguillon D., Lacroix C., Martin E., 2000. Interface debonding ahead of a primary crack, *J. Mech. Phys. Solids*, 48, 2137–2161.
[47] Martin E., Leguillon D., Lacroix C., 2002. An energy criterion for the initiation of interface failure ahead of a matrix crack in brittle matrix composites, *Compos. Interf.*, 9, 143–156.
[48] Cook J., Gordon J.E., 1964. A mechanism for the control of crack propagation in all brittle systems, *Proc. Roy. Soc A.*, 282, 508–520.
[49] Quesada D., Picard D., Putot C., Leguillon D., 2009. The role of the interbed thickness on the step-over fracture under overburden pressure, *Int. J. Rock Mech. Min. Sci.*, 46, 281–288.
[50] Leguillon D., Quesada D., Putot C., Martin E., 2007. Size effects for crack initiation at blunt notches or cavities, *Eng. Frac. Mech.*, 74, 2420–2436.
[51] Dundurs J., 1967. Effect of elastic constants on stress in a composite under plane deformations, *J. Comp. Mater.*, 1, 310–322.
[52] Rice J.R., 1988. Elastic fracture mechanics concept for interfacial cracks, *ASME J. Appl. Mech.*, 55, 98–103.
[53] Dundurs J., Comninou M., 1979. Some consequences of the inequality conditions in contact and crack problems, *J. Elast.*, 9, 71–82.
[54] Geubelle P.H., Knauss W.G., 1995. A note related to energy release rate computations for kinking interface cracks, *ASME J. Appl. Mech.*, 62, 266–267.
[55] Wang J.S., Suo Z., 1990. Experimental determination of interfacial toughness curves using brazil-nut-sandwiches, *Acta Metall. Mater.*, 38, 1279–1290.
[56] Liechti K.M., Chai Y.S., 1991. Biaxial loading experiments for determining interfacial fracture toughness, *ASME J. Appl. Mech.*, 58, 681–687.
[57] Leguillon D., Siruguet K., 2002. Finite fracture mechanics — Application to the onset of a crack at a bimaterial corner, in B.L. Karihaloo (ed.), *IUTAM Symp. Analytical and Computational Fracture Mechanics of Non-Homogeneous Materials*, Kluwer Academic Publishers, Dordrecht, pp. 11–18.
[58] Blanks K.S., Kristofferson A., Carlström E., Clegg W.J., 1998. Crack deflection in ceramic laminates using porous interlayers, *J. Eur. Ceram. Soc.*, 18, 1945–1951.
[59] Clegg W.J., Blanks K.S., Davis J.B., Lanckmans F., 1997. Porous interfaces as crack deflecting interlayers in ceramic laminates, *Key Engng. Mater.*, 132–136, 1866–1869.
[60] Yosibash Z., Adnan O., Schneck R., Atlas H., 2003. Thermo-mechanical failure criterion at the micron scale in electronic devices. *Int. J. Frac.*, 122, 47–64.
[61] Pahn L.O., Earmme Y.Y., 2000. Analysis of a short interfacial crack from the corner of a rectangular inclusion, *Int. J. Frac.*, 106, 341–356.
[62] Reedy E.D. Guess T.R., 2001. Rigid square inclusion embedded within an epoxy disk: asymptotic stress analysis, *Int. J. Solids Struct.*, 38, 1281–1293.

[63] Barroso A., Vicentini D., Mantič V., Paris F., 2012. Determination of generalized fracture toughness in composite multimaterial closed corners with two singular terms — Part I: Test proposal and numerical analysis. *Eng. Frac. Mech.*, 89, 1–14.

[64] Vicentini D., Barroso, A., Justo J., Mantič V., Paris F., 2012. Determination of generalized fracture toughness in composite multimaterial closed corners with two singular terms — Part II: Experimental results, *Eng. Frac. Mech.*, 89, 15–23.

[65] Mittelman B., Yosibash Z., 2014. Asymptotic analysis of the potential energy difference because of a crack at a V-notch edge in a 3D domain, *Eng. Frac. Mech.*, 131, 232–256.

[66] Mittelman B., Yosibash Z., 2015. Energy release rate cannot predict crack initiation orientation in domains with a sharp V-notch under mode III loading, *Eng. Frac. Mech.*, 141, 230–241.

https://doi.org/10.1142/9781800611887_0013

Chapter 13

Computational Procedure for Singularity Analysis of Anisotropic Elastic Multimaterial Corners — Applications to Composites and Their Joints

Vladislav Mantič*, Alberto Barroso†, and Federico París‡
Escuela Técnica Superior de Ingeniería, Universidad de Sevilla, Camino de los Descubrimientos s/n, 41092, Sevilla, Spain
**mantic@us.es*
†*abc@us.es*
‡*fparis@us.es*

Abstract

An approach to the evaluation of linear elastic solutions in anisotropic multimaterial corners under generalized plane-strain or plane-stress conditions is developed. This approach works for quite general configurations of piecewise homogeneous multimaterial corners covering discontinuities in geometry, materials and boundary conditions. Open and closed (periodic) corners including any finite number of single-material wedges converging at the corner tip are considered. General homogeneous boundary conditions and sliding friction contact can be imposed at corner boundaries and perfect adhesion or sliding friction contact can be imposed at corner interfaces. An anisotropic dry friction model is generally assumed, representing contact between surfaces with a strongly oriented surface topography or texture. A semi-analytic approach to the corner singularity analysis based on the Lekhnitskii–Stroh formalism of anisotropic elasticity, a transfer matrix concept for single-material wedges and a matrix formalism for boundary and interface conditions, is developed and implemented in a symbolic computation tool. A least-squares fitting technique for extracting generalized stress intensity factors (GSIFs) from finite element and boundary element results is proposed and implemented. Singularity analysis of a crack terminating at a ply interface in a laminate and of a bimaterial corner in a double-lap joint between a composite laminate and a metal layer is carried out as an application of the developed theory. A criterion for failure initiation at a closed corner tip based on GSIFs and the associated generalized fracture toughnesses is proposed, and a novel experimental procedure for the determination of the corresponding failure envelope is introduced and accomplished. Finally, the procedure developed is applied to removing or reducing stress singularities in some bimaterial joints.

Contents

13.1. Introduction

Composites are heterogeneous materials, which on microscales and mesoscales are usually considered as piecewise homogeneous materials. Therefore, models of composites and their adhesive joints with other composites or metals on micro-, meso- or macroscale often include interfaces between dissimilar materials, sometimes with potential cracks either along interfaces or terminating at interfaces [1]. Singular points, where there are discontinuities in the idealized geometry, material, interface or boundary conditions, can easily be identified in these models. We refer to a neighborhood of a singular point as a *corner*, and to the singular point itself as the *corner tip* (having in mind a 2D model). In particular, we refer to a neighborhood of a singular point where several materials meet as a *multimaterial corner*, see Figs. 13.1 and 13.2, and definitions in Section 13.3.

Linear elastic solutions for models of composites and their adhesive joints under mechanical and thermal loads may involve unbounded stresses at a corner tip, called *singular stresses* [2–8]. After the first fundamental

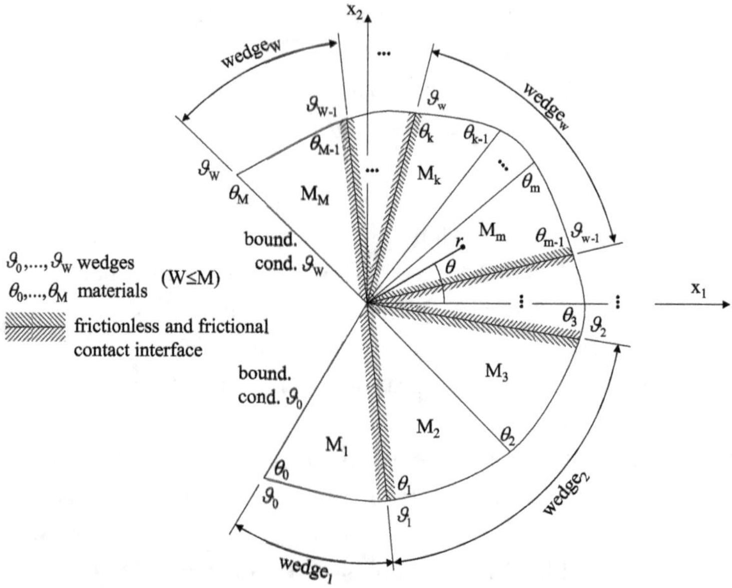

Fig. 13.1. A multimaterial corner (2D view).

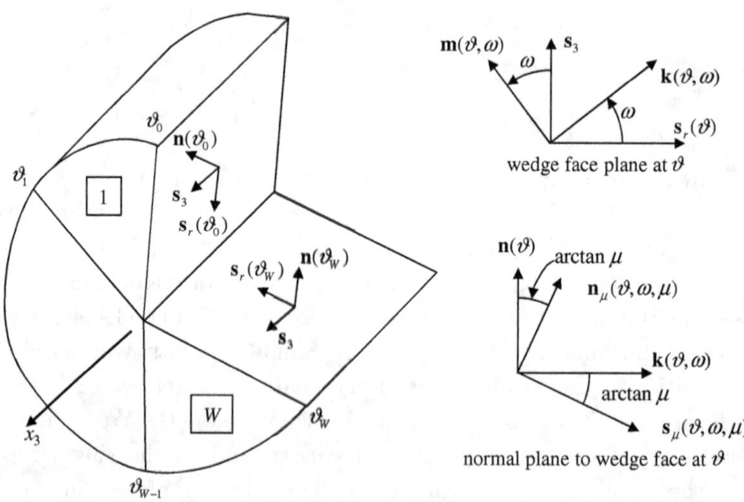

Fig. 13.2. Local Cartesian coordinate systems at a multimaterial corner (3D view).

contributions analysing singular stresses at homogeneous isotropic elastic corners in [9, 10], see also [11], considerable effort has been made by many authors to analyze singular stresses at multimaterial isotropic and anisotropic elastic corners [12–25]. A rigorous mathematical analysis of these solutions can be found in [26, 27].

The present work is focused on linear elastic anisotropic materials subjected to a *generalized plane-strain* state [28, 29] where three-dimensional (3D) displacements depend on only two Cartesian coordinates (or on the corresponding polar coordinates), namely $u_i = u_i(x_1, x_2) = u_i(r, \theta)$ with $i = 1, 2, 3$ or $i = r, \theta, 3$. The results obtained can easily be adapted to *generalized plane-stress* states [29] if required. The majority of practical problems found in composites can be studied under these hypotheses.

In general, a singular elastic solution in a neighborhood of the corner tip can be represented by the following asymptotic series expansion (a rigorous mathematical justification can be found in the fundamental works [30, 31], see also [2–5, 26, 27]), given by products of power-logarithmic terms in radial coordinate and angular functions:

$$u_i(r,\theta) \cong \sum_{n=1}^{N}\sum_{q=0}^{Q}\sum_{l=0}^{L} K_{nql} r^{\lambda_n+q} \log^l r g_i^{(nql)}(\theta) + \cdots , \tag{13.1}$$

where K_{nql} are *generalized stress intensity factors* (*GSIFs*), λ_n are *characteristic* or *singular exponents* (*eigenvalues*), and $g_i^{(nql)}(\theta)$ are *characteristic* or *singular angular shape functions* (*eigenfunctions*), $N \geq 1$ and $Q \geq 0$ can be finite or infinite numbers whereas $L \geq 0$ is a finite number.

The terms with the exponent $q = 0$ are referred to as principal terms, because the terms with $q \geq 1$ are due to the curvature of the corner boundaries and interfaces, the material non-homogeneity in the radial direction, non-homogeneous boundary or interface conditions or body forces in the neighborhood of the corner tip. In the present work, for the sake of simplicity, only straight boundaries and interfaces, piecewise homogeneous materials (in particular, those which are homogeneous in the radial direction), homogeneous boundary and interface conditions and vanishing body forces are considered in the neighborhood of the corner tip. Thus, $Q = 0$, the index q being omitted hereinafter. The so-called *characteristic exponents* λ_n are defined by the roots of a complex analytic (holomorphic) function given as the determinant of a matrix, referred to as a *characteristic matrix*, whose elements are also complex analytic functions.

The null space of this characteristic matrix determines the *characteristic angular shape functions* $g_i^{(nql)}(\theta)$. The vanishing determinant condition defines a transcendental equation referred to as a *characteristic equation* (*eigenequation*) *of the corner*, which depends on the problem configuration in the neighborhood of the corner tip: geometry, material properties, and boundary and interface conditions. Under the above assumptions, $L \geq 1$ is associated only with special cases with repeated roots of the characteristic equation of the corner, whose algebraic multiplicity is larger than the geometric multiplicity [7, 28, 32, 33]. For the sake of simplicity, $L = 0$ is assumed, and the index l is omitted as well. In view of the above assumptions, the displacements and stresses in the neighborhood of the corner tip can be represented by the following asymptotic series expansions:

$$u_i(r,\theta) \cong \sum_{n=1}^{N} K_n r^{\lambda_n} g_i^{(n)}(\theta) + \cdots, \tag{13.2}$$

$$\sigma_{ij}(r,\theta) \cong \sum_{n=1}^{N} K_n r^{\lambda_n - 1} f_{ij}^{(n)}(\theta) + \cdots. \tag{13.3}$$

If the number of terms in the series N is finite, a regular remainder term vanishing at the corner tip should be added to these series to obtain the complete solution in the corner. The terms in these series, referred to as *power-type singularities*, solve the elliptic system of three partial differential equations of generalized plane strain and satisfy the boundary and interface conditions in the neighborhood of the corner tip. All power type singularities are defined by three elements: the *characteristic exponent* λ_n, the *characteristic angular shape functions* $g_i^{(n)}(\theta)$ for displacements and $f_{ij}^{(n)}(\theta)$ for stresses, and the *generalized stress intensity factor* (GSIF) K_n. The characteristic angular functions are smooth functions inside each homogeneous material in the corner tip neighborhood, but may be non-smooth or even discontinuous at material interfaces. In an elastic boundary-value problem with a domain including one or more corners, the associated characteristic exponents and functions depend only on the local problem configuration (geometry, material properties, and boundary and interface conditions) in a neighborhood of the corner tip, whereas the GSIFs depend on the global problem configuration. If the boundary-value problem with vanishing body forces is linear, the GSIFs are linear functionals of the boundary conditions.

The rigid body motions are included in (13.2) and (13.3) for $\lambda_n = 0$ (translations) and $\lambda_n = 1$ (small rotations) with appropriate definitions of $g_i^{(n)}(\theta)$ and the corresponding $f_{ij}^{(n)}(\theta) = 0$. The terms with $0 < \lambda_n < 1$ give rise to unbounded (singular) stresses with a finite elastic strain energy. It is assumed that λ_n are naturally ordered satisfying $\mathrm{Re}\,\lambda_n \leq \mathrm{Re}\,\lambda_{n+1}$ (Re denotes the real part of a complex number).

Many different approaches to the corner singularity analysis, the evaluation of the characteristic exponents and functions, have been proposed in the past. In particular, with reference to the singularity analysis of a linear elastic anisotropic multimaterial corner in a generalized plane-strain state, several analytical, semi-analytical and numerical approaches are available at present. A numerical approach [2, 13–15] is expected to be more general and capable of analyzing corner singularity problems not tractable by analytical or semi-analytical approaches. The advantages of an analytical or semi-analytical approach, which provides an explicit closed-form expression of the characteristic equation of the corner, are its essentially arbitrarily high precision, fast computation and excellent possibilities for parametric studies and for understanding the influence of different parameters of the local corner configuration on the values and nature of the characteristic exponents and the behavior of characteristic functions.

Following the original proposal by Ting [16], several authors [18–21] have shown that for the development of a general semi-analytical approach to the singularity analysis of linear elastic anisotropic multimaterial corners in a generalized plane-strain state (assuming perfect adhesion between materials) it is very advantageous to employ the powerful Lekhnitskii–Stroh formalism of anisotropic elasticity [34–36], see also [28, 29, 37], together with a transfer matrix for a single-material angular sector (single-material wedge) in the corner. In fact, this methodology is essentially fully analytical except for the numerical evaluation of the complex roots of the Lekhnitskii–Stroh sextic polynomial for each anisotropic material in the corner. In the case of homogenized unidirectional fiber-reinforced composite laminas, represented by transversely isotropic materials, the roots of this sextic polynomial can be evaluated analytically leading to a fully analytical approach for corner singularity analysis [25]. It is worth pointing out that the application of the transfer matrix concept for all the single-material wedges in the corner allows the size of the matrix in the vanishing determinant condition to be made as small as possible. It should also be mentioned that the expressions introduced in [19, 21] can be applied directly to the corner singularity analysis for any kind of anisotropic material, namely non-degenerate, degenerate or extraordinary degenerate with a lower geometric than algebraic multiplicity of the repeated roots of the sextic polynomial, according to the classification in [38].

The above-described methodology was suitably adapted by Mantič *et al.* [22] to the simpler case of the singularity analysis of linear elastic anisotropic multimaterial corners in antiplane strain, governed by a scalar linear second-order elliptic equation (a generalization of the Laplace equation).

A general semi-analytical approach to the corner singularity analysis for linear elliptic systems of second-order partial differential equations in the plane was developed by Costabel and Dauge [17, 39], employing their previous fundamental mathematical results [31]. This general mathematical framework covers, as a particular case, linear elastic anisotropic materials under generalized plane-strain conditions. Their approach was implemented in a computational tool capable of carrying out the singularity analysis of multimaterial corners including anisotropic non-degenerate materials [17]. It is interesting to observe that, although not apparent at first sight, the approach due to Costabel and Dauge is in essence closely related to the above-described approaches based on the Lekhnitskii–Stroh formalism. In their approach, first, the complex roots of the symbol determinant of the partial differential system of linear elasticity in the plane are found, and then, the solution basis in the form of terms in the series expansion (13.2) is constructed analytically. This symbol determinant is, in fact, given by the Lekhnitskii–Stroh sextic polynomial.

Whereas, as discussed above, an arbitrarily high accuracy can be achieved in the evaluation of characteristic exponents and functions, the accuracy in the evaluation of GSIFs is, in general, substantially worse because a numerical solution of the global elastic problem (typically by means of finite or boundary element methods, FEM or BEM), or experimental tests (e.g., using photoelasticity), and usually also some post-processing of the results are needed [6, 8, 40, 41], see Section 13.6 for other references. Nevertheless, a few fast and highly accurate methods for the evaluation of stresses in the presence of crack and corner singularities in isotropic elastic materials are already available [41, 42].

In real composite structures, only high values of stresses (high stress concentrations), instead of singular stresses, are expected at these corners; first because strict discontinuities are hardly present in a real structure (e.g., a sharp corner tip in the model is usually rounded in a real structure), and second because a zone of nonlinear behavior (due to large strains, plasticity, damage, etc.) usually appears in the neighborhood of the corner tip.

The failure of a composite specimen, structure or joint may be initiated at a corner due to these high stress concentration there. Nevertheless, if the size of the nonlinear zone is sufficiently small with respect to a geometrical characteristic length (adjacent layer thickness, crack length, corner side length, etc.) the linear elastic solution for the idealized model may satisfactorily represent the solution in almost the whole volume of the structure and it will essentially determine the solution behavior in the nonlinear zone. In this case, failure initiation at a corner is governed by the linear elastic solution, and in particular by its asymptotic series expansion in the neighborhood of the corner tip, given by GSIFs, which could then be used in predictions of strength. A typical form of failure initiation at a corner tip, also considered in the present study, is the onset of a crack [9, 43–54]. In fact, the ultimate aim of the present study is to contribute to improving the accuracy and reliability of the strength predictions of composites in cases where a failure initiates at a multimaterial corner in the form of a crack.

The overall objective of the present work is the development of: (i) a general semi-analytical procedure for the singularity analysis of linear elastic anisotropic multimaterial corners in generalized plane strain (i.e., the evaluation of characteristic exponents and functions) with frictionless or friction contact surfaces in the corner, (ii) a general and sufficiently accurate and reliable numerical procedure for the extraction of GSIFs in these corners from FEM or BEM results and (iii) a procedure for failure assessment of corners in composites and their joints.

For the first objective, the methodology developed in [19] is generalized to include boundary surfaces and interfaces with frictionless or friction contact in the neighborhood of the corner tip. Although several relevant studies of singular elastic solutions at interface cracks and corners with frictionless contact or sliding friction contact have been published for isotropic [55–61] and also for anisotropic materials [62–75], it seems that a fully general approach to the singularity analysis of elastic anisotropic multimaterial corners in generalized plane strain including sliding frictional surfaces has not been developed yet. One of the open issues is related to the angle between the friction shear stress vector and the relative tangential displacement vector in the case where the in-plane and antiplane displacements are coupled. This fact is the reason for including in the present work a general and powerful matrix formalism for imposing rather general boundary and interface conditions, in particular frictionless and frictional sliding contact, by generalizing the methodology introduced in [19, 71]. This matrix formalism, which is especially suitable for a straightforward computational implementation, allows the characteristic corner matrix to be assembled in a fully automatic way for any finite number of materials and contact surfaces in the corner.

With reference to the second objective, a least-squares procedure for the extraction of multiple GSIFs at this kind of corner is developed and implemented [40] as a computational tool for post-processing results of a BEM code [75–79] solving boundary-value problems with linear elastic anisotropic materials in generalized plane strain.

The above-described computation tools are used to study a crack terminating at the interface between two plies in a $[0/90]_S$ laminate subjected to longitudinal tension and a critical bimaterial corner in an adhesively bonded double-lap joint. An altered configuration of the corner in this double-lap joint including an interface crack with sliding friction contact is also studied.

With reference to the third objective, the results of the singularity analysis for the critical corner in the double-lap joint are further used when analysing experimental results obtained by testing a novel modified configuration of the Brazilian disc specimen including the same bimaterial corner [80, 81] with two stress singularities. These tests provide a rough approximation of the failure envelope in the plane of GSIFs normalized by the pertinent values of the generalized fracture toughnesses. A corner failure criterion based on this failure envelope is proposed. The results for real adhesively bonded double-lap joints show a satisfactory agreement with the proposed failure criterion. The procedure developed can be applied as a general methodology for the failure assessment of these kind of multimaterial corners.

13.2. Lekhnitskii–Stroh Formalism for Linear Elastic Anisotropic Materials

The Lekhnitskii–Stroh complex-variable formalism [34–36, 82], or simply Stroh formalism, is a powerful and efficient theoretical tool for the analysis of anisotropic elastic problems. This section summarizes the fundamentals of this formalism employed in the analysis of singular stresses at multilateral corners in *generalized plane strain*. A comprehensive explanation of the

Lekhnitskii–Stroh formalism and of its numerous applications can be found in [28, 29, 37]. The application of the theoretical framework presented here to generalized plane-stress problems is straightforward; see [29] for the pertinent conversions of material constants.

13.2.1. Basic equations

Let x_i $(i = 1, 2, 3)$ be a Cartesian coordinate system. The constitutive law of a linear elastic anisotropic material relating the Cartesian components of stresses σ_{ij} and displacements u_i has the following form, at small strains $\varepsilon_{ij} = \frac{1}{2}(u_{i,j} + u_{j,i})$:

$$\sigma_{ij} = C_{ijkl}\varepsilon_{kl} = C_{ijkl}u_{k,l}, \tag{13.4}$$

where C_{ijkl} is the positive definite and symmetric fourth-order tensor of elastic stiffnesses, satisfying the symmetry relations $C_{ijkl} = C_{jikl} = C_{klij}$ and the positivity condition for the strain energy density $\frac{1}{2}C_{ijkl}\varepsilon_{ij}\varepsilon_{kl} = \frac{1}{2}\sigma_{ij}\varepsilon_{ij} > 0$ for any non-zero ε_{ij}. Then, the equilibrium equations (in the absence of body forces) can be written in terms of displacements as

$$C_{ijkl}u_{k,lj} = 0. \tag{13.5}$$

Under generalized plane-strain conditions the displacement field depends only on the plane coordinates x_1 and x_2, i.e., $u_i = u_i(x_1, x_2)$ $(i = 1, 2, 3)$. Hence, $\varepsilon_{33} = 0$. In this case (13.5) represents a linear elliptic system of three second-order partial differential equations in two dimensions for the 3D displacement vector field u_i. Let a solution of (13.5) be written as a function of a variable z defined by a linear combination of x_1 and x_2,

$$u_i = a_i f(z), \quad \text{with } z = x_1 + px_2, \tag{13.6}$$

where $f(z)$ is an arbitrary analytic function of z, while p and a_i are constants to be determined.

By differentiating the displacements in (13.6) twice with respect to x_l and x_j and substituting in (13.5), the following condition for the number p and vector $\mathbf{a}$ is obtained, taking into account that $f(z)$ is an arbitrary function:

$$[C_{i1k1} + p(C_{i1k2} + C_{i2k1}) + p^2 C_{i2k2}]a_k = 0. \tag{13.7}$$

Let the 3×3 matrices $\mathbf{Q}$, $\mathbf{R}$ and $\mathbf{T}$ be defined as

$$Q_{ik} = C_{i1k1}, \quad R_{ik} = C_{i1k2}, \quad T_{ik} = C_{i2k2}, \tag{13.8}$$

then, (13.7) can be written in the matrix form:

$$[\mathbf{Q} + p(\mathbf{R} + \mathbf{R}^T) + p^2\mathbf{T}]\mathbf{a} = \mathbf{0}. \tag{13.9}$$

It can be shown that $\mathbf{Q}$ and $\mathbf{T}$ are symmetric and positive definite [28, 29].

The components of the stress tensor σ_{i1} and σ_{i2} can be obtained by substituting the derivatives of the displacements $u_{k,l} = a_k f'(z)(\delta_{1l} + p\delta_{2l})$ and the definitions in (13.8) into (13.4),

$$\sigma_{i1} = (Q_{ik} + pR_{ik})a_k f'(z), \quad \sigma_{i2} = (R_{ki} + pT_{ik})a_k f'(z). \tag{13.10}$$

σ_{33} is determined using the constitutive law and the condition $\varepsilon_{33} = 0$. The relations (13.10) can be rewritten as

$$\sigma_{i1} = -pb_i f'(z), \quad \sigma_{i2} = b_i f'(z), \tag{13.11}$$

by defining

$$\mathbf{b} = (\mathbf{R}^T + p\mathbf{T})\mathbf{a} = -\frac{1}{p}(\mathbf{Q} + p\mathbf{R})\mathbf{a}. \tag{13.12}$$

By introducing the stress function vector $\boldsymbol{\varphi}$ as

$$\varphi_i = b_i f(z), \tag{13.13}$$

the stress tensor components in (13.10) can be expressed as

$$\sigma_{i1} = -\varphi_{i,2}, \quad \sigma_{i2} = \varphi_{i,1}. \tag{13.14}$$

The homogeneous linear system in (13.9) has a non-trivial solution if and only if its determinant is zero,

$$|\mathbf{Q} + p(\mathbf{R} + \mathbf{R}^T) + p^2\mathbf{T}| = 0. \tag{13.15}$$

The determinant in (13.15) is a polynomial of six degrees with real coefficients in a single variable p. The condition for a vanishing determinant in (13.15) is referred to as the *Lekhnitskii–Stroh sextic equation* of the anisotropic material in generalized plane strain. This polynomial has six complex roots (three pairs of complex conjugate values), called *eigenvalues*, p_α $(\alpha = 1, \ldots, 6)$. These eigenvalues are usually sorted by the sign of the imaginary part of p_α as follows (Im denotes the imaginary part and the overbar the complex conjugate value):

$$\operatorname{Im} p_\alpha > 0, \quad p_{\alpha+3} = \bar{p}_\alpha \; (\alpha = 1, 2, 3). \tag{13.16}$$

Let $\mathbf{a}_\alpha$ denote the *eigenvector* associated with p_α $(\alpha = 1, 2, 3)$ in (13.9). $\mathbf{b}_\alpha$ is obtained from (13.12). Then,

$$\mathbf{a}_{\alpha+3} = \bar{\mathbf{a}}_\alpha \quad \text{and} \quad \mathbf{b}_{\alpha+3} = \bar{\mathbf{b}}_\alpha \; (\alpha = 1, 2, 3). \tag{13.17}$$

13.2.2. Sextic eigen-relation: Stroh orthogonality and closure relations

The two equalities in (13.12) can be rewritten in the form:

$$\begin{bmatrix} -\mathbf{Q} & \mathbf{0}_{3\times3} \\ -\mathbf{R}^T & \mathbf{I}_{3\times3} \end{bmatrix} \begin{bmatrix} \mathbf{a}_\alpha \\ \mathbf{b}_\alpha \end{bmatrix} = p_\alpha \begin{bmatrix} \mathbf{R} & \mathbf{I}_{3\times3} \\ \mathbf{T} & \mathbf{0}_{3\times3} \end{bmatrix} \begin{bmatrix} \mathbf{a}_\alpha \\ \mathbf{b}_\alpha \end{bmatrix}, \tag{13.18}$$

where $\mathbf{I}_{3\times3}$ and $\mathbf{0}_{3\times3}$ are the 3×3 identity and zero matrices, respectively. Multiplying (13.18) by the inverse of the 6×6 matrix on the right-hand side of (13.18) leads to the *sextic eigen-relation*:

$$\begin{bmatrix} -\mathbf{T}^{-1}\mathbf{R}^T & \mathbf{T}^{-1} \\ \mathbf{R}\mathbf{T}^{-1}\mathbf{R}^T - \mathbf{Q} & -\mathbf{R}\mathbf{T}^{-1} \end{bmatrix} \begin{bmatrix} \mathbf{a}_\alpha \\ \mathbf{b}_\alpha \end{bmatrix} = p_\alpha \begin{bmatrix} \mathbf{a}_\alpha \\ \mathbf{b}_\alpha \end{bmatrix} \Rightarrow \mathbf{N}\boldsymbol{\xi}_\alpha = p_\alpha \boldsymbol{\xi}_\alpha, \tag{13.19}$$

where $\mathbf{N}$ is the 6×6 *fundamental elasticity matrix* [83], $\boldsymbol{\xi}_\alpha^T = (\mathbf{a}_\alpha^T, \mathbf{b}_\alpha^T)$ is the *right eigenvector* of $\mathbf{N}$ ($\mathbf{N}$ is non-symmetric) and p_α is the associated *eigenvalue*.

Equation (13.19) is valid for all α when $\mathbf{N}$ has three linearly independent eigenvectors $\boldsymbol{\xi}_\alpha$ $(\alpha = 1, 2, 3)$. Then, $\mathbf{N}$ is called *simple* if all the p_α are different and *semisimple* if there are repeated p_α but with three independent eigenvectors $\boldsymbol{\xi}_\alpha$ $(\alpha = 1, 2, 3)$. The associated materials are referred to as *non-degenerate materials*.

If $\mathbf{N}$ has less than three linearly independent eigenvectors $\boldsymbol{\xi}_\alpha$ associated with p_α $(\alpha = 1, 2, 3)$, i.e., the algebraic multiplicity of a repeated eigenvalue is larger than its geometric multiplicity, some expressions of the Stroh formalism have to be modified [28, 84–86].

When there are two linearly independent eigenvectors $\boldsymbol{\xi}_\alpha$ associated with p_α $(\alpha=1,2,3)$, $\mathbf{N}$ is called *non-semisimple* and the associated materials are known as *degenerate materials*. Then, for $p_1 = p_2$,

$$\mathbf{N}\boldsymbol{\xi}_1 = p_1\boldsymbol{\xi}_1, \quad \mathbf{N}\boldsymbol{\xi}_2 = p_1\boldsymbol{\xi}_2 + \boldsymbol{\xi}_1, \quad \mathbf{N}\boldsymbol{\xi}_3 = p_3\boldsymbol{\xi}_3, \tag{13.20}$$

where $\boldsymbol{\xi}_2$ is a *generalized eigenvector*.

When there is only one linearly independent eigenvector $\boldsymbol{\xi}_\alpha$ associated with p_α $(\alpha=1,2,3)$, $\mathbf{N}$ is called *extraordinary non-semisimple*, and the associated materials are known as *extraordinary degenerate materials*. Then, for $p_1 = p_2 = p_3 = p$,

$$\mathbf{N}\boldsymbol{\xi}_1 = p\boldsymbol{\xi}_1, \quad \mathbf{N}\boldsymbol{\xi}_2 = p\boldsymbol{\xi}_2 + \boldsymbol{\xi}_1, \quad \mathbf{N}\boldsymbol{\xi}_3 = p\boldsymbol{\xi}_3 + \boldsymbol{\xi}_2, \tag{13.21}$$

where $\boldsymbol{\xi}_2$ and $\boldsymbol{\xi}_3$ are *generalized eigenvectors*.

For each case, the relations satisfied by the eigenvalues p_α and the associated eigenvectors $\boldsymbol{\xi}_\alpha^T = (\mathbf{a}_\alpha^T, \mathbf{b}_\alpha^T)$ can be presented in the following explicit form:

- Non-degenerate case (three linearly independent eigenvectors). For $\alpha = 1, 2, 3$,

$$[\mathbf{Q} + (\mathbf{R} + \mathbf{R}^T)p_\alpha + \mathbf{T}p_\alpha^2]\mathbf{a}_\alpha = 0, \tag{13.22}$$

$$\mathbf{b}_\alpha = (\mathbf{R}^T + p_\alpha\mathbf{T})\mathbf{a}_\alpha = -\frac{1}{p_\alpha}(\mathbf{Q} + p_\alpha\mathbf{R})\mathbf{a}_\alpha. \tag{13.23}$$

- Degenerate case (two linearly independent eigenvectors). For $\alpha = 1,3$ Eq. (13.22) holds, and for the generalized eigenvector $\boldsymbol{\xi}_2^T = (\mathbf{a}_2^T, \mathbf{b}_2^T)$ with $p_2 = p_1 = p$.

$$\begin{aligned} -\{\mathbf{Q} + (\mathbf{R} + \mathbf{R}^T)p + \mathbf{T}p^2\}\mathbf{a}_2 &= -[2p\mathbf{T} + \mathbf{R} + \mathbf{R}^T]\mathbf{a}_1, \\ \mathbf{b}_2 &= \mathbf{T}\mathbf{a}_1 + [\mathbf{R}^T + p\mathbf{T}]\mathbf{a}_2. \end{aligned} \tag{13.24}$$

- Extraordinary degenerate case (one linearly independent eigenvector). Denoting $p_1 = p_2 = p_3 = p$, for $(p_1, \boldsymbol{\xi}_1)$ Eq. (13.22) holds with $\alpha = 1$, for $(p_2, \boldsymbol{\xi}_2)$ Eq. (13.24) holds, and for the generalized eigenvector $\boldsymbol{\xi}_3^T = (\mathbf{a}_3^T, \mathbf{b}_3^T)$:

$$\begin{aligned} -[\mathbf{Q} + (\mathbf{R} + \mathbf{R}^T)p + \mathbf{T}p^2]\mathbf{a}_3 &= -[2p\mathbf{T} + \mathbf{R} + \mathbf{R}^T]\mathbf{a}_2 - \mathbf{T}\mathbf{a}_1, \\ \mathbf{b}_3 &= \mathbf{T}\mathbf{a}_2 + [\mathbf{R}^T + p\mathbf{T}]\mathbf{a}_3. \end{aligned} \tag{13.25}$$

See [19, 38], for further details of the classification of $\mathbf{N}$.

The eigenvectors or generalized eigenvectors $\boldsymbol{\xi}_\alpha^T = (\mathbf{a}_\alpha^T, \mathbf{b}_\alpha^T)$ define the (3×3) complex matrices, $\mathbf{A} = [\mathbf{a}_1, \mathbf{a}_2, \mathbf{a}_3]$ and $\mathbf{B} = [\mathbf{b}_1, \mathbf{b}_2, \mathbf{b}_3]$, which are employed in representations of the displacement and stress function vectors.

Inasmuch as $\mathbf{N}$ is non-symmetric, the *left eigenvector* fulfills the relation:

$$\mathbf{N}^T \boldsymbol{\eta} = p\boldsymbol{\eta}, \tag{13.26}$$

where $\boldsymbol{\eta} = (\mathbf{b}^T, \mathbf{a}^T)$. It is easy to show that the right and left eigenvectors for the different eigenvalues are orthogonal:

$$\boldsymbol{\eta}_\alpha \cdot \boldsymbol{\xi}_\beta = 0, \quad \text{for } p_\alpha \neq p_\beta. \tag{13.27}$$

In general the following *Stroh orthogonality* and *closure relations* (written in a compact form) can be deduced, after a suitable normalization of the eigenvectors (in particular, the eigenvectors are normalized according to $\boldsymbol{\eta}_\alpha \cdot \boldsymbol{\xi}_\beta = \delta_{\alpha\beta}$):

$$\mathbf{X}\mathbf{X}^{-1} = \mathbf{X}^{-1}\mathbf{X} = \mathbf{I}_{6\times 6}, \tag{13.28}$$

where

$$\mathbf{X} = \begin{bmatrix} \mathbf{A} & \bar{\mathbf{A}} \\ \mathbf{B} & \bar{\mathbf{B}} \end{bmatrix}, \quad \mathbf{X}^{-1} = \begin{bmatrix} \boldsymbol{\Gamma}\mathbf{B}^T & \boldsymbol{\Gamma}\mathbf{A}^T \\ \boldsymbol{\Gamma}\bar{\mathbf{B}}^T & \boldsymbol{\Gamma}\bar{\mathbf{A}}^T \end{bmatrix}, \tag{13.29}$$

$\mathbf{I}_{6\times 6}$ is the 6×6 identity matrix and $\boldsymbol{\Gamma}$ is expressed as

$$\boldsymbol{\Gamma}^{ND} = \begin{bmatrix} 1 & 0 & 0 \\ 0 & 1 & 0 \\ 0 & 0 & 1 \end{bmatrix}, \quad \boldsymbol{\Gamma}^{D} = \begin{bmatrix} 0 & 1 & 0 \\ 1 & 0 & 0 \\ 0 & 0 & 1 \end{bmatrix}, \quad \boldsymbol{\Gamma}^{ED} = \begin{bmatrix} 0 & 0 & 1 \\ 0 & 1 & 0 \\ 1 & 0 & 0 \end{bmatrix}, \tag{13.30}$$

with superscripts *ND*, *D* and *ED* referring to the non-degenerate, degenerate and extraordinary degenerate cases, respectively.

Degenerate cases appear, strictly speaking, when a particular combination of the elastic stiffnesses of the material leads to the Lekhnitskii–Stroh sextic equation (13.15) with repeated roots (eigenvalues) whose algebraic multiplicity is larger than its geometric multiplicity, see (13.9). Thus, from an engineering point of view, the degenerate cases might be considered as very particular cases, and any degenerate case can be obtained as a limit of non-degenerate cases with respect to a continuous variation of the elastic stiffnesses. However, it should be stressed that in such a limit procedure the behavior of the eigenvectors $\boldsymbol{\xi}_\alpha$, and correspondingly of the matrices **A** and **B**, can be discontinuous, in particular their magnitudes can become infinite.

For composite materials reinforced with long fibers, which typically behave as transversely isotropic materials, the particular spatial orientations of these materials can lead to a non-semisimple **N**, cf. [25, 37, 87]. The fact that a spatial orientation of some materials, irrespective of the values of the elastic stiffnesses, can lead to a non-semisimple **N**, makes the analysis of degenerate cases relevant for applications of the Lekhnitskii–Stroh formalism to composite materials.

13.2.3. Representation of displacement and stress function vectors

The displacement and stress function vector solution of an anisotropic elastic problem under generalized plane-strain conditions can be expressed as a linear combination of terms written in compact (sextic) form as follows [19, 28, 84–86]:

$$\mathbf{w}(x_1,x_2) = \mathbf{X}\begin{bmatrix}\mathbf{F}(x_1,x_2) & \mathbf{0}_{3\times 3}\\ \mathbf{0}_{3\times 3} & \tilde{\mathbf{F}}(x_1,x_2)\end{bmatrix}\mathbf{v}, \quad \text{with}$$

$$\mathbf{w} = \begin{bmatrix}\mathbf{u}\\ \boldsymbol{\varphi}\end{bmatrix} \quad \text{and} \quad \mathbf{v} = \begin{bmatrix}\mathbf{q}\\ \tilde{\mathbf{q}}\end{bmatrix}, \tag{13.31}$$

where $\mathbf{q}=(q_1,q_2,q_3)^T$ and $\tilde{\mathbf{q}}=(\tilde{q}_1,\tilde{q}_2,\tilde{q}_3)^T$ are in general 3×1 constant vectors with real or complex components, and the elements of the 3×3 matrices **F** and $\tilde{\mathbf{F}}$ are defined by a complex analytic function f of complex variables $z_\alpha = x_1 + p_\alpha x_2$ and $\bar{z}_\alpha = x_1 + \bar{p}_\alpha x_2$, and by its first and second derivatives f' and f''. The structure of **F** and $\tilde{\mathbf{F}}$ depends on the number of linearly independent eigenvectors $\boldsymbol{\xi}_\alpha$, as follows,

- Non-degenerate case:

$$\mathbf{F} = \begin{bmatrix}f(z_1) & 0 & 0\\ 0 & f(z_2) & 0\\ 0 & 0 & f(z_3)\end{bmatrix}, \quad \tilde{\mathbf{F}} = \begin{bmatrix}f(\bar{z}_1) & 0 & 0\\ 0 & f(\bar{z}_2) & 0\\ 0 & 0 & f(\bar{z}_3)\end{bmatrix}. \tag{13.32}$$

- Degenerate case ($z_1 = z_2$):

$$\mathbf{F} = \begin{bmatrix} f(z_1) & x_2 f'(z_1) & 0 \\ 0 & f(z_1) & 0 \\ 0 & 0 & f(z_3) \end{bmatrix}, \quad \tilde{\mathbf{F}} = \begin{bmatrix} f(\bar{z}_1) & x_2 f'(\bar{z}_1) & 0 \\ 0 & f(\bar{z}_1) & 0 \\ 0 & 0 & f(\bar{z}_3) \end{bmatrix}. \tag{13.33}$$

- Extraordinary degenerate case ($z_1 = z_2 = z_3 = z$):

$$\mathbf{F} = \begin{bmatrix} f(z) & x_2 f'(z) & \frac{1}{2} x_2^2 f''(z) \\ 0 & f(z) & x_2 f'(z) \\ 0 & 0 & f(z) \end{bmatrix}, \quad \tilde{\mathbf{F}} = \begin{bmatrix} f(\bar{z}) & x_2 f'(\bar{z}) & \frac{1}{2} x_2^2 f''(\bar{z}) \\ 0 & f(\bar{z}) & x_2 f'(\bar{z}) \\ 0 & 0 & f(\bar{z}) \end{bmatrix}. \tag{13.34}$$

Finally, in view of (13.14), the traction vector $\mathbf{t}$ at a point (x_1, x_2) on a contour, with the unit normal vector $\mathbf{n} = (n_1, n_2)$ to this contour at (x_1, x_2), can be computed by the tangential derivative of $\boldsymbol{\varphi}$ with respect to this contour,

$$\mathbf{t}(x_1, x_2) = -\frac{\partial \boldsymbol{\varphi}}{\partial s}(x_1, x_2), \tag{13.35}$$

where $\mathbf{s} = (-n_2, n_1)$ is the unit tangential vector to this contour. Thus, a zero traction $\mathbf{t}$ along a contour corresponds to a constant (possibly zero) stress function vector $\boldsymbol{\varphi}$ along this contour.

13.3. Elastic Multimaterial Corner

13.3.1. Corner configuration

Consider an *elastic anisotropic multimaterial corner* composed of a finite number of *single-material wedges* with plane faces intersecting at a straight *corner edge* coincident with the x_3-axis of the Cartesian and cylindrical coordinate systems (x_1, x_2, x_3) and (r, θ, x_3), respectively. Let the corner be subjected to a generalized plane-strain state with zero body forces and constant *interface conditions* at the plane *interfaces* between the single-material wedges. *Perfect bonding* (traction equilibrium and displacement continuity), *friction* and *frictionless sliding* are considered at these interfaces. We distinguish between an *open corner* with two outer plane boundary faces, intersecting at the corner edge, where either constant *homogeneous orthogonal* or sliding friction *boundary conditions* are prescribed, and a *closed corner* (also called a periodic corner) with no such outer boundary faces. A homogeneous orthogonal boundary condition represents a general case of either displacement, traction or mixed homogeneous boundary conditions where the displacement and traction vectors are perpendicular to each other (including the case where one of them is zero), cf. [12, 19, 28, 71, 88].

Typical 2D and 3D representations of an open corner are shown in Figs. 13.1 and 13.2. Notice that the *corner tip* in Fig. 13.1 is a 2D view of the corner edge in Fig. 13.2. Let the corner contain M ($M \geq 1$) single-material wedges (in the following also referred to as materials), where material M_m ($m = 1, \ldots, M$) is defined by angles θ in the angular sector $\theta_{\mathrm{m}-1} < \theta < \theta_{\mathrm{m}}$. In an open corner, the boundary conditions are defined at angles θ_0 and θ_M, where $0° < \theta_M - \theta_0 \leq 360°$ and the interface conditions at $\theta_{\mathrm{m}}(m = 1, \ldots, M-1)$. In a closed corner, the interface conditions are prescribed at $\theta_m(m = 0, \ldots, M)$, where $\theta_M - \theta_0 = 360°$.

Let the sequence of all materials in the corner be partitioned to subsequences of the maximum length of consecutive materials with perfect bonding at the common interfaces. The materials in such a subsequence are grouped together and referred to as a *wedge*. The number of wedges in a corner is $W \geq 1$ ($W \leq M$), and the sequence of wedges is indexed by subscript w ($w = 1, \ldots, W$), see Fig. 13.1. Wedge w is defined by angles θ in the angular sector $\vartheta_{w-1} < \theta < \vartheta_w$. Thus, $\vartheta_0 = \theta_0$ and $\vartheta_W = \theta_M$. As shown in Section 13.5.1, the fact that materials in a wedge are perfectly bonded together allows us to work advantageously with a wedge as if it were a single entity, by defining a *transfer matrix* for the whole wedge. It follows from the wedge definition that friction or frictionless sliding is prescribed at the interface between two consecutive wedges. If there is no such interface in the corner the whole multimaterial corner is considered as a wedge, i.e., $W = 1$.

The present friction model can be described as a rate-independent dry Coulomb friction model with a linear variation of the limit shear traction with respect to the normal traction. Surface topography and texture can be isotropic or anisotropic. This sometimes requires a generalization of the standard isotropic friction model. Suitable anisotropic friction models with either an associated sliding rule (given by a version of the maximum dissipation principle) or a non-associated sliding rule have been studied in [89–92], see also [93]. Without loss of generality, we assume that suitable, physically based or experimentally determined, functions defining the angle of the sliding velocity ω^u and kinetic (or dynamic) friction coefficient μ in terms of the angle of the frictional shear traction ω are given at each friction surface by:

$$\begin{aligned} \omega^u &= \omega^u(\omega) \qquad \text{(sliding rule)}, \\ \mu &= \mu(\omega) > 0 \quad \text{(friction rule)}, \end{aligned} \tag{13.36}$$

where both angles, ω^u and ω, are measured with respect to the coordinate system in the wedge face.

Functions $\omega^u(\omega)$ and $\mu(\omega)$ are periodic with period 2π. In the case of an associated sliding rule, the polar diagram of the directionally dependent friction coefficient $\mu(\omega)$ is typically given by an ellipse with the following relationship for the sliding angle $\tan(\omega^u - \alpha) = (\mu(0° + \alpha)/\mu(90° + \alpha))^2 \tan(\omega - \alpha)$, where $\mu(0° + \alpha)$ and $\mu(90° + \alpha)$ define the major and minor semi-axes of this ellipse rotated by an angle α with respect to the coordinate system. In the usual isotropic friction model $\omega^u = \omega$ and μ is a constant independent of ω, the ellipse being replaced by a circle.

13.3.2. Boundary and interface conditions: Matrix formalism

In this section, a powerful matrix formalism for a compact representation of different boundary and interface conditions in the multimaterial corner, suitable for an efficient computer implementation, is introduced.

13.3.2.1. *Coordinate systems*

The following orthonormal vector basis attached to the wedge faces, cf. [19, 71], is employed to define the homogeneous orthogonal boundary and interface conditions:

$$(\mathbf{s}_r(\vartheta), \mathbf{s}_3, \mathbf{n}(\vartheta)), \tag{13.37}$$

with the Cartesian components of these vectors defined as

$$\mathbf{s}_r(\vartheta) = \begin{pmatrix} -\cos\vartheta \\ -\sin\vartheta \\ 0 \end{pmatrix}, \quad \mathbf{s}_3 = \begin{pmatrix} 0 \\ 0 \\ 1 \end{pmatrix}, \quad \mathbf{n}(\vartheta) = \begin{pmatrix} -\sin\vartheta \\ \cos\vartheta \\ 0 \end{pmatrix}, \tag{13.38}$$

where ϑ is the wedge face angle. Additionally, when defining the sliding friction condition, the following three orthonormal vector bases attached to the wedge faces are employed:

$$(\mathbf{k}(\vartheta,\omega),\mathbf{m}(\vartheta,\omega),\mathbf{n}(\vartheta)), \tag{13.39}$$

$$(\mathbf{k}(\vartheta,\omega^u),\mathbf{m}(\vartheta,\omega^u),\mathbf{n}(\vartheta)), \tag{13.40}$$

$$(\mathbf{n}_\mu(\vartheta,\omega,\mu),\mathbf{s}_\mu(\vartheta,\omega,\mu),\mathbf{m}(\vartheta,\omega)), \tag{13.41}$$

with

$$\mathbf{k}(\vartheta,\omega) = \cos\omega\mathbf{s}_r(\vartheta) + \sin\omega\mathbf{s}_3, \tag{13.42}$$

$$\mathbf{m}(\vartheta,\omega) = -\sin\omega\mathbf{s}_r(\vartheta) + \cos\omega\mathbf{s}_3, \tag{13.43}$$

$$\mathbf{n}_\mu(\vartheta,\omega,\mu) = \frac{\mathbf{n}(\vartheta)+\mu\mathbf{k}(\vartheta,\omega)}{\sqrt{1+\mu^2}}, \tag{13.44}$$

$$\mathbf{s}_\mu(\vartheta,\omega,\mu) = \frac{\mathbf{k}(\vartheta,\omega)-\mu\mathbf{n}(\vartheta)}{\sqrt{1+\mu^2}}, \tag{13.45}$$

where ω and ω^u, respectively, are the angles measured from the vector $\mathbf{s}_r(\vartheta)$ and are the directions of the friction shear and sliding in the wedge face, and μ is the corresponding kinetic (or dynamic) friction coefficient at the wedge face. Notice that, in the isotropic friction model the vector bases in (13.39) and (13.40) coincide. Subscript w has been omitted in the above vector definitions for the sake of simplicity.

13.3.2.2. *Boundary condition matrices*

The usual homogeneous orthogonal boundary conditions for the first and last corner faces, $w = 0$ and W, can be expressed formally, in view of (13.35) and assuming a zero stress function vector $\boldsymbol{\varphi}$ at the corner edge ($r = 0$), by a linear relation for $r > 0$:

$$\mathbf{D}_u(\vartheta_w)\mathbf{u}(r,\vartheta_w) + \mathbf{D}_\varphi(\vartheta_w)\boldsymbol{\varphi}(r,\vartheta_w) = \mathbf{0}, \tag{13.46}$$

where $\mathbf{D}_u(\vartheta_w)$ and $\mathbf{D}_\varphi(\vartheta_w)$ are 3×3 real matrices defined in Table 13.1, fulfilling the following orthogonality relations, cf. [19, 71]:

$$\mathbf{D}_u(\vartheta_w)\mathbf{D}_\varphi^T(\vartheta_w) = \mathbf{D}_\varphi(\vartheta_w)\mathbf{D}_u^T(\vartheta_w) = \mathbf{0}, \tag{13.47}$$

with superscript T denoting the transpose.

In general, the boundary conditions of friction contact are nonlinear, because of the unilateral Signorini conditions of impenetrability and non-adhesion, and the Amontons–Coulomb law for dry friction. Nevertheless,

Table 13.1. Boundary condition matrices $\mathbf{D}_u$ and $\mathbf{D}_\varphi$ for homogeneous orthogonal boundary conditions, with $\vartheta = \vartheta_w$ for $w = 0$ and W.

	Matrix definition	
Boundary condition	$\mathbf{D}_u(\vartheta)$	$\mathbf{D}_\varphi(\vartheta)$
Free	$\mathbf{0}_{3\times 3}$	$\mathbf{I}_{3\times 3}$
Fixed	$\mathbf{I}_{3\times 3}$	$\mathbf{0}_{3\times 3}$
Symmetry (only u_θ restricted)	$[\mathbf{n}(\vartheta), \mathbf{0}, \mathbf{0}]^T$	$[\mathbf{0}, \mathbf{s}_r(\vartheta), \mathbf{s}_3]^T$
Antisymmetry (only u_θ allowed)	$[\mathbf{s}_r(\vartheta), \mathbf{s}_3, \mathbf{0}]^T$	$[\mathbf{0}, \mathbf{0}, \mathbf{n}(\vartheta)]^T$
Only u_r restricted	$[\mathbf{s}_r(\vartheta), \mathbf{0}, \mathbf{0}]^T$	$[\mathbf{0}, \mathbf{n}(\vartheta), \mathbf{s}_3]^T$
Only u_r allowed	$[\mathbf{n}(\vartheta), \mathbf{s}_3, \mathbf{0}]^T$	$[\mathbf{0}, \mathbf{0}, \mathbf{s}_r(\vartheta)]^T$
Only u_3 restricted	$[\mathbf{s}_3, \mathbf{0}, \mathbf{0}]^T$	$[\mathbf{0}, \mathbf{s}_r(\vartheta), \mathbf{n}(\vartheta)]^T$
Only u_3 allowed	$[\mathbf{s}_r(\vartheta), \mathbf{n}(\vartheta), \mathbf{0}]^T$	$[\mathbf{0}, \mathbf{0}, \mathbf{s}_3]^T$

Table 13.2. Boundary condition matrices $\mathbf{D}_u$, $\mathbf{D}_\varphi$, $\tilde{\mathbf{D}}_u$ and $\tilde{\mathbf{D}}_\varphi$ for Coulomb sliding friction, with $\vartheta = \vartheta_w$, $\omega = \omega_w$ and $\omega^u = \omega^u_w(\omega_w)$ and $\mu = \mu_w(\omega_w)$ for $w = 0$ and W.

$\mathbf{D}_u(\vartheta, \omega, \omega^u) = [\mathbf{n}(\vartheta), \mathbf{0}, \mathbf{m}(\vartheta, \omega^u), \mathbf{0}]^T$	$\mathbf{D}_\varphi(\vartheta, \omega, \mu) = [\mathbf{0}, \mathbf{s}_\mu(\vartheta, \omega, \mu), \mathbf{0}, \mathbf{m}(\vartheta, \omega)]^T$
$\tilde{\mathbf{D}}_u(\vartheta, \omega, \omega^u) = [\mathbf{0}, \mathbf{k}(\vartheta, \omega^u)]^T$	$\tilde{\mathbf{D}}_\varphi(\vartheta, \omega, \mu) = [\mathbf{n}_\mu(\vartheta, \omega, \mu), \mathbf{0}]^T$

under the present hypothesis that the whole face is sliding in the same direction, i.e., ω, ω^u and μ are constant at the whole face, and assuming a monotonic loading from the unloaded state, the sliding friction boundary condition at the first and last corner faces, $w = 0$ and W, can also be expressed formally by a linear relation, cf. [71]:

$$\mathbf{D}_u(\vartheta_w, \omega_w, \omega^u_w(\omega_w))\mathbf{u}(r, \vartheta_w) + \mathbf{D}_\varphi(\vartheta_w, \omega_w, \mu_w(\omega_w))\boldsymbol{\varphi}(r, \vartheta_w) = \mathbf{0}, \tag{13.48}$$

where $\mathbf{D}_u(\vartheta_w, \omega_w, \omega^u_w(\omega_w))$ and $\mathbf{D}_\varphi(\vartheta_w, \omega_w, \mu_w(\omega_w))$ are 4×3 real matrices defined in Table 13.2. These matrices, however, do not fulfill orthogonality relations similar to (13.47). The fact that these matrices are rectangular, instead of the square matrices used in (13.46), implying four boundary conditions instead of three as would be expected, is associated with the fact that a certain direction of shear traction (or equivalently the sliding direction) is assumed here, although in general it is unknown. Recall, that the friction coefficient μ has only positive values. The fulfilment of the compression condition (the normal stresses are negative or vanish) and the dissipative character of friction (the friction shear stress is exerted in a direction that opposes sliding) should be checked at the wedge face after the problem is solved.

Notice that the boundary condition of frictionless sliding is referred to in Table 13.1 as a symmetry boundary condition, which requires a check after the problem is solved to ensure that the compression condition (the normal stresses are negative or vanish) is fulfilled at the wedge face.

In the following, a 6×6 real matrix $\mathbf{D}_{BC}$ including all the previously defined boundary condition matrices, is introduced and applied to partition the 6×1 vector $\mathbf{w}$ defined in (13.31) into the prescribed and unknown subvectors. We refer to $\mathbf{D}_{BC}$ as the *main boundary-condition-matrix.* $\mathbf{D}_{BC}$ is defined for homogeneous orthogonal boundary conditions and a sliding friction boundary condition, for a wedge face of angle ϑ_w, for $w = 0$ and W, respectively, as

$$\mathbf{D}_{BC}(\vartheta_w) = \begin{bmatrix} \mathbf{D}_u(\vartheta_w) & \mathbf{D}_\varphi(\vartheta_w) \\ \tilde{\mathbf{D}}_u(\vartheta_w) & \tilde{\mathbf{D}}_\varphi(\vartheta_w) \end{bmatrix} = \begin{bmatrix} \mathbf{D}_u(\vartheta_w) & \mathbf{D}_\varphi(\vartheta_w) \\ \mathbf{D}_\varphi(\vartheta_w) & \mathbf{D}_u(\vartheta_w) \end{bmatrix} \tag{13.49}$$

and

$$\mathbf{D}_{BC}(\vartheta_w, \omega_w, \mu_w(\omega_w), \omega_w^u(\omega_w)) = \begin{bmatrix} \mathbf{D}_u(\vartheta_w, \omega_w, \omega_w^u) & \mathbf{D}_\varphi(\vartheta_w, \omega_w, \mu_w) \\ \tilde{\mathbf{D}}_u(\vartheta_w, \omega_w, \omega_w^u) & \tilde{\mathbf{D}}_\varphi(\vartheta_w, \omega_w, \mu_w) \end{bmatrix}, \tag{13.50}$$

where the 2×3 real matrices $\tilde{\mathbf{D}}_u(\vartheta_w, \omega_w, \omega_w^u)$ and $\tilde{\mathbf{D}}_\varphi(\vartheta_w, \omega_w, \mu_w)$ are given in Table 13.2. It is straightforward to check that $\mathbf{D}_{BC}$ is an orthogonal matrix, i.e.,

$$\mathbf{D}_{BC}\mathbf{D}_{BC}^T = \mathbf{D}_{BC}^T\mathbf{D}_{BC} = \mathbf{I}_{6\times 6}, \tag{13.51}$$

where $\mathbf{I}_{6\times 6}$ is the 6×6 identity matrix. Evidently ω_w, μ_w and ω_w^u can be different for $w = 0$ and W. As shown in Section 13.5.2, the general orthogonality relation (13.51) is very useful in the application of different boundary conditions in the corner singularity analysis. Although an orthogonality relation analogous to (13.51) obtained for orthogonal boundary conditions can be found in [28] (Section 14.1) and [88] in a slightly different context, the orthogonality relation (13.51) appears to be new for Coulomb sliding friction for both isotropic and anisotropic friction models.

It is easy to show that if the vector $\mathbf{w}(r, \vartheta_w)$ in (13.31) is multiplied from the left by the matrix $\mathbf{D}_{BC}$, given either by (13.49) or (13.50), the prescribed and the unknown components of $\mathbf{w}(r, \vartheta_w)$ appear grouped in two separate blocks, $\mathbf{w}_P(r,\vartheta_w)$ and $\mathbf{w}_U(r,\vartheta_w)$, respectively,

$$\mathbf{D}_{BC}\mathbf{w}(r,\vartheta_w) = \begin{bmatrix} \mathbf{w}_P(r,\vartheta_w) \\ \mathbf{w}_U(r,\vartheta_w) \end{bmatrix}. \tag{13.52}$$

As follows from the above relations, $\mathbf{w}_P(r,\vartheta_w) = \mathbf{0}$ and $\mathbf{w}_U(r,\vartheta_w)$ are 3×1 vectors for orthogonal boundary conditions, whereas they are 4×1 and 2×1 vectors, respectively, for friction sliding. From the orthogonality relation (13.51) and the fact that $\mathbf{w}_P(r,\vartheta_w)$ is a zero vector, it is obtained, for $w = 0$ and W, that

$$\mathbf{w}(r,\vartheta_w) = \mathbf{D}_{BC}^T \begin{bmatrix} \mathbf{w}_P(r,\vartheta_w) \\ \mathbf{w}_U(r,\vartheta_w) \end{bmatrix} = \begin{bmatrix} \tilde{\mathbf{D}}_u^T \\ \tilde{\mathbf{D}}_\varphi^T \end{bmatrix} \mathbf{w}_U(r,\vartheta_w) = \tilde{\mathbf{D}}_{BC}^T \mathbf{w}_U(r,\vartheta_w). \tag{13.53}$$

Inasmuch as the order of the prescribed and unknown variables is not relevant, the definition of the main boundary-condition matrix $\mathbf{D}_{BC}$ is not unique, e.g. equivalent definitions can be obtained by permutations of rows corresponding to $\mathbf{w}_P(r,\vartheta_w)$ and $\mathbf{w}_U(r,\vartheta_w)$ separately.

13.3.2.3. *Interface condition matrices*

Similarly as in (13.46) and (13.48), the interface conditions between the wedges, for frictionless or frictional sliding, respectively, can be expressed, using (13.35) and assuming a zero stress function vector $\boldsymbol{\varphi}$ at the corner edge ($r = 0$), formally by linear relations for $r > 0$ and $1 \le w \le W - 1$:

$$\mathbf{D}_1(\vartheta_w)\mathbf{w}_w(r,\vartheta_w) + \mathbf{D}_2(\vartheta_w)\mathbf{w}_{w+1}(r,\vartheta_w) = \mathbf{0}, \tag{13.54}$$

and

$$\begin{aligned} &\mathbf{D}_1(\vartheta_w,\omega_w,\mu_w(\omega_w),\omega_w^u(\omega_w))\mathbf{w}_w(r,\vartheta_w) \\ &\quad + \mathbf{D}_2(\vartheta_w,\omega_w,\mu_w(\omega_w),\omega_w^u(\omega_w))\mathbf{w}_{w+1}(r,\vartheta_w) = \mathbf{0}, \end{aligned} \tag{13.55}$$

where $\mathbf{D}_i(\vartheta_w)$ and $\mathbf{D}_i(\vartheta_w,\omega_w,\mu_w,\omega_w^u)$ ($i = 1,2$), respectively, are real 6×6 and 7×6 matrices defined in Tables 13.3 and 13.4, $\mathbf{w}_w(r,\vartheta_w)$ and $\mathbf{w}_{w+1}(r,\vartheta_w)$, respectively, are 6×1 vectors of displacements and stress functions (13.31) associated with the wedges of number w and $w+1$ and defined at the wedge interface given by the angle ϑ_w.

A matrix analogous to the matrix $\mathbf{D}_{BC}$ in the case of boundary conditions is defined and applied for interface conditions as well. A 12×12

Table 13.3. Interface condition matrices $\mathbf{D}_1$, $\mathbf{D}_2$, $\tilde{\mathbf{D}}_1$ and $\tilde{\mathbf{D}}_2$ for frictionless sliding, with $\vartheta = \vartheta_w$ and $1 \leq w \leq W-1$.

$$\mathbf{D}_1(\vartheta) = \frac{1}{\sqrt{2}} \begin{bmatrix} -\mathbf{n}(\vartheta) & \mathbf{0}_{3\times1} & \mathbf{0}_{3\times3} & \mathbf{0}_{3\times1} \\ \mathbf{0}_{3\times1} & \mathbf{s}_r(\vartheta) & -\mathbf{I}_{3\times3} & \mathbf{s}_3 \end{bmatrix}^T$$

$$\mathbf{D}_2(\vartheta) = \frac{1}{\sqrt{2}} \begin{bmatrix} \mathbf{n}(\vartheta) & \mathbf{0}_{3\times1} & \mathbf{0}_{3\times3} & \mathbf{0}_{3\times1} \\ \mathbf{0}_{3\times1} & \mathbf{s}_r(\vartheta) & \mathbf{I}_{3\times3} & \mathbf{s}_3 \end{bmatrix}^T$$

$$\tilde{\mathbf{D}}_1(\vartheta) = \frac{1}{\sqrt{2}} \begin{bmatrix} \mathbf{n}(\vartheta) & \mathbf{0}_{3\times1} & \sqrt{2}\mathbf{s}_r(\vartheta) & \mathbf{0}_{3\times1} & \sqrt{2}\mathbf{s}_3 & \mathbf{0}_{3\times1} \\ \mathbf{0}_{3\times1} & \mathbf{n}(\vartheta) & \mathbf{0}_{3\times1} & \mathbf{0}_{3\times1} & \mathbf{0}_{3\times1} & \mathbf{0}_{3\times1} \end{bmatrix}^T$$

$$\tilde{\mathbf{D}}_2(\vartheta) = \frac{1}{\sqrt{2}} \begin{bmatrix} \mathbf{n}(\vartheta) & \mathbf{0}_{3\times1} & \mathbf{0}_{3\times1} & \sqrt{2}\mathbf{s}_r(\vartheta) & \mathbf{0}_{3\times1} & \sqrt{2}\mathbf{s}_3 \\ \mathbf{0}_{3\times1} & \mathbf{n}(\vartheta) & \mathbf{0}_{3\times1} & \mathbf{0}_{3\times1} & \mathbf{0}_{3\times1} & \mathbf{0}_{3\times1} \end{bmatrix}^T$$

Table 13.4. Interface condition matrices $\mathbf{D}_1$, $\mathbf{D}_2$, $\tilde{\mathbf{D}}_1$ and $\tilde{\mathbf{D}}_2$ for sliding friction, with $\vartheta = \vartheta_w$, $\omega = \omega_w$, $\omega^u = \omega^u_w(\omega_w)$, $\mu = \mu_w(\omega_w)$ and $1 \leq w \leq W-1$.

$$\mathbf{D}_1(\vartheta,\omega,\mu,\omega^u) = \frac{1}{\sqrt{2}} \begin{bmatrix} -\mathbf{n}(\vartheta) & \mathbf{0}_{3\times1} & \mathbf{0}_{3\times3} & -\mathbf{m}(\vartheta,\omega^u) & \mathbf{0}_{3\times1} \\ \mathbf{0}_{3\times1} & \mathbf{s}_\mu(\vartheta,\omega,\mu) & -\mathbf{I}_{3\times3} & \mathbf{0}_{3\times1} & \mathbf{m}(\vartheta,\omega) \end{bmatrix}^T$$

$$\mathbf{D}_2(\vartheta,\omega,\mu,\omega^u) = \frac{1}{\sqrt{2}} \begin{bmatrix} \mathbf{n}(\vartheta) & \mathbf{0}_{3\times1} & \mathbf{0}_{3\times3} & \mathbf{m}(\vartheta,\omega^u) & \mathbf{0}_{3\times1} \\ \mathbf{0}_{3\times1} & \mathbf{s}_\mu(\vartheta,\omega,\mu) & \mathbf{I}_{3\times3} & \mathbf{0}_{3\times1} & \mathbf{m}(\vartheta,\omega) \end{bmatrix}^T$$

$$\tilde{\mathbf{D}}_1(\vartheta,\omega,\mu,\omega^u) = \frac{1}{\sqrt{2}} \begin{bmatrix} \mathbf{n}(\vartheta) & \mathbf{0}_{3\times1} & \sqrt{2}\mathbf{k}(\vartheta,\omega^u) & \mathbf{0}_{3\times1} & \mathbf{m}(\vartheta,\omega^u) \\ \mathbf{0}_{3\times1} & \mathbf{n}_\mu(\vartheta,\omega,\mu) & \mathbf{0}_{3\times1} & \mathbf{0}_{3\times1} & \mathbf{0}_{3\times1} \end{bmatrix}^T$$

$$\tilde{\mathbf{D}}_2(\vartheta,\omega,\mu,\omega^u) = \frac{1}{\sqrt{2}} \begin{bmatrix} \mathbf{n}(\vartheta) & \mathbf{0}_{3\times1} & \mathbf{0}_{3\times1} & \sqrt{2}\mathbf{k}(\vartheta,\omega^u) & \mathbf{m}(\vartheta,\omega^u) \\ \mathbf{0}_{3\times1} & \mathbf{n}_\mu(\vartheta,\omega,\mu) & \mathbf{0}_{3\times1} & \mathbf{0}_{3\times1} & \mathbf{0}_{3\times1} \end{bmatrix}^T$$

real matrix $\mathbf{D}_I$ is defined for the frictionless and frictional sliding interface conditions, respectively, as

$$\mathbf{D}_I(\vartheta_w) = \begin{bmatrix} \mathbf{D}_1(\vartheta_w) & \mathbf{D}_2(\vartheta_w) \\ \tilde{\mathbf{D}}_1(\vartheta_w) & \tilde{\mathbf{D}}_2(\vartheta_w) \end{bmatrix} \tag{13.56}$$

and

$$\begin{aligned} &\mathbf{D}_I(\vartheta_w,\omega_w,\mu_w(\omega_w),\omega^u_w(\omega_w)) \\ &\quad = \begin{bmatrix} \mathbf{D}_1(\vartheta_w,\omega_w,\mu_w,\omega^u_w) & \mathbf{D}_2(\vartheta_w,\omega_w,\mu_w,\omega^u_w) \\ \tilde{\mathbf{D}}_1(\vartheta_w,\omega_w,\mu_w,\omega^u_w) & \tilde{\mathbf{D}}_2(\vartheta_w,\omega_w,\mu_w,\omega^u_w) \end{bmatrix}, \end{aligned} \tag{13.57}$$

where $\tilde{\mathbf{D}}_i(\vartheta_w)$ and $\tilde{\mathbf{D}}_i(\vartheta_w, \omega_w, \mu_w(\omega_w), \omega_w^u(\omega_w))$ $(i = 1, 2)$, respectively, are 6×6 and 5×6 real matrices, defined in Tables 13.3 and 13.4. We refer to $\mathbf{D}_I$ as the *main interface-condition matrix*. We can check by direct evaluation that $\mathbf{D}_I$ is an orthogonal matrix, i.e.,

$$\mathbf{D}_I \mathbf{D}_I^T = \mathbf{D}_I^T \mathbf{D}_I = \mathbf{I}_{12\times 12}, \tag{13.58}$$

where $\mathbf{I}_{12\times 12}$ is the 12×12 identity matrix. This general orthogonality relation will be useful in Section 13.5.2 when applying interface conditions in the corner singularity analysis.

Similarly, as for the boundary conditions, the fact that the matrices in (13.55) are rectangular, instead of the square matrices used in (13.54), in particular implying seven interface conditions instead of six as would be expected, is associated with the fact that some direction of friction shear traction (or equivalently sliding direction) is assumed here, although in general it is unknown. The fulfilment of the compression condition and the dissipative character of friction should be checked at the interface after the problem is solved.

The frictionless sliding condition also requires a check after the problem is solved to ensure the compression condition is fulfilled at the interface.

By multiplying the 12×1 vector $(\mathbf{w}_w^T(r, \vartheta_w), \mathbf{w}_{w+1}^T(r, \vartheta_w))^T$ from the left by the matrix $\mathbf{D}_I$, given either by (13.56) or (13.57), the prescribed and the unknown components of $\mathbf{w}(r, \vartheta_w)$ appear grouped in two separate blocks, $\mathbf{w}_P(r, \vartheta_w)$ and $\mathbf{w}_U(r, \vartheta_w)$, respectively,

$$\mathbf{D}_I \begin{bmatrix} \mathbf{w}_w(r, \vartheta_w) \\ \mathbf{w}_{w+1}(r, \vartheta_w) \end{bmatrix} = \begin{bmatrix} \mathbf{w}_P(r, \vartheta_w) \\ \mathbf{w}_U(r, \vartheta_w) \end{bmatrix}. \tag{13.59}$$

As follows from the above relations, $\mathbf{w}_P(r, \vartheta_w) = \mathbf{0}$ and $\mathbf{w}_U(r, \vartheta_w)$ are 6×1 vectors for frictionless sliding, whereas they are 7×1 and 5×1 vectors, respectively, for sliding friction. From the orthogonality relation (13.58) and the fact that $\mathbf{w}_P(r, \vartheta_w)$ is a zero vector, we obtain, for $1 \leq w \leq W - 1$, that

$$\begin{bmatrix} \mathbf{w}_w(r, \vartheta_w) \\ \mathbf{w}_{w+1}(r, \vartheta_w) \end{bmatrix} = \mathbf{D}_I^T \begin{bmatrix} \mathbf{w}_P(r, \vartheta_w) \\ \mathbf{w}_U(r, \vartheta_w) \end{bmatrix} = \begin{bmatrix} \tilde{\mathbf{D}}_1^T \\ \tilde{\mathbf{D}}_2^T \end{bmatrix} \mathbf{w}_U(r, \vartheta_w). \tag{13.60}$$

Similarly, as for the boundary conditions, the definition of the main interface-condition matrix $\mathbf{D}_I$ is not unique, e.g., equivalent definitions can be obtained by permutations of rows corresponding to $\mathbf{w}_P(r, \vartheta_w)$ and $\mathbf{w}_U(r, \vartheta_w)$ separately.

In the case of a closed corner, $\vartheta_W = \vartheta_0$ and $\mathbf{w}_{W+1}(r, \vartheta_W)$ should be replaced by $\mathbf{w}_1(r, \vartheta_0)$ in the relations (13.59) and (13.60), written for $w = W$. This is why a closed corner is sometimes referred to as periodic corner.

Finally, it should be mentioned that for a genuine 2D problem, where the in-plane-strain state can be uncoupled from the antiplane-strain state, the above vectors and matrices are correspondingly reduced. In particular, a significant reduction takes place in the sliding friction case, where a somewhat unpleasant feature with more boundary or interface conditions than usual, and consequently with rectangular boundary or interface condition matrices, is removed, as we know *a priori* that the sliding takes place in the x_1–x_2 plane, thus $\omega = \omega^u$ and ω equals either $0°$ or $\pm 180°$. Subsequently the vectors $\mathbf{s}_3$ and $\mathbf{m}$ and the angles ω and ω^u essentially disappear from the formulation, cf. [71].

13.4. Singular Elastic Solution in a Single-Material Wedge: Transfer Matrix

The particular geometrical configuration of the multimaterial corner shown in Fig. 13.1 allows us to take advantage of the *transfer matrix* concept similarly as in [16, 19]. The idea is explained briefly in the following. Let us assume a particular kind of elastic state in a single-material wedge M_m (defined by the angular sector between θ_{m-1} and θ_m), allowing separation of the variables in polar coordinates with the same radial dependence in all components of the displacement and stress function vectors. Then, we can relate the displacement and stress function vectors at both outer radial faces of the wedge, $\mathbf{w}_m(r, \theta_{m-1})$ and $\mathbf{w}_m(r, \theta_m)$, by a transfer matrix $\mathbf{E}_m$. It should be stressed that such a transfer matrix depends on the kind of the elastic state considered. If we can enforce continuity of the displacement and stress function vectors across interfaces between several perfectly bonded single-material wedges we arrive at a transfer matrix for the whole sequence of bonded wedges simply by sequentially multiplying the transfer matrices of all the single-material wedges in the sequence.

With reference to the representation of the displacement and stress function vector in (13.31) for the analysis of problems with stress singularities, in this section we will assume the following simple form for the complex analytic functions, in view of (13.2) and (13.3):

$$f(z_\alpha) = z_\alpha^\lambda \quad \text{and} \quad f(\bar{z}_\alpha) = \bar{z}_\alpha^\lambda \ (\alpha = 1, 2, 3), \tag{13.61}$$

where λ is a real or complex characteristic exponent. The complex variables z_α and $\bar{z}_\alpha$ in (13.61) can be expressed as

$$\begin{aligned} z_\alpha &= x_1 + p_\alpha x_2 = r(\cos\theta + p_\alpha \sin\theta) = r\zeta_\alpha(\theta), \\ \bar{z}_\alpha &= x_1 + \bar{p}_\alpha x_2 = r(\cos\theta + \bar{p}_\alpha \sin\theta) = r\bar{\zeta}_\alpha(\theta), \end{aligned} \tag{13.62}$$

considering the polar coordinate system (r, θ) centered at the *corner tip* $(r = 0)$.

13.4.1. Non-degenerate materials

Let us assume that the material in a single-material wedge M_m is non-degenerate, see Section 13.2. Then, by substituting (13.61) into (13.31) and taking into account (13.32), the displacement and stress function vector fields in this wedge can be written in the following compact form:

$$\mathbf{w}(r,\theta) = r^\lambda \mathbf{X}\mathbf{Z}^\lambda(\theta)\mathbf{v}, \quad \mathbf{w}(r,\theta) = \begin{bmatrix} \mathbf{u}(r,\theta) \\ \boldsymbol{\varphi}(r,\theta) \end{bmatrix}, \quad \mathbf{v} = \begin{bmatrix} \mathbf{q} \\ \tilde{\mathbf{q}} \end{bmatrix}, \tag{13.63}$$

where $\mathbf{X}$ is defined in (13.29), and $\mathbf{Z}^\lambda(\theta)$ is a diagonal matrix

$$\mathbf{Z}^\lambda(\theta) = \begin{bmatrix} \langle \zeta_*^\lambda(\theta) \rangle & \mathbf{0}_{3\times 3} \\ \mathbf{0}_{3\times 3} & \langle \bar{\zeta}_*^\lambda(\theta) \rangle \end{bmatrix}, \tag{13.64}$$

and

$$\begin{aligned} \langle \zeta_*^\lambda(\theta) \rangle &= \mathrm{diag}[\zeta_1^\lambda(\theta), \zeta_2^\lambda(\theta), \zeta_3^\lambda(\theta)], \\ \langle \bar{\zeta}_*^\lambda(\theta) \rangle &= \mathrm{diag}[\bar{\zeta}_1^\lambda(\theta), \bar{\zeta}_2^\lambda(\theta), \bar{\zeta}_3^\lambda(\theta)], \end{aligned}$$

with

$$\zeta_\alpha^\lambda(\theta) = (\cos\theta + p_\alpha \sin\theta)^\lambda \quad \text{and} \quad \bar{\zeta}_\alpha^\lambda(\theta) = (\cos\theta + \bar{p}_\alpha \sin\theta)^\lambda.$$

Thus, $\mathbf{F}$ and $\tilde{\mathbf{F}}$ in (13.32) are expressed as

$$\mathbf{F}(x_1, x_2) = r^\lambda \left\langle \zeta_*^\lambda(\theta) \right\rangle, \quad \tilde{\mathbf{F}}(x_1, x_2) = r^\lambda \left\langle \bar{\zeta}_*^\lambda(\theta) \right\rangle. \tag{13.65}$$

According to (13.63), if $0 < \lambda < 1$ (or $0 < \mathrm{Re}(\lambda) < 1$, if λ is a complex number), the associated stresses become singular at the origin of coordinates, i.e., they may become unbounded for $r \to 0^+$. If λ is a real number, then $\tilde{\mathbf{q}}$ is the complex conjugate of $\mathbf{q}$, and $\mathbf{u}$ and $\boldsymbol{\varphi}$ are also real functions. If λ is a complex number, then $\tilde{\mathbf{q}}$ is not necessarily the complex conjugate of $\mathbf{q}$, and $\mathbf{u}$ and $\boldsymbol{\varphi}$ are also complex functions. It can be deduced that if λ is a solution, then $\bar{\lambda}$ is also a solution, and the superposition of the solutions (13.63) for λ and $\bar{\lambda}$ leads to real-valued expressions of $\mathbf{u}$ and φ.

If (13.63) is evaluated for the single-material wedge M_m at $\theta = \theta_{m-1}$ and $\theta = \theta_m$, and $\mathbf{v}$ is eliminated, we obtain:

$$\begin{aligned} \mathbf{w}_m(r,\theta_{m-1}) &= r^\lambda \mathbf{X}\mathbf{Z}^\lambda(\theta_{m-1})\mathbf{v} \\ \mathbf{w}_m(r,\theta_m) &= r^\lambda \mathbf{X}\mathbf{Z}^\lambda(\theta_m)\mathbf{v} \end{aligned} \Rightarrow \mathbf{w}_m(r,\theta_m) = \mathbf{E}_m(\lambda,\theta_m,\theta_{m-1})\,\mathbf{w}_m(r,\theta_{m-1}), \tag{13.66}$$

where

$$\mathbf{E}_m(\lambda,\theta_m,\theta_{m-1}) = \mathbf{X}\mathbf{Z}^\lambda(\theta_m)[\mathbf{Z}^\lambda(\theta_{m-1})]^{-1}\mathbf{X}^{-1}. \tag{13.67}$$

The 3×3 complex matrix $\mathbf{E}_m(\lambda,\theta_m,\theta_{m-1})$, referred to as the *transfer matrix for the single-material wedge* M_m, depends on the wedge material properties, through the matrix $\mathbf{X}$ and its inverse $\mathbf{X}^{-1}$ defined in (13.29) and the eigenvalues p_α, on the wedge geometry, given by the angles θ_{m-1} and θ_m, and on the characteristic exponent λ. An explicit expression for $\mathbf{Z}^\lambda(\theta_m)[\mathbf{Z}^\lambda(\theta_{m-1})]^{-1}$ is obtained following [16, 94],

$$\begin{aligned} \mathbf{Z}^\lambda(\theta_m)[\mathbf{Z}^\lambda(\theta_{m-1})]^{-1} &= \mathbf{Z}^\lambda(\theta_m,\theta_{m-1}) \\ &= \begin{bmatrix} \langle\zeta_*^\lambda(\theta_m,\theta_{m-1})\rangle & \mathbf{0}_{3\times 3} \\ \mathbf{0}_{3\times 3} & \langle\bar{\zeta}_*^\lambda(\theta_m,\theta_{m-1})\rangle \end{bmatrix}, \end{aligned} \tag{13.68}$$

where

$$\langle\zeta_*^\lambda(\theta_m,\theta_{m-1})\rangle = \operatorname{diag}[\zeta_1^\lambda(\theta_m,\theta_{m-1}),\zeta_2^\lambda(\theta_m,\theta_{m-1}),\zeta_3^\lambda(\theta_m,\theta_{m-1})], \tag{13.69}$$

with

$$\zeta_\alpha(\theta_m,\theta_{m-1}) = \frac{\zeta_\alpha(\theta_m)}{\zeta_\alpha(\theta_{m-1})} = \cos(\theta_m - \theta_{m-1}) + p_\alpha(\theta_{m-1})\sin(\theta_m - \theta_{m-1}), \tag{13.70}$$

and

$$p_\alpha(\theta_{m-1}) = \frac{p_\alpha\cos(\theta_{m-1}) - \sin(\theta_{m-1})}{p_\alpha\sin(\theta_{m-1}) + \cos(\theta_{m-1})}. \tag{13.71}$$

13.4.2. Degenerate materials

Let us assume that the material in a single-material wedge M_m is degenerate, see Section 13.2. Then, by substituting (13.61) for $(\alpha = 1,3)$

into (13.31) and taking into account (13.33), the displacement and stress function vector fields in this wedge can be written in the following compact form:

$$\mathbf{w}(r,\theta) = r^{\lambda}\mathbf{X}\mathbf{Z}(\theta,\lambda)\,\mathbf{v}, \quad \mathbf{w}(r,\theta) = \begin{bmatrix}\mathbf{u}(r,\theta)\\ \varphi(r,\theta)\end{bmatrix}, \quad \mathbf{v} = \begin{bmatrix}\mathbf{q}\\ \tilde{\mathbf{q}}\end{bmatrix}, \tag{13.72}$$

where $\mathbf{X}$ is defined in (13.29), and $\mathbf{Z}(\theta,\lambda)$ is defined as

$$\mathbf{Z}(\theta,\lambda) = \begin{bmatrix}\boldsymbol{\Psi}(p_*,\theta,\lambda) & \mathbf{0}_{3\times 3}\\ \mathbf{0}_{3\times 3} & \boldsymbol{\Psi}(\bar{p}_*,\theta,\lambda)\end{bmatrix}, \tag{13.73}$$

with

$$\boldsymbol{\Psi}(p_*,\theta,\lambda) = \begin{bmatrix}\zeta_1^{\lambda}(\theta) & K(p_1,\theta,\lambda)\zeta_1^{\lambda}(\theta) & 0\\ 0 & \zeta_1^{\lambda}(\theta) & 0\\ 0 & 0 & \zeta_3^{\lambda}(\theta)\end{bmatrix},$$

$$K(p_1,\theta,\lambda) = \frac{\lambda\sin(\theta)}{\zeta_1(\theta)}. \tag{13.74}$$

Now, by applying a procedure similar to that of the previous section, first evaluating (13.72) at $\theta = \theta_{m-1}$ and $\theta = \theta_m$, and then eliminating $\mathbf{v}$, we again arrive at

$$\mathbf{w}_m(r,\theta_m) = \mathbf{E}_m\,(\lambda,\theta_m,\theta_{m-1})\mathbf{w}(r,\theta_{m-1}), \tag{13.75}$$

where

$$\mathbf{E}_m(\lambda,\theta_m,\theta_{m-1}) = \mathbf{X}\mathbf{Z}(\theta_m,\lambda)[\mathbf{Z}(\theta_{m-1},\lambda)]^{-1}\mathbf{X}^{-1} \tag{13.76}$$

is the *transfer matrix for the single-material wedge M_m* with a degenerate material. According to [19]

$$\mathbf{Z}(\theta_m,\lambda)[\mathbf{Z}(\theta_{m-1},\lambda)]^{-1} = \begin{bmatrix}\boldsymbol{\Psi}(p_*,\theta_m,\theta_{m-1},\lambda) & \mathbf{0}_{3\times 3}\\ \mathbf{0}_{3\times 3} & \boldsymbol{\Psi}(\bar{p}_*,\theta_m,\theta_{m-1},\lambda)\end{bmatrix}, \tag{13.77}$$

where

$$\mathbf{\Psi}(p_*,\theta_m,\theta_{m-1},\lambda) = \begin{bmatrix} \zeta_1^\lambda(\theta_m,\theta_{m-1}) & K(p_1,\theta_m,\theta_{m-1},\lambda)\zeta_1^\lambda(\theta_m,\theta_{m-1}) & 0 \\ 0 & \zeta_1^\lambda(\theta_m,\theta_{m-1}) & 0 \\ 0 & 0 & \zeta_3^\lambda(\theta_m,\theta_{m-1}) \end{bmatrix}, \tag{13.78}$$

with $\zeta_\alpha(\theta_m,\theta_{m-1})$ defined in (13.70) and

$$K(p_1,\theta_m,\theta_{m-1},\lambda) = \frac{\lambda\sin(\theta_m-\theta_{m-1})}{\zeta_1(\theta_m)\zeta_1(\theta_{m-1})}. \tag{13.79}$$

Isotropic materials are a typical example of degenerate anisotropic materials with a triple eigenvalue $p = i = \sqrt{-1}$ and two linearly independent eigenvectors. All the above expressions for these materials simplify due to the fact that $\zeta_\alpha(\theta) = \cos\theta + i\sin\theta = e^{i\theta}$. In particular, $\zeta_\alpha(\theta_m,\theta_{m-1})$ in (13.70) and K in (13.79) can be rewritten as

$$\zeta_\alpha(\theta_m,\theta_{m-1}) = e^{i\,(\theta_m-\theta_{m-1})}, \quad \text{and} \quad K(i,\theta_m,\theta_{m-1},\lambda) = \frac{\lambda\sin(\theta_m-\theta_{m-1})}{e^{i(\theta_m+\theta_{m-1})}}. \tag{13.80}$$

Expressions for the complex matrices **A** and **B** for an isotropic elastic material in [19, 28] define the complex matrix **X** in (13.29).

13.4.3. Extraordinary degenerate materials

Let us assume that the material in a single-material wedge M_m is extraordinary degenerate, see Section 13.2. Then, by substituting (13.61) with $p_\alpha = p$, $z_\alpha = z$ and $\zeta_\alpha(\theta) = \zeta(\theta) = \cos\theta + p\sin\theta$, into (13.31) and taking into account (13.34), the displacement and stress function vector fields in this wedge can be written in the form (13.72), where

$$\mathbf{Z}(\theta,\lambda) = \begin{bmatrix} \mathbf{\Psi}(p,\theta,\lambda) & \mathbf{0}_{3\times3} \\ \mathbf{0}_{3\times3} & \mathbf{\Psi}(\bar{p},\theta,\lambda) \end{bmatrix}, \tag{13.81}$$

with

$$\boldsymbol{\Psi}(p,\theta,\lambda) = \zeta^{\lambda}(\theta)\begin{bmatrix} 1 & K(p,\theta,\lambda) & \frac{1}{2}(1-\lambda^{-1})K^2(p,\theta,\lambda) \\ 0 & 1 & K(p,\theta,\lambda) \\ 0 & 0 & 1 \end{bmatrix},$$

$$K(p,\theta,\lambda) = \frac{\lambda\sin(\theta)}{\zeta(\theta)}. \tag{13.82}$$

By applying a procedure similar to that of the previous sections, we arrive at transfer expressions in the same form as in (13.75) and (13.76), where, according to [19],

$$\mathbf{Z}(\theta_m,\lambda)[\mathbf{Z}(\theta_{m-1},\lambda)]^{-1} = \begin{bmatrix} \boldsymbol{\Psi}(p,\theta_m,\theta_{m-1},\lambda) & \mathbf{0}_{3\times3} \\ \mathbf{0}_{3\times3} & \boldsymbol{\Psi}(\bar{p},\theta_m,\theta_{m-1},\lambda) \end{bmatrix}, \tag{13.83}$$

$$\begin{aligned} \boldsymbol{\Psi}(p,\theta_m,\theta_{m-1},\lambda) &= \zeta^{\lambda}(\theta_m,\theta_{m-1}) \\ &\times \begin{bmatrix} 1 & K(p,\theta_m,\theta_{m-1},\lambda) & K(p,\theta_m,\theta_{m-1},\lambda)\,Z(p,\theta_m,\theta_{m-1},\lambda) \\ 0 & 1 & K(p,\theta_m,\theta_{m-1},\lambda) \\ 0 & 0 & 1 \end{bmatrix}, \end{aligned} \tag{13.84}$$

and $\zeta(\theta_m,\theta_{m-1})$ is defined similarly as in (13.70), $K(p,\theta_m,\theta_{m-1},\lambda)$ is defined similarly as in (13.79), and $Z(p,\theta_m,\theta_{m-1},\lambda)$ is defined as

$$Z(p,\theta_m,\theta_{m-1},\lambda) = \frac{1}{2}\left(K(p,\theta_m,\theta_{m-1},\lambda) - \frac{\sin\theta_{m-1}}{\zeta(\theta_{m-1})} - \frac{\sin\theta_m}{\zeta(\theta_m)}\right). \tag{13.85}$$

13.5. Characteristic System for the Singularity Analysis of an Elastic Multimaterial Corner

Closed-form expressions of the transfer matrix for a single-material wedge made of any anisotropic linear elastic material were obtained in Section 13.4 by assuming a simple power law form $f(z_\alpha) = z_\alpha^\lambda$ for the complex analytic functions appearing in a general representation of any elastic solution in the Lekhnitskii–Stroh formalism (13.31). In the present section, this transfer matrix is used first to generate the transfer matrix of a sequence of perfectly bonded single-material wedges and then to assemble the corresponding characteristic system (also called an eigensystem) of the multimaterial corner with some boundary and/or interface conditions. The solution of this characteristic system gives the characteristic exponents of the corner problem and characteristic angular functions defining the singular elastic state in the corner. The formulation is quite general considering any finite number of single-material wedges bonded or with friction or frictionless contact between them, and any homogeneous orthogonal or contact boundary conditions for an open corner. Isotropic and also anisotropic friction contact conditions may be considered at the contact faces. The powerful matrix formalism introduced in the previous sections allows us to write general expressions in a compact form suitable for a straightforward computer implementation.

13.5.1. Transfer matrix for a multimaterial wedge

Let a multimaterial wedge w ($w = 1, \ldots, W$), see Fig. 13.1, be defined by a sequence of perfectly bonded single-material wedges with indices $m = i_w, i_w+1, \ldots, j_w-1, j_w$, where i_w and j_w are the indices of the first and last material in the wedge w. Notice that $\theta_{i_w-1} = \vartheta_{w-1}$ and $\theta_{j_w} = \vartheta_w$. Then, a wedge transfer matrix can be defined as follows. Using the continuity conditions corresponding to the hypothesis of perfect bonding between the materials $\mathbf{w}_m(r,\theta_m) = \mathbf{w}_{m+1}(r,\theta_m)$ ($i_w \leq m < m+1 \leq j_w$), and the 6×6 transfer matrix $\mathbf{E}_m(\lambda)$ for each single-material wedge, it is easy to arrive at the *transfer relation for the wedge* w, relating the elastic variables between the wedge external faces at angles ϑ_{w-1} and ϑ_w:

$$\mathbf{w}_w(r,\vartheta_w) = \mathbf{K}_w(\lambda)\mathbf{w}_w(r,\vartheta_{w-1})$$

or

$$\begin{bmatrix} \mathbf{u}_w(r,\vartheta_w) \\ \boldsymbol{\varphi}_w(r,\vartheta_w) \end{bmatrix} = \begin{bmatrix} \mathbf{K}_w^{(1)}(\lambda) & \mathbf{K}_w^{(2)}(\lambda) \\ \mathbf{K}_w^{(3)}(\lambda) & \mathbf{K}_w^{(4)}(\lambda) \end{bmatrix} \begin{bmatrix} \mathbf{u}_w(r,\vartheta_{w-1}) \\ \boldsymbol{\varphi}_w(r,\vartheta_{w-1}) \end{bmatrix}, \tag{13.86}$$

where the expression for *the* $6\times$ 6 *wedge transfer matrix* $\mathbf{K}_w$ is obtained by the sequential product of the transfer matrices $\mathbf{E}_m(\lambda)$ of all the single-material wedges in the multimaterial wedge w,

$$\mathbf{K}_w(\lambda) = \mathbf{E}_{j_w}(\lambda)\cdot\mathbf{E}_{j_w-1}(\lambda)\cdots\mathbf{E}_{i_w+1}(\lambda)\cdot\mathbf{E}_{i_w}(\lambda). \tag{13.87}$$

The transfer relation (13.86) can be rewritten in the following matrix form suitable for the easy assembly of the characteristic system of a multimaterial corner:

$$\begin{bmatrix}\mathbf{K}_w(\lambda) & -\mathbf{I}_{6\times6}\end{bmatrix}\begin{bmatrix}\mathbf{w}_w(r,\vartheta_{w-1})\\ \mathbf{w}_w(r,\vartheta_w)\end{bmatrix} = \mathbf{0}_{6\times1}. \tag{13.88}$$

13.5.2. Characteristic system assembly

The following linear system collects all the wedge transfer relations (13.88) for the corner,

$$\mathbf{K}_{\text{corner\_ext.}}(\lambda)\mathbf{w}_{\text{corner\_ext.}} = \mathbf{0}_{6W\times1}, \tag{13.89}$$

where the $6W \times 12W$ extended complex matrix of transfer relations of the multimaterial corner is defined as

$$\mathbf{K}_{\text{corner\_ext.}}(\lambda)$$
$$= \begin{bmatrix}\mathbf{K}_1(\lambda) & -\mathbf{I}_{6\times6} & \mathbf{0}_{6\times6} & \mathbf{0}_{6\times6} & \cdots & \cdots & \mathbf{0}_{6\times6}\\ \mathbf{0}_{6\times6} & \mathbf{0}_{6\times6} & \mathbf{K}_2(\lambda) & -\mathbf{I}_{6\times6} & \cdots & \cdots & \mathbf{0}_{6\times6}\\ \vdots & \vdots & \vdots & \vdots & \ddots & \vdots & \vdots\\ \mathbf{0}_{6\times6} & \mathbf{0}_{6\times6} & \mathbf{0}_{6\times6} & \mathbf{0}_{6\times6} & \cdots & \mathbf{K}_W(\lambda) & -\mathbf{I}_{6\times6}\end{bmatrix} \tag{13.90}$$

and the $12W \times 1$ vector of elastic variables at wedge faces in the corner is defined as

$$\mathbf{w}_{\text{corner\_ext.}} = \begin{bmatrix}\mathbf{w}_1(r,\vartheta_0)\\ \hline \mathbf{w}_1(r,\vartheta_1)\\ \mathbf{w}_2(r,\vartheta_1)\\ \hline \mathbf{w}_2(r,\vartheta_2)\\ \vdots\\ \mathbf{w}_W(r,\vartheta_{W-1})\\ \hline \mathbf{w}_W(r,\vartheta_W)\end{bmatrix}. \tag{13.91}$$

Let $\boldsymbol{\vartheta} = (\vartheta_0, \vartheta_1, \ldots, \vartheta_{W-1}, \vartheta_W)$ define the vector of polar angles of wedge faces in the whole multimaterial corner. Let the sliding friction condition be prescribed at F $(0 \leq F \leq W+1)$ boundary faces or interfaces (between wedges) whose polar angles are given by the sequence ϑ_{k_i} $(i = 1, \ldots, F)$ and $0 \leq k_1 < k_2 < \cdots < k_{F-1} < k_F \leq W$. The corresponding functions defining the friction coefficients $\mu_{k_i}(\omega_{k_i})$ and the angles of sliding direction $\omega^u_{k_i}(\omega_{k_i})$ in terms of *a priori* unknown angles of frictional shear traction ω_{k_i} are gathered in the following vectors of the assumed functions

$\boldsymbol{\mu} = (\mu_{k_1}, \mu_{k_2}, \ldots, \mu_{k_{F-1}}, \mu_{k_F})$ and $\boldsymbol{\omega}^u = (\omega^u_{k_1}, \omega^u_{k_2}, \ldots, \omega^u_{k_{F-1}}, \omega^u_{k_F})$. Additionally, we define the vector of unknown values of the angles of frictional shear traction $\boldsymbol{\omega} = (\omega_{k_1}, \omega_{k_2}, \ldots, \omega_{k_{F-1}}, \omega_{k_F})$. In the following expressions, for the simple case of isotropic friction, the vector $\boldsymbol{\omega}^u$ can be omitted as the angles of friction shear stress and sliding angles coincide and the vector $\boldsymbol{\mu}$ represents just friction coefficient values at each contact surface, independent of the angles of friction shear stress.

13.5.2.1. *Open multimaterial corner*

The following $12W \times 12W$ *extended matrix of boundary and interface conditions for an open multimaterial corner* collects all the matrices for the boundary and interface conditions of the multimaterial corner in a way that it is compatible with the definition of the vector $\mathbf{w}_{\text{corner\_ext.}}$ in (13.91):

$$\mathbf{D}_{\text{corner\_ext.}}(\vartheta, \omega, \mu, \omega^u) = \text{blocked\_diag}[\mathbf{D}_{BC}(\vartheta_0), \mathbf{D}_I(\vartheta_1), \ldots, \mathbf{D}_{BC}(\vartheta_W)]$$

$$= \begin{bmatrix} \mathbf{D}_{BC}(\vartheta_0) & \mathbf{0}_{6\times 12} & \mathbf{0}_{6\times 12} & \cdots & \mathbf{0}_{6\times 6} \\ \mathbf{0}_{12\times 6} & \mathbf{D}_I(\vartheta_1) & \mathbf{0}_{12\times 12} & \cdots & \mathbf{0}_{12\times 6} \\ \vdots & \vdots & \vdots & \ddots & \vdots \\ \mathbf{0}_{6\times 6} & \mathbf{0}_{6\times 12} & \mathbf{0}_{6\times 12} & \cdots & \mathbf{D}_{BC}(\vartheta_W) \end{bmatrix}. \tag{13.92}$$

Notice that some or all of the boundary or interface condition matrices, $\mathbf{D}_{BC}$ or $\mathbf{D}_I$, included in the definition of $\mathbf{D}_{\text{corner\_ext.}}(\vartheta, \omega, \mu, \omega^u)$ may additionally depend on the values of the angle of friction shear stress, and on the assumed functions for the friction coefficient and for the angle of sliding at the pertinent boundary surface or interface, which have not been explicitly indicated in the right-hand side of (13.92) for the sake of notation simplicity. When there are no sliding friction conditions in the corner, $\boldsymbol{\omega}$, μ and ω^u are omitted in (13.92). The matrix $\mathbf{D}_{\text{corner\_ext.}}(\vartheta, \omega, \mu, \omega^u)$ is orthogonal because of the orthogonality relations (13.51) and (13.58) of the diagonal submatrices.

By reordering the vector of elastic variables at the wedge faces in (13.91) into subvectors of prescribed and unknown variables according to the boundary and interface condition relations (13.52) and (13.59), respectively, the vector denoted as $\mathbf{w}_{\text{corner\_PU}}$ is obtained. This $12W \times 1$ vector is reduced to a $(6W - F) \times 1$ vector $\mathbf{w}_{\text{corner\_U}}$ by omitting the prescribed zero values of these variables,

$$\mathbf{w}_{\text{corner\_PU}} = \left[\begin{array}{c} \mathbf{w}_P(r,\vartheta_0) \\ \mathbf{w}_U(r,\vartheta_0) \\ \hline \mathbf{w}_P(r,\vartheta_1) \\ \mathbf{w}_U(r,\vartheta_1) \\ \hline \vdots \\ \hline \mathbf{w}_P(r,\vartheta_W) \\ \hline \mathbf{w}_U(r,\vartheta_W) \end{array}\right], \quad \mathbf{w}_{\text{corner\_U}} = \begin{bmatrix} \mathbf{w}_U(r,\vartheta_0) \\ \mathbf{w}_U(r,\vartheta_1) \\ \vdots \\ \mathbf{w}_U(r,\vartheta_W) \end{bmatrix}. \tag{13.93}$$

By collecting all the boundary and interface condition relations in the corner, (13.52) and (13.59), respectively, we can write, first,

$$\mathbf{w}_{\text{corner\_PU}} = \mathbf{D}_{\text{corner\_ext.}}(\vartheta, \omega, \mu, \omega^u)\mathbf{w}_{\text{corner\_ext.}} \tag{13.94}$$

and, then, taking into account that $\mathbf{D}_{\text{corner\_ext.}}(\vartheta, \omega, \mu, \omega^u)$ is an orthogonal matrix,

$$\mathbf{w}_{\text{corner\_ext.}} = \mathbf{D}^T_{\text{corner\_ext.}}(\vartheta, \omega, \mu, \omega^u)\mathbf{w}_{\text{corner\_PU}}. \tag{13.95}$$

In fact, the last relation collects all the relations (13.53) and (13.60) for the boundary surfaces and interfaces of an open multimaterial corner. By substituting this relation into (13.89) we obtain

$$\mathbf{K}_{\text{corner\_ext.}}(\lambda)\mathbf{D}^T_{\text{corner\_ext.}}(\vartheta, \omega, \mu, \omega^u)\mathbf{w}_{\text{corner\_PU}} = \mathbf{0}_{6W\times 1}. \tag{13.96}$$

Finally, by removing the columns of the matrix $\mathbf{K}_{\text{corner\_ext.}}(\lambda)$ $\mathbf{D}^T_{\text{corner\_ext.}}(\vartheta, \omega, \mathbf{m}, \omega^u)$ multiplied by the prescribed zero values of $\mathbf{w}(r, \vartheta_w)$ $(w = 0, W)$ the final form of the *characteristic system for the singularity analysis of an open multimaterial corner* (also called a corner eigensystem) is achieved:

$$\mathbf{K}_{\text{corner}}(\lambda, \omega)\mathbf{w}_{\text{corner\_U}} = \mathbf{0}_{6W\times 1}, \tag{13.97}$$

where only the unknown values of λ and ω remain as arguments of the *characteristic matrix of an open multimaterial corner*

$$\mathbf{K}_{\text{corner}}(\lambda, \omega) = \begin{bmatrix} \mathbf{K}_1\tilde{\mathbf{D}}^T_{BC}(\vartheta_0) & -\tilde{\mathbf{D}}^T_1(\vartheta_1) & \mathbf{0}_{6\times n_2} & \cdots & \mathbf{0}_{6\times n_{W-1}} & \mathbf{0}_{6\times n_W} \\ \mathbf{0}_{6\times n_0} & \mathbf{K}_2\tilde{\mathbf{D}}^T_2(\vartheta_1) & -\tilde{\mathbf{D}}^T_1(\vartheta_2) & \cdots & \mathbf{0}_{6\times n_{W-1}} & \mathbf{0}_{6\times n_W} \\ \vdots & \vdots & \vdots & \ddots & \vdots & \vdots \\ \mathbf{0}_{6\times n_0} & \mathbf{0}_{6\times n_1} & \mathbf{0}_{6\times n_2} & \cdots & \mathbf{K}_W\tilde{\mathbf{D}}^T_2(\vartheta_{W-1}) & -\tilde{\mathbf{D}}^T_{BC}(\vartheta_W) \end{bmatrix}, \tag{13.98}$$

n_w is the number of rows in matrices $\tilde{\mathbf{D}}_{BC}(\vartheta_w)$, for $w = 0, W$, and (defind in (13.53)) in the matrices $\tilde{\mathbf{D}}_i(\vartheta_w)$, for $w = 1, \ldots, W-1$ and $i = 1, 2$. Thus, according to Section 13.3.2, $n_w = 3$, for $w = 0, W$, and $n_w = 6$, for $w = 1, \ldots, W-1$, except for the faces where the sliding friction condition is prescribed at ϑ_w, then n_w is smaller by 1, i.e. $n_w = 2$ and $n_w = 5$, respectively. Therefore, the matrix $\mathbf{K}_{\text{corner}}(\lambda)$ is a $6W \times 6W$ square matrix if the sliding friction condition is not prescribed at any boundary surface or interface of the corner (i.e., $F = 0$), whereas with a sliding friction condition at one or more corner boundary surfaces or interfaces (i.e., $F > 0$) $\mathbf{K}_{\text{corner}}(\lambda, \omega)$ is a $6W \times (6W - F)$ rectangular matrix. In general, the elements of the matrix $\mathbf{K}_{\text{corner}}(\lambda, \omega)$ are transcendental complex analytic functions (holomorphic functions) of λ, including also real parameters ω for sliding friction conditions.

It is instructive to see how the characteristic matrix of the corner $\mathbf{K}_{\text{corner}}(\lambda, \omega)$ simplifies when there is only one wedge (i.e., $W = 1$) to a 6×6, 6×5 or 6×4 matrix, respectively, depending on whether there are none, or one or two boundary surfaces with a sliding friction condition:

$$\mathbf{K}_{\text{corner}}(\lambda, \omega) = [\mathbf{K}_1(\lambda)\tilde{\mathbf{D}}_{BC}^T(\vartheta_0) \quad -\tilde{\mathbf{D}}_{BC}^T(\vartheta_1)]. \tag{13.99}$$

When there are only orthogonal boundary conditions prescribed at both boundary surfaces of the wedge ($W = 1$), the characteristic system can further be reduced, as in [19] (see also [22]), to the form:

$$\mathbf{K}_{\text{corner\_reduced}}(\lambda)\mathbf{w}_U(r, \vartheta_0) = \mathbf{0}, \tag{13.100}$$

where

$$\begin{aligned}\mathbf{K}_{\text{corner\_reduced}}(\lambda) = {} & \mathbf{D}_u(\vartheta_1)\mathbf{K}_1^{(1)}(\lambda)\tilde{\mathbf{D}}_u^T(\vartheta_0) + \mathbf{D}_u(\vartheta_1)\mathbf{K}_1^{(2)}(\lambda)\tilde{\mathbf{D}}_\varphi^T(\vartheta_0) \\ & + \mathbf{D}_\varphi(\vartheta_1)\mathbf{K}_1^{(3)}(\lambda)\tilde{\mathbf{D}}_u^T(\vartheta_0) + \mathbf{D}_\varphi(\vartheta_1)\mathbf{K}_1^{(4)}(\lambda)\tilde{\mathbf{D}}_\varphi^T(\vartheta_0)\end{aligned} \tag{13.101}$$

is a 3×3 matrix defined by partitioning the wedge transfer matrix $\mathbf{K}_1(\lambda)$ to four 3×3 matrices in (13.86).

13.5.2.2. *Closed multimaterial corner (periodic corner)*

The following $12W \times 12W$ *extended rectangular matrix of interface conditions for a closed multimaterial corner* collects all the matrices of interface conditions of the corner in such a way that it is compatible with the definition of vector $\mathbf{w}_{\text{corner\_ext.}}$ in (13.91):

$$\mathbf{D}_{\text{corner\_ext.}}(\vartheta, \omega, \mu, \omega^u) = \begin{bmatrix} \mathbf{D}_2(\vartheta_0) & \mathbf{0}_{n_0'\times 12} & \mathbf{0}_{n_0'\times 12} & \cdots & \mathbf{0}_{n_0'\times 12} & \mathbf{D}_1(\vartheta_W) \\ \tilde{\mathbf{D}}_2(\vartheta_0) & \mathbf{0}_{n_0\times 12} & \mathbf{0}_{n_0\times 12} & \cdots & \mathbf{0}_{n_0\times 12} & \tilde{\mathbf{D}}_1(\vartheta_W) \\ \mathbf{0}_{12\times 6} & \mathbf{D}_I(\vartheta_1) & \mathbf{0}_{12\times 12} & \cdots & \mathbf{0}_{12\times 12} & \mathbf{0}_{12\times 6} \\ \vdots & \vdots & \vdots & \ddots & \vdots & \vdots \\ \mathbf{0}_{12\times 6} & \mathbf{0}_{12\times 12} & \mathbf{0}_{12\times 12} & \cdots & \mathbf{D}_I(\vartheta_{W-1}) & \mathbf{0}_{12\times 6} \end{bmatrix}, \tag{13.102}$$

where $n_0 = n_W$ is the number of rows in the matrices $\tilde{\mathbf{D}}_2(\vartheta_0)$ and $\tilde{\mathbf{D}}_1(\vartheta_W)$, and $n_0 + n_0' = 12$.

Comments similar to those given after Eq. (13.92) are valid here as well. In particular, the matrix $\mathbf{D}_{\text{corner\_ext.}}(\vartheta, \omega, \mu, \omega^u)$ is orthogonal. Recall that $\vartheta_W = \vartheta_0 + 360°$.

By reordering the vector of elastic variables for the wedge faces in (13.91) into subvectors of prescribed and unknown variables according to the interface condition relation (13.59), the vector denoted as $\mathbf{w}_{\text{corner\_PU}}$ is obtained. This $12W\times 1$ vector is reduced to a $(6W - F)\times 1$ vector $\mathbf{w}_{\text{corner\_U}}$ by omitting the prescribed zero values of these variables,

$$\mathbf{w}_{\text{corner\_PU}} = \left[\begin{array}{c} \mathbf{w}_P(r,\vartheta_0) \\ \mathbf{w}_U(r,\vartheta_0) \\ \hline \mathbf{w}_P(r,\vartheta_1) \\ \mathbf{w}_U(r,\vartheta_1) \\ \hline \vdots \\ \hline \mathbf{w}_P(r,\vartheta_{W-1}) \\ \mathbf{w}_U(r,\vartheta_{W-1}) \end{array}\right], \quad \mathbf{w}_{\text{corner\_U}} = \begin{bmatrix} \mathbf{w}_U(r,\vartheta_0) \\ \mathbf{w}_U(r,\vartheta_1) \\ \vdots \\ \mathbf{w}_U(r,\vartheta_{W-1}) \end{bmatrix}. \tag{13.103}$$

Several differences with respect to the similar vectors in (13.93) should be noted. In particular, while $\mathbf{w}_P(r,\vartheta_0)$ and $\mathbf{w}_U(r,\vartheta_0)$ are 3×1 vectors (or 4×1 and 2×1, respectively, for a sliding friction condition) in (13.93), representing the prescribed boundary conditions and unknowns at the corner boundary at ϑ_0, and similarly for ϑ_W, they are 6×1 vectors (or 7×1 and 5×1, respectively, for a sliding friction condition) in (13.103), representing the prescribed interface conditions and interface unknowns.

By collecting all the interface condition relations in the corner (13.59), we can write, first,

$$\mathbf{w}_{\text{corner\_PU}} = \mathbf{D}_{\text{corner\_ext.}}(\vartheta, \omega, \mu, \omega^u)\mathbf{w}_{\text{corner\_ext.}} \tag{13.104}$$

and, then, taking into account that $\mathbf{D}_{\text{corner\_ext.}}(\vartheta, \omega, \mu, \omega^u)$ is an orthogonal matrix,

$$\mathbf{w}_{\text{corner\_ext.}} = \mathbf{D}^T_{\text{corner\_ext.}}(\vartheta, \omega, \mu, \omega^u)\mathbf{w}_{\text{corner\_PU}}. \tag{13.105}$$

In fact, the last relation collects all the relations (13.60) for the interfaces of a closed multimaterial corner. By substituting this relation into (13.89) we obtain

$$\mathbf{K}_{\text{corner\_ext.}}(\lambda)\mathbf{D}^T_{\text{corner\_ext.}}(\vartheta, \omega, \mu, \omega^u)\mathbf{w}_{\text{corner\_PU}} = \mathbf{0}_{6W\times 1}. \tag{13.106}$$

Finally, by removing the columns of the matrix $\mathbf{K}_{\text{corner\_ext.}}(\lambda)$ $\mathbf{D}^T_{\text{corner\_ext.}}(\vartheta, \omega, \mu, \omega^u)$ multiplied by the prescribed zero values of $\mathbf{w}_P(r, \vartheta_w)$ ($w = 0, W-1$) the final form of *the characteristic system for the singularity analysis of a closed multimaterial corner* (also called a corner eigensystem) is achieved:

$$\mathbf{K}_{\text{corner}}(\lambda, \omega)\mathbf{w}_{\text{corner\_U}} = \mathbf{0}_{6W\times 1}, \tag{13.107}$$

where only the unknown values of λ and $\boldsymbol{\omega}$ remain as arguments of *the characteristic matrix of a closed multimaterial corner*

$$\mathbf{K}_{\text{corner}}(\lambda, \omega) = \begin{bmatrix} \mathbf{K}_1\tilde{\mathbf{D}}_2^T(\vartheta_0) & -\tilde{\mathbf{D}}_1^T(\vartheta_1) & \mathbf{0}_{6\times n_2} & \cdots & \mathbf{0}_{6\times n_{W-2}} & \mathbf{0}_{6\times n_{W-1}} \\ \mathbf{0}_{6\times n_0} & \mathbf{K}_2\tilde{\mathbf{D}}_2^T(\vartheta_1) & -\tilde{\mathbf{D}}_1^T(\vartheta_2) & \cdots & \mathbf{0}_{6\times n_{W-2}} & \mathbf{0}_{6\times n_{W-1}} \\ \vdots & \vdots & \vdots & \ddots & \vdots & \vdots \\ \mathbf{0}_{6\times n_0} & \mathbf{0}_{6\times n_1} & \mathbf{0}_{6\times n_2} & \cdots & \mathbf{K}_{W-1}\tilde{\mathbf{D}}_2^T(\vartheta_{W-2}) & -\tilde{\mathbf{D}}_1^T(\vartheta_{W-1}) \\ -\tilde{\mathbf{D}}_1^T(\vartheta_0) & \mathbf{0}_{6\times n_1} & \mathbf{0}_{6\times n_2} & \cdots & \mathbf{0}_{6\times n_{W-2}} & \mathbf{K}_W\tilde{\mathbf{D}}_2^T(\vartheta_{W-1}) \end{bmatrix}, \tag{13.108}$$

n_w is the number of rows in the matrices $\tilde{\mathbf{D}}_i(\vartheta_w)$, for $w = 0, \ldots, W-1$ and $i = 1, 2$. Thus, according to Section 13.3.2, $n_w = 6$ for $w = 0, \ldots, W-1$, except for the interfaces where the sliding friction condition is prescribed at ϑ_w, n_w then being smaller by 1, i.e., $n_w = 5$. Therefore, the matrix $\mathbf{K}_{\text{corner}}(\lambda)$ is a $6W \times 6W$ square matrix if the sliding friction condition is not prescribed at any corner interface (i.e., $F = 0$), whereas with the sliding friction condition at one or more corner interfaces (i.e., $F > 0$) $\mathbf{K}_{\text{corner}}(\lambda, \omega)$ is a $6W \times (6W - F)$ rectangular matrix.

13.5.3. Solution of the characteristic system — Singular elastic solution

We are looking for non-trivial solutions $\mathbf{w}_{\text{corner\_U}} \neq \mathbf{0}$ of the homogeneous linear system (13.97) or (13.107), respectively, for an open or closed corner, which define *the characteristic (singular) elastic solutions of the corner* verifying the prescribed boundary and/or interface conditions. Any non-trivial solution $\mathbf{w}_{\text{corner\_U}} \neq \mathbf{0}$ of (13.97) or (13.107) is a (right) null vector of the matrix $\mathbf{K}_{\text{corner}}(\lambda, \omega)$. Thus, first we need to find *the characteristic (singular) values of* λ and ω, which will provide a rank deficient matrix $\mathbf{K}_{\text{corner}}(\lambda, \omega)$ (rank $\mathbf{K}_{\text{corner}}(\lambda, \omega) < 6W - F$) with a non-trivial (right) null space.

In the most usual case without friction contact conditions, i.e., when $F = 0$, the linear system (13.97) or (13.107), with a $6W \times 6W$ square matrix, is a kind of *nonlinear eigenvalue problem of the multimaterial corner*

$$\mathbf{K}_{\text{corner}}(\lambda)\mathbf{w}_{\text{corner\_U}} = \mathbf{0}_{6W\times 1}. \tag{13.109}$$

The usual procedure for solving this nonlinear eigenvalue problem begins by finding the roots λ of the matrix determinant, which are the solutions of the *characteristic equation of the corner* (also called the corner eigenequation),

$$\det \mathbf{K}_{\text{corner}}(\lambda) = 0. \tag{13.110}$$

These roots, referred to as the *characteristic (singular) exponents* (also called eigenvalues), are fundamental in the corner singularity analysis. From the above, in general $\det \mathbf{K}_{\text{corner}}(\lambda)$ is a transcendental complex analytic function (a holomorphic function) of λ. Characteristic exponents usually form an infinite discrete set in the complex plane defining an infinite set of *characteristic elastic solutions* for the considered corner configuration. It can be shown that if a complex λ is root of (13.110), its complex conjugate $\bar{\lambda}$ is a root of (13.110) as well. While characteristic exponents λ with $\operatorname{Re}\lambda > 0$ lead to elastic solutions in the corner with a finite elastic strain energy, $\operatorname{Re}\lambda < 0$ correspond to elastic solutions with infinite strain energy, due to a non-integrable singularity at the corner tip. In particular, characteristic exponents with $0 < \operatorname{Re}\lambda < 1$ correspond to singular elastic solutions in the corner with unbounded stresses and strains at the corner tip but a finite elastic strain energy.

With friction contact conditions, i.e., when $F \geq 1$, linear system (13.97) or (13.107), with a $6W \times (6W - F)$ rectangular matrix, is an overdetermined homogeneous system. An efficient and reliable way to determine if it has a non-trivial solution for some particular values of λ and ω, and to evaluate such a non-trivial solution, is to compute the singular value decomposition (SVD) of the matrix $\mathbf{K}_{\text{corner}}(\lambda, \omega)$ [95, 96], by evaluating singular values $\sigma_i \geq 0$ $(i = 1, \ldots, 6W - F)$. Let $\boldsymbol{\sigma}_{\min}(\mathbf{K}_{\text{corner}}(\lambda, \omega))$ denote the smallest singular value. Then, *the characteristic exponents* λ and *characteristic friction-angles* ω (which in the isotropic friction case coincide with *the characteristic sliding angles* ω^u) are determined by solving the following *characteristic equation of the corner* with sliding friction contact:

$$\sigma_{\min}(\mathbf{K}_{\text{corner}}(\lambda, \omega)) = 0. \tag{13.111}$$

Due to the periodicity of the functions of ω_{k_i} $(i = 1, \ldots F)$ involved in the definition of $\mathbf{K}_{\text{corner}}(\lambda, \omega)$, see (13.36) and (13.42)–(13.45), we typically search for characteristic friction angles in the interval $-180° \leq \omega_{k_i} < 180°$.

It may be useful to write the characteristic equation of the corner as a condition for a vanishing determinant, which is (at least theoretically) equivalent to (13.111), by taking into account that the squares of the singular values σ_i^2 are eigenvalues of the matrix of the least-squares system for (13.97) or (13.107),

$$\det(\mathbf{K}_{\text{corner}}^{\mathrm{T}}(\lambda, \omega)\mathbf{K}_{\text{corner}}(\lambda, \omega)) = 0. \tag{13.112}$$

An advantage of this explicit form of the characteristic equation of the corner, in comparison with (13.111), is that we can search for roots of a transcendental complex analytic function of λ and ω_{k_i} (providing the functions in (13.36) are also analytic functions) similarly as in (13.110). However, a disadvantage of (13.112) may be that the algebraic multiplicity of the roots λ is doubled.

Previous formulations for a corner singularity with sliding friction contact are usually limited to corner configurations with uncoupled plane-strain and antiplane-strain states [68, 70, 71, 73]. The novelty of the present formulation is that in a corner singularity problem with friction under generalized plane-strain conditions we find not only the characteristic exponents λ as usual, but also the characteristic friction angles ω, which solve the characteristic equation of the corner, (13.111) or (13.112).

It should be stressed that the present procedure leads to a closed-form analytic expression of the matrix $\mathbf{K}_{\text{corner}}(\lambda, \omega)$, where the only numerically computed values are the roots p_α of the Lekhnitskii–Stroh sextic equation for anisotropic materials (13.15), if analytic expressions of these roots are not available. With the exception of very simple corner configurations, a computational tool for symbolic computations should be used to evaluate a closed-form expression of $\mathbf{K}_{\text{corner}}(\lambda, \omega)$. Furthermore, such a tool for symbolic computation can be used to evaluate a closed-form analytic expression of the corner eigenequation (13.110) or (13.112). In the present work, *Mathematica* [97] has successfully been used for these purposes.

Finding the real or complex roots λ of nonlinear eigenequations in the form of a vanishing determinant condition in (13.110) (and similarly in (13.112)), which are in general given by complex analytic functions (holomorphic functions), can be efficiently carried out by Muller's method [98, 99] or by more sophisticated algorithms [100]. Nevertheless, special care should be taken in finding all the roots in a region of interest, usually defined by an interval of the real part of λ, e.g., $0 < \text{Re}\,\lambda \leq 1$. For this purpose, the argument principle (e.g., [101]) is an excellent tool for identifying the number of roots of a holomorphic function, e.g., $h(\lambda) = \det \mathbf{K}_{\text{corner}}(\lambda)$, in a particular region of the complex plane. According to the argument principle, the following integral along a closed contour C (without self-intersections) in the complex plane:

$$J = \frac{1}{2\pi i} \oint_C \frac{h'(\lambda)}{h(\lambda)} d\lambda, \tag{13.113}$$

with $h'(\lambda)$ denoting the complex derivative of $h(\lambda)$, gives the change in the argument of $h(\lambda)$ around this contour (the difference between the final and initial values of the continuously varying polar angle of the complex number $h(\lambda)$) divided by 2π,

$$J = \frac{1}{2\pi} [\arg(h(\lambda))]_C. \tag{13.114}$$

It can be shown that J equals the number of zeros of $h(\lambda)$ (which has no poles), including their multiplicity, inside the domain defined by C. This statement is valid if there are no zeros of $h(\lambda)$ on the contour C itself. The integral in (13.113) can be evaluated by a numerical quadrature, usually giving a number very close to an integer representing the number of zeros of $h(\lambda)$. Nevertheless, in the present work we employ the representation (13.114). A couple of examples of applications of the argument principle are given in Sections 13.7.1 and 13.7.2.

With sliding friction conditions at some corner surfaces, and taking into account that a singular value of a matrix is always a real non-negative number, it appears that an efficient way of finding the solutions of the characteristic equation of the corner (13.111) is to apply a global minimization procedure to $\sigma_{\min}(\mathbf{K}_{\text{corner}}(\lambda, \omega))$ as an objective function of λ and ω from a feasible region of interest, e.g., defined by $0 < \text{Re}\lambda < 1$ and $-180° \leq \omega_{k_i} < 180°$. The region of interest should be thoroughly explored by an automatic and reliable minimization procedure capable of finding all the minimizer pairs λ and ω for which the objective function vanishes, i.e., $\sigma_{\min}(\mathbf{K}_{\text{corner}}(\lambda, \omega)) = 0$. Note that the minimization procedure requires the computation of the SVD of $\mathbf{K}_{\text{corner}}(\lambda, \omega)$ for all pairs λ and ω considered in the minimization iterations.

Once a particular value of the characteristic exponent λ, and with friction contact also of the vector of the characteristic friction angles ω, is obtained as a solution of the characteristic equation of the corner, the corresponding behavior of the displacements and stresses inside the corner can easily be computed. The procedure starts by computing a corresponding non-trivial solution $\mathbf{w}_{\text{corner\_U}} \neq \mathbf{0}$ of the homogeneous linear systems (13.97) or (13.107) considering a fixed value of the radial coordinate r, e.g., $r = 1$. The values of $\mathbf{w}_{\text{corner\_U}}$ with the pertinent radial dependence are obtained by multiplying it by r^λ, $\mathbf{w}_{\text{corner\_U}}(r) = r^\lambda \mathbf{w}_{\text{corner\_U}}(r = 1)$. Completing $\mathbf{w}_{\text{corner\_U}}$ with zero values of $\mathbf{w}_{\text{P}}(r, \vartheta_w)$, see (13.93) and (13.103), leads to the corresponding $\mathbf{w}_{\text{corner\_PU}} \neq \mathbf{0}$. Then, $\mathbf{w}_{\text{corner\_ext.}}$ in (13.91) is evaluated by means of (13.95) or (13.105). Now, from $\mathbf{w}_w(r, \vartheta_{w-1})$, known for any $w = 1, \ldots, W$, we can compute $\mathbf{w}_m(r, \theta_{m-1})$ for all single-material wedges in a multimaterial wedge w, with indices $m = i_w, \ldots, j_w$, by sequentially employing the transfer relation (13.66). Finally, we compute the Cartesian components of the displacement and stress function vectors, $\mathbf{u}(r, \theta)$ and $\varphi(r, \theta)$, inside a single-material wedge m by a transfer relation analogous to (13.66):

$$\mathbf{w}_m(r, \theta) = \mathbf{E}_m\,(\lambda, \theta, \theta_{m-1})\mathbf{w}_m(r, \theta_{m-1}) \quad \text{for} \quad \theta_{m-1} \leq \theta \leq \theta_m. \tag{13.115}$$

Then, the displacement vector and stress tensor in cylindrical coordinates are obtained as follows [28, Sec. 7.3]:

$$u_r = -\mathbf{s}_r^T(\theta)\mathbf{u}(r, \theta), \quad u_\theta = \mathbf{n}^T(\theta)\mathbf{u}(r, \theta), \quad u_3 = \mathbf{s}_3^T \mathbf{u}(r, \theta), \tag{13.116}$$

and

$$\begin{aligned}
\sigma_{rr} &= \mathbf{s}_r^T(\theta)\boldsymbol{\varphi}_{,\theta}(r,\theta)/r, \quad \sigma_{\theta\theta} = \mathbf{n}^T(\theta)\boldsymbol{\varphi}_{,r}(r,\theta),\\
\sigma_{r\theta} &= -\mathbf{n}^T(\theta)\boldsymbol{\varphi}_{,\theta}(r,\theta)/r = -\mathbf{s}_r^T(\theta)\boldsymbol{\varphi}_{,r}(r,\theta),\\
\sigma_{r3} &= -\mathbf{s}_3^T(\theta)\boldsymbol{\varphi}_{,\theta}(r,\theta)/r, \quad \sigma_{\theta 3} = \mathbf{s}_3^T(\theta)\boldsymbol{\varphi}_{,r}(r,\theta),
\end{aligned} \tag{13.117}$$

and σ_{33} is evaluated from the condition $\varepsilon_{33} = 0$. Taking into account that if $\mathbf{w}_{\text{corner\_U}} \neq \mathbf{0}$ is a solution of the homogeneous linear system (13.97) or (13.107) then also $c\mathbf{w}_{\text{corner\_U}}$, where $c \neq 0$, solves this system as well, it is useful to standardize the corresponding singular elastic solution obtained in the corner. A practical approach is to make a stress component at a suitably defined position in the corner equal to a given constant, e.g., requiring $\sigma_{\theta\theta}(r=1,\theta^*) = (2\pi)^{\text{Re}\,\lambda-1}$ [102] where θ^* is a specific angle in the corner (the angle of a wedge face, angle of the corner symmetry or antisymmetry plane if it exists, etc.). In this way, the standardized characteristic angular functions $g_i^{(n)}(\theta)$ and $f_{ij}^{(n)}(\theta)$ introduced in (13.2) and (13.3) are defined for a particular characteristic exponent $\lambda = \lambda_n$.

With reference to the corner singularity analysis with friction contact conditions, and as explained in Section 13.3.2, each solution of the characteristic equation of the corner (13.111) obtained should be checked to see if it satisfies the compression condition ($\sigma_{\theta\theta}(r,\vartheta_w) \leq 0$) and the friction dissipation condition (although the relative tangential displacement obtained by the solution of the characteristic system is parallel to $\mathbf{k}(\vartheta_w,\omega^u)$, we still should check its orientation with respect to the friction shear stress to guarantee the dissipative character of the friction) at all the friction surfaces of the corner. It appears that oscillatory solutions for a complex characteristic exponent λ are not compatible with these compression and friction dissipation conditions; consequently only real characteristic exponents λ are essentially admissible. Nevertheless, it is well known that complex characteristic exponents λ may appear in cracks at the straight interface between two anisotropic bodies with a frictionless contact zone at the crack tip [65–70]. Thus, assuming solution continuity with respect to a vanishing friction coefficient, it can be expected that similar complex characteristic exponents λ can also appear for sliding friction contact in a corner singularity analysis, a hypothesis that should still be checked numerically. Although, strictly speaking, such solutions are not admissible, in the global problem solution the portion of the contact zone adjacent to the corner tip, where these compression and friction dissipation conditions can be violated, could be of a very small size with respect to other characteristic lengths of the whole problem. Then, such a solution might be accepted, in the same way that crack face overlapping in an oscillatory solution of an open model of interface cracks is accepted with the hypothesis of small scale contact [103].

13.6. Evaluation of GSIFs

When the characteristic exponents λ_n and the characteristic angular shape functions $g_i^{(n)}(\theta)$ and $f_{ij}^{(n)}(\theta)$ in the asymptotic series expansions of the elastic solution in a corner in (13.2) and (13.3) are known, the only unknowns in these series to be determined are the *generalized stress intensity factors* K_n (GSIFs), which are coefficients of power-type singularities. The evaluation of the GSIFs usually requires a numerical model of the whole problem. Techniques for the evaluation of GSIFs can roughly be divided into four basic groups, see Table 13.5, according to their local or global character and whether they are implemented in the solution or post-processing stage of FEM or BEM analysis. Local techniques are usually sensitive to the accuracy of the numerical solution for stresses or displacements close to the corner tip, while global techniques, working also, or only, with the elastic solution far from the corner tip, are less sensitive to the solution accuracy at the corner tip. Techniques for extracting GSIFs from a numerical FEM or BEM solution in the post-processing stage do not need to be incorporated into the FEM or BEM codes, but do not typically have as good accuracy as methods which directly incorporate the singularity shape functions into the problem discretization, usually requiring a modification of these codes. References for techniques in

Table 13.5. Classification of procedures for the evaluation of GSIFs.

	Local/Global techniques	Global techniques
Extraction of GSIFs from FEM or BEM solution in the post-processing stage	Least-squares fitting [40, 108–111]	Conservative integrals [2, 105, 107, 112–117]
	Local techniques	**Global techniques**
Incorporation of singularity shape functions in problem discretization	Quarter point elements [118–120] and other singularity elements [121]	Functions in the whole domain or boundary [41, 122]

these groups are included in Table 13.5; further information can be found in [6, 8, 41]. Examples of the evaluation of GSIFs involving anisotropic materials can be found in [104, 105] by means of the H-integral, and in [106, 107] by means of the M-integral along with other techniques.

13.6.1. Least-squares fitting technique

The technique presented in this section is based on least-squares fitting of the finite asymptotic series expansions in (13.2) and (13.3) to the numerical results for the displacements and/or stresses in a multimaterial corner [40]. This is a reliable, accurate and easy-to-use technique for the extraction of GSIFs with no need to modify the FEM or BEM code applied. It has no limits for the number of power stress singularities considered in the analysis and shows an acceptable robustness when employing numerical results relatively far from the corner tip.

The technique minimizes an error function J

$$J(K_1,\ldots,K_N) = aJ_u(K_1,\ldots,K_N) + bJ_\sigma(K_1,\ldots,K_N) \quad (a, b \geq 0), \tag{13.118}$$

with

$$J_u(K_1,\ldots,K_N) = \sum_{\alpha=r,\theta,3}\sum_{i=1}^{N_r}\sum_{j=1}^{N_\theta} a_\alpha \times[u_\alpha^{\text{series}}(r_i,\theta_j,K_1,\ldots,K_N) - u_\alpha^{BEM}(r_i,\theta_j)]^2 \quad (a_\alpha \geq 0), \tag{13.119}$$

$$J_\sigma(K_1,\ldots,K_N) = \sum_{\alpha=r,\theta,3}\sum_{i=1}^{N_r}\sum_{j=1}^{N_\theta} b_\alpha \times[t_\alpha^{\text{series}}(r_i,\theta_j,K_1,\ldots,K_N) - t_\alpha^{BEM}(r_i,\theta_j)]^2 \quad (b_\alpha \geq 0), \tag{13.120}$$

where J_u and J_σ compute the sums of squares of differences between the numerical and analytical solution in terms of displacements and stresses, respectively, at a number of points (usually nodes of a mesh) given by polar coordinates (r_i, θ_j). The numerical and analytical solutions are denoted with the superscripts '*BEM*' and '*series*', as the first is obtained from a BEM model in the present work and the second is given by the asymptotic series expansion in (13.2) and (13.3). In (13.118), a and b are weighting factors, which allow us to consider only displacements $(a, b) = (1, 0)$, only stresses $(a, b) = (0, 1)$ or both $(a, b) = (l^{-2}, \sigma^{-2})$, l and σ being some characteristic length and stress values so that the terms given by the displacements and stresses have values of the same order of magnitude. Similarly, the dimensionless weighting factors in (13.119) and (13.120), a_α and b_α, allow the isolated components of the nodal displacements and tractions to be used. The points used for the evaluation of J in (13.118) are placed in the present work, without loss of generality, along radial lines defined by the corner boundaries and interfaces. This is a natural option when a BEM model is used as any BEM mesh has nodes at these locations. Thus, N_r and N_θ in (13.119) and (13.120) denote the number of BEM nodes at each radial line and the number of radial lines employed, respectively, as schematically illustrated in Fig. 13.3. Nevertheless, when using an FEM model, J can be evaluated at an arbitrary set of nodes in the corner.

The set of GSIFs K_n $(n = 1, \ldots, N)$ which minimizes J is obtained by solving the following linear system of equations:

$$\frac{\partial J(K_1, \ldots, K_N)}{\partial K_j} = 0 \quad (j = 1, \ldots, N). \tag{13.121}$$

The present technique admits solutions with complex values of GSIFs. If a characteristic exponent λ is a complex number, as in the open model of interface cracks, then its complex conjugate $\bar{\lambda}$ is also a characteristic exponent. The associated GSIF is also a complex number $K = K_{\mathrm{R}} + iK_{\mathrm{I}}$, where K_{R} and K_{I} are real numbers. In this case, two real terms can be included in (13.2) and (13.3) instead of two complex terms. In the representation of displacements, one term would be equal to $K_R \ \mathrm{Re}\lfloor r^\lambda g_i(\theta)\rfloor$ and the other to $K_I \ \mathrm{Im}\lfloor r^\lambda g_i(\theta)\rfloor$, while in the

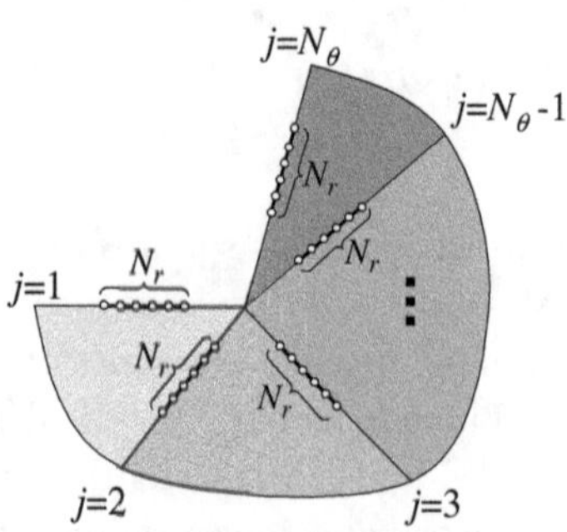

Fig. 13.3. BEM nodes for least-squares fitting.

representation of stresses, one term would be equal to $K_R \operatorname{Re}\lfloor r^{\lambda-1} f_{ij}(\theta)\rfloor$ and the other to $K_I \operatorname{Im}\lfloor r^{\lambda-1} f_{ij}(\theta)\rfloor$.

13.6.2. Implementation, accuracy and robustness

We will now briefly discuss additional issues regarding the implementation, accuracy and robustness of the above technique. First, the linear system arising from (13.121) is calculated explicitly. For the sake of simplicity and without loss of generality, let only the radial displacement component u_r be included in the error function J, by choosing $a_r = 1$, $a_\theta = a_3 = 0$, $a = 1$ and $b = 0$. The displacement u_r at a point p (usually a BEM mesh node) defined by the radius r_i and polar angle θ_j is approximated by using N terms of the asymptotic series expansion representation (13.2) as

$$u_r^{\text{series}}(r_i, \theta_j) \cong \sum_{n=1}^{N} K_n r_i^{\lambda_n} g_r^{(n)}(\theta_j) = \sum_{n=1}^{N} a_{pn} K_n, \qquad (13.122)$$

where $a_{pn} = r_i^{\lambda_n} g_r^{(n)}(\theta_j)$ is the coefficient for the K_n term of the series expansion of u_r in (13.2) evaluated at the point p with the polar coordinates (r_i, θ_j). The derivative of the error function J with respect to K_j evaluated at P points $(p = 1, \ldots, P)$ is a linear function of K_n $(n = 1, \ldots, N)$, which can be expressed as follows:

$$\begin{aligned}\frac{\partial J}{\partial K_j} &= 2\sum_{p=1}^{P}\left\{\sum_{n=1}^{N} a_{pn} K_n - u_r^{BEM}(p)\right\} a_{pj} \\ &= 2\sum_{p=1}^{P}\left\{\sum_{n=1}^{N} a_{pn} K_n a_{pj} - a_{pj} u_r^{BEM}(p)\right\} = 0, \qquad (13.123)\end{aligned}$$

where $u_r^{\text{BEM}}(p) = u_r^{\text{BEM}}(r_i, \theta_j)$ denotes the displacement u_r at the point p. Equation (13.123) is written in matrix notation as

$$\mathbf{A}^T \cdot \mathbf{A} \cdot \mathbf{K} = \mathbf{A}^T \cdot \mathbf{b}, \qquad (13.124)$$

where $\mathbf{A}$ is a $P \times N$ matrix, P and N being respectively the number of points (nodes) used for building the error function J, and the number of terms in the series (13.2) expansion (the number of GSIF values). As long as the number of points P is greater than the number of terms considered for the displacement representation N, $\mathbf{A}$ is a rectangular matrix with more rows than columns and thus expected to have full rank N. $\mathbf{K}$ is the $N \times 1$

vector of unknowns (GSIF values) and $\mathbf{b}$ is the $P \times 1$ vector of numerical results for u_r at the chosen points. Hence,

$$\mathbf{A} = \begin{bmatrix} a_{11} & \cdots & a_{1N} \\ \vdots & & \vdots \\ a_{P1} & \cdots & a_{PN} \end{bmatrix}, \quad \mathbf{K} = \begin{bmatrix} K_1 \\ \vdots \\ K_N \end{bmatrix}, \quad \mathbf{b} = \begin{bmatrix} b_1 \\ \vdots \\ b_P \end{bmatrix}. \tag{13.125}$$

It is clear that the $N \times N$ square matrix $\mathbf{A}^T \cdot \mathbf{A}$ can have rank $= N$ if and only if the number of points P is equal to or greater than the number N of GSIFs to be calculated. Only in that case can the inverse of $\mathbf{A}^T \cdot \mathbf{A}$ exist and be computed.

Equation (13.124) is a typical matrix expression that appears when solving an overdetermined linear system

$$\mathbf{A} \cdot \mathbf{K} = \mathbf{b}, \tag{13.126}$$

using the 2-norm minimization $\min_{\mathbf{K}} \|\mathbf{A} \cdot \mathbf{K} - \mathbf{b}\|_2$, see [95, 96]. Notice that the fulfilment of (13.126) corresponds to the vanishing differences in the error function J, i.e., $J = 0$.

The solution $\mathbf{b}$ of the full rank least-squares problem is theoretically unique. Nevertheless, due to the nature of the matrix components $a_{pn} = r_i^{\lambda_n} g_r^{(n)}(\theta_j)$, the evaluation of the matrix $\mathbf{A}^T \cdot \mathbf{A}$ using only points very close to the corner tip has been shown to give rise to ill-conditioned matrices with numerically computed rank $(\mathbf{A}^T \cdot \mathbf{A}) < N$. This ill-conditioning includes the cases in which the number of points P exceeds the number of GSIF terms N and the matrix $\mathbf{A}$ has numerically computed full rank N. The conditioning number for the 2-matrix norm [95, 96] $\kappa(\mathbf{A}) = \|\mathbf{A}\|_2 \|\mathbf{A}^+\|_2 = \sigma_{\max}/\sigma_{\min}$, where $\mathbf{A}^+$ is the pseudoinverse of $\mathbf{A}$, and $\sigma_{\max}$ and $\sigma_{\min}$, respectively, are the maximum and minimum singular values of $\mathbf{A}$, gives an indication of the conditioning of the problem. It has numerically been verified [40] that $\kappa(\mathbf{A})$ in these cases (in which the number of points P exceeds only slightly the number of GSIF terms N) may be around 10^3 times higher than the conditioning number obtained in those cases where P is much greater than N.

Also the relative proximity between the nodes chosen for the evaluation of the error function J or, equivalently, matrix $\mathbf{A}$ has been shown to affect the numerical conditioning of $\mathbf{A}^T \cdot \mathbf{A}$. When consecutive nodes are chosen for the evaluation of J, the number of points needed for $\mathbf{A}^T \cdot \mathbf{A}$ to have numerically computed full rank has been shown to be significantly greater than when non-consecutive nodes are chosen.

Taking all these considerations into account, the least-squares solution has been computed in the present work, solving the system in (13.126) by means of the QR decomposition of matrix $\mathbf{A}$, which is known to be more accurate than directly solving the system in (13.124) with a possibly ill-conditioned matrix $\mathbf{A}^T \cdot \mathbf{A}$.

13.7. Examples of Singularity Analysis

Two problems regarding multimaterial corners with composite materials are studied applying the computational tools developed in the previous sections and implemented in *Mathematica* [97] and numerical BEM codes [78, 79] with the aim of showing the capabilities of these tools and validating them. Both problems are of unquestionable engineering interest. The first is related to failure initiation in cross-ply laminates $[0/90]_\mathrm{s}$ under tension. The second deals with the failure analysis of an adhesively bonded double-lap joint (between a composite laminate $[0_\mathrm{n}]$ and an aluminium plate using an epoxy adhesive layer) subjected to tension load.

13.7.1. Transverse crack terminating at the interface in a $[0/90]_\mathrm{S}$ laminate

A transverse crack in the inner 90° ply of a cross-ply $[0/90]_\mathrm{S}$ laminate under tensile loading is considered, see [79] and Fig. 13.4(a). The transverse crack terminates perpendicularly at the interface with the outer 0° plies and the crack faces are assumed to be free. The neighborhood of the crack tip can be considered as a trimaterial corner, Fig. 13.4(b). The elastic properties of the unidirectional fiber-reinforced plastic ply (AS4/8552) considered as an orthotropic material are $E_{11} = 141.3\,\mathrm{GPa}$, $E_{22} = E_{33} = 9.58\,\mathrm{GPa}$, $G_{12} = G_{13} = 5.0\,\mathrm{GPa}$, $G_{23} = 3.5\,\mathrm{GPa}$, $\nu_{12} = \nu_{13} = 0.3$, $\nu_{23} = 0.32$, where subscript 1 denotes the fiber direction. The material in both plies is the same but they have a different spatial orientation. Due to the material symmetries and corner configuration, the in-plane and antiplane displacements are uncoupled.

In this study, first a complete singularity analysis of the trimaterial corner shown in Fig. 13.4(b) is carried out by solving (13.110) by finding characteristic exponents with $0 < \mathrm{Re}\,\lambda < 2$. The influence of the treatment of material degeneracy in the Lekhnitskii–Stroh formalism on the accuracy of the singularity analysis results is discussed. Then, the whole problem shown in Fig. 13.4(a) is solved by means of a BEM model such as that

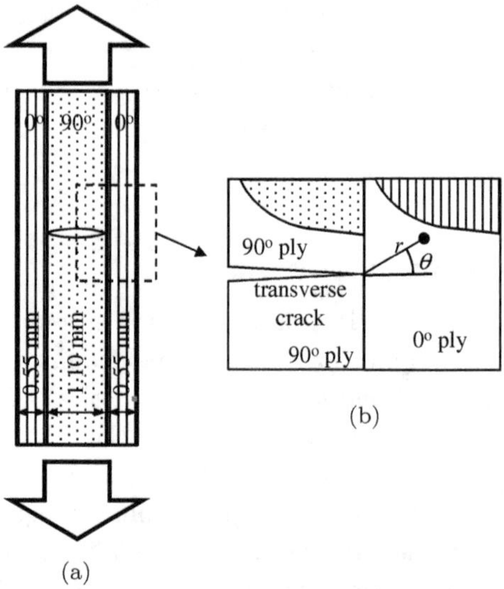

Fig. 13.4. (a) Cross-ply $[0/90]_S$ laminate under tensile loading with a transverse crack terminated at the interface. (b) Neighborhood of the transverse crack tip.

used in [79], a few GSIFs are evaluated by a least-squares fitting procedure (Section 13.6), and finally, to check the computational tools developed in the present work, the stresses computed by BEM at interior points of the corner are compared with those derived from the series approximation (13.3).

The roots of the characteristic equation of the corner (13.110) are found, by means of the argument principle (13.114), inside a rectangular contour in the complex plane defined by its corners $0.1 - i0.9$, $1.95 - i0.9$, $1.95 + i0.9$ and $0.1 + i0.9$, see Fig. 13.5; no roots exist along this contour. The argument principle indicates that inside the contour, in the domain with $0.1 < \mathrm{Re}\,\lambda < 1.95$ and $-0.9 < \mathrm{Im}\lambda < 0.9$, there are 11 roots of (13.110), including their multiplicities. The characteristic exponents found are: $\lambda_1 = 0.471654$ (antiplane), $\lambda_2 = 0.521510$ (antisymmetric), $\lambda_3 = 0.669888$ (symmetric), $\lambda_{4,5,6} = 1$ (a root of multiplicity three, one for rigid body rotation, one is an antiplane term and the third is a symmetric term), $\lambda_7 = 1.52834$ (antiplane), $\lambda_8 = 1.73425$ (symmetric), $\lambda_{9,10} = 1.84194 \pm 0.308109i$ (both antisymmetric) and $\lambda_{11} = 1.89369$ (symmetric). The labels symmetric and antisymmetric refer to plane-strain elastic solutions corresponding to particular terms in (13.2) and (13.3), which are symmetric or antisymmetric with respect to the transverse crack plane, whereas the label antiplane refers to an antiplane elastic solution

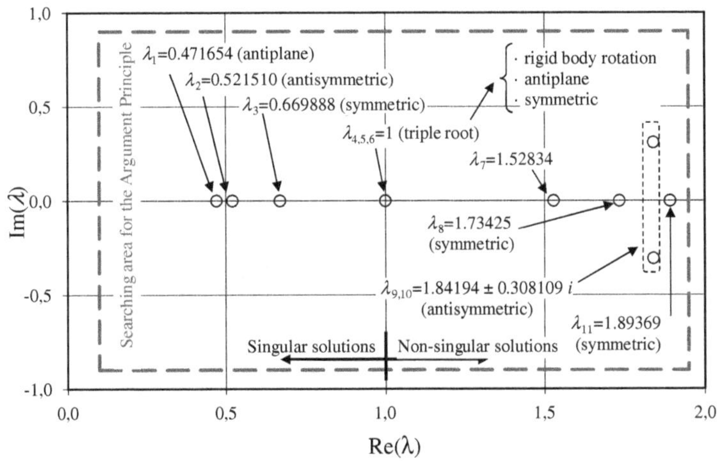

Fig. 13.5. Characteristic exponents for the open trimaterial corner in Fig. 13.4(b).

corresponding to a term in (13.2) and (13.3). In the case of complex roots, special care should be taken to identify them, especially those having $\mathrm{Re}\,\lambda > 1$, because of typically large variations in the determinant of the characteristic matrix there, making it difficult to find roots using standard algorithms such as Muller's method [98, 99], and thus more sophisticated algorithms may be required [100].

In many previous works on corner singularity analysis with anisotropic materials, only non-degenerate materials (in the Lekhnitskii–Stroh formalism of anisotropic elasticity) are treated explicitly, so it seems useful to analyse briefly the possible influence of an approximation of a degenerate material by a non-degenerate one on the results of the singularity analysis. For instance, let the unidirectional fiber-reinforced plastic ply (AS4/8552) be modeled as an elastic transversely isotropic material (instead of an orthotropic one) defined by five independent elastic constants (instead of nine): $E = 9.58\,\mathrm{GPa}$ and $\nu = 0.32$ define its isotropic behavior in the 2–3 plane, and $E_{11} = 141.3\,\mathrm{GPa}$, $\nu_{12} = \nu_{13} = 0.3$ and $G_{12} = G_{13} = 5.0\,\mathrm{GPa}$. Notice that the shear modulus in the 2–3 plane is $G = E/2(1+\nu) = 3.629\,\mathrm{GPa}$ (instead of $G_{23} = 3.5\,\mathrm{GPa}$ as given by the manufacturer considering this ply as an orthotropic material). A change in the spatial orientation of a transversely isotropic material can lead to a mathematical degeneracy in the Lekhnitskii–Stroh formalism [25, 36, 87] and if explicit expressions of the $\mathbf{A}$ and $\mathbf{B}$ matrices for the corresponding degenerate material are not available, only approximate

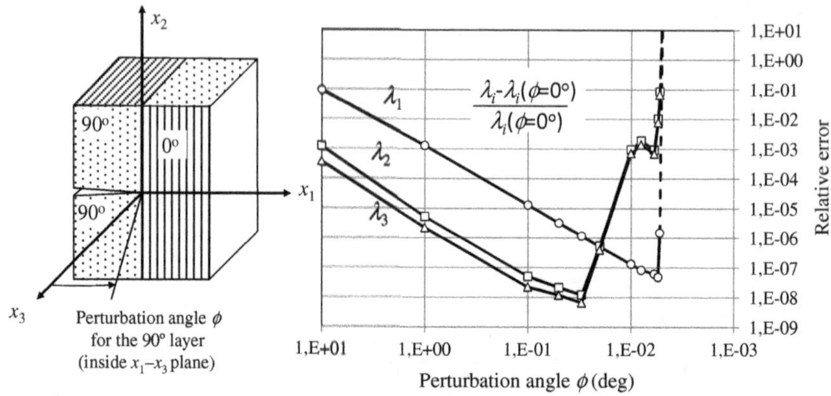

Fig. 13.6. Influence of the fiber-angle perturbation ϕ in the 90° ply on the characteristic exponents for the open trimaterial corner in Fig. 13.4(b).

singularity analysis of a corner including this material can be carried out through using a small perturbation of the elastic constants or spatial orientation of the material leading to a non-degenerate case. However, the accuracy of the results obtained by such a procedure is not known *a priori*. To illustrate these facts, Fig. 13.6 shows the numerical results for the characteristic exponents λ_1 (antiplane), λ_2 (antisymmetric) and λ_3 (symmetric), considering small perturbations of the spatial orientation of the transversely isotropic material. These characteristic exponents were obtained using a specific *Mathematica* [97] code developed for this kind of material in [25], employing standard machine precision for floating point computation, which is about 15–16 decimal digits. In particular, the influence of the fiber-angle perturbation ϕ of the 90° ply, which is a degenerate case [25], is studied. The relative error of the actual value of λ_i with respect to the solution for a vanishing fiber-angle perturbation $\phi = 0°$ is plotted as a function of ϕ. Numerical instabilities are seen for small fiber-angle perturbations. Consequently, the characteristic exponents obtained by approximating a degenerate material ($\phi = 0°$) with a non-degenerate one ($\phi \neq 0°$) with a sufficiently small fiber-angle perturbation may lead to significant numerical errors, as the threshold angle below which numerical instabilities appear is *a priori* unknown, and depends on the corner configuration. Therefore, in the present work all the mathematically degenerate cases are dealt with using the corresponding expressions introduced in the previous sections, which provide high accuracy in the evaluation of characteristic exponents and functions.

The plane-strain solution of the elastic problem defined in Fig. 13.4(a) is symmetric with respect to the transverse crack plane, and the crack is expected to open because of the tensile load applied perpendicularly to this plane. Thus, only the characteristic exponents marked as symmetric in Fig. 13.5, namely $\lambda_3 = 0.669888$ for singular stresses and $\lambda_6 = 1$, $\lambda_8 = 1.73425$ and $\lambda_{11} = 1.89369$ for finite stresses, are included in the 2D asymptotic series expansions (13.2) and (13.3) in the neighborhood of the transverse crack tip. The GSIFs extracted from the BEM results using the least-squares fitting technique, standardized following [102] in such a way that the stress component $\sigma_{\theta\theta|\theta=0^\circ} = K/(2\pi r)^{1-\lambda}$, are $K_{r(\lambda=1)} = -0.0000203291\,\text{MPa}$, $K_{3(\lambda=0.669888)} = 0.561634\,\text{MPa}\cdot\text{mm}^{0.330112}$, $K_{6(\lambda=1)} = 0.694945\,\text{MPa}$, $K_{8(\lambda=1.73425)} = 0.522112\,\text{MPa}\cdot\text{mm}^{-0.73425}$ and $K_{11(\lambda=1.89369)} = -0.288119\,\text{MPa}\cdot\text{mm}^{-0.89369}$, the K_r term being associated with rigid body rotation. The BEM results used in the least-squares fitting are the two displacement components (u_r, u_θ) along the radial lines emerging from the corner tip at $\theta = 90^\circ$ and 180°, computed at the nodes of the BEM mesh in the range $10^{-5}\,\text{mm} < r < 0.3\,\text{mm}$. Figure 13.7 shows the results for the stress components as functions of the angular coordinate θ, for a fixed radial coordinate $r = 0.1\,\text{mm}$, obtained by both the BEM model and the asymptotic series approximation (13.3) using the extracted GSIFs

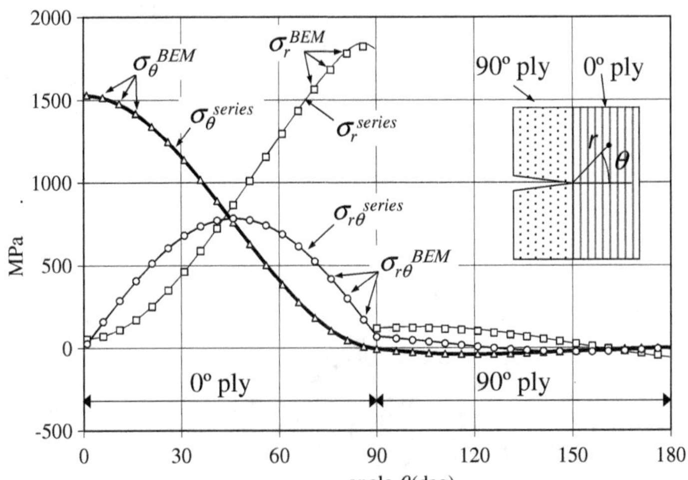

Fig. 13.7. Stress components evaluated by BEM and by the asymptotic series expansion at a distance $r = 0.1\,\text{mm}$ from the corner tip.

values. The fitting of both approximations for all three stress components is excellent along the whole range of θ. As could be expected, in view of the relative stiffnesses of the plies in the load direction parallel to the interface (the 0° ply is much stiffer than 90° ply), much higher stresses are observed in the 0° than in 90° ply. Notice the discontinuity of the σ_r stress component at the interface ($\theta = 90°$) between the 0° and 90° plies.

13.7.2. Bimaterial corner in an adhesively bonded double-lap joint

13.7.2.1. *Singularity analysis of a closed corner*

In an adhesively bonded double-lap joint between an aluminium plate and a composite laminate $[0_n]$ with an epoxy adhesive layer as shown in Fig. 13.8, several multimaterial corners can be identified. We will look at corner B, a closed bimaterial corner (the two materials are perfectly bonded at both interfaces), because it is a critical point at which failure typically initiates in this type of joint [123, 124]. The elastic properties of the unidirectional fiber-reinforced plastic ply are the same as in the first example in Section 13.7.1; the elastic properties of the adhesive are $E = 3.0$ GPa and $\nu = 0.35$ and of the aluminium $E = 68.67$ GPa and $\nu = 0.33$. Thermal stresses, which could arise in the curing process, are not considered [125].

In this study, first a comprehensive singularity analysis of the closed bimaterial corner B, Fig. 13.8, is carried out solving (13.110) by finding characteristic exponents with $0 < \operatorname{Re} \lambda < 1.5$. Then, the double-lap joint problem shown in Fig. 13.8 is solved by means of a BEM model such as that

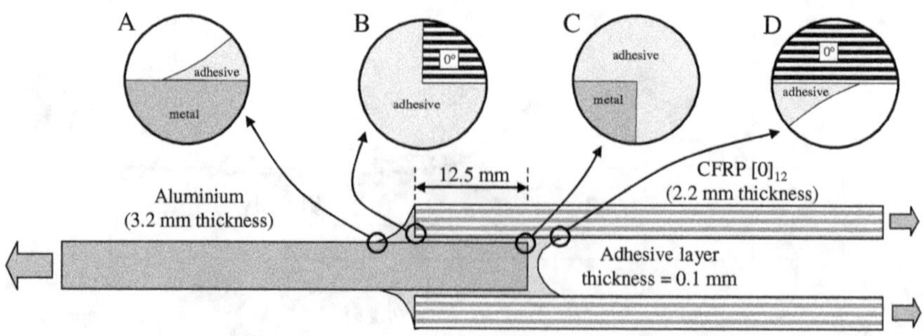

Fig. 13.8. Multimaterial corners in an adhesively bonded double-lap joint between an aluminium plate and two carbon-fiber-reinforced polymer (CFRP) laminates.

used in [124]. Three GSIFs for elastic plane-strain solutions are evaluated by a least-squares fitting procedure (Section 13.6), and finally, in order to check the computational tools developed in the present work, a displacement component computed by BEM at interior points of the corner is compared with the series approximation (13.2), identifying the contribution of each term in the series expansion.

The roots of the characteristic equation of the corner (13.110) are found, using the argument principle (13.114), inside a rectangular contour in the complex plane defined by its corners $0.1 - i0.9$, $1.5 - i0.9$, $1.5 + i0.9$ and $0.1 + i0.9$, see Fig. 13.9; there are no roots along this contour. The argument principle indicates that inside the contour, in the domain with $0.1 < \operatorname{Re}\lambda < 1.5$ and $-0.9 < \operatorname{Im}\lambda < 0.9$, there are six roots of (13.110), including their multiplicities. The characteristic exponents found are all real numbers: $\lambda_1 = 0.763236$, $\lambda_2 = 0.813696$, $\lambda_3 = 0.889389$, $\lambda_4 = 1$, $\lambda_5 = 1.106980$, $\lambda_6 = 1.185066$, all having a multiplicity equal to 1. The roots λ_2 and λ_6 correspond to antiplane solutions, while λ_4 is for rigid body rotation.

The elastic problem of the double-lap joint defined in Fig. 13.8 is solved under plane-strain conditions. Thus, only two characteristic exponents $\lambda_1 = 0.763236$ and $\lambda_3 = 0.889389$ for singular stresses and one characteristic exponent $\lambda_5 = 1.106980$ for finite stresses are considered in the following series approximations of the displacements (where the rigid body rotation

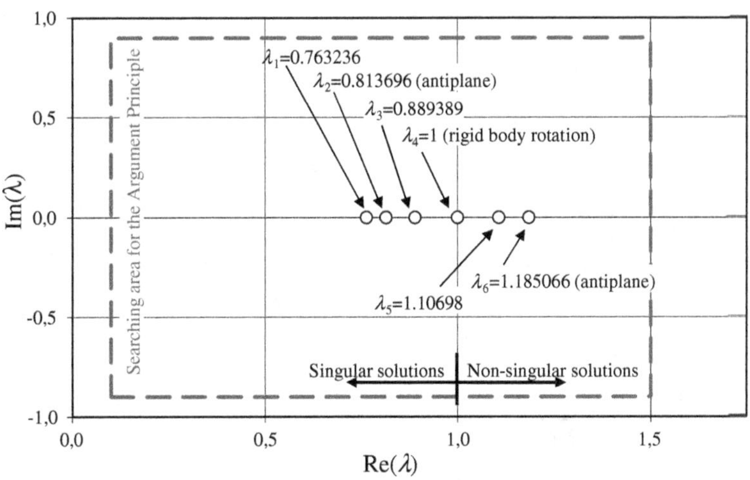

Fig. 13.9. Characteristic exponents for the closed bimaterial corner B in Fig. 13.8.

term is also included) and stresses in the neighborhood of the corner tip:

$$u_i(r,\theta) \cong K_1 r^{0.763236} g_i^{(1)}(\theta) + K_3 r^{0.889389} g_i^{(3)}(\theta) + K_4 r^1 g_i^{(4)}(\theta) + K_5 r^{1.106980} g_i^{(5)}(\theta), \quad (13.127)$$

$$\sigma_{ij}(r,\theta) \cong \frac{K_1}{r^{0.236764}} f_{ij}^{(1)}(\theta) + \frac{K_3}{r^{0.110611}} f_{ij}^{(3)}(\theta) + K_5 r^{0.106980} f_{ij}^{(5)}(\theta). \quad (13.128)$$

As an example, Fig. 13.10 shows the angular shape functions for the first singular term with $\lambda_1 = 0.763236$. Recall that $f_{ij}^{(n)}$ are dimensionless functions whereas the dimension of $g_i^{(n)}$ (with the exception of the dimensionless $g_i^{(4)}$ for rigid body rotation) is $F^{-1}L^2$ where F and L denote force and length, respectively. Notice that all the characteristic angular shape functions shown in Fig. 13.10 are continuous except for $f_{rr}^{(1)}$, which suffers jumps at the interfaces ($\theta = 0°$ and $90°$).

The plane-strain solution of the elastic problem shown in Fig. 13.8 can be obtained analysing only half of the aluminium plate and one CFRP laminate because of the problem symmetry. The right end of the CFRP laminate is fixed while the left end of the aluminium plate has a tensile stress of 125 MPa. Figure 13.11 shows a detail (for the overlap area) of the deformed shape calculated by the BEM model employed.

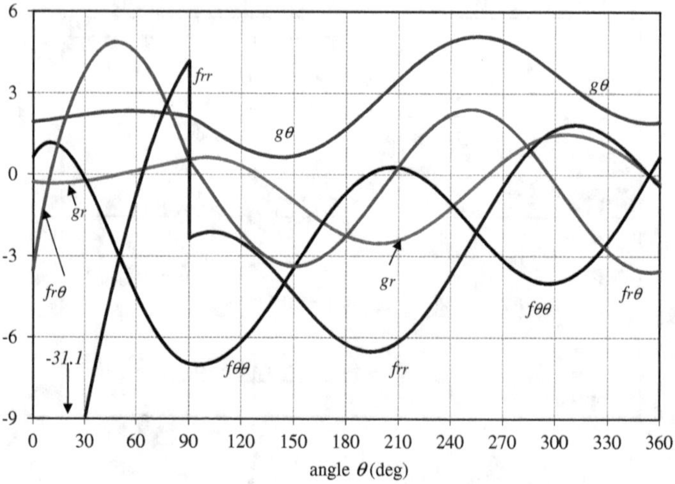

Fig. 13.10. Angular shape functions for λ_1: $g_r^{(1)}(\theta)$, $g_\theta^{(1)}(\theta)$, $f_{\theta\theta}^{(1)}(\theta)$, $f_{r\theta}^{(1)}(\theta)$ and $f_{rr}^{(1)}(\theta)$.

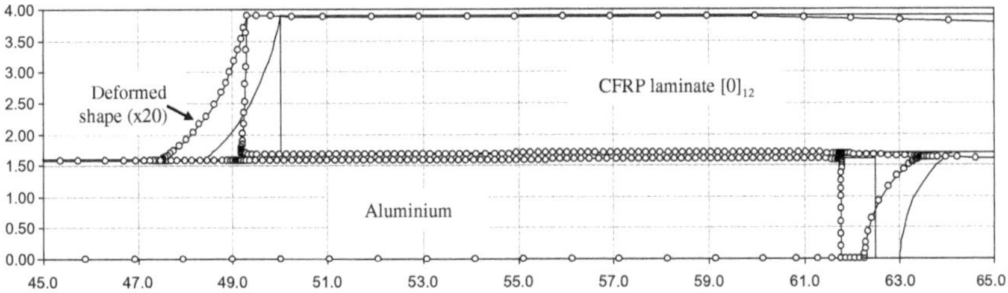

Fig. 13.11. Deformed shape of the BEM mesh and undeformed boundaries, detail of the overlap zone (in mm).

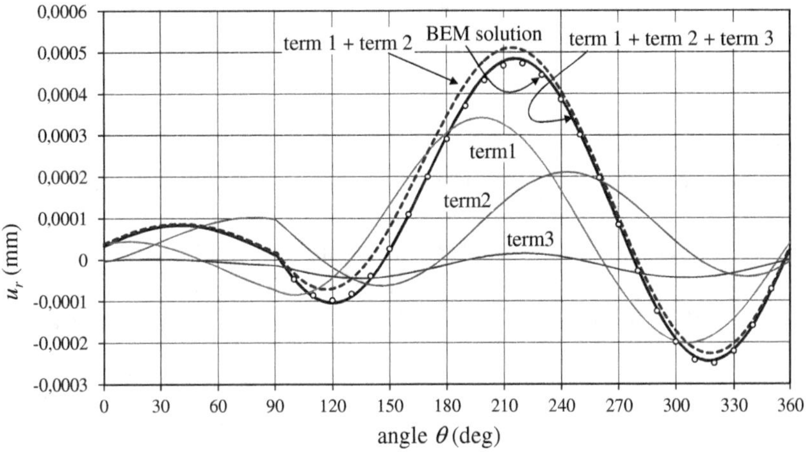

Fig. 13.12. Displacement component u_r evaluated by BEM and by the asymptotic series expansion at $r = 0.0194$ mm from the corner B tip.

The GSIFs extracted from the BEM results using the least-squares fitting technique [40], standardized as proposed by [102] in such a way that the stress component $\sigma_{\theta\theta|\theta=0^\circ} = K/(2\pi r)^{1-\lambda}$, are $K_r = K_{4(\lambda=1)} = -\ 0.00356242\,\text{MPa}$, $K_{1(\lambda=0.763236)} = -0.00275036\,\text{MPa}\cdot\text{mm}^{0.236764}$, $K_{2(\lambda=0.889389)} = 0.0273839\,\text{MPa}\cdot\text{mm}^{0.110611}$, $K_{3(\lambda=1.106980)} = -0.0114328\,\text{MPa}\cdot\text{mm}^{-0.10698}$. The BEM results used in the least-squares fitting are the two displacement components (u_r, u_θ) along the radial lines emerging from the corner tip at $\theta = 0^\circ$ and 90°. The BEM mesh has a progressive mesh refinement towards the corner tip, with the smallest element having a length of 10^{-8} mm; the selected range of the nodes is from 10^{-6} mm to 0.025 mm. Figure 13.12 shows the results for the u_r displacement component as a function of the angular coordinate θ, for a fixed radial coordinate $r = 0.0194\,\text{mm}$, obtained by the BEM model and by the asymptotic series approximation (13.2) particularized in (13.127) using the extracted GSIFs values. The contribution of each term in (13.127) is also indicated, showing that even the approximation given by the two-term series (term 1 + term 2) is reasonably close to the BEM results while the fitting of BEM results by the three-term series approximation (term 1 + term 2 + term 3) is excellent along the whole range of θ.

13.7.2.2. *Singularity analysis of a corner including an interface crack with sliding friction contact*

This section shows some results of a singularity analysis of the corner B in Fig. 13.8 altered by the presence of a crack at the interface between the composite lamina and adhesive at $\theta = 90°$. Such cracks were observed in experimental tests of double-lap joints [123, 124]. During the service life of this kind of joint and in testing under cyclic loading, the crack faces can make contact with each other and slide in any direction. To show the capability of the computational tools developed for the singularity analysis of sliding friction contact with coupled in-plane and antiplane displacements, the angle of the fibers in the unidirectional composite lamina was changed from $\phi = 0°$ (parallel to the load in the original configuration in Fig. 13.8) to $\phi = 90°$ (perpendicular to the load), while the fibers keep their horizontal position indicated in Fig. 13.8. An isotropic Coulomb friction model, with $\omega = \omega^u$ in (13.36), is considered with a relatively large value for the friction coefficient $\mu = 1$, in order to have a noteworthy influence of friction on the solution behavior.

Solutions of the characteristic equation of the corner (13.111) are given by pairs of characteristic exponents λ and characteristic friction angles ω. We searched for $0 < \lambda < 1$ and $-180° \leq \omega < 180°$. Plots of the contours of the function $\sigma_{\min}(\mathbf{K}_{\text{corner}}(\lambda, \omega))$ from (13.111) for $\phi = 0°$ and $40°$ are shown in Fig. 13.13 as examples. Some of the zeros (global minima) are indicated by arrows. For the case $\phi = 0°$, Fig. 13.13(a), besides the expected values of the characteristic friction angles $\omega = 0°, \pm 90°, \pm 180°$, characteristic values at $\omega = \pm 79.7°$ are unexpectedly found, which represent singular elastic solutions with coupled in-plane and antiplane displacements, in spite of the symmetry of this corner configuration with respect to the plane (x_1–x_2). For the case $\phi = 40°$, Fig. 13.13(b), all the singular solutions found have coupled in-plane and antiplane displacements as expected due to the non-symmetric corner configuration. For all the solutions of (13.111) the energy

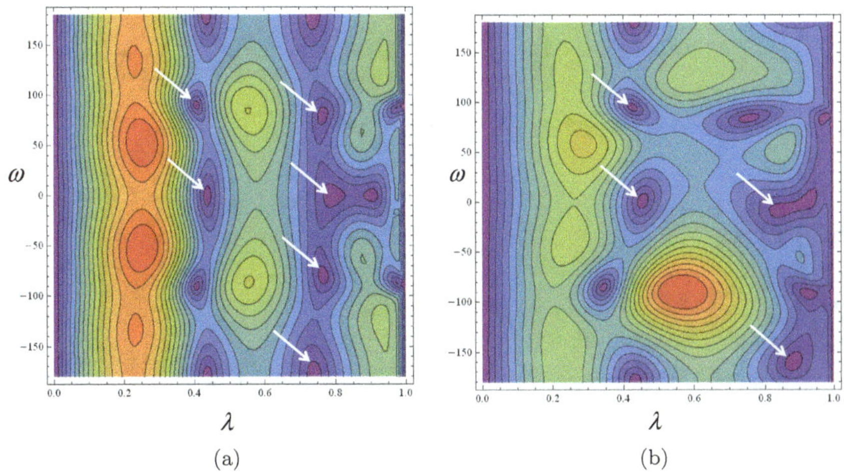

Fig. 13.13. Contour plots of $\sigma_{\min}(\mathbf{K}_{\text{corner}}(\lambda,\omega))$ from (13.111) for fiber angles (a) $\phi = 0°$ and (b) $\phi = 40°$.

dissipation condition under proportional loading (Section 13.5.3) is checked *a posteriori*, the arrows in Fig. 13.13 showing only those solutions (global minima) which satisfy this condition.

The solutions of (13.111) are plotted in Fig. 13.14 as functions of the fiber angle ϕ. Only solutions satisfying the energy dissipation condition are shown and the same symbols are used in both plots for the corresponding values of λ and ω. Only slight variations in the values of λ and ω are observed. Some of the series of characteristic values shown in Fig. 13.14 are shorter than others because, surprisingly, the singular elastic solutions with the remaining characteristic values violate the energy dissipation condition.

13.8. Failure Criterion for a Multimaterial Closed Corner Based on Generalized Fracture Toughness

This section introduces a quite general criterion for failure initiation at a multimaterial corner based on generalized stress intensity factors (GSIFs) and associated generalized fracture toughnesses. It is necessary to have an experimental procedure capable of generating the corresponding failure envelope covering all fracture mode mixities.

Consider a multimaterial corner under a plane strain with two stress singularities, represented by two terms (typically referred to as singular modes) in the asymptotic series expansion (13.3) defined by characteristic

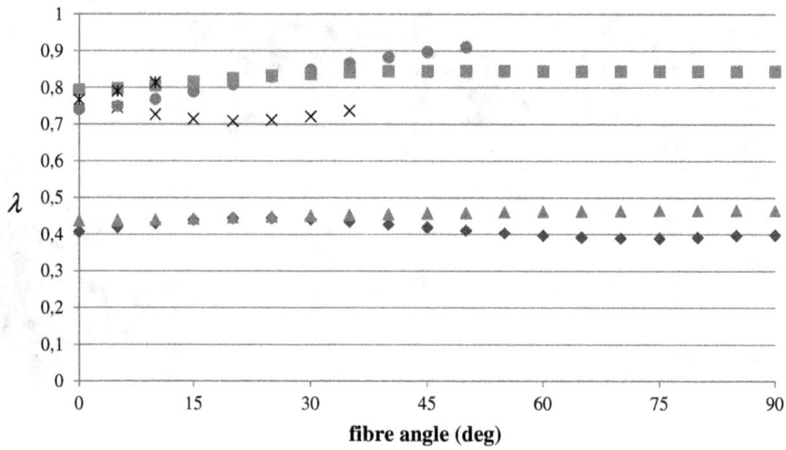

(a)

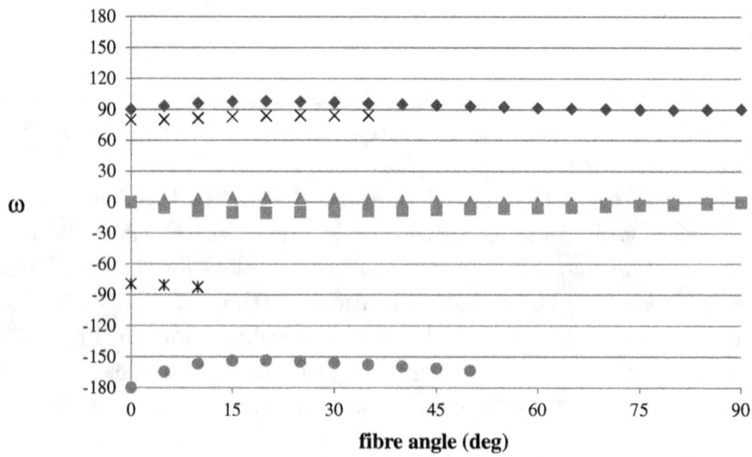

(b)

Fig. 13.14. Plots of (a) characteristic exponents λ and (b) characteristic friction angles ω for corner B, Fig. 13.8, for a crack in sliding friction contact with friction coefficient $\mu = 1$. Only characteristic values with singular elastic solutions satisfying the energy dissipation condition are shown.

exponents λ_1 and λ_2, which are assumed to govern the failure initiation at the corner tip through the associated GSIFs K_1 and K_2. Then, a general corner failure criterion can be expressed as

$$K = \kappa_C(\psi), \tag{13.129}$$

where

$$K = \sqrt{\left(\frac{K_1}{K_{1C}}\right)^2 + \left(\frac{K_2}{K_{2C}}\right)^2} \tag{13.130}$$

is a normalized GSIF modulus (a dimensionless magnitude), ψ is a normalized fracture-mode-mixity angle,

$$\tan\psi = \frac{K_2/K_{2C}}{K_1/K_{1C}}, \tag{13.131}$$

and $\kappa_C(\psi)$ is a dimensionless function of ψ giving the critical value of K. K_{1C} and K_{2C} are the generalized fracture-toughness values for pure singular modes, i.e., when either K_2 or K_1 equals zero, respectively. In particular, they can be chosen so that $\kappa_C(0°) = \kappa_C(90°) = 1$. $\psi = 0°$ corresponds to $K_1 > 0$ and $K_2 = 0$ while $\psi = 90°$ to $K_2 > 0$ and $K_1 = 0$. Unlike the traditional fracture-mode-mixity definition for a crack, $\tan\psi = (K_2/K_1)$, the inclusion of the generalized fracture toughnesses in (13.131) is due to the different dimensions of K_1 and K_2 in the present case. The parameterization $(\psi, \kappa_C(\psi))$ defines a hypothetical failure envelope curve in the plane $(K_1/K_{1C}, K_2/K_{2C})$.

The size-scale effect should be taken into account when evaluating GSIFs in geometrically similar specimens. From dimensional analysis, a GSIF can be expressed as [80]

$$K_k = \sigma_{\text{nom}} R^{1-\lambda_k} A_k, \tag{13.132}$$

where σ_{nom} is the nominal stress in the specimen, R is a characteristic length of the specimen and A_k is a shape factor that takes into account the geometry and material properties of the specimen.

Whereas a numerical procedure for the evaluation of GSIFs was proposed in Section 13.6 and applied in Section 13.7, the determination of particular values of K_{1C} and K_{2C}, and of the whole failure envelope curve $\kappa_C(\psi)$, requires the experimental testing of samples including the corner under several loading conditions. For the particular case of the bimaterial closed corner B in the adhesively bonded joint in Fig. 13.8, a suitable procedure proposed in [80,81] can be used to experimentally determine K_{1C}, K_{2C} and $\kappa_C(\psi)$. This procedure is based on a novel modified configuration of the Brazilian disc specimen, including the bimaterial corner tip at the centre of the disc (Fig. 13.15(a)) and subjected to a diametrical compression P at an angle α (Fig. 13.15(b)).

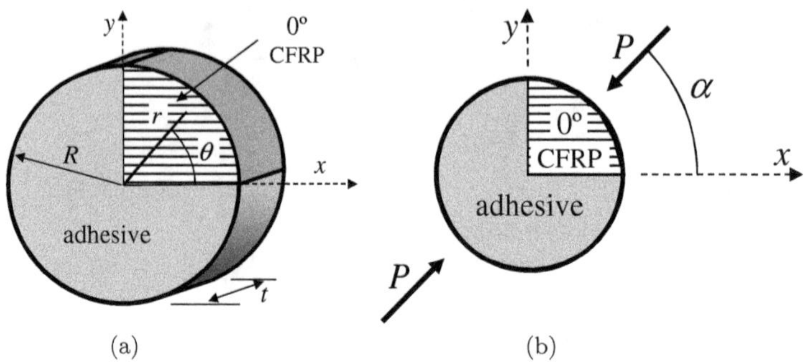

Fig. 13.15. Brazilian disc specimen with the bimaterial corner. (a) Geometry. (b) Loading.

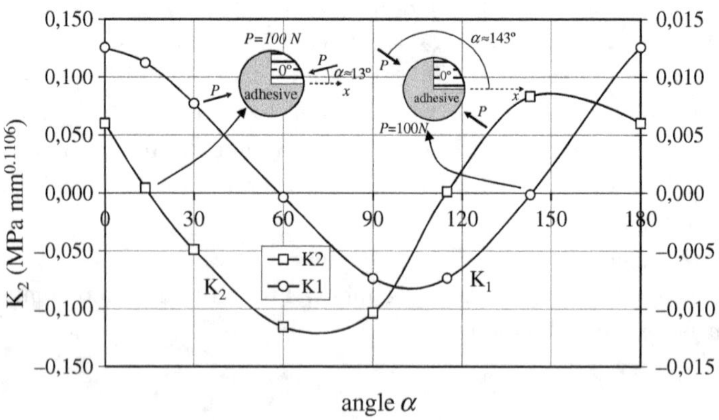

Fig. 13.16. Standardized values of K_1 and K_2 in the Brazilian disc specimen under diametrical compression.

GSIFs K_1 and K_2 were extracted from FEM results in [80] by employing the least-squares fitting technique presented in Section 13.6. The numerical results for a particular configuration ($t = 1\,\text{mm}$, $R = 1\,\text{mm}$ and $P = 100\,\text{N}$) are depicted in Fig. 13.16. The diametrical compression orientations providing (approximately) pure singular modes are $\alpha \approx 13°$ and $\alpha \approx 115°$ for $K_2 \approx 0$, and $\alpha \approx 60°$ and $\alpha \approx 143°$ for $K_1 \approx 0$. Angles $\alpha \approx 13°$ and $\alpha \approx 143°$, giving positive values of the non-vanishing GSIFs, were chosen to determine the generalized fracture toughnesses K_{1C} and K_{2C}, respectively. Let P^{exp} denote the experimental load for which failure initiation in the corner is observed, t^{exp} the real specimen thickness and R^{exp} the real specimen radius. Then the following equation, obtained from

(13.132), is used to determine K_{1C} and K_{2C}:

$$K_{kC} = K_k^{\text{FEM}} \frac{t^{\text{FEM}}}{t^{\exp}} \frac{R^{\text{FEM}}}{R^{\exp}} \frac{P^{\exp}}{P^{\text{FEM}}} \left(\frac{R^{\exp}}{R^{\text{FEM}}}\right)^{1-\lambda_k}$$

$$= K_k^{FEM} \frac{\sigma_{\text{nom}}^{\exp}}{\sigma_{\text{nom}}^{\text{FEM}}} \left(\frac{R^{\exp}}{R^{\text{FEM}}}\right)^{1-\lambda_k}, \tag{13.133}$$

where superscript 'FEM' denotes values from the above defined FEM model, and $\sigma_{\text{nom}}^{\text{FEM}}$ and $\sigma_{\text{nom}}^{\exp}$ are nominal stresses in the FEM model and the real specimen, respectively.

Figure 13.17 shows a real sample after failure and a scheme for the failure path for $\alpha \approx 13°$, the angle used to determine K_{1C}.

Experimental tests corresponding to any orientation of the diametrical compression define critical pairs of GSIFs for which failure initiates in the corner. All the critical pairs of GSIFs obtained can be represented in the plane $(K_1/K_{1C}, K_2/K_{2C})$ as shown in Fig. 13.18; see [81] for the details of these experiments. In Fig. 13.18, light (blue) circles are experimental results of single tests for particular load orientations (the load orientation α is indicated on the plot), and black circles are the average values at the same load orientation ($\alpha = 150°$ has only one specimen). The continuous line interpolates the black circles and represents an approximation of the failure envelope curve, which can be used to define the corner failure criterion in (13.129). Representations of the loading angle range leading to results in each quadrant in Fig. 13.18 are plotted schematically to visualize the correspondence between the loading angle α and the critical pairs of GSIFs.

In addition to the experimental results obtained by the Brazilian disc specimen, Fig. 13.18 shows another set of experimental results obtained

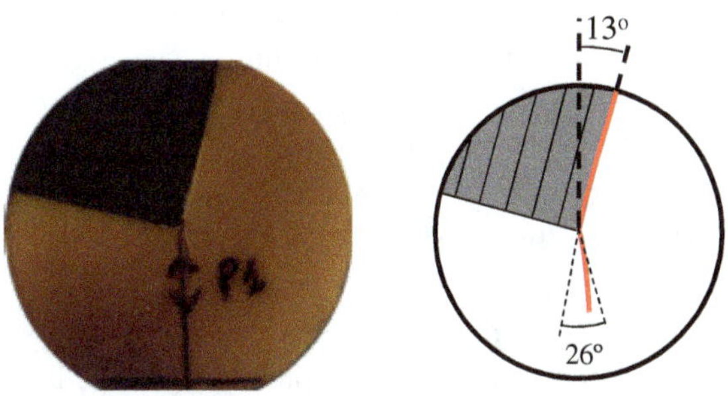

Fig. 13.17. Brazilian disc specimen tested (left) and failure path scheme (right) for a loading angle $\alpha \approx 13°$.

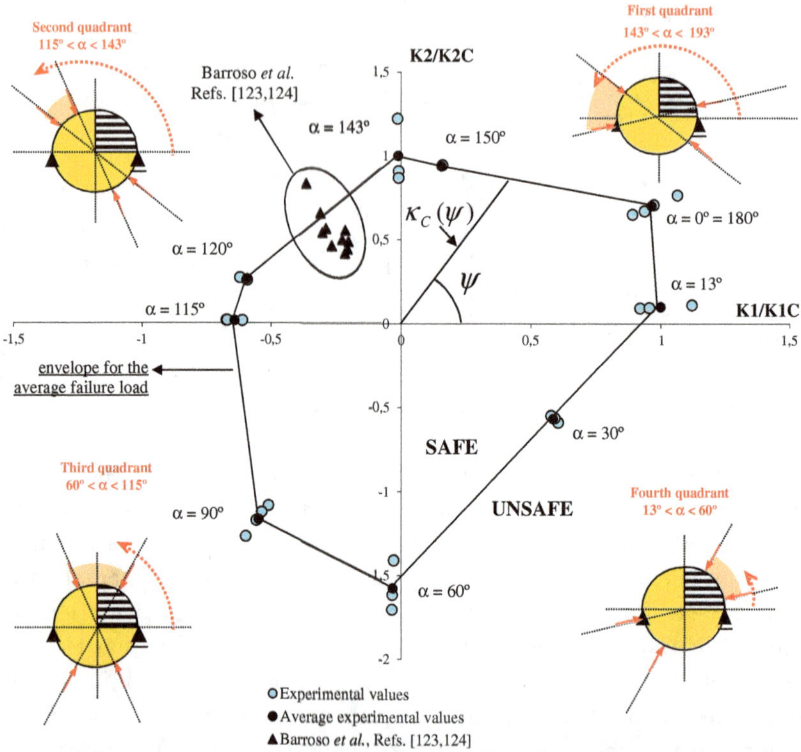

Fig. 13.18. Experimental results for the critical pairs of GSIFs in the Brazilian disc specimens (circle marks) and double-lap joint specimens (triangular marks) and an approximation of the corner failure envelope.

previously by Barroso and co-workers [123,124]. These results, indicated by black triangles in Fig. 13.18, represent the critical pairs of GSIFs for which complete failure was observed in real double-lap joints between the unidirectional CFRP laminate and aluminium, see Fig. 13.8, subjected to tension. Notice that both specimens, the Brazilian disc specimen and the double-lap joint, have the same corner configuration at the corner tip with reference to geometry and materials. The critical pairs of GSIFs shown in Fig. 13.18 for both sets of experiments for the same fracture-mode-mixity are very close. The observed agreement between both sets of experiments is quite significant as the double-lap joints, although having the same local corner configuration, are completely different in size, geometry and manufacturing process from the Brazilian disc specimens. While the Brazilian disc specimens were manufactured in an autoclave and their characteristic length (the radius) is 17 mm, the double-lap joint specimens were manufactured in a hot plate press and their characteristic length is 0.1 mm (the adhesive layer thickness). Additionally, the Brazilian disc specimens were tested only a few days after manufacture whereas the double-lap joints were tested a long time after manufacture (about half a year); this delay might lead to strength degradation due to moisture absorption.

It is important to notice that the novel experimental procedure employing the Brazilian disc specimen can directly be applied to the generation of failure criteria based on generalized fracture toughness for other multimaterial closed corners with two stress singularities under plane strain. In fact, placing a very thin strip of Teflon® at the interface so that it emerges from the corner tip can imitate a crack, as in [126] for a straight interface, allowing the present procedure to be applied to open multimaterial corners with a corner angle $\vartheta_W - \vartheta_0 = 360°$.

13.9. Removal of Stress Singularities in Bimaterial Joints

An interesting application of the singularity analysis developed in this chapter is the so-called *stress singularity removal* which consists in the determination of an appropriate local geometry configuration, for each particular problem, which eliminates the stress singularities present in the original structure or specimen, or at least substantially reduces their severity. In view of (13.3), it means that the real part of the first characteristic exponent becomes $\mathrm{Re}\,\lambda_1 \geq 1$, or equivalently, considering the order of stress singularity $\delta = 1 - \lambda$, $\mathrm{Re}\delta_1 \leq 0$. In configurations of samples with multimaterial corners generated by tabs bonded at the sample ends or by the presence of a bimaterial joint, the determination of the strength values may be greatly affected by the stress singularities arising at these points. Eliminating the stress singularity (or substantially relieving its severity) allows the determination of the strength values not affected (or less affected) by the presence of these singular points. This idea has been successfully applied by the authors and co-workers to various problems [127, 128, 134], two of them will be briefly presented in the following sections.

13.9.1. Tensile and shear strength in bimaterial samples

Tensile and shear strength in homogeneous materials can easily be determined by using standard tensile and Iosipescu specimens. In both

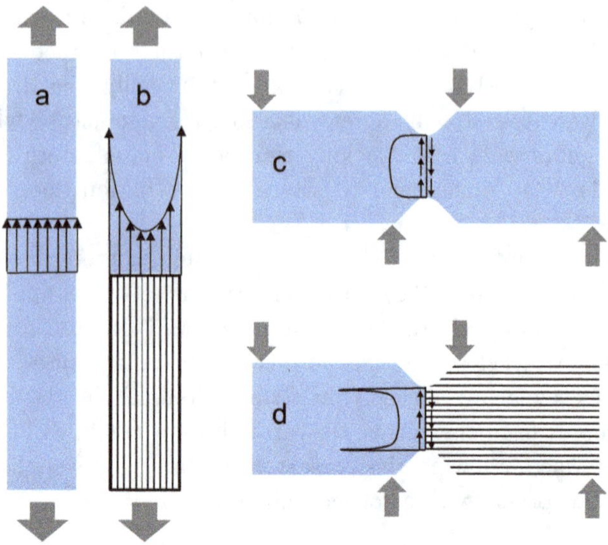

Fig. 13.19. Tensile and Iosipescu specimens with one or two materials.

configurations, the normal and shear stresses are quite uniform within the failure area, see Figs. 13.19(a) and (c), their values at the instant of failure being accepted as strength values. When trying to use these test configurations in the strength determination of bimaterial samples, the stresses at the interface are far from being uniform, due to the presence of stress singularities at the free edge of the samples, see a schematic representation in Figs. 13.19(b) and (d).

With such stress profiles, at the instant of failure, a strength value based on the (resultant) failure load divided by the failure area of interface would not be representative of the real strength of interface. In such cases, the evaluation of the characteristic exponents can help in modifying the local bimaterial-corner configuration to remove the stress singularity.

The characteristic exponents depend on the local material properties, local geometry and local boundary conditions. Obviously, the material properties cannot be modified, but the local geometry can be modified by changing, for instance, one single-material wedge angle, as schematically depicted in Fig. 13.20 [127, 128].

For a practical illustration of the *stress singularity removal* procedure, consider the bimaterial system formed by a structural epoxy adhesive FM-73M0.6 ($E = 3.0\,\text{GPa}$, $\nu = 0.35$) and a unidirectional carbon fiber laminate

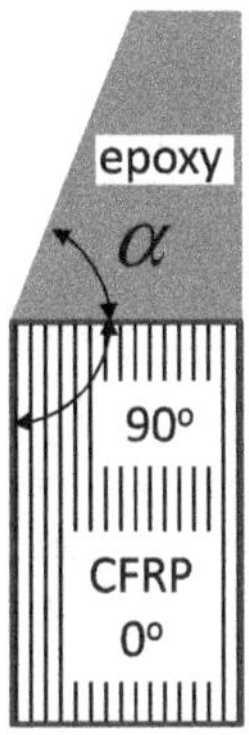

Angle α (degrees)	Order of stress singularity (CFRP 0°)
90	0.219697
80	0.152230
75	0.112593
70	0.067840
65	0.016722
60	**– 0.042452**

Fig. 13.20. Characteristic exponents for the adhesive-CFRP bimaterial system.

CFRP, AS4-8552 ($E_{11} = 141.3\,\text{GPa}$, $E_{22} = E_{33} = 9.58\,\text{GPa}$, $G_{12} = G_{13} = 5.0\,\text{GPa}$, $G_{23} = 3.5\,\text{GPa}$, $\nu_{12} = \nu_{13} = 0.3$, $\nu_{23} = 0.32$, where subindex 1 defines the fiber direction) with the fiber direction perpendicular to the bimaterial interface. With such material and geometrical configuration, the characteristic exponents associated to the bimaterial corner, and computed under the assumption of generalized plane strain, are shown in Fig. 13.20.

From the results shown in Fig. 13.20 it is clear that for a local geometrical configuration close to $\alpha = 65°$, the stress singularity vanishes. This fact can be easily verified by a numerical analysis of the joint, whose results are depicted in Fig. 13.21. The normal stresses perpendicular to the interface (normalized by the remote applied tension), Fig. 13.21(a), and shear stresses, Fig. 13.21(b), along the interface are shown for both the unmodified and modified bimaterial-corner configurations (both in generalized plane strain and plane stress). In the modified configurations, the stresses are quite uniform along the interface, providing in this way a new more reliable and accurate methodology for the measurement of the joint strength calculated as the failure load divided by the cross sectional area of the failure surface.

In both modified samples (for tensile and shear strength measurements) a notch was made at the epoxy side to get a wedge of angle 65°, which is approximately the angle for which the stress singularity vanishes, i.e., $\text{Re}\,\lambda_1 \simeq 1$ and equivalently $\text{Re}\delta_1 \simeq 0$. Figure 13.22 shows the load-displacement results of the tensile tests. Four samples from each configuration (4 without and 4 with notch) were tested. The samples with the notch showed a 92% increase of tensile strength value compared with

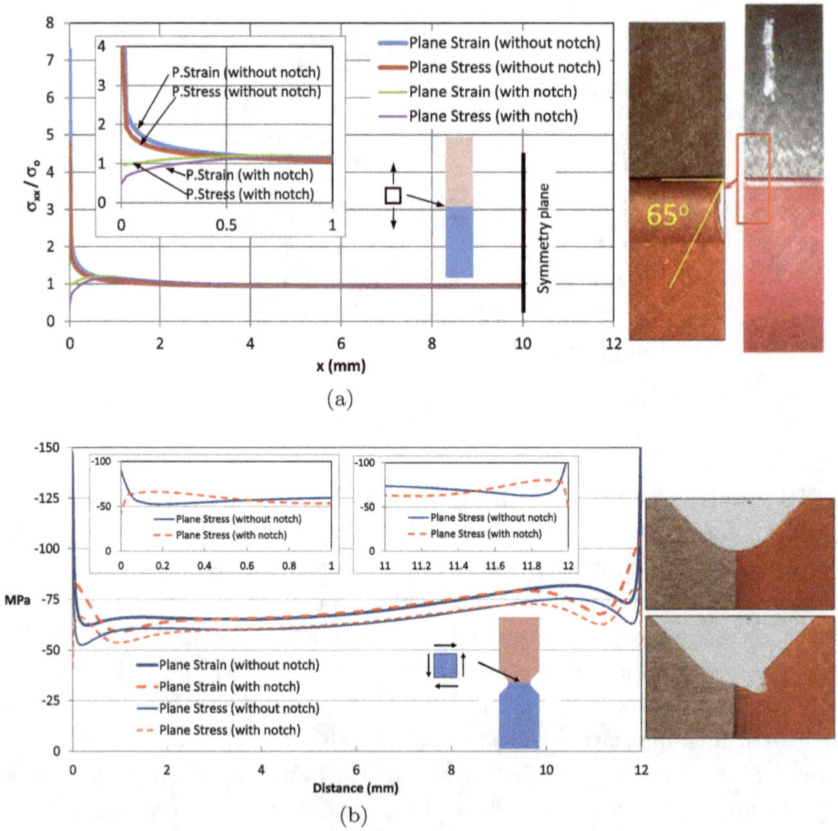

Fig. 13.21. (a) Normal stresses and (b) shear stresses along the specimen interface.

the original samples without notch. For the Iosipescu test, results were not as clear as in the tensile case, and the shear strength for the samples without and with notch were found to be quite similar. This result can be associated with the fact that in the tensile test configuration the notch is made all around the perimeter of the sample, at the bimaterial interface border line eliminating a lot of potential places for failure initiation with singularities. However, in the shear case, the notch is only performed at the thickness side, see Fig. 13.21(b), were the shear stress singularity appears, which is a small length compared to the whole perimeter of the interface. Moreover, the plasticity effects in Iosipescu shear specimen are more important than in tensile specimen, at least according to the von Misses plasticity criterion.

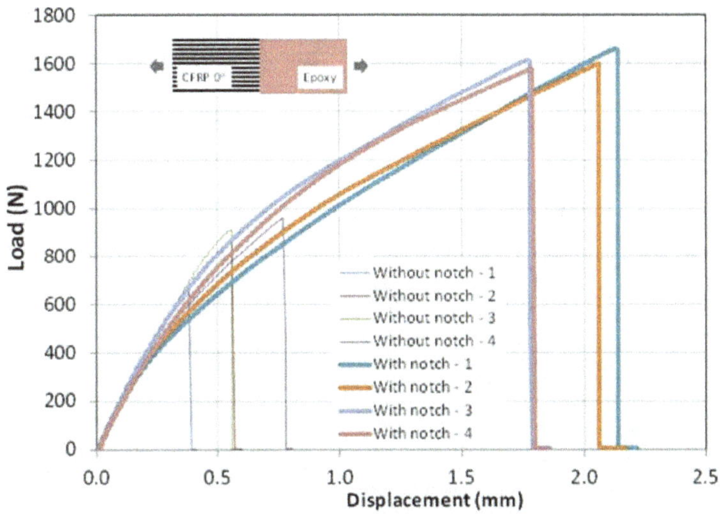

Fig. 13.22. Tensile test results for the bimaterial samples without and with a notch.

13.9.2. Removal or reduction of stress singularities associated to tabs bonding in standard composite testing

In the testing of standard coupons in composite materials, specifically designed tabs are bonded to specimens to achieve a smooth load transfer from the testing machine to the specimen. Nevertheless, the usage of these tabs creates critical points at tab locations, namely multimaterial corners, where premature failure can take place due to stress singularities appearing there. Two examples are studied in what follows, both related with geometrical modifications of the local tab geometry to remove the stress singularities at these locations or to reduce their severity. The first example is related to the compression testing of thick composite laminates, while the second deals with the intralaminar shear stiffness/strength determination in composite materials using the off-axis test.

13.9.2.1. *Compression test of thick composite laminates*

It is well known and documented in literature that thick composite laminates exhibit less compression strength values than the same materials with less thickness [129]. A reason for this effect is the difficulty in

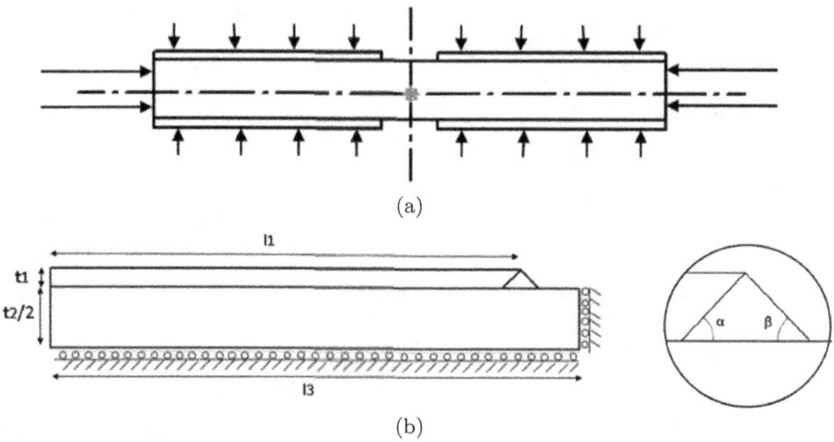

Fig. 13.23. (a) Original and (b) modified configurations of the compression sample geometry.

applying the compressive load to the thick specimen by means of tabs and the secondary compressive loads perpendicular to the thickness. Various devices have been proposed to try to solve this problem, see [129] for a comprehensive review.

As an example, Fig. 13.23(a) shows the standard sample geometry, with tabs designed for the ICSTM fixture [130] developed at Imperial College London. A local modification was proposed in [129] and carried out at the tab end, see Fig. 13.23(b) (showing due to symmetry only one quarter of the sample), which led to a higher compression strength.

The compression strength increment was associated to a benign stress profile at the tab ends. This fact was observed by means of numerical analysis in [129] and is briefly analyzed by the stress singularity approach in the following.

The characteristic exponents are evaluated for different mechanical and geometrical parameters defining the local corner configuration. The mechanical properties of the composite laminate, the composite tab and the adhesive are: Composite Laminate T300/914 ($E_{11} = 129\,\text{GPa}$, $E_{22} = E_{33} = 8.4\,\text{GPa}$, $G_{12} = G_{31} = 4.2\,\text{GPa}$, $G_{23} = 3.0\,\text{GPa}$, $\nu_{21} = \nu_{31} = 0.02$, $\nu_{32} = 0.40$), composite tab Woven EGlass/epoxy ($E_{11} = 27\,\text{GPa}$, $E_{22} = 29\,\text{GPa}$, $E_{33} = 7\,\text{GPa}$, $G_{12} = 7.4$, $G_{31} = G_{23} = 4.0\,\text{GPa}$, $\nu_{21} = 0.2$, $\nu_{31} = \nu_{32} = 0.30$), adhesive 3M Scotchweld ($E = 2.87\,\text{GPa}$, $\nu = 0.37$).

Figure 13.24(a) shows the order of stress singularity for the original corner configuration in which the tab angle varies between 0° (no tab, no

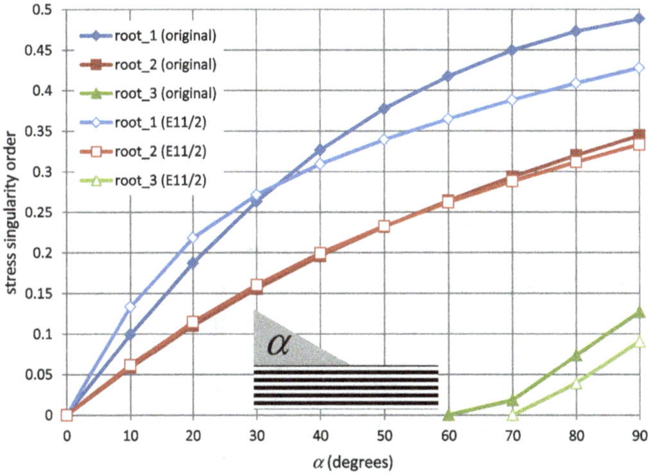

(a)

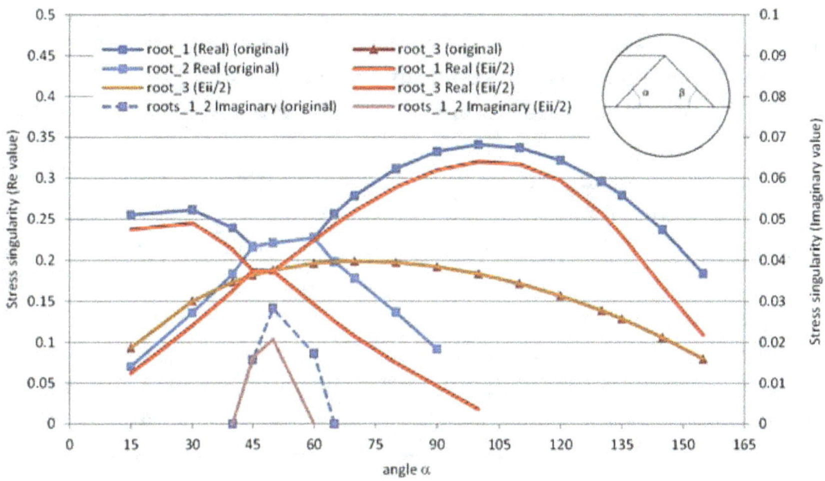

(b)

Fig. 13.24. Order of stress singularities for the (a) original and (b) modified configurations of the tab-end geometry.

singularity) and 90° (original configuration). The stiffness of the tab has also been modified to see its influence in the stress singularity evaluation.

Figure 13.24(b) shows the order of stress singularity for the most critical corner of the proposed modified configuration, the three-material corner between the composite laminate, the tab and the adhesive. In this case also the tab-adhesive angle is moved between 15° and 155° and the mechanical properties of the tab material (E_{11}, E_{22} and E_{33}) were modified in the parametrical study. For some angle values (around $\alpha = 50°$) the orders of stress singularity become complex conjugates. It is clear from the results shown in Fig. 13.24(b) that the modification proposed in [129] was an accurate choice, as the order of stress singularity for the original configuration (Fig. 13.24(a), $\alpha = 90°$) which is near 0.5, decreases to 0.2 (Fig. 13.24(b), $\alpha = 45°$). Any other choice would lead to higher values of the order of stress singularity, i.e., more severe stress singularities. This fact, which was observed numerically in [129] is confirmed here, for all possible angles, by the stress singularity analysis of the corner.

What is additionally observed in the present analysis is that a softer tab material ($E_{ii}/2$, $i = 1, 2, 3$) would give a lower order of stress singularity for all geometrical configurations studied, see Fig 13.24(b).

13.9.2.2. *Off-axis test of unidirectional composite materials*

The off-axis test is a standard test to determine the intralaminar stiffness and strength of unidirectional composite materials [131]. The test consists in a tensile test of a unidirectional composite material, but with a fiber orientation different than 0° (the load orientation), typically at 10° with respect to the load orientation. It is well known that to make the shear stress field more uniform and to avoid a premature failure at the tab locations due to the stress concentrations, some modifications have been proposed in literature, like pinned specimens or large aspect ratios of the specimens, there are also proposals that use particular tab angles, which depend on the fiber orientation [132, 133], which are easier to implement. In particular, the idea of using tabs with a certain inclination, it proposes different inclinations for different off-axis fiber orientations (Fig. 13.25), is also explored in the present work by means of stress singularities removal. The hypothesis for the present analysis is that clamped conditions can be assumed at the tab location.

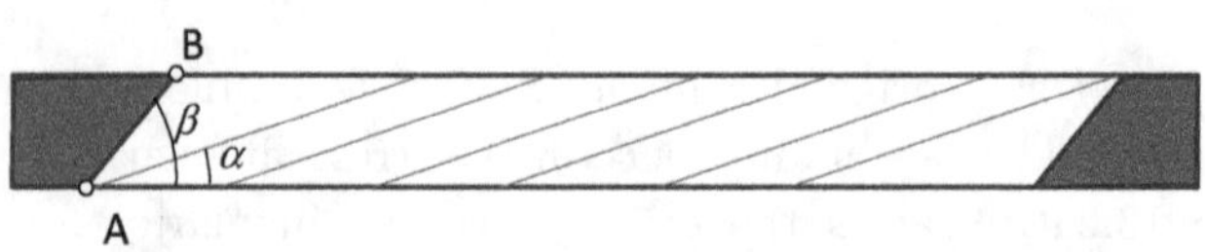

Fig. 13.25. Off-axis sample with oblique tabs.

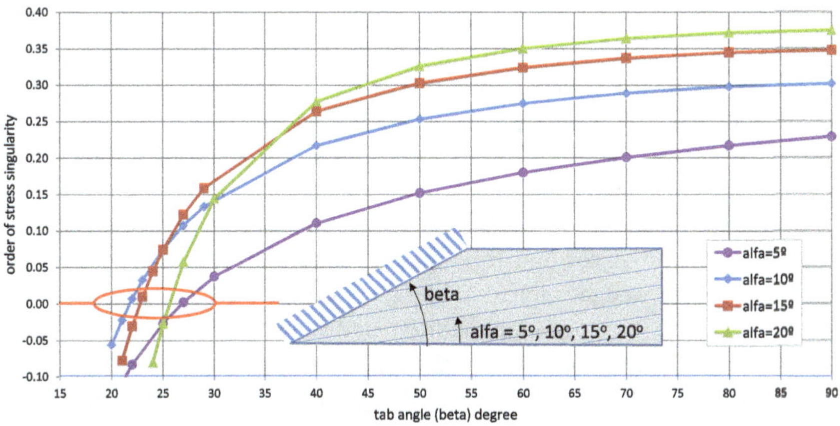

Fig. 13.26. Stress singularity orders for the off-axis sample with oblique tabs.

The stress singularity order δ_1 for corner A (Fig. 13.25), which is more critical than corner B, are plotted in Fig. 13.26, for generalized plane strain, for different fiber angle α and tab angle β. The optimum tab angles for each fiber orientation, those leading to the stress singularity order $\delta_1 = 0$, are: $\beta = 22°$ (for $\alpha = 10°$), $\beta = 23°$ (for $\alpha = 15°$), $\beta = 26°$ (for $\alpha = 20°$) and $\beta = 27°$ (for $\alpha = 5°$). These tab-fiber combinations should be used as an optimum configuration in the determination of the intralaminar shear strength of composite materials when using this type of test. Actually, the stress singularity orders for the standard straight tabs ($\beta = 90°$) are the highest ones, in the range $0° < \beta < 90°$ for all α values considered, thus leading to the most severe stress singularities there.

13.10. Conclusions

A novel procedure for the singularity analysis of linear elastic anisotropic multimaterial (piecewise homogeneous) corners in generalized plane strain including sliding friction contact surfaces has been developed and implemented. This semi-analytic approach, based on the Lekhnitskii–Stroh formalism of anisotropic materials, handles any kind of linear elastic material. The approach avoids numerical uncertainties and instabilities, which may appear when degenerate materials are treated as limit cases of non-degenerate materials. Specifically in the generation of the characteristic matrix of the corner, only the complex roots of the Lekhnitskii–Stroh sextic polynomial for an anisotropic material must, in general, be evaluated numerically. Nevertheless, for specific kinds of material, for example, transversely isotropic materials representing homogenized unidirectional fiber-reinforced composites, even these roots can be evaluated analytically for any orientation of the material axes. The application of the transfer matrix concept for homogeneous single-material wedges along with a powerful matrix formalism for prescribed boundary and interface (transmission) conditions leads to a fully automatic generation of the characteristic system for open and closed (periodic) corners including any finite number of wedges and frictionless or frictional contact surfaces and any homogeneous orthogonal boundary conditions.

A general anisotropic friction model was considered. It is useful for modeling contact between anisotropic materials, in particular for unidirectional composite laminas. If the angles of friction shear stresses and relative tangential displacements in a contact surface are not known *a priori*, which is the case where the in-plane and antiplane displacements are coupled, these angles should be determined when solving the corner characteristic equation. In this case, the corner characteristic equation includes, in addition to the characteristic exponent, one or more extra unknowns — the angles of friction shear stresses, depending on the number of friction contact surfaces in the corner. This fact appears to be a new feature of a corner characteristic equation to the best knowledge of the authors. The general procedure developed for the corner singularity analysis is very suitable for straightforward computational implementation.

A general post-processing procedure to extract multiple generalized stress intensity factors for multimaterial corners from a numerical solution obtained by FEM or BEM has also been presented, implemented and tested. The procedure employs least-squares fitting of a finite asymptotic series expansion of the singular solution at the corner to the numerical results in displacements and/or stresses along the boundaries and interfaces in the corner. The procedure has no special requirements regarding the accuracy of the numerical results close to the corner tip, where larger errors in the numerical solution are expected.

Singularity corner analyses have been carried out for a couple of examples of multimaterial corners in composites specimens: a transverse crack terminating at the interface between plies in a $[0/90]_S$ laminate, and a closed corner in a double-lap joint between a composite and a metal. The computational tool developed has been validated by comparing the BEM solution with the computed finite series approximation at the corner at a set distance to the corner tip, whereas their values along the corner boundaries were used in the fitting procedure. Furthermore, the corner in the latter example was altered to include a friction interface crack, an additional parametric study having been carried out for this particular case.

A novel experimental procedure for the determination of generalized fracture-toughness values and the subsequent generation of a failure criterion for closed anisotropic multimaterial corners under plane strain with two singularities has been proposed and tested. The procedure is based on a kind of Brazilian disc specimen with the corner tip at its centre and loaded in compression at any position along the external perimeter. A failure envelope curve in the plane of the generalized stress intensity factors normalized by the generalized fracture-toughness values has been determined experimentally, and used in the formulation of a failure criterion for corners of this kind. Satisfactory agreement has been observed between the predictions of this failure criterion and the failure loads of a real double-lap joint between a composite and a metal including such a corner. Finally, some examples of applications of the singularity analysis procedure presented here for the successful elimination or reduction of stress singularities in bimaterial joints are shown.

Acknowledgments

The authors are grateful to Dr Antonio Blázquez and Dr Enrique Graciani for their BEM codes and to Mr Israel G. García for carrying out calculations in some of the numerical examples. The authors also thank Dr Enrique Graciani for a correction of formula (13.133) in the first version of this chapter. Dr Daniane Vicentini has carried out and analysed tests with the new Brazilian disc specimen. V. Mantič thanks Prof Dominique Leguillon for his support and encouragement in analysing singularities with a sliding friction contact. This work was supported by Junta de Andalucía and the European Social Fund (Proyectos de Excelencia P08-TEP-4051 and P08-TEP-4071), by Ministerio de Ciencia e Innovación de España (Proyecto MAT2009-14022), by Ministerio de Ciencia, Innovación y Universidades de España and European Regional Development Fund (Proyecto PGC2018-099197-B-I00), and by Consejería de Economía y Conocimiento, Junta de Andalucía and European Regional Development Fund (Proyectos P18-FR-1928 and US-1266016).

References

[1] Hull D., Clyne T.W., 1996. *An Introduction to Composite Materials*, 2nd edn., Cambridge University Press, Cambridge.

[2] Leguillon D., Sanchez-Palencia E., 1987. *Computation of Singular Solutions in Elliptic Problems and Elasticity*, Masson, Paris.
[3] Grisvard P., 1992. *Singularities in Boundary Value Problems*. Springer-Verlag, London.
[4] Nazarov S.E., Plamenevskii B.A., 1994. *Elliptic Problems in Domains with Piecewise Smooth Boundaries*, Walter de Gruyter, Berlin.
[5] Kozlov V.A., Maz'ya V.G., Rossmann J., 2001. *Spectral Problems Associated with Corner Singularities of Solutions to Elliptic Equations*, American Mathematical Society, Providence, RI.
[6] Sinclair G.B., 2004. Stress singularities in classical elasticity — I: Removal, interpretation and analysis, *Appl. Mech. Rev.*, 57, 251–297.
[7] Sinclair G.B., 2004. Stress singularities in classical elasticity — II: Asymptotic identification, *Appl. Mech. Rev.*, 57, 385–439.
[8] Yosibash Z., 2012. *Singularities in Elliptic Boundary Value Problems and Elasticity and their Connection with Failure Initiation*, Springer, New York.
[9] Wieghardt K., 1907. Über Das Spalten und Zerreissen Elastischer Körper, *Z. Math. Phys.*, 55, 60–103.
[10] Williams M.L., 1952. Stress singularities resulting from various boundary conditions in angular corners of plates in extension, *J. Appl. Mech.*, 19, 526–528.
[11] Vasilopoulos D., 1988. On the determination of higher order terms of singular elastic stress fields near corners, *Numer. Math.*, 53, 51–95.
[12] Dempsey J.P., Sinclair G.B., 1979. On the stress singularities in the plane elasticity of the composite wedge, *J. Elasticity*, 9, 373–391.
[13] Papadakis P.J., Babuška I., 1995. A numerical procedure for the determination of certain quantities related to the stress intensity factors in two-dimensional elasticity, *Comput. Meth. Appl. Mech. Eng.*, 122, 69–92.
[14] Yosibash Z., Szabó B., 1995. Numerical analysis of singularities in two-dimensions Part 1: Computation of eigenpairs, *Int. J. Numer. Meth. Eng.*, 38, 2055–2082.
[15] Pageau S.S., Biggers Jr. S.B., 1996. A finite element approach to three-dimensional singular stress states in anisotropic multi-material wedges and junctions, *Int. J. Solids Struct.*, 33, 33–47.
[16] Ting T.C.T., 1997. Stress singularities at the tip of interfaces in polycrystals, in H.-P. Rossmanith, (ed.), *Damage and Failure of Interfaces* Balkema, Rotterdam, pp. 75–82.
[17] Costabel M., Dauge M., Lafranche Y., 2001. Fast semi-analytic computation of elastic edge singularities, *Comput. Meth. Appl. Mech. Eng.*, 190(24), 2111–2134.
[18] Wu K.-C., 2001. Near-tip field and the associated path-independent integrals for anisotropic composite wedges, *Chinese J. Mech.*, 17, 21–28.
[19] Barroso A., Mantič V., París F., 2003. Singularity analysis of anisotropic multimaterial corners, *Int. J. Fracture*, 119, 1–23.
[20] Hwu C., Omiya M., Kishimoto K., 2003. A key matrix N for the stress singularity of the anisotropic elastic composite wedges, *JSME Int. J., Series A*, 46, 40–50.
[21] Yin W.L., 2003. Anisotropic elasticity and multi-material singularities, *J. Elasticity*, 71, 263–292.

[22] Mantič V., París F., Berger J., 2003. Singularities in 2D anisotropic potential problems in multi-material corners. Real variable approach, *Int. J. Solids Struct.*, 40, 5197–5218.

[23] Linkov A.M., Koshelev V.F., 2006. Multi-wedge points and multi-wedge elements in computational mechanics: Evaluation of exponent and angular distribution, *Int. J. Solids Struct.*, 43, 5909–5930.

[24] Paggi M., Carpinteri A., 2008. On the stress singularities at multimaterial interfaces and related analogies with fluid dynamics and diffusion, *Appl. Mech. Rev.*, 61, 020801-1–020801-22.

[25] Barroso A., Mantič V., París F., 2009. Computing stress singularities in transversely isotropic multimaterial corners by means of explicit expressions of the orthonormalized Stroh-eigenvectors, *Eng. Frac. Mech.*, 76, 250–268.

[26] Nicaise S., Sändig A.-M., 1994. General interface problems I/II, *Math. Method Appl. Sci.*, 17, 395–450.

[27] Nicaise S., Sändig A.-M., 1999. Transmission problems for the Laplace and elasticity operators: Regularity and boundary integral formulation, *Math. Mod. Meth. Appl. Sci.*, 9, 855–898.

[28] Ting T.C.T., 1996. *Anisotropic Elasticity: Theory and Applications*, Oxford University Press, New York.

[29] Hwu C., 2010. *Anisotropic Elastic Plates*, Springer, New York.

[30] Kondratiev V.A., 1967. Boundary-value problems for elliptic equations in domains with conical or angular points, *Trans. Moscow Math. Soc.*, 16, 227–313.

[31] Costabel M., Dauge M., 1993. Construction of corner singularities for Agmon–Douglis–Niremberg elliptic systems, *Matematische Nachr.*, 162, 209–237.

[32] Dempsey J.P., 1995. Power-logarithmic stress singularities at bi-material corners and interface cracks, *J. Adhes. Sci. Technol.*, 9, 253–265.

[33] Sinclair G.B., 1999. Logarithmic stress singularities resulting from various boundary conditions in angular corners of plates in extension, *J. Appl. Mech.*, 66, 556–559.

[34] Lekhnitskii S.G., 1938. Some cases of the elastic equilibrium of a homogeneous cylinder with arbitrary anisotropy, *Appl. Math. Mech. (Prikladnaya Matematrikai Mekhanika)*, 2, 345–367 (in Russian).

[35] Stroh A.N., 1958. Dislocations and cracks in anisotropic elasticity, *Philos. Mag.* 3, 625–646.

[36] Stroh A.N., 1962. Steady state problems in anisotropic elasticity, *J. Math. Phys.*, 41, 77–103.

[37] Tanuma K., 2007. *Stroh Formalism and Rayleigh Waves*, Springer, New York.

[38] Ting T.C.T., 1999. A modified Lekhnitskii formalism à la Stroh for anisotropic elasticity and classifications of the 6×6 matrix N, *Proc. R. Soc. London A* 455, 69–89.

[39] Costabel M., Dauge M., 1995. Computation of corner singularities in linear elasticity, in (M. Costabel, M. Dauge, S. Nicaise eds), *Boundary Value Problems and Integral Equations in Nonsmooth Domains* Marcel Dekker, New York, pp. 59–68.
[40] Barroso A., Graciani E., Mantič V., París F., 2012. A least squares procedure for the evaluation of multiple generalized stress intensity factors at 2D multimaterial corners by BEM, *Eng. Anal. Bound. Elem.*, 36, 458–470.
[41] Helsing J., Jonsson A., 2002. On the computation of stress fields on polygonal domains with V-notches, *Int. J. Numer. Meth. Eng.*, 53, 433–454.
[42] Helsing J., Ojala R., 2009. Elastostatic computations on aggregates of grains with sharp interfaces, corners and triple-junctions, *Int. J. Solids Struct.*, 46, 4437–4450.
[43] Novozhilov V., 1969. On a necessary and sufficient criterion for brittle strength, *J. Appl. Math. Mech.*, 33, 212–222.
[44] Carpinteri A., 1987. Stress-singularity and generalized fracture toughness at the vertex of re-entrant corners, *Eng. Fract. Mech.*, 26, 143–155.
[45] Reedy Jr. E., Guess T., 1993. Comparison of butt tensile strength data with interface corner stress intensity factor prediction, *Int. J. Solids Struct.*, 30, 2929–2936.
[46] Seweryn A., 1994. Brittle fracture criterion for structures with sharp notches, *Eng. Fract. Mech.*, 47, 673–681.
[47] Dunn M., Suwito W., Cunningham S., 1997. Fracture initiation at sharp notches: Correlation using critical stress intensities, *Int. J. Solids Struct.*, 34, 3873–3883.
[48] Leblond J.B., Mouro P., 2000. Crack propagation from a pre-existing flaw at a notch root: I. Introduction and general form of the stress intensity factors at the initial crack tip, II Detailed form of the stress intensity factors at the initial crack tip and conclusion, *Int. J. Fracture*, 104, 211–239.
[49] Leguillon D., 2002. Strength or toughness? A criterion for crack onset at a notch, *Eur. J. Mech. A/Solid*, 21, 61–72.
[50] Yosibash Z., Priel E., Leguillon D., 2006. A failure criterion for brittle elastic materials under mixed mode loading, *Int. J. Fracture*, 141, 289–310.
[51] Taylor D., 2007. *The Theory of Critical Distance: A New Perspective in Fracture Mechanics*, Elsevier, Oxford.
[52] Carpinteri A., Cornetti P., Pugno N., Sapora A., Taylor D., 2008. A finite fracture mechanics approach to structures with sharp V-notches, *Eng. Fract. Mech.*, 75, 1736–1752.
[53] Andersons J., Tarasovs S., Sparnins E., 2010. Finite fracture mechanics analysis of crack onset at a stress concentration in a UD glass/epoxy composite in off-axis tension, *Compos. Sci. Technol.*, 70, 1380–1385.
[54] García I.G., Leguillon D., 2012. Mixed-mode crack initiation at a V-notch in presence of an adhesive joint, *Int. J. Solids Struct*, 49, 2138–2149.
[55] Dundurs J., Lee M.-S., 1972. Stress concentration at a sharp edge in contact problems, *J. Elasticity*, 2, 109.

[56] Gdoutos E.E., Theocaris P.S., 1975. Stress concentrations at the apex of a plane indenter acting on an elastic half plane, *J. Appl. Mech.*, 42, 688–692.
[57] Comninou M., 1976. Stress singularity at a sharp edge in contact problems with friction, *Z. Angew. Math. Phys. ZAMP*, 27, 493–499.
[58] Comninou M., 1977. The interface crack, *J. Appl. Mech.*, 44, 631–636.
[59] Comninou M., 1977. The interface crack with friction in the contact zone, *J. Appl. Mech.*, 44, 780–781.
[60] Audoly B., 2000. Asymptotic study of the interfacial crack with friction, *J. Mech. Phys. Solids*, 48, 1851–1864.
[61] Bui H.D., Oueslati A., 2005. The sliding interface crack with friction between elastic and rigid bodies, *J. Mech. Phys. Solids*, 53, 1397–1421.
[62] Wang S.S., Choi I., 1983. The interface crack between dissimilar anisotropic composite materials, *J. Appl. Mech.*, 50, 169–178.
[63] Wang S.S., Choi I., 1983. The interface crack behavior in dissimilar anisotropic composites under mixed-mode loading, *J. Appl. Mech.*, 50, 179–183.
[64] Wu K.-C., Hwang S.-J., 1990. Correspondence relations for the interface crack in monoclinic composites under mixed loading, *J. Appl. Mech.*, 57, 894–900.
[65] Wu K.-C., 1992. On the contact zone model for interface cracks, *J. Appl. Mech.*, 57, 92–94.
[66] Ni L., Nemat-Nasser S., 1992. Interface cracks in anisotropic dissimilar materials: General case, *Q. Appl. Math. L*, 305–322.
[67] Deng X., 1993. On stationary and moving interface cracks with frictionless contact in anisotropic biomaterials, *Proc. R. Soc. Lond. A*, 443, 563–572.
[68] Deng X., 1994. An asymptotic analysis of stationary and moving cracks with frictional contact along bimaterial interfaces and in homogeneous solids, *Int. J. Solids Struct.*, 31, 2407–2429.
[69] Lee J., Gao H., 1994. A generalized Comninou contact model for interface cracks in anisotropic elastic solids, *Int. J. Fracture*, 67, 53–68.
[70] Deng X., 1995. Mechanics of debonding and delamination in composites: Asymptotic studies, *Compos. Eng.*, 5, 1299–1315.
[71] Mantič V., París F., Cañas J., 1997. Stress singularities in 2D orthotropic corners, *Int. J. Fracture*, 83, 67–90.
[72] Leguillon D., 1999. Interface crack tip singularity with contact and friction, *C.R. Acad. Sci. Paris*, Ser. IIb, 327, 437–442.
[73] Poonsawat P., Wijeyewickrema A.C., Karasudhi P., 2001. Singular stress fields of angle-ply and monoclinic bimaterial wedges, *Int. J. Solids Struct.*, 38, 91–113.
[74] Guo C., Sun C.T., 2003. Stationary and moving interface cracks with contact in anisotropic bimaterials, *Chin. J. Mech.-Ser. A*, 19, 31–39.
[75] Mantič V., Blázquez A., Correa E., París F., 2006. Analysis of interface cracks with contact in composites by 2D BEM, in (M. Guagliano, M.H. Aliabadi (eds), *Fracture and Damage of Composites* WIT Press, Southampton, pp. 189–248.

[76] París F., Cañas J., 1997. *Boundary Element Method, Fundamentals and Applications*, Oxford University Press, Oxford.
[77] Mantič V., París F., 2004. Advanced formulation of the BIEs for elastic anisotropic materials in 2D, [V. Leitao, M.H. Aliabadi, (eds)], in *Proc. 5th Int. Conf. BETEQ-04, Advances in Boundary Element Techniques V* Engineering Conferences, Eastleigh, pp. 57–62.
[78] Graciani E., Mantič V., París F., Blázquez A., 2005. Weak formulation of axi-symmetric frictionless contact problems with boundary elements: Application to interface cracks, *Comput. Struct.*, 83, 836–55.
[79] Blázquez A., Mantič V., París F., McCartney L.N., 2008. Stress state characterization of delamination cracks in [0/90] symmetric laminates by BEM, *Int. J. Solids Struct.*, 45, 1632–1662.
[80] Barroso A., Vicentini D., Mantič V., París F., 2012. Determination of generalized fracture toughness in multimaterial closed corners with two singular terms. Part I: Test procedure and numerical analysis, *Eng. Fract. Mech.*, 89, 1–14.
[81] Vicentini D., Barroso A., Justo J., Mantič V., París F., 2012. Determination of generalized fracture toughness in multimaterial closed corners with two singular terms. Part II: Experimental results. *Eng. Fract. Mech.*, 89, 15–23.
[82] Eshelby J.D., Read W.T., Shockley W., 1953. Anisotropic elasticity with applications to dislocation theory, *Acta. Metall. Mater.*, 1, 251–259.
[83] Ingebrigtsen K.A., Tonning A., 1969. Elastic surface waves in crystals, *Phys. Rev.*, 184, 942–951.
[84] Ting T.C.T., Hwu C., 1988. Sextic formalism in anisotropic elasticity for almost non-semisimple matrix N, *Int. J. Solids Struct.*, 24, 65–76.
[85] Ting T.C.T., 1996. Existence of an extraordinary degenerate matrix N for anisotropic materials, *Q. J. Mech. Appl. Math.*, 49, 405–417.
[86] Wang Y.M., Ting T.C.T., 1997. The Stroh formalism for anisotropic materials that possess an almost extraordinary degenerate matrix N, *Int. J. Solids Struct.*, 34, 401–413.
[87] Tanuma K., 1996. Surface impedance tensors of transversely isotropic materials, *Q. J. Mech. Appl. Math.*, 49, 29–48.
[88] Ting T.C.T., Wang M.Z., 1992. Generalized Stroh formalism for anisotropic elasticity for general boundary conditions, *Acta Mech. Sinica*, 8, 193–207.
[89] Michalowski R., Mróz Z., 1978. Associated and non-associated sliding rules in contact friction problems, *Arch. Mech.*, 30, 259–276.
[90] Curnier A., 1984. A theory of friction, *Int. J. Solids Struct.*, 20, 637–647.
[91] Mróz Z., Stupkiewicz S., 1994. An anisotropic friction and wear model, *Int. J. Solids Struct.*, 31, 1113–1131.
[92] Hijiaj M., de Saxcé G., Mróz Z., 2002. A variational inequality-based formulation of the frictional contact law with a non-associated sliding rule, *Eur. J. Mech. A/Solids*, 21, 49–59.
[93] Wriggers P., 2006. *Computational Contact Mechanics*, 2nd edn., Springer, Berlin.

[94] Hwu C., Ting T.C.T., 1989. Two-dimensional problems of the anisotropic elastic solid with an elliptic inclusion, *Q. J. Mech. Appl. Math.*, 42, 553–572.

[95] Golub G.H., van Loan C.F., 1996. *Matrix Computations*, 3rd edn., The Johns Hopkins University Press, Baltimore.

[96] Trefethen L.N., Bau D. 1997. *Numerical Linear Algebra*, SIAM, Philadelphia.

[97] Wolfram S., 1991. *A System for Doing Mathematics by Computer*, Addison-Wesley, Redwood City.

[98] Muller D.E., 1956. A method for solving algebraic equations using an automatic computer, *Math. Tables Comput.*, 10, 208–215.

[99] Portela A., Aliabadi M.H., Rooke D.P., 1991. Efficient boundary element analysis of sharp notched plates, *Int. J. Numer. Meth. Eng.*, 32, 445–470.

[100] Johnson T., Tucker W., 2009. Enclosing all zeros of an analytic function — a rigorous approach, *J. Comput. Appl. Math.*, 228, 418–442.

[101] Shaw W.T., 2006. *Complex Analysis with Mathematica*, Cambridge University Press, Cambridge.

[102] Pageau S.P., Gadi K.S., Biggers Jr. S.B., Joseph P.F., 1996. Standardized complex and logarithmic eigensolutions for n-material wedges and junctions, *Int. J. Fracture*, 77, 51–76.

[103] Rice J.R., 1988. Elastic fracture mechanics concepts for interfacial cracks, *J. Appl. Mech.*, 55, 98–103.

[104] Nomura Y., Ikeda T., Miyazaki N., 2009. Stress intensity factor analysis at an interfacial corner between anisotropic bimaterials under thermal stress, *Eng. Fract. Mech.*, 76, 221–235.

[105] Hwu C., Kuo T.L., 2007. A unified definition for stress intensity factors of interface corners and cracks, *Int. J. Solids Struct.*, 44, 6340–6359.

[106] Banks-Sills L., Hershkovitz I., Wawrzynek P.A., Eliasi R., Ingraffea A.R., 2005. Methods for calculating stress intensity factors in anisotropic materials: Part I $z=0$ is a symmetric plane, *Eng. Fract. Mech.*, 72, 2328–2358.

[107] Banks-Sills L., Wawrzynek P.A., Carter B., Ingraffea A.R., Hershkovitz I., 2007. Methods for calculating stress intensity factors in anisotropic materials: Part II Arbitrary geometry, *Eng. Fract. Mech.*, 74, 1293–1307.

[108] Chan S.K., Tuba I.S., Wilson W.K., 1970. On the finite element method in linear fracture mechanics, *Eng. Fract. Mech.*, 2, 1–17.

[109] Munz D., Yang Y.Y., 1993. Stresses near the edge of bonded dissimilar materials described by two stress intensity factors, *Int. J. Fracture*, 60, 169–177.

[110] Liu Y., Wu Z., Liang Y., Liu X., 2008. Numerical methods for determination of stress intensity factors of singular stress field, *Eng. Fract. Mech.*, 75, 4793–803.

[111] Ju S.H., 2010. Finite element calculation of stress intensity factors for interface notches, *Comput. Meth. Appl. Mech. Eng.*, 199, 2273–2280.

[112] Sinclair G.B., Okajima M., Griffin J.H., 1984. Path independent integrals for computing stress intensity factors at sharp corners in elastic plates, *Int. J. Numer. Meth. Eng.*, 20, 999–1008.

[113] Choi N.Y., Earmme Y.Y., 1992. Evaluation of stress intensity factors in a circular arc-shaped interfacial crack using *L*-integral, *Mech. Mater.*, 14, 141–153.

[114] Banks-Sills L., 1997. A conservative integral for determining stress intensity factors of a bimaterial strip, *Int. J. Fracture*, 86, 385–398.

[115] Im S., Kim K.S., 2000. An application of two-state M-integral for computing the intensity of the singular near-tip field for a generic wedge, *J. Mech. Phys. Solids*, 48, 129–151.

[116] Omer N., Yosibash Z., 2005. On the path independency of the point-wise J integral in three-dimensions, *Int. J. Fracture*, 136, 1–36.

[117] Ortiz J.E., Mantič V., París F., 2006. A domain-independent integral for computation of stress intensity factors along three-dimensional crack front and edges by BEM, *Int. J. Solids Struct.*, 43, 5593–5612.

[118] Henshell R.D., Shaw K.G., 1975. Crack tip finite elements are unnecessary, *Int. J. Numer. Meth. Eng.*, 9, 495–507.

[119] Barsoum R.S., 1976. On the use of isoparametric finite element in linear fracture mechanics, *Int. J. Numer. Meth. Eng.*, 10, 25–37.

[120] Gray L., Phan A.V., Paulino G.H., Kaplan T., 2003. Improved quarter-point crack tip element, *Eng. Fract. Mech.*, 70, 269–283.

[121] Wait R., 1978. Finite element methods for elliptic problems with singularities, *Comput. Meth. Appl. Mech. Eng.*, 13, 141–150.

[122] Marin L., Lesnic D., Mantič V., 2004. Treatment of singularities in Helmholtz type equations using the boundary element method, *J. Sound Vib.*, 278, 39–62.

[123] Barroso A., 2007. *Characterization of Singular Stress Fields in Multimaterial Corners. Application to Adhesive Joints with Composite Materials*, Ph.D. thesis (in Spanish), University of Seville.

[124] Barroso A., París F., Mantič V., 2009. Representativity of the singular stress state in the failure of adhesively bonded joints between metals and composites, *Compos. Sci. Technol.*, 69, 1746–1755.

[125] Barroso A., Vicentini D., París F., Mantič V., 2011. Representativity of thermal stresses in designing composite joints based on singular stress states at multimaterial corners, *Compos. Part A Appl. S.*, 42, 1084–1092.

[126] Banks-Sills L., Konovalov N., Fliesher A., 2010. Comparison of two- and three-dimensional analyses of interface fracture data obtained from Brazilian disk specimens, *Int. J. Struct. Integrity*, 1, 20–42.

[127] Lauke B., Barroso A., 2011. Notched-butt test for the determination of adhesion strength at bimaterial interfaces, *Compos. Interfaces*, 18, 661–669.

[128] Barroso A., Lauke B., Mantič V., París F., 2016. Tensile and shear strength of bimaterial interfaces within composite materials, *Compos. Sci. Technol.*, 124, 81–88.

[129] Dogra J., Hodkinson J.M., Robinson P., Pinho S.T., 2007. Development of a compression test for thick composite laminates: Finite element analysis, in *Proc. 16th Int. Conf. Composite Materials, ICCM16*, Kyoto, Japan.

[130] Häberle J., 1992. *Strength and Failure Mechanisms of Unidirectional Carbon Fiber-Reinforced Plastics Under Axial Compression*, Ph.D. Thesis, Imperial College London.

[131] Chamis C.C., Sinclair J.H., 1976. 10° *Off-Axis Tensile Test for Intralaminar Shear Characterization of Fiber Composites*, Report No. NASA TND-8215, NASA Lewis Research Center, Cleveland.

[132] Sun C.T., Berreth S.P., 1988. A new end tab design for off-axis tension test of composite materials, *J. Compos. Mater.* 22, 766–799.

[133] Sun C.T., Chung I., 1993. An oblique end-tab design for testing off-axis composite specimens, *Composites* 24, 619-623.

[134] Barroso A., Marín J.C., Mantič V., París F., 2020. Premature failures in standard test specimens with composite materials induced by stress singularities in adhesive joints, *Int. J. Adhes. Adhes.*, 97, 102478.

https://doi.org/10.1142/9781800611887_bmatter

Index

B

C

D

E

F

G

M

P

T

V

W

Y

Z

Computational and Experimental Methods in Structures

(Continued from page ii)

www.ingramcontent.com/pod-product-compliance
Lightning Source LLC
LaVergne TN
LVHW010321220926
848484LV00004BB/824
* 9 7 8 1 8 0 0 6 1 1 8 7 0 *